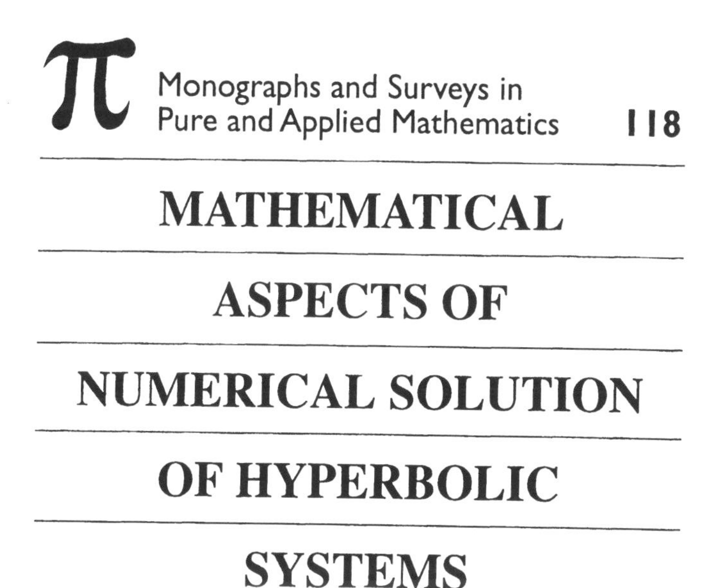

Monographs and Surveys in Pure and Applied Mathematics 118

MATHEMATICAL ASPECTS OF NUMERICAL SOLUTION OF HYPERBOLIC SYSTEMS

Monographs and Surveys in Pure and Applied Mathematics

Main Editors

H. Brezis, *Université de Paris*
R.G. Douglas, *Texas A&M University*
A. Jeffrey, *University of Newcastle upon Tyne (Founding Editor)*

Editorial Board

H. Amann, *University of Zürich*
R. Aris, *University of Minnesota*
G.I. Barenblatt, *University of Cambridge*
H. Begehr, *Freie Universität Berlin*
P. Bullen, *University of British Columbia*
R.J. Elliott, *University of Alberta*
R.P. Gilbert, *University of Delaware*
R. Glowinski, *University of Houston*
D. Jerison, *Massachusetts Institute of Technology*
K. Kirchgässner, *Universität Stuttgart*
B. Lawson, *State University of New York*
B. Moodie, *University of Alberta*
S. Mori, *Kyoto University*
L.E. Payne, *Cornell University*
D.B. Pearson, *University of Hull*
I. Raeburn, *University of Newcastle*
G.F. Roach, *University of Strathclyde*
I. Stakgold, *University of Delaware*
W.A. Strauss, *Brown University*
J. van der Hoek, *University of Adelaide*

MATHEMATICAL ASPECTS OF NUMERICAL SOLUTION OF HYPERBOLIC SYSTEMS

ANDREI G. KULIKOVSKII
NIKOLAI V. POGORELOV
ANDREI YU. SEMENOV

CRC Press is an imprint of the
Taylor & Francis Group, an informa business
A CHAPMAN & HALL BOOK

CRC Press
Taylor & Francis Group
6000 Broken Sound Parkway NW, Suite 300
Boca Raton, FL 33487-2742

First issued in paperback 2019

© 2001 by Taylor & Francis Group, LLC
CRC Press is an imprint of Taylor & Francis Group, an Informa business

No claim to original U.S. Government works

ISBN-13: 978-0-8493-0608-2 (hbk)
ISBN-13: 978-0-367-39773-9 (pbk)

This book contains information obtained from authentic and highly regarded sources. Reasonable efforts have been made to publish reliable data and information, but the author and publisher cannot assume responsibility for the validity of all materials or the consequences of their use. The authors and publishers have attempted to trace the copyright holders of all material reproduced in this publication and apologize to copyright holders if permission to publish in this form has not been obtained. If any copyright material has not been acknowledged please write and let us know so we may rectify in any future reprint.

Except as permitted under U.S. Copyright Law, no part of this book may be reprinted, reproduced, transmitted, or utilized in any form by any electronic, mechanical, or other means, now known or hereafter invented, including photocopying, microfilming, and recording, or in any information storage or retrieval system, without written permission from the publishers.

For permission to photocopy or use material electronically from this work, please access www.copyright.com (http://www.copyright.com/) or contact the Copyright Clearance Center, Inc. (CCC), 222 Rosewood Drive, Danvers, MA 01923, 978-750-8400. CCC is a not-for-profit organization that provides licenses and registration for a variety of users. For organizations that have been granted a photocopy license by the CCC, a separate system of payment has been arranged.

Trademark Notice: Product or corporate names may be trademarks or registered trademarks, and are used only for identification and explanation without intent to infringe.

Visit the Taylor & Francis Web site at
http://www.taylorandfrancis.com

and the CRC Press Web site at
http://www.crcpress.com

Library of Congress Cataloging-in-Publication Data

Kulikovskii, A.G. (Andrei Gennadievich)
Mathematical aspects of numerical solution of hyperbolic systems /
A.G. Kulikovskii, N.V. Pogorelov, A. Yu. Semenov.
p. cm.—(Chapman & Hall/CRC monographs and surveys in pure and applied mathematics ; 118)
Includes bibliographical references and index.
ISBN 0-8493-0608-6 (alk. paper)
1. Differential equations, Hyperbolic—Numerical solutions. I. Pogorelov, N.V. II. Semenov, A. Yu. (Andrei Yurievich), 1955- III. Title. IV. Series.
QA377 .K844 2000
515′.353—dc21 00-047568

Library of Congress Card Number 00-047568

Preface

This work gives a comprehensive description of various mathematical aspects of the problems originating in numerical solution of hyperbolic systems of partial differential equations. The material is presented in close relation with the important mechanical applications of such systems. They include both the Euler equations of gas dynamics and comparatively new fields such as magnetohydrodynamics (MHD), shallow water, and mechanics of solids. When considering the equations of gas dynamics, we mainly dwell on their applications to media with a complicated wide-range equation of state. Historically, high-resolution numerical schemes for hyperbolic conservation laws were first applied to purely gas dynamic problems. This can be explained by the fact that, due to the "convexity" of the Euler system, the Riemann problem for it generally has a unique solution. This is not true for more complicated MHD and solid dynamics equations. Although the solution of the MHD Riemann problem exists, it is too complicated to be used in regular calculations. In this book we give a collection of recipes for application of high-order nonoscillatory shock-capturing schemes to MHD modelling of complicated physical phenomena. Of great importance is also the problem of physical admissibility of solutions that are nonevolutionary if we solve the ideal MHD system. We discuss the current state of this problem and state our views on the stability of nonevolutionary shock waves.

The book deals with a number of new original problems called *nonclassical*. Among them are such problems as shock wave propagation in elastic rods and composite materials, ionization fronts in plasma, electromagnetic shocks in magnets, etc. We show that if a small-scale higher-order mathematical model results in oscillations of the discontinuity structure, the variety of admissible discontinuities can exhibit disperse behaviour. This variety includes discontinuities with additional boundary conditions not following from the hyperbolic conservation laws. Nonclassical problems are accompanied by a multiple nonuniqueness of solutions. An example of the selection rule is given that in certain cases permits one to easily make a correct, physically realizable choice.

The book is divided into seven chapters. For the reader's convenience, in Chapter 1 we introduce the main notions and definitions that make the book self-sufficient. The definitions are followed by mechanical examples that illustrate the essence of the subjects to be considered in the subsequent chapters. General properties of solutions are discussed. In Chapter 2 we formulate the approaches to numerical solution of quasilinear hyperbolic systems both in the conservative and nonconservative forms. The methods are subdivided into two classes: shock-fitting and shock-capturing schemes. Among shock-capturing schemes we choose only those belonging to the Godunov type, that is, which are based on the solution of the Riemann problem to determine fluxes through the computational cell interfaces. Since exact solutions are frequently not available, we describe also the methods based on approximate solutions and solutions of the linearized problem which always exists. The methods of increasing the order of accuracy are given, which include both the application of the generalized Riemann problem and various reconstruction procedures. Chapter 3 is devoted to the equations of gas dynamics. We present the exact solution of the Riemann problem for gases possessing complicated wide-range equations of state. The Courant–Isaacson–Rees, Roe, and Osher–Solomon methods are described. The cases are emphasized of fairly arbitrary equations of state. Genuine shock-fitting and floating shock-fitting techniques are discussed, including the self-adjusting grid approach. The applications include the problem of the chemically reacting airflow around blunt body at high angle of attack, modelling of shock-induced phenomena, jet-like structures in laser plasma, etc. Separately, both in Chapter 3 and Chapter 5, the problem is considered of the solar wind interaction with the magnetized interstellar medium. While investigating this problem, different aspects of the application of the introduced methods are comprehensively discussed. In addition to nonstationary hyperbolic systems, in Chapters 3 and 4 we dwell on the application of high-resolution methods to stationary, supersonic or supercritical, gas dynamic and shallow water equations. Chapter 4 also describes different Godunov-type methods for hyperbolic shallow water equations and is accompanied by a number of examples. Chapter 5 deals with MHD equations. For the reader's convenience we outline the assumptions adopted in ideal MHD and give the classification of discontinuities. The evolutionary property of MHD shock is discussed, emphasizing the degenerate (parallel and perpendicular) and singular (switch-on and switch-off) cases. The approaches to solving MHD system by nearly all modern high-resolution numerical methods are summarized. The problem of physical admissibility of nonevolutionary solutions is investigated, as well as its interaction with the application of shock-capturing numerical methods whose numerical dissipation can substantially exceed the physical dissipation existing in space plasma. Chapter 6 represents an attempt to outline the problems of solid dynamics that are governed by hyperbolic systems. For these problems, Courant–Isaacson–Rees type methods are formulated with the application to spallation phenomena, dynamics of Timoshenko-type shells, etc. Chapter 7 introduces the notion of nonclassical discontinuity and formulates evolutionary conditions for them. The correlation between the evolutionary conditions for discontinuities and the existence of their structure is investigated. The behavior of classical discontinuities near Jouget points on the shock adiabatic curves is explained. Further on, various examples are presented illustrating the application of the introduced theoretical basis to important physical phenomena.

The choice of the material naturally reflects the scientific interests of the authors and several important aspects arising in hyperbolic problems were discussed only briefly or were not discussed at all. We believe, however, that this book will substantially supplement existing literature devoted to this subject, since it concerns new areas of application and formulates new notions in order to clear out unexpected difficulties that may be encountered when one deals with nonconvex hyperbolic systems.

The book can be useful to graduate and postgraduate students majoring in the field of numerical, engineering, and applied mathematics and mechanics of continuous media. It is also aimed at the attention of specialists in pure and applied mathematics, and various fields of physics and mechanics where hyperbolic systems of partial differential equations find their application.

The authors would like to express their gratitude to V. B. Baranov, A. A. Barmin, A. A. Charakhch'yan, T. Hanawa, S. A. Ivanenko, I. E. Ivanov, V. D. Ivanov, A. S. Kholodov, V. I. Kondaurov, I. K. Krasyuk, I. A. Kryukov, T. Matsuda, G. P. Prokopov, E. A. Pushkar, N. Satofuka, K. Shibata, E. I. Sveshnikova, and A. V. Zabrodin for valuable discussions and helpful comments. Special thanks to N. I. Gvozdovskaya and A. I. Zhurov for their help in the preparation of the manuscript.

Contents

CREDITS

Figures 3.30 and 3.31: From Pogorelov, N.V., Periodic stellar wind/interstellar medium interaction, *Astron. Astrophys.*, 297, 835-840, 1995. With permission.

Figures 5.21 to 5.25: From Barmin, A.A., Kulikovskii, A.G., and Pogorelov, N.V., Shock-capturing approach and nonrevolutionary solutions in magnetohydrodynamics, *J. Comput. Phys.*, 126, 77-90, 1996. With permission.

Figure 5.26: From Pogorelov, N.V. and Semenov, A. Yu., Solar wind interaction with the magnetized interstellar medium: shock-capturing modeling, *Astron. Astrophys.*, 321, 330-337, 1997. With permission.

Figures 5.28 to 5.34: From Pogorelov, N.V. and Matsuda, T., Nonrevolutionary MHD shocks in the solar wind and interstellar medium interaction, *Astron. Astrophys.*, 354, 697-702, 2000. With permission.

Figures 5.38 to 5.43: From Pogorelov, N.V. and Matsuda, T., Influence of the interstellar magnetic field on the shape of the global heliopause, *J. Geophys. Res.*, 103, A1, 237-245, 1997. With permission.

Chapter 1
Hyperbolic Systems of Partial Differential Equations

1.1 Quasilinear systems

Let us define the system of first-order partial differential equations for the unknown vector-function $\mathbf{u}$ of the independent variables $\mathbf{x}$ and t as a system of relations

$$F_i\left(\mathbf{x}, t, \mathbf{u}, \frac{\partial \mathbf{u}}{\partial x_1}, \ldots, \frac{\partial \mathbf{u}}{\partial x_m}, \frac{\partial \mathbf{u}}{\partial t}\right) = 0, \qquad i = 1, \ldots, N. \tag{1.1.1}$$

Here

$$\mathbf{u} = [u_1, \ldots, u_n]^{\mathrm{T}}, \quad \mathbf{x} = [x_1, \ldots, x_m]^{\mathrm{T}}, \quad \frac{\partial \mathbf{u}}{\partial x_j} = \left[\frac{\partial u_1}{\partial x_j}, \ldots, \frac{\partial u_n}{\partial x_j}\right]^{\mathrm{T}}$$

are vector-columns, $j = 1, \ldots, m$.

This system is called determined if $N = n$. Later we shall consider only determined systems.

The system of (1.1.1) is called a system of quasilinear equations if the functions F_i are linear with respect to the derivatives of $\mathbf{u}$ occurring in (1.1.1) as arguments. If F_i are also linear with respect to $\mathbf{u}$, the system is called linear.

The system of the first-order quasilinear equations can be represented in the form

$$\tilde{A}\frac{\partial \mathbf{u}}{\partial t} + \sum_{j=1}^{m} \tilde{B}_j \frac{\partial \mathbf{u}}{\partial x_j} = \tilde{\mathbf{c}}, \tag{1.1.2}$$

where the coefficient matrices $\tilde{A}$ and $\tilde{B}_j$ and the source term vector $\tilde{\mathbf{c}}$ depend on t, $\mathbf{x}$, and $\mathbf{u}$.

The vector-row $\mathbf{l} = [l_1, \ldots, l_n]$ and the number λ are called a left eigenvector and an eigenvalue of a matrix A, respectively, if

$$\mathbf{l}A = \lambda \mathbf{l}, \quad ||\mathbf{l}|| \neq 0. \tag{1.1.3}$$

As a norm of a vector $\mathbf{a}$ we adopt the quantity $||\mathbf{a}|| = \sqrt{\mathbf{a} \cdot \mathbf{a}} = \sqrt{\sum_{k=1}^{n} a_k^2}$. The dot sign here stands for the scalar product of two vectors.

Similarly, the vector-column $\mathbf{r} = [r_1, \ldots, r_n]^{\mathrm{T}}$ is called a right eigenvector of a matrix A if

$$A\mathbf{r} = \lambda \mathbf{r}, \quad ||\mathbf{r}|| \neq 0. \tag{1.1.4}$$

Owing to Eqs. (1.1.3)–(1.1.4), the eigenvalue λ of the matrix A is the root of the characteristic equation

$$\det(A - \lambda I) = 0, \tag{1.1.5}$$

where $I = \mathrm{diag}[1, \ldots, 1]$ is the $n \times n$ identity matrix.

Suppose all the eigenvalues λ of the matrix A are real. Let us enumerate them in increasing order, that is,

$$\lambda_1 \leq \ldots \leq \lambda_k \leq \ldots \leq \lambda_n. \tag{1.1.6}$$

The equality signs in Eq. (1.1.6) correspond to the case of the eigenvalue multiplicity.

If for any eigenvalue λ of multiplicity α the rank of the matrix $A - \lambda I$ is equal to $n - \alpha$, then both the right and left eigenvectors corresponding to all eigenvalues form a basis in the Euclidean space $\mathbf{E}^n(\mathbf{u})$.

It is easy to check that if $\lambda_k \neq \lambda_p$, the vectors $\mathbf{l}^k$ and $\mathbf{r}^p$ are mutually orthogonal. Indeed,

$$\begin{aligned} \mathbf{l}^k A = \lambda_k \mathbf{l}^k \quad &\Longrightarrow \quad \mathbf{l}^k A \cdot \mathbf{r}^p = \lambda_k \mathbf{l}^k \cdot \mathbf{r}^p, \\ A\mathbf{r}^p = \lambda_p \mathbf{r}^p \quad &\Longrightarrow \quad \mathbf{l}^k \cdot A\mathbf{r}^p = \lambda_p \mathbf{l}^k \cdot \mathbf{r}^p. \end{aligned}$$

Hence, $(\lambda_k - \lambda_p)\mathbf{l}^k \cdot \mathbf{r}^p = 0$.

Unfortunately, in a variety of mechanical applications of quasilinear systems (Euler gas dynamic equations, magnetohydrodynamic equations (MHD), solid dynamics equations, etc.) the multiplicity of eigenvalues can be greater than one and the choice of the set of independent nondegenerate eigenvectors requires additional analysis.

A matrix A is called nonsingular if $\lambda = 0$ is not its eigenvalue. Since in the applications to be studied in what follows the matrix $\tilde{A}$ in Eq. (1.1.2) is nonsingular, we can resolve the system for $\partial \mathbf{u}/\partial t$

$$\frac{\partial \mathbf{u}}{\partial t} + \sum_{j=1}^{m} A_j \frac{\partial \mathbf{u}}{\partial x_j} = \mathbf{b}, \tag{1.1.7}$$

where $A_j = \tilde{A}^{-1}\tilde{B}_j$ and $\mathbf{b} = \tilde{A}^{-1}\tilde{\mathbf{c}}$.

1.2 Hyperbolic systems of quasilinear differential equations

1.2.1 Definitions. Let us define a matrix P associated with Eq. (1.1.7) by the formula

$$P = P(\boldsymbol{\alpha}) = \sum_{j=1}^{m} \alpha_j A_j, \quad -\infty < \alpha_j < \infty. \tag{1.2.1}$$

The quasilinear system of (1.1.7) is called hyperbolic at the point $(\mathbf{x}, t, \mathbf{u})$ if there exists a nonsingular matrix $\Omega(\boldsymbol{\alpha})$ diagonalizing P,

$$\Omega^{-1} P \Omega = \Lambda = \mathrm{diag}[\lambda_1, \ldots, \lambda_n], \tag{1.2.2}$$

where all eigenvalues λ_k of the matrix P are real and the norms of Ω and Ω^{-1} are uniformly bounded in $\boldsymbol{\alpha} = [\alpha_1, \ldots, \alpha_m]^{\mathrm{T}}$. If all the eigenvalues are distinct, the system is called strictly hyperbolic. The matrix Λ is here a diagonal matrix whose entries are the eigenvalues of the matrix P.

The hyperbolicity condition implies that there exists an independent basis $\{\mathbf{l}^1, \ldots, \mathbf{l}^n\}$ composed of the left eigenvectors of the matrix P. It is clear that there also exists a basis composed of the right eigenvectors. Note that the following relations hold:

$$\Omega^{-1} P = \Lambda \Omega^{-1}, \quad P\Omega = \Omega\Lambda.$$

Thus, the kth row of Ω^{-1} consists of the components of the kth left eigenvector $\mathbf{l}^k$ corresponding to the eigenvalue λ_k, see Eq. (1.1.3). In addition, the kth column of Ω consists of the corresponding components of the kth right eigenvector $\mathbf{r}^k$, see Eq. (1.1.4). In what follows we also use the notation $\Omega_L = \Omega^{-1}$ and $\Omega_R = \Omega$, where $\Omega_L \Omega_R = I$.

Thus, relations (1.2.2) can be rewritten as

$$\Omega_L P \Omega_R = \Lambda, \quad P = \Omega_R \Lambda \Omega_L.$$

As a norm of a matrix A we adopt the square root of the spectral radius of the matrix $A A^T$, where the spectral radius is equal to the largest eigenvalue of this matrix.

Since numerical solutions of the systems of hyperbolic quasilinear equations occurring in mechanical applications are frequently constructed on the basis of the system with two independent variables, we shall consider the essence of hyperbolicity for the system

$$\frac{\partial \mathbf{u}}{\partial t} + A \frac{\partial \mathbf{u}}{\partial x} = \mathbf{b}. \tag{1.2.3}$$

Multiplying this system by the left eigenvector $\mathbf{l}^k$, we can transform it into the form

$$\mathbf{l}^k \cdot \left(\frac{\partial \mathbf{u}}{\partial t} + \lambda_k \frac{\partial \mathbf{u}}{\partial x} \right) = f_k, \qquad k = 1, \ldots, n; \tag{1.2.4}$$

where $f_k = \mathbf{l}^k \cdot \mathbf{b}$.

The system of (1.2.4) can be rewritten as

$$\mathbf{l}^k \cdot \left(\frac{d\mathbf{u}}{dt} \right)_k = f_k, \qquad k = 1, \ldots, n; \tag{1.2.5}$$

where $(d\mathbf{u}/dt)_k$ is the derivative of $\mathbf{u}$ with respect to t in the direction $dx/dt = \lambda_k$. This direction is called a characteristic direction and Eq. (1.2.5) is called a characteristic form of Eq. (1.2.3). If the system of (1.2.3) is linear and its coefficient matrix is constant, then the eigenvalues λ_k, which are also called characteristic velocities, are constant and the characteristic lines in the x–t plane become straight lines,

$$x = \lambda_k t + \text{const}. \tag{1.2.6}$$

In certain cases the system of (1.2.4), or (1.2.5), can be further simplified. If A does not depend on $\mathbf{u}$, Eq. (1.2.4) can be rewritten for Riemann invariants (Riemann 1860) w_k as

$$\left(\frac{dw_k}{dt} \right)_k = g_k, \quad k = 1, \ldots, n; \tag{1.2.7}$$

where

$$w_k = \mathbf{l}^k \cdot \mathbf{u}, \quad g_k = f_k + \sum_{j=1}^{m} u_j \left(\frac{dl_j^k}{dt} \right)_k = f_k + \sum_{j=1}^{m} u_j \left[\frac{\partial l_j^k}{\partial t} + \lambda_k \frac{\partial l_j^k}{\partial x} \right].$$

Choosing w_k as components of a new unknown vector, we can rewrite the system in the form such that each equation involves the derivatives of only one function of x, t, and $\mathbf{u}$, that is,

$$\frac{\partial w_k}{\partial t} + \lambda_k \frac{\partial w_k}{\partial x} = g_k(x, t, \mathbf{w}), \qquad k = 1, \ldots, n. \tag{1.2.8}$$

Although a quasilinear hyperbolic system generally cannot be written out for the Riemann invariants, these invariants play an important role in the construction of numerical solutions to these systems.

Note that the system of hyperbolic equations with two independent variables can be extended in a peculiar way. If we introduce the notation

$$\frac{\partial \mathbf{u}}{\partial x} = \mathbf{p}, \quad \frac{\partial \mathbf{u}}{\partial t} = \mathbf{q}, \tag{1.2.9}$$

then the system of (1.2.4) acquires the form

$$\mathbf{l}^k \cdot (\mathbf{q} + \lambda_k \mathbf{p}) = f_k, \qquad k = 1, \ldots, n. \tag{1.2.10}$$

By differentiating each equation of (1.2.10) with respect to t and x and taking into account the integrability condition $\partial \mathbf{q}/\partial x = \partial \mathbf{p}/\partial t$, we obtain

$$\mathbf{l}^k \cdot \left(\frac{\partial \mathbf{q}}{\partial t} + \lambda_k \frac{\partial \mathbf{q}}{\partial x} \right) = E_k, \quad \mathbf{l}^k \cdot \left(\frac{\partial \mathbf{p}}{\partial t} + \lambda_k \frac{\partial \mathbf{p}}{\partial x} \right) = G_k, \tag{1.2.11}$$

where E_k and G_k are the function x, t, $\mathbf{u}$, $\mathbf{p}$, and $\mathbf{q}$.

Thus, since $\mathbf{l}^k$ do not depend on $\mathbf{q}$ and $\mathbf{p}$, the extended system (1.2.11) can always be written out for the Riemann invariants (Courant and Lax 1949).

1.2.2 Systems of conservation laws.

Let us consider a system

$$\frac{\partial \mathbf{U}(\mathbf{x}, t, \mathbf{u})}{\partial t} + \sum_{j=1}^{m} \frac{\partial \mathbf{F}_j(\mathbf{x}, t, \mathbf{u})}{\partial x_j} = \mathbf{c}(\mathbf{x}, t, \mathbf{u}) \tag{1.2.12}$$

that is the consequence of the quasilinear system (1.1.7) for any of its solutions. The systems describing a number of mechanical problems (gas dynamics, MHD, shallow water equations, thermoelasticity and elastoviscoplasticity equations, etc.) can be written in the form that reflects the conservation of such fundamental physical properties as mass, momentum, energy, etc. and expressed by Eq. (1.2.12). If the number of such fundamental conservation laws is equal to the number of equations in (1.1.7), we say that this system is written in the conservation-law, or conservative, form. Most of hyperbolic systems are solved numerically on the basis of their conservative rather than quasilinear form. Moreover, in a Cartesian coordinate system they acquire the simplest form

$$\frac{\partial \mathbf{U}(\mathbf{u})}{\partial t} + \sum_{j=1}^{m} \frac{\partial \mathbf{F}_j(\mathbf{u})}{\partial x_j} = \mathbf{c}(\mathbf{x}, t, \mathbf{u}). \tag{1.2.13}$$

In this case $\mathbf{U}$ and $\mathbf{F}$ do not depend on the independent variables t and $\mathbf{x}$.

If $\mathbf{c} \equiv \mathbf{0}$, the system is said to be written in a strictly conservative, or divergent, form. The source term $\mathbf{c}$ can be both of physical (volume production of mass, momentum, and energy) and geometrical origin. This issue will be discussed in subsequent chapters. We shall assume $\mathbf{c} \equiv \mathbf{0}$ in the general discussion of hyperbolic systems, although the source term may appear in applications.

If the system is written in the conservative form, the quantities $\mathbf{U}$ and $\mathbf{F}_j$ are called a vector of conservative variables and a flux vector, respectively. The sense of these notions is readily understood if we consider a bounded region $V \subset \mathbf{E}^m(\mathbf{x})$ and let $\mathbf{n} = [n_1, \ldots, n_m]$ be the outward unit normal to the boundary ∂V of V. Then it follows from (1.2.13) that

$$\frac{d}{dt} \int_V \mathbf{U}\, dV + \sum_{j=1}^{m} \int_{\partial V} n_j \mathbf{F}_j(\mathbf{U})\, dS = \mathbf{0}, \tag{1.2.14}$$

that is, the time variation of the quantity $\mathbf{U}$ in the volume V is equal to its losses through the boundary.

1.3 Mechanical examples

Let us consider several examples of hyperbolic systems of partial differential equations that are frequently encountered in mechanical applications. The formulas to be presented below are useful for the implementation of modern high-resolution numerical methods.

1.3.1 Nonstationary equations of gas dynamics.

The system of nonstationary Euler equations for primitive variables in Cartesian coordinates can be written as follows:

$$\frac{\partial \mathbf{u}}{\partial t} + A_i \frac{\partial \mathbf{u}}{\partial x_i} = \mathbf{0}, \qquad \mathbf{u} = [\rho, u, v, w, p]^{\mathrm{T}}. \tag{1.3.1}$$

Here and further on in this section $i, j = 1, 2, 3$ and summation is adopted over repeated indices. If we introduce the Kronecker delta δ_{ij}, then the coefficient matrices can be written out as

$$A_i = \begin{bmatrix} v_i & \delta_{i1}\rho & \delta_{i2}\rho & \delta_{i3}\rho & 0 \\ 0 & v_i & 0 & 0 & \delta_{i1}/\rho \\ 0 & 0 & v_i & 0 & \delta_{i2}/\rho \\ 0 & 0 & 0 & v_i & \delta_{i3}/\rho \\ 0 & \delta_{i1}\rho c^2 & \delta_{i2}\rho c^2 & \delta_{i3}\rho c^2 & v_i \end{bmatrix}. \tag{1.3.2}$$

The entries of the coefficient matrices are constituted by the density ρ, the pressure p, the components $v_1 = u$, $v_2 = v$, $v_3 = w$ of the velocity vector $\mathbf{v}$, and the speed of sound $c = c(p, \rho)$.

It follows from the definition of hyperbolicity that the system remains hyperbolic if we perform an arbitrary differentiable self-invertible transformation of independent variables $\xi = \xi(x, y, z)$, $\eta = \eta(x, y, z)$, $\zeta = \zeta(x, y, z)$. In this case the system of (1.1.7) acquires the form

$$\frac{\partial \mathbf{u}}{\partial t} + A \frac{\partial \mathbf{u}}{\partial \xi} + B \frac{\partial \mathbf{u}}{\partial \eta} + C \frac{\partial \mathbf{u}}{\partial \zeta} = \mathbf{0}, \tag{1.3.3}$$

where $A = \xi_{x_i} A_i$, $B = \eta_{x_i} A_i$, and $C = \zeta_{x_i} A_i$. Here $\xi_{x_i} = \partial \xi / \partial x_i$, $\eta_{x_i} = \partial \eta / \partial x_i$, and $\zeta_{x_i} = \partial \zeta / \partial x_i$.

Let us introduce the vectors

$$\boldsymbol{\alpha} = [\xi_x, \xi_y, \xi_z], \quad \boldsymbol{\beta} = [\eta_x, \eta_y, \eta_z], \quad \boldsymbol{\gamma} = [\zeta_x, \zeta_y, \zeta_z]. \tag{1.3.4}$$

If the system of (1.3.3) is t-hyperbolic at the point $(\mathbf{x}, t, \mathbf{u})$, there must exist nonsingular matrices $\Omega_R^A(\boldsymbol{\alpha})$, $\Omega_R^B(\boldsymbol{\beta})$, and $\Omega_R^C(\boldsymbol{\gamma})$ such that

$$\Omega_L^A A \Omega_R^A = \mathrm{diag}[\lambda_k^A], \quad \Omega_L^B B \Omega_R^B = \mathrm{diag}[\lambda_k^B], \quad \Omega_L^C C \Omega_R^C = \mathrm{diag}[\lambda_k^C], \tag{1.3.5}$$

where λ_k^A, λ_k^B, and λ_k^C are all real, $k = 1, \ldots, 5$. The norms of $\Omega_{L,R}^A$, $\Omega_{L,R}^B$, and $\Omega_{L,R}^C$ must be uniformly bounded in $\boldsymbol{\alpha}$, $\boldsymbol{\beta}$, and $\boldsymbol{\gamma}$, respectively.

Without loss of generality we can consider only the diagonalization of the matrix

$$A = \begin{bmatrix} U & \alpha_1\rho & \alpha_2\rho & \alpha_3\rho & 0 \\ 0 & U & 0 & 0 & \alpha_1/\rho \\ 0 & 0 & U & 0 & \alpha_2/\rho \\ 0 & 0 & 0 & U & \alpha_3/\rho \\ 0 & \alpha_1\rho c^2 & \alpha_2\rho c^2 & \alpha_3\rho c^2 & U \end{bmatrix}. \tag{1.3.6}$$

In this formula $U = \alpha_i v_i$ is the contravariant component of the vector $\mathbf{v}$ along the curvilinear coordinate ξ.

The eigenvalues of the matrix A can be easily determined as

$$\lambda_1 = \lambda_2 = \lambda_3 = U, \quad \lambda_{4,5} = U \pm c\,(\alpha_i \times \alpha_i)^{1/2}. \tag{1.3.7}$$

We see that they are obviously real.

Although the matrix A has the eigenvalue of multiplicity three, it has a complete set of linearly independent eigenvectors. They constitute a matrix

$$\Omega_R^A = \begin{bmatrix} \tilde\alpha_1 & \tilde\alpha_2 & \tilde\alpha_3 & \rho/(c\sqrt{2}) & \rho/(c\sqrt{2}) \\ 0 & -\tilde\alpha_3 & \tilde\alpha_2 & \tilde\alpha_1/\sqrt{2} & -\tilde\alpha_1/\sqrt{2} \\ \tilde\alpha_3 & 0 & -\tilde\alpha_1 & \tilde\alpha_2/\sqrt{2} & -\tilde\alpha_2/\sqrt{2} \\ -\tilde\alpha_2 & \tilde\alpha_1 & 0 & \tilde\alpha_3/\sqrt{2} & -\tilde\alpha_3/\sqrt{2} \\ 0 & 0 & 0 & \rho c/\sqrt{2} & \rho c/\sqrt{2} \end{bmatrix}, \tag{1.3.8}$$

where $\tilde\alpha_j = \alpha_j/(\alpha_i \times \alpha_i)^{1/2}$. The kth column of Ω_R^A is the right eigenvector corresponding to the eigenvalue λ_k. The inverse of Ω_R^A can be constructed from the left eigenvectors of A as

$$\Omega_L^A = \begin{bmatrix} \tilde\alpha_1 & 0 & \tilde\alpha_3 & -\tilde\alpha_2 & -\tilde\alpha_1/c^2 \\ \tilde\alpha_2 & -\tilde\alpha_3 & 0 & \tilde\alpha_1 & -\tilde\alpha_2/c^2 \\ \tilde\alpha_3 & \tilde\alpha_2 & -\tilde\alpha_1 & 0 & -\tilde\alpha_3/c^2 \\ 0 & \tilde\alpha_1/\sqrt{2} & \tilde\alpha_2/\sqrt{2} & \tilde\alpha_3/\sqrt{2} & (\rho c\sqrt{2})^{-1} \\ 0 & -\tilde\alpha_1/\sqrt{2} & -\tilde\alpha_2/\sqrt{2} & -\tilde\alpha_3/\sqrt{2} & (\rho c\sqrt{2})^{-1} \end{bmatrix}. \tag{1.3.9}$$

The determinants of Ω_R^A and Ω_L^A are

$$\det \Omega_R^A = (\det \Omega_L^A)^{-1} = \rho c.$$

The spectral norm of the matrix Ω_R^A can be calculated from the formula

$$\|\Omega_R^A\| = \sqrt{r\left(\Omega_R^A\,(\Omega_R^A)^{\mathrm{T}}\right)},$$

where r is the spectral radius of a corresponding matrix. The eigenvalues σ_k of $\Omega_{\mathrm{R}}^A(\Omega_{\mathrm{R}}^A)^{\mathrm{T}}$ are

$$\sigma_1 = \sigma_2 = \sigma_3 = 1, \quad \sigma_{4,5} = \frac{\phi \pm \sqrt{\phi^2 - 4\rho^2 c^6}}{2c^2},$$

where $\phi = \rho^2 + c^2 + \rho^2 c^4$, and consequently

$$\|\Omega_{\mathrm{R}}^A\|^2 = \max \sigma_k = \frac{\phi + \sqrt{\phi^2 - 4\rho^2 c^6}}{2c^2} > 1.$$

Since the norm of Ω_{R}^A and, hence, the norm of Ω_{L}^A are independent of the real parameters α_i, they are uniformly bounded in $\boldsymbol{\alpha}$. This is required for hyperbolicity of the considered system.

As shown by Warming, Beam, and Hyett (1975), the similarity transformation based on the matrix Ω_{R}^A not only diagonalizes A but also symmetrizes the matrices A_i and, hence, the matrices B and C. Note that such a presentation cannot be obtained for an arbitrary linear combination of noncommuting matrices.

Suppose we are solving the system in the conservation-law form

$$\frac{\partial \mathbf{U}}{\partial t} + \frac{\partial \mathbf{F}_j}{\partial x_j} = \mathbf{0}, \tag{1.3.10}$$

where

$$\mathbf{U} = \begin{bmatrix} \rho \\ \rho v_1 \\ \rho v_2 \\ \rho v_3 \\ e \end{bmatrix}, \quad \mathbf{F}_j = \begin{bmatrix} \rho v_j \\ \rho v_1 v_j + p\delta_{1j} \\ \rho v_2 v_j + p\delta_{2j} \\ \rho v_3 v_j + p\delta_{3j} \\ (e+p)v_j \end{bmatrix},$$

$$e = \frac{p}{\gamma - 1} + \frac{\rho \mathbf{v}^2}{2}, \quad \mathbf{v}^2 = v_1^2 + v_2^2 + v_3^2.$$

Here e is the total gas energy per unit volume and γ is the specific heat ratio, or adiabatic index.

Equation (1.3.1) can be expressed in the form

$$\frac{\partial \mathbf{U}}{\partial t} + \tilde{A}_j \frac{\partial \mathbf{U}}{\partial x_j} = \mathbf{0}, \tag{1.3.11}$$

where $\tilde{A}_j$ are the Jacobian matrices $\partial \mathbf{F}_j / \partial \mathbf{U}$.

The matrices A_j of the nonconservative form (1.3.1) and the matrices $\tilde{A}_j$ of the conservative form (1.3.11) are related via the similarity transformation $A_j = M^{-1}\tilde{A}_j M$, where M is the Jacobian matrix $\partial \mathbf{U}/\partial \mathbf{u}$. It is easy to find that

$$M = \begin{bmatrix} 1 & 0 & 0 & 0 & 0 \\ v_1 & \rho & 0 & 0 & 0 \\ v_2 & 0 & \rho & 0 & 0 \\ v_3 & 0 & 0 & \rho & 0 \\ \frac{1}{2}\mathbf{v}\cdot\mathbf{v} & \rho v_1 & \rho v_2 & \rho v_3 & \beta^{-1} \end{bmatrix}, \quad M^{-1} = \begin{bmatrix} 1 & 0 & 0 & 0 & 0 \\ -v_1/\rho & 1/\rho & 0 & 0 & 0 \\ -v_2/\rho & 0 & 1/\rho & 0 & 0 \\ -v_3/\rho & 0 & 0 & 1/\rho & 0 \\ \frac{1}{2}\beta(\mathbf{v}\cdot\mathbf{v}) & -\beta v_1 & -\beta v_2 & -\beta v_3 & \beta \end{bmatrix}, \tag{1.3.12}$$

where $\beta = \gamma - 1$.

Thus, the eigenvectors $\tilde{\mathbf{r}}^k$ and $\tilde{\mathbf{l}}^k$ of the matrix $\tilde{A}$ can be calculated by the formulas

$$\tilde{\mathbf{r}}^k = M\mathbf{r}^k, \quad \tilde{\mathbf{l}}^k = \mathbf{l}^k M^{-1}.$$

1.3.2 Stationary Euler equations. Let now the system of gas dynamic equations (1.3.3) be stationary and assume that the matrix A is nonsingular. Then we obtain the system

$$\frac{\partial \mathbf{u}}{\partial \xi} + \mathcal{B}\frac{\partial \mathbf{u}}{\partial \eta} + \mathcal{C}\frac{\partial \mathbf{u}}{\partial \zeta} = \mathbf{0}, \tag{1.3.13}$$

where $\mathcal{B} = A^{-1}(\eta_{x_i} A_i)$ and $\mathcal{C} = A^{-1}(\zeta_{x_i} A_i)$.

If this system is ξ-hyperbolic at the point $(\mathbf{x}, \mathbf{u})$, there must exist nonsingular matrices $\Omega_{\mathrm{R}}^{\mathcal{B}}(\boldsymbol{\alpha}, \boldsymbol{\beta})$ and $\Omega_{\mathrm{R}}^{\mathcal{C}}(\boldsymbol{\alpha}, \boldsymbol{\gamma})$ such that

$$\Omega_{\mathrm{R}}^{\mathcal{B}}\,\mathcal{B}\,\Omega_{\mathrm{L}}^{\mathcal{B}} = \mathrm{diag}[\lambda_k^{\mathcal{B}}], \quad \Omega_{\mathrm{R}}^{\mathcal{C}}\,\mathcal{C}\,\Omega_{\mathrm{L}}^{\mathcal{C}} = \mathrm{diag}[\lambda_k^{\mathcal{C}}]$$

and $\lambda_k^{\mathcal{B}}$ and $\lambda_k^{\mathcal{C}}$ are all real, $k = 1, \ldots, 5$. The norms of the matrices $\Omega_{\mathrm{R}}^{\mathcal{B}}$ and $\Omega_{\mathrm{R}}^{\mathcal{C}}$ must be uniformly bounded in $\mathbf{x}$, $\boldsymbol{\alpha}$, $\boldsymbol{\beta}$, and $\boldsymbol{\gamma}$.

Without loss of generality we can consider (Pogorelov 1987) only the diagonalization of the matrix

$$\mathcal{B} = \frac{1}{Q^2 U}\begin{bmatrix} VQ^2 & -\rho d_1 U & -\rho d_2 U & -\rho d_3 U & \alpha_1 d_1 + \alpha_2 d_2 + \alpha_3 d_3 \\ 0 & Q^2 V + \alpha_1 c^2 d_1 & \alpha_1 c^2 d_2 & \alpha_1 c^2 d_3 & (-U d_1 + r_1 c^2)/\rho \\ 0 & \alpha_2 c^2 d_1 & Q^2 V + \alpha_2 c^2 d_2 & \alpha_2 c^2 d_3 & (-U d_2 + r_2 c^2)/\rho \\ 0 & \alpha_3 c^2 d_1 & \alpha_3 c^2 d_2 & Q^2 V + \alpha_3 c^2 d_3 & (-U d_3 + r_3 c^2)/\rho \\ 0 & -\rho c^2 d_1 U & -\rho c^2 d_2 U & -\rho c^2 d_3 U & (UV - qc^2)U \end{bmatrix}.$$

Here $U = \alpha_i v_i$ and $V = \beta_i v_i$ are the contravariant components of the vector $\mathbf{v}$ along the curvilinear coordinates ξ and η, respectively. Besides,

$$\begin{aligned} &Q^2 = U^2 - (sc)^2, \quad s^2 = \alpha_1^2 + \alpha_2^2 + \alpha_3^2, \\ &q = \alpha_1\beta_1 + \alpha_2\beta_2 + \alpha_3\beta_3, \quad d_i = \alpha_i V - \beta_i U, \\ &r_1 = (\alpha_1\beta_2 - \alpha_2\beta_1)\alpha_2 + (\alpha_1\beta_3 - \alpha_3\beta_1)\alpha_3, \\ &r_2 = (\alpha_2\beta_3 - \alpha_3\beta_2)\alpha_3 + (\alpha_2\beta_1 - \alpha_1\beta_2)\alpha_1, \\ &r_3 = (\alpha_3\beta_1 - \alpha_1\beta_3)\alpha_1 + (\alpha_3\beta_2 - \alpha_2\beta_3)\alpha_2. \end{aligned}$$

The eigenvalues of the matrix $\mathcal{B}$ are the following:

$$\lambda_1^{\mathcal{B}} = \lambda_2^{\mathcal{B}} = \lambda_3^{\mathcal{B}} = \frac{V}{U}, \quad \lambda_{4,5}^{\mathcal{B}} = \frac{UV - qc^2 \mp cT}{Q^2}, \tag{1.3.14}$$

where

$$T = \left[D^2 - (s^2\tilde{s}^2 - q^2)c^2\right]^{1/2}, \quad \tilde{s}^2 = \beta_1^2 + \beta_2^2 + \beta_3^2, \quad D^2 = d_1^2 + d_2^2 + d_3^2.$$

It is easy to check that they are real for $U^2 > (sc)^2$.

The matrices $\Omega_R^{\mathcal{B}}$ and $\Omega_L^{\mathcal{B}}$ are constructed using the right and the left eigenvectors of the matrix $\mathcal{B}$ as follows:

$$\Omega_R^{\mathcal{B}} = \begin{bmatrix} -d_3/D & d_2/D & d_1/D & 1/(2c^2\delta T) & -1/(2c^2\varepsilon T) \\ -d_2/D & -d_3/D & 0 & (d_1T + cb_1)/(2\delta\rho c D^2 T) & (d_1T - cb_1)/(2\varepsilon\rho c D^2 T) \\ d_1/D & 0 & d_3/D & (d_2T + cb_2)/(2\delta\rho c D^2 T) & (d_2T - cb_2)/(2\varepsilon\rho c D^2 T) \\ 0 & d_1/D & -d_2/D & (d_3T + cb_3)/(2\delta\rho c D^2 T) & (d_3T - cb_3)/(2\varepsilon\rho c D^2 T) \\ 0 & 0 & 0 & 1/(2\delta T) & -1/(2\varepsilon T) \end{bmatrix}$$

and

$$\Omega_L^{\mathcal{B}} = \begin{bmatrix} -d_3/D & -d_2/D & d_1/D & 0 & [(\alpha_1\beta_2 - \alpha_2\beta_1)c^2 + \rho d_3]/(\rho c^2 D) \\ d_2/D & -d_3/D & 0 & d_1/D & [(\alpha_1\beta_3 - \alpha_3\beta_1)c^2 - \rho d_2]/(\rho c^2 D) \\ d_1/D & 0 & d_3/D & -d_2/D & [(\alpha_3\beta_2 - \alpha_2\beta_3)c^2 - \rho d_1]/(\rho c^2 D) \\ 0 & \delta\rho c d_1 & \delta\rho c d_2 & \delta\rho c d_3 & \delta T \\ 0 & \varepsilon\rho c d_1 & \varepsilon\rho c d_2 & \varepsilon\rho c d_3 & -\varepsilon T \end{bmatrix},$$

where

$$b_1 = (\alpha_1\beta_2 - \alpha_2\beta_1)d_2 + (\alpha_1\beta_3 - \alpha_3\beta_1)d_3,$$
$$b_2 = (\alpha_2\beta_3 - \alpha_3\beta_2)d_3 + (\alpha_2\beta_1 - \alpha_1\beta_2)d_1,$$
$$b_3 = (\alpha_3\beta_1 - \alpha_1\beta_3)d_1 + (\alpha_3\beta_2 - \alpha_2\beta_3)d_2.$$

The quantities δ and ε in these formulas are arbitrary nonzero real constants.

After proper normalization of the columns and rows of $\Omega_R^{\mathcal{B}}$ and $\Omega_L^{\mathcal{B}}$, respectively, these matrices will not only diagonalize the matrix $\mathcal{B}$ but also symmetrize all the matrices X_i of the linear combination forming $\mathcal{B}$ and $\mathcal{C}$

$$X_i = A^{-1}A_i, \quad \mathcal{B} = \eta_{x_i}X_i, \quad \mathcal{C} = \zeta_{x_i}X_i.$$

For this purpose we must choose

$$\delta^2 = \frac{cf + UT}{2\rho^2c^2D^2UT}, \quad \varepsilon^2 = \frac{UT - cf}{2\rho^2c^2D^2UT}, \quad f = s^2V - qU.$$

Note that from $U^2 > (sc)^2$ it follows that $|U|T > c|f|$.

In this case we obtain

$$\Omega_L^{\mathcal{B}} X_i \Omega_R^{\mathcal{B}} = \begin{bmatrix} \mathcal{D}_i & \mathcal{B}_i \\ \mathcal{B}_i^{\mathrm{T}} & \mathcal{C}_i \end{bmatrix},$$

where $\mathcal{D}_i$ and $\mathcal{C}_i$ are the diagonal and the symmetric matrix, respectively equal to

$$\mathcal{D}_i = \begin{bmatrix} v_i/U & 0 & 0 \\ 0 & v_i/U & 0 \\ 0 & 0 & v_i/U \end{bmatrix},$$

$$\mathcal{C}_i = \begin{bmatrix} \dfrac{cd_iQ^2 + (v_iU - \alpha_ic^2)(UT - cf)}{UQ^2T} & c_i \\ c_i & \dfrac{-cd_iQ^2 + (v_iU - \alpha_ic^2)(UT + cf)}{UQ^2T} \end{bmatrix},$$

$$c_1 = \frac{(\alpha_3\beta_2 - \alpha_2\beta_3)\tilde{c}}{QDT}, \quad c_2 = \frac{(\alpha_1\beta_3 - \alpha_3\beta_1)\tilde{c}}{QDT}, \quad c_3 = \frac{(\alpha_2\beta_1 - \alpha_1\beta_2)\tilde{c}}{QDT},$$

$$\tilde{c} = [(\alpha_2\beta_3 - \alpha_3\beta_2)u + (\alpha_3\beta_1 - \alpha_1\beta_3)v + (\alpha_1\beta_2 - \alpha_2\beta_1)w]\, c^2,$$

$$\mathcal{B}_1 = \frac{\alpha_3\beta_2 - \alpha_2\beta_3}{2\rho UDT}\mathcal{B}_c, \quad \mathcal{B}_2 = \frac{\alpha_1\beta_3 - \alpha_3\beta_1}{2\rho UDT}\mathcal{B}_c, \quad \mathcal{B}_3 = \frac{\alpha_2\beta_1 - \alpha_1\beta_2}{2\rho UDT}\mathcal{B}_c,$$

$$\mathcal{B}_c = \begin{bmatrix} w/\delta & -w/\varepsilon \\ -v/\delta & v/\varepsilon \\ -u/\delta & u/\varepsilon \end{bmatrix}.$$

Besides, such normalization results in a very sparse form of the matrix

$$(\Omega_{\mathrm{L}}^{\mathcal{B}})^{\mathrm{T}}\,\Omega_{\mathrm{L}}^{\mathcal{B}} = \begin{bmatrix} 1 & 0 & 0 & 0 & -c^{-2} \\ 0 & 1 & 0 & 0 & \alpha_1(\rho U)^{-1} \\ 0 & 0 & 1 & 0 & \alpha_2(\rho U)^{-1} \\ 0 & 0 & 0 & 1 & \alpha_3(\rho U)^{-1} \\ -c^{-2} & \alpha_1(\rho U)^{-1} & \alpha_2(\rho U)^{-1} & \alpha_3(\rho U)^{-1} & (c^2 + \rho^2)\rho^{-2}c^{-4} \end{bmatrix},$$

whose eigenvalues σ_k are

$$\sigma_1 = \sigma_2 = \sigma_3 = 1, \quad \sigma_{4,5} = \frac{U\varphi \pm \sqrt{(U\varphi)^2 - 4[U^2 - (sc)^2]\rho^2c^6}}{2\rho^2c^4U},$$

where $\varphi = \rho^2c^4 + c^2 + \rho^2$.

The determinants of the matrices $\Omega_{\mathrm{R}}^{\mathcal{B}}$ and $\Omega_{\mathrm{L}}^{\mathcal{B}}$ are the following:

$$\det \Omega_{\mathrm{R}}^{\mathcal{B}} = (\det \Omega_{\mathrm{L}}^{\mathcal{B}})^{-1} = -\frac{\rho c U}{Q}.$$

The spectral norm of $\Omega_{\mathrm{L}}^{\mathcal{B}}$ is equal to $\sqrt{r}$, where $r = \max|\sigma_k|$. It is easy to see that for $U^2 > (sc)^2$ it is uniformly bounded in $\mathbf{x}$, $\boldsymbol{\alpha}$, and $\boldsymbol{\beta}$. The same is true for $\Omega_{\mathrm{R}}^{\mathcal{B}}$. Thus, the system of (1.3.13) is hyperbolic if the flow is supersonic with respect to ξ.

If we pass to the conservative variables $\mathbf{F}$, the system of (1.3.13) acquires the form

$$\frac{\partial \mathbf{F}}{\partial \xi} + \tilde{\mathcal{B}}\frac{\partial \mathbf{F}}{\partial \eta} + \tilde{\mathcal{C}}\frac{\partial \mathbf{F}}{\partial \zeta} = \mathbf{0}, \tag{1.3.15}$$

where

$$\tilde{A}_i = MA_iM^{-1}, \quad \tilde{A} = \xi_{x_i}\tilde{A}_i, \quad \tilde{\mathcal{B}} = \tilde{A}^{-1}(\eta_{x_i}\tilde{A}_i), \quad \tilde{\mathcal{C}} = \tilde{A}^{-1}(\zeta_{x_i}\tilde{A}_i)$$

and M is the Jacobian matrix $\partial\mathbf{F}/\partial\mathbf{U}$.

It is clear that $\tilde{\mathcal{B}}$ and $\tilde{\mathcal{C}}$ are diagonalized with the use of $\tilde{\Omega}_{\mathrm{R}}^{\mathcal{B}}$ and $\tilde{\Omega}_{\mathrm{R}}^{\mathcal{C}}$ such that

$$\tilde{\Omega}_{\mathrm{L}}^{\mathcal{B}}\,\tilde{\mathcal{B}}\,\tilde{\Omega}_{\mathrm{R}}^{\mathcal{B}} = \Omega_{\mathrm{L}}^{\mathcal{B}}\,M^{-1}\,\tilde{\mathcal{B}}\,M\,\Omega_{\mathrm{R}}^{\mathcal{B}} = \Omega_{\mathrm{L}}^{\mathcal{B}}\,\mathcal{B}\,\Omega_{\mathrm{R}}^{\mathcal{B}},$$
$$\tilde{\Omega}_{\mathrm{L}}^{\mathcal{C}}\,\tilde{\mathcal{C}}\,\tilde{\Omega}_{\mathrm{R}}^{\mathcal{C}} = \Omega_{\mathrm{R}}^{\mathcal{C}}\,M^{-1}\,\tilde{\mathcal{C}}\,M\,\Omega_{\mathrm{R}}^{\mathcal{C}} = \Omega_{\mathrm{L}}^{\mathcal{C}}\,\mathcal{C}\,\Omega_{\mathrm{R}}^{\mathcal{C}}.$$

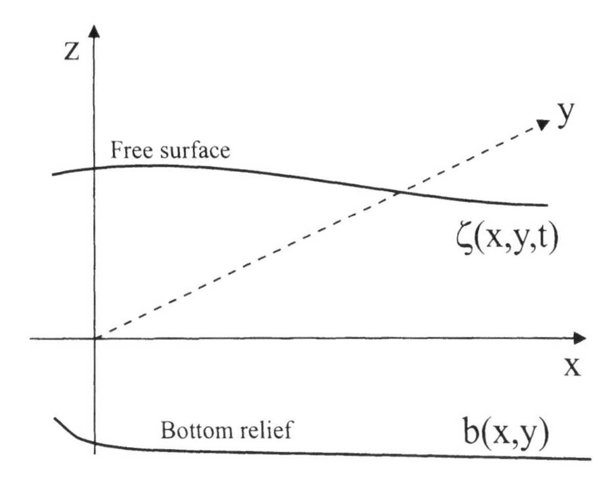

Figure 1.1 Shallow water layer.

1.3.3 Shallow water equations. If we consider the flow of an incompressible fluid, the mass conservation equation in Eq. (1.3.1) reduces to div $\mathbf{v} = 0$. Let the gravitation force act along the z-axis of the Cartesian coordinate system shown in Fig. 1.1. In this case the condition of the divergence-free velocity field must be accompanied by the momentum equation in the form

$$\frac{\partial \mathbf{v}}{\partial t} + (\mathbf{v} \cdot \nabla)\mathbf{v} = -\frac{\nabla p}{\rho} + \mathbf{F}_g, \tag{1.3.16}$$

where $\mathbf{F}_g = [0, 0, -g]^{\mathrm{T}}$ and g is the free-fall acceleration.

In the shallow water approximation the substantive time derivative along the z-axis is assumed negligibly small and we obtain (Stoker 1957)

$$p = \rho g(\zeta - z). \tag{1.3.17}$$

Here we assumed that pressure is equal to zero at the free-surface level $z = \zeta(x, y, t)$. Substituting expression (1.3.17) into Eq. (1.3.16) and adding the mass conservation equation, we arrive at the system governing the shallow water behavior in the form

$$\begin{aligned}
\frac{\partial h}{\partial t} + \frac{\partial uh}{\partial x} + \frac{\partial vh}{\partial y} &= 0, \\
\frac{\partial u}{\partial t} + u\frac{\partial u}{\partial x} + v\frac{\partial u}{\partial y} &= -g\frac{\partial \zeta}{\partial x}, \\
\frac{\partial v}{\partial t} + u\frac{\partial v}{\partial x} + v\frac{\partial v}{\partial y} &= -g\frac{\partial \zeta}{\partial y}.
\end{aligned} \tag{1.3.18}$$

Here $h(x, y, t) = \zeta - b(x, y)$, where $b(x, y)$ describes the shape of the bottom relief.

This system can be rewritten in the conservation-law form

$$\mathbf{U}_t + \mathbf{E}(\mathbf{U})_x + \mathbf{G}(\mathbf{U})_y = \mathbf{S}(\mathbf{U}) \tag{1.3.19}$$

with

$$\mathbf{U} = \begin{bmatrix} h \\ hu \\ hv \end{bmatrix}, \quad \mathbf{E} = \begin{bmatrix} hu \\ hu^2 + \frac{1}{2}gh^2 \\ huv \end{bmatrix}, \quad \mathbf{G} = \begin{bmatrix} hv \\ huv \\ hv^2 + \frac{1}{2}gh^2 \end{bmatrix}, \quad \mathbf{S} = \begin{bmatrix} 0 \\ -ghb_x \\ -ghb_y \end{bmatrix}.$$

Note that if we formally put $\rho = h$ and $p = \frac{1}{2}gh^2$, for $b(x, y) = \text{const}$ we obtain the gas dynamic system for barotropic gas with the specific heat ratio equal to 2. This system preserves the hyperbolic properties of the Euler equations. The methods of its solution will be discussed in Chapter 4.

1.3.4 Equations of ideal magnetohydrodynamics.

The equations of ideal magnetohydrodynamics generalize the Euler gas dynamic equations in the presence of an electromagnetic field. In this case appropriate terms must be added to the momentum and energy equations of the Euler system and it must be supplemented by the Maxwell equations (see Landau and Lifshitz 1984; Jeffrey and Taniuti 1964; Kulikovskii and Lyubimov 1965; Akhiezer et al. 1975),

$$\frac{\partial \rho}{\partial t} + \operatorname{div} \rho \mathbf{v} = 0, \tag{1.3.20}$$

$$\frac{\partial \mathbf{v}}{\partial t} + (\mathbf{v} \cdot \nabla)\mathbf{v} = -\frac{\nabla p}{\rho} - \frac{\mathbf{B} \times \operatorname{curl} \mathbf{B}}{4\pi\rho}, \tag{1.3.21}$$

$$\frac{dp}{dt} = c_e^2 \frac{d\rho}{dt}, \tag{1.3.22}$$

$$\frac{\partial \mathbf{B}}{\partial t} = \operatorname{curl}(\mathbf{v} \times \mathbf{B}). \tag{1.3.23}$$

Here $\mathbf{B}$ is the magnetic field strength vector and c_e is the acoustic speed of sound. For the sake of simplicity, we substituted the energy equation, which will be written out in Chapter 5, by the equation expressing the definition of the speed of sound $(\partial p/\partial \rho)_S = c_e^2$, where S is entropy. In the corresponding equation (1.3.22) we also introduced the total, or substantive, time derivative

$$\frac{df}{dt} = \frac{\partial f}{\partial t} + (\mathbf{v} \cdot \nabla) f.$$

If we take into account the vector analysis formula

$$\operatorname{curl}(\mathbf{v} \times \mathbf{B}) = (\mathbf{B} \cdot \nabla)\mathbf{v} - (\mathbf{v} \cdot \nabla)\mathbf{B} + \mathbf{v} \operatorname{div} \mathbf{B} - \mathbf{B} \operatorname{div} \mathbf{v},$$

the quasilinear form of the MHD system acquires the form

$$\frac{\partial \mathbf{u}}{\partial t} + A_1 \frac{\partial \mathbf{u}}{\partial x} + A_2 \frac{\partial \mathbf{u}}{\partial y} + A_3 \frac{\partial \mathbf{u}}{\partial z} = \mathbf{0}, \tag{1.3.24}$$

where

$$\mathbf{u} = [\rho,\ u,\ v,\ w,\ p,\ B_x,\ B_y,\ B_z]^{\mathrm{T}},$$

$$A_1 = \begin{bmatrix} u & \rho & 0 & 0 & 0 & 0 & 0 & 0 \\ 0 & u & 0 & 0 & \frac{1}{\rho} & 0 & \frac{B_y}{4\pi\rho} & \frac{B_z}{4\pi\rho} \\ 0 & 0 & u & 0 & 0 & 0 & -\frac{B_x}{4\pi\rho} & 0 \\ 0 & 0 & 0 & u & 0 & 0 & 0 & -\frac{B_x}{4\pi\rho} \\ 0 & \rho c_e^2 & 0 & 0 & u & 0 & 0 & 0 \\ 0 & 0 & 0 & 0 & 0 & u & 0 & 0 \\ 0 & B_y & -B_x & 0 & 0 & 0 & u & 0 \\ 0 & B_z & 0 & -B_x & 0 & 0 & 0 & u \end{bmatrix}, \quad A_2 = \begin{bmatrix} v & 0 & \rho & 0 & 0 & 0 & 0 & 0 \\ 0 & v & 0 & 0 & 0 & -\frac{B_y}{4\pi\rho} & 0 & 0 \\ 0 & 0 & v & 0 & \frac{1}{\rho} & \frac{B_x}{4\pi\rho} & 0 & \frac{B_z}{4\pi\rho} \\ 0 & 0 & 0 & v & 0 & 0 & 0 & -\frac{B_y}{4\pi\rho} \\ 0 & 0 & \rho c_e^2 & 0 & v & 0 & 0 & 0 \\ 0 & -B_y & B_x & 0 & 0 & v & 0 & 0 \\ 0 & 0 & 0 & 0 & 0 & 0 & v & 0 \\ 0 & 0 & B_z & -B_y & 0 & 0 & 0 & v \end{bmatrix}$$

$$A_3 = \begin{bmatrix} w & 0 & 0 & \rho & 0 & 0 & 0 & 0 \\ 0 & w & 0 & 0 & 0 & -\frac{B_z}{4\pi\rho} & 0 & 0 \\ 0 & 0 & w & 0 & 0 & 0 & -\frac{B_z}{4\pi\rho} & 0 \\ 0 & 0 & 0 & w & \frac{1}{\rho} & \frac{B_x}{4\pi\rho} & \frac{B_y}{4\pi\rho} & 0 \\ 0 & 0 & 0 & \rho c_e^2 & w & 0 & 0 & 0 \\ 0 & -B_z & 0 & B_x & 0 & w & 0 & 0 \\ 0 & 0 & -B_z & B_y & 0 & 0 & w & 0 \\ 0 & 0 & 0 & 0 & 0 & 0 & 0 & w \end{bmatrix}$$

The system of (1.3.20)–(1.3.23) must be supplemented by the equation

$$\operatorname{div} \mathbf{B} = 0 \tag{1.3.25}$$

expressing the absence of magnetic charge. It is not important for our current consideration, since if $\operatorname{div} \mathbf{B} = 0$ initially at $t = 0$, then the Faraday equation (1.3.23) will preserve the absence of magnetic charge later in time. This can be seen if we apply the divergence operator to the both parts of Eq. (1.3.23),

$$\frac{\partial \operatorname{div} \mathbf{B}}{\partial t} = \operatorname{div} \operatorname{curl} (\mathbf{v} \times \mathbf{B}) \equiv 0.$$

Note that we did not use Eq. (1.3.25) when deriving the system of (1.3.24).

For simplicity, we shall consider in this section only the one-dimensional system postponing the detailed description of the MHD equations to Chapter 5. Solution of the characteristic equation

$$\det(A_1 - \lambda I) = 0$$

gives the following eigenvalues:

$$\lambda_{1,2} = u, \quad \lambda_{3,4} = u \pm a_{\mathrm{A}}, \quad \lambda_{5,6} = u \pm a_{\mathrm{f}}, \quad \lambda_{7,8} = u \pm a_{\mathrm{s}}$$

with

$$a_{\mathrm{A}} = \frac{|B_x|}{\sqrt{4\pi\rho}}, \quad a_{f,s} = \frac{1}{2}\left[\left(c_e^2 + \frac{\mathbf{B}^2}{4\pi\rho} + \frac{|B_x| c_e}{\sqrt{\pi\rho}}\right)^{1/2} \pm \left(c_e^2 + \frac{\mathbf{B}^2}{4\pi\rho} - \frac{|B_x| c_e}{\sqrt{\pi\rho}}\right)^{1/2}\right].$$

The velocity a_A is called the Alfvén, or rotational, velocity; a_f and a_s are called the fast and slow magnetosonic velocities, respectively. All eigenvalues are clearly real.

Note that the equation for B_x in the one-dimensional treatment reduces to

$$\frac{\partial B_x}{\partial t} + u\frac{\partial B_x}{\partial x} = 0, \tag{1.3.26}$$

that is, to the one-dimensional convection equation for B_x. Of course, in the genuinely one-dimensional problem, with all quantities depending only on the space variable x, one can simply put $B_x \equiv \text{const}$ and omit the corresponding equation. By omitting Eq. (1.3.26), we reduce the system dimension to 7×7 (seven equations for seven dependent variables). Both the extended 8×8 system and the reduced one have real eigenvalues and a nondegenerate set of eigenvectors. It is apparent that the eigenvalues of the truly one-dimensional system are the same as those of the extended system. Only the eigenvalue $\lambda = u$ has the multiplicity 1.

Below we present a complete set of right and left eigenvectors for the extended system:

$$\mathbf{r}^1 = [1,\ 0,\ 0,\ 0,\ 0,\ 0,\ 0,\ 0]^{\mathrm{T}},$$
$$\mathbf{r}^2 = [0,\ 0,\ 0,\ 0,\ 0,\ 1,\ 0,\ 0]^{\mathrm{T}},$$
$$\mathbf{r}^3 = \left[0,\ 0,\ -\frac{b_z}{\sqrt{2}},\ \frac{b_y}{\sqrt{2}},\ 0,\ 0,\ b_z\sqrt{2\pi\rho}\,\mathrm{sgn}\,B_x,\ -b_y\sqrt{2\pi\rho}\,\mathrm{sgn}\,B_x\right]^{\mathrm{T}},$$
$$\mathbf{r}^4 = \left[0,\ 0,\ -\frac{b_z}{\sqrt{2}},\ \frac{b_y}{\sqrt{2}},\ 0,\ 0,\ -b_z\sqrt{2\pi\rho}\,\mathrm{sgn}\,B_x,\ b_y\sqrt{2\pi\rho}\,\mathrm{sgn}\,B_x\right]^{\mathrm{T}},$$
$$\mathbf{r}^5 = [\rho\alpha_f,\ \alpha_f a_f,\ -\alpha_s a_s b_y\,\mathrm{sgn}\,B_x,\ -\alpha_s a_s b_z\,\mathrm{sgn}\,B_x,\ \alpha_f\rho c_e^2,\ 0,\ 2\alpha_s\sqrt{\pi\rho}c_e b_y,\ 2\alpha_s\sqrt{\pi\rho}c_e b_z]^{\mathrm{T}},$$
$$\mathbf{r}^6 = [\rho\alpha_f,\ -\alpha_f a_f,\ \alpha_s a_s b_y\,\mathrm{sgn}\,B_x,\ \alpha_s a_s b_z\,\mathrm{sgn}\,B_x,\ \alpha_f\rho c_e^2,\ 0,\ 2\alpha_s\sqrt{\pi\rho}c_e b_y,\ 2\alpha_s\sqrt{\pi\rho}c_e b_z]^{\mathrm{T}},$$
$$\mathbf{r}^7 = [\rho\alpha_s,\ \alpha_s a_s,\ \alpha_f a_f b_y\,\mathrm{sgn}\,B_x,\ \alpha_f a_f b_z\,\mathrm{sgn}\,B_x,\ \alpha_s\rho c_e^2,\ 0,\ -2\alpha_f\sqrt{\pi\rho}c_e b_y,\ -2\alpha_f\sqrt{\pi\rho}c_e b_z]^{\mathrm{T}},$$
$$\mathbf{r}^8 = [\rho\alpha_s,\ -\alpha_s a_s,\ -\alpha_f a_f b_y\,\mathrm{sgn}\,B_x,\ -\alpha_f a_f b_z\,\mathrm{sgn}\,B_x,\ \alpha_s\rho c_e^2,\ 0,\ -2\alpha_f\sqrt{\pi\rho}c_e b_y,\ -2\alpha_f\sqrt{\pi\rho}c_e b_z]^{\mathrm{T}},$$

$$\mathbf{l}^1 = \left[1,\ 0,\ 0,\ 0,\ -\frac{1}{c_e^2},\ 0,\ 0,\ 0\right],$$
$$\mathbf{l}^2 = [0,\ 0,\ 0,\ 0,\ 0,\ 1,\ 0,\ 0],$$
$$\mathbf{l}^3 = \left[0,\ 0,\ -\frac{b_z}{\sqrt{2}},\ \frac{b_y}{\sqrt{2}},\ 0,\ 0,\ \frac{b_z\,\mathrm{sgn}\,B_x}{2\sqrt{2\pi\rho}},\ -\frac{b_y\,\mathrm{sgn}\,B_x}{2\sqrt{2\pi\rho}}\right],$$
$$\mathbf{l}^4 = \left[0,\ 0,\ -\frac{b_z}{\sqrt{2}},\ \frac{b_y}{\sqrt{2}},\ 0,\ 0,\ -\frac{b_z\,\mathrm{sgn}\,B_x}{2\sqrt{2\pi\rho}},\ \frac{b_y\,\mathrm{sgn}\,B_x}{2\sqrt{2\pi\rho}}\right],$$
$$\mathbf{l}^5 = \left[0,\ \frac{\alpha_f a_f}{2c_e^2},\ -\frac{\alpha_s a_s}{2c_e^2}b_y\,\mathrm{sgn}\,B_x,\ -\frac{\alpha_s a_s}{2c_e^2}b_z\,\mathrm{sgn}\,B_x,\ \frac{\alpha_f}{2\rho c_e^2},\ 0,\ \frac{\alpha_s b_y}{4c_e\sqrt{\pi\rho}},\ \frac{\alpha_s b_z}{4c_e\sqrt{\pi\rho}}\right],$$
$$\mathbf{l}^6 = \left[0,\ -\frac{\alpha_f a_f}{2c_e^2},\ \frac{\alpha_s a_s}{2c_e^2}b_y\,\mathrm{sgn}\,B_x,\ \frac{\alpha_s a_s}{2c_e^2}b_z\,\mathrm{sgn}\,B_x,\ \frac{\alpha_f}{2\rho c_e^2},\ 0,\ \frac{\alpha_s b_y}{4c_e\sqrt{\pi\rho}},\ \frac{\alpha_s b_z}{4c_e\sqrt{\pi\rho}}\right],$$
$$\mathbf{l}^7 = \left[0,\ \frac{\alpha_s a_s}{2c_e^2},\ \frac{\alpha_f a_f}{2c_e^2}b_y\,\mathrm{sgn}\,B_x,\ \frac{\alpha_f a_f}{2c_e^2}b_z\,\mathrm{sgn}\,B_x,\ \frac{\alpha_s}{2\rho c_e^2},\ 0,\ -\frac{\alpha_f b_y}{4c_e\sqrt{\pi\rho}},\ -\frac{\alpha_f b_z}{4c_e\sqrt{\pi\rho}}\right],$$

$$\mathbf{l}^8 = \left[0,\ -\frac{\alpha_s a_s}{2c_e^2},\ -\frac{\alpha_f a_f}{2c_e^2} b_y \operatorname{sgn} B_x,\ -\frac{\alpha_f a_f}{2c_e^2} b_z \operatorname{sgn} B_x,\ \frac{\alpha_s}{2\rho c_e^2},\ 0,\ -\frac{\alpha_f b_y}{4c_e\sqrt{\pi\rho}},\ -\frac{\alpha_f b_z}{4c_e\sqrt{\pi\rho}}\right].$$

In these equations

$$\alpha_f = \sqrt{\frac{c_e^2 - a_s^2}{a_f^2 - a_s^2}}, \qquad \alpha_s = \sqrt{\frac{a_f^2 - c_e^2}{a_f^2 - a_s^2}}.$$

Also, by definition, we put $\operatorname{sgn} 0 \equiv 1$.

We also normalized the tangential components of the magnetic field vector

$$b_y = \frac{B_y}{\sqrt{B_y^2 + B_z^2}}, \qquad b_z = \frac{B_z}{\sqrt{B_y^2 + B_z^2}}.$$

The quantities b_y and b_z clearly degenerate in the absence of the tangential magnetic field. If one notices, however, that the only important relationship to be preserved is $b_y^2 + b_z^2 = 1$, the following regularization can be chosen:

$$b_y = \sin\psi, \qquad b_z = \cos\psi.$$

The value of ψ is arbitrary. The compact form of the MHD eigenvector normalization presented here is based on the paper by Brio and Wu (1988) (see also Roe and Balsara 1996). Besides, they used $\psi = \pi/4$, though this does not seem to be important, and $\psi = 0$ or $\pi/2$ can also be a good choice. It is easy to check that $\det \Omega_R = 16\pi\rho^2 c^5$.

The formulas for the eigenvectors obviously become degenerate if $a_f = a_s$. This occurs for $B_y^2 + B_z^2 = 0$ and $c_e = a_A$. If this rare occasion happens, we can simply put $B_x = B_x(1+\varepsilon)$, where ε is a small constant.

1.3.5 Elasticity equations.

Let us consider as an example a linear equation describing the vibration of a uniform rod in the Timoshenko theory (Grigolyuk and Selezov 1973). A Timoshenko-type equation has the form

$$\frac{\partial^2 w}{\partial t^2} + a_1 \frac{\partial^4 w}{\partial x^4} - a_2 \frac{\partial^4 w}{\partial t^2 \partial x^2} + a_3 \frac{\partial^4 w}{\partial t^4} = 0, \tag{1.3.27}$$

where a_1, a_2, and a_3 are positive constants depending on the rod density, its geometrical parameters and elastic coefficients, and w is a deflection. In particular,

$$a_1 = \frac{EI}{\mu F l^2}, \qquad a_2 = \frac{I}{F l^2} + \frac{EI}{k\mu F l^2}, \qquad a_3 = \frac{I}{kF l^2},$$

where F is the area of the cross-section of the rod, l is its length, μ is the shear modulus, I is the moment of inertia of the cross-section about the axis passing through the center of mass, EI is the bending rigidity, E is Young's modulus, and k is the shear coefficient. The variables x and w are normalized by l, and t is normalized by l/c, where $c = \sqrt{\mu/\rho}$, ρ is the density of material.

If we introduce

$$M = \frac{\partial^2 w}{\partial t^2}, \qquad N = \frac{\partial^2 w}{\partial x^2},$$

Eq. (1.3.27) can be written as

$$a_3\frac{\partial^2 M}{\partial t^2} - a_2\frac{\partial^2 M}{\partial x^2} + a_1\frac{\partial^2 N}{\partial x^2} + M = 0, \tag{1.3.28}$$

$$\frac{\partial^2 N}{\partial t^2} - \frac{\partial^2 M}{\partial x^2} = 0. \tag{1.3.29}$$

If we also introduce

$$P = \frac{\partial M}{\partial t}, \quad Q = \frac{\partial N}{\partial t}, \quad R = \frac{\partial M}{\partial x}, \quad S = \frac{\partial N}{\partial x},$$

Eqs. (1.3.28)–(1.3.29) acquire the form

$$\frac{\partial P}{\partial t} - \frac{a_2}{a_3}\frac{\partial R}{\partial x} + \frac{a_1}{a_3}\frac{\partial S}{\partial x} = -\frac{1}{a_3}M,$$

$$\frac{\partial Q}{\partial t} - \frac{\partial R}{\partial x} = 0, \quad \frac{\partial R}{\partial t} - \frac{\partial P}{\partial x} = 0, \quad \frac{\partial S}{\partial t} - \frac{\partial Q}{\partial x} = 0,$$

$$\frac{\partial M}{\partial t} = P, \quad \frac{\partial N}{\partial t} = Q.$$

This system of the first-order equations can be rewritten in the form

$$\frac{\partial \mathbf{U}}{\partial t} + A\frac{\partial \mathbf{U}}{\partial x} = \mathbf{f},$$

where

$$A = \begin{bmatrix} 0 & 0 & -\alpha_2 & \alpha_1 & 0 & 0 \\ 0 & 0 & -1 & 0 & 0 & 0 \\ -1 & 0 & 0 & 0 & 0 & 0 \\ 0 & -1 & 0 & 0 & 0 & 0 \\ 0 & 0 & 0 & 0 & 0 & 0 \\ 0 & 0 & 0 & 0 & 0 & 0 \end{bmatrix}, \quad \alpha_1 = \frac{a_1}{a_3}, \quad \alpha_2 = \frac{a_2}{a_3};$$

$$\mathbf{U} = [P, Q, R, S, N, M]^{\mathrm{T}}, \quad \mathbf{f} = \left[-\frac{M}{a_3}, 0, 0, 0, P, Q\right]^{\mathrm{T}}.$$

The matrix A can be diagonalized,

$$A = \Omega_{\mathrm{R}}\Lambda\Omega_{\mathrm{L}},$$

where

$$\Lambda = \operatorname{diag}[\beta_1, -\beta_1, \beta_2, -\beta_2, 0, 0],$$

$$\beta_1 = \frac{1}{\sqrt{2}}\sqrt{\alpha_2 + \sqrt{\alpha_2^2 - 4\alpha_1}} = \sqrt{\frac{E}{\mu}}, \quad \beta_2 = \frac{1}{\sqrt{2}}\sqrt{\alpha_2 - \sqrt{\alpha_2^2 - 4\alpha_1}} = \sqrt{k};$$

and Ω_{R}, Ω_{L} have the block form

$$\Omega_{\mathrm{R}} = \begin{bmatrix} B & 0 \\ 0 & I \end{bmatrix}, \quad \Omega_{\mathrm{L}} = \begin{bmatrix} B^{-1} & 0 \\ 0 & I \end{bmatrix},$$

where $I = \operatorname{diag}[1, 1]$ is the 2×2 identity matrix, and the 4×4 matrices B and B^{-1} have the following structure:

$$B = \begin{bmatrix} -\beta_1^3 & \beta_1^3 & -\beta_2^3 & \beta_2^3 \\ -\beta_1 & \beta_1 & -\beta_2 & \beta_2 \\ \beta_1^2 & \beta_1^2 & \beta_2^2 & \beta_2^2 \\ 1 & 1 & 1 & 1 \end{bmatrix}, \quad B^{-1} = \frac{1}{2(\beta_2^2 - \beta_1^2)\beta_1\beta_2} \begin{bmatrix} \beta_2 & -\beta_2^3 & -\beta_2\beta_1 & \beta_1\beta_2^3 \\ -\beta_2 & \beta_2^3 & -\beta_2\beta_1 & \beta_1\beta_2^3 \\ -\beta_1 & \beta_1^3 & \beta_2\beta_1 & -\beta_1^3\beta_2 \\ \beta_1 & -\beta_1^3 & \beta_2\beta_1 & -\beta_1^3\beta_2 \end{bmatrix},$$

$$\det B = -4(\beta_2 - \beta_1)^2(\beta_2 + \beta_1)^2\beta_1\beta_2.$$

1.4 Properties of solutions

The purpose of this section is to provide a brief discussion of several mathematical properties of solutions of quasilinear hyperbolic equations. We restrict ourselves to those properties that are essential for further presentation. Detailed information can be found in a number of monographs and textbooks. The incomplete list includes those by Jeffrey and Taniuti (1964), Lax (1972), Rozhdestvenskii and Yanenko (1983), Jeffrey (1976), Le Veque (1992), Kulikovskii and Sveshnikova (1995), Godlewski and Raviart (1996), Serre (1996), Kröner (1997), and Toro (1997).

1.4.1 Classical solutions.

We assume for simplicity that the system is homogeneous with the flux vector, depending only on the unknown vector itself but not on the independent variables

$$\frac{\partial \mathbf{U}}{\partial t} + \frac{\partial \mathbf{F}(\mathbf{U})}{\partial x} = \mathbf{0}. \tag{1.4.1}$$

As shown in Section 1.3, this form of the system is frequently encountered in mechanical applications.

As is apparent from Eq. (1.1.7), the unknown vector must at least be differentiable. To pass from the integral form (1.2.14) to the corresponding differential form (1.2.13), we must also assume a definite smoothness of $\mathbf{U}$. If a sufficiently smooth solution does not exist, one must use the integral form.

Let us consider a Cauchy problem by specifying

$$\mathbf{U}(x, 0) = \mathbf{U}_0(x) \tag{1.4.2}$$

for Eq. (1.4.1).

The vector function $\mathbf{U}$ is called a classical solution of the Cauchy problem (1.4.1)–(1.4.2) if $\mathbf{U}$ is a continuously differentiable function that satisfies these equations pointwise.

Existence of smooth solutions

Before giving definition to generalized solutions of a hyperbolic system, we shall show that a classical solution, in fact, can sometimes exist only within a finite time interval even for a smooth initial distribution given by the function $\mathbf{U}_0$. For this purpose we shall consider the simplest case of the one-dimensional scalar equation

$$\frac{\partial U}{\partial t} + \frac{\partial F}{\partial x} = 0 \tag{1.4.3}$$

with the initial condition

$$U(x,0) = U_0(x), \quad -\infty < x < \infty. \tag{1.4.4}$$

Note that Eq. (1.4.3) can also be written in a quasilinear form

$$\frac{\partial U}{\partial t} + a(U)\frac{\partial U}{\partial x} = 0 \tag{1.4.5}$$

by putting $a(U) = \partial F(U)/\partial U$. Hence, the characteristic line (1.2.6) passing through the point $(x_0, 0)$ in the x–t plane becomes

$$x = x_0 + ta(U_0(x_0)). \tag{1.4.6}$$

It is apparent from Eq. (1.4.5) that

$$\frac{dU}{dt} = 0$$

along the characteristic line. This allows us to find a smooth solution using a so-called *method of characteristics* by specifying

$$U(x,t) = U_0(x_0),$$

where x_0 should be found from formula (1.4.6) describing the characteristic line.

For linear systems, when matrix A is independent of $\mathbf{U}$, characteristic lines $x = x_0 + at$ are straight lines that never intersect each other. This means that we can find the exact solution in the whole half-plane $\{-\infty < x < \infty,\ t > 0\}$. This solution has the form of a travelling wave

$$U(x,t) = U_0(x - at). \tag{1.4.7}$$

On the contrary, if $a(U) \neq \text{const}$ and for a certain $x_1 < x_2$,

$$\frac{1}{a(U_0(x_1))} < \frac{1}{a(U_0(x_2))},$$

then the characteristics that issue from $x = x_1$ and $x = x_2$ inevitably intersect each other. The classical solution no longer exists beyond the intersection point, since it must become discontinuous at it. This feature also outlines the limits of the method of characteristics.

An important notice must be given at this stage. Though we shall mainly deal with nonlinear systems in this monograph, linear equations are frequently encountered in numerical methods when a linearized system is used in the computational procedure to advance the solution within a small time interval.

If we linearize the system

$$\frac{\partial \mathbf{u}}{\partial t} + A(\mathbf{u})\frac{\partial \mathbf{u}}{\partial x} = \mathbf{0} \tag{1.4.8}$$

in a small vicinity of some constant value $\mathbf{u}_0$, that is, assume $\mathbf{u} = \mathbf{u}_0 + \tilde{\mathbf{u}}(x,t)$, then $\tilde{\mathbf{u}}$ must satisfy the linear system

$$\frac{\partial \tilde{\mathbf{u}}}{\partial t} + A(\mathbf{u}_0)\frac{\partial \tilde{\mathbf{u}}}{\partial x} = \mathbf{0} \tag{1.4.9}$$

to the second order of accuracy. The characteristic velocities λ_k and the eigenvectors $\mathbf{r}^k$ of the coefficient matrix A are constant in this case. Since A is diagonalizable, that is, $A = \Omega_R \Lambda \Omega_L$, on multiplying the system of (1.4.9) by Ω_L and introducing the Riemann variables $\mathbf{w} = \Omega_L \tilde{\mathbf{u}}$ we obtain

$$\frac{\partial \mathbf{w}}{\partial t} + \Lambda \frac{\partial \mathbf{w}}{\partial x} = \mathbf{0}, \tag{1.4.10}$$

where $\Lambda = \mathrm{diag}[\lambda_1, \ldots, \lambda_n]$.

Thus, the system splits into separate equations whose solutions w_k are travelling waves

$$w_k = w_k(x - \lambda_k t), \qquad k = 1, \ldots, n. \tag{1.4.11}$$

In this case the functions w_k are called Riemann invariants of the system (1.4.9) with constant coefficients a_{ij} constituting the matrix A. Travelling waves (1.4.11) propagate at constant velocities λ_k preserving their shape.

The general solution of the system (1.4.9) is the sum of n travelling waves propagating at the corresponding characteristic velocities,

$$\tilde{\mathbf{u}} = \sum_{k=1}^{n} \mathbf{r}^k w_k(x - \lambda_k t). \tag{1.4.12}$$

If the eigenvectors $\mathbf{r}^k$ are normalized so that $|\mathbf{r}^k| = 1$, the quantities w_k can be considered as the amplitudes of the corresponding linear waves.

Riemann waves

We now proceed to the properties of the nonlinear system (1.4.1). Let us consider an important special class of solutions to the Cauchy problem (1.4.1)–(1.4.2) that are called Riemann, or simple, waves. The latter notion seems to be more general, since it also includes such steady two-dimensional solution of the gas dynamic system (1.3.13) as the Prandtl–Meyer waves and similar waves in MHD and elasticity.

It is assumed that in simple waves the unknown vector $\mathbf{U}$ depends on a certain combination $\theta(x, t)$ of the independent variables, that is, $\mathbf{U} = \mathbf{U}(\theta(x, t))$. In this case from Eq. (1.4.1) we obtain

$$(A - \lambda I)\frac{d\mathbf{U}}{d\theta} = \mathbf{0} \tag{1.4.13}$$

with

$$\lambda = -\frac{\partial \theta}{\partial t} \bigg/ \frac{\partial \theta}{\partial x}. \tag{1.4.14}$$

To obtain nontrivial solutions, we must demand

$$\det(A - \lambda I) = 0, \tag{1.4.15}$$

that is, λ must be the eigenvalue of the matrix A, thus coinciding with one of the characteristic velocities $\lambda_k(\mathbf{U})$ of the system (1.4.1). The increments dU_k coincide in this case with the corresponding increments of small disturbances propagating through the uniform distribution U_k. It is clear that the derivative $d\mathbf{U}/d\theta$ is parallel in this case to the corresponding right eigenvector of the matrix A

$$\frac{d\mathbf{U}}{d\theta} = \alpha \mathbf{r}, \tag{1.4.16}$$

where the eigenvector number is omitted.

Equation (1.4.16) defines a family of integral curves tangent at each of their points to the right eigenvector of the matrix A. The number of the simple wave solutions is equal to the number of linearly independent eigenvectors $\mathbf{r}$. Since the system is hyperbolic, this number is equal to n.

Let us consider one of the Riemann waves corresponding to a simple root of the characteristic equation. On the integral curve, the vector $\mathbf{U}$ is a function of a single parameter θ. This parameter can be chosen arbitrarily depending on our needs and tastes. This can be an arc length in the $\mathbf{U}$-space measured from the initial value $\theta_0 = \theta(x, 0)$, one of the components U_k of the vector $\mathbf{U}$ or a characteristic velocity λ. The only requirement is that the chosen parameter must be monotone along the part of the integral curve under consideration. On choosing θ, we can determine its value from Eq. (1.4.14),

$$\frac{\partial \theta}{\partial t} + \lambda(\mathbf{U}(\theta)) \frac{\partial \theta}{\partial x} = 0. \tag{1.4.17}$$

A scalar function of the vector here is regarded as a function of its components.

On the other hand, the characteristics

$$\frac{dx}{dt} = \lambda(\mathbf{U}(\theta))$$

of Eq. (1.4.17) coincide with the chosen characteristic family of the system (1.4.1). Along each of the characteristic lines of (1.4.17), we have $d\theta/dt = 0$, that is, $\theta = \text{const}$ and, hence, all the components U_k of the vector $\mathbf{U}$ and λ itself are constant. This means that in the x–t plane the characteristics are represented by straight lines and their slopes can be determined even at $t = 0$. The remaining families of characteristics are generally curvilinear.

If we wish the solution to be in the form of a Riemann wave, initial conditions for $\mathbf{U}$ must also be represented by some function of θ. This means that only one arbitrary function $\theta(x, 0) = \theta_0(x)$ must occur in them. In addition, the solution contains $n - 1$ constants, which are necessary to single out the integral curve of Eq. (1.4.13). As shown above, the solution in the form of a Riemann wave can be constructed uniquely only in the part of the x–t plane where characteristics do not intersect each other.

The above consideration shows that simple waves generalize small-perturbation waves governed by Eq. (1.4.9). In fact, each element $d\mathbf{U} = (d\mathbf{U}/d\theta)d\theta$ of the Riemann wave varies proportionally to the right eigenvector of the coefficient matrix of the system, exactly in the same way as in the case of propagation of a small perturbation. The velocities of propagation are also the same. A simple wave can therefore be represented as a series of small perturbations, each moving in a wake behind the foregoing one. Depending on the character of the characteristic velocity variation along the integral curve, the profile of the wave can suffer deformations.

If $\lambda(\theta) = \text{const}$ on the integral curve, the characteristics in the x–t plane are straight lines (see Fig. 1.2b) and never intersect each other. In this case the Riemann wave is a travelling wave with

$$\mathbf{U} = \mathbf{U}(\theta), \quad \theta = \theta_0(x - \lambda t).$$

If λ is not constant, in the monotonicity intervals of the characteristic velocity we can choose it as a parameter θ. Thus, we obtain

$$\frac{\partial \lambda}{\partial t} + \lambda \frac{\partial \lambda}{\partial x} = 0$$

with the initial condition $\lambda(x, 0) = \lambda_0(x)$.

Let the initial profile be represented by the function shown in Fig. 1.2. As shown above for the scalar equation (1.4.3), on the interval with $d\lambda/dx > 0$, the solution can be uniquely determined. This is caused by the fact that preceding parts of the profile move faster than those following them. On the other hand, on the interval with $d\lambda/dx < 0$ the elements closer to the maximum of $\lambda_0(x)$ move faster than foregoing ones and sooner or later catch them. This process is known by the name of the *wave steepening*. It finally results in the wave breaking (see Fig. 1.2e). Since nonunique solutions are mostly disregarded in continuum mechanics, it is adopted that at the moments of the characteristic intersection the classical solution ceases to exist and a discontinuity originates.

There is a class of equations possessing solutions in the form of Riemann waves in which the characteristic velocity λ is constant. This occurs if λ is constant along each integral curve of the wave. In this case the Riemann wave is a traveling wave, that is, $U_k = U_k(x - \lambda t)$, $\lambda = \text{const}$. Such waves that propagate without changing their shape will be called nondeforming waves. If the function θ is initially discontinuous, it will remain discontinuous for all subsequent t. Each discontinuity of this type has a counterpart discontinuity with the opposite quantity variation. Discontinuities that are at the same time Riemann waves will be called reversible. As examples of nondeforming Riemann waves and reversible discontinuities one can indicate rotational (Alfvén) Riemann waves and rotational discontinuities in magnetohydrodynamics (Landau and Lifshitz 1984).

Let us note an important particular case of Riemann waves, namely, self-similar Riemann waves. Initial condition at $t = 0$ for such waves is a piecewise-constant function θ with a discontinuity at the point taken as the origin $x = 0$. If $\lambda|_{x>0} > \lambda|_{x<0}$, a fan of rectilinear characteristics $x/t = \lambda(\theta)$ corresponding to the considered wave starts from the origin. Since the quantities U_k are constant along the characteristics, the characteristic velocity $\lambda = x/t$ itself can be taken as a parameter θ.

The properties of classical solutions described in this subsection show that, since the general system of quasilinear hyperbolic equations cannot be written for the Riemann invariants, the domain of definition of the solution can only be found simultaneously with the solution itself. In addition, a classical solution and its derivatives do not remain bounded.

1.4.2 Generalized solutions. The above reasoning leads us to introduce the notion of a generalized solution. We shall call the vector function *a generalized solution* of the system (1.4.1) if it satisfies the system of (1.2.14) for an arbitrary piecewise-smooth contour ∂V bounding the volume V. It is clear that all classical solutions form a subset of generalized solutions. On the other hand, $\mathbf{U}(x, t)$ can only be piecewise-continuous with continuous first derivatives within each continuity interval.

The surface on which the function $\mathbf{U}(x, t)$ is discontinuous is called a *surface of strong discontinuity*, or a *shock surface*. If only the first derivatives of $\mathbf{U}$ are discontinuous, then we have a *weak discontinuity*.

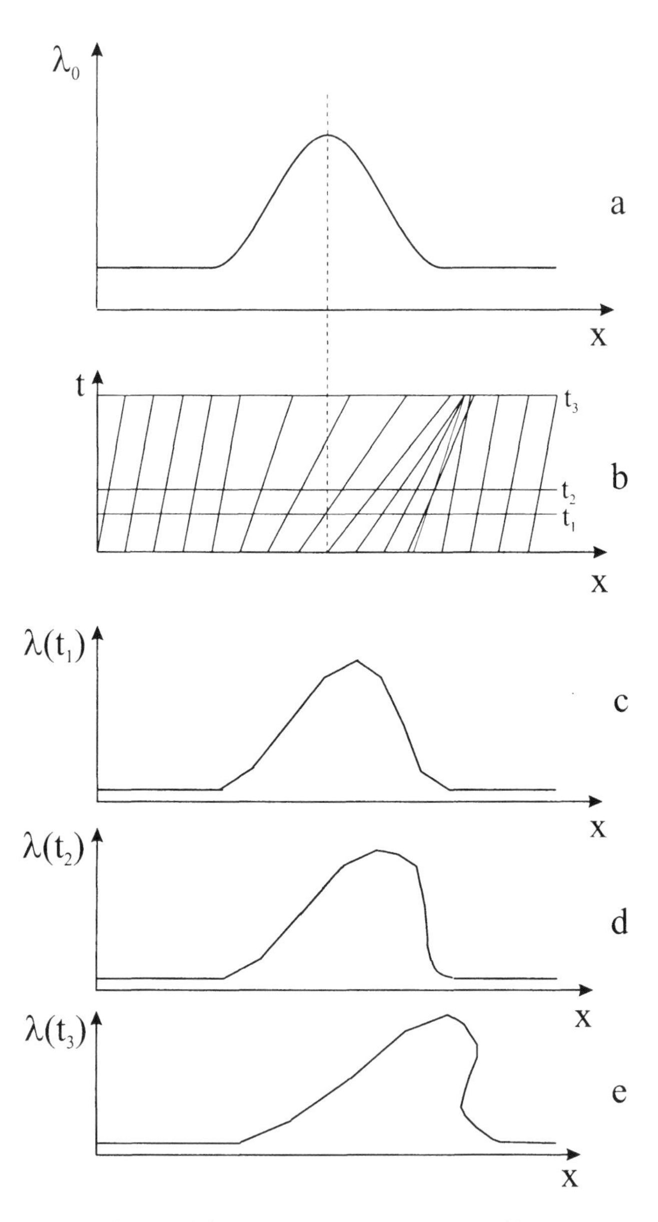

Figure 1.2 Breaking of the smooth profile.

Shock relations

Let us find the formulas relating unknown functions on the shock. Although such relations can be obtained directly in the multidimensional form (see, e.g., Godlewski and Raviart 1996), for the sake of simplicity we shall present their derivation for the system of the two variables x and t. In this case the system (1.2.14) transforms into

$$\frac{d}{dt}\int_{x_1}^{x_2} U_i(\mathbf{u})dx + F_i(\mathbf{u})|_{x=x_2} - F_i(\mathbf{u})|_{x=x_1} = 0. \tag{1.4.18}$$

Let the vector $\mathbf{u}$ be discontinuous on the line $x = X(t)$ in the x–t plane while remaining continuous on both sides of this line. Here we assumed that the shock moves from the left to the right and its velocity is

$$W = \frac{dX}{dt}. \tag{1.4.19}$$

If we denote as $u_i^{\mathrm{R}} = u_i(X+0,t)$ and $u_i^{\mathrm{L}} = u_i(X-0,t)$ the values of the functions u_i ahead of and behind the discontinuity, respectively, then for the fixed time instant from Eq. (1.4.18) we can easily obtain

$$\frac{dX}{dt}(U_i^{\mathrm{R}} - U_i^{\mathrm{L}}) + \int_{x_1}^{X} \frac{\partial U_i}{\partial t}dx + \int_{X}^{x_2} \frac{\partial U_i}{\partial t}dx + F_i(\mathbf{U}(x_2)) - F_i(\mathbf{U}(x_1)) = 0, \tag{1.4.20}$$

where $U_i^{\mathrm{R}} = U_i(\mathbf{u}^{\mathrm{R}})$, $U_i^{\mathrm{L}} = U_i(\mathbf{u}^{\mathrm{L}})$, and x_1 and x_2 are constant.

If we let $x_1 \to X+0$ and $x_2 \to X-0$, the integrals in Eq. (1.4.20) vanish and we obtain

$$W\{U_i\} = \{F_i\}. \tag{1.4.21}$$

Here by definition

$$\{f\} = f^{\mathrm{R}} - f^{\mathrm{L}}.$$

Note that $W = 0$ in the coordinate system attached to the shock and, hence, $\{F_i\} = 0$. This means that the flux vector remains constant in this frame. Relations (1.4.21), by analogy with gas dynamics, are called the Hugoniot relations.

Uniqueness of generalized solutions

Once we know that generalized solutions of hyperbolic systems can be discontinuous, the question arises whether solutions of this type are unique. It is well known (see, e.g., Rozhdestvenskii and Yanenko 1983) that satisfaction of the conservation equations and initial conditions is not sufficient to determine a unique solution of the hyperbolic system. This can be seen from the simple model equation

$$\frac{\partial u}{\partial t} + \frac{\partial}{\partial x}\left(\frac{u^2}{2}\right) = 0 \tag{1.4.22}$$

with the initial condition

$$u(x,0) = u_0(x) = \begin{cases} u^{\mathrm{L}} & \text{for } x < 0, \\ u^{\mathrm{R}} & \text{for } x > 0. \end{cases} \tag{1.4.23}$$

Since we are interested in the generalized solution, we seek it in the class of piecewise-continuous functions satisfying the integral equation

$$\oint_{\partial V} u\,dx - \tfrac{1}{2}u^2\,dt = 0 \tag{1.4.24}$$

and the initial condition (1.4.23).

The straightforward solution involving a single shock is

$$u_1(x,t) = \begin{cases} u^{\mathrm{L}} & \text{for } x < Wt, \\ u^{\mathrm{R}} & \text{for } x > Wt \end{cases} \tag{1.4.25}$$

with the shock speed $W = \frac{1}{2}(u^{\mathrm{L}} + u^{\mathrm{R}})$.

Let $u^{\mathrm{L}} < u^{\mathrm{R}}$. Then we can construct another solution to the problem (1.4.22)–(1.4.23) in the form

$$u_2(x,t) = \begin{cases} u^{\mathrm{L}} & \text{for} \quad x \le u^{\mathrm{L}}t, \\ \dfrac{x}{t} & \text{for } u^{\mathrm{L}}t \le x \le u^{\mathrm{R}}t, \\ u^{\mathrm{R}} & \text{for} \quad x \ge u^{\mathrm{R}}t. \end{cases} \tag{1.4.26}$$

Thus, we encountered the fact of the solution nonuniqueness, whereas one would expect a unique solution of the Cauchy problem in the class of discontinuous functions.

To determine the unique solution, the following assumptions (see Rozhdestvenskii and Yanenko 1983) can be adopted:

- any classical solution, if exists, is also a solution in the generalized sense;
- limits of classical solutions are also solutions of the integral conservation laws in the class of discontinuous functions.

While the first assumption is fairly natural, the second one implies continuous dependence of the Cauchy problem solutions on initial data. Thus, the above assumptions are based on the well-posedness of the Cauchy problem. Later on we shall discuss the requirements to generalized solutions in detail. The above two assumptions, however, allow us to make a choice between the solutions u_1 and u_2 of the Cauchy problem (1.4.22)–(1.4.23).

In Figs. 1.3a and 1.3b the behavior of characteristics $x = x_0 + u(x,t)\,t$ for the solutions u_1 and u_2, respectively, are presented. If we smear the initial data (1.4.23) in the vicinity of the shock and use the obtained function $u_\delta(x)$ as a new initial profile (see Fig. 1.4a) that coincides with $u_0(x)$ outside the interval $|x| \le \delta$, the characteristic behavior represented in Fig. 1.4b will tend to that shown in Fig. 1.3b as $\delta \to 0$. Thus, in accordance with our assumptions, only u_2 is a valid solution to the Cauchy problem under consideration. In fact, the first solution is completely artificial, since the shock inherent in it occurs at the intersection points of characteristics that originate at infinity rather than on the axis $t = 0$ of the initial data. If on the contrary $u_{\mathrm{L}} > u_{\mathrm{R}}$, then the shock will be characteristically consistent with the initial data and the solution u_1 will be valid. The principles formulated by Rozhdestvenskii and Yanenko (1983) are of great importance in view of the shock-capturing numerical approach, which is frequently used to obtain discontinuous solutions of hyperbolic systems. The discontinuities in this approach are represented by sharp gradients of appropriate functions on the computational mesh. We can hope that the solutions obtained

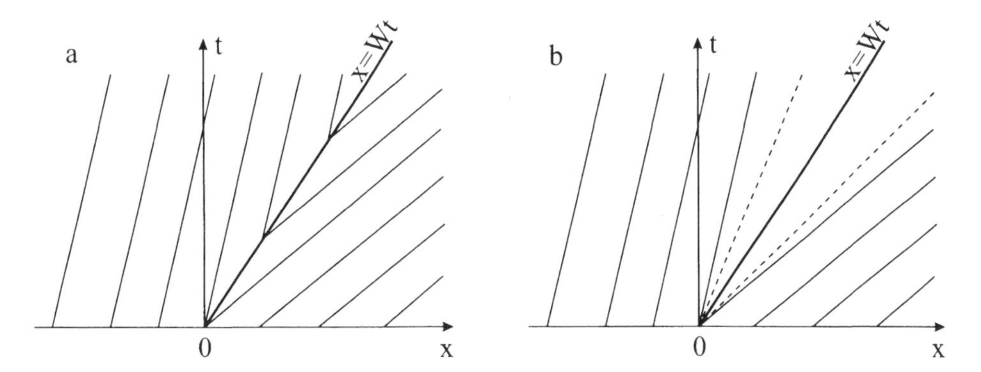

Figure 1.3 Characteristics corresponding to the generalized solutions u_1 (a) and u_2 (b).

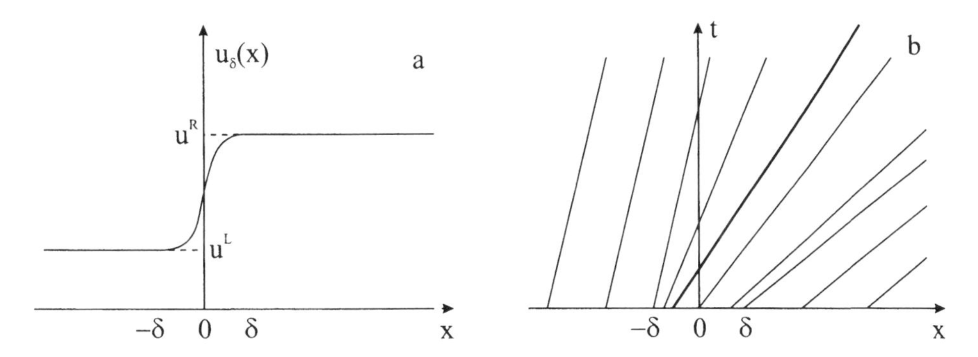

Figure 1.4 A smeared initial profile (a) and corresponding characteristics (b).

by this approach will tend to truly discontinuous solutions of the hyperbolic system as we refine the mesh. On the other hand, the zones of high gradients occurring in these solutions are obviously governed by numerical viscosity. Thus, we can say that a shock is admissible if it can be obtained by steepening of a corresponding viscous profile.

1.4.3 Small-amplitude shocks.

We shall now consider the discontinuities satisfying the Hugoniot relations (1.4.21) and assume that

$$U^{\mathrm{L,R}} = \lim_{\epsilon \to 0} U(X(t) \mp \epsilon, t).$$

In what follows the superscript L will be omitted. The equations describing continuous smooth solutions have a standard quasilinear form resulting from Eq. (1.4.1),

$$\frac{\partial U_i}{\partial t} + F_{ij}\frac{\partial U_j}{\partial x} = 0, \qquad F_{ij} = \frac{\partial F_i(\mathbf{U})}{\partial U_j}. \tag{1.4.27}$$

For any fixed initial values of U_i^{R}, Eq. (1.4.21) determines a curve in the U_i-space and the value of W on it. This curve passes through the initial point U_i^{R} and is called a shock adiabatic curve, or a Hugoniot curve. Consider small-amplitude discontinuities for which the quantities $\{U_i\}$ are small (Lax 1957). Expanding $\{F_i\}$ in a power series of $\{U_j\}$ and retaining only the first two terms, we obtain

$$\{F_i\} = F_{ij}^{\mathrm{R}}\{U_j\} + \frac{1}{2}\left(\frac{\partial^2 F_i}{\partial U_j \partial U_k}\right)^{\mathrm{R}} \{U_j\}\{U_k\}. \tag{1.4.28}$$

The expansion coefficients are calculated ahead of the shock.

In accordance with (1.4.28) we can rewrite Eq. (1.4.21) in the form

$$(F_{ij}^* - W\delta_{ij})\{U_j\} = 0, \tag{1.4.29}$$

where

$$F_{ij}^* = F_{ij}^{\mathrm{R}} + \frac{1}{2}\left(\frac{\partial F_{ij}}{\partial U_k}\right)^{\mathrm{R}} \{U_k\}.$$

The matrix F_{ij}^* is calculated at the point $U_i^{\mathrm{R}} + \frac{1}{2}\{U_i\}$, which is the middle of the chord connecting the initial point U_i^{R} with a point U_i on the Hugoniot curve.

It follows from Eq. (1.4.29) that (i) W coincides with the characteristic velocity at the middle of the above-mentioned chord and (ii) the chord direction coincides with that of the corresponding eigenvector $\mathbf{r}^*$ of the matrix F_{ij} at the chord middlepoint.

Let $\{U_i\} \to 0$. Then from Eq. (1.4.29) we obtain that the velocity W of an infinitely weak shock is equal to the characteristic velocity λ^{R} and the vector $\mathbf{t}^{\mathrm{R}}$ tangent to the Hugoniot curve at the initial point coincides with the right eigenvector $\mathbf{r}^{\mathrm{R}}$ of the matrix F_{ij}^{R}. Assuming that in a small vicinity of the initial point U_i^{R} the characteristic velocity λ up to small correction terms is a linear function of U_i, from (ii) we obtain that

$$W = \lambda^* = \frac{1}{2}(\lambda^{\mathrm{R}} + \lambda) + O(\{\lambda\}^2). \tag{1.4.30}$$

Let us show that the curvature of the shock curve at the initial point coincides with the curvature of the integral curve of the Riemann wave passing through this point. For the unit vectors $\mathbf{t}$ and $\mathbf{r}$ tangent to the Hugoniot curve and to the integral curve of the Riemann wave, respectively, the following formulas are valid:

$$\mathbf{t}\left(\frac{l}{2}\right) = \mathbf{t}^{\mathrm{R}} + \left(\frac{\partial \mathbf{t}}{\partial l}\right)^{\mathrm{R}} \frac{l}{2} + O(l^2), \tag{1.4.31}$$

$$\mathbf{r}\left(\frac{l}{2}\right) = \mathbf{r}^{\mathrm{R}} + \left(\frac{\partial \mathbf{r}}{\partial l}\right)^{\mathrm{R}} \frac{l}{2} + O(l^2), \tag{1.4.32}$$

where l is the distance from the initial point U_i^{R} to U_i along the corresponding curve. The derivatives $(\partial\mathbf{t}/\partial l)^{\mathrm{R}}$ and $(\partial\mathbf{r}/\partial l)^{\mathrm{R}}$ are the curvatures of the corresponding curves at the initial point. The tangent to the Hugoniot curve at the point $l/2$ with an accuracy to $O(l^2)$ is directed along the chord of the arc of length l. According to (ii), this direction to the same accuracy is defined by the eigenvector $\mathbf{r}^*$ of the matrix $F_{ij}^* = F_{ij}(U_k^{\mathrm{R}} + \frac{1}{2}\{U_k\})$, that is, by the matrix F_{ij} evaluated at the middle of the chord.

As the integral curve of the Riemann wave and the Hugoniot curve are tangent to each other at the initial point, the points corresponding to $l/2$ on both curves and the middle of the chord of the Hugoniot curve with the arc length equal to l are separated by the distance of the order of $O(l^2)$. Thus, we can conclude that the left-hand side of (1.4.32) also coincides with $\mathbf{r}^*$ to the same order of accuracy.

Since the left-hand sides of Eqs. (1.4.31)–(1.4.32) and the first terms on their right-hand sides coincide, the curvatures of the Hugoniot curve and the Riemann wave integral curve are also the same at the initial point, that is,

$$\left(\frac{\partial \mathbf{t}}{\partial l}\right)^{\mathrm{R}} = \left(\frac{\partial \mathbf{r}}{\partial l}\right)^{\mathrm{R}}$$

As apparent from (1.4.30), the variation of W and λ occurs in the same direction both along the shock curve and along the corresponding Riemann wave. In particular, the segment of the Hugoniot curve with growing W and the segment of the Riemann wave with decreasing λ can be combined at the initial point into one curve with a continuous tangent and curvature. This will be used later for the construction of solutions of certain self-similar problems.

1.4.4 Evolutionary conditions for shocks. In the general case, discontinuities of solutions are surfaces on which the conditions are imposed that relate the quantities on both sides of the discontinuities. These conditions usually involve the discontinuity velocity W. For hyperbolic systems in the conservation-law form, these relations have the form (1.4.21). Note that $\mathbf{U} = [U_i]$ and $\mathbf{F} = [F_i]$ are the conservative variables and the fluxes of them through a unit area of the discontinuity surface.

The evolutionary conditions are necessary conditions for resolvability of the problem of the discontinuity interaction with small disturbances depending on the x-coordinate normal to the discontinuity surface. Consider small disturbances $\delta U_i^{\mathrm{L,R}}$ propagating through the states $U_i^{\mathrm{L,R}}$ behind and ahead of the discontinuity. Linearizing Eq. (1.4.21), we obtain n relations for δU_i^{L}, δU_i^{R}, and the disturbance δW of the shock velocity. For hyperbolic equations, linear one-dimensional disturbances can be represented as a superposition of n waves, each one being a travelling wave propagating at the characteristic velocity $\lambda_i^{\mathrm{L,R}}$. This allows us to subdivide all these waves into incoming and outgoing ones, depending on the sign of the difference $\lambda_i^{\mathrm{L,R}} - W$. Incoming waves are fully determined by the initial conditions, while outgoing ones must be determined from the linearized boundary conditions on the shock.

Each of the linear waves is described by a single quantity w_i called the amplitude, see Eq. (1.4.12). The disturbances of all quantities can be expressed in terms of these amplitudes. It is obvious that w_i and δU_i are related by a linear invertible transformation.

Performing the same transformation in the linearized relations on the discontinuity, we obtain n linear equations relating $2n + 1$ quantities w_i^{L}, w_i^{R}, and δW with the coefficients depending on W, U_i^{L}, and U_i^{R}. According to the above considerations, only δW and the amplitudes w_i^{L} and w_i^{R}, which correspond to outgoing waves, are to be determined from this system.

Let s^{R} and s^{L} be the numbers of rightward and leftward outgoing waves, respectively. The number of quantities to be determined from the linearized system of the Hugoniot relations is thus equal to $s^{\mathrm{R}} + s^{\mathrm{L}} + 1$. If this number is equal to the number of equations, that is,

$$s^{\mathrm{R}} + s^{\mathrm{L}} + 1 = n, \tag{1.4.33}$$

then, in the general case of a nonzero determinant of the coefficient matrix, the problem is uniquely resolvable. This means that small incoming disturbances generate small outgoing disturbances and small δW. Equality (1.4.33) is called the Lax condition (Lax 1957).

If (1.4.33) is satisfied, that is, if the number of weak outgoing disturbances of different types is equal but one to the number of boundary conditions at the discontinuity, the discontinuity is called *evolutionary* (Landau and Lifshitz 1987). Otherwise, it is nonevolutionary.

If

$$s^{\mathrm{R}} + s^{\mathrm{L}} + 1 > n, \tag{1.4.34}$$

that is, if the number of unknown quantities to be determined is greater than the number of boundary conditions, these quantities cannot be found uniquely and depend on one or more arbitrary functions of time. This implies that such discontinuities do not exist or the conditions on them are underdetermined and there are physical reasons to impose additional, independent of (1.4.21), boundary conditions that make the discontinuity evolutionary (see Chapter 7).

If

$$s^{\mathrm{R}} + s^{\mathrm{L}} + 1 < n, \tag{1.4.35}$$

then the linearized boundary conditions cannot be satisfied in the general case by means of the quantities to be determined. Thus, the problem of the discontinuity interaction with small disturbances has no solution in the linear approximation. Since we expect a well-posed physical problem to have a solution, this means that finite (not small) deviations from the initial state must occur. Previous studies of various physical problems show that interaction of nonevolutionary discontinuities with small disturbances results in their disintegration into two or more evolutionary discontinuities (see Chapter 7).

The evolutionary condition (1.4.33) can be rewritten in the form of inequalities relating the shock velocity W and the velocities $\lambda_i^{\mathrm{L,R}}$ of small disturbances. Let us enumerate the characteristic velocities on both sides of the discontinuity as follows:

$$\lambda_1 \leq \lambda_2 \leq \ldots \leq \lambda_n. \tag{1.4.36}$$

Condition (1.4.33) can then be rewritten in the form

$$\lambda_k^{\mathrm{R}} \leq W \leq \lambda_{k+1}^{\mathrm{R}}, \qquad \lambda_{k-1}^{\mathrm{L}} \leq W \leq \lambda_k^{\mathrm{L}}, \tag{1.4.37}$$

where $k = 1, \ldots, n$. To avoid misinterpreting, we must put here $\lambda_0^{\mathrm{L}} = -\infty$ and $\lambda_{n+1}^{\mathrm{R}} = \infty$.

Inequalities (1.4.37) allow us to divide all evolutionary shocks into n types depending on the value of k. The shock satisfying the relation (1.4.37) is called k-shock. The above relations can be represented as an evolutionary diagram (Akhiezer, Lyubarskii and Polovin 1958). The values of W and λ_k^{R} on the horizontal axis of this diagram are represented in the real scale, whereas the values of W and λ_k^{L} on the vertical axis are arbitrarily scaled with retaining the inequalities between the quantities. The straight lines parallel to the axes and passing through the points λ_k^{L} and λ_k^{R} divide the plane into several rectangles (see Fig. 1.5). If the point (W, W) lies in one of the outlined rectangles, the evolutionary inequalities (1.4.37) are satisfied.

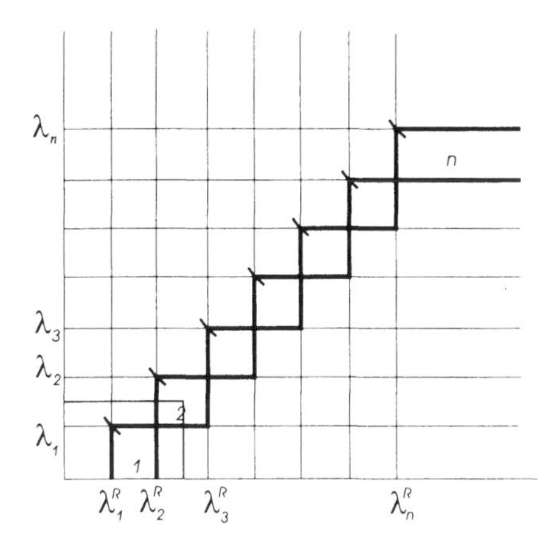

Figure 1.5 Evolutionary diagram.

Let us check whether small-amplitude shocks are evolutionary. According to Section 1.4.3, the k-shock velocity in this case is

$$W = \frac{1}{2}(\lambda_k^{\mathrm{L}} + \lambda_k^{\mathrm{R}}), \tag{1.4.38}$$

under the assumption that the difference $\lambda_k^{\mathrm{L}} - \lambda_k^{\mathrm{R}}$ is small for a chosen k. The differences $\lambda_j - W$ for $j \neq k$ are not considered small and therefore cannot change the sign when crossing the discontinuity. According to (1.4.38), the inequality $\lambda_k^+ > W$ holds for $W > \lambda_k^-$ and the characteristics of the same (kth) family approach the discontinuity from both its sides. The remaining $n-1$ characteristics arrive at the discontinuity from one side and leave it from another. Thus, there are $n-1$ outgoing characteristics and such discontinuity is evolutionary.

If $W < \lambda_k^{\mathrm{R}}$, then we have $W > \lambda_k^{\mathrm{L}}$ and the characteristics of the kth type leave the discontinuity on both its sides. We have $n+1$ outgoing characteristics and this discontinuity is nonevolutionary. In Fig. 1.5 weak shock waves are shown as the segments of the curves passing through the points $(\lambda_k^{\mathrm{R}}, \lambda_k^{\mathrm{L}})$, $k = 1, \ldots, n$.

The shocks whose velocity coincides with one of the characteristic velocities can also be evolutionary. Evolutionarity of these limiting shocks (we shall call them Jouget discontinuities, or Jouget shocks), however, must be checked separately in each individual case. For example, weak shocks for which

$$\lambda_k^{\mathrm{R}} = W = \lambda_k^{\mathrm{L}}$$

are evolutionary. This a reason for using the equality signs in the relations (1.4.37).

1.4.5 Entropy behavior on discontinuities.

The concept of entropy plays an important role in continuum mechanics. Godunov (1961, 1978) introduced an important class of hyperbolic partial differential equations expressing the conservation laws for which the notion of entropy is defined. It is assumed that the conservative system (1.4.1) results in one additional conservation equation for entropy in the form

$$\frac{\partial S(\mathbf{U})}{\partial t}+\frac{\partial F(\mathbf{U})}{\partial x}=0. \tag{1.4.39}$$

It can be obtained from (1.4.1) if we multiply the latter by factors q_i and perform summation over i. Choosing q_i as independent variables and introducing the functions

$$T=S-q_iU_i, \qquad H=F-q_iF_i, \tag{1.4.40}$$

we can rewrite Eqs. (1.4.1) and (1.4.39) in the form

$$\frac{\partial}{\partial t}\left(\frac{\partial T}{\partial q_i}\right)+\frac{\partial}{\partial x}\left(\frac{\partial H}{\partial q_i}\right)=0, \tag{1.4.41}$$

$$\frac{\partial}{\partial t}\left(T-q_i\frac{\partial T}{\partial q_i}\right)-\frac{\partial}{\partial x}\left(H-q_i\frac{\partial H}{\partial q_i}\right)=0. \tag{1.4.42}$$

Equations (1.4.41)–(1.4.42) represent a canonical form of the Godunov system. It is apparent that Eq. (1.4.39) can result from Eq. (1.4.1) only if

$$dS=q_idU_i, \qquad dF=q_idF_i.$$

We see that in the one-dimensional case only two unknown functions T and S occur in it. The function $T(\mathbf{q})$ is a convex function if the function $S(\mathbf{q})$ is convex, since $-S$ and T are related via the Legendre transform

$$S=T-q_i\frac{\partial T}{\partial q_i}, \qquad F=H-q_i\frac{\partial H}{\partial q_i}.$$

In mechanics of continuous media S is interpreted as entropy and F as its flux.

Let us introduce the entropy production at the discontinuity as the difference between the entropy inflow and outflow

$$\{P(q_i,W)\}=\{WS-F\}=W\{T-q_iT_i\}-\{H-q_iH_i\}, \tag{1.4.43}$$

where

$$T_i=\frac{\partial T}{\partial q_i}, \qquad H_i=\frac{\partial H}{\partial q_i}. \tag{1.4.44}$$

According to the second law of thermodynamics, $\{P\}$ must be nonnegative.

Let us estimate the variation of $\{P\}$ along the Hugoniot curve. This curve for a given state q_i^- ahead of the discontinuity is described by the equation

$$\{H_i\}-W\{T_i\}=0 \tag{1.4.45}$$

corresponding to the conservation law (1.4.41).

Differentiating Eqs. (1.4.43) and (1.4.45) under assumption $q_i^{\mathrm{R}}=\text{const}$ and eliminating dq_i, we obtain

$$d\{P\}=[T-T^{\mathrm{R}}-T_i^{\mathrm{R}}(q_i-q_i^{\mathrm{R}})]\,dW. \tag{1.4.46}$$

If the function $T(\mathbf{q})$ is convex, the sign of the right-hand side of Eq. (1.4.46) coincides with the sign of dW. Discontinuities with $W\geq\lambda^{\mathrm{R}}$, which correspond to the Hugoniot curve points adjacent to the initial point and belonging to the segment with nondecreasing W, do not in this case contradict the second law of thermodynamics.

1.5 Disintegration of a small arbitrary discontinuity

Let us consider the Riemann problem describing the disintegration of an arbitrary initial discontinuity, or simply the Riemann problem, for brevity. This is the Cauchy problem for the system

$$\frac{\partial \mathbf{U}}{\partial t} + \frac{\partial \mathbf{F}(\mathbf{U})}{\partial x} = \mathbf{0}$$

with special initial conditions in the form (Riemann 1860)

$$\mathbf{U}(x,0) = \begin{cases} \mathbf{U}^{\mathrm{R}} & \text{for } x > 0, \\ \mathbf{U}^{\mathrm{L}} & \text{for } x < 0. \end{cases} \tag{1.5.1}$$

Let us find the form of the solution of this problem for $t > 0$ and for small differences $U_i^{\mathrm{R}} - U_i^{\mathrm{L}}$, where U_i are the components of the vector of conservative variables $\mathbf{U}$. The Riemann problem is self-similar and its solution must depend on x/t and consist of the Riemann waves, regions of constant U_i, and discontinuities.

Examine, first, the solution of a linearized problem assuming that all characteristic velocities λ_i are different. Since we have a discontinuity at $t = 0$ and $x = 0$, the solution consists of n characteristic waves (1.4.12). Each of them is a half-infinite step profile and has a jump at its right end. The discontinuities propagate at their own velocities λ_i. The variation of parameters in the wave is proportional to the right eigenvector $\mathbf{r}^i$ of the matrix

$$F_{lm} = \frac{\partial F_l}{\partial U_m},$$

which is assumed constant in the linear approximation. The proportionality factors w_i will be called wave amplitudes, provided that after proper normalization the eigenvectors have unit lengths. The problem of determining the wave amplitudes reduces to decomposition of the vector $\mathbf{U}^{\mathrm{L}} - \mathbf{U}^{\mathrm{R}}$ in terms of the eigenvectors $\mathbf{r}^i$,

$$\mathbf{U}^{\mathrm{L}} - \mathbf{U}^{\mathrm{R}} = \sum_i w_i \mathbf{r}^i.$$

Since eigenvectors corresponding to different eigenvalues are linearly independent, the problem is resolvable. The amplitudes w_i can be considered as a coordinate system in the vicinity of the initial point. As follows from the above discussion, the Jacobian of the coordinate transformation matrix is not equal to zero. For $t = \text{const} > 0$, the motion from the right to the left along the x-axis corresponds in the U_i-space to a sequence of jumps, distributed in the order of decreasing λ_i and located along the coordinate lines w_i. They form a broken line connecting the points U_i^{R} and U_i^{L}

If nonlinearity must be taken into account, we still seek the solution of the Riemann problem as a sequence of n waves. The rightmost wave corresponds to the highest characteristic velocity λ_n. Depending on the direction of the quantity variation in the wave, it can be either an expanding Riemann wave with λ_n decreasing from the forward front to the backward one, or an evolutionary shock with $\lambda_n^{\mathrm{R}} < W < \lambda_n^{\mathrm{L}}$. The latter inequality excludes coexistence of the n-shock and the n-Riemann wave. The state behind the nth wave is the point in the U_i-space on the curve composed of the part of the Riemann wave integral curve with decreasing λ_n and of the evolutionary segment of the Hugoniot curve along which the shock velocity grows in the direction from the initial point. As shown earlier, this is the curve with the continuous tangent and curvature at the initial point.

The next change occurs in the wave of the $(n-1)$th type, which can also be either a Riemann or a shock wave. If n- and $(n-1)$-shocks are weak, their velocities are close to the corresponding characteristic velocities, and these shocks turn out to be separated on the x-axis by the region of parameters independent of x. The state behind the $(n-1)$-wave belongs to the curve in the U_i-space composed of the evolutionary part of the Hugoniot curve and the segment of the Riemann wave integral curve corresponding to an expanding wave.

Proceeding with the construction of the solution, we obtain a broken line whose ith segment is the segment of a curve corresponding to the ith wave. The lengths and directions of the segments must be chosen in a way such that the broken line connects the points U_i^{R} and U_i^{L}. The lengths of the segments together with the signs determining their directions can be considered as new coordinates in the vicinity of the point U_i^{R} in the U_i-space.

In an infinitely small vicinity of the initial point this coordinate transformation reduces to that considered above when constructing the solution of the linearized problem (the segments of the curve in this case are replaced by the elements of their tangents and the variation in the eigenvector directions is disregarded). Thus, the Jacobian of the transformation from U_i to the new coordinate system is not equal to zero at the initial point. By continuity, this means that it is not equal to zero in a vicinity of this point. If U_i^{L} belongs to this vicinity, the Riemann problem can be uniquely resolved.

It is worth noting that the described classical behavior of the solution to the Riemann problem can be violated for non-small $\mathbf{U}^{\mathrm{L}} - \mathbf{U}^{\mathrm{R}}$. There are a few reasons for this violation. One of them is in the nonuniqueness of the transformation from U_i to the variables characterizing the wave amplitudes. Another one is in the appearance of new types of discontinuities with additional relations to be satisfied on them. These cases will be considered in detail in Chapter 7.

Chapter 2
Numerical Solution of Quasilinear Hyperbolic Systems

In this chapter we describe basic approaches to constructing shock-capturing and shock-fitting methods for solving multidimensional quasilinear hyperbolic systems of general form. Among numerical methods we mainly select those that are based on the exact or approximate solution of the corresponding one-dimensional Riemann problem of disintegration of an arbitrary discontinuity or can be interpreted as based on this solution. Such methods are called Godunov methods. They proved to be extremely fruitful in numerous applications. This is due to the fact that the Godunov-type methods are based on the fundamental properties of hyperbolic systems.

The numerical algorithms described below can adequately predict the propagation of discontinuities, which are common for quasilinear hyperbolic systems, and simulate monotone profiles of grid variables in the vicinity of discontinuities.

A method for solving the Riemann problem will be referred to as a Riemann problem solver or simply a solver. We present numerical schemes using both the exact solver and several approximate Riemann problem solvers. The approximate Riemann problem solvers include the Courant–Isaacson–Rees (CIR), Roe and Osher solvers. In the Osher solver the solution is constructed of the Riemann waves only. The CIR and Roe solvers are based on the solution of the Riemann problem for a linearized hyperbolic system of equations. In this case the solution contains only travelling discontinuities, since in the linearized problem there is no difference between a Riemann wave and a travelling discontinuity, see Chapter 1. Indeed, all Riemann waves are represented in this problem by step-functions dividing the regions of constant parameters. The solvers to be described permit one to construct finite-difference and finite-volume schemes for both conservative and nonconservative hyperbolic systems.

We describe also some specific issues of reconstruction of discrete mesh functions, the generalized Riemann problem, additional monotonization procedures, algorithms for selecting physically admissible solutions, and others.

The basic methods presented in this chapter are written for the general hyperbolic system. In Chapters 3–6 they will be applied to the construction of specific numerical algorithms and Riemann problem solvers for gas dynamic equations, shallow water equations, and equations of magnetohydrodynamics and solid dynamics.

2.1 Introduction

The construction of new numerical methods and modification of known methods in order to improve their efficiency have always been topical problems of computational sciences. This is connected both with practical demand to obtain numerical solutions of new complicated problems and the logic of development of numerical methods as a theoretical branch of mathematical sciences.

It is well known that solutions of various problems of mathematical physics described by hyperbolic systems can be smooth in one subdomain and discontinuous in another (Rozhdestvenskii and Yanenko 1983; Petrovskii 1991; Godlewski and Raviart 1996), see Chapter 1. Note that discontinuous solutions can arise even from smooth initial data. Such properties of solutions result in contradictory requirements on algorithms of numerical calculations. The algorithms must preserve the monotonicity of the unknown functions in subdomains where these functions have large gradients and simultaneously ensure high order of accuracy in subdomains where the solution varies smoothly. Godunov's theorem (1959) states that within the framework of linear finite-difference schemes, these two requirements cannot be met simultaneously.

To overcome this difficulty, *shock-fitting* finite-difference methods can be applied, which are based on a direct fitting of discontinuities in the solution. This fitting is produced by appropriate generation of a discrete mesh associated with discontinuities. In particular, the method of characteristics can be used here, see Zhukov (1960) and Richardson (1964). As far as the shock-fitting methods are concerned, we can subdivide them into several groups. One of them is represented by genuinely shock-fitting methods. They are applied if the internal structure of the solution, as well as the number and type of each discontinuity are known in advance. The location and velocity of the discontinuities are to be determined. In this case, one makes an initial guess about the location of a discontinuity and organizes a numerical process so that in the calculation of the derivatives, using finite differences crossing the discontinuity is not allowed. This implies that the numerical grid must be adjusted to the discontinuity surface. This can be done if we always have grid points on this surface. Alternatively, in the finite volume methods computational cell surfaces must coincide with discontinuities. Note that the initial conditions may not satisfy the discontinuity relations; nevertheless, in this case the discontinuities will move in order to finally adjust themselves to these relations. The steady-state solution is obtained if all discontinuities have zero velocity.

It is obvious that approximation of derivatives in the vicinity of a discontinuity must only involve one-sided differences. This may require invoking the characteristic properties of the hyperbolic system to choose the correct direction of the wave propagation. The relations on discontinuities occurring in solutions of hyperbolic equations are satisfied exactly in shock-fitting methods. Note also that, since we perform numerical approximation of derivatives only in smooth regions, requirements on the choice of a particular numerical scheme are not so strict as in the case of uniform methods, known also as *shock-capturing* methods. These methods smear all discontinuities over a length scale determined by the numerical dissipation of the scheme and transform the discontinuities into narrow domains with large gradients. The widths of these domains are smaller for higher-order numerical schemes. On the other hand, spurious oscillations, inevitable in this case, manifest themselves mainly in the vicinity of discontinuities and must be damped by an artificial viscosity. For example, either a linear or a quadratic viscosity (von Neumann and Richtmyer 1950) can be introduced; for details see Richtmyer and Morton (1967), Roache (1976), and Wilkins (1980). It is worth mentioning that the use of the artificial viscosity may essentially change the solution (Latter 1955), and the numerical results must be thoroughly verified. In smooth regions nonmonotone low-viscosity central numerical schemes can also be applied (Lax and Wendroff 1960, 1964; MacCormack 1969). Some applications of genuinely shock-fitting methods will be described in Chapter 3, which deals with equations of gas dynamics.

Theoretically, shock-fitting methods fit all discontinuities, although this seems to be possible only in the one-dimensional case. As far as the two- and three-dimensional cases are concerned, the fitting of all discontinuities can encounter substantial difficulties. In these cases, one can fit only a few main surfaces of discontinuities. The numerical modelling in domains between these surfaces can be carried out by a uniform (shock-capturing) finite-difference or finite-volume scheme. Such an approach with partial fitting of discontinuities is widely used, see Moretti (1963), Moretti and Abbett (1966), Moretti and Bleich (1967), Richtmyer and Morton (1967), Lyubimov and Rusanov (1970), and Roache (1976).

Another group of shock-fitting methods will be referred to as *floating shock-fitting* methods. These methods are designed to fit all discontinuities that originate with time. This requires the development of algorithms for their detection and further tracking as boundaries of smooth subregions of numerical calculation. Algorithms of this sort are becoming more and more complicated if the number of discontinuities to be fitted increases. In Sections 2.9 and 3.5 we outline this approach and give appropriate references.

It is clear that in order to avoid spurious oscillations near discontinuities, one should add viscosity in their vicinity. On the other hand, a higher-order approximation is preferable in smooth regions. Another approach to the numerical investigation of a solution to a hyperbolic system of equations with different properties in different subdomains is the use of *hybrid* shock-capturing schemes, or schemes of varying order of accuracy. The hybridity means that the numerical scheme can locally change its properties, for example, the order of accuracy. In particular, the hybridity permits one to carry out shock-capturing calculations within the framework of the second or higher order of accuracy of the scheme in subdomains with a smooth solution and within the framework of a first-order monotone scheme in subdomains where the solution has large gradients. This approach permits one to combine the positive properties of different methods in a shock-capturing algorithm.

One can simultaneously apply shock-fitting and shock-capturing approaches. A combination of a discontinuity-fitting technique with shock-capturing schemes in domains between the discontinuity surfaces can also be quite useful. Also it seems favorable to use shock-capturing numerical schemes based on moving (dynamic) adaptive meshes (McRae and Lafli 1999; Zegeling 1999; Ivanenko 1999; Azarenok and Ivanenko 1999, 2000).

The basic methods of the fitting technique were created 30 to 40 years ago and are topical to the present day. Today, new fitting methods appear very rarely. In contrast, the shock-capturing numerical schemes are under constant development.

Primarily, shock-capturing numerical methods of a fixed order of accuracy were devised. First-order methods were developed by Courant, Isaacson, and Rees (1952), Lax and Friedrichs (Lax 1954), and Godunov (1959). Subsequently, methods of the second order were suggested by Lax and Wendroff (1960, 1964), MacCormack (1969), and Kutler, Lomax, and Warming (1973). The methods of third order of accuracy were developed by Rusanov (1968, 1970), Burstein and Mirin (1970), Abarbanel and Zwas (1971), and Kutler, Lomax, and Warming (1973). Schemes of the fourth order of accuracy were created by Abarbanel and Zwas (1971), Abarbanel and Gottlieb (1973), and Abarbanel, Gottlieb, and

Turkel (1975). In parallel with the practical application of the above numerical methods, comparative reviews of different methods were published, see Emery (1968), Taylor, Ndefo, and Masson (1972), and Anderson (1974). Note also the reviews of reviews published by Srinivas, Gururaja, and Krishra (1976) and Sod (1978). Those papers and practical requirements stimulated creation of the hybrid schemes, or schemes with the variable order of accuracy.

The hybrid difference schemes were the first stage in the development of the shock-capturing schemes of variable order of accuracy. In the simplest case, a hybrid scheme is a combination of two schemes. This combination has the form $gS_1 + (1 - g)S_2$, where S_1 is a first-order scheme, S_2 a second-order scheme, and g the hybridity coefficient, $0 \leq g \leq 1$. The first hybrid scheme was presented by Fedorenko (1962) for an advection equation. He suggested a hybrid difference scheme and a rule of local switching between two basic schemes S_1 and S_2 on the basis of an analysis of the ratio between the second difference of the solution and its first difference. Gol'din, Kalitkina, and Shishova (1965) developed several hybrid numerical schemes for linear and nonlinear advection equations with smooth switching between two schemes of the first and second order of accuracy. Their hybridity coefficient g depended on the gradient of the solution. The first hybrid scheme for a system of equations was presented by Harten and Zwas (1972a, 1972b). In particular, Harten and Zwas (1972a) combined the Lax–Friedrichs scheme (Lax 1954) of the first order of accuracy with the second-order scheme by Lax and Wendroff (1960, 1964). van Leer (1973, 1974) presented a special algorithm for the monotonization of the Lax–Wendroff method, see also a generalization of the monotonization by van Leer (1977a, 1977b). A hybridization of the Godunov method (1959) was first presented by Kolgan (1972, 1975). For the hybridization he used piecewise linear functions and was the first to apply a version of the minmod limiter. Kutler, Lomax, and Warming (1972), and Beam and Warming (1976) hybridized a symmetric and a nonsymmetric scheme. In their scheme the hybridity coefficient depended on the Mach number.

Boris and Book (1973, 1975, 1976) developed a hybrid method that permits one to increase the order of accuracy by a special procedure of flux corrected transport (FCT). At the first stage a numerical solution is calculated by a monotone first-order scheme. The second stage must modify the numerical solution and provide the second order of accuracy in time and space. This stage must not generate any new extrema in numerical solution and must not lead to an increase (or decrease) in the maxima (or minima) that already exist. Note that these requirements are equivalent to the condition of boundedness of the total variation of the numerical solution. Thus, the FCT method contains elements of the total variation diminishing (TVD) schemes (Harten 1983). To make the solution satisfy the TVD property, an instrument of piecewise linear (polynomial) function reconstruction was developed. The slopes of the function being reconstructed are limited by special functions called limiters. The limiters depend on finite differences. A detailed analysis of the properties of modern limiters was given by Sweby (1984) and, on a different basis, by Roe (1985), see also Yee (1989), Hirsch (1990), and Toro (1997).

Today the term *hybridity* is rarely used in the context of difference schemes. However, schemes of variable order of accuracy still exist and represent the main instrument of numerical simulation. The hybrid difference schemes have been transformed into modern schemes of variable order of accuracy by eliminating from consideration some formal and semiempirical, although interesting and sophisticated, approaches. In particular, the modern schemes of variable order of accuracy are based mainly on the piecewise-polynomial reconstruction of discrete mesh functions. However the simplest hybrid difference schemes based on taking into account specific features of hyperbolic systems may be useful as a first step in the numerical investigation of the hyperbolic system. Some of such schemes will be described below. Note that these schemes can be used for both conservative and nonconservative forms of equations.

Our consideration is based on a special selection of numerical schemes that allow us to give a clear physical interpretation of the schemes and establish their connection with the solution of the Riemann problem. This class of schemes is called Godunov methods. They are currently under intensive development. This is due to their efficiency in a number of numerical applications. This is accounted for by the fact that the Godunov-type methods are based on the fundamental properties of hyperbolic systems.

It should be noted that the knowledge of the solution to the Riemann problem may be very important by itself. The use of the Riemann problem solution may add to the reliability and accuracy of known numerical methods for hyperbolic systems. Recent examples of this are the works by Monaghan (1997) and Parshikov (1999), who used some elements of the linearized gas dynamic Riemann problem solution to improve the smooth particle hydrodynamics (SPH) method (Monaghan and Gingold 1983; Benz 1988; Monaghan 1989; Stellingwerf and Wingate 1993; Chow and Monaghan 1997; and V. D. Ivanov et al. 1999).

Hyperbolic systems of equations can also be applied in grid generation problems, see Steger and Chaussee (1980), Thompson, Warsi, and Mastin (1985), and Chan (1999). For the marching noniterative generation of orthogonal grids, Semenov (1995c, 1995d, 1996) applied a system which is hyperbolic only in the extended form (Courant and Lax 1949); see also Section 1.2.1. Matsuno (1999) developed a high-order accurate TVD upwind grid-generation method for both two- and three-dimensional grid generation problems.

2.2 Methods based on the exact solution of the Riemann problem

We shall consider the one-dimensional Riemann problem (see, for details, Section 1.4) for a quasilinear hyperbolic conservation law of the form

$$\frac{\partial \mathbf{U}}{\partial t} + \frac{\partial \mathbf{F}(\mathbf{U})}{\partial x} = \mathbf{0}, \quad A = \frac{\partial \mathbf{F}}{\partial \mathbf{U}}. \tag{2.2.1}$$

Here $\mathbf{U} = \mathbf{U}(t,x) = [U_1, \dots, U_n]^{\mathrm{T}}$, $\mathbf{F}(\mathbf{U}) = [F_1, \dots, F_n]^{\mathrm{T}}$, $t \geq 0$, $-\infty < x < \infty$, and $\mathbf{U}(0,x) = \mathbf{U}_0(x)$. The vector of the initial data, $\mathbf{U}_0$, is a vector step function,

$$\mathbf{U}_0(x) = \begin{cases} \mathbf{Q}_+, & x > 0, \\ \mathbf{Q}_-, & x < 0, \end{cases} \tag{2.2.2}$$

where $\mathbf{Q}_+$ and $\mathbf{Q}_-$ are constant vectors.

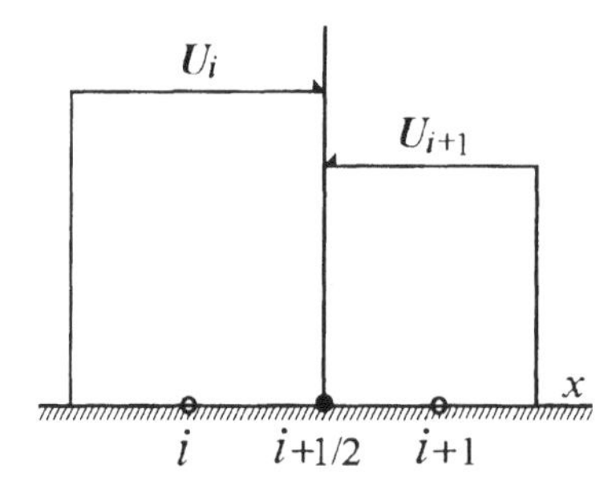

Figure 2.1 Piecewise constant distribution of **U**.

2.2.1 The Godunov method of the first order. In 1959, Godunov suggested a numerical scheme for solution of gas dynamic equations. This scheme is essentially based on either the exact or approximate solution of the Riemann problem (2.2.1)–(2.2.2). For the system (2.2.1) this method can be formulated as follows.

Let us introduce a discrete uniform mesh with size Δx. Let $\mathbf{U}_i$ be the mesh function values, where the integer subscript $i = 1, 2, \ldots$ refers to the center of the ith computational cell, see Fig. 2.1. The half-integer subscript $i \pm 1/2$ refers to the boundary between the cells with the numbers i and $i \pm 1$. Assume that all mesh functions are constant inside each space cell. Let the integer superscript $k = 0, 1, 2, \ldots$ indicate the time layer and Δt be the time increment. Then for the cell boundary $i + 1/2$ and for each time step we can solve the Riemann problem with the following initial data: $\mathbf{U}_i^k = \text{const}$ for $x < x_{i+1/2}$ and $\mathbf{U}_{i+1}^k = \text{const}$ for $x > x_{i+1/2}$. Let $\mathbf{U}_{i+1/2}$ be a solution of this problem. In the same way we can calculate $\mathbf{U}_{i-1/2}$ for the boundary $i - 1/2$. The Godunov finite-volume explicit scheme has the form

$$\frac{\mathbf{U}_i^{k+1} - \mathbf{U}_i^k}{\Delta t} + \frac{\mathbf{F}_{i+1/2} - \mathbf{F}_{i-1/2}}{\Delta x} = \mathbf{0}, \qquad \mathbf{F}_{i\pm 1/2} = \mathbf{F}(\mathbf{U}_{i\pm 1/2}). \tag{2.2.3}$$

This scheme is of the first order of accuracy with respect to the space variable and time.

The spectral analysis (Richtmyer and Morton 1967) of the linearized equations (2.2.3) leads to the stability condition

$$\max_p |C_p| \leq 1, \quad C_p = \lambda_p \frac{\Delta t}{\Delta x}, \tag{2.2.4}$$

where λ_p are the eigenvalues of Jacobian matrix A of (2.2.1). Inequality (2.2.4) is known as the CFL condition (Courant, Friedrichs, and Lewy 1928) and C_p is the CFL number corresponding to λ_p.

Scheme (2.2.3) can easily be extended to a nonuniform space mesh, self-adjusting and moving grids, shock-fitting calculations, two- or three-dimensional cases, etc. Let us consider, for instance, the two-dimensional hyperbolic system

$$\frac{\partial \mathbf{U}}{\partial t} + \frac{\partial \mathbf{F}(\mathbf{U})}{\partial x} + \frac{\partial \mathbf{E}(\mathbf{U})}{\partial y} = \mathbf{0}. \tag{2.2.5}$$

The scheme for this system for a uniform Cartesian space mesh can be written as

$$\frac{\mathbf{U}_{i,j}^{k+1}-\mathbf{U}_{i,j}^{k}}{\Delta t}+\frac{\mathbf{F}_{i+1/2,\,j}-\mathbf{F}_{i-1/2,\,j}}{\Delta x}+\frac{\mathbf{E}_{i,\,j+1/2}-\mathbf{E}_{i,\,j-1/2}}{\Delta y}=\mathbf{0}, \tag{2.2.6}$$
$$\mathbf{F}_{i\pm1/2,\,j}=\mathbf{F}(\mathbf{U}_{i\pm1/2,\,j}),\quad \mathbf{E}_{i,\,j\pm1/2}=\mathbf{E}(\mathbf{U}_{i,\,j\pm1/2}),$$

where Δx and Δy are the mesh sizes in the x- and y-direction, respectively. The double integer subscripts (i,j) refer to the centers of two-dimensional space cells and the half-integer subscripts refer to the corresponding boundaries of the cells. The quantities $\mathbf{U}_{i+1/2,\,j}$, $\mathbf{U}_{i-1/2,\,j}$, $\mathbf{U}_{i,\,j+1/2}$, and $\mathbf{U}_{i,\,j-1/2}$ are solutions of the corresponding Riemann problem. That is, we must solve the one-dimensional Riemann problem for each cell boundary. The scheme thus constructed is a two-dimensional finite-volume Godunov scheme for a uniform mesh.

It is not difficult to generalize the Godunov scheme (2.2.6) to an arbitrary space grid. Let us rewrite Eq. (2.2.5) in integral form as

$$\frac{d}{dt}\left(\iint_G \mathbf{U}\,dG\right)+\oint_S(\mathbf{F}\,dy-\mathbf{E}\,dx)=\mathbf{0}. \tag{2.2.7}$$

Here G is a domain in the two-dimensional (x,y) space, $dG = dx\,dy$ is the area element, and S is the boundary of G.

Let us construct the explicit Godunov finite-volume scheme for equations written in the integral form (2.2.7). We discretize the computational domain by constructing a grid of arbitrary convex polygons having areas G_i, $i = 1,2,\ldots$, with $m = m(i)$ sides having lengths S_j, $j = 1,\ldots,m(i)$; $\mathbf{S}_j = \mathbf{n}_j S_j \equiv [S_x,S_y]_j^{\mathrm{T}} = [S_{xj},S_{yj}]^{\mathrm{T}}$, where $\mathbf{n}_j$ is the outward normal to S_j. On each polygon the above integral equations can be approximated as follows:

$$G_i\frac{\mathbf{U}_i^{k+1}-\mathbf{U}_i^{k}}{\Delta t}+\sum_{j=1}^{m(i)}(\mathbf{F}_jS_{xj}^k+\mathbf{E}_jS_{yj}^k)=\mathbf{0}. \tag{2.2.8}$$

The integer subscript i in Eq. (2.2.8) denotes the values of grid variables calculated at the center of mass of the ith polygon, and the subscript j denotes their values on the middle of jth side of the polygon. The quantities $\mathbf{F}_j$ and $\mathbf{E}_j$ are calculated by solving the corresponding Riemann problem in the direction of the jth outward normal.

The Godunov finite-volume scheme in the case of general moving grids can be written out as

$$\frac{(G\mathbf{U})_i^{k+1}-(G\mathbf{U})_i^{k}}{\Delta t}+\sum_{j=1}^{m(i)}\left[(\mathbf{F}_j-D_{xj}\mathbf{U}_j)S_{xj}^{k+1/2}+(\mathbf{E}_j-D_{yj}\mathbf{U}_j)S_{yj}^{k+1/2}\right]=\mathbf{0}, \tag{2.2.9}$$

$$\frac{G_i^{k+1}-G_i^{k}}{\Delta t}-\sum_{j=1}^{m(i)}(\mathbf{D}_j\cdot\mathbf{S}_j^{k+1/2})=0. \tag{2.2.10}$$

Here $\mathbf{D}_j = [D_x,D_y]_j^{\mathrm{T}} = [D_{xj},D_{yj}]^{\mathrm{T}}$ is the velocity at the jth cell side center and $\mathbf{a}\cdot\mathbf{b}$ stands for the scalar product of the vectors $\mathbf{a}$ and $\mathbf{b}$. The quantities $\mathbf{F}_j$, $\mathbf{E}_j$, and $\mathbf{U}_j$ are defined by solving the corresponding Riemann problem and are calculated for the cell side moving with the velocity $\mathbf{D}_j$. The vectors $\widetilde{\mathbf{F}} = \mathbf{F}-D_x\mathbf{U}$ and $\widetilde{\mathbf{E}} = \mathbf{E}-D_y\mathbf{U}$ are the modified function fluxes along the x- and y-direction, respectively. We see that now the quantities

G_i and $\mathbf{S}_j$ are functions of time. The half-integer subscript $k+1/2$ in Eqs. (2.2.9)–(2.2.10) denotes the values for the time instant $t+\frac{1}{2}\Delta t$. Equation (2.2.10) is a discrete equation that describes the evolution of the computational cell volume G_i. The important property of this approximation is that the uniform-flow solutions $\mathbf{U} = \mathbf{U}_0 = \text{const}$ to Eq. (2.2.5) are also solutions to the discrete equations (2.2.9)–(2.2.10). This condition is of great importance in using arbitrary moving curvilinear coordinate systems. For given approximation of $\mathbf{S}$ in time one can use Eq. (2.2.10) to calculate the areas G_i^{k+1}.

The numerical scheme (2.2.8) is stable on a uniform Cartesian grid if

$$\max(|C_x| + |C_y|) \leq \max|C_x| + \max|C_y| \leq 1, \tag{2.2.11}$$

where C_x and C_y are the CFL numbers that correspond to the x-axis and y-axis and depend on the eigenvalues of the matrices $\partial\mathbf{F}/\partial\mathbf{U}$ and $\partial\mathbf{E}/\partial\mathbf{U}$, respectively.

By analogy we can construct the Godunov method for multidimensional cases.

2.2.2 Exact solution of the Riemann problem.

In this section we describe the main approaches and algorithms for constructing the general exact solution of the Riemann problem. One must first construct two basic elementary solutions. Then the general solution—particularly, in gas dynamics, shallow water equations, etc.—can be represented as a combination of these solutions.

Elementary solution 1: Strong discontinuity

The first elementary solution is a moving discontinuity. For obtaining the discontinuity relations let us integrate of Eq. (2.2.1) over t and x and consider their integral form

$$\oint_L (\mathbf{U}\,dx - \mathbf{F}\,dt) = \mathbf{0}, \tag{2.2.12}$$

where L is the boundary of a region in the (t,x) plane. Let us search for a solution of Eq. (2.2.1) in the form of a travelling wave $f(t,x) = f(\zeta) \equiv f(x - Wt)$, where $W = \text{const}$ is a wave velocity. Consider Eq. (2.2.12) in orthogonal coordinates (ζ, τ) associated with discontinuity, where $\zeta = x - Wt$, and $\tau = Wx + t$. The coordinate ζ is normal and coordinate τ is tangential with respect to the discontinuity. Using transformation

$$x = \frac{\zeta + W\tau}{1+W^2}, \qquad t = \frac{-W\zeta + \tau}{1+W^2}$$

we can rewrite Eq. (2.2.12) as

$$\frac{1}{1+W^2}\oint_L [(W\mathbf{U} - \mathbf{F})\,d\tau + (\mathbf{U} + W\mathbf{F})\,d\zeta] = \mathbf{0}. \tag{2.2.13}$$

Let us integrate (2.2.13) over a rectangular region $\tau_0 - \delta\tau \leq \tau \leq \tau_0 + \delta\tau$, $\zeta_0 - \delta\zeta \leq \zeta \leq \zeta_0 + \delta\zeta$, where $\zeta = \zeta_0$ corresponds to the discontinuity. We can find

$$\begin{aligned} &(W\mathbf{U} - \mathbf{F})_1 2\delta\tau - (W\mathbf{U} - \mathbf{F})_2 2\delta\tau = \mathbf{0} \\ \Longrightarrow \quad &W\{\mathbf{U}\} - \{\mathbf{F}\} = \mathbf{0}, \end{aligned} \tag{2.2.14}$$

where $\{q\} \equiv q_1 - q_2$, and indices 1 and 2 denote the variables on the left- and right-hand side of the discontinuity.

Let the piecewise constant initial data $\mathbf{U}_0(x)$ be defined as follows: $\mathbf{U}_0(x) = \mathbf{U}_1 = \text{const}$ for $x < 0$ and $\mathbf{U}_0(x) = \mathbf{U}_2 = \text{const}$ for $x > 0$. Both $\mathbf{U}_1$ and $\mathbf{U}_2$ satisfy Eq. (2.2.1). Suppose that this initial strong discontinuity moves with a velocity of W and is a solution of the Riemann problem. Then, Eq. (2.2.14) relates $\mathbf{U}_1$, $\mathbf{U}_2$, and W. If these relations are satisfied, then the moving discontinuity is a formal solution of Eq. (2.2.1).

The above solution is self-similar with respect to the variable $\zeta = x - Wt$, where $W = \text{const}$. This solution is also self-similar with respect to the variable $\xi = x/t$. In fact, a moving discontinuity in the (t, x) coordinates is a straight line that satisfies the relation $\xi = W = \text{const}$.

Elementary solution 2: Riemann wave

There also exists a continuous elementary solution of the Riemann problem, which is called a Riemann wave. Let us seek a continuous solution in the form $f(t, x) = f(\xi) \equiv f(x/t)$ and consider the nonconservative form of Eq. (2.2.1),

$$\mathbf{U}_t + A\mathbf{U}_x = \mathbf{0}. \tag{2.2.15}$$

Equation (2.2.15) can also be rewritten in terms of some other variables $\mathbf{u}$ such that $\mathbf{U} = \mathbf{U}(\mathbf{u})$:

$$\mathbf{u}_t + B\mathbf{u}_x = \mathbf{0}, \tag{2.2.16}$$

where $B = M^{-1}AM$, $M = \partial\mathbf{U}/\partial\mathbf{u}$. The subscripts t and x in Eqs. (2.2.15)–(2.2.16) denote the respective partial derivatives. Further we will transform this system of equations into the characteristic form.

Consider the hyperbolic system (2.2.16) and multiply it by a matrix Ω_L composed of the left eigenvectors of the matrix B. Then the characteristic form of system (2.2.16) is given by

$$\Omega_L\mathbf{u}_t + \Lambda\Omega_L\mathbf{u}_x = \mathbf{0},$$

or in the expanded form,

$$\sum_{k=1}^{n} \Omega_{Lpk}\frac{\partial u_k}{\partial t} + \lambda_p \sum_{k=1}^{n} \Omega_{Lpk}\frac{\partial u_k}{\partial x} = 0, \qquad p = 1, \ldots, n;$$

where $\mathbf{u} = [u_1, \ldots, u_n]^{\mathrm{T}} = [u_k]^{\mathrm{T}}$, $k = 1, \ldots, n$; $\Omega_L = [\Omega_{Lpk}]$ and $\Lambda = [\lambda_p\delta_{pk}]$. Let us search for an exact continuous solution as a function of $\xi = x/t$. Substituting it in the characteristic form, we obtain

$$(\lambda_p - \xi)\left[\sum_{k=1}^{n} \Omega_{Lpk}\frac{\partial u_k}{\partial \xi}\right] = 0, \qquad p = 1, \ldots, n.$$

Thus, if there exists an exact solution, then it must satisfy one of the n systems of equations

$$\lambda_\alpha - \xi = 0, \tag{2.2.17}$$

$$(\lambda_\beta - \xi)\left[\sum_{k=1}^{n} \Omega_{L\beta k}\frac{\partial u_k}{\partial \xi}\right] = 0 \implies \sum_{k=1}^{n} \Omega_{L\beta k}\frac{\partial u_k}{\partial \xi} = 0 \text{ if } \lambda_\beta \neq \lambda_\alpha. \tag{2.2.18}$$

Here $\beta = 1, \ldots, n$; $\beta \neq \alpha$. If the variables $\mathbf{u}$ are chosen successfully, Eqs. (2.2.17)–(2.2.18) may be solved in a closed form, see Chapters 3 and 4. There can exist no more than n different exact solutions in the form of the Riemann waves.

The general self-similar solution, if it exists, can be described as a combination of strong discontinuities and Riemann waves divided by the regions of constant quantities.

General solution for the linear system

Consider the hyperbolic system (2.2.1) with $\mathbf{F} = A\mathbf{U}$, where A is a constant coefficient matrix. Equation (2.2.1) becomes

$$\mathbf{U}_t + \mathbf{F}_x = \mathbf{U}_t + (A\mathbf{U})_x = \mathbf{U}_t + A\mathbf{U}_x = \mathbf{0}. \tag{2.2.19}$$

Solution of the Riemann problem for the linear/linearized hyperbolic system was given in Section 1.4.1, see Eq. (1.4.12), and has the form

$$\mathbf{U}(t,x) = \sum_{k=1}^{n} \mathbf{r}^k w_k(x - \lambda_k t) \tag{2.2.20}$$

$$\Longrightarrow \quad U_p(t,x) = \sum_{k=1}^{n} \Omega_{\mathrm{R}pk} w_k(x - \lambda_k t), \qquad p = 1, \ldots, n, \tag{2.2.21}$$

where $\mathbf{w} = \Omega_{\mathrm{L}}\mathbf{U}$ and $\Omega_{\mathrm{R}} = [\Omega_{\mathrm{R}pk}]$. This solution is a combination of travelling waves with wave velocities λ_p.

The relation (2.2.14) for Eq. (2.2.19) has the form

$$(A - WI)\{\mathbf{U}\} = \mathbf{0}, \tag{2.2.22}$$

where I is the $n \times n$ identity matrix. Thus, $\lambda_p = W$ is an eigenvalue of A and $\{\mathbf{U}\}$ is a right eigenvector corresponding to $\lambda_p = W$. Considering Eq. (2.2.21) for k-discontinuity yields $\{\mathbf{U}\} = \mathbf{r}^k\{w_k\}$. Hence, $\{\mathbf{U}\}$ is the kth right eigenvector and (2.2.21) satisfies (2.2.14).

The relation (2.2.17)–(2.2.18) for Eq. (2.2.19) becomes

$$\lambda_\alpha - \xi = 0,$$
$$\sum_{k=1}^{n} \Omega_{\mathrm{L}\beta k} \frac{\partial U_k}{\partial \xi} = 0 \quad \text{if} \quad \lambda_\beta \neq \lambda_\alpha, \quad \beta = 1, \ldots, n; \quad \beta \neq \alpha.$$

Thus, for a discontinuity moving with the velocity λ_α we have

$$\sum_{k=1}^{n} \Omega_{\mathrm{L}\beta k}\{U_k\} = 0 \quad \text{if} \quad \lambda_\beta \neq \lambda_\alpha. \tag{2.2.23}$$

Here $\beta = 1, \ldots, n$; $\beta \neq \alpha$. Indeed, from Eq. (2.2.22) we have

$$(A - \lambda_\alpha I)\{\mathbf{U}\} = \Omega_{\mathrm{R}}(\Lambda - \lambda_\alpha I)\Omega_{\mathrm{L}}\{\mathbf{U}\} = \mathbf{0} \quad \Longrightarrow \quad (\Lambda - \lambda_\alpha I)\Omega_{\mathrm{L}}\{\mathbf{U}\} = \mathbf{0}.$$

This is equivalent to (2.2.23) if λ_α is an eigenvalue of A. Hence, (2.2.21) satisfies (2.2.17)–(2.2.18). The cases $\lambda_\beta = \lambda_\alpha$ can be considered analogously.

The solution of the Riemann problem for Eq. (2.2.19) consists of n jumps, or step functions, moving with the velocities λ_k. Since the initial function in the Riemann problem is represented by a step function, the elementary Riemann waves w_k will also be step functions. Thus, the difference between the strong discontinuities and Riemann waves in system (2.2.19) disappears.

2.3 Methods based on approximate Riemann problem solvers

In this section we describe Godunov-type methods for solving Eqs. (2.2.1) based on approximate solution of the Riemann problem. A method of solution of the Riemann problem will be called a Riemann problem solver, or simply Riemann solver. Then, an exact solution can be called an exact Riemann solver, and an approximate solution is an approximate Riemann solver. We shall describe three basic numerical methods based on approximate Riemann solvers, namely, the Courant–Isaacson–Rees, Roe, and Osher methods. These methods have the following characteristic features.

The Courant–Isaacson–Rees (CIR) method. This method is based on the solution of the locally linearized Riemann problem. In this case the solution consists of only discontinuities dividing the regions of constant quantities. For the linear/linearized hyperbolic system of equations this solution is exact. It can be constructed for an arbitrary hyperbolic systems in both conservative and nonconservative forms.

Roe's method. This method is also based on the solution of the linearized Riemann problem. The difference is that in this case the exact shock relations (2.2.14) are preserved on all elementary step functions. Obviously such a method must be constructed only for conservation law systems.

The Osher method. The approximate solution of the Riemann problem is constructed for the nonlinear system, but only solutions in the form of Riemann waves are used.

2.3.1 Courant–Isaacson–Rees-type methods.

Consider first a single advection equation with a constant coefficient λ:

$$u_t + \lambda u_x = 0. \tag{2.3.1}$$

The upwind finite-difference scheme of the first order in time and space can be written out for Eq. (2.3.1):

$$\frac{u_j^{k+1} - u_j^k}{\Delta t} + \lambda \frac{u_j^k - u_{j-1}^k}{\Delta x} = 0 \quad \text{if} \quad \lambda \geq 0, \tag{2.3.2}$$

$$\frac{u_j^{k+1} - u_j^k}{\Delta t} + \lambda \frac{u_{j+1}^k - u_j^k}{\Delta x} = 0 \quad \text{if} \quad \lambda < 0. \tag{2.3.3}$$

Here Δx is the discrete mesh size and Δt the time increment. Let the integer subscripts $j = 1, 2, \ldots$ refer to the centers of mesh cells, and half-integer ones to the boundaries. The integer superscript is the number of the layer with respect to time. The CFL number C, see (2.2.4), is defined as $C = \lambda \Delta t / \Delta x$.

For the stability analysis we can use the von Neumann spectral method, see Richtmyer and Morton (1967). Let us seek an exact solution of the difference scheme (2.3.2)–(2.3.3) in the spectral form

$$u_m^k = \sum_{\varphi} (q)^k u_m^0 \exp(im\varphi), \qquad i = \sqrt{-1}, \quad \varphi \in [0, 2\pi],$$

where q is an unknown function. Substituting u_m^k into Eqs. (2.3.2)–(2.3.3), we find that

$$q = q(C, \varphi) = \begin{cases} 1 - C + C\exp(-i\varphi) & \text{for } \lambda \geq 0, \\ 1 + C - C\exp(i\varphi) & \text{for } \lambda < 0. \end{cases}$$

A sufficient stability condition is $|q|^2 \leq 1$, whence it follows that $|C| \leq 1$. Note that a necessary and sufficient stability condition is $|q| \leq 1 + Q\tau$, where Q is a constant independent on Δt and Δx (Godunov and Ryabenkii 1987).

A uniform difference scheme for Eq. (2.3.1) can be introduced by a combination of schemes (2.3.2)–(2.3.3):

$$\frac{u_j^{k+1} - u_j^k}{\Delta t} + \lambda \frac{u_{j+1/2} - u_{j-1/2}}{\Delta x} = 0, \tag{2.3.4}$$

$$u_{m+1/2} = \tfrac{1}{2}(u_m^k + u_{m+1}^k) + \tfrac{1}{2}\,\mathrm{sgn}\,\lambda\,(u_m^k - u_{m+1}^k), \qquad m = j, j-1. \tag{2.3.5}$$

Equation (2.3.5) gives u at the interface $m+1/2$ between two discrete cells with numbers m and $m+1$. We can consider $u_{m+1/2}$ an approximate solution of the Riemann problem for Eq. (2.3.1) with the initial data u_{m+1}^k for $x > x_{m+1/2}$ and u_m^k for $x < x_{m+1/2}$. To be specific, we rewrite the solution in the local coordinate system where $x_{m+1/2} = 0$. The solution $u_{m+1/2}$ is determined by Eq. (2.3.5) only for $\xi = x/t = 0$. For arbitrary $\xi = x/t$, Eq. (2.3.5) takes the form

$$u_{m+1/2}(\xi) = \tfrac{1}{2}(u_m^k + u_{m+1}^k) + \tfrac{1}{2}\,\mathrm{sgn}(\lambda - \xi)\,(u_m^k - u_{m+1}^k), \qquad -\infty < \xi < +\infty. \tag{2.3.6}$$

Indeed, if $\xi < \lambda$, then $u_{m+1/2} = u_m^k$ and if $\xi > \lambda$, then $u_{m+1/2} = u_{m+1}^k$.

Consider a hyperbolic system with constant coefficients:

$$\mathbf{U}_t + \mathbf{F}_x = \mathbf{U}_t + (A\mathbf{U})_x = \mathbf{U}_t + A\mathbf{U}_x = \mathbf{0}, \tag{2.3.7}$$

where $\mathbf{F} = A\mathbf{U}$, A is a constant coefficient matrix, $\mathbf{U} = \mathbf{U}(t,x) = [U_1, \ldots, U_n]^{\mathrm{T}}$, $t \geq 0$, $-\infty < x < \infty$, and $\mathbf{U}(0,x) = \mathbf{U}_0(x)$. Multiplying Eq. (2.3.7) by the matrix $\mathbf{\Omega}_{\mathrm{L}}$ we obtain

$$\Omega_{\mathrm{L}}\mathbf{U}_t + \Omega_{\mathrm{L}}A\mathbf{U}_x = \Omega_{\mathrm{L}}\mathbf{U}_t + \Lambda\Omega_{\mathrm{L}}\mathbf{U}_x = (\Omega_{\mathrm{L}}\mathbf{U})_t + \Lambda(\Omega_{\mathrm{L}}\mathbf{U})_x = \mathbf{w}_t + \Lambda\mathbf{w}_x = \mathbf{0}, \tag{2.3.8}$$

where $\mathbf{w} = \Omega_{\mathrm{L}}\mathbf{U}$. In fact, this system is equivalent to a system of Eq. (2.3.1). Then for each equation we can write out the finite-difference scheme (2.3.4)–(2.3.5). By passing from the variables $\mathbf{w} = \Omega_{\mathrm{L}}\mathbf{U}$ to $\mathbf{U}$, we arrive at the scheme

$$\frac{\mathbf{U}_j^{k+1} - \mathbf{U}_j^k}{\Delta t} + A\frac{\mathbf{U}_{j+1/2} - \mathbf{U}_{j-1/2}}{\Delta x} = \mathbf{0}, \tag{2.3.9}$$

$$\mathbf{U}_{m+1/2} = \tfrac{1}{2}(\mathbf{U}_m^k + \mathbf{U}_{m+1}^k) + \tfrac{1}{2}S\,(\mathbf{U}_m^k - \mathbf{U}_{m+1}^k), \qquad m = j, j-1; \tag{2.3.10}$$

$$S = \Omega_{\mathrm{R}}[\mathrm{sgn}\,\lambda_p\,\delta_{pl}]\Omega_{\mathrm{L}}.$$

Multiplying Eq. (2.3.10) by A and using the fact that $\mathbf{F} = A\mathbf{U}$, we can rewrite this scheme as

$$\frac{\mathbf{U}_j^{k+1} - \mathbf{U}_j^k}{\Delta t} + \frac{\mathbf{F}_{j+1/2} - \mathbf{F}_{j-1/2}}{\Delta x} = \mathbf{0}, \tag{2.3.11}$$

$$\mathbf{F}_{m+1/2} = \tfrac{1}{2}(\mathbf{F}_m^k + \mathbf{F}_{m+1}^k) + \tfrac{1}{2}|A|\,(\mathbf{U}_m^k - \mathbf{U}_{m+1}^k), \qquad m = j, j-1; \tag{2.3.12}$$

$$|A| = \Omega_{\mathrm{R}}[\lambda_p\ \mathrm{sgn}\,\lambda_p\,\delta_{pl}]\Omega_{\mathrm{L}}.$$

A stability condition for schemes (2.3.9)–(2.3.10) and (2.3.11)–(2.3.12) is given by (2.2.4):

$$\max_p |\lambda_p| \frac{\Delta t}{\Delta x} \leq 1.$$

Similarly to Eq. (2.3.6), we can construct an approximate solution of the Riemann problem for Eqs. (2.3.9)–(2.3.12):

$$\mathbf{U}_{m+1/2}(\xi) = \tfrac{1}{2}(\mathbf{U}^k_m + \mathbf{U}^k_{m+1}) + \tfrac{1}{2}S(\xi)\,(\mathbf{U}^k_m - \mathbf{U}^k_{m+1}), \tag{2.3.13}$$

$$S(\xi) = \Omega_{\mathrm{R}}[\operatorname{sgn}(\lambda_p - \xi)\,\delta_{pl}]\Omega_{\mathrm{L}},$$

$$\mathbf{F}_{m+1/2}(\xi) = \tfrac{1}{2}(\mathbf{F}^k_m + \mathbf{F}^k_{m+1}) + \tfrac{1}{2}A(\xi)\,(\mathbf{U}^k_m - \mathbf{U}^k_{m+1}), \tag{2.3.14}$$

$$A(\xi) = \Omega_{\mathrm{R}}[\lambda_p \operatorname{sgn}(\lambda_p - \xi)\,\delta_{pl}]\Omega_{\mathrm{L}}, \qquad -\infty < \xi < +\infty.$$

The CIR scheme for a quasilinear system

We shall describe below CIR finite-difference and finite-volume schemes for quasilinear nonconservative and conservative hyperbolic systems represented, respectively, by

$$\mathbf{U}_t + A(\mathbf{U})\mathbf{U}_x = \mathbf{0}, \quad A(\mathbf{U}) = \Omega_{\mathrm{R}}\Lambda\Omega_{\mathrm{L}}, \tag{2.3.15}$$

$$\mathbf{U}_t + [\mathbf{F}(\mathbf{U})]_x = \mathbf{0}, \quad \frac{\partial \mathbf{F}}{\partial \mathbf{U}} = A(\mathbf{U}) = \Omega_{\mathrm{R}}\Lambda\Omega_{\mathrm{L}}. \tag{2.3.16}$$

We will use these two forms of hyperbolic systems independently and note that a matrix A of (2.3.15), in general, can be different from the Jacobian matrix. System (2.3.15) can be rewritten as

$$\mathbf{l}^p \cdot \mathbf{U}_t + \lambda_p \mathbf{l}^p \cdot \mathbf{U}_x = 0, \qquad p = 1, \ldots, n, \tag{2.3.17}$$

where a left eigenvector $\mathbf{l}^p$ corresponds to eigenvalue λ_p.

Various finite-difference schemes for the quasilinear system (2.3.15) can be obtained from Eqs. (2.3.9)–(2.3.10) by different approximations of the matrices $A(\mathbf{U})$ and $S(\mathbf{U})$. We refer to the following scheme as

Version 1:

$$\frac{\mathbf{U}^{k+1}_j - \mathbf{U}^k_j}{\Delta t} + A(\mathbf{U}^k_j)\frac{\mathbf{U}_{j+1/2} - \mathbf{U}_{j-1/2}}{\Delta x} = \mathbf{0}, \tag{2.3.18}$$

$$\mathbf{U}_{m+1/2} = \tfrac{1}{2}(\mathbf{U}^k_m + \mathbf{U}^k_{m+1}) + \tfrac{1}{2}S^k_j\,(\mathbf{U}^k_m - \mathbf{U}^k_{m+1}), \quad m = j, j-1; \tag{2.3.19}$$

$$S = \Omega_{\mathrm{R}}[\operatorname{sgn}\lambda_p \delta_{pl}]\Omega_{\mathrm{L}}.$$

In notations of (2.3.17) this scheme can be written as

$$\mathbf{l}^p_j \cdot \frac{\mathbf{U}^{k+1}_j - \mathbf{U}^k_j}{\Delta t} + (\lambda_p \mathbf{l}^p)_j \cdot \frac{\mathbf{U}^k_j - \mathbf{U}^k_{j-1}}{\Delta x} = 0, \quad \text{if} \quad (\lambda_p)_j \geq 0, \tag{2.3.20}$$

$$\mathbf{l}^p_j \cdot \frac{\mathbf{U}^{k+1}_j - \mathbf{U}^k_j}{\Delta t} + (\lambda_p \mathbf{l}^p)_j \cdot \frac{\mathbf{U}^k_{j+1} - \mathbf{U}^k_j}{\Delta x} = 0, \quad \text{if} \quad (\lambda_p)_j < 0. \tag{2.3.21}$$

In 1952, Courant, Isaacson, and Rees constructed a finite-difference scheme similar to (2.3.20)–(2.3.21) for a quasilinear hyperbolic system for n functions of two variables x and y in the nonconservative form and investigated its convergence. They extended the upwind difference scheme (2.3.2)–(2.3.3) to systems of equations. The schemes they constructed are equivalent in our notation to schemes (2.3.9)–(2.3.12) and (2.3.18)–(2.3.19) described above.

Another version of scheme (2.3.18)–(2.3.19) can be constructed by using another approximation for the matrix $S(\mathbf{U})$,

$$\frac{\mathbf{U}_j^{k+1}-\mathbf{U}_j^k}{\Delta t}+A(\mathbf{U}_j^k)\frac{\mathbf{U}_{j+1/2}-\mathbf{U}_{j-1/2}}{\Delta x}=\mathbf{0}, \tag{2.3.22}$$

$$\mathbf{U}_{m+1/2}=\tfrac{1}{2}(\mathbf{U}_m^k+\mathbf{U}_{m+1}^k)+\tfrac{1}{2}S_{m+1/2}^k(\mathbf{U}_m^k-\mathbf{U}_{m+1}^k), \quad m=j,j-1; \tag{2.3.23}$$
$$S=\Omega_{\mathrm{R}}[\operatorname{sgn}\lambda_p\delta_{pl}]\Omega_{\mathrm{L}}.$$

The scheme (2.3.22)–(2.3.23) is rather formal but it can be extended to the conservative system (2.3.16) as

$$\frac{\mathbf{U}_j^{k+1}-\mathbf{U}_j^k}{\Delta t}+\frac{\mathbf{F}(\mathbf{U}_{j+1/2})-\mathbf{F}(\mathbf{U}_{j-1/2})}{\Delta x}=\mathbf{0} \tag{2.3.24}$$

with $\mathbf{U}_{j\pm1/2}$ defined by (2.3.23). Based on the nonconservative form of the system, this scheme is not reliable in numerical calculations of shock wave propagation; see one-dimensional gas dynamic tests by Semenov (1987). Therefore we modify it as follows. Approximate the functions $\mathbf{F}(\mathbf{U}_{m+1/2})$ by two terms of their Taylor series expansions assuming that the differences $\mathbf{U}_m^k-\mathbf{U}_{m+1}^k$ are small. Then

$$\mathbf{F}(\mathbf{U}_{m+1/2})\simeq\mathbf{F}\left(\frac{\mathbf{U}_m^k+\mathbf{U}_{m+1}^k}{2}\right)+\frac{1}{2}\left(\frac{\partial\mathbf{F}}{\partial\mathbf{U}}S\right)_{m+1/2}^k(\mathbf{U}_m^k-\mathbf{U}_{m+1}^k)$$
$$\simeq\tfrac{1}{2}(\mathbf{F}_m^k+\mathbf{F}_{m+1}^k)+\tfrac{1}{2}|A|_{m+1/2}^k(\mathbf{U}_m^k-\mathbf{U}_{m+1}^k),$$
$$|A|=AS=\Omega_{\mathrm{R}}[\,\lambda_p\operatorname{sgn}\lambda_p\,\delta_{pl}]\Omega_{\mathrm{L}}.$$

Here, the sign $\simeq$ stands for the approximation of the first order. Then scheme (2.3.24) can be transformed into

$$\frac{\mathbf{U}_j^{k+1}-\mathbf{U}_j^k}{\Delta t}+\frac{\mathbf{F}_{j+1/2}-\mathbf{F}_{j-1/2}}{\Delta x}=\mathbf{0}, \tag{2.3.25}$$

$$\mathbf{F}_{m+1/2}=\tfrac{1}{2}(\mathbf{F}_m^k+\mathbf{F}_{m+1}^k)+\tfrac{1}{2}|A|_{m+1/2}^k(\mathbf{U}_m^k-\mathbf{U}_{m+1}^k), \quad m=j,j-1; \tag{2.3.26}$$
$$|A|=\Omega_{\mathrm{R}}[\,|\lambda_p|\,\delta_{pl}]\Omega_{\mathrm{L}}.$$

This reliable conservative form of the CIR scheme was developed and used by Kholodov (1978) for multidimensional calculations; see also Belotserkovskii, Kholodov, and Turchak (1986), and Magomedov and Kholodov (1988).

By analogy with (2.3.13)–(2.3.14), we can write out the CIR solver using in finite-difference schemes (2.3.22)–(2.3.23) and (2.3.25)–(2.3.26):

$$\mathbf{U}_{m+1/2}(\xi)=\tfrac{1}{2}(\mathbf{U}_m^k+\mathbf{U}_{m+1}^k)+\tfrac{1}{2}S(\xi)_{m+1/2}^k(\mathbf{U}_m^k-\mathbf{U}_{m+1}^k), \tag{2.3.27}$$
$$S(\xi)=\Omega_{\mathrm{R}}[\operatorname{sgn}(\lambda_p-\xi)\,\delta_{pl}]\Omega_{\mathrm{L}},$$

$$\mathbf{F}_{m+1/2}(\xi)=\tfrac{1}{2}(\mathbf{F}_m^k+\mathbf{F}_{m+1}^k)+\tfrac{1}{2}A(\xi)_{m+1/2}^k(\mathbf{U}_m^k-\mathbf{U}_{m+1}^k), \tag{2.3.28}$$
$$A(\xi)=\Omega_{\mathrm{R}}[\lambda_p\operatorname{sgn}(\lambda_p-\xi)\,\delta_{pl}]\Omega_{\mathrm{L}}, \qquad -\infty<\xi<+\infty.$$

Thus, Eqs. (2.3.27)–(2.3.28) represent solution of the linearized Riemann problem for Eq. (2.2.1) as a combination of regions with constant parameters separated by discontinuities. Just as in the case of the exact solution, this approximate solution is self-similar in the variable ξ. The flux vector $\widetilde{\mathbf{F}} = \mathbf{F} - \xi\mathbf{U}$ is calculated using expressions (2.3.27) and (2.3.28); we have

$$\begin{aligned} \widetilde{\mathbf{F}}_{m+1/2} &= \mathbf{F}_{m+1/2}(\xi) - \xi\mathbf{U}_{m+1/2}(\xi) \\ &= \tfrac{1}{2}(\mathbf{F}^k_m - \xi\mathbf{U}^k_m + \mathbf{F}^k_{m+1} - \xi\mathbf{U}^k_{m+1}) + \tfrac{1}{2}|A(\xi)|^k_{m+1/2}\,(\mathbf{U}^k_m - \mathbf{U}^k_{m+1}), \quad (2.3.29) \\ |A(\xi)| &= \Omega_{\mathrm{R}}[|\lambda_p - \xi|\,\delta_{pl}]\Omega_{\mathrm{L}}. \end{aligned}$$

One can use $\widetilde{\mathbf{F}}$ instead of $\mathbf{F}$ in moving grid calculations; see Eqs. (2.2.9)–(2.2.10).

In addition, the quantities $\mathbf{U}_{m+1/2}$ and $\mathbf{F}_{m+1/2}$ behind the p-discontinuity and those ahead of the discontinuity are related by the equation $\{\mathbf{F}\} = \lambda_p\{\mathbf{U}\}$. In particular, for $\lambda = \lambda_1$ we have

$$\{\mathbf{U}\} = \lim_{\xi\to+\lambda_1}\mathbf{U}(\xi) - \lim_{\xi\to-\lambda_1}\mathbf{U}(\xi) = \tfrac{1}{2}\Omega_{\mathrm{R}}\,\mathrm{diag}[2, 0, \ldots, 0]\Omega_{\mathrm{L}}\Delta\mathbf{U}, \quad (2.3.30)$$

$$\{\mathbf{F}\} = \lim_{\xi\to+\lambda_1}\mathbf{F}(\xi) - \lim_{\xi\to-\lambda_1}\mathbf{F}(\xi) = \tfrac{1}{2}\Omega_{\mathrm{R}}\,\mathrm{diag}[2\lambda_1, 0, \ldots, 0]\Omega_{\mathrm{L}}\Delta\mathbf{U}, \quad (2.3.31)$$

$$\Delta\mathbf{U} = \mathbf{U}^k_m - \mathbf{U}^k_{m+1}.$$

Comparing Eqs. (2.3.30) and (2.3.31) we see that $\{\mathbf{F}\} = \lambda_1\{\mathbf{U}\}$. Similarly, the relation $\{\mathbf{F}\} = \lambda_p\{\mathbf{U}\}$ can be proven for any λ_p.

We can derive Eq. (2.3.26) from Eq. (2.3.23). Indeed, premultiplying system (2.3.16) by the matrix A, we obtain the system

$$A\mathbf{U}_t + A\mathbf{F}_x = \mathbf{F}_t + A\mathbf{F}_x = \mathbf{0}, \quad A = \frac{\partial\mathbf{F}}{\partial\mathbf{U}}. \quad (2.3.32)$$

Applying Eq. (2.3.23) to this system, we arrive at the relations

$$\begin{aligned} \mathbf{F}_{m+1/2} &= \tfrac{1}{2}(\mathbf{F}^k_m + \mathbf{F}^k_{m+1}) + \tfrac{1}{2}S^k_{m+1/2}\,(\mathbf{F}^k_m - \mathbf{F}^k_{m+1}), \quad (2.3.33) \\ S &= \Omega_{\mathrm{R}}[\operatorname{sgn}\lambda_p\,\delta_{pl}]\Omega_{\mathrm{L}}. \end{aligned}$$

Substituting the first-order accurate relation

$$\mathbf{F}^k_m - \mathbf{F}^k_{m+1} = A^k_{m+1/2}(\mathbf{U}^k_m - \mathbf{U}^k_{m+1})$$

into Eq. (2.3.33), we obtain (2.3.26).

Note that for the construction of the Godunov method it suffices to use formulas (2.3.28) for $\mathbf{F}$. However, formula (2.3.27) for $\mathbf{U}$ may also be useful. Sometimes one may need to calculate a vector $\mathbf{V}(\mathbf{U})$ for cell boundary $m + 1/2$, i.e., $\mathbf{V}_{m+1/2}$. Let us derive formulas for this case. Substituting $\mathbf{U} = \mathbf{U}(\mathbf{V})$ into system (2.3.15) yields

$$\begin{aligned} &U_V\mathbf{V}_t + AU_V\mathbf{V}_x = 0, \quad U_V = \frac{\partial\mathbf{U}}{\partial\mathbf{V}}, \quad V_U = \frac{\partial\mathbf{V}}{\partial\mathbf{U}}, \\ &\mathbf{V}_t + V_U A U_V \mathbf{V}_x = 0. \end{aligned}$$

Applying Eq. (2.3.23) and the identity $U_V \mathbf{V}_x = \mathbf{U}_x$, we arrive at the formulas

$$\mathbf{V}_{m+1/2} = \tfrac{1}{2}(\mathbf{V}^k_m + \mathbf{V}^k_{m+1}) + \tfrac{1}{2}(V_U S)^k_{m+1/2}(\mathbf{U}^k_m - \mathbf{U}^k_{m+1}), \qquad (2.3.34)$$
$$S = \Omega_{\mathrm{R}}[\operatorname{sgn}\lambda_p\,\delta_{pl}]\Omega_{\mathrm{L}}.$$

Below we present some other versions of the CIR scheme. These versions are modifications of (2.3.18)–(2.3.19) that preserve the first order of accuracy in space and time. These schemes show the flexibility of the CIR approach and provide a basic instrument for the construction of finite-difference schemes for nonconservative or mixed systems with conservative and nonconservative terms. The necessity to solve systems of this kind arises, for example, in solid dynamics (Kondaurov and Nikitin 1990), magnetohydrodynamics (Zachary, Malagoli, and Colella 1994), etc.

Substituting (2.3.19) into (2.3.18), we rewrite Version 1 in the form of

Version 2:

$$\frac{\mathbf{U}^{k+1}_j - \mathbf{U}^k_j}{\Delta t} + (\Omega_{\mathrm{R}}\Lambda^-\Omega_{\mathrm{L}})^k_j \frac{\mathbf{U}^k_{j+1} - \mathbf{U}^k_j}{\Delta x} + (\Omega_{\mathrm{R}}\Lambda^+\Omega_{\mathrm{L}})^k_j \frac{\mathbf{U}^k_j - \mathbf{U}^k_{j-1}}{\Delta x} = \mathbf{0}, \qquad (2.3.35)$$
$$\Lambda^\pm = [\tfrac{1}{2}(1 \pm \operatorname{sgn}\lambda_p)\,\lambda_p\,\delta_{pl}] = [\tfrac{1}{2}(\lambda_p \pm |\lambda_p|)\,\delta_{pl}].$$

Version 2 permitted Kamenetskii and Semenov (1989) to develop an algorithm for an implicit approximation of boundary conditions, particularly on a solid wall. Let us find solution in the region $x \ge 0$ where the point $x = 0$ corresponds to the solid wall. Assuming further that the point $x = 0$ corresponds to the j_0th mesh point, we can write out scheme (2.3.35) at this point in the implicit–explicit form

$$\frac{\mathbf{U}^{k+1}_{j_0} - \mathbf{U}^k_{j_0}}{\Delta t} + (\Omega_{\mathrm{R}}\Lambda^-\Omega_{\mathrm{L}})^{k+1}_{j_0} \frac{\mathbf{U}^k_{j_0+1} - \mathbf{U}^{k+1}_{j_0}}{\Delta x} + (\Omega_{\mathrm{R}}\Lambda^+\Omega_{\mathrm{L}})^{k+1}_{j_0} \frac{\mathbf{U}^{k+1}_{j_0} - \mathbf{U}^k_{j_0-1}}{\Delta x} = \mathbf{0}, \qquad (2.3.36)$$
$$\Lambda^\pm = [\tfrac{1}{2}(\lambda_p \pm |\lambda_p|)\,\delta_{pl}].$$

In other points $j = j_0 + 1, j_0 + 2, \ldots$ one can use both the explicit scheme defined by (2.3.35) and other versions of the explicit Godunov method without any implicit approximation. Further, from system (2.3.36) we must eliminate the equations dependent on the finite differences $\mathbf{U}_{j_0} - \mathbf{U}_{j_0-1}$, since they are not defined. Instead of the eliminated equations we will use the boundary conditions. Note that in a correct boundary problem the number of conditions is equal to the number of positive eigenvalues, see Section 2.8. The described approach can easily be generalized for multidimensional cases (Kamenetskii and Semenov 1989). To solve (2.3.36) at the point j_0 one can use the method of simple iterations. This correction can be important for simulation of essentially nonstationary problems, such as strong jet interaction with a solid wall. Its application allows one to preserve the adequate approximation of boundary conditions in time.

From Version 2 we can obtain relations for

Version 3:

$$\frac{\mathbf{U}^{k+1}_j - \mathbf{U}^k_j}{\Delta t} + (\Omega_{\mathrm{R}}\Lambda^-\Omega_{\mathrm{L}})^k_{j+1/2} \frac{\mathbf{U}^k_{j+1} - \mathbf{U}^k_j}{\Delta x} + (\Omega_{\mathrm{R}}\Lambda^+\Omega_{\mathrm{L}})^k_{j-1/2} \frac{\mathbf{U}^k_j - \mathbf{U}^k_{j-1}}{\Delta x} = \mathbf{0}. \qquad (2.3.37)$$

Equation (2.3.37) differs from Eq. (2.3.35) by the space approximation of the matrices $\Omega_{\mathrm{R}}\Lambda^\pm\Omega_{\mathrm{L}}$. In comparison with (2.3.35) Version 3 permits us to reduce the amount of calculations. Version 3 can be rewritten in the form of

Version 4:

$$\frac{\mathbf{U}_j^{k+1}-\mathbf{U}_j^k}{\Delta t}+\frac{A_{j+1/2}^k(\mathbf{U}_{j+1}^k-\mathbf{U}_j^k)+A_{j-1/2}^k(\mathbf{U}_j^k-\mathbf{U}_{j-1}^k)}{2\Delta x} \tag{2.3.38}$$
$$+\tfrac{1}{2}|A|_{j+1/2}^k\frac{\mathbf{U}_j^k-\mathbf{U}_{j+1}^k}{\Delta x}-\tfrac{1}{2}|A|_{j-1/2}^k\frac{\mathbf{U}_{j-1}^k-\mathbf{U}_j^k}{\Delta x}=\mathbf{0},$$
$$|A|=\Omega_\mathrm{R}[\lambda_p \ \mathrm{sgn}\,\lambda_p \ \delta_{pl}]\Omega_\mathrm{L}.$$

The last scheme permits us to construct a finite-volume scheme for the conservative system (2.3.16). We have the following sequence of formulas:

$$A_{j+1/2}\frac{\mathbf{U}_{j+1}-\mathbf{U}_j}{\Delta x}\simeq(A\mathbf{U}_x)_{j+1/2}\simeq(\mathbf{F}_x)_{j+1/2}\simeq\frac{\mathbf{F}_{j+1}-\mathbf{F}_j}{\Delta x}, \tag{2.3.39}$$
$$A_{j-1/2}\frac{\mathbf{U}_j-\mathbf{U}_{j-1}}{\Delta x}\simeq(A\mathbf{U}_x)_{j-1/2}\simeq(\mathbf{F}_x)_{j-1/2}\simeq\frac{\mathbf{F}_j-\mathbf{F}_{j-1}}{\Delta x}. \tag{2.3.40}$$

Substituting (2.3.39)–(2.3.40) into Version 4, we obtain the familiar scheme (2.3.25)–(2.3.26):

$$\frac{\mathbf{U}_j^{k+1}-\mathbf{U}_j^k}{\Delta t}+\frac{\mathbf{F}_{j+1/2}-\mathbf{F}_{j-1/2}}{\Delta x}=\mathbf{0},$$
$$\mathbf{F}_{m+1/2}=\tfrac{1}{2}(\mathbf{F}_m^k+\mathbf{F}_{m+1}^k)+\tfrac{1}{2}|A|_{m+1/2}^k\,(\mathbf{U}_m^k-\mathbf{U}_{m+1}^k),\quad m=j,j-1;$$
$$|A|=\Omega_\mathrm{R}[\,|\lambda_p|\ \delta_{pl}]\Omega_\mathrm{L}.$$

Magomedov and Kholodov developed and used several numerical schemes for a multidimensional quasilinear hyperbolic system in both nonconservative (Magomedov and Kholodov 1969; Magomedov 1971; Belotserkovskii et al. 1974a) and conservative form (Kholodov 1978), see also Magomedov and Kholodov (1988). In our notation, these schemes in one-dimensional case are equivalent to schemes (2.3.25)–(2.3.26), (2.3.35), (2.3.37) and (2.3.38) described above.

Let us write out a finite-difference scheme for a mixed system, or a system with both conservative and nonconservative terms,

$$\mathbf{U}_t+\mathbf{G}(\mathbf{U})_x+B(\mathbf{U})\mathbf{U}_x=\mathbf{0},\qquad D=\frac{\partial\mathbf{G}}{\partial\mathbf{U}}+B=\Omega_\mathrm{R}\Lambda\Omega_\mathrm{L}. \tag{2.3.41}$$

The scheme can be represented as

$$\frac{\mathbf{U}_j^{k+1}-\mathbf{U}_j^k}{\Delta t}+\frac{\mathbf{G}_{j+1/2}^k-\mathbf{G}_{j-1/2}^k}{\Delta x} \tag{2.3.42}$$
$$+\frac{B_{j+1/2}^k(\mathbf{U}_{j+1}^k-\mathbf{U}_j^k)+B_{j-1/2}^k(\mathbf{U}_j^k-\mathbf{U}_{j-1}^k)}{2\Delta x}+\frac{R_{j+1/2}-R_{j-1/2}}{\Delta x}=\mathbf{0},$$
$$\mathbf{G}_{m+1/2}^k=\tfrac{1}{2}(\mathbf{G}_m^k+\mathbf{G}_{m+1}^k),\quad B_{m+1/2}^k=B(\mathbf{U}_{m+1/2}^k),\qquad m=j,j-1;$$
$$\mathbf{U}_{m+1/2}^k=\tfrac{1}{2}(\mathbf{U}_m^k+\mathbf{U}_{m+1}^k),\quad R_{m+1/2}^k=\tfrac{1}{2}|D(\mathbf{U}_{m+1/2}^k)|(\mathbf{U}_m^k-\mathbf{U}_{m+1}^k),$$
$$|D|=\Omega_\mathrm{R}[\,|\lambda_p|\ \delta_{pl}]\Omega_\mathrm{L}.$$

Scheme (2.3.42) is a combination of schemes (2.3.25)–(2.3.26) and (2.3.38). Note that one can use a combination of other versions of CIR schemes.

The classification of the CIR schemes described here follows the analysis by Semenov (1984, 1987, 1988) and Kamenetskii and Semenov (1989).

Monotonization for discontinuous Jacobian matrices

We will describe a special modification of CIR-type schemes that increases their monotone properties at the points where the Jacobian matrix $A = \partial\mathbf{F}/\partial\mathbf{U}$ is discontinuous, whereas the left and right limits of $\partial\mathbf{F}/\partial\mathbf{U}$ exist. In particular, such cases may occur in gas dynamic flow of two gases with essentially different equations of state, in solid dynamics with large gradients of elastic and plastic moduli, and so on. In dynamics of solids, cases are encountered where the properties of the material can substantially change from point to point. For example, this occurs in studying intensive fracture phenomena and/or phase transformations. The use of standard numerical methods for simulating such processes can result in nonphysical oscillations of mesh functions. To avoid these oscillations, a monotonization of the numerical scheme should be used. In particular, a monotonization procedure for CIR-type schemes was developed by V. D. Ivanov (private communication) for studying problems of solid dynamics; see Ivanov, Petrov, and Suvorova (1989), Ivanov et al. (1990), Ivanov and Petrov (1992), Blazhevich et al. (1999), and V. D. Ivanov et al. (1999).

Below we discuss Ivanov's monotonization for the general case of hyperbolic systems. Consider first a linear hyperbolic system of the form (2.2.1) with $\mathbf{F} = A\mathbf{U}$, where $A = \Omega_{\mathrm{R}}\Lambda\Omega_{\mathrm{L}}$ is a matrix of constant coefficients for $x < 0$ and for $x > 0$. This system can be reduced to a system of n advection equations for the Riemann invariants, see Eq. (2.3.8):

$$\frac{\partial w_p}{\partial t} + \lambda_p \frac{\partial w_p}{\partial x} = 0, \qquad \lambda_p = \text{const}, \quad p = 1, \dots, n. \tag{2.3.43}$$

Here $\mathbf{w} = \Omega_{\mathrm{L}}\mathbf{U}$ is a vector of the n Riemann invariants, $\mathbf{w} = [w_1, \dots, w_n]^{\mathrm{T}}$, $w_p = \mathbf{l}^p \cdot \mathbf{U}$, where $\mathbf{l}^p$ is the pth left eigenvector of the matrix A.

Suppose that matrix A has a discontinuity at the point $x = 0$, i.e., $A = A_1 = \Omega_{\mathrm{R}1}\Lambda_1\Omega_{\mathrm{L}1}$ for $x < 0$ and $A = A_2 = \Omega_{\mathrm{R}2}^{-1}\Lambda_2\Omega_{\mathrm{L}2}$ for $x > 0$. Assume, in addition, that the signs of entries of the matrices Λ_1 and Λ_2 remain the same. This often arises, in particular, in using Lagrangian coordinates. Assuming that the system of equations (2.3.43) has a continuous solution, we can write out the finite-difference scheme (2.3.2) for each advection equation:

$$\frac{(w_\alpha)_j^{k+1} - (w_\alpha)_j^k}{\Delta t} + \lambda_\alpha \frac{(w_\alpha)_j^k - (w_\alpha)_{j-1}^k}{\Delta x} = 0 \quad \text{if} \quad \lambda_\alpha \geq 0, \tag{2.3.44}$$

$$\frac{(w_\beta)_j^{k+1} - (w_\beta)_j^k}{\Delta t} + \lambda_\beta \frac{(w_\beta)_{j+1}^k - (w_\beta)_j^k}{\Delta x} = 0 \quad \text{if} \quad \lambda_\beta < 0. \tag{2.3.45}$$

Assuming further that the point $x = 0$ corresponds to the jth mesh point, we can rewrite scheme (2.3.44)–(2.3.45) with respect to the variables $\mathbf{U}$ satisfying the relation $w_p = \mathbf{l}^p \cdot \mathbf{U}$:

$$\mathbf{l}_1^\alpha \cdot \frac{\mathbf{U}_j^{k+1} - \mathbf{U}_j^k}{\Delta t} + \lambda_\alpha \mathbf{l}_1^\alpha \cdot \frac{\mathbf{U}_j^k - \mathbf{U}_{j-1}^k}{\Delta x} = 0 \quad \text{if} \quad \lambda_\alpha \geq 0, \tag{2.3.46}$$

$$\mathbf{l}_2^\beta \cdot \frac{\mathbf{U}_j^{k+1} - \mathbf{U}_j^k}{\Delta t} + \lambda_\beta \mathbf{l}_2^\beta \cdot \frac{\mathbf{U}_{j+1}^k - \mathbf{U}_j^k}{\Delta x} = 0 \quad \text{if} \quad \lambda_\beta < 0. \tag{2.3.47}$$

In constructing this scheme, we used the fact that the difference $\mathbf{U}_j^k - \mathbf{U}_{j-1}^k$ is calculated for the points in the domain $x \leq 0$, where $A = A_1$ and $\mathbf{l}^\alpha = \mathbf{l}_1^\alpha$, and the difference $\mathbf{U}_{j+1}^k - \mathbf{U}_j^k$ is calculated for the points in the domain $x \geq 0$, where $A = A_2$ and $\mathbf{l}^\beta = \mathbf{l}_2^\beta$. Passing to the variables $\mathbf{U}$, we can rewrite Eqs. (2.3.46)–(2.3.47) for an arbitrary jth point in the following way (cf. (2.3.9)–(2.3.10)):

$$\frac{\mathbf{U}_j^{k+1}-\mathbf{U}_j^k}{\Delta t}+A_j\frac{\mathbf{U}_{j+1/2}-\mathbf{U}_{j-1/2}}{\Delta x}=\mathbf{0}, \tag{2.3.48}$$

$$\mathbf{U}_{m+1/2}=\tfrac{1}{2}(\mathbf{U}_m^k+\mathbf{U}_{m+1}^k)+\tfrac{1}{2}S_j\,(\mathbf{U}_m^k-\mathbf{U}_{m+1}^k),\quad m=j,j-1; \tag{2.3.49}$$

$$S_j=(\Omega_{\mathrm{R}})_j[\operatorname{sgn}(\lambda_p)_j\,\delta_{pl}](\Omega_{\mathrm{L}})_j,\quad A_j=(\Omega_{\mathrm{R}})_j\Lambda_j(\Omega_{\mathrm{L}})_j,\quad \Omega_{\mathrm{R}}=\Omega_{\mathrm{L}}^{-1},$$

$$(\Omega_{\mathrm{L}})_j=(\Omega_{\mathrm{L}}^{+})_{j-1/2}+(\Omega_{\mathrm{L}}^{-})_{j+1/2},\quad \Lambda_j=\Lambda_{j-1/2}^{+}+\Lambda_{j+1/2}^{-}. \tag{2.3.50}$$

The superscript in Ω_{L}^{+} and Λ^{+} indicates that the matrix contains only the coefficients corresponding to positive λ's, with the other coefficients equal to zero. The superscript in Ω_{L}^{-} and Λ^{-} indicates that the matrix contains only the coefficients corresponding to negative λ's, with the other coefficients equal to zero. The approximation in Eq. (2.3.50) provides the equality $A=A_1$ on the left of the discontinuity and the equality $A=A_2$ on the right of the discontinuity and gives a special approximation at the point $x=0$.

We can easily generalize scheme (2.3.48)–(2.3.50) for the quasilinear nonconservative system of equations (2.2.15) by substituting

$$\frac{\mathbf{U}_j^{k+1}-\mathbf{U}_j^k}{\Delta t}+A(\mathbf{U}_j^k)\frac{\mathbf{U}_{j+1/2}-\mathbf{U}_{j-1/2}}{\Delta x}=\mathbf{0}, \tag{2.3.51}$$

$$\mathbf{U}_{m+1/2}=\tfrac{1}{2}(\mathbf{U}_m^k+\mathbf{U}_{m+1}^k)+\tfrac{1}{2}S(\mathbf{U}_j^k)\,(\mathbf{U}_m^k-\mathbf{U}_{m+1}^k),\quad m=j,j-1;$$

$$S(\mathbf{U})=\Omega_{\mathrm{R}}[\operatorname{sgn}\lambda_p\,\delta_{pl}]\Omega_{\mathrm{L}},\quad A(\mathbf{U})=\Omega_{\mathrm{R}}\Lambda\Omega_{\mathrm{L}},\quad \Omega_{\mathrm{R}}=\Omega_{\mathrm{L}}^{-1},$$

$$(\Omega_{\mathrm{L}})_j=(\Omega_{\mathrm{L}}^{+})_{j-1/2}+(\Omega_{\mathrm{L}}^{-})_{j+1/2},\quad \Lambda_j=\Lambda_{j-1/2}^{+}+\Lambda_{j+1/2}^{-}.$$

Suppose further that the matrix A has discontinuities only at boundaries of computational cells. We can do this, since we work now with the first-order schemes. Assuming that the system of equations (2.3.43) has a continuous solution, we can write out the finite-difference scheme (2.3.2) for each advection equation in the form (2.3.44)–(2.3.45). Then we can rewrite this scheme in terms $w_p=\mathbf{l}^p\cdot\mathbf{U}$:

$$\mathbf{l}_j^{\alpha}\cdot\frac{\mathbf{U}_j^{k+1}-\mathbf{U}_j^k}{\Delta t}+\lambda_{\alpha}\frac{\mathbf{l}_j^{\alpha}\cdot\mathbf{U}_j^k-\mathbf{l}_{j-1}^{\alpha}\cdot\mathbf{U}_{j-1}^k}{\Delta x}=0\quad\text{if}\quad\lambda_{\alpha}\geq 0,$$

$$\mathbf{l}_j^{\beta}\cdot\frac{\mathbf{U}_j^{k+1}-\mathbf{U}_j^k}{\Delta t}+\lambda_{\beta}\frac{\mathbf{l}_{j+1}^{\beta}\cdot\mathbf{U}_{j+1}^k-\mathbf{l}_j^{\beta}\cdot\mathbf{U}_j^k}{\Delta x}=0\quad\text{if}\quad\lambda_{\beta}<0.$$

Assuming that the signs of the entries of $\Lambda_{j\pm1}$ and Λ_j remain the same, otherwise, it can be used refined mesh and/or method (2.3.51), we can rewrite this scheme as

$$\frac{\mathbf{U}_j^{k+1}-\mathbf{U}_j^k}{\Delta t}+(\Omega_{\mathrm{R}})_j\,\Lambda_j\frac{\mathbf{w}_{j+1/2}-\mathbf{w}_{j-1/2}}{\Delta x}=\mathbf{0}, \tag{2.3.52}$$

$$\mathbf{w}_{m+1/2}=\tfrac{1}{2}(\mathbf{w}_m^k+\mathbf{w}_{m+1}^k)+\tfrac{1}{2}S_{m+1/2}(\mathbf{w}_m^k-\mathbf{w}_{m+1}^k),\quad m=j,j-1.$$

Introducing numerical fluxes $\mathbf{F}$ we can obtain

$$\frac{\mathbf{U}_j^{k+1}-\mathbf{U}_j^k}{\Delta t}+\frac{\mathbf{F}_{j+1/2}-\mathbf{F}_{j-1/2}}{\Delta x}+\mathbf{H}_j=\mathbf{0}, \tag{2.3.53}$$

$$\mathbf{F}_{m+1/2} = \tfrac{1}{2}(\mathbf{F}^k_m + \mathbf{F}^k_{m+1}) + \tfrac{1}{4}(|A|_m + |A|_{m+1})(\mathbf{U}^k_m - \mathbf{U}^k_{m+1}), \quad \mathbf{F} = A\mathbf{U}, \quad m = j, j-1;$$
$$\mathbf{H}_j = \frac{1}{\Delta x}[(\Omega_{\rm R})_j \Lambda_j(\mathbf{w}_{j+1/2} - \mathbf{w}_{j-1/2}) - (\mathbf{F}_{j+1/2} - \mathbf{F}_{j-1/2})],$$
$$\mathbf{w}_{m+1/2} = \tfrac{1}{2}(\mathbf{w}^k_m + \mathbf{w}^k_{m+1}) + \tfrac{1}{2}S_{m+1/2}(\mathbf{w}^k_m - \mathbf{w}^k_{m+1}), \quad \mathbf{w} = \Omega_{\rm L}\mathbf{U}, \quad m = j, j-1.$$

Similar to (2.3.51), this scheme can be generalized for quasilinear conservative system of equations (2.3.16). The scheme obtained permits one to modify finite-difference schemes (2.3.25)–(2.3.26). The schemes (2.3.51) and (2.3.53) permit one to modify also scheme (2.3.42). These modifications make it possible to describe more adequately the rare specific case where $A_j \neq A_{j\pm1}$ for $\mathbf{U}_j = \mathbf{U}_{j\pm1}$.

The described monotonization permits the CIR solver to preserve the properties of exact solutions. Therefore, such an approach gives more adequate results of computational simulations in regions with large gradients of material parameters. The most significant improvement of numerical results was achieved in calculations of multilayered media under the action of a strong pressure pulse with the aid of Lagrangian coordinates; see Ivanov and Petrov (1992). The results were verified by shock-fitting techniques.

The realization of this monotonization method requires analytical calculation of the matrix $\Omega_{\rm R} = \Omega_{\rm L}^{-1}$ in the general case and can be cumbersome. Analytical formulas for $\Omega_{\rm R}$ can only be obtained for relatively simple systems of equations. In multidimensional cases, $\Omega_{\rm R}$ and, hence, A can be calculated numerically. Note that if the global monotonization is not sufficiently fast, then it should be applied locally only in regions with large gradients of material parameters. It is worth emphasizing that such monotonization can be important even if we use monotone first-order numerical schemes.

Hybrid schemes

The "hybrid" difference schemes represent an intermediate stage of development of numerical methods in 1960–1985. Some hundreds of works were devoted to hybrid schemes. In fact, by hybridity we mean the possibility of a numerical scheme to change its properties, for instance, the order of approximation; see early studies by Fedorenko (1962), Gol'din et al. (1965), Harten and Zwas (1972a, 1972b), van Leer (1973), Kutler, Lomax, and Warming (1973), and Boris and Book (1973). The introduction of hybridity is aimed at guaranteeing, within the framework of shock-capturing numerical methods, the second order of accuracy in the domains of smoothness of the solution, as well as the possibility to carry out calculations in domains with strong gradients according to a scheme of the first order of accuracy possessing good smoothing properties. A number of achievements of hybrid numerical schemes are included in modern numerical methods. The simplest hybrid difference scheme can be stated in the operator form as

$$gS_1 + (1-g)S_2,$$

where S_1 is a scheme of the first order, S_2 is a scheme of the second order, and g is the coefficient of hybridity, $0 \leq g \leq 1$.

In particular, Boris and Book (1973, 1976), Boris, Book, and Hain (1975), Zalesak (1979), and Knorr and Mond (1980) developed the flux corrected transport (FCT) method, see also Oran and Boris (1987). The FCT method is a two-stage hybrid numerical method of the form

$$gS_1 + (1-g)S_2 = S_1 + (1-g)(S_2 - S_1)$$

developed for a quasilinear multidimensional advection equation. A hyperbolic system in FCT approach is considered as a set of advection equations with additional source terms calculated explicitly and/or implicitly (Oran and Boris 1987). This approach can be also formally generalized for an arbitrary hyperbolic system of conservation laws, see Zalesak (1979).

At the first stage a numerical solution is calculated by the monotone scheme S_1 of the first order of accuracy. The second, "anti-diffusion" stage $(1-g)(S_2 - S_1)$ must modify the numerical solution and provide the second order of accuracy in time and space. This modification, that is, the choice of g, must satisfy the following condition. The anti-diffusion stage must not generate any new extrema in the numerical solution and must not lead to an increase (or decrease) in the maxima (or minima) that already exist. Note that these requirements are equivalent to the condition of boundedness of the total variation of the numerical solution. Thus, the FCT method belongs to the total variation diminishing (TVD) schemes (Harten 1983), see Section 2.7.2 below.

A disadvantage of the hybrid scheme approach is that the hybridization is often formal or semi-empirical. In particular, the hybrid schemes often do not take into account the characteristic properties of the hyperbolic systems.

Below we briefly outline the hybrid method suggested by Semenov (1984, 1987). The distinguishing feature of this method is the use of characteristic directions in constructing explicit hybrid schemes for n-dimensional hyperbolic systems. For specific choices of the hybridity coefficients the method reverts to the famous, namely, to the CIR method and the Lax–Wendroff method. We outline this hybrid method, since some of the ideas contained in it will be used later in Section 2.10.

Consider the difference scheme (2.3.4)–(2.3.5) with the following modification:

$$\frac{u_j^{k+1} - u_j^k}{\Delta t} + \lambda \frac{u_{j+1/2} - u_{j-1/2}}{\Delta x} = 0, \tag{2.3.54}$$

$$u_{m+1/2} = \tfrac{1}{2}(u_m^k + u_{m+1}^k) + \tfrac{1}{2}g\,(u_m^k - u_{m+1}^k), \quad m = j, j-1, \tag{2.3.55}$$

where g is the hybridity coefficient. Specifically,

$$g = C = \lambda \frac{\Delta t}{\Delta x} \tag{2.3.56}$$

in the domains of smooth solution and

$$g = \operatorname{sgn} \lambda \tag{2.3.57}$$

in the domains of nonsmooth solution. Using formula (2.3.56) yields the Lax–Wendroff scheme of the second order of accuracy in space and time, and Eq. (2.3.57) leads to a CIR-type scheme of the first order.

We can easily generalize this approach by incorporating all numerical schemes for the system of equations (2.3.11)–(2.3.12), (2.3.35), (2.3.37), and (2.3.42). In particular, we reduce the matrix $|A|$ that occurs in (2.3.12) to the form

$$|A| = \Omega_{\mathrm{R}}[\lambda_p\, g_p\, \delta_{pl}]\Omega_{\mathrm{L}}, \tag{2.3.58}$$

where

$$g_p = C_p = \lambda_p \frac{\Delta t}{\Delta x} \tag{2.3.59}$$

in the domains of smoothness and

$$g_p = \operatorname{sgn} \lambda_p \tag{2.3.60}$$

in the domains of nonsmoothness of the solution. Note that for each of the characteristic directions the coefficient of hybridity can be chosen independently. We adopt the following criteria to distinguish between smooth and nonsmooth domains (Evseev and Semenov 1990): if the inequalities

$$\begin{aligned} &(\lambda_p)_{m+1} \geq (\lambda_p)_m, \quad (\lambda_p)_m (\lambda_p)_{m+1} > 0, \\ &\big||w_p|_{m+1} - |w_p|_m\big| \leq 0.3(|w_p|_{m+1} + |w_p|_m), \quad m = j, j-1, \end{aligned} \tag{2.3.61}$$

hold in some domain, then it is a domain of smoothness. Otherwise it is a domain of nonsmoothness. The quantity w_p is the pth Riemann invariant of the linearized system corresponding to the eigenvalue λ_p, $w_p = (\Omega_{\mathrm{L}}\mathbf{U})_p = \mathbf{l}^p \cdot \mathbf{U}$, see Eq. (2.3.8). Thus, a solution is considered smooth if the relative drops of w_p are small and the eigenvalues in the neighboring computational cells of the same sign do not "overtake" each other; see Section 1.4. We shall return to this criterion in Section 2.10. For particular values of the hybridity coefficient the method transforms into the Lax–Wendroff and CIR methods. If for all p we choose g_p on the basis of (2.3.60), then we obtain a CIR-type scheme of the first order of accuracy in t and x. If for all p we use Eq. (2.3.59), then we obtain the Lax–Wendroff scheme of the second order of accuracy in x and t. A condition for the stability of this scheme can be obtained by spectral method and is expressed as

$$C_p^2 \leq g_p C_p \leq 1. \tag{2.3.62}$$

It is clear that the CFL condition $\max |C_p| \leq 1$ provides a stability condition for the hybrid scheme with g_p of (2.3.59) and/or (2.3.60). Note that one can use the continuous choice of the hybridity coefficient:

$$g_p = (1 - \theta_p) \operatorname{sgn} \lambda_p + \theta_p C_p,$$

where $0 \leq \theta_p \leq 1$ depends on gradient of w_p such that $\theta_p \to 0$ in the domains of nonsmoothness of the solution; see, for example, Gol'din et al. (1965) and Harten and Zwas (1972a, 1972b).

For the construction of a hybrid scheme on the basis the CIR schemes (2.3.35) and (2.3.37), one should rewrite the matrices $\Lambda^{\pm}$ in the form depending on g_p:

$$\Lambda^{\pm} = \left[\tfrac{1}{2}(1 \pm g_p)\, \lambda_p\, \delta_{pl}\right] = \left[\tfrac{1}{2}(\lambda_p \pm \lambda_p g_p)\, \delta_{pl}\right].$$

This is a simple way of increasing the order of accuracy and can be used as a first step in the numerical investigation of the hyperbolic system in question.

The Lax–Friedrichs scheme

The Lax–Friedrichs scheme can be interpreted as a simplified CIR numerical scheme. Let us replace all $|\lambda_p|$ in the matrix $|A|$ occurring in Eqs. (2.3.12), (2.3.26), and (2.3.38) by $|\lambda| = \max_p |\lambda_p|$. Then $|A|$ transforms into a simple form:

$$|A| = \Omega_R[\,|\lambda_p|\,\delta_{pl}]\Omega_L = |\lambda|\Omega_R\Omega_L = |\lambda| I,$$

where I is the $n \times n$ identity matrix. In particular, system (2.3.25)–(2.3.26) then reduces to

$$\frac{\mathbf{U}_j^{k+1} - \mathbf{U}_j^k}{\Delta t} + \frac{\mathbf{F}_{j+1/2} - \mathbf{F}_{j-1/2}}{\Delta x} = \mathbf{0}, \tag{2.3.63}$$

$$\mathbf{F}_{m+1/2} = \tfrac{1}{2}(\mathbf{F}_m^k + \mathbf{F}_{m+1}^k) + \tfrac{1}{2}|\lambda|_{m+1/2}^k(\mathbf{U}_m^k - \mathbf{U}_{m+1}^k), \quad m = j, j-1; \tag{2.3.64}$$

$$|\lambda| = \max_p |\lambda_p|.$$

The scheme for mixed system (2.3.42) acquires in this case the following form:

$$\begin{aligned} &\frac{\mathbf{U}_j^{k+1} - \mathbf{U}_j^k}{\Delta t} + \frac{\mathbf{G}_{j+1/2}^k - \mathbf{G}_{j-1/2}^k}{\Delta x} \\ &\quad + \frac{B_{j+1/2}^k(\mathbf{U}_{j+1}^k - \mathbf{U}_j^k) + B_{j-1/2}^k(\mathbf{U}_j^k - \mathbf{U}_{j-1}^k)}{2\Delta x} + \frac{R_{j+1/2} - R_{j-1/2}}{\Delta x} = \mathbf{0}, \\ &\mathbf{G}_{m+1/2}^k = \tfrac{1}{2}(\mathbf{G}_m^k + \mathbf{G}_{m+1}^k), \quad B_{m+1/2}^k = B(\mathbf{U}_{m+1/2}^k), \quad m = j, j-1; \\ &\mathbf{U}_{m+1/2}^k = \tfrac{1}{2}(\mathbf{U}_m^k + \mathbf{U}_{m+1}^k), \quad R_{m+1/2}^k = \tfrac{1}{2}|\lambda|_{m+1/2}^k(\mathbf{U}_m^k - \mathbf{U}_{m+1}^k). \end{aligned} \tag{2.3.65}$$

The Lax–Friedrichs scheme is much simpler than the CIR scheme and can be used as a first simplest step in the numerical investigation of the system of equations in question. We write out the Lax–Friedrichs scheme based on the local maxima of eigenvalues (Rusanov 1961). The original Lax–Friedrichs method (Lax 1954) used the $|\lambda|$ maximal on the whole computational region

$$|\lambda^k| = \max_m |\lambda_{m+1/2}^k|,$$

and $\Delta t = \Delta x / |\lambda^k|$.

2.3.2 Roe's scheme.

Roe's numerical scheme is based on an approximate Riemann problem solver for Eqs. (2.2.1)–(2.2.2). This solver is similar to the CIR Riemann solver (2.3.28) with a special choice of the matrix A, denoted here by $\mathcal{A}$. The exact formulas for determining the matrix $\mathcal{A}$ are the following:

$$\begin{aligned} &\mathbf{F}(\mathbf{U}_1) - \mathbf{F}(\mathbf{U}_2) = \mathcal{A}(\mathbf{U}_1, \mathbf{U}_2)(\mathbf{U}_1 - \mathbf{U}_2), \\ &\mathcal{A}(\mathbf{U}, \mathbf{U}) = A(\mathbf{U}), \\ &\mathbf{U}_1 = \mathbf{U}_m, \quad \mathbf{U}_2 = \mathbf{U}_{m+1}, \end{aligned}$$

or

$$\begin{aligned} &\Delta\mathbf{F} = \mathcal{A}\Delta\mathbf{U}, \quad \Delta\mathbf{F} = \mathbf{F}_1 - \mathbf{F}_2, \quad \Delta\mathbf{U} = \mathbf{U}_1 - \mathbf{U}_2, \\ &\mathbf{F}_1 = \mathbf{F}(\mathbf{U}_1), \quad \mathbf{F}_2 = \mathbf{F}(\mathbf{U}_2). \end{aligned}$$

Here $\mathcal{A}$ must have only real eigenvalues and a complete set of eigenvectors. Consider a nondegenerate nonlinear transformation $\mathbf{U} = \mathbf{U}(\mathbf{Y})$. From the exact relations $\Delta\mathbf{F} = \mathcal{A}_F\Delta\mathbf{Y}$ and $\Delta\mathbf{U} = \mathcal{A}_U\Delta\mathbf{Y}$ it follows that $\mathcal{A} = \mathcal{A}_F\mathcal{A}_U^{-1}$. Since $\mathbf{F}$ is generally a nonlinear function of $\mathbf{U}$, the obtained matrix $\mathcal{A}$ is not unique.

For the scalar case we can illustrate this nonuniqueness as follows. Assuming that $F = U^3$ and taking $U = Y$, and hence $F = Y^3$, we obtain

$$\frac{\Delta F}{\Delta U} = U_2^2 + U_2U_1 + U_1^2.$$

Assuming $F = U^3$ and taking $U = Y^2$, and hence $F = Y^6$, we find that

$$\frac{\Delta F}{\Delta U} = \frac{(U_2 + \sqrt{U_2U_1} + U_1)\left(\sqrt{U_2^3} + \sqrt{U_1^3}\right)}{\sqrt{U_2} + \sqrt{U_1}}.$$

Therefore, the entries of $\mathcal{A}$ can depend on the choice of $\mathbf{Y}$. The difference between the matrices $\mathcal{A}$ corresponding to different $\mathbf{Y}$ is of the order of $O(|\Delta\mathbf{U}|^2)$.

The exact analytical formulas for $\mathcal{A}_F$ and $\mathcal{A}_U$ can be written out explicitly if $\mathbf{U}$ and $\mathbf{F}$ are polynomials or rational functions in the components of $\mathbf{Y}$. To derive the entries of $\mathcal{A}_F$ and $\mathcal{A}_U$, one can use the following exact identities:

$$\Delta(ab) = \bar{b}\Delta a + \bar{a}\Delta b; \tag{2.3.66}$$

$$\begin{aligned}\Delta(abc) &= \bar{b}\bar{c}\Delta a + \bar{a}\,\bar{c}\Delta b + \bar{a}\bar{b}\Delta c + \tfrac{1}{4}\Delta a\Delta b\Delta c \\ &= (\bar{b}\bar{c} + \theta_1\Delta b\Delta c)\Delta a + (\bar{a}\,\bar{c} + \theta_2\Delta a\Delta c)\Delta b + (\bar{a}\bar{b} + \theta_3\Delta a\Delta b)\Delta c, \\ &\quad \theta_1 + \theta_2 + \theta_3 = \tfrac{1}{4};\end{aligned} \tag{2.3.67}$$

$$\begin{aligned}\Delta(a^2b) &= (2\bar{a}\bar{b} + \theta_1\Delta a\Delta b)\Delta a + (\bar{a}^2 + \theta_2(\Delta a)^2)\Delta b, \\ &\quad \theta_1 + \theta_2 = \tfrac{1}{4};\end{aligned} \tag{2.3.68}$$

$$\begin{aligned}\Delta(abcd) &= \bar{b}\bar{c}\bar{d}\Delta a + \bar{a}\,\bar{c}\bar{d}\Delta b + \bar{a}\bar{b}\,\bar{d}\Delta c + \bar{a}\bar{b}\bar{c}\Delta d \\ &\quad + \tfrac{1}{4}\bar{d}\Delta a\Delta b\Delta c + \tfrac{1}{4}\bar{a}\Delta b\Delta c\Delta d + \tfrac{1}{4}\bar{b}\Delta a\Delta c\Delta d + \tfrac{1}{4}\bar{c}\Delta a\Delta b\Delta d \\ &= (\bar{b}\bar{c}\bar{d} + \theta_1\bar{d}\Delta b\Delta c + \kappa_1\bar{b}\Delta c\Delta d + \pi_1\bar{c}\Delta b\Delta d)\Delta a \\ &\quad + (\bar{a}\,\bar{c}\bar{d} + \theta_2\bar{d}\Delta a\Delta c + \eta_1\bar{a}\Delta c\Delta d + \pi_2\bar{c}\Delta a\Delta d)\Delta b \\ &\quad + (\bar{a}\bar{b}\,\bar{d} + \theta_3\bar{d}\Delta a\Delta b + \eta_2\bar{a}\Delta b\Delta d + \kappa_2\bar{b}\Delta a\Delta d)\Delta c \\ &\quad + (\bar{a}\bar{b}\bar{c} + \pi_3\bar{c}\Delta a\Delta b + \eta_3\bar{a}\Delta b\Delta c + \kappa_3\bar{b}\Delta a\Delta c)\Delta d, \\ &\quad \theta_1+\theta_2+\theta_3 = \tfrac{1}{4}, \quad \eta_1+\eta_2+\eta_3 = \tfrac{1}{4}, \quad \pi_1+\pi_2+\pi_3 = \tfrac{1}{4}, \quad \kappa_1+\kappa_2+\kappa_3 = \tfrac{1}{4}.\end{aligned} \tag{2.3.69}$$

Here $\Delta f = f_1 - f_2$ and $\bar{f} = \frac{1}{2}(f_1 + f_2)$ is the arithmetic average, where $f = a$, b, c, and d. Relation (2.3.66) determines the coefficients of the differences Δa and Δb uniquely for a quadratic polynomial. It is apparent from Eqs. (2.3.67) and (2.3.68) that, for the cubic polynomials, the coefficients of Δa, Δb, and Δc are not unique and belong to a one- or two-parameter family of coefficients. For polynomials of higher degree, the nonuniqueness of the coefficients in the differences is even more complicated, see (2.3.69). Thus, examples (2.3.67)–(2.3.69) demonstrate the cause of nonuniqueness in constructing entries of the matrix $\mathcal{A}$. The matrix $\mathcal{A}$ approximates the matrix $A = \partial\mathbf{F}/\partial\mathbf{U}$ and must preserve its hyperbolic properties. The main problem is choosing an appropriate $\mathbf{Y}$.

We can interpret Roe's method as a special case of the CIR method (2.3.27)–(2.3.28) for which

$$\mathbf{U}(\xi) = \tfrac{1}{2}(\mathbf{U}_1 + \mathbf{U}_2) + \tfrac{1}{2}S(\xi)\Delta\mathbf{U}, \quad S(\xi) = \Omega_\mathrm{R}[\operatorname{sgn}(\lambda_p - \xi)\,\delta_{pl}]\Omega_\mathrm{L}, \tag{2.3.70}$$

$$\mathbf{F}(\xi) = \tfrac{1}{2}(\mathbf{F}_1 + \mathbf{F}_2) + \tfrac{1}{2}\mathcal{A}(\xi)\Delta\mathbf{U}, \quad \mathcal{A}(\xi) = \Omega_\mathrm{R}[\lambda_p \operatorname{sgn}(\lambda_p - \xi)\,\delta_{pl}]\Omega_\mathrm{L}, \tag{2.3.71}$$

where $\xi = x/t$, $-\infty < \xi < +\infty$, and $\Delta\mathbf{U} = \mathbf{U}_1 - \mathbf{U}_2$, $\mathbf{U}_1 = \mathbf{U}_m$, $\mathbf{U}_2 = \mathbf{U}_{m+1}$. The corresponding flux vector $\tilde{\mathbf{F}}$ of (2.3.29) for moving grid calculations is given by

$$\tilde{\mathbf{F}} = \mathbf{F}(\xi) - \xi\mathbf{U}(\xi) = \tfrac{1}{2}(\mathbf{F}_1 - \xi\mathbf{U}_1 + \mathbf{F}_2 - \xi\mathbf{U}_2) + \tfrac{1}{2}|\mathcal{A}(\xi)|\Delta\mathbf{U}, \tag{2.3.72}$$

$$|\mathcal{A}(\xi)| = \Omega_\mathrm{R}[|\lambda_p - \xi|\,\delta_{pl}]\Omega_\mathrm{L}.$$

Equations (2.3.70)–(2.3.71) define piecewise constant functions $\mathbf{U}(\xi)$ and $\mathbf{F}(\xi)$ that relate the right values $\mathbf{U}_2$, $\mathbf{F}_2$ to the left values $\mathbf{U}_1$, $\mathbf{F}_1$ through a set of discontinuities. Note that for a system with constant coefficients this solution is exact and unique.

Let the right quantities $\mathbf{U}_2$, $\mathbf{F}_2$ and left quantities $\mathbf{U}_1$, $\mathbf{F}_1$ be related in accordance with the shock relations (2.2.14):

$$\{\mathbf{F}\} = W\{\mathbf{U}\} \quad \Longleftrightarrow \quad \Delta\mathbf{F} = W\Delta\mathbf{U},$$

where W is the velocity of a discontinuity. Then W is an eigenvalue of $\mathcal{A}$. Indeed,

$$\Delta\mathbf{F} - W\Delta\mathbf{U} = (\mathcal{A}_F - W\mathcal{A}_U)\Delta\mathbf{Y} = (\mathcal{A} - WI)\mathcal{A}_U\Delta\mathbf{Y} = (\mathcal{A} - WI)\Delta\mathbf{U} = \mathbf{0}.$$

Hence, $\det[\mathcal{A} - WI] = 0$ and $\Delta\mathbf{U}$ is an eigenvector of $\mathcal{A}$. Thus, Eq. (2.3.71) describes the jump condition exactly. This is an important feature of Roe's scheme distinguishing it from the CIR family where the jump conditions are satisfied generally with the order of accuracy of the scheme.

For the construction of the Godunov method, only formula (2.3.71) will suffice. Indeed, the numerical scheme (2.2.3) contains only the quantities $\mathbf{F}$. The formulas for $\mathbf{U}$ may be useful, for instance, for the approximation of the right-hand sides in the hyperbolic system. In addition, $\mathbf{F}(\xi)$ and $\mathbf{U}(\xi)$ ahead of and behind the p-discontinuity are related by $\{\mathbf{F}\} = \lambda_p\{\mathbf{U}\}$, see Eqs. (2.3.30)–(2.3.31).

For schemes using stationary grids one can obtain $\mathbf{F}$ and $\mathbf{U}$ by setting $\xi = 0$ in Eqs. (2.3.70)–(2.3.71). Thus,

$$\mathbf{U}(0) = \tfrac{1}{2}(\mathbf{U}_1 + \mathbf{U}_2) + \tfrac{1}{2}S\Delta\mathbf{U}, \qquad S = S(0) = \Omega_\mathrm{R}[\operatorname{sgn}\lambda_p\,\delta_{pl}]\Omega_\mathrm{L},$$

$$\mathbf{F}(0) = \tfrac{1}{2}(\mathbf{F}_1 + \mathbf{F}_2) + \tfrac{1}{2}|\mathcal{A}|\Delta\mathbf{U}, \qquad |\mathcal{A}| = \mathcal{A}(0) = \Omega_\mathrm{R}[\,|\lambda_p|\,\delta_{pl}]\Omega_\mathrm{L}.$$

Roe's method was first constructed for gas dynamics (Roe 1981), see also the review by Roe (1986) and Section 3.4.4. Roe applied special variables $\mathbf{Y}(\mathbf{U})$, known as Roe's variables, which permitted him to obtain a unique matrix $\mathcal{A}$, in the sense that $\mathcal{A} = \mathcal{A}(\mathbf{U}_1, \mathbf{U}_2) = \mathcal{A}(\overline{\mathbf{Y}})$, where $\overline{\mathbf{Y}}$ stands for the arithmetic average of the vector $\mathbf{Y}$ at the boundary,

$$\overline{\mathbf{Y}} = \tfrac{1}{2}[\mathbf{Y}(\mathbf{U}_1) + \mathbf{Y}(\mathbf{U}_2)].$$

Roe's matrix $\mathcal{A}$ preserves the hyperbolic properties of the matrix A. In addition, the entries of $\mathcal{A}$ are functions of some variables that are averaged in the same manner, see Section 3.4.4.

Roe's Riemann problem solver for gas dynamic equations is presented in Chapter 3, and for shallow water equations it is presented in Chapter 4. Roe's Riemann problem solver for magnetohydrodynamics (MHD) also exists, see Chapter 5, though it is not unique. This Riemann problem solver is much simpler, particularly, than an exact one, and preserves the shock relations.

Computational practice shows that the solution given by (2.3.71) is often a reasonable compromise between the exact solution and the approximate solution of the CIR type. Although rather simple, these solutions ensure exact treatment of the shock relations and the corresponding velocities, thus partially preserving the properties of the starting nonlinear equations. As a result, shocks and contact discontinuities can be captured exactly, for example, by the self-adjusting methods described in Harten and Hyman (1983), Kamenetskii and Semenov (1994), see Section 3.5.4. If Eulerian grids are employed, the usage of approximate Riemann problem solvers (apart from low costs) is not obvious, because the discontinuity velocity W, which is calculated exactly in the analytical solution, would generally be inexact at the next time step. Comparative shock-capturing numerical calculations of some two-dimensional problems of gas dynamics demonstrated, however, that Roe's method provides a more consistent treatment of the interaction between shock waves. In particular, with Roe's method, the shock waves are smeared over fewer computational cells as compared with the CIR method.

2.3.3 The Osher numerical scheme. As mentioned above, Roe's numerical scheme is based on the "shock-only" approximation and rarefaction waves in the Riemann problem solution are replaced by expansion shocks. This is a dangerous procedure, since spurious expansion shocks can truly originate. Special entropy fix procedures invented to avoid this phenomenon are represented in Section 2.10.

Since we are interested in numerical schemes based on the approximate solution of the Riemann problem, we might want to obtain such a scheme constructed of simple waves. As shown in Section 1.4, such solution can, in principle, be obtained in that part of the (x, t)-plane where the characteristics of the same family do not intersect each other. This allows us to construct a local Riemann problem solution on the basis of simple waves only. Such a scheme was proposed by Engquist and Osher (1981) for a single conservation law and extended by Osher (1981) to strictly hyperbolic systems.

Prior to describing the Osher scheme for hyperbolic systems, we shall explain its essence using a single conservation law

$$\frac{\partial u}{\partial t} + \frac{\partial f(u)}{\partial x} = 0 \tag{2.3.73}$$

as an example.

In full agreement with the idea of introducing the upwind approximation of Eq. (2.3.73), the Engquist–Osher (EO) flux is written out as

$$f^{\mathrm{EO}}(u^{\mathrm{L}}, u^{\mathrm{R}}) = f(u^*) - \int_{u^{\mathrm{L}}}^{u^*} \max(a(s), 0)\, ds + \int_{u^*}^{u^{\mathrm{R}}} \min(a(s), 0)\, ds, \tag{2.3.74}$$

where $a(u) = \partial f/\partial u$ and integration is performed in the u-space along the integral curve connecting u^{L} and u^{R} (see Section 1.5). The quantity u^* occurring in Eq. (2.3.73) is an arbitrary reference state on the integral curve and f^{EO} is clearly independent of u^*.

Being nonpositive, the integral terms in this flux provide dissipation to the numerical flux. Osher (1984) showed that the flux (2.3.74) can be represented as the Godunov numerical flux plus a dissipation term.

Consider the analogy of the EO flux with the Godunov flux

$$f^{G}_{j+1/2} = f(u_j, u_{j+1}) = f(\bar{u}(x_{j+1/2}, t^*)) \tag{2.3.75}$$

for the scalar conservation law. Here $\bar{u}(x_{j+1/2}, s)$ for $s > t$ is the unique entropy-consistent solution of the Riemann problem for Eq. (2.3.73) with the initial data

$$\bar{u}(x, t) = \begin{cases} u_j & \text{if } x < x_{j+1/2}, \\ u_{j+1} & \text{if } x > x_{j+1/2} \end{cases}$$

and

$$\bar{u}(x_{j+1/2}, t) = \lim_{s \to t} \bar{u}(x_{j+1/2}, s).$$

Note that the solution to this Riemann problem for a single conservation law with a convex $f(u)$ exists for arbitrary u_j and u_{j+1}.

The Godunov flux can be represented in the form (Osher 1984)

$$f^{G}_{j+1/2} = \begin{cases} \min\limits_{u_j \le u \le u_{j+1}} f(u) & \text{for } u_j < u_{j+1}, \\ \max\limits_{u_j \ge u \ge u_{j+1}} f(u) & \text{for } u_j > u_{j+1}. \end{cases} \tag{2.3.76}$$

Now if we assign $u^{\mathrm{L}} = u_j \le u^{\mathrm{R}} = u_{j+1}$ and assume that f is a monotone increasing function of u, then

$$f^{G}_{j+1/2} = f(u^{\mathrm{L}}).$$

Otherwise, if f is a monotone decreasing function, then

$$f^{G}_{j+1/2} = f(u^{\mathrm{R}}).$$

If an extremum of f exists inside the interval, a proper choice must be made between this extremum and the end point values.

Similarly, in the case of the monotone behavior of the flux f, we choose an appropriate end point value for u^* in the EO scheme, namely,

$$f^{EO}_{j+1/2} = \begin{cases} f(u^{\mathrm{L}}) & \text{for } \sigma \ge 0, \\ f(u^{\mathrm{R}}) & \text{for } \sigma \le 0 \end{cases}$$

with

$$\sigma = \frac{f(u^{\mathrm{R}}) - f(u^{\mathrm{L}})}{u^{\mathrm{R}} - u^{\mathrm{L}}}.$$

This means that we choose $u^* = u^{\mathrm{L}}$ for $\sigma \ge 0$ and $u^* = u^{\mathrm{R}}$ otherwise. Thus, similarly to the Godunov scheme, if f is monotone between u^{L} and u^{R}, we choose the appropriate end point value. The Osher and Godunov fluxes coincide in this case. Otherwise, we must scan the interval to record the values of f at the critical points and use Eq. (2.3.74). It is easy to see that the dissipation in the EO scheme for Eq. (2.3.73) with the convex downward flux f

is nonzero only for two states that describe a transonic compression, that is, $a(u^{\mathrm{L}}) > 0$ and $a(u^{\mathrm{R}}) < 0$. Note that this occurs only if $u^{\mathrm{R}} < u^{\mathrm{L}}$.

The important feature of the EO scheme is in its retrieving the sonic points of $\partial f/\partial u$, that is, the points where

$$\frac{\partial f}{\partial u} = a = 0.$$

These points are then used in the construction of the numerical flux. The addition of numerical viscosity near sonic transitions helps to select the entropy-consistent solution to the problem.

The advantage of the EO scheme is also revealed when we solve the hyperbolic equation (2.3.73) with a nonconvex $f(u)$. Provided that the extremum points are known, addition of extra numerical viscosity near these points allows us to achieve an efficient determination of physically admissible solutions that otherwise can be mistreated if $f(u)$ is concave. On the other hand, this will not suffice if the small-scale, higher-order differential model includes not only second- but also third-order derivatives, that is, not only dissipation but also dispersion must participate in the selection of an admissible solution (Kulikovskii 1984).

The value of a can be approximated using the values of the first and second derivatives of f at u^{L} and u^{R}. Since the linear interpolation of a that corresponds to the parabolic interpolation of f cannot represent inflection points of f, Bell, Colella, and Trangenstein (1989) suggested approximating the function a on the interval between u^{L} and u^{R} by the Hermitian cubic interpolant $\tilde{a}$, which, in turn, is represented by a piecewise-linear approximation $\bar{a}$ passing through u^{L}, u^{R}, and the extremum points of $\tilde{a}$. Thus the final form of the numerical flux acquires the form

$$f^{\mathrm{EO}}(u^{\mathrm{L}}, u^{\mathrm{R}}) = f(u^*) - \int_{u^{\mathrm{L}}}^{u^*} \max(\bar{a}(s), 0)\, ds + \int_{u^*}^{u^{\mathrm{R}}} \min(\bar{a}(s), 0)\, ds. \tag{2.3.77}$$

Another form can be obtained by introducing nonnegative quantities

$$\alpha^+ = \frac{1}{u^* - u^{\mathrm{L}}} \int_{u^{\mathrm{L}}}^{u^*} \max(\bar{a}(s), 0)\, ds, \quad \alpha^- = \frac{1}{u^{\mathrm{R}} - u^*} \int_{u^*}^{u^{\mathrm{R}}} |\min(\bar{a}(s), 0)|\, ds$$

and

$$\beta = \frac{u^* - u^{\mathrm{L}}}{u^{\mathrm{R}} - u^{\mathrm{L}}}.$$

Then,

$$f^{\mathrm{EO}}(u^{\mathrm{L}}, u^{\mathrm{R}}) = f(u^*) - [\alpha^+\beta + \alpha^-(1-\beta)](u^{\mathrm{R}} - u^{\mathrm{L}}).$$

Note that $0 \le \beta \le 1$ for $u^* \in [u^{\mathrm{L}}, u^{\mathrm{R}}]$. Thus, the additional integral terms in Eq. (2.3.74) represent dissipation.

A convenient form of the EO scheme that allows one to avoid any ambiguity in the choice of u^* is in writing the flux as an arithmetic average of those for $u^* = u^{\mathrm{L}}$ and $u^* = u^{\mathrm{R}}$,

$$f^{\mathrm{EO}}(u^{\mathrm{L}}, u^{\mathrm{R}}) = \frac{1}{2}\left[f(u^{\mathrm{L}}) + f(u^{\mathrm{R}}) - \int_{u^{\mathrm{L}}}^{u^{\mathrm{R}}} |a(s)|\, ds\right].$$

In the case of a system, the numerical flux of the Osher scheme can be determined from the formula

$$\mathbf{F}^{\mathrm{O}}_{j+1/2} = \frac{1}{2}\left(\left[\mathbf{F}(\mathbf{U}_{j+1}) - \int_{\mathbf{U}_j}^{\mathbf{U}_{j+1}} A^{+}(\mathbf{U})\, d\mathbf{U}\right] + \left[\mathbf{F}(\mathbf{U}_j) + \int_{\mathbf{U}_j}^{\mathbf{U}_{j+1}} A^{-}(\mathbf{U})\, d\mathbf{U}\right]\right) \tag{2.3.78}$$

or

$$\mathbf{F}^{\mathrm{O}}_{j+1/2} = \frac{1}{2}\left[\mathbf{F}(\mathbf{U}_{j+1}) + \mathbf{F}(\mathbf{U}_j) - \int_{\mathbf{U}_j}^{\mathbf{U}_{j+1}} |A(\mathbf{U})|\, d\mathbf{U}\right]. \tag{2.3.79}$$

Here

$$A = \frac{\partial \mathbf{F}}{\partial \mathbf{U}} = \Omega_{\mathrm{R}}\Lambda\Omega_{\mathrm{L}}, \quad \Lambda = \mathrm{diag}[\lambda_1, \ldots, \lambda_n], \quad |\Lambda| = \mathrm{diag}[|\lambda_1|, \ldots, |\lambda_n|],$$
$$A^{+} = \Omega_{\mathrm{R}}\Lambda^{+}\Omega_{\mathrm{L}}, \quad A^{-} = \Omega_{\mathrm{R}}\Lambda^{-}\Omega_{\mathrm{L}}, \quad \Lambda^{+} = \tfrac{1}{2}(\Lambda + |\Lambda|), \quad \Lambda^{-} = \tfrac{1}{2}(\Lambda - |\Lambda|).$$

For strictly hyperbolic systems the eigenvalues can be ordered as $\lambda_1 < \ldots < \lambda_n$. Let the corresponding eigenvectors be $\mathbf{r}^1, \ldots, \mathbf{r}^n$. Since we are supposed to construct a solution on the basis of simple waves, consider the system of ordinary differential equations in the n-dimensional phase space (see Section 1.4.1),

$$\frac{d\mathbf{U}}{ds} = \mathbf{r}^k\,(\mathbf{U}(s))\,. \tag{2.3.80}$$

Along the subpath Γ_k (s is an arc length) we have

$$\left|\frac{\partial \mathbf{F}(\mathbf{U})}{\partial \mathbf{U}}\right| d\mathbf{U} = \left|\frac{\partial \mathbf{F}}{\partial \mathbf{U}}\right| \frac{d\mathbf{U}}{ds}\, ds = |\lambda_k|\, \mathbf{r}^k\, ds.$$

System (2.2.1) is called "convex" if each of its characteristic fields is either genuinely nonlinear, that is,

$$\frac{\partial \lambda_k}{\partial \mathbf{U}} \cdot \mathbf{r}^k \neq 0,$$

or linearly degenerate, that is,

$$\frac{\partial \lambda_k}{\partial \mathbf{U}} \cdot \mathbf{r}^k \equiv 0.$$

If $\mathbf{U}^{\mathrm{R}}$ belongs to a vicinity of $\mathbf{U}^{\mathrm{L}}$, the solution of the Riemann problem for such systems is known to exist. This is always true for the Euler gas dynamic system and many other physical systems. For genuinely nonlinear fields the eigenvectors $\mathbf{r}^k$ can be normalized so that

$$\frac{\partial \lambda_k}{\partial \mathbf{U}} \cdot \mathbf{r}^k \equiv 1.$$

We see that for a linearly degenerate field λ_k is constant along the path Γ_k, which is always parallel to $\mathbf{r}^k$ and, hence,

$$\int_{\Gamma_k} \left|\frac{\partial \mathbf{F}(\mathbf{U})}{\partial \mathbf{U}}\right| d\mathbf{U} = \int_{\Gamma_k} |\lambda_k| \mathbf{r}^k ds = |\lambda_k| \int_{\Gamma_k} \mathbf{r}^k ds$$
$$= (\mathrm{sgn}\,\lambda_k) \times \int_{\Gamma_k} \lambda_k \mathbf{r}^k\, ds = (\mathrm{sgn}\,\lambda_k) \times \int_{\Gamma_k} \frac{\partial \mathbf{F}}{\partial \mathbf{U}} d\mathbf{U} = (\mathrm{sgn}\,\lambda_k) \times \mathbf{F}(\mathbf{U})\Big|_{\mathbf{U}^k}^{\mathbf{U}^{k+1}}.$$

Here $\mathbf{U}^k$ and $\mathbf{U}^{k+1}$ are the quantities at the beginning and at the end of the arc Γ_k.

For a genuinely nonlinear field, λ_k is a monotone function along the orbit (2.3.80). This means that the corresponding eigenvalue can change sign only once along the path Γ_k, that is, there can be only a unique $\tilde{\mathbf{U}}$ for which $\lambda_k(\mathbf{U}) = 0$ with $\mathbf{U}$ on the orbit Γ_k connecting $\mathbf{U}^k$ and $\mathbf{U}^{k+1}$. In this case it is easy to see that

$$\int_{\Gamma_k} \left| \frac{\partial \mathbf{F}}{\partial \mathbf{U}} \right| d\mathbf{U} = 2\mathbf{F}(\mathbf{B}) - 2\mathbf{F}(\mathbf{A}) + \mathbf{F}(\mathbf{U}^{k+1}) - \mathbf{F}(\mathbf{U}^k),$$

where

$$\mathbf{B} = \begin{cases} \mathbf{U}^{k+1} & \text{if } \lambda_k(\mathbf{U}^{k+1}) > 0, \\ \tilde{\mathbf{U}} & \text{if } \lambda_k(\mathbf{U}^{k+1}) \le 0, \end{cases} \qquad \mathbf{A} = \begin{cases} \mathbf{U}^k & \text{if } \lambda_k(\mathbf{U}^k) > 0, \\ \tilde{\mathbf{U}} & \text{if } \lambda_k(\mathbf{U}^k) \le 0. \end{cases}$$

The path connecting $\mathbf{U}_j$ and $\mathbf{U}_{j+1}$ consists of a continuous union of segments Γ_k^j. There remains a choice to start with $k = n$ in Eq. (2.3.78) emanating from $\mathbf{U}_j$ and continue until, for $k = 1$, the last segment Γ_1^j ends at $\mathbf{U}_{j+1}$, or to begin with $k = 1$ emanating from $\mathbf{U}_j$ and continue until, for $k = n$, the last segment Γ_n^j ends at $\mathbf{U}_{j+1}$ (see the discussion by Pandolfi 1984). The original Osher scheme used the former approach. The latter ordering of the subpaths seems, however, more physical, since it uses increasing order of the wave speeds.

The form (2.3.78)–(2.3.79) of the Osher flux allowed us to avoid the choice of the reference state $\mathbf{U}^*$. Similarly to the scalar case, we can introduce the mean speed

$$\bar{\sigma} = \frac{[\mathbf{F}(\mathbf{U}^{\mathrm{L}}) - \mathbf{F}(\mathbf{U}^{\mathrm{R}})] \cdot (\mathbf{U}^{\mathrm{L}} - \mathbf{U}^{\mathrm{R}})}{|\mathbf{U}^{\mathrm{L}} - \mathbf{U}^{\mathrm{R}}|^2}$$

and define $\mathbf{U}^* = \mathbf{U}^{\mathrm{L}}$ if $\bar{\sigma} \ge 0$ and $\mathbf{U}^* = \mathbf{U}^{\mathrm{R}}$ if $\bar{\sigma} < 0$. Note that $\mathbf{U}^{\mathrm{L}} \equiv \mathbf{U}_j$ and $\mathbf{U}^{\mathrm{R}} \equiv \mathbf{U}_{j+1}$.

Suppose we chose $\mathbf{U}^* = \mathbf{U}^{\mathrm{L}}$, then

$$\mathbf{F}^{\mathrm{O}}_{j+1/2} = \mathbf{F}(\mathbf{U}^{\mathrm{L}}) + \int_{\mathbf{U}^{\mathrm{L}}}^{\mathbf{U}^{\mathrm{R}}} A^-(\mathbf{U})\, d\mathbf{U},$$

or, using Eq. (2.3.80),

$$\mathbf{F}^{\mathrm{O}}_{j+1/2} = \mathbf{F}(\mathbf{U}^{\mathrm{L}}) + \sum_{k=1}^{n} \int_{\mathbf{U}^k}^{\mathbf{U}^{k+1}} A^-(\mathbf{U})\mathbf{r}^k\, ds = \mathbf{F}(\mathbf{U}^{\mathrm{L}}) + \sum_{k=1}^{n} \int_{\mathbf{U}^k}^{\mathbf{U}^{k+1}} \lambda_k^-(\mathbf{U})\mathbf{r}^k\, ds.$$

It is obvious that ds is equal to

$$ds = \mathbf{l}^k \cdot d\mathbf{U} = dw_k.$$

In general, the last relation cannot be integrated to obtain the Riemann invariant w_k, and we must apply the approximate integration from $\mathbf{U}^k$ to $\mathbf{U}^{k+1}$ along the subpath Γ_k parallel to $\mathbf{r}^k$. It gives the upper limit of the considered integral in the form

$$\bar{w}_k = \bar{\mathbf{l}}^k \cdot (\mathbf{U}^{k+1} - \mathbf{U}^k).$$

The kth left eigenvector is approximated here at some point between $\mathbf{U}^{\mathrm{R}}$ and $\mathbf{U}^{\mathrm{L}}$, e. g., at the arithmetic average of these quantities.

Thus, we obtain

$$\mathbf{F}^{\mathrm{O}}_{j+1/2} = \mathbf{F}(\mathbf{U}^{\mathrm{L}}) + \sum_{k=1}^{n} \int_0^{\bar{w}_k} \lambda_k^- \mathbf{r}^k \, dw_k.$$

Now it is only natural to use the same approximate value for the right eigenvector, and we arrive at the formula

$$\mathbf{F}^{\mathrm{O}}_{j+1/2} = \mathbf{F}(\mathbf{U}^{\mathrm{L}}) + \sum_{k=1}^{n} \left(\int_0^{\bar{w}_k} \min(\lambda_k, 0) \, dw_k \right) \times \bar{\mathbf{r}}^k, \qquad (2.3.81)$$

similar to Eq. (2.3.77). We implicitly assumed in deriving Eq. (2.3.81) that $\bar{w}_k > 0$. This always can be done if we admit that eigenvectors are defined up to a constant multiplier.

Note that generally there is no need to know the solution of the problem along the path connecting $\mathbf{U}_j$ and $\mathbf{U}_{j+1}$. It is sufficient to determine the zero points of λ_k. This can be done, as mentioned earlier for the case of a single equation, by approximating the eigenvalues on the interval under consideration using cubic polynomials. In this case the method can also be applied to nonstrictly hyperbolic systems with coinciding eigenvalues and/or a nonconvex flux $\mathbf{F}$.

Suppose the system is not strictly hyperbolic. As seen from the above considerations, this is not very important if the eigenvalue multiplicity does not depend on the solution of the problem, that is, if the multiplicity is always the same. Otherwise, even if we assume that the characteristic fields are in general either genuinely nonlinear or linearly degenerate, it may happen that the coincidence of the two or more eigenvalues at some phase space point results in the degeneracy of an initially nonlinear wave field. If we encounter the local linear degeneracy of the characteristic field, we must know the behavior of the wave speed at all points along the wave curve connecting the pre-wave and post-wave states. Besides, as shown by Osher (1984), the only schemes for nonlinear *nonconvex* scalar conservation law in one space dimension that are known to converge to weak solutions satisfying appropriate entropy conditions are monotone schemes that are only first-order accurate. The addition of dissipation near transonic expansions, though permitting us to separate entropy-consistent solutions for a strictly hyperbolic system, can become insufficient in a degenerate case. If some eigenvalues are close one to another, the corresponding eigenvectors may turn out to be nearly parallel. A proper modification of the Osher flux that takes into account the deficiency of the eigenvectors was suggested by Bell et al. (1989).

The eigenvalues λ_l and λ_m are considered to be close if

$$|\lambda_l^e - \lambda_m^e| < \varepsilon \sum_{k=1}^{n} \bar{w}_k |\mathcal{K}_{lk} - \mathcal{K}_{mk}|,$$

where $\lambda_{l,m}^e$ are calculated at the point $\frac{1}{2}(\mathbf{U}^{\mathrm{R}} + \mathbf{U}^{\mathrm{L}})$ and

$$\mathcal{K}_{lk} = \frac{\partial \lambda_l}{\partial \mathbf{U}} \cdot \bar{\mathbf{r}}^k.$$

The coefficient ε is assumed to be 0.1.

The actual modification depends on whether the eigenvector deficiency is associated with the change in sign of the wave speeds during the transition from $\mathbf{U}^{\mathrm{L}}$ to $\mathbf{U}^{\mathrm{R}}$. Let us introduce

$$\lambda_{lm}^{\min} = \min(\lambda_l^{\mathrm{L}}, \lambda_l^{\mathrm{R}}, \lambda_m^{\mathrm{L}}, \lambda_m^{\mathrm{R}}), \quad \lambda_{lm}^{\max} = \max(\lambda_l^{\mathrm{L}}, \lambda_l^{\mathrm{R}}, \lambda_m^{\mathrm{L}}, \lambda_m^{\mathrm{R}}).$$

If $\lambda_{lm}^{\min}$ and $\lambda_{lm}^{\max}$ have the same sign, then we do not have a transonic wave. In this case, the variation of $\mathbf{U}$ in the sum of characteristic waves $\bar{w}_l\,\bar{\mathbf{r}}^l + \bar{w}_m\,\bar{\mathbf{r}}^m$ can be substituted as a the single jump by introducing

$$\bar{\mathbf{r}}^{lm} = \frac{\bar{w}_l\bar{\mathbf{r}}^l + \bar{w}_m\bar{\mathbf{r}}^m}{\left|\bar{w}_l\bar{\mathbf{r}}^l + \bar{w}_m\bar{\mathbf{r}}^m\right|}.$$

If $\lambda_{lm}^{\min}$ and $\lambda_{lm}^{\max}$ have opposite signs, extra dissipation is introduced for deficient modes by approximating them on the basis of the local Lax–Friedrichs-type scheme by Rusanov (1961), see Eqs. (2.3.63)–(2.3.64).

Let us have a closer look at the formula (2.3.77) in the case of a reference state fixed at $u = u^{\mathrm{L}}$. It can be rewritten in the following form convenient for numerical implementation:

$$f^{\mathrm{EO}} = f(u^{\mathrm{L}}) + \int_{u^{\mathrm{L}}}^{u^{\mathrm{R}}} \min(a(u), 0)\,du = f(u^{\mathrm{L}}) + \int_{u^{\mathrm{L}}}^{u^{\mathrm{R}}} \chi(u)a(u)\,du,$$

where

$$\chi(u) = \begin{cases} 1 & \text{if } a(u) < 0, \\ 0 & \text{if } a(u) > 0. \end{cases}$$

The total integral can be represented as the sum of terms of the form

$$\int_{u^k}^{u^{k+1}} a(u)\,du = f(u^{k+1}) - f(u^k),$$

where $a < 0$ for all $u \in \left[u^k, u^{k+1}\right]$. The left end of the interval is defined either by the relation $a(u^k) = 0$ or $u^k = u^{\mathrm{L}}$. The right end of the interval is defined either by the relation $a(u^{k+1}) = 0$ or $u^{k+1} = u^{\mathrm{R}}$.

Thus, we can write out

$$f^{\mathrm{EO}} = f(u^{\mathrm{L}}) + \sum_{s=1}^{S} (-1)^s f(u^s), \tag{2.3.82}$$

where $qu^1 < qu^2 < \cdots < qu^S$ ($q = \operatorname{sgn}(u^{\mathrm{R}} - u^{\mathrm{L}})$) and u^s satisfies either $a(u^s) = 0$ or $u^1 = u^{\mathrm{L}}$ with $a(u^1) \le 0$ or $u^S = u^{\mathrm{R}}$ with $a(u^S) \le 0$.

The modification of Eq. (2.3.81) can be written out as (see Zachary and Colella 1992)

$$\mathbf{F}^{\mathrm{O}} = \mathbf{F}(\mathbf{U}^{\mathrm{L}}) + \sum_{k=1}^{n}\sum_{s=1}^{S(k)} (-1)^s \mathbf{F}(\mathbf{U}_k^s), \tag{2.3.83}$$

where

$$\mathbf{U}_k = \mathbf{U}^{\mathrm{L}} + \sum_{k'<k} \bar{w}_{k'}\bar{\mathbf{r}}^{k'}, \quad \mathbf{U}_k^s = \mathbf{U}_k + w_k^s\bar{\mathbf{r}}^k.$$

Here $qw_k^1 < qw_k^2 < \cdots < qw_k^s$ ($q = \operatorname{sgn}\bar{w}_k$). Besides, w_i^s satisfies

$$\begin{aligned} &\bar{\lambda}_k(w_k^s) = 0, \quad 2 \le s \le S(k) - 1; \\ &\bar{\lambda}_k(w_k^1) \le 0, \quad w_k^1 = 0; \\ &\bar{\lambda}_k(w_k^{S(k)}) \le 0, \quad w_k^{S(k)} = \bar{w}_k. \end{aligned}$$

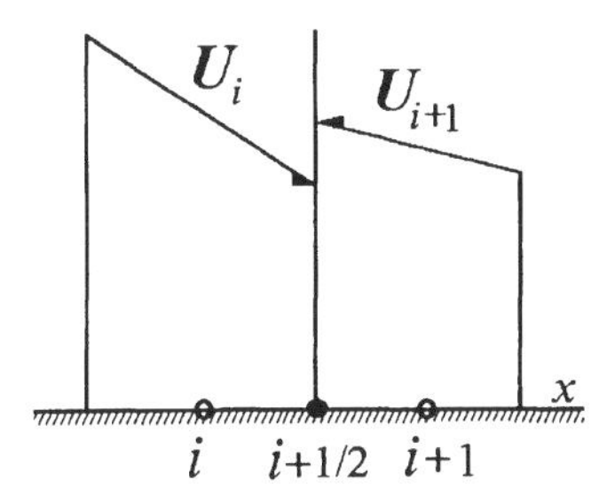

Figure 2.2 Piecewise linear distribution of **U**.

The modifications of this flux, which may turn out to be necessary for nonstrictly hyperbolic systems, are similar to those outlined earlier for another form of the EO flux.

Thus, summarizing, we can admit that, since the Osher numerical scheme approximates shocks by compression waves, it rules out expansion shocks. If the system is nonconvex, its solution may be nonunique. The EO algorithm can be interpreted in this case as averaging of the multivalued solution manifold with each elementary solution consisting of rarefaction, compression, and contact waves only (Osher 1984). Application of this scheme to the Euler, shallow water, and MHD equations will be described in Chapters 3, 4, and 5, respectively.

2.4 Generalized Riemann problem

The generalized Riemann problem is a special case of the Cauchy problem for a one-dimensional quasilinear hyperbolic system in the conservative form with piecewise-linear initial data. Consider

$$\frac{\partial \mathbf{U}}{\partial t} + \frac{\mathbf{F}(\mathbf{U})}{\partial x} = \mathbf{0}, \qquad A = \frac{\partial \mathbf{F}}{\partial \mathbf{U}} = \Omega_{\mathrm{R}} \Lambda \Omega_{\mathrm{L}}, \tag{2.4.1}$$

where $\mathbf{U} = \mathbf{U}(t, x) = [U_1, \ldots, U_n]^{\mathrm{T}}, \mathbf{F}(\mathbf{U}) = [F_1, \ldots, F_n]^{\mathrm{T}}$, $t \geq 0$, $-\infty < x < \infty$, and $\mathbf{U}(0, x) = \mathbf{U}_0(x)$. The vector $\mathbf{U}_0$ has the following form (see Fig. 2.2):

$$\mathbf{U}_0(x) = \begin{cases} \mathbf{Q}_{+0} + \mathbf{Q}_{+1} x, & x > 0, \\ \mathbf{Q}_{-0} + \mathbf{Q}_{-1} x, & x < 0, \end{cases} \tag{2.4.2}$$

where $\mathbf{Q}_{+0}$, $\mathbf{Q}_{+1}$, $\mathbf{Q}_{-0}$, and $\mathbf{Q}_{-1}$ are constant vectors. We can also consider a piecewise parabolic or more general piecewise polynomial initial value problem.

The generalized Riemann problem is stated in order to construct the Godunov method of the second order accuracy in time and space. The term *generalized Riemann problem* was first introduced by Ben-Artzi and Falcovitz (1984). They were first to study analytically the Riemann problem with piecewise linear initial data. For example, they obtained particular exact relations at gas dynamic discontinuities. It should be emphasized that the generalized Riemann problem is not self-similar. Men'shov (1990, 1991) presented a complete solution of this problem by the method of asymptotic expansions.

We shall seek a solution of Eqs. (2.4.1)–(2.4.2) in the form of a series

$$\mathbf{U}(t,x) \equiv \mathbf{U}(\xi,\theta) = \sum_{m=0}^{+\infty} \mathbf{V}_m(\xi)\theta^m = \mathbf{V}_0(\xi) + \mathbf{V}_1(\xi)\theta + O(\theta^2), \tag{2.4.3}$$

$$\xi = x/t, \quad \theta = \sqrt{x^2+t^2}; \qquad t = \theta/\sqrt{1+\xi^2}, \quad x = \theta\xi/\sqrt{1+\xi^2}.$$

Equation (2.4.3) agrees with the initial value of (2.4.2) if we have

$$\lim_{t\to 0} \mathbf{V}_0(\xi) = \begin{cases} \mathbf{Q}_{+0}, & x>0, \\ \mathbf{Q}_{-0}, & x<0; \end{cases} \quad \lim_{t\to 0} \mathbf{V}_1(\xi) = \begin{cases} \mathbf{Q}_{+1}, & x>0, \\ \mathbf{Q}_{-1}, & x<0. \end{cases}$$

Substituting $\mathbf{U}(\xi,\theta)$ from (2.4.3) into Eq. (2.4.1), we obtain

$$\frac{\partial \mathbf{U}}{\partial \xi}\frac{\partial \xi}{\partial t} + \frac{\partial \mathbf{U}}{\partial \theta}\frac{\partial \theta}{\partial t} + A\left(\frac{\partial \mathbf{U}}{\partial \xi}\frac{\partial \xi}{\partial x} + \frac{\partial \mathbf{U}}{\partial \theta}\frac{\partial \theta}{\partial x}\right)$$

$$= -\frac{\partial \mathbf{U}}{\partial \xi}\frac{x}{t^2} + \frac{\partial \mathbf{U}}{\partial \theta}\frac{t}{\sqrt{x^2+t^2}} + A\left(\frac{\partial \mathbf{U}}{\partial \xi}\frac{1}{t} + \frac{\partial \mathbf{U}}{\partial \theta}\frac{x}{\sqrt{x^2+t^2}}\right)$$

$$= -\frac{\partial \mathbf{U}}{\partial \xi}\frac{\xi\sqrt{1+\xi^2}}{\theta} + \frac{\partial \mathbf{U}}{\partial \theta}\frac{1}{\sqrt{1+\xi^2}} + A\left(\frac{\partial \mathbf{U}}{\partial \xi}\frac{\sqrt{1+\xi^2}}{\theta} + \frac{\partial \mathbf{U}}{\partial \theta}\frac{\xi}{\sqrt{1+\xi^2}}\right) = \mathbf{0};$$

whence,

$$\left(-\xi\frac{\partial \mathbf{U}}{\partial \xi} + A\frac{\partial \mathbf{U}}{\partial \xi}\right)\sqrt{1+\xi^2} + \left(\frac{\partial \mathbf{U}}{\partial \theta} + \xi A\frac{\partial \mathbf{U}}{\partial \theta}\right)\frac{\theta}{\sqrt{1+\xi^2}} = \mathbf{0}. \tag{2.4.4}$$

Thus, the zeroth approximation term $\mathbf{V}_0$ can be found from the equation

$$-\xi\frac{\partial \mathbf{V}_0}{\partial \xi} + A(\mathbf{V}_0)\frac{\partial \mathbf{V}_0}{\partial \xi} = (A_0 - \xi I)\frac{\partial \mathbf{V}_0}{\partial \xi} = \mathbf{0}, \tag{2.4.5}$$

where I is the $n \times n$ identity matrix and $A_0 = A(\mathbf{V}_0)$. The first approximation term $\mathbf{V}_1$ can be obtained from Eq. (2.4.4) as follows. Substituting series (2.4.3) into Eq. (2.4.4) and taking into account the terms of the order of θ only, we obtain the following equation for $\mathbf{V}_1$:

$$-\frac{\partial \mathbf{V}_1}{\partial \xi}\xi\sqrt{1+\xi^2} + A_0\frac{\partial \mathbf{V}_1}{\partial \xi}\sqrt{1+\xi^2} + \left(\frac{\partial A}{\partial \mathbf{U}}\right)_0 : \left(\mathbf{V}_1\frac{\partial \mathbf{V}_0}{\partial \xi}\right)\sqrt{1+\xi^2}$$

$$+ \mathbf{V}_1\frac{1}{\sqrt{1+\xi^2}} + A_0\mathbf{V}_1\frac{\xi}{\sqrt{1+\xi^2}} = \mathbf{0},$$

where the colon stands for the contraction of tensors and/or matrices over two indices. The notation $\mathbf{v}_1 = \mathbf{V}_1\sqrt{1+\xi^2}$ permits us to represent the equation in a more compact form,

$$(A_0 - \xi I)\frac{\partial \mathbf{v}_1}{\partial \xi} + \mathbf{v}_1 + \left(\frac{\partial A}{\partial \mathbf{U}}\right)_0 : \left(\mathbf{v}_1\frac{\partial \mathbf{V}_0}{\partial \xi}\right) = \mathbf{0}. \tag{2.4.6}$$

In addition, for each eigenvalue λ of the hyperbolic system we can write out the expansion

$$\lambda = \lambda_0 + \lambda_1\theta + O(\theta^2),$$

where λ_0 and λ_1 are constants. The solution $\mathbf{V}_0$ of Eq. (2.4.5) is an exact solution of the Riemann problem in terms of the self-similar variable ξ. Equation (2.4.6) is linear in $\mathbf{v}_1$, and if $\mathbf{V}_0$ is known, we can always determine $\mathbf{V}_1$ from the solution of Eq. (2.4.6). In particular, let the initial data be equal to $\mathbf{Q}_0 + \mathbf{Q}_1 x$, $-\infty < x < +\infty$, where $\mathbf{Q}_0$ and $\mathbf{Q}_1$ are constant vectors. Then we obtain $\mathbf{V}_0(\xi) = \mathbf{Q}_0$ and $\partial\mathbf{V}_0/\partial\xi = 0$. Hence from Eq. (2.4.6) it follows that $\mathbf{v}_1 = \mathbf{Q}_1\xi - A_0\mathbf{Q}_1$, or $\mathbf{V}_1 = (\mathbf{Q}_1\xi - A\mathbf{Q}_1)/\sqrt{1+\xi^2}$. Thus, with $O(\theta^2)$ error we have

$$\mathbf{U}(t,x) = \mathbf{Q}_0 + (\mathbf{Q}_1\xi - A\mathbf{Q}_1)\frac{\theta}{\sqrt{1+\xi^2}} = \mathbf{Q}_0 + \mathbf{Q}_1 x - A_0\mathbf{Q}_1 t. \tag{2.4.7}$$

Expansions similar to (2.4.3) were constructed for one-dimensional time-dependent gas dynamic equations by Men'shov (1990, 1991). In particular, the formulas of $\mathbf{V}_1(\xi)$ for the case where the term $\mathbf{V}_0(\xi)$ of the series is a rarefaction wave were also determined. They turned out to be rather complicated. Besides, Men'shov (1992) devised similar expansions for stationary two-dimensional gas dynamic equations in terms of $\xi = y/x$ and $\theta = \sqrt{x^2+y^2}$. These papers contain exact formulas for boundary fluxes, see Eq. (2.2.8), which can be used for the construction of the Godunov method of the second order of accuracy.

2.5 The Godunov method of the second order

In this section we describe the Godunov method of the second order. Let us consider a uniform discrete space mesh with step Δx and let Δt be the time increment. Then we can integrate Eq. (2.2.1) in space and time in the domain $x \in [x, x+\Delta x]$, $t \in [t, t+\Delta t]$ to obtain

$$\begin{aligned}&\int_t^{t+\Delta t}\int_x^{x+\Delta x}\left(\frac{\partial\mathbf{U}}{\partial t} + \frac{\partial\mathbf{F}(\mathbf{U})}{\partial x}\right) dt\, dx\\ &= \int_x^{x+\Delta x}\mathbf{U}(t+\Delta t, x)dx - \int_x^{x+\Delta x}\mathbf{U}(t,x)dx\\ &+ \int_t^{t+\Delta t}\mathbf{F}[\mathbf{U}(t, x+\Delta x)]dt - \int_t^{t+\Delta t}\mathbf{F}[\mathbf{U}(t,x)]dt = \mathbf{0}.\end{aligned} \tag{2.5.1}$$

Let us approximate all integrals in (2.5.1) by the simple one-point Gauss quadrature formula

$$\begin{aligned}&[\mathbf{U}(t+\Delta t, x+\tfrac{1}{2}\Delta x) - \mathbf{U}(t, x+\tfrac{1}{2}\Delta x)]\Delta x\\ &+ [\mathbf{F}(\mathbf{U}(t+\tfrac{1}{2}\Delta t, x+\Delta x)) - \mathbf{F}(\mathbf{U}(t+\tfrac{1}{2}\Delta t, x))]\Delta t = \mathbf{0},\end{aligned} \tag{2.5.2}$$

or

$$\begin{aligned}&\frac{\mathbf{U}(t+\Delta t, x+\frac{1}{2}\Delta x) - \mathbf{U}(t, x+\frac{1}{2}\Delta x)}{\Delta t}\\ &+ \frac{\mathbf{F}[\mathbf{U}(t+\frac{1}{2}\Delta t, x+\Delta x)] - \mathbf{F}[\mathbf{U}(t+\frac{1}{2}\Delta t, x)]}{\Delta x} = \mathbf{0}.\end{aligned} \tag{2.5.3}$$

Then the Godunov finite-volume explicit scheme acquires the form of (2.2.3):

$$\frac{\mathbf{U}_j^{k+1} - \mathbf{U}_j^k}{\Delta t} + \frac{\mathbf{F}_{j+1/2} - \mathbf{F}_{j-1/2}}{\Delta x} = \mathbf{0}, \tag{2.5.4}$$
$$\mathbf{F}_{j+1/2} = \mathbf{F}[\mathbf{U}(t+\tfrac{1}{2}\Delta t, x+\Delta x)] = \mathbf{F}(t+\tfrac{1}{2}\Delta t, x+\Delta x),$$
$$\mathbf{F}_{j-1/2} = \mathbf{F}[\mathbf{U}(t+\tfrac{1}{2}\Delta t, x)] = \mathbf{F}(t+\tfrac{1}{2}\Delta t, x),$$
$$\mathbf{U}_j^{k+1} = \mathbf{U}(t+\Delta t, x+\tfrac{1}{2}\Delta x), \quad \mathbf{U}_j^k = \mathbf{U}(t, x+\tfrac{1}{2}\Delta x).$$

Here the integer subscripts $i = 1, 2, \ldots$ refer to the centers of space cells and the half-integer ones to boundaries. The superscript k is the index of the layer with respect to time. The boundary values $\mathbf{F}_{j\pm1/2}$ are calculated from the solution of the corresponding Riemann problem.

Suppose that the Riemann problem of (2.2.1)–(2.2.2) with piecewise constant initial data, see Fig. 2.1, has both exact and approximate self-similar solutions in the variable $\xi = x/t$. Then the functions $\mathbf{F}_{j\pm1/2}$ are constant for fixed ξ and $t \in [t, t+\Delta t]$. In the generalized Riemann problem of (2.4.1)–(2.4.2) we must additionally take into account the dependence of $\mathbf{F}_{j\pm1/2}$ from Eq. (2.5.4) on ξ and θ, see Eq. (2.4.3), or on ξ and time. Then, we shall have a scheme of the second order of accuracy with respect to space and time. In particular, analytical formulas for determining of $\mathbf{F}_{j\pm1/2}$ in the generalized Riemann problem for gas dynamics were derived by Men'shov (1990, 1991, 1992). However, they are rather complicated. Therefore it may be easier to construct a second-order scheme by using higher-order time-approximation procedures, e.g., Runge–Kutta predictor–corrector methods (Hairer, Norsett, and Wanner 1987).

The specific features of a predictor–corrector algorithm can be illustrated by the second-order scheme of Borrel and Montagne (1985) and Rodionov (1987a). The scheme has a clear interpretation at all stages of calculations. Applications of this scheme proved its reliability and efficiency and were used to simulate the unsteady two-dimensional gas dynamics with transonic and supersonic flows (Borrel and Montagne 1985) and stationary gas dynamics flows with chemical reactions (Rodionov 1987a, 1987b).

Predictor: first step

Let us assume that the following piecewise linear distributions of the mesh function inside the cell are given:

$$\mathbf{U}(t^k, x) = \mathbf{U}_j^k + \mathbf{Q}_j^k\,(x - x_j), \qquad x \in [x_j - \tfrac{1}{2}\Delta x,\ x_j + \tfrac{1}{2}\Delta x], \tag{2.5.5}$$

where x_j is the space coordinate of the jth cell center and $\mathbf{Q}_j^k$ is the vector of slopes for $\mathbf{U}$ inside this cell.

Let us determine the time variation of $\mathbf{U}$ at the cell center. It follows immediately from Eq. (2.2.1), see also (2.4.7), that

$$\mathbf{U}_t = -A\mathbf{U}_x = -A\mathbf{Q}_j^k. \tag{2.5.6}$$

Then, we can write out the equation for $\mathbf{U}$ as

$$\frac{\hat{\mathbf{U}}_j^{k+1} - \mathbf{U}_j^k}{\Delta t} + \frac{\mathbf{F}(\mathbf{U}_j^k + \tfrac{1}{2}\Delta x\,\mathbf{Q}_j^k) - \mathbf{F}(\mathbf{U}_j^k - \tfrac{1}{2}\Delta x\,\mathbf{Q}_j^k)}{\Delta x} = \mathbf{0}. \tag{2.5.7}$$

Predictor: second step

We calculate the values of $\mathbf{U}$ at $t+\frac{1}{2}\Delta t$, or $\mathbf{U}_j^{k+1/2}$, in the following manner:

$$\mathbf{U}_j^{k+1/2} = \tfrac{1}{2}(\hat{\mathbf{U}}_j^{k+1} + \mathbf{U}_j^k). \tag{2.5.8}$$

Corrector step

At the final, corrector stage we use the finite-volume scheme

$$\frac{\mathbf{U}_j^{k+1} - \mathbf{U}_j^k}{\Delta t} + \frac{\mathbf{F}_{j+1/2} - \mathbf{F}_{j-1/2}}{\Delta x} = \mathbf{0}, \tag{2.5.9}$$

where all values of $\mathbf{F}_{j+1/2}$ are determined from the solution of the Riemann problem with the following piecewise constant initial data:

$$\mathbf{U}_0(x) = \begin{cases} \mathbf{U}_{j+1}^{k+1/2} - \frac{1}{2}\Delta x\,\mathbf{Q}_{j+1}^k, & x_{j+1/2} > 0, \\ \mathbf{U}_j^{k+1/2} + \frac{1}{2}\Delta x\,\mathbf{Q}_j^k, & x_{j+1/2} < 0. \end{cases} \tag{2.5.10}$$

The values of $\mathbf{F}_{j-1/2}$ are determined similarly. For the solution of these Riemann problems we can use either the exact or an approximate Riemann solver (Borrel and Montagne 1985; Rodionov 1987a). This completes the construction of scheme (2.5.5)–(2.5.10). Note that another scheme for the predictor step can be used, for example, similar to (2.5.9)–(2.5.10) with

$$\mathbf{U}_0(x) = \begin{cases} \mathbf{U}_{j+1}^k - \frac{1}{2}\Delta x\,\mathbf{Q}_{j+1}^k, & x_{j+1/2} > 0, \\ \mathbf{U}_j^k + \frac{1}{2}\Delta x\,\mathbf{Q}_j^k, & x_{j+1/2} < 0; \end{cases} \quad \text{or} \quad \mathbf{U}_0(x) = \begin{cases} \mathbf{U}_{j+1}^k, & x_{j+1/2} > 0, \\ \mathbf{U}_j^k, & x_{j+1/2} < 0. \end{cases}$$

For the construction of a higher-order scheme, one can use more accurate quadrature formulas and multilevel algorithms of recalculations of the predictor–corrector type, see Harten and Osher (1987) and Shu and Osher (1988).

One of the simplest ways to determine slopes Q_m in the mth point for mesh function U_j ($j = 1, 2, \ldots$) is to use the minmod function

$$Q_m = \operatorname{minmod}\left(\frac{U_{m+1} - U_m}{\Delta x}, \frac{U_m - U_{m-1}}{\Delta x}\right),$$
$$\operatorname{minmod}(a, b) = \tfrac{1}{2}(\operatorname{sgn} a + \operatorname{sgn} b)\min(|a|, |b|). \tag{2.5.11}$$

The minmod function selects the slope with the minimum absolute value if the signs of both arguments are similar and returns zero if the signs are opposite; for more detailed consideration see Section 2.7.

Let us prove that if we use the CIR Riemann problem solver such a predictor–corrector scheme provides the second order of accuracy in time and space.

For the initial piecewise linear distributions (2.5.10) the approximate solution is not constant in time and changes in accordance with Eq. (2.5.6), see also Eq. (2.4.7). Thus, we can rewrite Eqs. (2.3.27)–(2.3.28) in the form

$$\mathbf{U}_{m+1/2}(\xi,t) = \tfrac{1}{2}(\mathbf{U}_m + \mathbf{U}_{m+1}) + \tfrac{1}{2}S(\xi,t)_{m+1/2}\,(\mathbf{U}_m - \mathbf{U}_{m+1}), \tag{2.5.12}$$
$$\mathbf{U}_m = \mathbf{U}_m^k + \tfrac{1}{2}\Delta x\,\mathbf{Q}_m^k + x\,\mathbf{Q}_m^k - tA_m^k\mathbf{Q}_m^k = \mathbf{U}_m^k + \tfrac{1}{2}\Delta x\,\mathbf{Q}_m^k + t\xi\mathbf{Q}_m^k - tA_m^k\mathbf{Q}_m^k,$$
$$\mathbf{U}_{m+1} = \mathbf{U}_{m+1}^k - \tfrac{1}{2}\Delta x\,\mathbf{Q}_{m+1}^k + x\,\mathbf{Q}_{m+1}^k - tA_{m+1}^k\mathbf{Q}_{m+1}^k$$
$$= \mathbf{U}_{m+1}^k - \tfrac{1}{2}\Delta x\,\mathbf{Q}_{m+1}^k + t\xi\mathbf{Q}_{m+1}^k - tA_{m+1}^k\mathbf{Q}_{m+1}^k,$$
$$S(\xi,t) = \Omega_{\mathrm{R}}(t)\,[\mathrm{sgn}(\lambda_p(t)-\xi)\,\delta_{pl}]\,\Omega_{\mathrm{L}}(t),$$
$$\mathbf{F}_{m+1/2}(\xi,t) = \tfrac{1}{2}(\mathbf{F}_m + \mathbf{F}_{m+1}) + \tfrac{1}{2}A(\xi,t)_{m+1/2}\,(\mathbf{U}_m - \mathbf{U}_{m+1}), \tag{2.5.13}$$
$$\mathbf{F}_m = \mathbf{F}(\mathbf{U}_m), \quad \mathbf{F}_{m+1} = \mathbf{F}(\mathbf{U}_{m+1}),$$
$$A(\xi,t) = \Omega_{\mathrm{R}}(t)[\lambda_p(t)\,\mathrm{sgn}(\lambda_p(t)-\xi)\,\delta_{pl}]\Omega_{\mathrm{L}}(t).$$

Here, $x=0$ corresponds to boundary $m+1/2$, and $\mathbf{U}_m^k + \frac{1}{2}\Delta x\,\mathbf{Q}_m^k$ and $\mathbf{U}_{m+1}^k - \frac{1}{2}\Delta x\,\mathbf{Q}_{m+1}^k$ are the left and right boundary values for $t=0$, respectively. We took into account the variation of all matrices, eigenvalues, and the values $\mathbf{U}_m$ and $\mathbf{U}_{m+1}$ in time and space. It is necessary to assume that $\det\Omega_{\mathrm{R}}(t)\neq 0$ and $\det\Omega_{\mathrm{L}}(t)\neq 0$ for $0\le t\le\Delta t$. Equations (2.5.12)–(2.5.13) determine an approximate solution of the Riemann problem for Eq. (2.2.1) as a combination of the discontinuities that move with velocities $\lambda_p(t)$ and regions with quantities linearly dependent on t and x. In addition, the values $\mathbf{U}_{m+1/2}$ and $\mathbf{F}_{m+1/2}$ behind the p-discontinuity and those ahead of it are related by the equation $\{\mathbf{F}\} = \lambda_p(t)\{\mathbf{U}\}$, see Eqs. (2.3.30)–(2.3.31). It is obvious that this solution generalizes the one obtained by the CIR solver (2.3.27)–(2.3.28). The CIR solver provides the solution, which consists of regions with only constants quantities that are separated by p-discontinuities $\xi=\lambda_p=\mathrm{const}$, $p=1,\ldots,n$.

To construct the second-order Godunov scheme in the form (2.5.4) we calculate $\mathbf{F}$ for $t+\frac{1}{2}\Delta t$ and $\xi=0$. Substituting $t=\frac{1}{2}\Delta t$ and $\xi=0$ into Eq. (2.5.13), we obtain

$$\mathbf{F}_{m+1/2} = \tfrac{1}{2}(\mathbf{F}_m + \mathbf{F}_{m+1}) + \tfrac{1}{2}\,|A|_{m+1/2}^{k+1/2}\,(\mathbf{U}_m - \mathbf{U}_{m+1}), \tag{2.5.14}$$
$$\mathbf{U}_m = \mathbf{U}_m^k + \tfrac{1}{2}\Delta x\,\mathbf{Q}_m^k - \tfrac{1}{2}\Delta t\,A_m^k\mathbf{Q}_m^k,$$
$$\mathbf{U}_{m+1} = \mathbf{U}_{m+1}^k - \tfrac{1}{2}\Delta x\,\mathbf{Q}_{m+1}^k - \tfrac{1}{2}\Delta t\,A_{m+1}^k\mathbf{Q}_{m+1}^k,$$
$$\mathbf{F}_m = \mathbf{F}(\mathbf{U}_m), \quad \mathbf{F}_{m+1} = \mathbf{F}(\mathbf{U}_{m+1}),$$
$$|A| = \Omega_{\mathrm{R}}|\Lambda|\Omega_{\mathrm{L}} = \Omega_{\mathrm{R}}[|\lambda_p|\,\delta_{pl}]\Omega_{\mathrm{L}}.$$

Here we assumed that $x=0$ corresponds to boundary $m+1/2$ and $\mathbf{U}_m^k + \frac{1}{2}\Delta x\,\mathbf{Q}_m^k$ and $\mathbf{U}_{m+1}^k - \frac{1}{2}\Delta x\,\mathbf{Q}_{m+1}^k$ are the left and right boundary values for $t=0$, respectively.

The equation obtained by linearization of Eq. (2.5.14) provides the same results as the corrector step (2.5.9). Thus, the predictor–corrector scheme (2.5.7)–(2.5.10) is of the second order.

Note that the above predictor–corrector approach can be applied also to equations in both nonconservative (2.3.15) and mixed form (2.3.41). In particular, for the CIR scheme (2.3.37) the predictor–corrector steps are the following.

Predictor: first and second steps

$$\frac{\hat{\mathbf{U}}_j^{k+1} - \mathbf{U}_j^k}{\Delta t} + A_j^k\mathbf{Q}_j^k = \mathbf{0},$$

$$\mathbf{U}_j^{k+1/2} = \tfrac{1}{2}(\hat{\mathbf{U}}_j^{k+1} + \mathbf{U}_j^k) = \mathbf{U}_j^k - \tfrac{1}{2}\Delta t\, A_j^k \mathbf{Q}_j^k.$$

Corrector

At the corrector step we use the finite-volume scheme (2.3.37) modified with the piecewise constant initial data (2.5.10):

$$\frac{\mathbf{U}_j^{k+1} - \mathbf{U}_j^k}{\Delta t} + (\Omega_{\mathrm{R}}\Lambda^-\Omega_{\mathrm{L}})_{j+1/2}^{k+1/2} \frac{\mathbf{U}_{j+1}^{k+1/2} - \frac{1}{2}\Delta x\, \mathbf{Q}_{j+1}^k - \mathbf{U}_j^{k+1/2} - \frac{1}{2}\Delta x\, \mathbf{Q}_j^k}{\Delta x}$$

$$+ (\Omega_{\mathrm{R}}\Lambda^+\Omega_{\mathrm{L}})_{j-1/2}^{k+1/2} \frac{\mathbf{U}_j^{k+1/2} - \frac{1}{2}\Delta x\, \mathbf{Q}_j^k - \mathbf{U}_{j-1}^{k+1/2} - \frac{1}{2}\Delta x\, \mathbf{Q}_{j-1}^k}{\Delta x} = \mathbf{0}.$$

By analogy one can increase the order of accuracy for other versions of CIR-type schemes (2.3.38), (2.3.42), etc.

Let us derive the stability condition for scheme (2.5.5)–(2.5.10) in the case of one advection equation (2.3.1) with $\lambda = 1$. Here the finite-difference solution has the form

$$u_{m+1}^{k+1} = (1-C)u_{m+1}^k + Cu_m^k - \tfrac{1}{2}C(1-C)\alpha_{m+1}\Delta x + \tfrac{1}{2}C(1-C)\alpha_m \Delta x, \quad C = \frac{\Delta t}{\Delta x}.$$

By setting

$$\alpha_{m+1} = \frac{u_{m+1}^k - u_m^k}{\Delta x}\psi_{m+1}, \quad \alpha_m = \frac{u_{m+1}^k - u_m^k}{\Delta x}\psi_m, \tag{2.5.15}$$

where $\psi_{m+1} \geq 0$ and $\psi_m \geq 0$ are some coefficients limiting the values of the function slopes, we obtain

$$\begin{aligned} u_{m+1}^{k+1} &= r_{m+1}u_{m+1}^k + r_m u_m^k, \\ r_{m+1} &= 1 - C - \tfrac{1}{2}C(1-C)(\psi_{m+1} - \psi_m), \\ r_m &= C + \tfrac{1}{2}C(1-C)(\psi_{m+1} - \psi_m). \end{aligned} \tag{2.5.16}$$

The above scheme can be written out in matrix form as

$$\mathbf{u}^{k+1} = R\mathbf{u}^k, \tag{2.5.17}$$

where $R = [R_{ij}]$ is a two-diagonal matrix and $\mathbf{u} = [\ldots, u_{m+1}, u_m, \ldots]^{\mathrm{T}}$. A sufficient stability condition for scheme (2.5.17) is $\|R\|^k \leq 1$ and, hence, $\max|\lambda_R| \leq 1$, where λ_R is an eigenvalue of the matrix R. Using the estimate $|\lambda_R| \leq \sum_i |R_{ij}|$ for an eigenvalue (Korn and Korn 1968), we find that

$$|\lambda_R| \leq |r_{m+1}| + |r_m| \leq 1.$$

Thus,

$$0 \leq C + \tfrac{1}{2}C(1-C)(\psi_{m+1} - \psi_m) \leq 1, \tag{2.5.18}$$

or

$$-\frac{2C}{1-C} \leq C(\psi_{m+1} - \psi_m) \leq 2. \tag{2.5.19}$$

Assuming that $0 \le \psi_{m+1} \le \psi$ and $0 \le \psi_m \le \psi$, we obtain

$$-C\psi \le C(\psi_{m+1} - \psi_m) \le C\psi. \tag{2.5.20}$$

Inequalities (2.5.19) and (2.5.20) hold simultaneously if

$$C \le 1, \quad 0 \le \psi = \max\left(\frac{\alpha_{m+1}\Delta x}{\Delta_m}, \frac{\alpha_m \Delta x}{\Delta_m}\right) \le 2, \quad \Delta_m = u^k_{m+1} - u^k_m. \tag{2.5.21}$$

Thus, the stability condition for scheme (2.5.5)–(2.5.10) consists of the condition $C \le 1$ supplemented by the constraint $0 \le \psi \le 2$. The constraint imposed on ψ implies that the slopes α must be somehow limited. One of the simplest ways to do so is to determine α_{m+1} and α_m using the minmod function (2.5.11), see also Section 2.7. In this case, $\psi = 1$ and the stability condition (2.5.21) holds.

2.6 Multidimensional schemes and their stability conditions

So far, we have considered only one-dimensional numerical schemes. All these schemes can be written in operator form as

$$\frac{\mathbf{U}^{k+1} - \mathbf{U}^k}{\Delta t} + \mathbf{B}\mathbf{U}^k = \mathbf{0}, \quad \mathbf{U}^{k+1} = (I - \Delta t\,\mathbf{B})\mathbf{U}^k, \tag{2.6.1}$$

where I is the $n \times n$ identity matrix and $I - \Delta t\,\mathbf{B}$ can be the first- or the second-order accurate operator of transition to the upper time level. In particular, for the scheme (2.5.5)–(2.5.10) the operator $\mathbf{B}$ has the form

$$\mathbf{B} = \mathcal{A}\,(I - \tfrac{1}{2}\Delta t\,\tilde{\mathcal{A}}), \tag{2.6.2}$$

where $I - \frac{1}{2}\Delta t\,\tilde{\mathcal{A}}$ corresponds to the predictor steps (2.5.6)–(2.5.8) and the operator $\mathcal{A}\,(I - \frac{1}{2}\Delta t\,\tilde{\mathcal{A}})$ corresponds to the final corrector step (2.5.9). Here, the operators $\mathcal{A}$ and $\tilde{\mathcal{A}}$ correspond to some finite-difference approximations of $A\mathbf{U}_x$ or $\mathbf{F}_x$. In this notation the schemes of the first order of accuracy correspond to $\mathbf{B} = \mathcal{A}$. In our discussion we will consider the particular case $\tilde{\mathcal{A}} = \mathcal{A}$. The cases $\tilde{\mathcal{A}} \ne \mathcal{A}$ must be investigated separately.

Let us construct a numerical scheme for the two-dimensional system (2.2.5), or

$$\frac{\partial \mathbf{U}}{\partial t} + A_1 \frac{\partial \mathbf{U}}{\partial x} + A_2 \frac{\partial \mathbf{U}}{\partial y} = \mathbf{0}.$$

One of the methods of construction involves using the one-dimensional schemes (2.6.1), which are written for each space direction (2.2.6):

$$\frac{\mathbf{U}^{k+1} - \mathbf{U}^k}{\Delta t} + \mathbf{B}_1\mathbf{U}^k + \mathbf{B}_2\mathbf{U}^k = \mathbf{0}, \quad \mathbf{U}^{k+1} = (I - \Delta t\,\mathbf{B}_1 - \Delta t\,\mathbf{B}_2)\mathbf{U}^k. \tag{2.6.3}$$

Here $\mathbf{B}_1$ corresponds to the one-dimensional scheme in the x-direction and $\mathbf{B}_2$ to that in the y-direction.

The difference scheme (2.6.3) is always of the first order with respect to time. A scheme of the second order of accuracy with respect to time and space can be written out as

$$\frac{\mathbf{U}^{k+1}-\mathbf{U}^k}{\Delta t}+\mathbf{B}_1(I-\tfrac{1}{2}\Delta t\,\mathbf{B}_2)\mathbf{U}^k+\mathbf{B}_2(I-\tfrac{1}{2}\Delta t\,\mathbf{B}_1)\mathbf{U}^k=\mathbf{0}. \tag{2.6.4}$$

Then the operator of transition to the upper level in (2.6.4) coincides with that investigated by Strang (1963),

$$\tfrac{1}{2}(I-\Delta t\,\mathbf{B}_1)(I-\Delta t\,\mathbf{B}_2)+\tfrac{1}{2}(I-\Delta t\,\mathbf{B}_2)(I-\Delta t\,\mathbf{B}_1). \tag{2.6.5}$$

Substituting the expressions for $\mathbf{B}_1$ and $\mathbf{B}_2$ of (2.6.2) into (2.6.4) we rewrite the scheme based on the Strang splitting with $O(\Delta t^2)$ accuracy in the form

$$\frac{\mathbf{U}^{k+1}-\mathbf{U}^k}{\Delta t}+(\mathcal{A}_1+\mathcal{A}_2)[I-\tfrac{1}{2}\Delta t(\mathcal{A}_1+\mathcal{A}_2)]\mathbf{U}^k=\mathbf{0}. \tag{2.6.6}$$

Relation (2.6.6) represents the simpler scheme for the two-dimensional case with second-order accuracy in time. In particular, the operator $I-\frac{1}{2}\Delta t(\mathcal{A}_1+\mathcal{A}_2)$ is equivalent to the predictor step (2.5.5)–(2.5.8) along x- and y-directions simultaneously. Here, this predictor step uses a piecewise linear two-dimensional space distribution. The final, corrector step also uses the same space distribution of the mesh function.

Strang (1963) generalized the second-order scheme (2.6.4) for an arbitrary n-dimensional hyperbolic system

$$\frac{\partial \mathbf{U}}{\partial t}+\frac{\partial \mathbf{F}_1(\mathbf{U})}{\partial x_1}+\cdots+\frac{\partial \mathbf{F}_n(\mathbf{U})}{\partial x_n}=\mathbf{0}$$

or

$$\frac{\partial \mathbf{U}}{\partial t}+A_1\frac{\partial \mathbf{U}}{\partial x_1}+\cdots+A_n\frac{\partial \mathbf{U}}{\partial x_n}=\mathbf{0}.$$

In this case the transition operator has the following form:

$$\frac{1}{n!}\sum_{i_1,\ldots,i_n}(I-\Delta t\,\mathbf{B}_{i_1})\cdots(I-\Delta t\,\mathbf{B}_{i_n}), \tag{2.6.7}$$

where the summation is performed over all $n!$ different transpositions of the numbers $1,\ldots,n$. Operator (2.6.7) can be simplified by neglecting the $O(\Delta t^2)$ terms, thus resulting in the scheme

$$\frac{\mathbf{U}^{k+1}-\mathbf{U}^k}{\Delta t}+(\mathcal{A}_1+\cdots+\mathcal{A}_n)\,[I-\tfrac{1}{2}\Delta t(\mathcal{A}_1+\cdots+\mathcal{A}_n)]\mathbf{U}^k=\mathbf{0}. \tag{2.6.8}$$

Stability conditions

The stability condition for scheme (2.6.7), as well as for (2.6.5), reduces to the following stability condition for all one-dimensional operators $I-\Delta t\,\mathbf{B}_i$:

$$\|I-\Delta t\,\mathbf{B}_i\|=\left\|I-\Delta t\,\mathcal{A}_i(I-\tfrac{1}{2}\Delta t\,\mathcal{A}_i)\right\|\le 1. \tag{2.6.9}$$

Assume that the initial operator of the first order of accuracy, $I - \Delta t\,\mathcal{A}_i$, is stable and a stability condition for it has the form

$$\|\mathcal{A}_i\|\Delta t \le \varepsilon_i, \tag{2.6.10}$$

where $\varepsilon_i > 0$ is a constant. Suppose $\varepsilon_i = 1$ and

$$\|\mathcal{A}_i\| = \frac{\|A_i\|}{\Delta x_i},$$

where Δx_i is the mesh step along the x_i-direction.

From the identity

$$I - \Delta t\,\mathcal{A}_i\left(I - \tfrac{1}{2}\Delta t\,\mathcal{A}_i\right) = \tfrac{1}{2}[I + (I - \Delta t\,\mathcal{A}_i)^2]$$

one can find that a sufficient condition for inequality (2.6.9) to be satisfied is $\|I - \Delta t\,\mathcal{A}_i\| \le 1$, or (2.6.10). Hence, scheme (2.6.7) does not change one-dimensional stability conditions.

Consider scheme (2.6.8). It can be rewritten in the form

$$\begin{aligned}\mathbf{U}^{k+1} &= \left[I - (\mathcal{A}_1 + \cdots + \mathcal{A}_n)\Delta t + \tfrac{1}{2}(\mathcal{A}_1 + \cdots + \mathcal{A}_n)^2(\Delta t)^2\right]\mathbf{U}^k\\ &= \tfrac{1}{2}[I + (I - \mathcal{A}_1\Delta t - \cdots - \mathcal{A}_n\Delta t)^2]\mathbf{U}^k.\end{aligned} \tag{2.6.11}$$

A stability condition for scheme (2.6.11) is

$$\tfrac{1}{2}\|I + (I - \mathcal{A}_1\Delta t - \cdots - \mathcal{A}_n\Delta t)^2\| \le 1.$$

It follows that a sufficient stability condition is

$$\|I - \mathcal{A}_1\Delta t - \cdots - \mathcal{A}_n\Delta t\| \le 1.$$

Then we have

$$\begin{aligned}&\|I - \mathcal{A}_1\Delta t - \cdots - \mathcal{A}_n\Delta t\| = \|(a_1 I - \mathcal{A}_1\Delta t) + \cdots + (a_n I - \mathcal{A}_n\Delta t)\|\\ &\le a_1\left\|I - \mathcal{A}_1\frac{\Delta t}{a_1}\right\| + \cdots + a_n\left\|I - \mathcal{A}_n\frac{\Delta t}{a_n}\right\| \le 1\\ &\Longrightarrow \quad \left\|I - \mathcal{A}_1\frac{\Delta t}{a_1}\right\| \le 1, \ldots, \left\|I - \mathcal{A}_n\frac{\Delta t}{a_n}\right\| \le 1,\\ &a_1 + \cdots + a_n = 1, \qquad a_i > 0, \quad i = 1, \ldots, n.\end{aligned} \tag{2.6.12}$$

Inequality (2.6.12) is equivalent to the inequalities

$$\|\mathcal{A}_i\|\frac{\Delta t}{a_i} \le \varepsilon_i, \quad a_1 + \cdots + a_n = 1, \quad a_i > 0,$$

or

$$\Delta t \le \max_{a_1+\cdots+a_n=1} \min_i \left(\frac{\varepsilon_i a_i}{\|\mathcal{A}_i\|}\right).$$

The maxmin is attained if all $\varepsilon_i a_i/||\mathcal{A}_i||$ are the same. Hence,

$$\Delta t \leq \left(\sum_{i=1}^{n} \frac{||\mathcal{A}_i||}{\varepsilon_i}\right)^{-1}, \quad \text{or} \quad \left(\sum_{i=1}^{n} \frac{||\mathcal{A}_i||}{\varepsilon_i}\right) \Delta t \leq 1. \tag{2.6.13}$$

For $\varepsilon_i = 1$, condition (2.6.13) is equivalent to

$$\left(\sum_{i=1}^{n} ||\mathcal{A}_i||\right) \Delta t = \left(\sum_{i=1}^{n} \frac{||A_i||}{\Delta x_i}\right) \Delta t \leq 1. \tag{2.6.14}$$

Condition (2.6.14) is a sufficient stability condition. For specific methods other conditions may be obtained. In particular, for the two-dimensional Lax–Wendroff scheme, Turkel (1977) proved the unusual sufficient stability condition

$$\Delta t \left(|\mathcal{A}_1\mathbf{q}|^{2/3} + |\mathcal{A}_2\mathbf{q}|^{2/3}\right) \leq 1, \quad |\mathbf{q}| = 1. \tag{2.6.15}$$

Note that condition (2.6.14) is weaker than (2.6.15), but not better. In particular, there are matrices A_1 and A_2 such that $||A_1|| = ||A_2||$ for which condition (2.6.15) gives $C \leq 1/\sqrt{8}$, condition (2.6.13) gives $C \leq 1/2$; however, the upper stability bound is determined by the condition $C \leq 1/\sqrt{2}$ (Clifton 1967; Turkel 1977).

From (2.6.14) one can obtain stronger stability conditions, for example,

$$\max_i ||\mathcal{A}_i||\Delta t \leq \frac{1}{n},$$

and some others.

The general stability condition for scheme (2.6.3) is

$$||I - \Delta t \mathbf{B}_1 - \Delta t \mathbf{B}_2|| \leq 1.$$

We have

$$\begin{aligned} ||I - \Delta t \mathbf{B}_1 - \Delta t \mathbf{B}_2|| &= ||(b_1 I - \mathbf{B}_1 \Delta t) + (b_2 I - \mathbf{B}_2 \Delta t)|| \\ &\leq b_1 \left\|I - \mathbf{B}_1 \frac{\Delta t}{b_1}\right\| + b_2 \left\|I - \mathbf{B}_2 \frac{\Delta t}{b_n}\right\| \leq 1, \\ \Longrightarrow \quad &\left\|I - \mathbf{B}_1 \frac{\Delta t}{b_1}\right\| \leq 1, \quad \left\|I - \mathbf{B}_2 \frac{\Delta t}{b_2}\right\| \leq 1, \\ b_1 + b_2 = 1, \quad &b_1 > 0, \quad b_2 > 0. \end{aligned} \tag{2.6.16}$$

Let us derive stability conditions for scheme (2.6.3) in the case where the operators $\mathbf{B}_1$ and $\mathbf{B}_2$ are one-dimensional CIR operators of the first order of accuracy. Inequality (2.6.16) is equivalent to the following stability conditions:

$$\frac{|C_p|}{b_1} \leq 1, \quad \frac{|C_q|}{b_2} \leq 1, \quad b_1 + b_2 = 1,$$

where C_p and C_q are CFL numbers along the x- and y-direction, respectively. Hence, we obtain condition (2.2.11):

$$\max(|C_p| + |C_q|) \leq \max|C_p| + \max|C_q| \leq 1.$$

Let us derive stability conditions for scheme (2.6.3) in the case where the operators $\mathbf{B}_1$ and $\mathbf{B}_2$ are hybrid one-dimensional CIR operators; see, for example, scheme (2.3.12) with hybrid modification (2.3.58)–(2.3.60). Inequality (2.6.16) is equivalent to the following two one-dimensional stability conditions (2.3.62):

$$\frac{C_p^2}{b_1^2} \leq g_p \frac{C_p}{b_1} \leq 1, \quad \frac{C_q^2}{b_2^2} \leq g_q \frac{C_q}{b_2} \leq 1, \quad b_1 + b_2 = 1,$$

where C_p and g_p are CFL numbers and hybridity coefficients along the x-direction, and C_q and g_q are those along the y-direction. Thus we arrive at the stability conditions (Semenov 1984, 1987)

$$C_p g_p + C_q g_q \leq 1, \quad \frac{C_p}{g_p} + \frac{C_q}{g_q} \leq 1, \qquad C_p g_p \geq 0, \quad C_q g_q \geq 0. \qquad (2.6.17)$$

In particular, these inequalities can be satisfied if

$$g_p = \operatorname{sgn}\lambda_p, \quad g_q = \operatorname{sgn}\lambda_q$$

in the domains of nonsmoothness and

$$g_p = 2C_p, \quad g_q = 2C_q$$

or

$$g_p = (|C_p| + \max|C_q|)\operatorname{sgn}\lambda_p, \quad g_q = (\max|C_p| + |C_q|)\operatorname{sgn}\lambda_q$$

in the domains of smoothness of the solution. This leads to the sufficient stability condition $\max(|C_p|, |C_q|) \leq \frac{1}{2}$ or $\max|C_p| + \max|C_q| \leq 1$, respectively. By analogy one can determine a stability condition in the n-dimensional case.

2.7 Reconstruction procedures and slope limiters

In the construction of the Godunov-type methods of higher order in space, piecewise-linear (Section 2.5) or polynomial distributions of functions should be used (Harten et al. 1986; Harten and Osher 1987). In this section we describe methods for the determination of piecewise-linear distributions in discrete cells.

2.7.1 Preliminary remarks. Consider a uniform discrete space mesh with step Δx. Let a discrete function u_m, $m = 1, \ldots, N$, or $\{u_m\}$, be defined on this mesh. The integer subscript m refers to the centers of space cells and the half-integer subscript to boundaries. In order to determine the values of u on the cell sides, we must determine a reconstruction procedure. Suppose for simplicity that we deal with piecewise-linear distributions inside the cells:

$$u(x) = u_m + \alpha_m x, \quad x \in [-\tfrac{1}{2}\Delta x, \tfrac{1}{2}\Delta x]. \tag{2.7.1}$$

We use here the local coordinates, with $x = 0$ corresponding to the center of the mth cell. For simplicity we assume that $\Delta x = 1$. Note that the simplest determination of slopes, for example, by formulas $\alpha_m = \frac{1}{2}(u_{m+1} - u_{m-1})$ and/or $\alpha_m = u_{m+1} - u_m$, can lead to unstable calculations and spurious oscillations in numerical results. The slopes must satisfy the stability condition and the natural condition requiring that the numerical solution must have properties similar to those of the initial function $\{u_m\}$. That condition imposes some constraints on the slope values, for example, in the form $\alpha_m = (u_{m+1} - u_m)\psi_m$, where $0 \leq \psi_m \leq 2$, see (2.5.15), (2.5.21) and so on. The slopes α_m are constrained by the so-called limiters ψ_m, which are defined by a particular slope-limiting algorithm. In the simplest case a limiter is a function that determines and simultaneously limits the slopes α on the basis of the analysis of given values $\{u_m\}$ or their differences $\{u_{m+1} - u_m\}$.

To study the relations between a numerical solution and its reconstruction, we consider the single advection equation

$$\frac{\partial u}{\partial t} + \frac{\partial u}{\partial x} = 0, \qquad u(0, x) = u_0(x), \quad a \leq x \leq b. \tag{2.7.2}$$

The exact solution $u(t, x)$ of this equation is $u = u_0(x - t)$, see Eqs. (2.2.20). This solution preserves the monotonicity of the initial function u_0. Consider further the numerical solution obtained by using the schemes (2.3.4)–(2.3.5) of the first order of accuracy. The numerical solution can be represented as

$$u_m^{k+1} = (1 - C)u_m^k + Cu_{m-1}^k, \tag{2.7.3}$$

where $C = \Delta t/\Delta x$ is the CFL number and $C \leq 1$. Then, for the finite difference $\Delta_m = u_{m+1} - u_m$, from Eq. (2.7.3) we find that

$$\Delta_m^{k+1} = (1 - C)\Delta_m^k + C\Delta_{m-1}^k. \tag{2.7.4}$$

One can easily verify that the solution (2.7.4) preserves the monotonicity of the initial data. Indeed, if $\Delta_j^k \geq 0$ in some domain, then $\Delta_j^{k+1} \geq 0$, an so on.

This result holds for the general case of linear finite-difference schemes written in the form

$$u_m^{k+1} = \sum_{j \in J(m)} \gamma_j u_j^k, \quad \gamma_i \geq 0, \tag{2.7.5}$$

where the summation is performed over the stencil $J(m)$ involving the point of number m. It can be similarly proved that this scheme is also monotone.

Consider the finite-difference scheme of the second order of accuracy. Godunov (1959) proved that there are no monotone linear schemes (2.7.5) of second or higher order of accuracy. Therefore, we must look for other classes of schemes that permit one to carry out calculations with a higher order of accuracy without spurious oscillations.

The first approach involves the consideration of nonlinear schemes instead of the linear schemes (2.7.5). In particular, the hybrid difference schemes belong to the class of nonlinear schemes, see Section 2.3.1, page 52.

Another approach is the construction of a scheme of the first order of accuracy with minimal numerical viscosity (Kholodov 1978, 1980; Magomedov and Kholodov 1988; Belotserkovskii and Kholodov 1999). The methods of linear programming were used for this purpose. The schemes can be treated as points in a Euclidean space, with the coordinates of a point specified by the coefficients of the corresponding scheme. For example, from this point of view, scheme (2.7.5) is a point with coordinates γ_j in the Euclidean space whose dimension is equal to the number of the points in the stencil. The first-order schemes with minimal numerical viscosity are the schemes nearest to a scheme of a higher order of accuracy in the sense of a distance in the Euclidean space of coefficients. Analogously, among higher-order schemes a scheme with the best monotonicity can be selected, see the above citations. Such a scheme is closest to the monotone first-order scheme in the Euclidean space of coefficients.

In particular, the consideration of schemes as points in a Euclidean space allows one to perform a natural classification of several tens of known explicit and implicit schemes of the first, second, and third orders of accuracy with respect to their viscous properties (Magomedov and Kholodov 1988).

2.7.2 TVD schemes. The next approach is associated with schemes that diminish (or preserve) the total variation of a function, instead of monotonicity. The TVD (total variation diminishing) principle is the condition weaker that the monotonicity requirement. The total variation $\mathrm{TV}[u]$ of a bounded function $u = u(x)$ is defined, see Kolmogorov and Fomin (1975), as

$$\mathrm{TV}[u] = \sup_M \sum_{m=1}^{M} |u(x_m) - u(x_{m-1})|, \quad a = x_0 < x_1 < \cdots < x_M = b,$$

or, for a differentiable function $u = u(x)$, as

$$\mathrm{TV}[u] = \int_a^b \left|\frac{\partial u}{\partial x}\right| dx.$$

One can easily verify that the solution u of Eq. (2.7.2) is a function of bounded variation if $\mathrm{TV}[u]$ exists. A piecewise constant function with a finite number of jumps belongs to the class of functions with the bounded total variation. In particular, the exact solution of the Riemann problem for a linear system, see Eqs. (2.2.19)–(2.2.20), is also a function with bounded total variation.

The total variation can also be defined for a discrete function $u = \{u_m^k\}$. Assume first that the function must be reconstructed in accordance with (2.7.1). If all $\alpha_m = 0$, $\mathrm{TV}[u]$ is equal to

$$\mathrm{TV}_0^k = \sum_m |u_{m+1}^k - u_m^k| = \sum_m |\Delta_m^k|, \quad \Delta_m^k = u_{m+1}^k - u_m^k, \tag{2.7.6}$$

where the summation is performed over all discrete points of the computational interval $[a,b]$ and the zero subscript indicates the reconstruction with zero α's. For an arbitrary piecewise linear reconstruction of (2.7.1), TV$[u]$ has the form

$$\mathrm{TV}_\alpha^k = \sum_m \left(|\Delta_m^k - \tfrac{1}{2}(\alpha_{m+1} + \alpha_m)| + \tfrac{1}{2}|\alpha_{m+1}| + \tfrac{1}{2}|\alpha_m|\right). \tag{2.7.7}$$

If we introduce $\alpha = \max_m |\alpha_m|$, then $\lim_{\alpha\to 0} \mathrm{TV}_\alpha^k = \mathrm{TV}_0^k$.

Equation (2.7.2) can be solved with the help of scheme (2.5.5)–(2.5.10). In this case the finite-difference solution has the form

$$\begin{aligned} u_{m+1}^{k+1} &= (1-C)u_{m+1}^k + Cu_m^k - \tfrac{1}{2}C(1-C)\alpha_{m+1} + \tfrac{1}{2}C(1-C)\alpha_m, \\ u_m^{k+1} &= (1-C)u_m^k + Cu_{m-1}^k - \tfrac{1}{2}C(1-C)\alpha_m + \tfrac{1}{2}C(1-C)\alpha_{m-1}. \end{aligned} \tag{2.7.8}$$

Subtracting u_m^{k+1} from u_{m+1}^{k+1}, we obtain

$$\Delta_m^{k+1} = (1-C)\Delta_m^k + C\Delta_{m-1}^k - \tfrac{1}{2}C(1-C)\alpha_{m+1} + C(1-C)\alpha_m - \tfrac{1}{2}C(1-C)\alpha_{m-1}. \tag{2.7.9}$$

Using the inequality $|p+q| \le |p| + |q|$, we arrive at the following inequality for TV_0^{k+1}:

$$\begin{aligned} \mathrm{TV}_0^{k+1} &\le \sum_m \left(|\Delta_m^k - R_m| + |R_m|\right), \\ R_m &= C\Delta_m^k + \tfrac{1}{2}C(1-C)\alpha_{m+1} - \tfrac{1}{2}C(1-C)\alpha_m. \end{aligned} \tag{2.7.10}$$

This inequality will be used to limit the total variation. A numerical scheme satisfying the property

$$\mathrm{TV}_0^{k+1} \le \mathrm{TV}_0^k \tag{2.7.11}$$

is called a total variation diminishing (TVD) scheme (Harten 1983), or a total variation non-increasing (TVNI) scheme. Further, we will study only TVD schemes.

Scheme (2.5.5)–(2.5.9) is a TVD scheme if for all m the following inequality holds:

$$|\Delta_m^k - R_m| + |R_m| \le |\Delta_m^k|.$$

Thus, the TVD property is equivalent to the inequalities

$$0 \le R_m \operatorname{sgn} \Delta_m^k \le |\Delta_m^k|,$$

and, hence, to the general sufficient relations

$$\begin{aligned} &0 \le R_m \operatorname{sgn} \Delta_m^k = C|\Delta_m^k| + \tfrac{1}{2}C(1-C)(\alpha_{m+1} - \alpha_m) \operatorname{sgn} \Delta_m^k \le |\Delta_m^k|, \\ &\alpha_{m+1}\Delta_m^k \ge 0, \quad \alpha_{m+1}\Delta_m^k \ge 0. \end{aligned} \tag{2.7.12}$$

Assuming $\alpha_{m+1} = \Delta_m^k \psi_{m+1}$ and $\alpha_m = \Delta_m^k \psi_m$, where $\psi_{m+1} \ge 0$ and $\psi_m \ge 0$ are some coefficients, we rewrite the above TVD condition in the form

$$0 \le C + \tfrac{1}{2}C(1-C)(\psi_{m+1} - \psi_m) \le 1,$$

or

$$-\frac{2}{1-C} \leq \psi_{m+1} - \psi_m \leq \frac{2}{C}.$$

It is apparent that the TVD condition is equivalent to the stability condition (2.5.19). If we use a CFL number such that $0 \leq C \leq C_{\max} < 1$, then

$$-2 \leq \psi_{m+1} - \psi_m \leq \frac{2}{C_{\max}}. \tag{2.7.13}$$

If $0 \leq C \leq 1$, then the sufficient condition becomes

$$-2 \leq \psi_{m+1} - \psi_m \leq 2. \tag{2.7.14}$$

For inequality (2.7.14) to hold, it is sufficient that

$$0 \leq \psi_{m+1} \leq 2, \quad 0 \leq \psi_m \leq 2. \tag{2.7.15}$$

The inequality (2.7.12) is satisfied for all $0 \leq C \leq 1$ if

$$-2|\Delta^k_m| \leq (\alpha_{m+1} - \alpha_m)\,\mathrm{sgn}\,\Delta^k_m \leq 2|\Delta^k_m|, \qquad \alpha_{m+1}\Delta^k_m \geq 0, \quad \alpha_m\Delta^k_m \geq 0,$$

and, hence,

$$\begin{aligned} &\Delta^k_m - |\Delta^k_m| \leq \tfrac{1}{2}(\alpha_{m+1} + \alpha_m) \leq \Delta^k_m + |\Delta^k_m|, \\ &\alpha_{m+1}\Delta^k_m \geq 0, \quad \alpha_m\Delta^k_m \geq 0. \end{aligned} \tag{2.7.16}$$

For the advection equation

$$\frac{\partial u}{\partial t} - \frac{\partial u}{\partial x} = 0, \quad u(0,x) = u_0(x), \quad a \leq x \leq b, \tag{2.7.17}$$

the TVD property (2.7.13) becomes

$$-2 \leq \psi_m - \psi_{m+1} \leq \frac{2}{C_{\max}}. \tag{2.7.18}$$

In what follows, we will study the reconstruction of functions only for TVD schemes. Among all reconstructions providing the TVD property, two types of reconstruction can be singled out. The first is a "monotone" reconstruction, which provides the TVD property and, additionally, preserves the total variation of the function being reconstructed. The second reconstruction provides only the TVD property.

2.7.3 Monotone and limiting reconstructions. Let us find α_m such that the additional condition

$$\mathrm{TV}^k_\alpha = \mathrm{TV}^k_0 \tag{2.7.19}$$

holds, see Eqs. (2.7.6)–(2.7.7). Relation (2.7.19) expresses the invariance condition for the total variation of the initial function and reconstructed function. Comparing Eq. (2.7.6) and Eq. (2.7.7), we see that relation (2.7.19) is equivalent to the following sufficient conditions:

$$\begin{aligned} &\tfrac{1}{2}(\Delta_m - |\Delta_m|) \leq \tfrac{1}{2}(\alpha_{m+1} + \alpha_m) \leq \tfrac{1}{2}(\Delta_m + |\Delta_m|), \\ &\alpha_{m+1}\Delta_m \geq 0, \quad \alpha_m\Delta_m \geq 0, \quad \Delta_m = u_{m+1} - u_m. \end{aligned} \tag{2.7.20}$$

If all α's satisfy inequalities (2.7.20), one can easily verify that condition (2.7.19) also holds.

It can be proved that conditions (2.7.20) also provide the TVD condition (2.7.11). To prove this, for simplicity we set in (2.7.10) $\Delta_m^k = 1$, $\alpha_m = \epsilon$, and $\alpha_{m+1} = 2 - \epsilon$, where $0 \leq \epsilon \leq 2$. Then,

$$R_m = C + C(1-C)(1-\epsilon), \quad 0 \leq \epsilon \leq 2, \quad 0 \leq C \leq 1.$$

Since for $\epsilon = 0$, we have $R_m = 2C - C^2 \leq 1$ and for $\epsilon = 2$, $R_m = C^2 \leq 1$, the inequality $R_m \leq 1$ holds for all admissible C and ϵ. Then we have $|\Delta_m^k - R_m| + |R_m| \leq |\Delta_m^k|$, which proves the TVD condition (2.7.11). We considered above the values of α_{m+1} and α_m such that $\alpha_{m+1} + \alpha_m = 2$. If the neighboring α's satisfy the inequality $\alpha_{m+1} + \alpha_m < 2$, the corresponding scheme is certainly a TVD scheme.

We can prove also the TVD property by using inequalities (2.7.20). Indeed,

$$\psi = \max\left(\frac{\alpha_{m+1}}{\Delta_m^k}, \frac{\alpha_m}{\Delta_m^k}\right) \leq 2,$$

and hence the sufficient condition of (2.7.15) holds.

Note that to check whether the scheme satisfying Eq. (2.7.19) actually possesses the TVD property one can carry out direct conclusive calculations. In particular, we checked the validity of inequality (2.7.11) for 2×10^6 uniform distributed values of α_{m+1} and α_m satisfying inequality (2.7.20) with C from the interval $0 \leq C \leq 1$.

Thus, scheme (2.5.5)–(2.5.10) with the reconstruction defined by (2.7.1) and satisfying (2.7.20) is a TVD scheme. We will refer to this reconstruction as the monotone TVD reconstruction, since the slopes provided by this reconstruction preserve the monotonicity of the initial function u and its total variation. Specifically, if the inequality $u_m \leq u_{m+1}$ holds, then, for the monotonicity preservation, after the reconstruction the following inequalities must hold:

$$u_m \leq u_m + \tfrac{1}{2}\alpha_m \leq u_{m+1} - \tfrac{1}{2}\alpha_{m+1} \leq u_{m+1}, \tag{2.7.21}$$

or

$$u_m \leq u_{m+1/2}^{\mathrm{L}} \leq u_{m+1/2}^{\mathrm{R}} \leq u_{m+1},$$

where the values $u_{m+1/2}^{\mathrm{L}} = u_m + \frac{1}{2}\alpha_m$ and $u_{m+1/2}^{\mathrm{R}} = u_{m+1} - \frac{1}{2}\alpha_{m+1}$ are the left and right values of u for the boundary $m + 1/2$, see Fig. 2.3. Hence,

$$0 \leq \tfrac{1}{2}(\alpha_{m+1} + \alpha_m) \leq u_{m+1} - u_m, \quad \alpha_{m+1} \geq 0, \quad \alpha_m \geq 0. \tag{2.7.22}$$

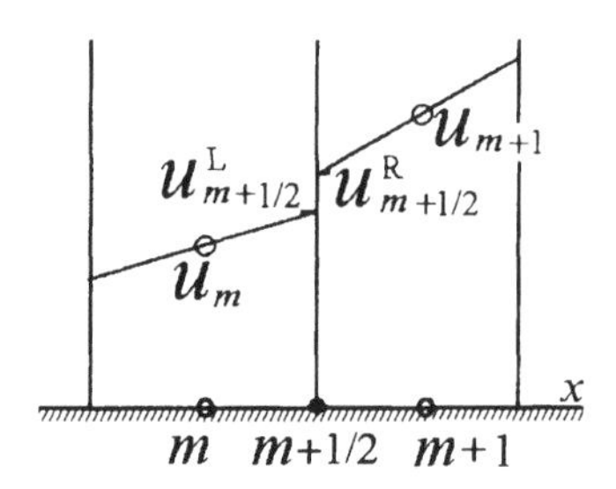

Figure 2.3 Piecewise constant distribution of **u**.

For the case $u_m \geq u_{m+1}$ we have

$$u_m \geq u_m + \tfrac{1}{2}\alpha_m \geq u_{m+1} - \tfrac{1}{2}\alpha_{m+1} \geq u_{m+1}. \tag{2.7.23}$$

Thus,

$$0 \geq \tfrac{1}{2}(\alpha_{m+1} + \alpha_m) \geq u_{m+1} - u_m, \quad \alpha_{m+1} \leq 0, \quad \alpha_m \leq 0. \tag{2.7.24}$$

Inequalities (2.7.22) or (2.7.24) must be satisfied for all cell boundaries. The set of the slopes α is not empty, since at least all $\alpha_m = 0$ satisfy (2.7.21)–(2.7.22) or (2.7.23)–(2.7.24). We can combine inequalities (2.7.22) and (2.7.24) to represent them in a unified form as (2.7.20). Relations (2.7.20) can be rewritten as

$$\begin{gathered} \Delta_m^- \leq \tfrac{1}{2}(\alpha_{m+1} + \alpha_m) \leq \Delta_m^+, \\ \alpha_{m+1}\Delta_m \geq 0, \quad \alpha_m \Delta_m \geq 0, \quad \Delta_m = u_{m+1} - u_m, \\ \Delta^- = \tfrac{1}{2}(\Delta - |\Delta|), \quad \Delta^+ = \tfrac{1}{2}(\Delta + |\Delta|), \end{gathered} \tag{2.7.25}$$

where $\Delta_m^- \leq 0$, $\Delta_m^+ \geq 0$, and $\Delta_m^- \Delta_m^+ = 0$.

For the inequalities (2.7.25) to be satisfied automatically, an algorithm for selection of α's should be suggested. Such an algorithm is provided by the introduction of limiting functions (limiters). The simplest limiter is a function of two variables, $L(a, b)$. This function serves to limit the values of α as follows:

$$\alpha_m = L\left(\frac{u_{m+1} - u_m}{\Delta x}, \frac{u_m - u_{m-1}}{\Delta x}\right). \tag{2.7.26}$$

In addition, the function L possesses the property of symmetry, $L(a, b) = L(b, a)$, and the compatibility condition, $L(a, a) = a$. More generally, the limiter can be a function of many finite differences $L(a, b, \ldots, c)$. The compatibility condition provides in this case the determination of the exact slope for a linear function: $L(a, a, \ldots, a) \equiv a$.

The minmod limiter

Consider two consecutive inequalities of the monotonicity condition (2.7.25):

$$\begin{gathered} \Delta_m^- \leq \tfrac{1}{2}(\alpha_{m+1} + \alpha_m) \leq \Delta_m^+, \\ \Delta_{m-1}^- \leq \tfrac{1}{2}(\alpha_m + \alpha_{m-1}) \leq \Delta_{m-1}^+. \end{gathered} \tag{2.7.27}$$

To close system (2.7.27) it is convenient to set $\alpha_{m+1} = \alpha_m$ and $\alpha_{m-1} = \alpha_m$; thus, we obtain

$$\Delta_m^- \le \alpha_m \le \Delta_m^+, \qquad (2.7.28)$$
$$\Delta_{m-1}^- \le \alpha_m \le \Delta_{m-1}^+,$$

or

$$\max(\Delta_m^-, \Delta_{m-1}^-) \le \alpha_m \le \min(\Delta_m^+, \Delta_{m-1}^+).$$

Using the fact that $\Delta_m^- \Delta_m^+ = 0$, we obtain

$$\alpha_m = \max(\Delta_m^-, \Delta_{m-1}^-) + \min(\Delta_m^+, \Delta_{m-1}^+) \equiv \operatorname{minmod}(\Delta_m, \Delta_{m-1}), \qquad (2.7.29)$$
$$\Delta_n = u_{n+1} - u_n, \quad n = m, m-1;$$
$$\Delta^+ = \tfrac{1}{2}(\Delta + |\Delta|), \quad \Delta^- = \tfrac{1}{2}(\Delta - |\Delta|).$$

One can verify that the constructed α's satisfy (2.7.25). Another form of the minmod limiter is given by

$$\alpha_m = \operatorname{minmod}\left(\frac{u_{m+1} - u_m}{\Delta x}, \frac{u_m - u_{m-1}}{\Delta x}\right),$$
$$\operatorname{minmod}(a, b) = \tfrac{1}{2}(\operatorname{sgn} a + \operatorname{sgn} b) \min(|a|, |b|). \qquad (2.7.30)$$

The minmod limiter preserves the monotonicity of the initial function u. Kolgan (1972) was the first to suggest a limiting function for modelling two-dimensional stationary gas dynamic flows by the Godunov method with the exact Riemann solver. Kolgan used the following limiter:

$$L_{\text{Kolgan}}(a, b) = \begin{cases} a & \text{if } |a| = \min\left[|a|, |b|, \tfrac{1}{2}|a+b|\right], \\ b & \text{if } |b| = \min\left[|a|, |b|, \tfrac{1}{2}|a+b|\right], \\ \tfrac{1}{2}(a+b) & \text{if } \tfrac{1}{2}|a+b| = \min\left[|a|, |b|, \tfrac{1}{2}|a+b|\right]. \end{cases}$$

Note that if a and b have opposite signs, the slope can be nonzero.

The minmod limiter (2.7.29)–(2.7.30) preserves the monotonicity of reconstructed function and the TVD property of scheme (2.5.5)–(2.5.10). The stability condition for a scheme with the minmod limiter reduces to the CFL condition $C \le 1$, see (2.5.21). This limiter provides stable numerical simulation in many problems, but it is not free of disadvantages. It is important to observe that the scheme becomes first-order accurate in space in the neighborhood of a local extremum of $u(x)$ where the minmod slope is zero. This leads to an increase in the dissipation near extrema. Switching to zero slopes at the extrema obviously results in their "plateauing" (flatting) and extrema distortion. Another drawback can arise in regions where the second derivative of $u(x)$ changes in sign, and the scheme can change its stencil. When this occurs, the approximation error changes so that the oscillations of curvature may develop on smooth portions of the curve $u(x)$. Excessive smearing of contact discontinuities is yet another disadvantage. To eliminate them, finer limiters have been proposed for the slope calculation.

The minmod function is not the only limiter that ensures a monotone reconstruction. Later we shall show that the minmod reconstruction is a special case of a more general procedure of "limiting" reconstruction, see Semenov (1992), Kamenetskii and Semenov (1994), and Semenov (1995a).

The minmod limiting reconstruction

Here we discuss a more general case of reconstruction for a nonuniform mesh. Consider a point distribution x_m, where $m = 0, 1, \ldots, N$ and $x_m < x_{m+1}$. Each interval $[x_m, x_{m+1}]$ of the obtained mesh will be assigned a value u_m which is the integral average of a function $u(x)$ on this interval. We would like to reconstruct the function $u(x)$ as

$$u(x) = u_m + \alpha_m \left[x - \tfrac{1}{2}(x_m+x_{m+1})\right], \quad x \in [x_m, x_{m+1}] \tag{2.7.31}$$

in a way such that the resulting reconstruction is monotone in the regions where the discrete function u is monotone and the sum of the moduli of the differences in the values of the function at the cell boundaries has a minimum. Thus, it is required that the sum of the boundary jumps

$$J_b = \sum_{m=0}^{N-1} |\Delta_m^\alpha|, \quad \Delta_m^\alpha = \left[u_{m+1} - \tfrac{1}{2}(x_{m+1}-x_m)\alpha_{m+1}\right] - \left[u_m + \tfrac{1}{2}(x_m - x_{m-1})\alpha_m\right], \tag{2.7.32}$$

is minimal. We call this a "limiting" reconstruction (Semenov 1992). The limiting reconstruction can be considered as a natural generalization of the monotone reconstruction.

A solution of the problem always exists, since the upper limit of J_b is its value for $\alpha_m = 0$, which corresponds to a piecewise constant reconstruction; in this case $\Delta_m^\alpha \equiv u_{m+1} - u_m$. If $J_b = 0$, then we obtain a continuous piecewise-linear reconstruction.

From the monotonicity condition (2.7.20) it follows that for nonzero α_m the sign of Δ_m^α remains the same as the sign of Δ_m. Then we can find the limiting reconstruction by determining the values of α_m that satisfy the system of the following inequalities, see Eq. (2.7.20):

$$\tfrac{1}{2}\alpha_{m+1}(x_{m+1}-x_m) + \tfrac{1}{2}\alpha_m(x_m - x_{m-1}) \leq u_{m+1} - u_m = \Delta_m \quad \text{if} \quad \Delta_m \geq 0. \tag{2.7.33}$$

If $\Delta_m < 0$, the sign of the inequality in (2.7.33) is reversed. Note that if for certain α_m inequalities (2.7.33) become equalities, we obtain an exact piecewise-linear continuous reconstruction.

We will now describe an iterative algorithm for obtaining α_m with continuously increasing moduli that ensures, however, that inequalities (2.7.33) are always satisfied.

At first we represent Δ_m in the form $\Delta_m = s_m|\Delta_m|$, where $s_m = \operatorname{sgn}\Delta_m$. Note that, in order to diminish $|\Delta_m|$ it is necessary that the signs of both α_m and α_{m+1} in (2.7.32) are the same as that of Δ_m. Let $\alpha_m = s_m|\alpha_m|$. On the other hand, we have $\alpha_m = s_{m+1}|\alpha_m|$, since α_m occurs in two inequalities of the type (2.7.33). Then it is convenient to write α_m in the form $\alpha_m = \frac{1}{2}(s_m + s_{m+1})|\alpha_m|$. In the case $s_m + s_{m+1} = 0$, we automatically obtain $\alpha_m = 0$ similarly to the case of using the minmod function. Then we represent (2.7.33) in the form

$$\tfrac{1}{4}(1+s_m s_{m+1})|\alpha_m|(x_m - x_{m-1}) + \tfrac{1}{4}(1+s_m s_{m+1})|\alpha_{m+1}|(x_{m+1}-x_m) \leq |\Delta_m|. \tag{2.7.34}$$

Here $m = 1, \ldots, N-1$ and $\alpha_0 = \alpha_N = 0$. It is not difficult to see that

$$|\alpha_m^{(0)}| = 2\min\left(\frac{|\Delta_m|}{x_{m+1}-x_{m-1}}, \frac{|\Delta_{m-1}|}{x_m - x_{m-2}}\right) \tag{2.7.35}$$

is a solution of the system of (2.7.34). This can be verified by the direct substitution of (2.7.35) into (2.7.34).

This result seems quite reasonable, since it gives the same values of α_m as those obtained with the minmod function. However, in general the α_m provided by (2.7.35) are not the largest possible ones. We shall therefore try to refine the values of $|\alpha_m|$. Setting $|\alpha_m| = |\alpha_m^{(0)}| + |\alpha_m^{(1)}|$ and substituting this expression into inequality (2.7.34), we obtain

$$\frac{1}{4}(1+s_m s_{m+1})|\alpha_m^{(1)}|(x_m - x_{m-1}) + \frac{1}{4}(1+s_m s_{m+1})|\alpha_{m+1}^{(1)}|(x_{m+1} - x_m) \le |\Delta_m^{(1)}|, \tag{2.7.36}$$
$$|\Delta_m^{(1)}| = |\Delta_m| - \frac{1}{4}(1+s_m s_{m+1})|\alpha_m^{(0)}|(x_m - x_{m-1}) - \frac{1}{4}(1+s_m s_{m+1})|\alpha_{m+1}^{(0)}|(x_{m+1} - x_m).$$

The system of inequalities (2.7.36) has the same form as (2.7.34). Solutions of this system are therefore given by (2.7.35), if we replace Δ_m by $\Delta_m^{(1)}$. Repeating this procedure, we obtain a method of successive iterations for refining the values of $|\alpha_m|$. It can be proved that the process converges after $N-1$ iterations. (However, $N-1$ iterations are realized only if $\Delta_m^{(n)}$, $n = 1, 2, \ldots$, has the same sign at each iteration. In practice, this situation occurs extremely rarely.) The iterative process is rapidly convergent and requires often no more than three to five iterations, see investigation by Malyshev (1996). Thus, the limiting reconstruction can be equivalent to the triple minmod application in accordance with a special rule. The realization of this algorithm is compact and convenient for use in numerical codes.

The minmod limiting reconstruction preserves the monotonicity of a reconstructed function and the TVD property for scheme (2.5.5)–(2.5.10). According to (2.5.21), (2.7.15) the stability condition for the scheme is the CFL condition $C \le 1$, since

$$\psi = \max\left[\frac{\alpha_{m+1}(x_{m+1} - x_{m-1})}{2\Delta_m}, \frac{\alpha_m(x_{m+1} - x_{m-1})}{2\Delta_m}\right] \le 1 < 2.$$

Note that the limiting reconstruction involves, in general, the variable number of finite differences of the function u. Therefore, a numerical scheme based on the limiting reconstruction has no *a priori* fixed stencil.

We can suggest other versions of the limiting reconstruction. For example, we can replace the quantity Δ_m by $\theta_m \Delta_m$, $0 \le \theta_m \le 1$. In this case the mth boundary jump of (2.7.32) will be no less than $(1 - \theta_m)\Delta_m$. Otherwise, we can also replace Δ_m by δ_m such that $\operatorname{sgn} \Delta_m = \operatorname{sgn} \delta_m$ and $|\Delta_m| \ge |\delta_m|$. In this case the mth boundary jump will be no less than $\Delta_m - \delta_m$. These special cases may be necessary for obtaining a prescribed jump of the function at some boundaries.

Finally, we can restrict the number of iterations. If it is equal to l, then the reconstruction in general has a fixed stencil with $2l + 1$ points. The same result can be obtained if in the iterative process for each mth point we set $\Delta_M^0 = 0$ for all M such that $M \ge m + l + 1$ and $M \le m - l - 1$.

All these procedures also provide a monotone reconstruction that preserves the TVD property of the scheme (2.5.7)–(2.5.9).

Let us illustrate some advantages of the limiting reconstruction. Figures 2.4 and 2.5 demonstrate the numerical results of a one-dimensional gas dynamic test performed by the scheme (2.5.7)–(2.5.10) based on Roe's method and using the limiting reconstruction. We considered the interaction of the two blast waves. It involves a number of strong shocks,

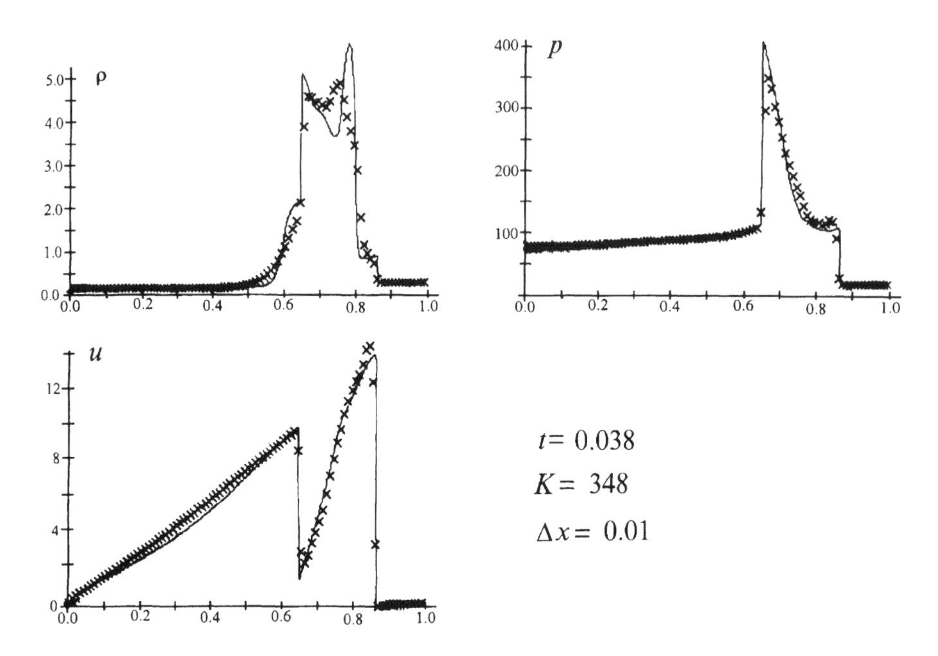

Figure 2.4 Blast wave interaction (rough mesh).

rarefaction waves, and contact discontinuities. The initial data at $t = 0$ are the following:

$$\rho(0 \le x \le 1) = 1, \quad u(0 \le x \le 1) = 0,$$
$$p(0 \le x < 0.1) = 1000, \quad p(0.1 \le x < 0.9) = 0.01, \quad p(0.9 \le x \le 1) = 100,$$

where ρ, u, and p are the density, velocity, and pressure, respectively. The adiabatic index is equal to 1.4. The boundaries $x = 0$ and $x = 1$ are solid walls. This test problem was studied by Woodward and Colella (1984).

In Figs. 2.4 and 2.5 the markers show the numerical solution at $t = 0.038$, and the solid line is the "exact" solution calculated on a fine grid with $N = 800$ (K is the time layer number.) Figure 2.4 presents our results for $N = 100$. One can check that these results are very close to those obtained with the second-order ENO (essentially nonoscillatory) scheme for $N = 200$, see Harten et al. (1987). Figure 2.5 presents our results for $N = 200$; they nearly coincide with the results obtained with the fourth-order ENO scheme for $N = 200$, see Harten et al. (1987). This shows a high efficiency of the limiting reconstruction. Thus, this permits us to use in calculations only half the number of the mesh points without any loss of accuracy. Note that the use of a limiting reconstruction requires calculations with smaller CFL numbers than those for using the only minmod limiter. Results of simulation of different problems can be found in Malyshev (1996).

Treating the reconstruction procedure as a separate stage, we note that the limiting reconstruction permits one to diminish the losses of information on the behavior of function under reconstruction. Such a reconstruction can be useful in many practical applications

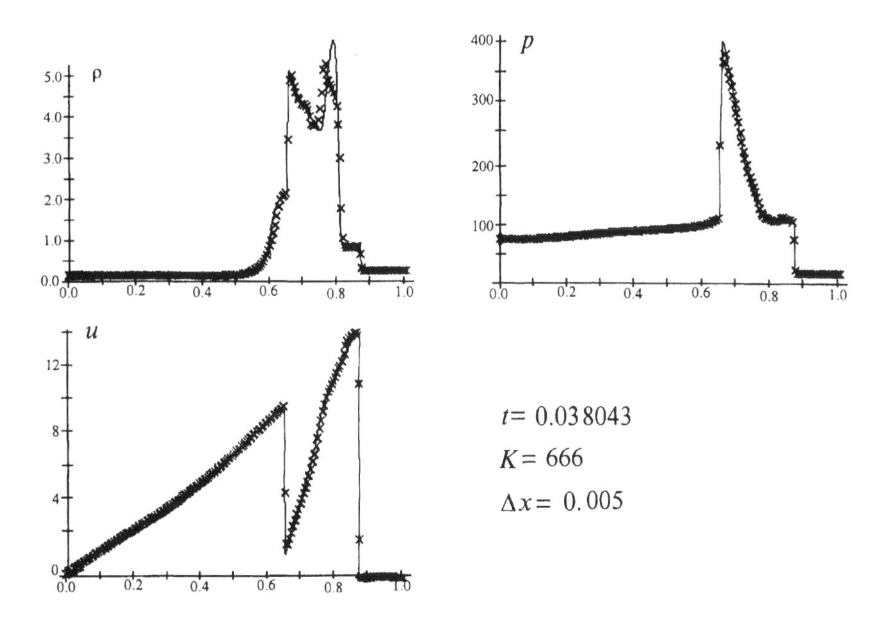

Figure 2.5 Blast wave interaction (fine mesh).

(Kamenetskii and Semenov 1994; Malyshev 1996). In particular, Kamenetskii and Semenov (1994) used the limiting reconstruction as a substage for the realization of a self-adjusting grid algorithm for bow shock fitting in a two-dimensional supersonic blunt body flow problem by the shock-capturing Roe method, see Section 3.5.4. The similar algorithm for one-dimensional case was created by Harten and Hyman (1983).

2.7.4 Genuine TVD and TVD limiting reconstructions. The TVD reconstruction is a reconstruction that preserves only the TVD property (2.7.11) of the scheme. It does not preserve the monotonicity property (2.7.19) of the function under reconstruction. We will refer to this reconstruction as a TVD reconstruction. Assume for simplicity that $\Delta x = 1$. We found previously that the TVD property (2.7.11) imposes the following sufficient condition (2.7.16) for piecewise-linear reconstruction:

$$\begin{gathered}\Delta_m - |\Delta_m| \leq \tfrac{1}{2}(\alpha_{m+1} + \alpha_m) \leq \Delta_m + |\Delta_m|, \\ \alpha_{m+1}\Delta_m \geq 0, \quad \alpha_m\Delta_m \geq 0, \quad \Delta_m = u_{m+1} - u_m.\end{gathered} \tag{2.7.37}$$

The reconstruction satisfying the property of (2.7.11) has the following interpretation: preserves and/or diminishes the sum of the boundary jumps of (2.7.32),

$$J_b = \sum_{m=0}^{N-1} |\Delta_m^\alpha|, \quad \Delta_m^\alpha = [u_{m+1} - \tfrac{1}{2}\alpha_{m+1}] - [u_m + \tfrac{1}{2}\alpha_m]. \tag{2.7.38}$$

Indeed, to satisfy the TVD property, the α's must ensure that

$$|u^{\mathrm{R}}_{m+1/2} - u^{\mathrm{L}}_{m+1/2}| \leq |u_{m+1} - u_m|,$$

$$u^{\mathrm{L}}_{m+1/2} = u_m + \tfrac{1}{2}\alpha_m, \quad u^{\mathrm{R}}_{m+1/2} = u_{m+1} - \tfrac{1}{2}\alpha_{m+1},$$

or

$$|\Delta^{\alpha}_{m}| \leq |\Delta_m|,$$
$$\Delta_m = u_{m+1} - u_m,$$
$$\Delta^{\alpha}_{m} = (u_{m+1} - \tfrac{1}{2}\alpha_{m+1}) - (u_m + \tfrac{1}{2}\alpha_m).$$

Hence, $-|\Delta_m| \leq -\Delta^{\alpha}_{m} \leq |\Delta_m|$, or

$$\Delta_m - |\Delta_m| \leq \tfrac{1}{2}(\alpha_{m+1} + \alpha_m) \leq \Delta_m + |\Delta_m|, \tag{2.7.39}$$

and

$$2\Delta^{-}_{m} \leq \tfrac{1}{2}(\alpha_{m+1} + \alpha_m) \leq 2\Delta^{+}_{m}, \tag{2.7.40}$$
$$\Delta^{-} = \tfrac{1}{2}(\Delta - |\Delta|), \quad \Delta^{+} = \tfrac{1}{2}(\Delta + |\Delta|).$$

Inequalities (2.7.20) for a monotone reconstruction are more restrictive than those of (2.7.39). Therefore, a TVD reconstruction can break the monotonicity of the reconstructed function.

TVD limiters

We describe below how to determine the coefficients α for the reconstruction of (2.7.37). First we consider limiters for a TVD reconstruction in the form of the function $L(a, b)$, see (2.7.26), and investigate their properties and monotonicity. Assuming for definiteness that $\Delta_m > 0$ and $\Delta_{m+1} > 0$, we consider two inequalities of (2.7.40):

$$0 \leq \tfrac{1}{2}(\alpha_{m+1} + \alpha_m) \leq 2\Delta_m, \tag{2.7.41}$$
$$0 \leq \tfrac{1}{2}(\alpha_m + \alpha_{m-1}) \leq 2\Delta_{m-1}. \tag{2.7.42}$$

Assuming also that $\alpha_k = \Psi_k\Delta_k \geq 0$, where $k = m{+}1, m, m{-}1$, we can rewrite (2.7.41)–(2.7.42) as

$$0 \leq \Psi_{m+1}\frac{\Delta_{m+1}}{\Delta_m} + \Psi_m \leq 4, \tag{2.7.43}$$

$$0 \leq \Psi_m\frac{\Delta_m}{\Delta_{m-1}} + \Psi_{m-1} \leq 4. \tag{2.7.44}$$

To satisfy inequalities (2.7.43)–(2.7.44), we adopt the necessary conditions

$$0 \leq \Psi_k \leq 2, \quad 0 \leq \Psi_k \leq 2\frac{\Delta_{k-1}}{\Delta_k}$$

for all k and all ratios $r_k = \Delta_{k-1}/\Delta_k$. Let $\Psi_k \equiv \Psi(r_k)$. Then, inequality (2.7.40) holds if $\Psi(r)$ satisfies the inequality

$$0 \leq \Psi(r) \leq \min(2, 2r). \tag{2.7.45}$$

By analogy, we can consider the case of negative differences. For $r \leq 0$ we set $\Psi(r) \equiv 0$. All slopes given by

$$\alpha_m = \Delta_m \Psi\left(\frac{\Delta_{m-1}}{\Delta_m}\right),$$

where Ψ satisfies (2.7.45), define a TVD reconstruction. Hence, the function $\Psi(r_m)$ is a TVD limiter. In particular, the minmod limiter corresponds to $\Psi(r) = \min(1, r)$ if $r > 0$, and $\Psi(r) = 0$, otherwise.

In addition, the function $\Psi(r)$ possesses the property

$$\frac{\Psi(r)}{r} = \Psi\left(\frac{1}{r}\right).$$

Indeed, from the symmetry of the function $L(a, b)$ it follows that

$$L(a, b) = b\Psi\left(\frac{a}{b}\right) = b\Psi(r) = L(b, a) = a\Psi\left(\frac{b}{a}\right) = rb\Psi\left(\frac{1}{r}\right), \quad r = \frac{a}{b}$$

$$\Longrightarrow \quad \left[\frac{\Psi(r)}{r} - \Psi\left(\frac{1}{r}\right)\right] b \equiv 0.$$

A detailed analysis of the properties of limiters represented in the form of $\Psi(r)$ was given by Sweby (1984) and from another viewpoint by Roe (1985); see also Hirsch (1990).

The above TVD limiters preserve the TVD property of the scheme represented by (2.5.5)–(2.5.10) and the stability condition $C \leq 1$, since

$$\psi = \max \Psi(r) \leq 2,$$

see (2.5.21), (2.7.15).

Van Leer (1977a) suggested a TVD limiter in the form

$$L(a, b) = \begin{cases} \dfrac{2ab}{a+b} & \text{if } \ ab > 0, \\ 0 & \text{if } \ ab \leq 0. \end{cases} \tag{2.7.46}$$

This limiter corresponds to $\Psi(r) = 2r/(1 + r)$. Let us prove that the function of (2.7.46) is a TVD limiter, though nonmonotone. Assume for definiteness that $\Delta_k = u_{k+1} - u_k > 0$, $k = m-1, m, m+1$. Then, in accordance with (2.7.39),

$$R = \frac{1}{2}(\alpha_{m+1} + \alpha_m) = \frac{1}{2}\left[\frac{2\Delta_{m+1}\Delta_m}{\Delta_{m+1} + \Delta_m} + \frac{2\Delta_m\Delta_{m-1}}{\Delta_m + \Delta_{m-1}}\right]$$
$$= \left[2 - \frac{\Delta_m}{\Delta_{m+1} + \Delta_m} - \frac{\Delta_m}{\Delta_m + \Delta_{m-1}}\right]\Delta_m \leq 2\Delta_m.$$

This proves inequality (2.7.39). Hence, Eq. (2.7.46) defines a TVD limiter.

Consider the discrete mesh function

$$u_m = \sum_{n=1}^{m} q^{(n-1)^2}, \quad 1 < q < 2.$$

We have $\Delta_k = q^{k^2}$ and

$$R = \frac{1}{2}(\alpha_{m+1} + \alpha_m) = \frac{1}{2}\left[\frac{2q^{(m+1)^2}q^{m^2}}{q^{(m+1)^2} + q^{m^2}} + \frac{2q^{m^2}q^{(m-1)^2}}{q^{m^2} + q^{(m-1)^2}}\right]$$
$$= \left[1 + \frac{q^{2m+1} - q^{2m-1}}{(q^{2m+1} - 1)(q^{2m-1} - 1)}\right] q^{m^2} > q^{m^2} = \Delta_m.$$

Thus, the inequality (2.7.20) is not satisfied, and hence the limiter defined by (2.7.46) is not monotone.

Another limiter suggested by van Leer (1979) can be written in the form

$$L(a, b) = \text{minmod}[\tfrac{1}{2}(a+b), 2\,\text{minmod}(a, b)\,]. \tag{2.7.47}$$

This is also a TVD limiter but not monotone. Another limiter that corresponds to $\Psi(r) = (r + |r|)/(1 + r)$ was studied by van Leer (1974).

Van Albada, van Leer, and Roberts (1982) suggested a limiter in the form

$$L(a, b) = \frac{(a^2 + \epsilon)b + (b^2 + \epsilon)a}{a^2 + b^2 + 2\epsilon}, \tag{2.7.48}$$

where ϵ is a small constant ranging between 10^{-7} and 10^{-5} that is used to avoid division by zero. This limiter corresponds to $\Psi(r) = (r^2 + r)/(1 + r^2)$ and has smooth derivatives with respect to a and b. Let us check that the function of (2.7.48) represents a nonmonotone TVD limiter. For simplicity we set $\epsilon = 0$. Assume also that $\Delta_k = u_{k+1} - u_k > 0$, $k = m-1, m, m+1$. Then, in accordance with (2.7.39),

$$R = \frac{1}{2}(\alpha_{m+1} + \alpha_m) = \frac{1}{2}\left[\frac{\Delta_{m+1}^2\Delta_m + \Delta_{m+1}\Delta_m^2}{\Delta_{m+1}^2 + \Delta_m^2} + \frac{\Delta_m^2\Delta_{m-1} + \Delta_m\Delta_{m-1}^2}{\Delta_m^2 + \Delta_{m-1}^2}\right]$$
$$= \frac{\Delta_m}{2}\left[\frac{p^2 + p}{p^2 + 1} + \frac{q^2 + q}{q^2 + 1}\right],$$

where $\Delta_{m+1} = p\Delta_m$ and $\Delta_{m-1} = q\Delta_m$, $0 < p < +\infty$, $0 < q < +\infty$. The function R attains a maximum at $p = q = 1 + \sqrt{2}$. Hence,

$$R \le R_{\max} = \frac{1 + \sqrt{2}}{2}\Delta_m \quad \Longrightarrow \quad \Delta_m < R_{\max} < 2\Delta_m.$$

This proves the TVD property (2.7.39), but inequality (2.7.20) is not always satisfied and hence the limiter is not monotone.

Consider now the one-parameter limiter

$$L_k(a, b) = \begin{cases} 0 & \text{if } \; ab \le 0, \\ k\,\text{minmod}(a, b) & \text{if } \; ab > 0, \quad \max\left(\dfrac{|a|}{|b|}, \dfrac{|b|}{|a|}\right) > k, \\ k\,\text{sgn}(a)\max(|a|, |b|) & \text{if } \; ab > 0, \quad \max\left(\dfrac{|a|}{|b|}, \dfrac{|b|}{|a|}\right) \le k, \end{cases} \tag{2.7.49}$$

where $1 \le k \le 2$. Function (2.7.49) can be rewritten in a more compact form as

$$L_k(a,b) = \tfrac{1}{2}(\operatorname{sgn} a + \operatorname{sgn} b)\max[\,|\operatorname{minmod}(ka,b)|, |\operatorname{minmod}(a,kb)|\,].$$

For $k = 2$ we obtain a so-called "superbee" limiter by Roe and Pike (1984). It corresponds to $\Psi(r) = \max[0, \min(2r, 1), \min(r, 2)]$. To check the TVD property, one has to consider all ratios of Δ_{m-1}, Δ_m, and Δ_{m+1}. This limiter turns out to be not monotone. For instance, assume for simplicity that $\Delta x = 1$ and define the mesh discrete function u_m as

$$u_m = \sum_{n=1}^{m} q^{n-1}, \qquad 1 < q < 2.$$

Then $\Delta_k = q^k$, $\alpha_k = 2q^k$, and

$$R = \tfrac{1}{2}(\alpha_{m+1} + \alpha_m) = \tfrac{1}{2}(2q^m + 2q^{m+1}) = (1+q)q^m > q^m = \Delta_m.$$

Thus, inequality (2.7.20) is not satisfied, and hence the superbee limiter is not monotone.

TVD limiting reconstruction

Note that the TVD limiter is a contraction operator with respect to the boundary jumps (2.7.32), (2.7.38). Hence, we can construct an iterative limiting reconstruction using a TVD limiter. It is required that the sum

$$J_b = \sum_{m=0}^{N-1} |\Delta_m^\alpha|, \quad \Delta_m^\alpha = [u_{m+1} - \tfrac{1}{2}\alpha_{m+1}\Delta x] - [u_m + \tfrac{1}{2}\alpha_m \Delta x]$$

attains a minimum. Consider for simplicity a uniform space mesh with size Δx. Using a TVD limiter of the form (2.7.26), we can write out the following iterative process, where k is the number of the iteration:

$k = 0$:

$$\alpha_m^{(0)} = L\left(\frac{\Delta_m}{\Delta x}, \frac{\Delta_{m-1}}{\Delta x}\right), \qquad \Delta_m = u_{m+1} - u_m,$$

$$\Delta_m^{(1)} = \Delta_m - \tfrac{1}{2}(\alpha_{m+1}^{(0)} + \alpha_m^{(0)})\Delta x, \qquad m = 1, \ldots, N-1; \quad \alpha_0^{(0)} = \alpha_N^{(0)} = 0;$$

$k = 1, \ldots, K$:

$$\alpha_m^{(k)} = L\left(\frac{\Delta_m^{(k)}}{\Delta x}, \frac{\Delta_{m-1}^{(k)}}{\Delta x}\right),$$

$$\Delta_m^{(k+1)} = \Delta_m^{(k)} - \tfrac{1}{2}(\alpha_{m+1}^{(k)} + \alpha_m^{(k)})\Delta x, \qquad m = 1, \ldots, N-1; \quad \alpha_0^{(k)} = \alpha_N^{(k)} = 0.$$

Then the final slope α_m is expressed as

$$\alpha_m = \sum_{k=0}^{K} \alpha_m^{(k)}, \qquad m = 1, \ldots, N-1.$$

Calculations demonstrate that after three to five iterations the piecewise linear function reconstructed in this way is practically the same for all limiters defined by (2.7.46)–(2.7.49) and, except for small nonmonotonicities, is similar to the result obtained by the limiting minmod reconstruction (Semenov, unpublished).

We consider here a more complex limiter that originates from the UNO (uniformly nonoscillatory) schemes, see Harten and Osher (1987). The UNO schemes are nonoscillatory in the sense that the number of extrema of the discrete solution is not increasing in time. This UNO limiter involves the analysis not only of the first-order finite differences Δ that approximate the first derivatives of the function u, but also of the second-order finite differences δ that approximate the second derivatives of u. The stencil for this limiter consists of five points. In what follows we assume for definiteness that $\Delta x = 1$. Then,

$$\begin{aligned} \alpha_m &= \operatorname{minmod}\left(\Delta_m - \tfrac{1}{2}\delta_{m+1/2},\ \Delta_{m-1} + \tfrac{1}{2}\delta_{m-1/2}\right), \\ \Delta_k &= u_{k+1} - u_k, \quad k = m, m-1, \\ \delta_{k+1/2} &= \operatorname{minmod}\left(u_{k+1} - 2u_k + u_{k-1},\ u_{k+2} - 2u_{k+1} + u_k\right). \end{aligned} \tag{2.7.50}$$

Note that α_m can be nonzero if $\Delta_m = 0$. It follows that the limiter defined by the functions of (2.7.50) is neither a monotone nor a TVD limiter; see Eqs. (2.7.20) and (2.7.37). This is not surprising, since the UNO and ENO schemes by Harten et al. (1987) allow the increase of the total variation within the accuracy of the numerical scheme. The discussion of various limiting procedures can be found, e.g., in Yee (1989), Hirsch (1990), and Toro (1997). In Chapter 5 we shall test different limiters for the application to MHD problems. A comparison of different limiters has recently been given for several one- and two-dimensional gas dynamic problems, in particular, by Velichko, Lifshitz, and Solntsev (1999). They arranged five limiters in the order of increasing compressive properties (Yee 1989), i.e., in the order of increasing the accuracy and decreasing the smearing of discontinuities, as follows: the minmod limiter defined by (2.7.30), the UNO limiter (2.7.50), the van Leer limiter (2.7.47), which is numerically equivalent to (2.7.49) with $k = 1.4$, and, finally, the superbee limiter, see (2.7.49) for $k = 2$. This classification is not general and is problem-dependent. One may easily design and use his or her own limiters.

Since the locally linearized hyperbolic system can be interpreted as a set of independent advection equations for the Riemann invariants, it seems reasonable to limit the slopes (or increments) of the characteristic variables

$$\delta \mathbf{w} = \Omega_{\mathrm{L}} \delta \mathbf{U}. \tag{2.7.51}$$

If this method of calculation turns out to be insufficiently fast, then it should be applied only in the regions with large gradients.

2.7.5 TVD limiters of nonsymmetric stencil.

The limiters described above use only symmetric stencil in analysis of finite differences, since the function $L(a, b)$ of limiter definition (2.7.26) is symmetric. The construction of limiters with nonsymmetric stencils also makes sense. A nonsymmetric limiter permits one to take into account the sign of the flow velocity in the reconstruction of the function.

The basic properties (2.7.25) of a monotone reconstruction in the form of inequalities permit us to construct nonsymmetric TVD limiters. (By analogy, one can use also Eq. (2.7.37).)

Unlike (2.7.27)–(2.7.29), we now consider three consecutive inequalities of (2.7.25) for α_{m+1}, α_m, α_{m-1}, and α_{m-2}:

$$\begin{aligned}
&\Delta_m^- \le \tfrac{1}{2}(\alpha_{m+1} + \alpha_m) \le \Delta_m^+,\\
&\Delta_{m-1}^- \le \tfrac{1}{2}(\alpha_m + \alpha_{m-1}) \le \Delta_{m-1}^+,\\
&\Delta_{m-2}^- \le \tfrac{1}{2}(\alpha_{m-1} + \alpha_{m-2}) \le \Delta_{m-2}^+.
\end{aligned} \tag{2.7.52}$$

To close this system of inequalities it is convenient for us to set $\alpha_{m+1} = \alpha_m$ and $\alpha_{m-2} = \alpha_{m-1}$. Thus, we obtain

$$\begin{aligned}
&\Delta_m^- \le \alpha_m \le \Delta_m^+,\\
&\Delta_{m-1}^- \le \tfrac{1}{2}(\alpha_m + \alpha_{m-1}) \le \Delta_{m-1}^+,\\
&\Delta_{m-2}^- \le \alpha_{m-1} \le \Delta_{m-2}^+.
\end{aligned}$$

Whence,

$$\begin{aligned}
&\Delta_m^- \le \alpha_m \le \Delta_m^+,\\
&2\Delta_{m-1}^- - \alpha_{m-1} \le \alpha_m \le 2\Delta_{m-1}^+ - \alpha_{m-1},\\
&\alpha_{m-1} \le \Delta_{m-2}^+, \quad \Delta_{m-2}^- \le \alpha_{m-1}.
\end{aligned}$$

Hence,

$$\begin{aligned}
&\Delta_m^- \le \alpha_m \le \Delta_m^+,\\
&2\Delta_{m-1}^- \le \alpha_m + \Delta_{m-2}^+, \quad \alpha_m + \Delta_{m-2}^- \le 2\Delta_{m-1}^+.
\end{aligned}$$

Finally,

$$\max(\Delta_m^-,\ 2\Delta_{m-1}^- - \Delta_{m-2}^+) \le \alpha_m \le \min(\Delta_m^+,\ 2\Delta_{m-1}^+ - \Delta_{m-2}^-). \tag{2.7.53}$$

Using the fact that $\Delta_m^- \Delta_m^+ = 0$, we have

$$\begin{aligned}
&\alpha_m = \max(\Delta_m^-,\ 2\Delta_{m-1}^- - \Delta_{m-2}^+) + \min(\Delta_m^+,\ 2\Delta_{m-1}^+ - \Delta_{m-2}^-),\\
&\Delta_k = u_{k+1} - u_k, \quad k = m, m-1, m-2;\\
&\Delta^+ = \tfrac{1}{2}(\Delta + |\Delta|), \quad \Delta^- = \tfrac{1}{2}(\Delta - |\Delta|).
\end{aligned} \tag{2.7.54}$$

To obtain a TVD limiter, one should modify (2.7.54) as follows:

$$\begin{aligned}
\alpha_m = {}& \max[\Delta_m^-,\ 3\Delta_{m-1}^- - \min(0, \Delta_{m-1}^- + \Delta_{m-2}^+)]\\
&+ \min[\Delta_m^+,\ 3\Delta_{m-1}^+ - \max(0, \Delta_{m-1}^+ + \Delta_{m-2}^-)].
\end{aligned} \tag{2.7.55}$$

Considering three inequalities (2.7.25) for α_{m+2}, α_{m+1}, α_m, and α_{m-1} permits us to generate a three-difference TVD limiter:

$$\begin{aligned}
\alpha_m = {}& \max[\Delta_{m-1}^-,\ 3\Delta_m^- - \min(0, \Delta_m^- + \Delta_{m+1}^+)]\\
&+ \min[\Delta_{m-1}^+,\ 3\Delta_m^+ - \max(0, \Delta_m^+ + \Delta_{m+1}^-)].
\end{aligned} \tag{2.7.56}$$

By analogy, we can consider four consecutive inequalities of (2.7.25) for α_{m+1}, α_m, α_{m-1}, α_{m-2}, and α_{m-3} and obtain in a similar manner the following four-difference nonsymmetric TVD limiter:

$$\begin{aligned}\alpha_m = \max[\Delta_m^-,\ 3\Delta_{m-1}^- - \min(0, \Delta_{m-1}^- + 2\Delta_{m-2}^+ - \Delta_{m-3}^-)]\\ + \min[\Delta_m^+,\ 3\Delta_{m-1}^+ - \max(0, \Delta_{m-1}^+ + 2\Delta_{m-2}^- - \Delta_{m-3}^+)].\end{aligned} \tag{2.7.57}$$

Considering four consecutive inequalities of (2.7.25) for α_{m+3}, α_{m+2}, α_{m+1}, α_m, and α_{m-1} yields analogously

$$\begin{aligned}\alpha_m = \max[\Delta_{m-1}^-,\ 3\Delta_m^- - \min(0, \Delta_m^- + 2\Delta_{m+1}^+ - \Delta_{m+2}^-)]\\ + \min[\Delta_{m-1}^+,\ 3\Delta_m^+ - \max(0, \Delta_m^+ + 2\Delta_{m+1}^- - \Delta_{m+2}^+)].\end{aligned} \tag{2.7.58}$$

The limiters of (2.7.55) and (2.7.57) result in

$$-1 \le \psi_{m+1} - \psi_m \le 3, \qquad \psi_{m+1} = \frac{\alpha_{m+1}}{\Delta_m}, \qquad \psi_m = \frac{\alpha_m}{\Delta_m}.$$

Comparing these inequalities with (2.7.13), we arrive at the following stability condition for the scheme: $C \le \frac{2}{3}$. The limiters of (2.7.56) and (2.7.58) for advection equation (2.7.17), see Eq. (2.7.18), satisfy the same stability condition (Semenov, unpublished).

If we consider four consecutive inequalities of (2.7.25) for α_{m+2}, α_{m+1}, α_m, α_{m-1}, and α_{m-2}, we can obtain two four-difference symmetric TVD limiters with the stencil similar to that of the UNO limiter (2.7.50). For the advection equation (2.7.2) the limiter becomes

$$\begin{aligned}\alpha_m = \max[\Delta_m^-,\ 3\Delta_{m-1}^- - \min(0, \Delta_{m-1}^- + \Delta_{m-2}^+),\ 2\Delta_m^- - \Delta_{m+1}^+]\\ + \min[\Delta_m^+,\ 3\Delta_{m-1}^+ - \max(0, \Delta_{m-1}^+ + \Delta_{m-2}^-),\ 2\Delta_m^+ - \Delta_{m+1}^-]\end{aligned}$$

and the limiter for Eq. (2.7.17) becomes

$$\begin{aligned}\alpha_m = \max[\Delta_{m-1}^-,\ 2\Delta_{m-1}^- - \Delta_{m-2}^+,\ 3\Delta_m^- - \min(0, \Delta_m^- + \Delta_{m+1}^+)]\\ + \min[\Delta_{m-1}^+,\ 2\Delta_{m-1}^+ - \Delta_{m-2}^-,\ 3\Delta_m^+ - \max(0, \Delta_m^+ + \Delta_{m+1}^-)].\end{aligned}$$

2.7.6 Multidimensional reconstruction. So far, we have only considered the linear reconstruction for function of one space variable. Here, we will discuss some specific approaches for the 2D or 3D cases.

Consider the two-dimensional case. Suppose a discrete space mesh consisting of convex polygons $G_0, G_1, G_2, \dots$ is specified. Let a discrete function $\{u_m\}$ be defined on this mesh. Let the integer subscript m denote the values of grid variables calculated at the center of mass of G_m. Let G_n ($n = n_1, \dots, n_N$) be the neighbors of G_m. Then $\{u_{mn}\}$ denotes the values of grid variables at the middle of the boundary between the G_m and G_n; see Fig. 2.6. Figure 2.6 presents two cases: (a) a general convex quadrangular cell G_0 with four neighbors G_1, G_2, G_3, and G_4, and (b) a triangular cell G_0 with three neighbors G_1, G_2, and G_3.

In order to determine the values of the function u at the sides of the polygon, we must first define a reconstruction procedure. Suppose we deal with piecewise linear distributions inside the cells (polygons)

$$u(x, y) = u_m + (x - x_m)\alpha_m + (y - y_m)\beta_m, \tag{2.7.59}$$

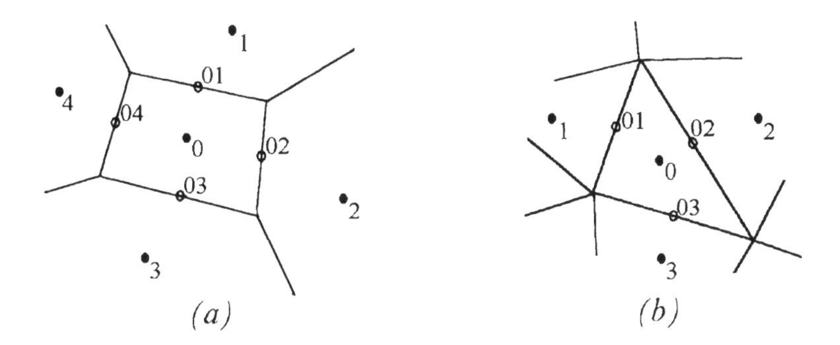

Figure 2.6 Quadrangular and triangular cells.

where the α_m and β_m are some coefficients and the coordinates (x_m, y_m) correspond to the center of mass of the mth cell. Hence,

$$u_m = \frac{1}{|G_m|} \iint_{G_m} u(x, y)\, dx\, dy, \quad |G_m| = \iint_{G_m} dx\, dy,$$

where $|G_m|$ is the area of G_m. The value of the grid variable at the middle of the mnth boundary may be determined as

$$u_{mn} = u(x_{mn}, y_{mn}) = u_m + (x_{mn} - x_m)\alpha_m + (y_{mn} - y_m)\beta_m, \tag{2.7.60}$$

where the coordinates (x_{mn}, y_{mn}) correspond to the middle of the boundary between G_m and G_n.

Consider how one can determine the coefficients α_m and β_m. The following conditions must be satisfied:

- $\alpha_m = \alpha$ and $\beta_m = \beta$ for a linear function $u(x, y) \equiv u_0 + \alpha x + \beta y$;
- the coefficients α_m and β_m must be determinable for arbitrary meshes;
- the coefficients α_m and β_m or the function increments, which are determined by these coefficients, must be limited so that in the one-dimensional case we obtain a TVD scheme.

In particular, Barth and Jespersen (1989) suggested limiting α_m and β_m so as to provide the minmax inequalities

$$u_{\min} \le u_m + (x - x_m)\alpha_m + (y - y_m)\beta_m \le u_{\max}, \quad (x, y) \in G_m, \tag{2.7.61}$$
$$u_{\min} = \min\left(u_m, \min_n u_n\right), \quad u_{\max} = \max\left(u_m, \max_n u_n\right), \quad n = n_1, \ldots, n_N.$$

For linear reconstruction, extrema in $u(x, y)$ occur at the vertices of the faces and sufficient conditions can be easily obtained. Compute the values u_i for each ith vertex ($i = 1, \ldots N$) of the mth polygon G_m and determine

$$\psi_i = \begin{cases} \min\left(1, \dfrac{u_{\max} - u_m}{u_i - u_m}\right) & \text{if} \quad u_i - u_m > 0, \\ \min\left(1, \dfrac{u_{\min} - u_m}{u_i - u_m}\right) & \text{if} \quad u_i - u_m < 0, \\ 1 & \text{if} \quad u_i - u_m = 0, \end{cases}$$

and $\psi = \min(\psi_1, \ldots, \psi_N)$. Then the limited coefficients $\psi\alpha_m$ and $\psi\beta_m$ satisfy inequalities (2.7.61).

For two- or three-dimensional cases, there are various definitions of the variation of the function, for example, the variations by Arzelá, Frechet, Hahn, Tonelli, Vitali and others, see Vitushkin (1955), and Golubov and Vitushkin (1977). This is accounted for by the the fact that multivariable functions can be characterized by several independent variations (Vitushkin 1955). In particular, a function $u(x, y)$ can be characterized by two variations, the so-called linear and plane variations. No sufficient conditions ensuring the boundedness of all variations of the numerical solution for an arbitrary mesh exist by now. For this reason, the methods of multidimensional linear reconstruction known so far, including those outlined here, do not have rigorous backgrounds. Hence, the applicability of these methods should be checked for each particular problem.

Note that for a regular mesh, all one-dimensional algorithms of linear function reconstruction described above can be used for each coordinate axis independently.

Tillyaeva (1986) suggested the following method of determining the slopes α and β in the two-dimensional case. Consider for simplicity a curvilinear quadrangular cell G_0 with four neighbors (Fig. 2.6a). Let us solve the following four systems of linear equations for α_0 and β_0:

$$u_0 + (x_i - x_0)\alpha_0^{(i)} + (y_i - y_0)\beta_0^{(i)} = u_i, \tag{2.7.62}$$

$$u_0 + (x_{i+1} - x_0)\alpha_0^{(i)} + (y_{i+1} - y_0)\beta_0^{(i)} = u_{i+1},$$

where $i = 1, 2, 3, 4$. For $i = 4$ we assume that $i + 1 \equiv 1$. Let

$$\delta u_{0n}^{(i)} = (x_{0n} - x_0)\alpha_0^{(i)} + (y_{0n} - y_0)\beta_0^{(i)}, \qquad i = 1, 2, 3, 4; \quad n = 1, 2, 3, 4.$$

In determining u_{0n}, the increments $\delta u_{0n}^{(i)}$ can be limited in the following way:

$$u_{0n} = u_0 + \operatorname{minmod}\left(\delta u_{0n}^{(1)}, \delta u_{0n}^{(2)}, \delta u_{0n}^{(3)}, \delta u_{0n}^{(4)}\right), \qquad n = 1, 2, 3, 4;$$

where

$$\operatorname{minmod}(a_1, \ldots, a_n) = \begin{cases} \min(|a_1|, \ldots, |a_n|) \operatorname{sgn} a_1 & \text{if } \operatorname{sgn} a_1 = \ldots = \operatorname{sgn} a_n, \\ 0 & \text{otherwise.} \end{cases}$$

For simplicity, we shall also use the notation

$$\operatorname{minmod}_n(a_n) = \operatorname{minmod}(a_1, \ldots, a_n).$$

In the two-dimensional gas dynamic simulation of stationary problems by Godunov's method with the exact solver, this approach permits one to preserve monotone profiles of the numerical solution and to provide weak sensitivity of numerical results to the geometry of the discrete space mesh (Tillyaeva 1986). For rectangular meshes this approach is equivalent to application of the minmod limiter along x- and y-direction independently. A more detailed analysis of the increments can be found in Rodionov (1996).

Consider some modifications of this approach. Note that the analysis of only two increments can be often sufficient for reliable gas dynamic simulation (Ivanov and Kryukov 1996), i.e.,

$$u_{0n} = u_0 + \text{minmod}\left(\delta u_{0n}^{(1)}, \delta u_{0n}^{(3)}\right), \quad \text{or} \quad u_{0n} = u_0 + \text{minmod}\left(\delta u_{0n}^{(2)}, \delta u_{0n}^{(4)}\right).$$

Instead of limiting the increments, one can limit the slopes; namely,

$$\begin{aligned}
&\text{(i) } \alpha_0 = \text{minmod}\left(\alpha_0^{(1)}, \alpha_0^{(2)}, \alpha_0^{(3)}, \alpha_0^{(4)}\right), \qquad && \beta_0 = \text{minmod}\left(\beta_0^{(1)}, \beta_0^{(2)}, \beta_0^{(3)}, \beta_0^{(4)}\right); \\
&\text{(ii) } \alpha_0 = \text{minmod}\left(\alpha_0^{(1)}, \alpha_0^{(3)}\right), && \beta_0 = \text{minmod}\left(\beta_0^{(1)}, \beta_0^{(3)}\right); \\
&\text{(iii) } \alpha_0 = \text{minmod}\left(\alpha_0^{(2)}, \alpha_0^{(4)}\right), && \beta_0 = \text{minmod}\left(\beta_0^{(2)}, \beta_0^{(4)}\right).
\end{aligned}$$

Note that other limiters can be used instead of the minmod limiter.

By analogy with the one-dimensional limiting reconstruction, a two-dimensional limiting reconstruction can be implemented. In this case the slopes can be sequentially increased in an iterative process analogous to (2.7.34)–(2.7.36). Suppose one can solve the following inequalities for α_m and β_m:

$$\begin{aligned}
u_m &\le u_m + (x_{mn} - x_m)\alpha_m + (y_{mn} - x_m)\beta_m \\
&\le u_n + (x_{mn} - x_n)\alpha_n + (y_{mn} - x_n)\beta_n \le u_n.
\end{aligned} \tag{2.7.63}$$

Note that they preserve the monotonicity of a function being reconstructed at the middle of the cell boundaries. If $u_m \ge u_n$, the sign of the inequality is reversed. From (2.7.63) one can find that

$$\begin{aligned}
u_m &\le u_m + \text{minmod}_n\left[(x_{mn} - x_m)\alpha_m + (y_{mn} - x_m)\beta_m\right] \\
&\le u_n + \text{minmod}_m\left[(x_{mn} - x_n)\alpha_n + (y_{mn} - x_n)\beta_n\right] \le u_n,
\end{aligned}$$

or

$$\begin{aligned}
&u_m \le \tilde{u}_m \le \tilde{u}_n \le u_n, \\
&\tilde{u}_k = u_k - \text{minmod}_i[(x_{ki} - x_k)\alpha_k + (y_{ki} - x_k)\beta_k], \qquad k = m, n.
\end{aligned}$$

The most accurate reconstruction is attained for the largest $|\alpha_m|$ and $|\beta_m|$ for which inequalities (2.7.63) are satisfied. We shall therefore try to refine the α_m and β_m. Assume that $\tilde{\alpha}_m = \alpha_m + \delta\alpha_m$ and $\tilde{\beta}_m = \beta_m + \delta\beta_m$ satisfy (2.7.63). Then $\delta\alpha$ and $\delta\beta$ satisfy

$$\begin{aligned}
\tilde{u}_m &\le \tilde{u}_m + (x_{mn} - x_m)\delta\alpha_m + (y_{mn} - x_m)\delta\beta_m \\
&\le \tilde{u}_n + (x_{mn} - x_n)\delta\alpha_n + (y_{mn} - x_n)\delta\beta_n \le \tilde{u}_n.
\end{aligned}$$

These inequalities are analogous to those of (2.7.63). Then, one can repeat the procedure for determining the slopes $\delta\alpha_m$ and $\delta\beta_m$ by using the values of $\tilde{u}$ and, hence, refine α_m and β_m. This sort of refining procedure can be used in other methods for slope determination.

Note that the slopes can be found also by using an approximation of contour integrals (Barth and Jespersen 1989) and/or the least squares method (Vankeirsbilck and Deconinck 1992). Consider the gradient theorem in the general case:

$$\iiint_G \nabla u \, dG = \oint_S u \, d\mathbf{S} \quad \Longrightarrow \quad \nabla u \approx \frac{1}{|G|} \oint_S u \, d\mathbf{S}, \tag{2.7.64}$$

where G is a domain in Euclidian space $\mathbf{E}^k$, $dG = dx_1 \dots dx_k$ is the volume element, $|G|$ is the volume of G, $\nabla = [\partial/\partial x_1, \dots, \partial/\partial x_k]^{\mathrm{T}}$, S is the boundary surface of G, $d\mathbf{S} = \mathbf{n}\, dS$ is the oriented element of S, where $\mathbf{n}$ is the outward normal to S, and dS is the area element. In particular, in the two-dimensional case for quadrangular cell G_0, we can obtain

$$\begin{aligned}
(u_x)_0 = \alpha_0^C &= \frac{1}{|G|} \oint_S u\, dy \approx \frac{1}{|G|} \big[\tfrac{1}{2}(u_1 + u_2)(y_1 - y_2) + \tfrac{1}{2}(u_4 + u_1)(y_4 - y_1) \\
&\quad + \tfrac{1}{2}(u_3 + u_4)(y_3 - y_4) + \tfrac{1}{2}(u_2 + u_3)(y_2 - y_3) \big] \\
&= \frac{1}{2|G|} [(u_1 - u_3)(y_4 - y_2) - (u_4 - u_2)(y_1 - y_3)], \\
(u_y)_0 = \beta_0^C &= -\frac{1}{|G|} \oint_S u\, dx \approx \frac{1}{2|G|} [(u_4 - u_2)(x_1 - x_3) - (u_1 - u_3)(x_4 - x_2)], \\
|G| &= \oint_S x\, dy = -\oint_S y\, dx = \tfrac{1}{2}[(x_1 - x_3)(y_4 - y_2) - (x_4 - x_2)(y_1 - y_3)],
\end{aligned}$$

where G is a quadrangular region with vertices at the points 1, 2, 3, and 4; see Fig. 2.6a. Otherwise, the slopes α_0^C and β_0^C can be chosen so that the sum

$$\sum_{n=1}^{4} \left[u_0 + (x_n - x_0)\alpha_0^C + (y_n - y_0)\beta_0^C - u_n \right]^2$$

attains minimum. Note that these two versions of the slope determination give similar results (Ivanov and Kryukov 1996). Other combinations of neighbors can also be used to evaluate the slopes.

In determining u_{0n}, we can limit the increments in the following way:

$$u_{0n} = u_0 + \operatorname{minmod}\left(2\delta u_{0n}^{(1)}, 2\delta u_{0n}^{(3)}, \delta u_{0n}^C \right),$$

where

$$\delta u_{0n}^C = (x_{0n} - x_0)\alpha_0^C + (y_{0n} - y_0)\beta_0^C.$$

For rectangular meshes this reconstruction was first suggested by van Leer (1977b).

The two-dimensional reconstruction for a general convex quadrangular cell with variables located at the nodes of the grid can be found in Borrel and Montagne (1985).

In a two-dimensional simulation based on triangular cells (see Fig. 2.6b) Durlofsky, Engquist, and Osher (1992) determined the slopes α_0 and β_0 by the solution of the three systems of linear equations (2.7.62)–(2.7.63). It was assumed that $i + 1 \equiv 1$ for $i = 3$. The final values $\alpha_0 = \alpha_0^{(i)}$ and $\beta_0 = \beta_0^{(i)}$ must be chosen so that $[\alpha_0^{(i)}]^2 + [\beta_0^{(i)}]^2$ be minimal. At extrema (i.e., when u_0 is maximal or minimal among u_0, u_1, u_2, and u_3) α_0 and β_0 become zero. Durlofsky et al. (1992) suggested also a more compressive limiter similar to the one-dimensional superbee limiter.

Note that we can also determine the slopes α_{0n} and β_{0n} at the middle of boundaries rather than the centers of cells. Then, one can set

$$\alpha_0 = \operatorname{minmod}(\alpha_{01}, \dots, \alpha_{0n}), \quad \beta_0 = \operatorname{minmod}(\beta_{01}, \dots, \beta_{0n}). \tag{2.7.65}$$

The two-dimensional approaches described above can be easily generalized to the three-dimensional case.

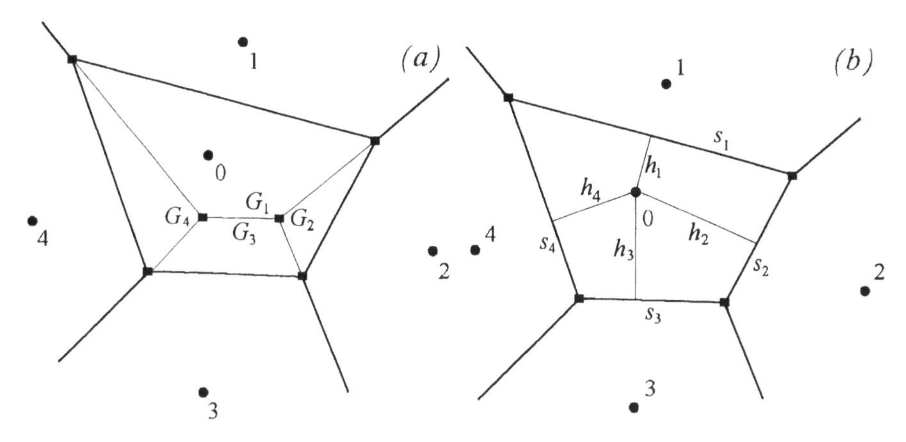

Figure 2.7 An example of a Dirichlet cell.

To determine a linear reconstruction of a function for both unstructured meshes of arbitrary convex polygons and arbitrary systems of points in accordance with (2.7.65), the Sibsonian (Sibson 1980, 1981) and harmonic (which is not Sibsonian) (Belikov et al. 1997; Belikov and Semenov 1997a) first-order interpolations can also be used. In particular, an arbitrary system of points arises in using the free-Lagrange method (Fritts, Crowley, and Trease 1988), particle methods (Brackbill and Monaghan 1988), meshless methods (Belytshko et al. 1996), and others. The two interpolations are rather sophisticated and based on a decomposition of the plane (space) into Dirichlet cells (polyhedra). However, these interpolations guarantee the uniqueness and continuity of the interpolation results and their stability with respect to small geometric perturbations. For a given node $\mathbf{x}_0$ the Dirichlet cell (Fig. 2.7) is the union of points from which the distance to the point $\mathbf{x}_0$ is less than to any other node. Just for the sake of illustration, consider one simple method for constructing these cells; unfortunately, it is too slow to be used in applications—the corresponding number of operations is $O(M^2)$, where M is the number of points. The approach is this: if we link the point $\mathbf{x}_0$ with all points $\mathbf{x}_i$ ($i \neq 0$) by segments and erect a perpendicular (or a hyperplane in $\mathbf{E}^k$) at the center of each segment on both sides, then the resulting convex polygon (polyhedron) is a Dirichlet cell. There are faster algorithms of the Dirichlet decomposition based on the Delaunay triangulation (George and Borouchaki 1998; Baker 1999); see, for example, Bowyer's method (Bowyer 1981). The Dirichlet cells divide the entire space into convex polygons (polyhedra), each containing only one of the nodes. The gradient theorem (2.7.64) permits one to determine the space derivatives at an arbitrary point $\mathbf{x}_0$. Specifically,

$$(\nabla u)_0 \approx \frac{1}{|G_0|} \oint_{S_0} u \, d\mathbf{S} = \frac{1}{|G_0|} \sum_{n=1}^{N} (\mathbf{x}_n - \mathbf{x}_0) \frac{s_n \widetilde{u}_n}{2h_n},$$

where $|G_0|$ is the volume of the Dirichlet cell corresponding to $\mathbf{x}_0$, and the $\mathbf{x}_n$, $n = 1, \ldots, N$, are the Dirichlet neighbors of $\mathbf{x}_0$. Here we denote the lengths of the sides (or the areas of the faces in the three-dimensional case) by s_n, $n = 1, \ldots, N$, and the lengths of the altitudes from $\mathbf{x}_0$ to the nth side (i.e., the distances from $\mathbf{x}_0$ to the nth face) by h_n (Fig. 2.7b).

The values $\widetilde{u}_n$ are calculated at the middle of the cell faces in accordance with the Sibsonian or harmonic non-Sibsonian interpolation. Such a procedure of determining the derivatives is exact for any linear function. Apart from second-order terms, we can also set $\widetilde{u}_n \approx \frac{1}{2}(u_n + u_0)$.

In the Sibsonian interpolation, the coefficients are set proportional to the areas (volumes) G_n of the polygons (polyhedra) that are cut out of the Dirichlet cell for $\mathbf{x}_0$ by the Dirichlet cells that are constructed for the corresponding neighbors of $\mathbf{x}_0$ in the absence of $\mathbf{x}_0$ (Fig. 2.7a):

$$u_0 = \sum_{n=1}^{N} \gamma_n u_n, \qquad \gamma_n = |G_n| \Big/ \sum_{j=1}^{N} |G_j|, \quad n = 1, \ldots, N.$$

A rigorous definition of the Sibsonian interpolation and discussion of its properties can be found in Sibson (1980, 1981) and Farin (1990).

The harmonic interpolation is based also on determining the neighbors through decomposition into Dirichlet cells,

$$u_0 = \sum_{n=1}^{N} \gamma_n u_n, \qquad \gamma_n = \frac{s_n}{h_n} \Big/ \sum_{j=1}^{N} \frac{s_j}{h_j}, \qquad n = 1, 2, \ldots, N.$$

This method of calculating γ_n is simpler and more economical than that employed in the Sibson method, since it does not require computing the intersection areas of polygons in the two-dimensional case or the intersection volumes of polyhedra in the three-dimensional case. A rigorous definition of the harmonic interpolation can be found in Belikov et al. (1997), and Belikov and Semenov (1997a), and several applications can be found in Belikov and Semenov (2000) and Sukumar et al. (2000).

Worth mentioning is the approach to determining the slopes that differs from the methods outlined above. This approach was used in the one- and two-dimensional shock-capturing and shock-fitting gas dynamic simulations by Grudnitskii and Prokhorchuk (1977), Belotserkovskii and Grudnitskii (1980), Belotserkovskii, Grudnitskii, and Prokhorchuk (1983). To describe the slope evolution, they used an extended system of equations (Courant and Lax 1949); see also Section 1.2.1. Consider the gas dynamic equations in the general form

$$\frac{\partial \mathbf{U}}{\partial t} + \frac{\partial \mathbf{F}(\mathbf{U})}{\partial x} + \frac{\partial \mathbf{E}(\mathbf{U})}{\partial y} = \mathbf{0}. \tag{2.7.66}$$

Differentiating Eqs. (2.7.66) with respect to x and y yields two systems that are also hyperbolic,

$$\frac{\partial \mathbf{U}_x}{\partial t} + \frac{\partial A\mathbf{U}_x}{\partial x} + \frac{\partial B\mathbf{U}_x}{\partial y} = \mathbf{0}, \qquad \mathbf{U}_x = \frac{\partial \mathbf{U}}{\partial x}, \quad \mathbf{U}_y = \frac{\partial \mathbf{U}}{\partial y}, \tag{2.7.67}$$

$$\frac{\partial \mathbf{U}_y}{\partial t} + \frac{\partial A\mathbf{U}_y}{\partial x} + \frac{\partial B\mathbf{U}_y}{\partial y} = \mathbf{0}, \tag{2.7.68}$$

where $A = \partial\mathbf{F}/\partial\mathbf{U}$ and $B = \partial\mathbf{E}/\partial\mathbf{U}$. Further, the nonstationary systems (2.7.66)–(2.7.68) are solved for the variables $\mathbf{U}$, $\mathbf{U}_x$, and $\mathbf{U}_y$. Note that for the solution of Eq. (2.7.66) a special second-order modification of the Lax–Friedrichs scheme (Lax 1954) taking into account the function slopes $\mathbf{U}_x$ and $\mathbf{U}_y$ was used (Grudnitskii and Prokhorchuk 1977). It is obvious that genuine Godunov schemes can be also be applied.

This approach proves to be fruitful in shock-fitting techniques. In particular, it permits one to capture newly arising discontinuities. Weak discontinuities can also be captured, since discontinuities in derivatives can be clearly seen in this approach (Belotserkovskii, Grudnitskii, and Rygalin 1983; Azarova et al. 1993a, 1993b). As regards the shock-capturing method for solving the equations of (2.7.67)–(2.7.68), one has to artificially diminish large values of $\mathbf{U}_x$ and/or $\mathbf{U}_y$ in the vicinity of a shock wave. However, near contact surfaces, rarefaction waves, and in subdomains of smooth flow, one need not use any artificial corrections in $\mathbf{U}_x$ and/or $\mathbf{U}_y$. By taking into account the system for $\mathbf{U}_z$, one can generalize the approach for the three-dimensional case.

2.8 Boundary conditions for hyperbolic systems

2.8.1 General notions. Consider an initial boundary value problem for the hyperbolic system in the form

$$\frac{\partial \mathbf{U}}{\partial t} + \frac{\partial \mathbf{F}}{\partial x} = \mathbf{H}, \tag{2.8.1}$$

where $\mathbf{U}$ is the vector of conservative variables and $\mathbf{F}$ is the flux vector. $\mathbf{H}$ is the source term that does not contain derivatives. This system can also be rewritten in the canonical quasilinear form

$$\frac{\partial \mathbf{U}}{\partial t} + A\frac{\partial \mathbf{U}}{\partial x} = \mathbf{H}, \tag{2.8.2}$$

where A is a $n \times n$ matrix. As the system of (2.8.1) is hyperbolic, we have

$$\mathbf{l}^i A = \lambda_i \mathbf{l}^i, \quad A\mathbf{r}^i = \lambda_i \mathbf{r}^i, \quad i = 1, \ldots, n,$$

where λ_i are the eigenvalues of A which are ordered so that

$$\lambda_1 \leq \lambda_2 \leq \ldots \leq \lambda_n.$$

The characteristic equations have the form

$$\Omega_{\mathrm{L}}\frac{\partial \mathbf{U}}{\partial t} + \Lambda\Omega_{\mathrm{L}}\frac{\partial \mathbf{U}}{\partial x} = \Omega_{\mathrm{L}}\mathbf{H} \tag{2.8.3}$$

or

$$\mathbf{l}^i \cdot \left(\frac{\partial \mathbf{U}}{\partial t} + \lambda_i \frac{\partial \mathbf{U}}{\partial x}\right) = \mathbf{l}^i \cdot \mathbf{H}. \tag{2.8.4}$$

In many cases we need to find the solution of Eq. (2.8.1) in a bounded region of the independent variable x. In this case proper physical boundary conditions must be stated. On the other hand, if discontinuity-fitting methods (to be described in Section 2.9) are used, the boundary conditions are also required on discontinuities that move with unknown velocities. Statement of boundary conditions at the fixed boundaries of the computational region usually does not represent mathematical difficulties, although their numerical implementation can be rather complicated. As such fixed boundaries we can consider, for example, solid walls and entrance or exit boundaries (Hirsch 1990).

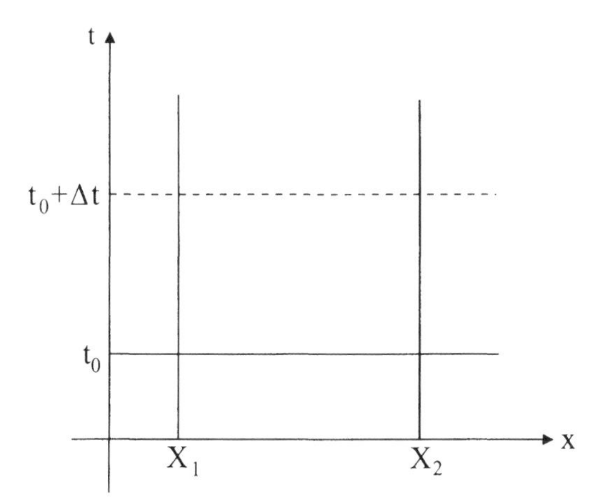

Figure 2.8 Computational region.

At first, let us determine the number of conditions necessary to obtain a unique solution in the vicinity of the fixed boundary $x = X_2$ (Fig. 2.8). We calculate the functions U_i for $X_1 < x < X_2$ and their values at $t = t_0$ are given. System (2.8.3), though not reducible to the Riemann invariants, can however be rewritten for the Riemann variable $\mathbf{w}$ defined as

$$\delta \mathbf{w} = \Omega_{\mathrm{L}} \delta \mathbf{U} - \Omega_{\mathrm{L}} \mathbf{H} dt. \tag{2.8.5}$$

Although the nonlinear system (2.8.5) cannot in general be resolved for $\mathbf{w}$, the obtained system

$$\frac{\partial \mathbf{w}}{\partial t} + \Lambda \frac{\partial \mathbf{w}}{\partial x} = \mathbf{0} \tag{2.8.6}$$

is very convenient for the characteristic analysis.

Let us linearize the system of (2.8.6) by freezing the entries of the diagonal matrix Λ at the point $(X_2, t_0 + \Delta t)$. The eigenvalues occurring in Λ can be both positive and negative at this point. The solution of Eq. (2.8.6) is a linear combination of travelling waves, and we can state that the equations corresponding to $\lambda_i \geq 0$ in (2.8.6) describe the waves arriving at the point under consideration from the interior region. Let the number of such equations be equal to s. Then, to obtain a unique solution at the point $(X_2, t_0 + \Delta t)$, we must specify $n - s$ boundary conditions. They are determined by the physical formulation of the problem.

2.8.2 Nonreflecting boundary conditions. In a number of cases we know the solution behavior only for $x \ll X_1$ or for $x \gg X_2$. In this case we must either find a way to transfer the quantities from infinity to our boundary or, if the above is not possible, to damp the waves incoming to the computational region. It is clear that the waves are incoming for $\lambda_i \geq 0$ at $x = X_1$ and for $\lambda_i \leq 0$ at $x = X_2$. The rest of the waves are outgoing.

The boundary conditions that damp or eliminate incoming waves are called nonreflecting, absorbing, or radiation conditions. Bayliss and Turkel (1982) and Engquist and Majda (1977) studied such conditions for linear hyperbolic systems. Hedstrom (1979) suggested

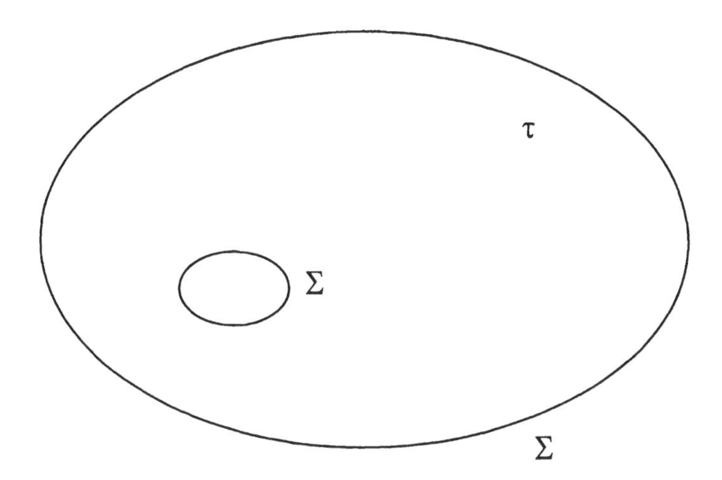

Figure 2.9 Multidimensional computational region.

the boundary conditions for a nonlinear system with two independent variables, which turned out to be very efficient for time-dependent problems. They were later generalized for multidimensional problems by Thompson (1987, 1990). Worth noting is also the application of Thompson's approach to MHD equations by Sun, Wu, and Dryer (1995). The essence of this approach can be stated as follows: the amplitudes of incoming waves are assumed to be constant in time at the boundaries. Mathematically this means that for incoming waves

$$\left.\frac{\partial \mathbf{w}}{\partial t}\right|_{x=X_1,X_2} = \mathbf{0} \tag{2.8.7}$$

or

$$\left.\left(\mathbf{l}^i \cdot \frac{\partial \mathbf{U}}{\partial t} - \mathbf{l}^i \cdot \mathbf{H}\right)\right|_{x=X_1,X_2} = 0. \tag{2.8.8}$$

The system of equations that combines Eqs. (2.8.4) and (2.8.8) reads

$$\left.\left(\mathbf{l}^i \cdot \frac{\partial \mathbf{U}}{\partial t} + \mathcal{L}_i - \mathbf{l}^i \cdot \mathbf{H}\right)\right|_{x=X_1,X_2} = 0, \tag{2.8.9}$$

where

$$\mathcal{L}_i = \begin{cases} \lambda_i \mathbf{l}^i \cdot \dfrac{\partial \mathbf{U}}{\partial x} & \text{for outgoing waves,} \\ 0 & \text{for incoming waves.} \end{cases}$$

In the numerical procedure we can approximate the x-derivative in Eq. (2.8.9) by backward differences.

In the multidimensional case we solve the initial boundary value problem for $\mathbf{U}(\mathbf{x}, t)$ in the volume τ bounded by the surface Σ (Fig. 2.9). The solution depends in this case (i) on the system of equations, (ii) on the initial values of $\mathbf{U}$ in τ, and (iii) on the time-dependent boundary conditions on Σ. If we rewrite (2.8.1) on the surface Σ in the form

$$\frac{\partial \mathbf{U}}{\partial t} + \frac{\partial \mathbf{F}_1}{\partial x_1} = \mathbf{H}, \tag{2.8.10}$$

or

$$\frac{\partial \mathbf{U}}{\partial t} + A\frac{\partial \mathbf{U}}{\partial x_1} = \mathbf{H}, \quad A = \frac{\partial \mathbf{F}_1}{\partial \mathbf{U}}, \tag{2.8.11}$$

where x_1 is directed perpendicular to Σ, then the procedure described earlier for two independent variables can be generalized. Note that in the multidimensional case $\mathbf{H}$ contains not only nondifferential terms but also derivatives in the direction tangent to the boundary. On diagonalizing the Jacobian matrix $\partial \mathbf{F}_1/\partial \mathbf{U}$ and multiplying Eq. (2.8.11) by the corresponding left eigenvector matrix Ω_{L}, we obtain

$$\Omega_{\mathrm{L}}\frac{\partial \mathbf{U}}{\partial t} + \Lambda\Omega_{\mathrm{L}}\frac{\partial \mathbf{U}}{\partial x_1} = \Omega_{\mathrm{L}}\mathbf{H}. \tag{2.8.12}$$

If we introduce the vector

$$\mathbf{L} = \Lambda\Omega_{\mathrm{L}}\frac{\partial \mathbf{U}}{\partial x_1}, \tag{2.8.13}$$

Eq. (2.8.12) acquires the form

$$\Omega_{\mathrm{L}}\frac{\partial \mathbf{U}}{\partial t} + \mathbf{L} = \Omega_{\mathrm{L}}\mathbf{H}.$$

Once the vector $\mathbf{L}$ is found on the boundary surface Σ, we can calculate the time derivative of the unknown vector on it:

$$\left(\frac{\partial \mathbf{U}}{\partial t}\right)_{\Sigma} = (\mathbf{H} - \Omega_{\mathrm{R}}\mathbf{L})_{\Sigma}. \tag{2.8.14}$$

For outgoing waves the components L_i are calculated by the formula (2.8.13) using backward differences, whereas for incoming waves they are put to zero, according to the assumption that the amplitudes of the latter waves remain constant in time.

Slightly different formulation of nonreflecting boundary conditions was suggested by Rizzi (1982) (see also Rizzi and Eriksson 1984). It is similarly based on the simplest local conditions that can be obtained from the theoretical nonlocal boundary conditions of Engquist and Majda (1977). In this approach the system is linearized at the point P of the boundary in a vicinity of the state known at the previous time level. Then the characteristic form is considered of the corresponding one-dimensional problem,

$$\frac{\partial \mathbf{U}}{\partial t} + A\frac{\partial \mathbf{U}}{\partial x_1} = \mathbf{0} \tag{2.8.15}$$

(note that x_1 is the coordinate perpendicular to the boundary at the point P). Since the system is linear, it can be rewritten for the Riemann invariants $\mathbf{w} = \Omega_{\mathrm{L}}\mathbf{U}$,

$$\frac{\partial \mathbf{w}}{\partial t} + \Lambda\frac{\partial \mathbf{w}}{\partial x_1} = \mathbf{0}, \quad \Lambda = \Omega_{\mathrm{L}}A\Omega_{\mathrm{R}}. \tag{2.8.16}$$

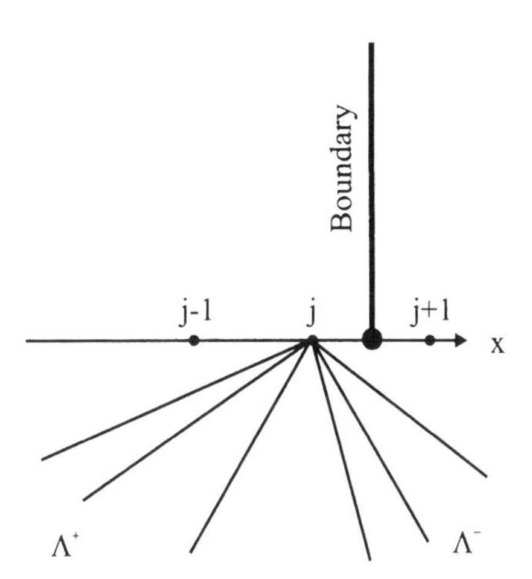

Figure 2.10 Calculation near the boundary.

Consider for definiteness the exit boundary. In this case the components of the vector $\mathbf{w}$ corresponding to the incoming waves are transferred to the point P from infinity. The components corresponding to the outgoing waves can be extrapolated to the boundary from the interior points. On determining $\mathbf{w}$ at the boundary, we can reconstruct the vector of unknown variables as $\mathbf{U} = \Omega_{\mathrm{R}}\mathbf{w}$.

We shall illustrate this approach using a simple upwind scheme of the first order of accuracy. Let us introduce the grid functions $\mathbf{w}_j^n = \mathbf{w}(x_j, t^n)$ with $x_j = j\,\Delta x$ and $t^n = n\,\Delta t$. On introducing the matrices

$$\Lambda^+ = \tfrac{1}{2}(\Lambda + |\Lambda|), \quad \Lambda^- = \tfrac{1}{2}(\Lambda - |\Lambda|), \tag{2.8.17}$$

the characteristics in the vicinity of the grid point $P\,(x_j, t^n)$ will be distributed as shown in Fig. 2.10. Hence, a characteristically consistent numerical scheme has the form

$$\mathbf{w}_j^{n+1} = \mathbf{w}_j^n - \frac{\Delta t}{\Delta x}\left[\Lambda^+\left(\mathbf{w}_j^n - \mathbf{w}_{j-1}^n\right) + \Lambda^-\left(\mathbf{w}_{j+1}^n - \mathbf{w}_j^n\right)\right]. \tag{2.8.18}$$

The matrix $|\Lambda|$ by definition is a diagonal matrix formed by the eigenvalue moduli.

If the boundary is located at the point $x_{j+1/2}$, then the values $\mathbf{w}_j$ and $\mathbf{w}_{j-1}$ refer to the interior points, while $\mathbf{w}_{j+1}$ corresponds to the exterior point. If we assume that the functions remain constant within a computational cell, which is common for the first-order schemes, we can put

$$\begin{aligned} (w_i)_{j+1} &= (w_i)_j && \text{for } \lambda_i \geq 0, \\ (w_i)_{j+1} &= (w_i)_\infty && \text{for } \lambda_i < 0. \end{aligned} \tag{2.8.19}$$

Here $(w_i)_\infty$ refers to the value of the Riemann invariant at infinity. Thus, we can simply put

$$\mathbf{w}_{j+1} = \mathbf{w}_\infty, \tag{2.8.20}$$

since our numerical scheme deals only with those components of $\mathbf{w}_{j+1}$ that correspond to negative eigenvalues.

As mentioned earlier, we must first linearize the system in the vicinity of the boundary point P to manipulate later with Riemann invariants. It is obvious that the application of formulas (2.8.19) to nonlinear systems can lead to large numerical errors unless the numerical boundary is placed at such a large distance from the interaction region that it can be considered as located at infinity (see, e. g., calculations of the solar wind interaction with the magnetized interstellar medium by Ratkiewicz et al. 1998).

The described approach was modified by Grin' et al. (1981), Mileshin and Tillyaeva (1982), and Sawada et al. (1986). The general idea is to freeze the matrix A in Eq. (2.8.15) not at the point P of the interior region but using the quantities obtained from the solution of the one-dimensional Riemann problem with initial conditions consisting of the values at infinity and at the cell adjacent to the boundary.

Another approach originates from the Osher numerical scheme, though it will become obvious later on that its implementation in gas dynamics and MHD problems does not necessarily require the application of Osher's formulas. The rarefaction wave solution in analytical form is, in fact, not available for MHD equations.

Suppose we want to formulate the boundary conditions for the right exit boundary (see Fig. 2.10). In the framework of finite-volume methods, we would like to find the numerical flux $\mathbf{F}_{j+1/2}$ at this boundary. The values of $\mathbf{U}$ at the point $j + 1$ are generally unknown. We know them, however, at infinity and can calculate the signs of the eigenvalues $(\lambda_i)_\infty$. Suppose $(\lambda_m)_\infty$ is the smallest positive eigenvalue for which $(\lambda_m)_j < 0$. Then we can evaluate the boundary flux by the formula (compare with Eq. (2.3.78))

$$\mathbf{F}_{j+1/2} = \mathbf{F}(\mathbf{U}_j) + \int_{\mathbf{U}_j}^{\mathbf{U}_b} A^-(\mathbf{U})d\mathbf{U}, \tag{2.8.21}$$

where $\mathbf{U}_b$ corresponds to the point on the segment Γ_m of the path connecting $\mathbf{U}_j$ and $\mathbf{U}_{j+1}$ for which $\lambda_m = 0$. This allows us to determine the flux on the boundary using the quantities in the inner cells. This approach provides the maximum possible information about the boundary values using only internal points. This formulation implies that we must use the variant of the Osher scheme with a natural choice of the segment distribution in which the segment Γ_1^j emanates from $\mathbf{U}_j$ and the path terminates by the segment Γ_n^j at $\mathbf{U}_{j+1}$ (see Section 2.3.3). What we need technically, for a first order in space scheme, is to spread the m-Riemann wave starting from $\mathbf{U}_j$ (or from $\mathbf{U}_{j+1/2}^{\mathrm{L}}$ for higher-order approximations) over the boundary in such a way that $\lambda_m = 0$ at it.

It is worth noting that the above approach can be considered as a supplement to the method based on the solution of the Riemann problem between the inner cell and parameters at infinity, since that method generally fails exactly in the case of the eigenvalue behavior we considered.

The implementation of the boundary conditions based on the solution in the form of Riemann waves for gas dynamic and MHD problems will be given later in the corresponding chapters.

It must be noted that we described only nonreflecting boundary conditions that are based on the local characteristic approach. On the other hand (see Engquist and Majda 1977), exact nonreflecting boundary conditions that eliminate errors caused by truncation of the computational region must necessarily be nonlocal, being expressed in terms of pseudodifferential operators. This makes them inconvenient for practical calculations. They were applied, however, for certain problems (see, e.g., Givoli 1991; Tsynkov 1995). In the last referenced paper a technique was developed for implementation of nonlocal nonreflecting boundary conditions for the problem of gas flow around airfoils. Sofronov (1993, 1998) suggested very efficient transparency conditions for time-dependent multidimensional flows. Worth mentioning also is the review article by Tsynkov (1998). Unfortunately, in this and other nonlocal approaches the artificial boundary is supposed to lie in the region of smooth flow. This does not allow us to apply such methods in the cases for which the boundary is crossed by discontinuities. The same seems to be true for the remote boundary conditions for unsteady multidimensional aerodynamic computations based on the bicharacteristic analysis (Roe 1989). That paper is very useful, since it describes some unsound practices in the implementation of the far-field boundary conditions.

2.8.3 Evolutionary boundary conditions.

Another type of boundary condition originates when we solve the system in the bounded region with unknown functions being governed by the same or different system outside the boundary. Such a boundary can be a shock, a tangential discontinuity, an ionization front, etc. In this case we must formulate the boundary conditions that allow us to obtain a unique solution. In contrast to the boundary conditions described in the previous subsection, we must also determine the velocity of the discontinuity surface.

Consider, first, the question of the necessary number of boundary conditions required to obtain a unique solution of the initial value problem for a linear system with constant coefficients

$$\frac{\partial \mathbf{u}}{\partial t} + A\frac{\partial \mathbf{u}}{\partial x} = \mathbf{0}. \tag{2.8.22}$$

Let the boundary be defined by its law of motion $x = X(t)$ or by its velocity $dX/dt = W(t)$. Let also the components u_j of the unknown vector $\mathbf{u}$ be given at $t = t_0$ in the region $x \geq X(t_0)$. We want to construct a solution to the system of (2.8.22) in the region $x > X(t)$. Assume initially that the velocity of the boundary does not coincide with any of the characteristic velocities of our system, that is,

$$W \neq \lambda_i, \qquad i = 1, \ldots, n.$$

In this case, some of the waves in the solution (1.4.12) are incoming, that is, they deliver the values of their amplitudes w_m from the region of initial conditions. The rest of the waves are outgoing, that is, they carry information that is not prescribed by the initial conditions. They originate at the boundary under consideration (see Fig. 2.11). Thus, we must know the amplitudes of these waves at the boundary. This states the initial boundary value problem for Eq. (2.8.22).

Let us assume that

$$\lambda_r > W, \qquad r = 1, \ldots, s$$

and

$$\lambda_l < W, \qquad l = s + 1, \ldots, n.$$

The values w_l of the Riemann invariants along the incoming characteristics are known, while s values of w_r must be prescribed on the basis of additional consideration. Since the

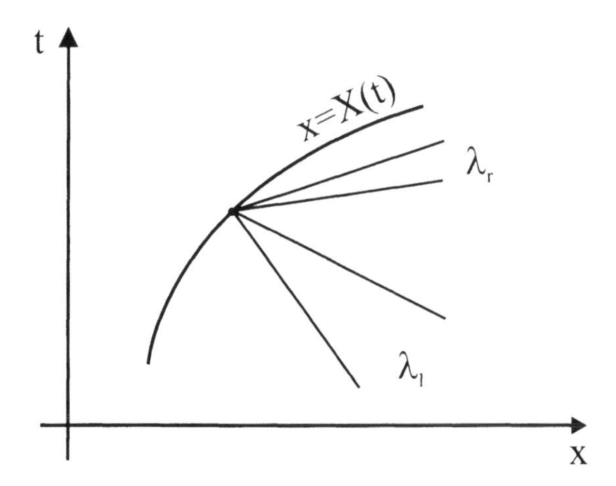

Figure 2.11 Behavior of characteristics near the boundary.

velocity W of the boundary is generally not known beforehand and must also be determined, the total number of boundary conditions to be stated becomes $s+1$.

It frequently happens in mechanical problems that even if the motion of the boundary is not prescribed we have an expression for W in terms of u_i. On the other hand, relationships between unknown functions on the boundary can include W. Thus we have $s+1$ equations,

$$\Phi_l(w_1, w_2, \ldots, w_n, W) = 0, \qquad l = 1, \ldots, s+1. \tag{2.8.23}$$

Relations (2.8.23) must uniquely determine $w_1, \ldots, w_s$ and W. In addition, no consequence of them can relate the amplitudes $w_{s+1}, \ldots, w_n$ of incoming waves. The latter can only occur if

$$\det\left(\frac{\partial \Phi_l}{\partial w_1 \ldots \partial w_r \partial W}\right) \equiv 0, \qquad r = 1, \ldots, s.$$

The summary of the above considerations can be formulated as follows: the initial boundary value problem for the linear system (2.8.22) can be uniquely resolved at the point x, t only if we have $s+1$ boundary relations in the form (2.8.23) with a nonzero Jacobian with respect to $w_1, \ldots, w_s$ and W, where s is the number of outgoing characteristics at this point. Such boundary conditions are called evolutionary. We can also say that the boundary itself is evolutionary at this point. If any expressions $W - \lambda_i$ changes its sign along the boundary, the necessary number of boundary conditions is obviously a function of the position of the boundary point.

If we solve the problem with a moving boundary dividing the computational region into the parts with the solution governed by the same or by another hyperbolic system, then s must be considered as a total number of outgoing characteristics leaving the boundary in both directions.

The boundary conditions for nonlinear problems are called evolutionary if they are evolutionary for a linearized problem, with linearization in the vicinity of $\mathbf{u} = \mathbf{u}_0 + \delta\mathbf{u}(x, t)$, where $\mathbf{u}_0 = \text{const}$ and $|\delta\mathbf{u}|$ is small. Thus, the evolutionary property means in this case the uniqueness of the solution of the interaction of the boundary with the small perturbations $\delta\mathbf{u}$ and δW (compare with Section 1.4.4). If the number of boundary conditions is excessive, the solution does not exist. If this number is less than $s+1$, the solution cannot be uniquely determined. This allows one, for example, to construct an arbitrarily growing solution.

It is worth noting that although a linear problem cannot be resolved if the number of boundary conditions is greater than $s+1$, the solution of the nonlinear problem can exist, only the magnitude of the disturbances becomes nonsmall. This can lead, for example, to disintegration of an arbitrary discontinuity.

If the velocity W of the boundary coincides at some point with one of the characteristic velocities λ_i, we shall say that this is a Jouget point (by analogy with the detonation wave theory, see Landau and Lifshitz 1987).

2.9 Shock-fitting methods

The quality of a numerical method applied to solution of a hyperbolic system strongly depends on its ability to reproduce shock profiles in the computational region. The best choice, of course, is to satisfy exactly the conservation relations on all shocks and approximate derivatives only in smooth subregions of the flow. Such a procedure, however, can result in a very complicated numerical algorithm. Besides, shocks can originate in the process of calculation, thus necessitating the algorithm for their effective detection. The methods based on this approach are called shock-fitting methods. Their application allowed Marconi, Salas, and Yaeger (1976) to model very complicated flow structure near realistic aircraft configurations. In the previous sections of this chapter we described the alternative, which lies in treating discontinuities as continuous, though sharp, profiles of parameters. In this case one must be careful to avoid possible monotonicity breakdown and to preserve as precisely as possible the Hugoniot relations. It is worth reminding in this connection that if we remain in the framework of linear schemes only the first order of space approximation allows one to preserve the monotonicity of grid functions (see Section 2.7.1).

As mentioned earlier, the Euler system of gas dynamic equations is a quasilinear hyperbolic system frequently encountered in mechanical applications. Moreover, owing to the importance of this system for various aerodynamic problems and taking into account its relative simplicity, numerical methods for solution of this system are rather well developed in comparison with those for more complicated hyperbolic systems of mechanical interest. For this reason in Chapter 3 we describe both genuinely shock-fitting methods and the elements of the floating shock-fitting method choosing the Euler system for ideal gas and for a chemically reacting mixture. Here we present some considerations outlining the general features of the floating shock-fitting method.

2.9.1 Floating shock fitting.

The genuinely shock-fitting method allows us to fit the discontinuities whose existence is known beforehand. We must only find their exact position in the computational region. The method is useful and widely applied for calculations of complicated problems with one external shock dividing the region of unknowns to be found from the region where all quantities are determined by the boundary conditions. The other discontinuities that may originate in the course of time are calculated in this case on the basis of the shock-capturing method. In principle, all discontinuities that exist in the computational region can be fitted. The fitting procedure, however, becomes very complicated and, which is even more important, time-consuming. We shall present here only the general approach to this problem based on the one-dimensional hyperbolic system

$$\frac{\partial \mathbf{U}}{\partial t} + \frac{\partial \mathbf{F}(\mathbf{U})}{\partial x} = \mathbf{0} \tag{2.9.1}$$

as an example. It can also be represented, as usual, in the quasilinear form

$$\frac{\partial \mathbf{U}}{\partial t} + A\frac{\partial \mathbf{U}}{\partial x} = \mathbf{0}, \quad A = \frac{\partial \mathbf{F}}{\partial \mathbf{U}}. \tag{2.9.2}$$

The procedure of exact fitting the discontinuities that propagate through the computational region and interact with other discontinuities (they can also appear and/or disappear in the course of time) is often called *floating shock fitting*.

The idea of this shock-fitting method lies in the application of a high order of accuracy finite-difference method to the system of (2.9.2) only in the smoothness regions. Approximation of derivatives across the discontinuities is not allowed. The method of characteristics is used to detect and track the discontinuities.

Once we decided to use shock fitting, it is not obligatory to approximate numerically the conservation laws (2.9.1) or the system for conservative variables (2.9.2). A quasilinear form of equations is simpler and can be written out for convenient dependent variables $\mathbf{u}$ (see Section 1.3.1),

$$\frac{\partial \mathbf{u}}{\partial t} + \mathcal{A}\frac{\partial \mathbf{u}}{\partial x} = \mathbf{0}, \tag{2.9.3}$$

where

$$\mathcal{A} = M^{-1}AM, \quad M = \frac{\partial \mathbf{U}}{\partial \mathbf{u}}.$$

For example, in gas dynamics the system written in terms of the pressure and density logarithms (see Moretti 1979) can be of great convenience for the calculation of strong rarefactions.

Suppose the system (2.9.3) is strictly hyperbolic. If we multiply it by the matrix Ω_L composed of left eigenvectors (vector-rows) $\mathbf{l}^i$ of $\mathcal{A}$, we obtain the system of compatibility relations along the characteristics C_i in the form

$$\mathbf{l}^i \cdot \left[\frac{\partial \mathbf{u}}{\partial t} + \lambda_i(\mathbf{u})\frac{\partial \mathbf{u}}{\partial x}\right] = 0. \tag{2.9.4}$$

The system (2.9.4) can also be written as a set of equations

$$\mathbf{l}^i \cdot \left(\frac{d\mathbf{u}}{dt}\right)_{C_i} = 0, \tag{2.9.5}$$

where the derivatives are taken along the characteristic directions $C_i = dx/dt = \lambda_i$, $i = 1, \ldots, n$.

Below we describe the elements of the shock-fitting procedure suggested by Henshaw (1987). Let us divide the part of the x-axis where we seek the solution of the system

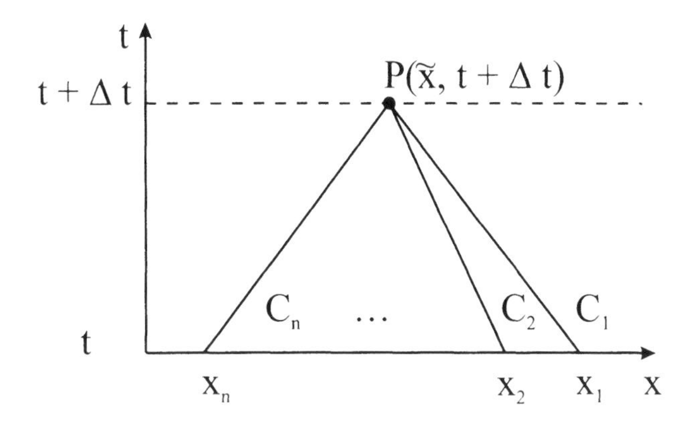

Figure 2.12 Method of characteristics for regular points.

into the intervals of equal size Δx with centers at x_j. The values of $\mathbf{u}$ at these points will be used for calculation only in smooth regions. To solve the characteristic equations, we need to introduce supplementary sets of points $x = x_c(k)$. The number k of such sets is equal to the number of discontinuities in the computational region. The points belonging to the introduced sets can move through the fixed grid in the course of calculation. The regular points lying beneath any set of characteristic points must not be used in the calculation. In general, characteristic points can be initially chosen among the points of the fixed grid in domains of large derivatives. One of the possible algorithms is the following. Let the point x_j be acceptable for inclusion into the set of characteristic points if

$$\max_{1 \le i \le n} \frac{|u_i(x_{j+1}) - 2u_i(x_j) + u_i(x_{j-1})|}{||u_i||} > \epsilon, \tag{2.9.6}$$

where $||u_i||$ is some global norm of u_i on the computational grid. As the above expression is of the order of $(\Delta x)^2$, in the smooth portion of the solution we can require that ϵ must be larger than this value. Similarly we can exclude the points from the characteristic set if it does not satisfy the above criterion. This criterion produces points both near strong and weak discontinuities.

In the smooth regions the characteristic equations can be solved using the following implicit numerical procedure. Suppose $\mathbf{q}(x, t)$ is a function that coincides with the solution at the grid points and varies linearly in between. This function is known at the starting time t. To find the solution at the point $P(\tilde{x}, t + \Delta t)$ (see Fig. 2.12) to the second order of accuracy, we have the system of $2n$ characteristic equations ($i = 1, \ldots, n$)

$$\mathbf{l}^i \left(\frac{\mathbf{q}(\tilde{x}, t + \Delta t) + \mathbf{q}(x_i, t)}{2} \right) \cdot (\mathbf{q}(\tilde{x}, t + \Delta t) - \mathbf{q}(x_i, t)) = 0, \tag{2.9.7}$$

$$\tilde{x} - x_i = \lambda_i \left(\frac{\mathbf{q}(\tilde{x}, t + \Delta t) + \mathbf{q}(x_i, t)}{2} \right) \times \Delta t. \tag{2.9.8}$$

They are written out for the characteristics emanating from the point $\tilde{x}$ at the time level $t + \Delta t$. Thus, both x_i and $\mathbf{q}$ at P are unknown.

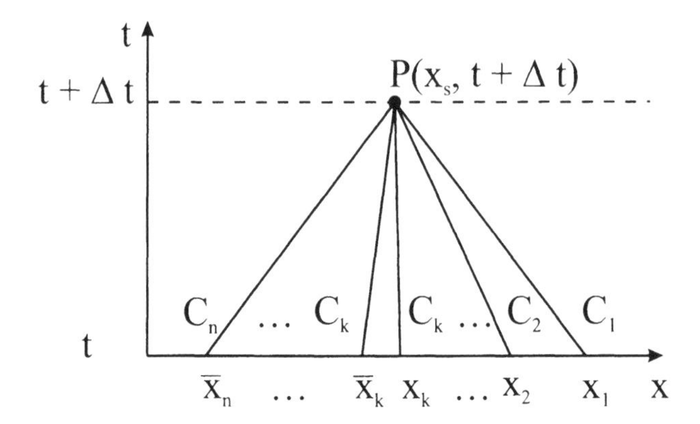

Figure 2.13 Method of characteristics for the shock points.

In order to advance the chosen set of characteristic points from t to $t + \Delta t$ we may also need to solve the system for unknown $\bar{x}$ and given one of the points x_i belonging to the characteristic group. Then we can spread n curves for each characteristic emanating from each of the points x_i and obtain n^2 points at the new time level. Each of them must pass the test (2.9.6) in order to decide whether it should be kept in the characteristic set.

The procedure of solving characteristic equations near discontinuities must incorporate the boundary conditions on them. For shocks, we have the Hugoniot relations

$$\mathbf{F}(\mathbf{U}^{\mathrm{R}}) - \mathbf{F}(\mathbf{U}^{\mathrm{L}}) = W(\mathbf{U}^{\mathrm{R}} - \mathbf{U}^{\mathrm{L}}), \tag{2.9.9}$$

with W being the shock velocity. The distinctive feature of the k-shock (see Section 1.4.4) is the intersection of characteristics of the kth family drawn from both sides of the shock.

For contact discontinuities Eq. (2.9.9) still holds, but the characteristics of the same family are becoming parallel one to another. In this case the additional relations are

$$W = \lambda_k^{\mathrm{R}} = \lambda_k^{\mathrm{L}}. \tag{2.9.10}$$

Prior to proceeding to the next time level, the characteristic points must be monitored for the presence of shocks. This is done by checking the possibility of intersection of the characteristics belonging to the same family. If the shock has already been detected among the point set under consideration, then the vectors $\mathbf{U}^{\mathrm{R}}$ and $\mathbf{U}^{\mathrm{L}}$ are not arbitrary and, if Eq. (2.9.9) is satisfied, we have an initial guess for the shock velocity. If Eq. (2.9.9) is not satisfied, than we have a more general case and all quantities must be found by solving the Riemann problem. This can happen, for example, if an arbitrary jump of parameters is prescribed at the initial time or if two or more discontinuities approach each other so close that the interaction between them must be considered. As an alternative, for some special systems like the Euler gas dynamic equations (Di Giacinto and Valorani 1989), appropriate analytical solutions describing such crossing can be used (Courant and Friedrichs 1976).

Suppose we found out that the k-shock lies beneath the set of characteristic points. We must determine its velocity and the quantities on its right- and left-hand sides. In this case the characteristics from 1 to k arrive at the shock position point x_s at $t + \Delta t$ from the

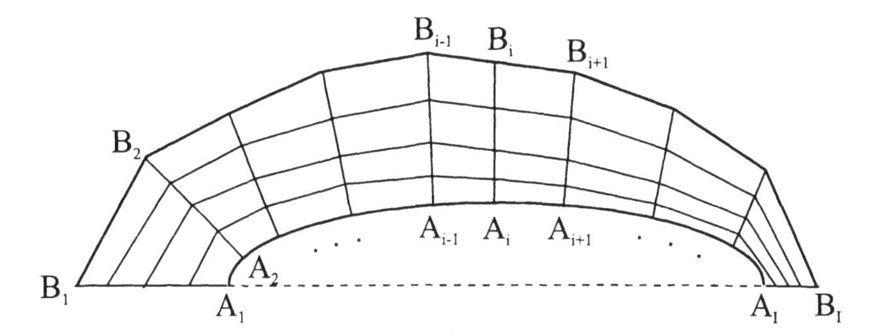

Figure 2.14 Distribution of grid points.

right-hand side of the shock, while the characteristics from k to n arrive at this point from the left-hand side (see Fig. 2.13). Thus, we have the system of $3n + 2$ equations:

$$
\begin{aligned}
&\mathbf{l}^i\left(\frac{\mathbf{u}^{\mathrm{R}} + \mathbf{q}(x_i, t)}{2}\right) \cdot \left(\mathbf{u}^{\mathrm{R}} - \mathbf{q}(x_i, t)\right) = 0, \qquad i = 1, \dots, k;\\
&\mathbf{l}^i\left(\frac{\mathbf{u}^{\mathrm{L}} + \mathbf{q}(\bar{x}_i, t)}{2}\right) \cdot \left(\mathbf{u}^{\mathrm{L}} - \mathbf{q}(\bar{x}_i, t)\right) = 0, \qquad i = k, \dots, n;\\
&\mathbf{F}(\mathbf{U}^{\mathrm{R}}) - \mathbf{F}(\mathbf{U}^{\mathrm{L}}) = W(\mathbf{U}^{\mathrm{R}} - \mathbf{U}^{\mathrm{L}});\\
&x_{\mathrm{s}} - x_i = \lambda_i\left(\frac{\mathbf{u}^{\mathrm{R}} + \mathbf{q}(x_i, t)}{2}\right)\Delta t, \qquad i = 1, \dots, k;\\
&x_{\mathrm{s}} - \bar{x}_i = \lambda_i\left(\frac{\mathbf{u}^{\mathrm{L}} + \mathbf{q}(\bar{x}_i, t)}{2}\right)\Delta t, \qquad i = k, \dots, n.
\end{aligned}
\tag{2.9.11}
$$

The unknowns are $\mathbf{u}^{\mathrm{R}}$, $\mathbf{u}^{\mathrm{L}}$, W, and the starting points $x_1, \dots, x_k, \bar{x}_k, \dots, \bar{x}_n$ of characteristics at the time level t. Using the shock velocity W_0 obtained at the previous time level as an initial guess, we can organize the iteration procedure by assuming

$$x_{\mathrm{s}} = x_{\mathrm{s}}^{(0)} + \tfrac{1}{2}(W + W_0)\Delta t.$$

2.9.2 Shock fitting on moving grids. In this subsection we describe another discontinuity-fitting method that is based on the calculation of unknown quantities and velocities of discontinuities using the solution of the corresponding Riemann problem (Godunov et al. 1979). In this case each discontinuity must coincide with the boundary of a computational cell. This obviously requires calculations on moving grids adjusted to discontinuities. For simplicity we shall outline this approach for the case of a nonstationary hyperbolic system with two space variables. We also consider the solution in the subdomain between some fixed boundary and one of the moving discontinuities (see Fig. 2.14). The shape of the computational cells can generally be arbitrary. The cell system is structured in the sense that the cells can be enumerated similarly to the enumeration of entries in a rectangular matrix. Let the points $B_0, \dots, B_I$ lie on the discontinuity whose initial position is known. The main problem now is to determine the new coordinates of the cell vertices, since the motion of the discontinuity will result in the deformation of the whole cell structure. We substitute the values of unknown functions inside the cells by some constant averaged values. The shock velocity is determined in the process of solution of the problem. This also determines the way of restructuring the grid system.

The solution of the problem consists of the following three steps. At the first step we calculate the motion of the boundary contour $B_0, \ldots, B_I$ within the time interval between t_0 and $t_0 + \Delta t$. At the second step we determine the positions of all internal cell vertices in accordance with the new position of the boundary. Finally we calculate the new values of unknowns at $t_0 + \Delta t$ on the basis of the conservation laws for each cell. The algorithms of this calculation will be discussed for the particular case of the Euler gas dynamic equations in the next chapter.

In the simplest case the rays $A_0B_0, \ldots, A_IB_I$ are fixed and we only need to find the method to calculate the new positions of the cell vertices lying on these lines. Let us assume that setting the coordinates of the boundary vertices on the lines $A_0, \ldots, A_I$ and $B_0, \ldots, B_I$ uniquely defines the positions of all remaining vertices. For this purpose it is sufficient to specify the law of their distribution along the rays. This can be done, for example, by simple formulas

$$x_{i,j} = x_i^A(1 - s_j) + x_i^B s_j, \qquad y_{i,j} = y_i^A(1 - s_j) + y_i^B s_j;$$

$$0 = s_0 < \cdots < s_j < \cdots < s_J = 1,$$

where x_i^A and y_i^A and x_i^B and y_i^B are the Cartesian coordinates of A_i and B_i, respectively. The grid construction is complete if we connect the points with the same index j and lying on the neighboring rays by straight lines. The choice of s_j allows us to manipulate with the cell size decreasing it in the regions of large gradients of the solution. Now we must determine the velocity of the discontinuity $B_0, \ldots, B_I$. This can be done in the following way. At each of the segments $B_{i-1}B_i$ of the discontinuity we solve the one-dimensional Riemann problem with initial conditions specified by the quantities inside and outside of the subdomain under consideration. The space coordinate for this purpose must be chosen perpendicular to the segment $B_{i-1}B_i$. The possibility of such an approach depends on the availability of the exact solution to the corresponding Riemann problem. As a velocity of the discontinuity we can choose that of the corresponding discontinuity originating in this exact solution. In this way we can determine the velocity that is precisely consistent with the Hugoniot relations. In pure gas dynamics the exact solution of the Riemann problem always exists. It is a combination of discontinuities and rarefaction waves separated by the regions of constant quantities. If a computational cell segment is a contact discontinuity, then its velocity coincides with the velocity of the contact discontinuity in the one-dimensional Riemann problem. If the segment is a part of the shock, then the solution will consist of the two shocks and a contact discontinuity and the sought velocity must be chosen equal to the velocity of the shock that faces the incoming flow (Godunov et al. 1979). It is clear that the velocities of neighboring segments are, generally speaking, nonconsistent. This can lead to some indeterminacy of the new position of B_i. To determine the speed of the point B_i one may use the linear interpolation between the velocities of the adjacent segments.

2.10 Entropy correction procedures

The entropy correction procedure is an algorithm permitting one to avoid the appearance in the numerical results of some nonphysical solutions, such as gas dynamic rarefaction shocks. This correction introduces a mechanism to select physically relevant solution. We will describe several algorithms of entropy correction to be used with the numerical schemes based on the CIR, Roe, and exact Riemann problem solvers.

Numerical effects of the rarefaction shock can originate in the region where some eigenvalue λ of the hyperbolic system changes in sign, for instance, $\lambda_m < 0$ and $\lambda_{m+1} > 0$, see Eqs. (2.3.25)–(2.3.26). In this case the calculation of $|\lambda|$ in the matrix $|A|$ in accordance with the averaging formula

$$|\lambda| = |\hat{\lambda}|, \quad \hat{\lambda} = \tfrac{1}{2}(\lambda_m + \lambda_{m+1})$$

can give $|\lambda| \to 0$ for $|\lambda_{m+1}| \neq 0$ and $|\lambda_m| \neq 0$. As a result, regions can appear with small (or zero) numerical dissipation (viscosity) proportional to $|\lambda|$. This can affect the numerical stability, and some abnormal and physically unstable solutions, such as rarefaction shocks can occur. As note earlier, the Osher scheme monitors all points of zero eigenvalues and automatically rules out rarefaction shocks.

Consider the entropy correction procedures for the Godunov-type schemes based on the exact, CIR, and Roe solvers. As described above, the mechanism of generation of extraneous solutions results is due to the lack of numerical dissipation in the vicinity of zero eigenvalues. Besides, in this case the numerical scheme does not distinguish, in particular, between the cases of flow with the same in moduli divergent velocities $\pm u$ and convergent velocities $\mp u$. In the former case, a rarefaction wave can appear. Otherwise, a shock wave can arise. However, both cases lead to the same equality $|\lambda| = |u| = 0$ at the line of symmetry. This example may explain some entropy wakes (nonphysical nonmonotonicities) that can arise in numerical results near the axis of symmetry and rigid walls, where the normal velocity is equal to zero.

In general, the entropy correction procedures add extra numerical dissipation in such domains by correcting the algorithms for the calculations of $|\lambda|$. In the notation of schemes (2.3.25)–(2.3.26) the entropy correction is equivalent to the transformation of the matrix

$$|A| = \Omega_{\mathrm{R}}[|\lambda_p|\, \delta_{pl}]\Omega_{\mathrm{L}} = \Omega_{\mathrm{R}}|\Lambda|\Omega_{\mathrm{L}}$$

into the matrix

$$|A| = \Omega_{\mathrm{R}}|\widetilde{\Lambda}|\Omega_{\mathrm{L}} = \Omega_{\mathrm{R}}(|\Lambda| + \Theta)\Omega_{\mathrm{L}},$$

where $\Theta = \mathrm{diag}[\theta_1, \ldots, \theta_n]$ is a diagonal matrix with nonnegative entries. In the general case, for each λ we can formulate the following entropy correction:

$$|\widetilde{\lambda}| = |\lambda| + \theta = |\hat{\lambda}| + \max\left(0, -|\hat{\lambda}| + \tfrac{1}{2}|\Delta\lambda|\right), \quad \Delta\lambda = \lambda_{m+1} - \lambda_m, \quad \hat{\lambda} = \tfrac{1}{2}(\lambda_m + \lambda_{m+1}).$$

It follows that

$$|\widetilde{\lambda}| = |\lambda| + \theta \geq \max\left(|\hat{\lambda}|, \tfrac{1}{2}|\Delta\lambda|\right)$$

and $|\widetilde{\lambda}| \to 0$ if and only if $\lambda_m \to 0$ and $\lambda_{m+1} \to 0$.

Huynh (1995), see also Velichko et al. (1999), used a procedure of this type in one- and two-dimensional gas dynamic calculations only to genuinely nonlinear characteristic fields. On the other hand, the necessity may appear to apply similar correction procedures to linearly degenerate characteristic fields. This is done in order to suppress a "carbuncle" phenomenon and numerical instabilities due to an odd–even decoupling at shocks slowly moving nearly parallel to the coordinate line (Quirk 1994; Pandolfi and D'Ambrosio 1998; Gressier and Moschetta 1998). This approach can very often stabilize time-convergent solutions (Pogorelov, Ohsugi, and Matsuda 2000). Of course, it is very difficult to make a decision in such cases whether we suppress the numerical or physical instability.

Let λ_m and λ_{m+1} be the eigenvalues that lie, respectively, on the left and right of cell boundary $m + 1/2$. In particular, for the two-dimensional gas dynamic equations the eigenvalues are equal to $u \pm c$, u and $v \pm c$, v. Then, the entropy correction by Harten (1983) can be written as

$$|\tilde{\lambda}| = \begin{cases} |\hat{\lambda}| & \text{if} \quad |\hat{\lambda}| \geq \delta, \\ \dfrac{\hat{\lambda}^2 + \delta^2}{2\delta} & \text{if} \quad |\hat{\lambda}| < \delta. \end{cases} \tag{2.10.1}$$

In the one-dimensional case, the quantity δ can be set equal to a constant, 0.1, 0.2, and 0.5 (Harten 1983). In two-dimensional calculations δ cannot be constant but must be chosen equal to $(|\hat{u}| + |\hat{v}|)\epsilon$, where ϵ is a small parameter that ranges between 0.1 and 0.5. One can also use $\delta = (|\hat{u}| + |\hat{v}| + c)\epsilon$, see Yee (1989). The "hat" means the average value. For instance, for the CIR method we can use the arithmetic average,

$$\hat{u} = \tfrac{1}{2}(u_m + u_{m+1}), \quad \hat{v} = \tfrac{1}{2}(v_m + v_{m+1}).$$

In Roe's method we can choose

$$\hat{u} = \frac{\sqrt{\rho_m}\, u_m + \sqrt{\rho_{m+1}}\, u_{m+1}}{\sqrt{\rho_m} + \sqrt{\rho_{m+1}}}, \quad \hat{v} = \frac{\sqrt{\rho_m}\, v_m + \sqrt{\rho_{m+1}}\, v_{m+1}}{\sqrt{\rho_m} + \sqrt{\rho_{m+1}}},$$

where ρ is the density. The averaging of other quantities in Roe's method is described in Section 3.4.4. If the inequality $|\hat{\lambda}| \ll \delta$ holds, we obtain $|\tilde{\lambda}| = \frac{1}{2}\delta$.

Instead of Eq. (2.10.1), its numerous versions can be used, see, e.g., Sazonov and Semenov (1993).

Version 1:

$$|\tilde{\lambda}| = |\lambda|_s = \frac{s_m|\lambda_m| + s_{m+1}|\lambda_{m+1}|}{s_m + s_{m+1}}, \tag{2.10.2}$$

where $s_m = s_{m+1} = 1$ for the CIR schemes and $s_m = \sqrt{\rho_m}$, $s_{m+1} = \sqrt{\rho_{m+1}}$ for Roe's scheme.

We can combine the methods of (2.10.1) and (2.10.2) and obtain

Version 2:

$$|\tilde{\lambda}| = \begin{cases} |\lambda|_s = \dfrac{s_m|\lambda_m| + s_{m+1}|\lambda_{m+1}|}{s_m + s_{m+1}} & \text{if} \quad |\lambda|_s \geq \delta, \\ \dfrac{\lambda_s^2 + \delta^2}{2\delta} & \text{if} \quad |\lambda|_s < \delta. \end{cases} \tag{2.10.3}$$

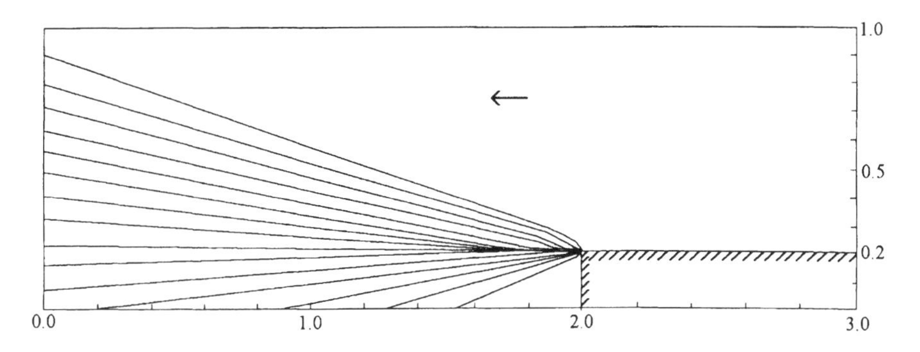

Figure 2.15 Wind tunnel flow.

In the entropy correction (2.10.3) we can incorporate the eigenvalue analysis in the form (2.3.61), see Evseev and Semenov (1990). This gives Version 3 of the entropy correction algorithm (2.10.1). In fact, if $\lambda_{m+1} > \lambda_m$, then the characteristics are divergent and a rarefaction shock can be generated locally. In this case we use the entropy correction (2.10.3) for the corresponding λ. Otherwise, if $\lambda_{m+1} < \lambda_m$, the local flow is convergent. Thus, a shock can be generated locally and we need not use an entropy correction.

All these versions of the entropy correction have a wide area of application. Version 3 proved to be very efficient and rather universal. The application of this correction permits one to smear the rarefaction shock and obtain more adequate flow in a rarefaction fan than that of Version 1. However, the best quality in eliminating rarefaction shocks in the wind tunnel problem (Woodward and Colella 1984) was demonstrated by Version 2. However, Version 3 is more universal and reliable as compared with Version 2.

Figure 2.15 shows the numerical results for the problem of a steady gas dynamic flow in a tunnel with a step with the flow velocity directed from right to left, unlike the wind tunnel problem (Woodward and Colella 1984), where the flow is directed from left to right. In addition, $x = 0$ is the free boundary. Figure 2.15 shows the density isolines with $\rho_{\max} = 1.3124$ and $\rho_{\min} = 0.0858$ and the number of isolines equal to 15. The calculations were made for $\Delta x = \Delta y = 0.05$. The initial data for $t = 0$ were taken as follows:

$$\rho(0, x, y{>}0.2) = 1.4, \quad \rho(0, x, y{<}0.2) = 0.14,$$
$$u(0, x, y{>}0.2) = -3, \quad u(0, x, y{<}0.2) = 0,$$
$$v(0, x, y) = 0, \quad p(0, x, y) = 1.$$

The adiabatic index is equal to 1.4. This problem provides a good test for checking the accuracy of a numerical method. In this problem the relation $\sin\alpha = 1/\mathrm{M}$ must hold, where M is the Mach number, and α is the slope angle of the upper density isoline in Fig. 2.15; see Eq. (3.6.63). The results were calculated for $\mathrm{M} = 3$.

In the numerical simulation of this problem the negative values of the density develop in a region near the step. The application of the entropy correction (2.10.1)–(2.10.3) permits one to delay the development of the flow breaking. The application of Version 3 permits one to remove any flow breaking and provides a correct simulation of the above problem, see Sazonov and Semenov (1993).

For the Godunov method based on the exact Riemann problem solver, an entropy correction suggested by Charakhch'yan (2000a) may be recommended. Consider his correction for a two-dimensional gas dynamic scheme of the form (2.2.6)

$$\frac{\mathbf{U}_{i,j}^{k+1}-\mathbf{U}_{i,j}^{k}}{\Delta t}+\frac{\mathbf{F}_{i+1/2,j}-\mathbf{F}_{i-1/2,j}}{\Delta x}+\frac{\mathbf{E}_{i,j+1/2}-\mathbf{E}_{i,j-1/2}}{\Delta y}=\mathbf{0},$$
$$\mathbf{F}_{i\pm1/2,j}=\mathbf{F}(\mathbf{U}_{i\pm1/2,j}),\quad \mathbf{E}_{i,j\pm1/2}=\mathbf{E}(\mathbf{U}_{i,j\pm1/2}),$$

where Δx and Δy are the mesh sizes in x- and y-direction, respectively. The double integer subscripts (i,j) refer to the centers of two-dimensional space cells and the half-integer subscripts refer to the corresponding boundaries of the cells. The quantities $\mathbf{U}_{i\pm1/2,j}$ and $\mathbf{U}_{i,j\pm1/2}$ are solutions of the corresponding Riemann problem.

The idea of the correction procedures is the following. For each time level and each computational cell (i,j) we calculate the quantity

$$S_{\max}^{k}=\max(S_{ij}^{k},S_{i+1,j}^{k},S_{i-1,j}^{k},S_{i,j+1}^{k},S_{i,j-1}^{k},S_{i+1/2,j}^{k},S_{i-1/2,j}^{k},S_{i,j+1/2}^{k},S_{i,j-1/2}^{k}),\qquad(2.10.4)$$

where S is the entropy. The values of S with half-integer subscripts must be taken into account in Eq. (2.10.4) only if a shock wave moves into the cell across the corresponding boundary. One need not use an entropy correction if $S_{ij}^{k+1}\le S_{\max}^{k}$. Otherwise, one must put $\tilde{S}_{ij}^{k+1}=S_{\max}^{k}$ and $\tilde{p}_{ij}^{k+1}=p_{ij}^{k+1}$, where p is the pressure and $\tilde{f}$ stands here for the corrected value of a function f. The internal energy $\tilde{\varepsilon}_{ij}^{k+1}$ and density $\tilde{\rho}_{ij}^{k+1}$ are determined by using the quantities $S_{\max}$ and p and the equation of state (EOS) of the medium. Other variables are not corrected.

If only rarefaction waves move into the cell across all cell boundaries, one must assume $\tilde{S}_{ij}^{k+1}=S_{ij}^{k}$ and $\tilde{p}_{ij}^{k+1}=p_{ij}^{k}$, and determine $\tilde{\varepsilon}_{ij}^{k+1}$ and $\tilde{\rho}_{ij}^{k+1}$ in the manner described above.

In the case of an arbitrary semi-empirical EOS that does not permit one to determine the entropy in the analytical form, in the entropy correction of the type (2.10.4) one can use any gas dynamic function that increases with S. For example, in the above entropy analysis one may use the entropy temperature $T_S=T(S,\rho)$. For an ideal perfect gas we have $T_S=T_{ij}^{k}\left(\rho/\rho_{ij}^{k}\right)^{\gamma-1}$, where T is the temperature, and γ the adiabatic index.

This algorithm of the entropy correction permits one to remove any entropy wakes near the solid walls and in the rarefaction waves. This algorithm was used for numerical simulation of two-dimensional cumulative jets and interaction of a strong shock wave with the metal–vacuum interface governed by wide-range EOSs for metals in Lagrangian coordinates (Charakhch'yan 2000a). Figure 2.16 shows the numerical results of the simulation of the impact of aluminum plate with a solid wall at $x=0$. The plate moves at the relative velocity of 5 km/s. The solution consists of a shock wave moving from the left to the right. The figure shows the entropy temperature for the 30th time layer after the collision. The open circles mark the results obtained by using the Godunov scheme (Godunov 1959) in the Lagrangian coordinates, and the triangles mark the results practically coincided with the exact solution and obtained using the scheme with the entropy correction (2.10.4).

Although Charakhch'yan used the entropy correction for gas dynamic equations in Lagrangian coordinates, his correction apparently can be used to remove stationary entropy wakes near solid walls and in rarefaction shocks also in Eulerian calculations.

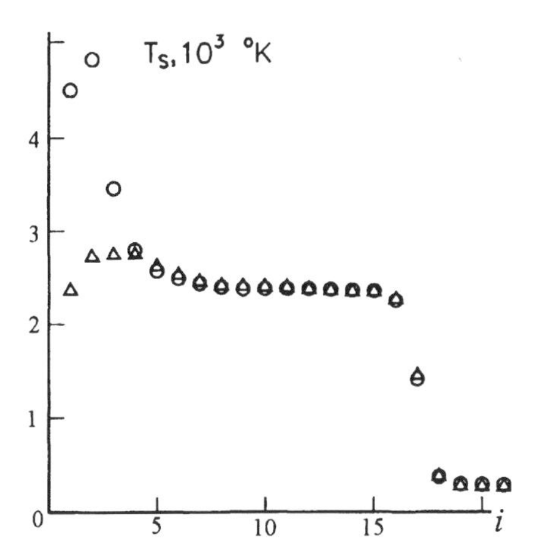

Figure 2.16 Entropy wake in temperature (o) near the left solid wall and its correction (Δ).

2.11 Final remarks

We described several basic elements used in the construction of high-resolution TVD methods for solution of hyperbolic systems. They include the description of the approaches to solving the Riemann problem, which is one of the main components of the Godunov-type methods. It is important to admit in this connection that we did not describe in detail various possible approaches of advancing the solution in time and increasing the space resolution on the basis of the application of higher-order polynomial interpolations. We mainly dwelled on the methods based on the MUSCL (monotone upstream schemes for conservation laws) approach, which involves the interpolation of grid variables to attain the higher order of accuracy, and paid little attention to non-MUSCL numerical methods. The discussion of these subjects can be found in Yee (1989), Hirsch (1990), and Toro (1997).

One must also mention the existence of flux-splitting methods, which represent the simplest way of introducing upwinding into systems by the representation of the flux $\mathbf{F}(\mathbf{U})$ as the sum $\mathbf{F}^-(\mathbf{U}) + \mathbf{F}^+(\mathbf{U})$ such that one can apply forward and backward differentiation to the Jacobian matrices $\partial\mathbf{F}^-(\mathbf{U})/\partial\mathbf{U}$ and $\partial\mathbf{F}^+(\mathbf{U})/\partial\mathbf{U}$, respectively. This approach was introduced by Mulder and van Leer (1983) and Anderson, Thomas, and van Leer (1985), who added the MUSCL approach to the Steger–Warming (1981) flux-vector splitting. Worth mentioning is also van Leer's (1982) splitting especially devised for perfect gas dynamic equations. In this connection we also refer to the advection upstream splitting method (AUSM) suggested by Liou and Steffen (1993) and its sequel by Liou (1996). These methods are based on the splitting that makes a distinction between convection and acoustic waves as two different processes. The devised technique turned out to preserve the exact resolution of one-dimensional contact and shock discontinuities and ensure the positivity of density for strong rarefactions. It allows one to avoid an unphysical "carbuncle" phenomenon (Quirk 1994). Very efficient scheme HLLE (for Harten, Lax, van Leer, and Einfeldt), which also improves the original method by Roe in the low density regions, was suggested by Einfeldt et al. (1991). We note also the implicit (Yee 1989) and symmetric (Yee 1987; Tadmor 1997) TVD schemes.

It is mentioned in the main body of the chapter that TVD schemes of any order of approximation degrade to the first order at discontinuities. In order to avoid this drawback, the TVD constraint must be weakened. For example, in the TVB (total variation bounded) methods it is required that $\mathrm{TV}[u^n] \leq B$, where B is positive and depends on u^0. Clearly, TVD implies TVB. Such methods were invented by Shu (1987) and proved to be uniformly high-order accurate in space. A further step is introducing ENO schemes by Harten (1987), Harten and Osher (1987), Harten et al. (1986, 1987), and Shu and Osher (1988, 1989), in which the TVD constraint is substituted by the requirement of nonincrease in the number of extrema. Technically the result is achieved by choosing the high-order polynomial interpolation stencil that ensures the least oscillation of all possible choices. Recently, high-order weighted essentially nonoscillatory (WENO) schemes have been developed. While ENO schemes are based on the smoothest among several stencils, WENO schemes take a weighted average of all candidates. The weights are adjusted by the local smoothness of the solution so that essentially zero weights are given to nonsmooth stencils while optimal weights are prescribed in smooth solutions. WENO schemes (Liu, Osher, and Chan 1994) act very similar to ENO schemes near discontinuities, but in the smooth regions they act more like an upstream centered scheme.

One must admit that modern shock-capturing methods (TVD, ENO, WENO) that are higher than third-order accurate are rather time-consuming and obviously require special treatment near the boundaries. On the other hand, higher order of accuracy is crucial for modelling complicated shock–turbulence interaction problems. Note also that the entropy fix procedures, applied to linearly degenerate fields, add too much viscosity to be suitable for correct viscous–inviscid interactions. One of the approaches is based on introducing a viscous flux limiter (Toro 1992). Yee, Sandham, and Djomehri (1999) suggested the application of the narrow grid stencil of high-order classical space differencing as base schemes. Further on, TVD, ENO, or WENO dissipations in conjunction with the artificial compression switch (Harten 1978) are used as characteristic filters. As a result, the final grid stencil substantially decreases. This approach can improve the resolution of fine scale flow structure when applied to existing numerical methods.

Chapter 3
Gas Dynamic Equations

In this chapter we describe the Godunov-type explicit shock-capturing and shock-fitting methods developed to compute flows governed by the multidimensional gas dynamic equations. These methods are based on the solution of the one-dimensional Riemann problem. We present numerical schemes using both the exact and several approximate Riemann problem solvers for gas dynamic equations. The approximate Riemann problem solvers involve the Courant–Isaacson–Rees (CIR), Roe, and Osher–Solomon solvers.

All developed methods permit us to manage calculations involving an arbitrary equation of state (EOS). For numerical modelling in many calculations of gas dynamic problems the EOS of the ideal perfect gas is used. There exist a number of physical problems that employ more complicated EOSs, for instance, those for metals, ceramics, polymers, and so on. Modern EOSs often have intricate behavior and properties. They are semi-empirical, wide-range, and multi-phase and include parameters of solid state, gas, liquid, plasma and so on; see Ionov and Selivanov (1987), Bushman, Lomonosov, and Fortov (1992), Kuropatenko (1992), Bushman, Kanel, Ni, and Fortov (1993), and Ross and Young (1993) In addition, these EOSs are sometimes given in a tabulated form. This leads to difficulties in adequate numerical incorporation of an EOS into known numerical algorithms. We describe some recipes for solving such problems. The numerical algorithms described below can adequately describe the propagation of discontinuities and preserve the monotone profiles of grid variables in the vicinity of shocks and contact discontinuities.

3.1 Systems of governing equations

The Euler equations governing the time-dependent three-dimensional flows of gas read

$$\frac{\partial \rho}{\partial t} + \operatorname{div}(\rho \mathbf{v}) = 0, \tag{3.1.1}$$

$$\frac{\partial \rho \mathbf{v}}{\partial t} + \operatorname{div}(\rho \mathbf{v}\mathbf{v} + p\hat{\mathbf{I}}) = \mathbf{0}, \tag{3.1.2}$$

$$\frac{\partial e}{\partial t} + \operatorname{div}[(e+p)\mathbf{v}] = 0. \tag{3.1.3}$$

Here $\rho = \rho(t,x,y,z)$ is the density, t is time, (x,y,z) are the Cartesian space coordinates, $\mathbf{v} = \mathbf{v}(t,x,y,z) = [u,v,w]^{\mathrm{T}}$ is the flow velocity, $\hat{\mathbf{I}} = \operatorname{diag}[1,1,1]$ is the 3×3 identity tensor, ε is the specific internal energy, $e = \rho\varepsilon + \frac{1}{2}\rho(u^2+v^2+w^2)$ is the total energy per unit volume, and $p = p(\rho,\varepsilon)$ is the pressure.

A relation between p and ε in the form $p = p(\rho, \varepsilon)$ or $\varepsilon = \varepsilon(\rho, p)$ is called an equation of state (EOS). In particular, the EOS of an ideal perfect gas is

$$p = (\gamma - 1)\rho\varepsilon, \quad \varepsilon = \frac{p}{(\gamma - 1)\rho}. \tag{3.1.4}$$

The quantity $T = (\gamma - 1)\varepsilon$ is the temperature in energy units and $\gamma > 1$ is the adiabatic index. A more complicated, so-called two-term EOS has the form

$$p = (\gamma - 1)\rho\varepsilon + (\rho - \rho_0)c_0^2, \quad \varepsilon = \frac{p + \gamma p_0}{(\gamma - 1)\rho} - \frac{c_0^2}{\gamma - 1}. \tag{3.1.5}$$

This EOS sometimes allows one to describe approximately the properties of solid state of materials. In (3.1.5) the quantities p_0 and c_0 are constants such that for $\varepsilon = 0$ the relation $p = 0$ must hold; hence $p_0 = \rho_0 c_0^2/\gamma$. The condition $p = 0$ for $\varepsilon = 0$ and $\rho = \rho_0$ provides the initial static equilibrium of the medium. Particularly, in modelling of explosive welding of metals (Godunov et al. 1970; Godunov et al. 1979) it was assumed $\rho_0 = 8.93\,\mathrm{g/cm^3}$ and $c_0 = 3.97\,\mathrm{km/s}$ for copper and $\rho_0 = 7.87\,\mathrm{g/cm^3}$ and $c_0 = 5.0\,\mathrm{km/s}$ for iron. Equation (3.1.4) can be obtained from Eq. (3.1.5) for $c_0 = 0$ and $p_0 = 0$.

Note that the equation of state (3.1.5), as well as more sophisticated EOSs, admit negative values of pressure. In particular, a negative pressure occurs in (3.1.5) for small values of ε and $\rho < \rho_0$. A negative pressure describes a stretching of a medium. If the negative pressure exceeds the strength limit of the material, then fracture of the material can occur. Therefore, the use of EOSs that admit negative pressures permits one to take into account the fracture processes.

Let us describe another, refined method of the continuous fracture modelling based on the consideration of the bulk porosity, ψ, where $0 \le \psi \le 1$. In this case the current density of the material, ρ, is equal to $(1 - \psi)\rho^\circ$, where ρ° is the density of the unfractured material. Then the governing system of equations becomes

$$\frac{\partial \rho}{\partial t} + \operatorname{div}(\rho\mathbf{v}) = 0, \tag{3.1.6}$$

$$\frac{\partial \rho\mathbf{v}}{\partial t} + \operatorname{div}(\rho\mathbf{v}\mathbf{v}) + (1 - \psi)\operatorname{div}(p\hat{\mathbf{I}}) = \mathbf{0}, \tag{3.1.7}$$

$$\frac{\partial e}{\partial t} + \operatorname{div}[e\mathbf{v} + (1 - \psi)p\mathbf{v}] = 0, \tag{3.1.8}$$

where $p = p(\rho^\circ, \varepsilon)$ and the current effective pressure is equal to $(1 - \psi)p$. If $\psi = 1$ the effective pressure is equal to zero and material is in a state of powder.

The above system should be complemented by an equation for ψ,

$$\frac{\partial \rho\psi}{\partial t} + \operatorname{div}(\rho\psi\mathbf{v}) = \rho\Psi, \tag{3.1.9}$$

where the right-hand side function Ψ describes the kinetics of the nucleation and growth of microvoids (pores) depending on the properties of the material; see Zukas et al. (1982), Bushman et al. (1993), and Kanel, Razorenov, Utkin, and Fortov (1996).

Equations (3.1.1)–(3.1.3) express the conservation of the mass, momentum, and energy of the medium, respectively. System (3.1.1)–(3.1.3) can be rewritten in expanded form as

$$\frac{\partial \rho}{\partial t} + \frac{\partial \rho u}{\partial x} + \frac{\partial \rho v}{\partial y} + \frac{\partial \rho w}{\partial z} = 0, \tag{3.1.10}$$

$$\frac{\partial \rho u}{\partial t} + \frac{\partial (\rho u^2 + p)}{\partial x} + \frac{\partial \rho uv}{\partial y} + \frac{\partial \rho uw}{\partial z} = 0, \tag{3.1.11}$$

$$\frac{\partial \rho v}{\partial t} + \frac{\partial \rho vu}{\partial x} + \frac{\partial (\rho v^2 + p)}{\partial y} + \frac{\partial \rho vw}{\partial z} = 0, \tag{3.1.12}$$

$$\frac{\partial \rho w}{\partial t} + \frac{\partial \rho wu}{\partial x} + \frac{\partial \rho wv}{\partial y} + \frac{\partial (\rho w^2 + p)}{\partial z} = 0, \tag{3.1.13}$$

$$\frac{\partial e}{\partial t} + \frac{\partial (e+p)u}{\partial x} + \frac{\partial (e+p)v}{\partial y} + \frac{\partial (e+p)w}{\partial z} = 0. \tag{3.1.14}$$

Flows of continuous media described by the gas dynamic equations are rather common. They include various gas flows, see Liepmann and Roshko (1957), Zel'dovich and Raizer (1967), Abramovich (1976), Courant and Friedrichs (1976), Rozhdestvenskii and Yanenko (1983), etc.

The general conservative form of Eqs. (3.1.10)–(3.1.14) in the one-dimensional case is

$$\frac{\partial \mathbf{U}}{\partial t} + \frac{\partial \mathbf{F}}{\partial x} = \mathbf{0}, \tag{3.1.15}$$

$$\mathbf{U} = [\rho, \rho u, \rho v, \rho w, e]^{\mathrm{T}}, \quad e = \rho\varepsilon + \tfrac{1}{2}\rho\,(u^2 + v^2 + w^2); \tag{3.1.16}$$

$$\mathbf{F} = [\rho u, \rho u^2 + p, \rho uv, \rho uw, (e+p)u]^{\mathrm{T}}. \tag{3.1.17}$$

In the nonconservative form Eqs. (3.1.15)–(3.1.17) acquire the form

$$\frac{\partial \mathbf{U}}{\partial t} + A\frac{\partial \mathbf{U}}{\partial x} = \mathbf{0}, \tag{3.1.18}$$

where the matrix $A = \partial\mathbf{F}/\partial\mathbf{U}$ is expressed as

$$A = \begin{bmatrix} 0 & 1 & 0 & 0 & 0 \\ -u^2+\theta b & 2u-ub & -vb & -wb & b \\ -uv & v & u & 0 & 0 \\ -uw & w & 0 & u & 0 \\ -uh+u\theta b & h-u^2 b & -uvb & -uwb & u+ub \end{bmatrix}, \tag{3.1.19}$$

$$h = \frac{e+p}{\rho}, \quad \theta = q^2 - \frac{e}{\rho} + \frac{\rho p_\rho}{p_\varepsilon}, \quad q^2 = u^2 + v^2 + w^2,$$

$$b = \frac{p_\varepsilon}{\rho}, \quad p_\varepsilon = \left(\frac{\partial p}{\partial \varepsilon}\right)_\rho, \quad p_\rho = \left(\frac{\partial p}{\partial \rho}\right)_\varepsilon.$$

The function h is called the total enthalpy. System (3.1.18) is hyperbolic and its coefficient matrix A has only real eigenvalues and a complete system of eigenvectors. Thus, the matrix A can be rewritten equivalently as

$$A = \Omega_{\mathrm{R}}\Lambda\Omega_{\mathrm{L}}. \tag{3.1.20}$$

It can be shown that

$$\Omega_{\mathrm{R}} = \begin{bmatrix} 1 & 0 & 0 & 1 & 1 \\ u-c & 0 & 0 & u & u+c \\ v & 1 & 0 & v & v \\ w & 0 & 1 & w & w \\ h-uc & v & w & h-c^2/b & h+uc \end{bmatrix}, \quad \det \Omega_{\mathrm{R}} = \frac{2c^3}{b}, \quad b = \frac{p_\varepsilon}{\rho}; \tag{3.1.21}$$

$$\Lambda = \mathrm{diag}[u-c,\ u,\ u,\ u,\ u+c], \tag{3.1.22}$$

$$\Omega_{\mathrm{L}} = \frac{b}{2c^2} \begin{bmatrix} \theta+uc/b & -u-c/b & -v & -w & 1 \\ -2vc^2/b & 0 & 2c^2/b & 0 & 0 \\ -2wc^2/b & 0 & 0 & 2c^2/b & 0 \\ 2h-2q^2 & 2u & 2v & 2w & -2 \\ \theta-uc/b & -u+c/b & -v & -w & 1 \end{bmatrix}, \tag{3.1.23}$$

$$c = \sqrt{p_\rho + \frac{p p_\varepsilon}{\rho^2}}. \tag{3.1.24}$$

The function c is called the speed of sound. Using (3.1.24) we can rewrite the quantity θ occurring in the matrices A and Ω_{L} as

$$\theta = q^2 - \frac{e}{\rho} + \frac{\rho p_\rho}{p_\varepsilon} = q^2 - h + \frac{c^2}{b}. \tag{3.1.25}$$

To solve the system of (3.1.1)–(3.1.3) we must specify appropriate initial and boundary conditions. At $t = 0$ conditions of the form $\mathbf{v}(0,x,y,z) = \mathbf{v}_0(x,y,z)$, $\rho(0,x,y,z) = \rho_0(x,y,z)$ and $\varepsilon(0,x,y,z) = \varepsilon_0(x,y,z)$ are prescribed. On the boundaries of the computational domain, boundary conditions are set, for example, in terms of fluxes or conditions on an impermeable wall. See also Section 2.8.

There exist a number of generalizations of the classical gas dynamic equations. More general models can involve the description of a set of reacting gases (components) that can have their own different densities, velocities, and energies, as well as equations of states, see Soo (1967), Sedov (1971), and Oran and Boris (1987). In the general case of n components, a new system of gas dynamics can have n equations of conservation of mass, n vector equations for the momentum conservation, and n energy equations. Sometimes, taking into account some physical relations and assumptions, the total number of these equations can be less than $3n$. All these models represent hyperbolic systems.

3.1.1 Two-temperature gas dynamic equations. Here we present a simple model of such a generalization that describes quasi-neutral plasma flows of electrons and ions. Let us consider the two-temperature gas dynamic model. Plasma flows described by the two-temperature gas dynamic equations are rather common in plasma physics, see Golant, Zhilinsky, and Sakharov (1980), Hora (1981), and Duderstadt and Moses (1982).

The electron mass is negligible in comparison with the ion mass and therefore the two-temperature gas dynamics takes into account only the ion density. In addition, in quasi-neutral plasma, the velocities of ions and electrons are the same, because the separation of charges is absent, and the model takes into account only the unified velocity. Both electrons and ions have their own energy equations and EOSs. For simplicity we consider a two-dimensional model.

The time-dependent equations of the two-dimensional, two-temperature gas dynamics in the Euler form read

$$\frac{\partial \rho}{\partial t} + \operatorname{div}(\rho \mathbf{v}) = 0, \tag{3.1.26}$$

$$\frac{\partial \rho \mathbf{v}}{\partial t} + \operatorname{div}(\rho \mathbf{v}\mathbf{v} + p\hat{\mathbf{I}}) = \mathbf{0}, \tag{3.1.27}$$

$$\frac{\partial E_1}{\partial t} + \operatorname{div}(E_1 \mathbf{v}) + p_1 \operatorname{div} \mathbf{v} = f_1, \tag{3.1.28}$$

$$\frac{\partial E_2}{\partial t} + \operatorname{div}(E_2 \mathbf{v}) + p_2 \operatorname{div} \mathbf{v} = f_2. \tag{3.1.29}$$

Here $\rho = \rho(t, x, y)$ is the density, t is time, (x, y) are the Cartesian space coordinates, $\mathbf{v} = \mathbf{v}(t, x, y) = [u, v]^{\mathrm{T}}$ is the flow velocity, $\hat{\mathbf{I}} = \operatorname{diag}[1, 1]^{\mathrm{T}}$ is the 2×2 identity tensor, $E_1 = \rho\varepsilon_1$ and $E_2 = \rho\varepsilon_2$ are the internal energies of electrons and ions per unit volume, respectively, where ε_1 and ε_2 are the specific internal energies of electrons and ions, $e = \rho\varepsilon_1 + \rho\varepsilon_2 + \frac{1}{2}\rho\,(u^2 + v^2)$ is the total energy per unit volume, $p_1 = p_1(\rho, E_1)$ is the electron pressure, $p_2 = p_2(\rho, E_2)$ is the ion pressure and, according to Pascal's law, $p = p(\rho, E_1, E_2) = p_1 + p_2$ is the total pressure. In general, Eqs. (3.1.28)–(3.1.29) can be complemented by right-hand sides f_1 and f_2 that take into account other processes, in particular, X-ray or laser radiation absorption, electron-ion temperature exchange, and electron and ion heat conduction (Brueckner and Jorna 1974; Duderstadt and Moses 1982).

The electron and ion gases have their own EOSs. In particular, the EOS of the perfect gas (3.1.4) can be used for electrons taking into account the phenomenon of electron gas degeneration, see Duderstadt and Moses (1982) and Zel'dovich and Raizer (1967):

$$\varepsilon_1 = \frac{T_1}{\gamma_1 - 1} + \varepsilon_{\mathrm{deg}}(\rho, T_1), \quad p_1 = \rho T_1 + p_{\mathrm{deg}}(\rho, T_1),$$

$$\varepsilon_{\mathrm{deg}} = \beta\rho^{-1/2}T_1^2, \quad p_{\mathrm{deg}} = \tfrac{1}{2}\rho\varepsilon_{\mathrm{deg}}.$$

Here T_1 is the electron gas temperature in energy units, γ_1 is the electron adiabatic index, and β is a constant depending on the substance. In particular, the EOS for ions of the perfect gas can be used with additional terms in the ion pressure and energy taking into account the elastic properties of the "cold" substance, or for $T_2 = 0$, see Ionov and Selivanov (1987), and Zel'dovich and Raizer (1967):

$$\varepsilon_2 = \frac{T_2}{\gamma_2 - 1} + \varepsilon_{\mathrm{cold}}(\rho), \quad p_2 = \rho T_2 + p_{\mathrm{cold}}(\rho), \tag{3.1.30}$$

$$p_{\mathrm{cold}} = -\frac{d\varepsilon_{\mathrm{cold}}}{d\eta} = \rho^2\frac{d\varepsilon_{\mathrm{cold}}}{d\rho}, \quad \eta = \frac{1}{\rho}. \tag{3.1.31}$$

Here T_2 is the ion temperature in energy units and γ_2 is the ion adiabatic index. Comparing Eqs. (3.1.5) and (3.1.30)–(3.1.31), we can express $\varepsilon_{\mathrm{cold}}$ and p_{cold} for EOS (3.1.5) as

$$\varepsilon_{\mathrm{cold}} = c_*\rho^{\gamma-1} + \frac{p_0}{\rho} - \frac{c_0^2}{\gamma - 1}, \quad p_{\mathrm{cold}} = (\gamma - 1)c_*\rho^{\gamma} - p_0, \quad c_* = \mathrm{const}.$$

Equations (3.1.26) and (3.1.27) express the conservation of the mass and momentum of the gas. Equations (3.1.28)–(3.1.29) express the evolution of the energies E_1 and E_2. In addition, Eq. (3.1.3) expressing the conservation of the total energy can be used instead of Eq. (3.1.28) or Eq. (3.1.29).

Analogously one can use the three-temperature gas dynamics that describe the motion of ions and electrons, and also the thermal radiation in matter; see, for example, Zabrodin and Prokopov (1998). In this case Eqs. (3.1.26)–(3.1.29) must be supplemented by an equation describing the evolution of the internal energy E_3 per unit volume of the thermal radiation gas. The last equation is similar to Eqs. (3.1.28)–(3.1.29). The thermal radiation EOS becomes

$$p_3 = \frac{1}{3}\rho\varepsilon_3 = \frac{1}{3}E_3; \qquad p_3 = \frac{\chi}{3}T_3^4, \quad \varepsilon_3 = \frac{\chi}{\rho}T_3^4, \quad \chi = \frac{4\sigma}{c_0}.$$

Here $p_3 = p_3(\rho, E_3)$ is the thermal radiation pressure, χ is a coefficient that depends on the light velocity c_0 and the Stefan–Boltzmann constant σ, T_3 is the thermal radiation temperature, and ε_3 is the specific internal energy of thermal radiation (Zel'dovich and Raizer 1967). In this case the total pressure in Eq. (3.1.27) is expressed as $p = p_1 + p_2 + p_3$.

In the one-dimensional case, Eqs. (3.1.26)–(3.1.29) have the following nonconservative form:

$$\frac{\partial \mathbf{U}}{\partial t} + A\frac{\partial \mathbf{U}}{\partial x} = \mathbf{0}, \quad \mathbf{U} = [\rho, \rho u, \rho v, E_1, E_2]^{\mathrm{T}}, \tag{3.1.32}$$

where

$$A = \frac{\partial \mathbf{F}}{\partial \mathbf{U}} = \begin{bmatrix} 0 & 1 & 0 & 0 & 0 \\ -u^2+a_0 & 2u & 0 & a_1 & a_2 \\ -uv & v & u & 0 & 0 \\ -uh_1 & h_1 & 0 & u & 0 \\ -uh_2 & h_2 & 0 & 0 & u \end{bmatrix}, \tag{3.1.33}$$

$$h_1 = \frac{E_1+p_1}{\rho}, \quad h_2 = \frac{E_2+p_2}{\rho}, \quad a_0 = \frac{\partial p}{\partial \rho}, \quad a_1 = \frac{\partial p}{\partial E_1}, \quad a_2 = \frac{\partial p}{\partial E_2}.$$

System (3.1.26)–(3.1.29) is hyperbolic and its coefficient matrix A has only real eigenvalues and a complete system of eigenvectors. Thus, the matrix A may be rewritten in the equivalent form (3.1.20) with

$$\Omega_{\mathrm{L}} = \begin{bmatrix} -v & 0 & 1 & 0 & 0 \\ -h_1 & 0 & 0 & 1 & 0 \\ -h_2 & 0 & 0 & 0 & 1 \\ a_0-uc & +c & 0 & a_1 & a_2 \\ a_0+uc & -c & 0 & a_1 & a_2 \end{bmatrix}, \quad \det\Omega_{\mathrm{L}} = -2c^3; \tag{3.1.34}$$

$$\Lambda = \mathrm{diag}[u,\ u,\ u,\ u+c,\ u-c], \tag{3.1.35}$$

$$\Omega_R = \frac{1}{2c^2}\begin{bmatrix} 0 & -2a_1 & -2a_2 & 1 & 1 \\ 0 & -2ua_1 & -2ua_2 & u+c & u-c \\ 2c^2 & -2va_1 & -2va_2 & v & v \\ 0 & 2(a_0+h_2a_2) & -2h_1a_2 & h_1 & h_1 \\ 0 & -2h_2a_1 & 2(a_0+h_1a_1) & h_2 & h_2 \end{bmatrix}, \tag{3.1.36}$$

$$c = \sqrt{a_0 + h_1a_1 + h_2a_2} = \sqrt{\frac{\partial p}{\partial \rho} + \frac{E_1+p_1}{\rho}\frac{\partial p}{\partial E_1} + \frac{E_2+p_2}{\rho}\frac{\partial p}{\partial E_2}}. \tag{3.1.37}$$

The function c is the speed of sound.

3.1.2 The mixture of ideal gases in chemical nonequilibrium.

It may be convenient to represent the system of governing equations in the curvilinear system of coordinates x^i $(i = 1, 2, 3)$. The vectors of the covariant basis ϵ_i, which are parallel to the coordinate lines of the system at any point defined by the radius-vector $\mathbf{r}$, are given by the formulas

$$\epsilon_i = \frac{\partial \mathbf{r}}{\partial x^i}.$$

Let $\hat{\mathbf{g}}$ be the metric tensor of this coordinate system and g_{ij} and g^{ij} its covariant and contravariant components. These components constitute the entries of mutually inverse matrices (see any textbook on the tensor calculus). Then, for any infinitesimal length increment ds we have

$$ds^2 = |d\mathbf{r}|^2 = d\mathbf{r}\cdot d\mathbf{r} = dx^i dx^j \epsilon_i \cdot \epsilon_j = g_{ij}dx^i dx^j.$$

From this it follows that the matrix $[g_{ij}]$ and, hence, the matrix $[g^{ij}]$ are symmetric.

By using g^{ij}, we can introduce a contravariant basis $\epsilon^i = g^{ij}\epsilon_j$. Then, the contravariant and covariant components A_i of the vector $\mathbf{A}$ can be introduced as

$$\mathbf{A} = A^i\epsilon_i = A_i\epsilon^i.$$

Similarly, a second-rank tensor $\hat{\mathbf{T}}$ can be represented as

$$\hat{\mathbf{T}} = T_{ij}\epsilon^i\epsilon^j = T^{ij}\epsilon_i\epsilon_j.$$

It is easy to see that

$$\epsilon^j \cdot \epsilon_p = g^{ij}\epsilon_i \cdot \epsilon_p = g^{ij}g_{ip} = \delta^j_p,$$

where δ^j_p is the Kronecker delta.

Since $g_{ij} = \epsilon_i \cdot \epsilon_j$ and $g^{ij} = \epsilon^i \cdot \epsilon^j$, we see that the lengths of the basis vectors ϵ_i and ϵ^i are $\sqrt{g_{ii}}$ and $\sqrt{g^{ii}}$, respectively. We can normalize the basis vectors and introduce physical components of vectors as

$$\mathbf{A} = A^i\epsilon_i \equiv A^{(i)}\frac{\epsilon_i}{\sqrt{g_{ii}}} = A^{(i)}\mathbf{e}_i;$$

$$\mathbf{A} = A_i\epsilon^i \equiv A_{(i)}\frac{\epsilon^i}{\sqrt{g^{ii}}} = A_{(i)}\mathbf{e}^i.$$

Consider the system

$$\frac{\partial \mathbf{U}'}{\partial t} + \frac{\partial \mathbf{E}'}{\partial x^1} + \frac{\partial \mathbf{F}'}{\partial x^2} + \frac{\partial \mathbf{G}'}{\partial x^3} = \mathbf{H}' \tag{3.1.38}$$

describing three-dimensional inviscid flows of a multicomponent reacting gas written for the curvilinear contravariant physical components $u = v^{(1)}$, $v = v^{(2)}$, and $w = v^{(3)}$ of the velocity vector $\mathbf{v}$ in the arbitrary nonorthogonal curvilinear coordinate system (Pogorelov and Shevelev 1981, Pogorelov 1988a, 1988b). Here

$$\mathbf{U}' = \sqrt{g}\begin{bmatrix} \rho \\ \rho u \\ \rho v \\ \rho w \\ e \\ \rho c_l \end{bmatrix}, \quad \mathbf{E}' = \sqrt{\frac{g}{g_{11}}}\begin{bmatrix} \rho u \\ \rho u^2 + pg^{11}g_{11} \\ \rho uv + pg^{12}\sqrt{g_{11}g_{22}} \\ \rho uw + pg^{13}\sqrt{g_{11}g_{33}} \\ (e+p)u \\ \rho u c_l \end{bmatrix},$$

$$\mathbf{F}' = \sqrt{\frac{g}{g_{22}}}\begin{bmatrix} \rho v \\ \rho uv + pg^{12}\sqrt{g_{11}g_{22}} \\ \rho v^2 + pg^{22}g_{22} \\ \rho vw + pg^{23}\sqrt{g_{22}g_{33}} \\ (e+p)v \\ \rho v c_l \end{bmatrix}, \quad \mathbf{G}' = \sqrt{\frac{g}{g_{33}}}\begin{bmatrix} \rho w \\ \rho uw + pg^{13}\sqrt{g_{11}g_{33}} \\ \rho vw + pg^{23}\sqrt{g_{22}g_{33}} \\ \rho w^2 + pg^{33}g_{33} \\ (e+p)w \\ \rho w c_l \end{bmatrix}$$

and $\mathbf{H}' = [0, H'^i, 0, \sqrt{g}\,\sigma_l]^{\mathrm{T}}$ with

$$H'^i = \frac{1}{2g_{ii}}\frac{\partial g_{ii}}{\partial x^j}\left[\sqrt{g}\left(\frac{\rho v^{(i)}v^{(j)}}{\sqrt{g_{jj}}} + pg^{ij}\sqrt{g_{ii}}\right)\right] - \Gamma^i_{mj}\left[\sqrt{gg_{ii}}\left(\frac{\rho v^{(m)}v^{(j)}}{\sqrt{g_{mm}g_{jj}}} + pg^{mj}\right)\right].$$

Here ρ, p, e, and $c_l = \rho_l/\rho$ are density, pressure, total energy per unit volume of the mixture, and mass fractions, or concentrations, of species ($l = 1, \dots, N$), respectively. Besides, g is the determinant of the metric tensor and Γ^i_{mj} are the Christoffel symbols. No summation must be made over the index i in the last formula. Note that passing to a curvilinear coordinate system, as a rule, results in the origin of a nondifferential source term $\mathbf{H}'$. This can sometimes complicate the numerical procedure. The appearance of Christoffel symbols is caused by the rule of the tensor differentiation in curvilinear coordinate systems:

$$\frac{\partial \mathbf{A}}{\partial x^i} = \frac{\partial A^k \boldsymbol{\epsilon}_k}{\partial x^i} = \frac{\partial A^k}{\partial x^i}\boldsymbol{\epsilon}_k + A^k\frac{\partial \boldsymbol{\epsilon}_k}{\partial x_i} \equiv \frac{\partial A^k}{\partial x^i}\boldsymbol{\epsilon}_k + A^k\Gamma^j_{ki}\boldsymbol{\epsilon}_j.$$

Note that the structure of the system has not changed much after introducing the continuity equations for individual species and it remains hyperbolic.

Consider a mixture with K chemical reactions between N components A_l,

$$\sum_{l=1}^{N} \nu_l^{(r)'} A_l \rightleftharpoons \sum_{l=1}^{N} \nu_l^{(r)''} A_l, \qquad r = 1, \dots, K;$$

with k_r^+ and k_r^- being the rate constants of direct and reverse reactions, respectively. The net rate of rth reaction is

$$\mathcal{R}_r = k_r^+ \prod_{l=1}^{N}[A_l]^{\nu_l^{(r)'}} - k_r^- \prod_{l=1}^{N}[A_l]^{\nu_l^{(r)''}},$$

where $[A_l] = \rho c_l/M_l = \rho_l/M_l$ is the molar density of lth component, with M_l and ρ_l being its molecular mass and mass density, respectively. Since rth reaction gives $\nu_l^{(r)''} - \nu_l^{(r)'}$ moles of A_l, the mass production rate of A_l is

$$\sigma_l^{(r)} = M_l[\nu_l^{(r)''} - \nu_l^{(r)'}]\mathcal{R}_r.$$

Performing summation over all chemical reactions, we obtain

$$\sigma_l = \sum_{r=1}^{K} \sigma_l^{(r)}. \tag{3.1.39}$$

3.2 The Godunov method for gas dynamic equations

The integral form of Eqs. (3.1.1)–(3.1.3) is

$$\frac{d}{dt}\left(\iiint_G \rho\, dG\right) + \oint_S \rho\mathbf{v}\cdot d\mathbf{S} = 0, \tag{3.2.1}$$

$$\frac{d}{dt}\left(\iiint_G \rho\mathbf{v}\, dG\right) + \oint_S (\rho\mathbf{v}\mathbf{v} + p\hat{\mathbf{I}})\cdot d\mathbf{S} = \mathbf{0}, \tag{3.2.2}$$

$$\frac{d}{dt}\left(\iiint_G e\, dG\right) + \oint_S (e+p)\mathbf{v}\cdot d\mathbf{S} = 0. \tag{3.2.3}$$

Here G is a domain in the three-dimensional space (x, y, z); $dG = dx\,dy\,dz$ is the volume element; S is the boundary surface of G, $d\mathbf{S} = \mathbf{n}\,dS$ is the oriented element of S, where $\mathbf{n}$ is the outward normal to S, and dS is the area element. In addition, $\mathbf{a}\cdot\mathbf{b}$ stands for the scalar product of the vectors $\mathbf{a}$ and $\mathbf{b}$.

Let us construct the explicit Godunov finite-volume scheme for equations (3.2.1)–(3.2.3) written in integral form. We discretize the computational domain by constructing a grid of arbitrary convex polyhedrons having volumes G_i, $i = 1, 2, \ldots$, with $m = m(i)$ faces having areas S_j, $j = 1, \ldots, m(i)$. For each polyhedron, the integral equations are approximated as follows:

$$G_i \frac{\rho_i^{k+1} - \rho_i^k}{\Delta t} + \sum_{j=1}^{m(i)} R_j\,(\mathbf{V}_j\cdot\mathbf{S}_j) = 0, \tag{3.2.4}$$

$$G_i \frac{(\rho\mathbf{v})_i^{k+1} - (\rho\mathbf{v})_i^k}{\Delta t} + \sum_{j=1}^{m(i)} (R\mathbf{V})_j\,(\mathbf{V}_j\cdot\mathbf{S}_j) + \sum_{j=1}^{m(i)} P_j\,\mathbf{S}_j = \mathbf{0}, \tag{3.2.5}$$

$$G_i \frac{e_i^{k+1} - e_i^k}{\Delta t} + \sum_{j=1}^{m(i)} (E_j + P_j)(\mathbf{V}_j\cdot\mathbf{S}_j) = 0, \tag{3.2.6}$$

where $\mathbf{S}_j = \mathbf{n}_j S_j$. The integer subscript i in Eqs. (3.2.4)–(3.2.6) denotes the values of grid variables calculated at the center of mass of the ith polyhedral cell, and the subscript j denotes their values on the jth face of the mesh cell. The integer superscript k indicates the time layer number and Δt is the time increment. The corresponding capital letters denote the density R, velocity $\mathbf{V}$, pressure P, and total energy E at the cell faces. Their values are calculated by solving the Riemann problem for the gas dynamic equations.

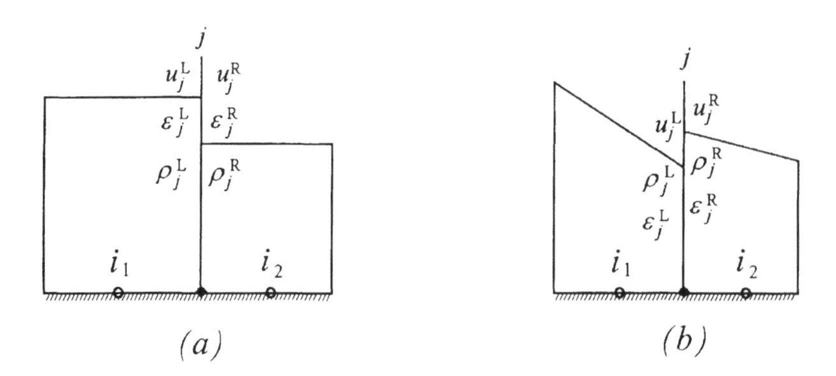

Figure 3.1 Choice of function slopes.

The explicit Godunov finite-volume scheme in the general case of moving grids can be approximated as follows:

$$\frac{(\rho G)_i^{k+1} - (\rho G)_i^k}{\Delta t} + \sum_{j=1}^{m(i)} R_j\left([\mathbf{V} - \mathbf{D}] \cdot \mathbf{S}^{k+1/2}\right)_j = 0, \tag{3.2.7}$$

$$\frac{(\rho \mathbf{v} G)_i^{k+1} - (\rho \mathbf{v} G)_i^k}{\Delta t} + \sum_{j=1}^{m(i)} (R\mathbf{V})_j\left([\mathbf{V} - \mathbf{D}] \cdot \mathbf{S}^{k+1/2}\right)_j + \sum_{j=1}^{m(i)} P_j \mathbf{S}_j^{k+1/2} = \mathbf{0}, \tag{3.2.8}$$

$$\frac{(eG)_i^{k+1} - (eG)_i^k}{\Delta t} + \sum_{j=1}^{m(i)} E_j\left([\mathbf{V} - \mathbf{D}] \cdot \mathbf{S}^{k+1/2}\right)_j + \sum_{j=1}^{m(i)} P_j\left(\mathbf{V} \cdot \mathbf{S}^{k+1/2}\right)_j = 0, \tag{3.2.9}$$

$$\frac{G_i^{k+1} - G_i^k}{\Delta t} - \sum_{j=1}^{m(i)} \left(\mathbf{D} \cdot \mathbf{S}^{k+1/2}\right)_j = 0; \tag{3.2.10}$$

see also Eqs. (2.2.9)–(2.2.10). Here $\mathbf{D}_j$ is the velocity of the center of the jth face. The quantities R_j, $\mathbf{V}_j$, E_j, and P_j are defined by solving the gas dynamic Riemann problem and are calculated for the jth face moving with velocity $\mathbf{D}_j$. In addition, the volumes G_i and oriented surface elements $\mathbf{S}_j$ are now functions of time. The half-integer subscript $k + 1/2$ in Eqs. (3.2.7)–(3.2.10) denotes the corresponding values for the time $t + \frac{1}{2}\Delta t$. Equation (3.2.10) is a discrete condition of consistent approximation for G and $\mathbf{S}$, or the equation for the time evolution of the cell volume G_i. The important property of this approximation is that the constant-flow solutions $\mathbf{U} = \mathbf{U}_0 = \text{const}$ to the gas dynamic equations are also solutions to the difference equations (3.2.7)–(3.2.9). This condition is essential when one uses arbitrary curvilinear and/or moving coordinate systems. For a given approximation of $\mathbf{S}$, one can use Eq. (3.2.10) to calculate the volumes G_i^{k+1}.

For $\mathbf{D} = \mathbf{0}$ we obtain Eqs. (3.2.4)–(3.2.6) in the Euler form and for $\mathbf{D} = \mathbf{V}$ we obtain the representation of the gas dynamic equations in the Lagrangian form with zero mass flux across all cell boundaries.

The idea of the Godunov method can be explained by the following one-dimensional considerations. Consider two adjacent cells, with numbers i_1 and i_2, separated by the jth boundary (Fig. 3.1a). Treating the density, velocity, and energy inside each cell as piecewise constant functions on G, we calculate the values of the density, ρ_j^L and ρ_j^R, velocity, u_j^L and u_j^R, and internal energy, ε_j^L and ε_j^R, at the left- and right-hand sides of the jth stationary or moving boundary. Then, we use these values to solve the Riemann problem and find the density R, total energy E, and velocity U in the direction normal to the boundary.

In the three-dimensional problem, the vector $\mathbf{V} = [U, V, W]^T$ is represented as the sum of the components that are normal (U) and tangential (V and W) to the cell boundary. The Riemann problem is solved for the normal component U, and the tangential components V and W are transferred in the direction of the velocity U obtained by solving the Riemann problem.

The linearized finite-volume scheme (3.2.4)–(3.2.6) is stable on the uniform Cartesian grid if

$$\max(|C_x| + |C_y| + |C_z|) \leq \max |C_x| + \max |C_y| + \max |C_z| \leq 1, \qquad (3.2.11)$$

where C_x, C_y, and C_z are the CFL numbers that correspond to the x-, y-, and z-axis, respectively,

$$C_x = (|u| + c)\frac{\Delta t}{\Delta x}, \quad C_y = (|v| + c)\frac{\Delta t}{\Delta y}, \quad C_z = (|w| + c)\frac{\Delta t}{\Delta z},$$

where Δx, Δy, and Δz are the mesh sizes in the respective directions and c is the sound speed; see Section 2.6. For an arbitrary mesh constructed of convex polyhedra, the time increment can be estimated in the following way (Belikov and Semenov 1997c):

$$\max_i \frac{1}{2G_i}\left[\sum_{j=1}^{m(i)}(|V_n| - V_n + c)_j S_j\right]\Delta t \leq 1, \qquad V_n = \mathbf{V}\cdot\mathbf{n}.$$

For the uniform Cartesian grid and constant flow velocity this criterion transforms into $\max(|C_x| + |C_y| + |C_z|) \leq 1$.

The scheme described above is first-order accurate in time and space. The accuracy of these schemes can be increased by adopting the piecewise polynomial distribution of functions inside computational cells, see Section 2.5. Higher-order accuracy may be required for the following reasons: (i) if the grid is highly nonuniform in space, a first-order scheme may be unacceptable, since the error can be of order unity; (ii) first-order schemes have considerable numerical viscosity, which may significantly flatten the profiles of grid variables in the domains of smooth solution. In particular, the order of accuracy with respect to spatial coordinates can be increased by applying the following algorithm. We assume that ρ, u, and ε are linear functions within a cell, see Fig. 3.1b, and calculate the initial values of ρ^L, ρ^R, u^L, u^R, ε^L, and ε^R at the centers of the cell faces. We construct linear distributions of the density, velocity, and energy inside the cell. This procedure is called piecewise-linear function reconstruction. To satisfy the stability condition the function slopes must be bounded by special algorithms referred to as "limiters." For description of limiters, reconstruction procedures, and limiting reconstructions, one can refer to Section 2.7. Higher order of accuracy with respect to time can be attained by using predictor–corrector methods, see Sections 2.5 and 2.7.

3.3 Exact solution of the Riemann problem

Let us construct an exact solution of the Riemann problem for the gas dynamic equations with two different two-term EOSs (3.1.5). The importance of this exact solution stems from the fact that it permits one to use an arbitrary EOS by a special approach. In constructing the solution below we follow the logical scheme of Godunov et al. (1979) for the same two-term EOSs. We use the relations at discontinuities in the form of Kotchine (1926).

First we shall construct the basic elementary solutions. Then an exact solution will be described using the superposition principle.

3.3.1 Elementary solution 1: Shock wave. The first elementary solution of the gas dynamic equations is a moving discontinuity (a shock wave or a contact discontinuity). For obtaining of the discontinuity relations let us integrate Eq. (3.1.15) over t and x and consider their integral form

$$\oint_L (\mathbf{U}\,dx - \mathbf{F}\,dt) = \mathbf{0}, \tag{3.3.1}$$

where L is the boundary of a region in the (t, x) plane. Let us seek the discontinuous solution of Eqs. (3.1.15)–(3.1.17) in the form of a travelling wave $f(t,x) = f(\zeta) \equiv f(x - Wt)$, where $W = \text{const}$ is the wave velocity. Consider Eq. (3.3.1) in orthogonal coordinates (ζ, τ) associated with discontinuity, where $\zeta = x - Wt$ and $\tau = Wx + t$. The coordinate ζ is normal and coordinate τ is tangential with respect to the discontinuity. Using transformation

$$x = \frac{\zeta + W\tau}{1 + W^2}, \qquad t = \frac{-W\zeta + \tau}{1 + W^2}$$

we can rewrite Eq. (3.3.1) as

$$\frac{1}{1 + W^2} \oint_L [(W\mathbf{U} - \mathbf{F})\,d\tau + (\mathbf{U} + W\mathbf{F})\,d\zeta] = \mathbf{0}. \tag{3.3.2}$$

Let us integrate (3.3.2) for the rectangular region $\tau_0 - \delta\tau \le \tau \le \tau_0 + \delta\tau$ and $\zeta_0 - \delta\zeta \le \zeta \le \zeta_0 + \delta\zeta$, where $\zeta = \zeta_0$ corresponds to the discontinuity. We can find

$$\begin{aligned} &(W\mathbf{U} - \mathbf{F})_1 2\delta\tau - (W\mathbf{U} - \mathbf{F})_2 2\delta\tau = \mathbf{0} \\ &\Longrightarrow \quad W\{\mathbf{U}\} - \{\mathbf{F}\} = \mathbf{0}, \end{aligned} \tag{3.3.3}$$

where $\{q\} \equiv q_1 - q_2$, and indices 1 and 2 denote the variables on the left- and right-hand side of the discontinuity.

Consider the Riemann problem with the following initial data: $\mathbf{U}_1 = [\rho_1, u_1, v_1, w_1, e_1]^{\mathrm{T}}$, $p_1 = p(\rho_1, \varepsilon_1)$ for $x < 0$, and $\mathbf{U}_2 = [\rho_2, u_2, v_2, w_2, e_2]^{\mathrm{T}}$, $p_2 = p(\rho_2, \varepsilon_2)$ for $x > 0$. The two-term EOS has the parameters γ, p_0 and c_0 of (3.1.5). Both $\mathbf{U}_1$ and $\mathbf{U}_2$ satisfy Eq. (3.1.15). Assume that this initial discontinuity moves with a velocity of W and is a solution of the Riemann problem. Then we find from Eq. (3.3.3) the desired formulas relating $\mathbf{U}_1$, $\mathbf{U}_2$, and W:

$$W\{\rho\} - \{\rho u\} = 0, \tag{3.3.4}$$

$$W\{\rho u\} - \{\rho u^2 + p\} = 0, \tag{3.3.5}$$
$$W\{\rho v\} - \{\rho u v\} = 0, \tag{3.3.6}$$
$$W\{\rho w\} - \{\rho u w\} = 0, \tag{3.3.7}$$
$$W\{e\} - \{eu + pu\} = 0. \tag{3.3.8}$$

If these relations are satisfied, then this discontinuity is a solution of Eq. (3.1.15). To describe all discontinuity configurations it is convenient to rewrite relations (3.3.4)–(3.3.8) in the coordinate system attached to the discontinuity. Let us introduce a new variable $V = u - W$. Using the fact that $\{W\} = 0$, we obtain the more compact relations

$$\{\rho V\} = 0, \tag{3.3.9}$$
$$\{\rho V^2 + p\} = 0, \tag{3.3.10}$$
$$\{\tfrac{1}{2}\rho V^3 + pV + \rho\varepsilon V + \tfrac{1}{2}\rho(v^2 + w^2)V\} = 0, \tag{3.3.11}$$
$$\{\rho v V\} = 0, \quad \{\rho w V\} = 0. \tag{3.3.12}$$

In particular, let us verify formulas (3.3.11). Indeed,

$$\begin{aligned}
&\{\tfrac{1}{2}\rho V^3 + pV + \rho\varepsilon V + \tfrac{1}{2}\rho(v^2 + w^2)V\} \\
&\quad = \{\tfrac{1}{2}\rho(u^2 - 2uW + W^2)(u - W) + (u - W)p + (u - W)[\rho\varepsilon + \tfrac{1}{2}\rho(v^2 + w^2)]\} \\
&\quad = \{\tfrac{1}{2}\rho(-2uW + W^2)(u - W) + pu - pW + (u - W)e\} \\
&\quad = -[W\{\rho\} - \{\rho u\}]\tfrac{1}{2}W^2 - [W\{e\} - \{eu + pu\}] + [W\{\rho u\} - \{\rho u^2 + p\}]W = 0
\end{aligned}$$

in accordance with relations (3.3.4), (3.3.8), and (3.3.5).

Suppose that the mass flux $m = \rho V = \rho_1 V_1 = \rho_2 V_2$ through the discontinuity is not equal to zero. In this case we have a shock wave that holds

$$\{v\} = v_1 - v_2 = 0, \quad \{w\} = w_1 - w_2 = 0, \tag{3.3.13}$$

see Eqs. (3.3.12). The case $\rho V = 0$ corresponding a contact discontinuity will be considered separately.

In the case of $\rho V \neq 0$, or $\{v\} = \{w\} = 0$, Eqs. (3.3.9)–(3.3.11) permit us to find compact formulas for u_2, ε_2, and ρ_2 as functions of only p_2, p_1, ρ_1, and W, see Kotchine (1926) and Godunov et al. (1979).

From (3.3.9) we have

$$V_2 = \frac{\rho_1 V_1}{\rho_2}. \tag{3.3.14}$$

Using this relation and Eq. (3.3.10), we obtain

$$V_1 = \pm\sqrt{\frac{p_2 - p_1}{\rho_2 - \rho_1}\frac{\rho_2}{\rho_1}} \quad \Longrightarrow \quad m = \rho_1 V_1 = \pm\sqrt{\frac{p_2 - p_1}{1/\rho_1 - 1/\rho_2}}. \tag{3.3.15}$$

Substituting V_1 and V_2 into Eq. (3.3.11) yields

$$\varepsilon_2 - \varepsilon_1 = \frac{p_2 + p_1}{2}\left(\frac{1}{\rho_1} - \frac{1}{\rho_2}\right). \tag{3.3.16}$$

Using the EOS (3.1.5) to eliminate ε from Eq. (3.3.16), we arrive at the so-called Rankine–Hugoniot relation:

$$\rho_2 = \rho_1 \frac{(p_2 + p_0)(\gamma + 1) + (p_1 + p_0)(\gamma - 1)}{(p_2 + p_0)(\gamma - 1) + (p_1 + p_0)(\gamma + 1)}.$$

Thus, V_1 can be expressed in terms of p_2, p_1, and ρ_1:

$$V_1^2 = \frac{(p_2 + p_0)(\gamma + 1) + (p_1 + p_0)(\gamma - 1)}{2\rho_1}.$$

Let $m = \rho V$ denote the mass flux across the discontinuity. From Eqs. (3.3.9)–(3.3.10) it follows that $\{m\} = 0$ and $\{mV + p\} = 0$. Then,

$$m_1 = \rho_1 V_1 = \pm\sqrt{\frac{\rho_1}{2}\,[(p_2 + p_0)(\gamma + 1) + (p_1 + p_0)(\gamma - 1)]}.$$

It is necessary to decide which sign must be chosen in this formula. The rule can be stated as follows: there only exist compression shock waves. The pressure behind the shock must be larger than that ahead of the shock. This rule is confirmed by various experiments and by the stability analysis of shock wave. It must be used to select physically relevant solutions (Rozhdestvenskii and Yanenko 1983).

Let us describe all shock wave configurations.

Left shock wave

We say that the shock wave is *left* if the gas flow through it is directed along the x-direction from left to right (Fig. 3.2a), i.e., $m = \rho V = (u - W)\rho > 0$. Denote the variables on the left- and right-hand side of the shock wave by ρ_{I}, u_{I}, ε_{I}, and p_{I} and R_{I}, U, E_{I}, and P, respectively, see Fig. 3.2a. The EOS is characterized by the parameters γ_{I}, $p_{0\text{I}}$, and $c_{0\text{I}}$. Using the condition $\{mV + p\} = 0$, we obtain the following equation for U and P, where $P \geq p_{\text{I}}$,

$$U - u_{\text{I}} + \frac{P - p_{\text{I}}}{|m_{\text{I}}|} = 0, \tag{3.3.17}$$
$$m_{\text{I}} = \sqrt{\frac{\rho_{\text{I}}}{2}[(P + p_{0\text{I}})(\gamma_{\text{I}} + 1) + (p_{\text{I}} + p_{0\text{I}})(\gamma_{\text{I}} - 1)]}.$$

Right shock wave

We say that the shock wave is *right* if the gas flow through it is directed from right to left, i.e., $m = \rho V = (u - W)\rho < 0$. Denote the variables on the left- and right-hand side of the shock wave by R_{II}, U, E_{II}, and P and ρ_{II}, u_{II}, ε_{II}, and p_{II}, respectively, see Fig. 3.2b. The EOS involves the parameters $\gamma_{\text{II}}, p_{0\text{II}}$, and $c_{0\text{II}}$. Using the condition $\{mV + p\} = 0$, we arrive at the following equation for U and P, where $P \geq p_{\text{II}}$,

$$U - u_{\text{II}} - \frac{P - p_{\text{II}}}{|m_{\text{II}}|} = 0, \tag{3.3.18}$$
$$m_{\text{II}} = \sqrt{\frac{\rho_{\text{II}}}{2}[(P + p_{0\text{II}})(\gamma_{\text{II}} + 1) + (p_{\text{II}} + p_{0\text{II}})(\gamma_{\text{II}} - 1)]}.$$

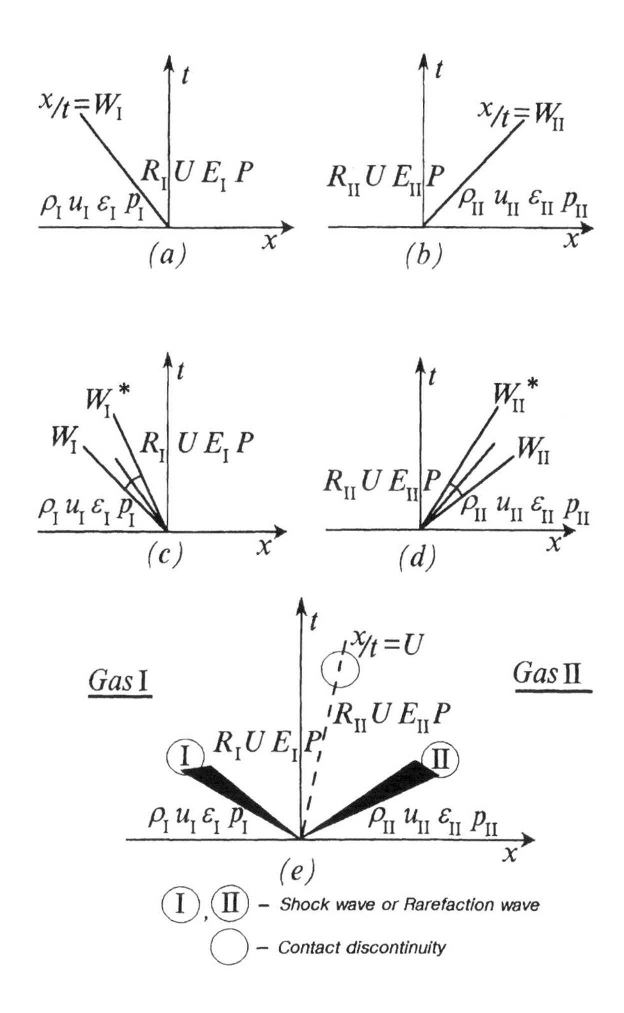

Figure 3.2 Scheme of flow configurations.

A shock wave is a self-similar solution with respect to the variable $\zeta = x - Wt$, $W = \text{const}$. This solution is also self-similar with respect to $\xi = x/t$. In fact, a moving shock wave in the (t, x)-coordinates is a straight line defined by the equation $\xi = W = \text{const}$.

3.3.2 Elementary solution 2: Contact discontinuity. Another elementary solution of the gas dynamic equations is the so-called contact discontinuity, which is a special case of a discontinuity moving with a velocity $W = U = \text{const}$ and having zero mass flux across the discontinuity, i.e., $m = \rho V = (u - U)\rho = \rho_1 V_1 = \rho_2 V_2 = 0$. Denote the variables on the left- and right-hand side of the contact discontinuity by p_1, u_1, v_1, and w_1 and p_2, u_2, v_2, and w_2, respectively. Then from formulas (3.3.9)–(3.3.10) we obtain $u_1 = u_2 = U$ and $p_1 = p_2 = P$, and from relations (3.3.12) we obtain that v_1, v_2, w_1, and w_2 can be arbitrary.

We would like to emphasize that the term "contact discontinuity" is used here only by analogy with the general case of a hyperbolic system, where it refers to discontinuities related to linearly degenerate characteristic fields. In gas dynamics, strictly speaking, we must call them *tangential* discontinuities. From this viewpoint, the contact discontinuity is just a particular case of tangential discontinuities. It separates two gases with different densities and/or EOSs with no relative motion occurring between them. In more complicated fluid models, such as magnetohydrodynamics, the difference between tangential and contact discontinuities is more fundamental (see Chapter 5).

3.3.3 Elementary solution 3: Rarefaction wave. The third elementary solution of the gas dynamic equations is a simple rarefaction wave. This solution is continuous. Let us seek the solution in the self-similar form $f(x, t) = f(\xi) \equiv f(x/t)$. Since we seek a continuous solution we can consider the equations in a nonconservative form, particularly,

$$\rho_t + u\rho_x + \rho u_x = 0, \tag{3.3.19}$$

$$u_t + uu_x + \frac{p_x}{\rho} = 0, \quad v_t + uv_x = 0, \quad w_t + uw_x = 0, \tag{3.3.20}$$

$$p_t + up_x + \gamma(p + p_0)u_x = 0. \tag{3.3.21}$$

To solve this system we first transform it to the characteristic form.

If we introduce the state vector $\mathbf{W} = [\rho, u, v, w, p]^{\mathrm{T}}$ the system (3.3.19)–(3.3.21) acquires the form

$$\mathbf{W}_t + B\mathbf{W}_x = \mathbf{0}, \tag{3.3.22}$$

where

$$B = \begin{bmatrix} u & \rho & 0 & 0 & 0 \\ 0 & u & 0 & 0 & 1/\rho \\ 0 & 0 & u & 0 & 0 \\ 0 & 0 & 0 & u & 0 \\ 0 & \gamma(p + p_0) & 0 & 0 & u \end{bmatrix},$$

and the matrix Ω_L of the left eigenvectors of B is

$$\Omega_L = \begin{bmatrix} \gamma(p+p_0) & 0 & 0 & 0 & -\rho \\ 0 & 0 & 1 & 0 & 0 \\ 0 & 0 & 0 & 1 & 0 \\ 0 & \gamma(p+p_0) & 0 & 0 & -c \\ 0 & \gamma(p+p_0) & 0 & 0 & +c \end{bmatrix}, \quad \det \Omega_L = 2\rho^2 c^5, \quad c = \sqrt{\frac{\gamma(p+p_0)}{\rho}}.$$

If we introduce a diagonal matrix Λ with the entries equal to the eigenvalues of B, $\Lambda = \text{diag}[u, u, u, u-c, u+c]$, we can write out the initial system in the characteristic form

$$\Omega_L \mathbf{W}_t + \Lambda \Omega_L \mathbf{W}_x = \mathbf{0}.$$

This equation is said to be the characteristic form of system (3.3.22). In the expanded form it reads

$$\sum_{k=1}^{5} \Omega_{Lpk} \frac{\partial W_k}{\partial t} + \lambda_p \sum_{k=1}^{5} \Omega_{Lpk} \frac{\partial W_k}{\partial x} = 0, \qquad p = 1, \ldots, 5;$$

where $\Omega_L = [\Omega_{Lpk}]$ and $\Lambda = [\lambda_p \, \delta_{pk}]$. Let us seek an exact continuous solution as a function of $\xi = x/t$. Then we find that

$$(\lambda_p - \xi) \sum_{k=1}^{5} \Omega_{Lpk} \frac{\partial W_k}{\partial \xi} = 0, \qquad p = 1, \ldots, 5.$$

Thus, an exact solution must satisfy one of the following five systems of equations ($\alpha = 1, \ldots, 5$):

$$\lambda_\alpha - \xi = 0, \tag{3.3.23}$$

$$(\lambda_\beta - \xi) \sum_{k=1}^{5} \Omega_{L\beta k} \frac{\partial W_k}{\partial \xi} = 0, \qquad \beta = 1, \ldots, 5; \quad \beta \neq \alpha. \tag{3.3.24}$$

Thus, we must consider three cases of Eqs. (3.3.23)–(3.3.24) corresponding to the three different eigenvalues u, $u-c$, and $u+c$.

Case I:

$$u - \xi = 0,$$
$$\gamma(p+p_0)u_\xi - cp_\xi = 0, \quad \gamma(p+p_0)u_\xi + cp_\xi = 0.$$

The exact, however, particular solution of this system is $u = \xi$, $p = -p_0 = \text{const}$. The functions v and w are arbitrary.

Case II:

$$u - c - \xi = 0, \tag{3.3.25}$$

$$\gamma(p+p_0)\rho_\xi - \rho p_\xi = 0, \quad \gamma(p+p_0)u_\xi + cp_\xi = 0, \tag{3.3.26}$$

$$v_\xi = 0, \quad w_\xi = 0 \quad \Longrightarrow \quad v = \text{const}, \quad w = \text{const}. \tag{3.3.27}$$

The second equation is equivalent to $S_\xi = 0$, or $S = \text{const}$, where $S = \gamma(p + p_0)/\rho^\gamma$ is the entropy. From the third equation and the condition $S = \text{const}$ it follows that

$$\left(u + \frac{2c}{\gamma - 1}\right)_\xi = 0,$$

or $w_\mathrm{L} = u + 2c/(\gamma - 1) = \text{const}$, where w_L is a so-called left Riemann invariant.

Case III:

$$u + c - \xi = 0, \tag{3.3.28}$$

$$\gamma(p + p_0)\rho_\xi - \rho p_\xi = 0, \quad \gamma(p + p_0)u_\xi - c p_\xi = 0, \tag{3.3.29}$$

$$v_\xi = 0, \quad w_\xi = 0 \quad \Longrightarrow \quad v = \text{const}, \quad w = \text{const}. \tag{3.3.30}$$

These equations give $S = \text{const}$ and $w_\mathrm{R} = u - 2c/(\gamma - 1) = \text{const}$, where w_R is a so-called right Riemann invariant.

Left rarefaction wave

We say that the rarefaction wave is *left* if the gas flow through it is directed along the x-direction from left to right, i.e., $(u - \xi)\rho > 0$. It implies that we deal with the relations

$$u - c = \xi,$$

$$\{w_\mathrm{L}\} = \left\{u + \frac{2c}{\gamma - 1}\right\} = 0, \tag{3.3.31}$$

$$\{S\} = \left\{\frac{\gamma(p + p_0)}{\rho^\gamma}\right\} = 0. \tag{3.3.32}$$

In this case one can obtain $(u - \xi)\rho = \rho c > 0$. Denote the variables on the left- and on the right-hand side of the rarefaction wave by ρ_I, u_I, ε_I, and p_I and R_I, U, E_I, and P, respectively, see Fig. 3.2c. The EOS involves the parameters γ_I, $p_{0\mathrm{I}}$, and $c_{0\mathrm{I}}$. Using Eq. (3.3.31), we obtain the following equation for U and P:

$$U - u_\mathrm{I} + \frac{2}{\gamma_\mathrm{I} - 1}(c - c_\mathrm{I}) = 0,$$

or, using Eq. (3.3.32),

$$U - u_\mathrm{I} + \frac{2c_\mathrm{I}}{\gamma_\mathrm{I} - 1}\left[\left(\frac{P + p_{0\mathrm{I}}}{p_\mathrm{I} + p_{0\mathrm{I}}}\right)^\chi - 1\right] = 0, \quad \chi = \frac{\gamma_\mathrm{I} - 1}{2\gamma_\mathrm{I}}. \tag{3.3.33}$$

Right rarefaction wave

We say that the rarefaction wave is *right* if the gas flow through it is directed from right to left, i.e., $(u - \xi)\rho < 0$. It implies that we deal with the relations

$$u + c = \xi,$$

$$\{w_\mathrm{R}\} = \left\{u - \frac{2c}{\gamma - 1}\right\} = 0, \tag{3.3.34}$$

$$\{S\} = \left\{\frac{\gamma(p + p_0)}{\rho^\gamma}\right\} = 0. \tag{3.3.35}$$

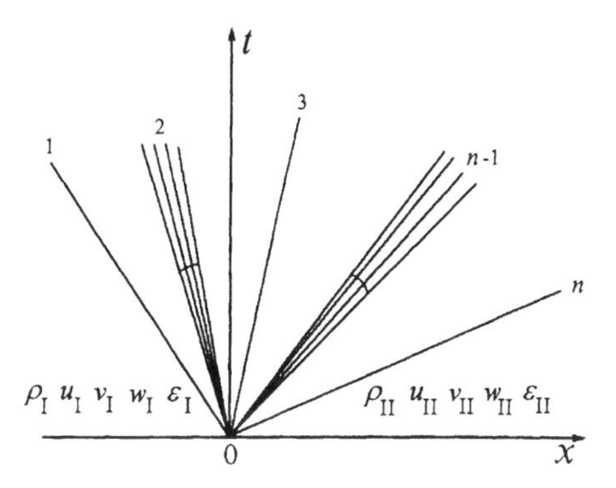

Figure 3.3 Construction of the general solution for gas dynamic equations.

In this case one can obtain $(u - \xi)\rho = -\rho c < 0$. Denote the variables on the left- and on the right-hand side of the rarefaction wave by R_{II}, U, E_{II}, and P and ρ_{II}, u_{II}, ε_{II}, and p_{II}, respectively, see Fig. 3.2d. The EOS involves the parameters γ_{II}, $p_{0\text{II}}$, and $c_{0\text{II}}$. Using Eq. (3.3.34), we arrive at the following equation for U and P:

$$U - u_{\text{II}} - \frac{2}{\gamma_{\text{II}} - 1}(c - c_{\text{II}}) = 0,$$

or, using Eq. (3.3.35),

$$U - u_{\text{II}} - \frac{2c_{\text{II}}}{\gamma_{\text{II}} - 1}\left[\left(\frac{P + p_{0\text{II}}}{p_{\text{II}} + p_{0\text{II}}}\right)^{\chi} - 1\right] = 0, \quad \chi = \frac{\gamma_{\text{II}} - 1}{2\gamma_{\text{II}}}. \tag{3.3.36}$$

3.3.4 General exact solution. The general exact solution of the Riemann problem consists of the elementary solutions separated by regions with constant quantities and can be represented on the plane (t, x) in the way shown in Fig. 3.3. The solution consists of, in total, n shock waves (SW), contact discontinuities (CD), and rarefaction waves (RW). The RWs are represented by a bundle of straight lines, or by a rarefaction fan. The elementary solutions are separated by regions with constant quantities, since a constant flow is also an exact solution of the equations. The n basic solutions provide $3n$ relations for the constant flow parameters. These $3n$ relations must relate $3(n - 1)$ unknowns ρ, u, and ε located in $(n - 1)$ inner regions with constant quantities, and n unknown velocities of the SWs, CDs and right boundaries of the RWs. The number of unknowns must be equal to the number of relations. Thus, we have $3(n - 1) + n = 3n$ and $n = 3$. In 1926 Kotchine proved that the general solution is a combination of only two waves (SW+SW, SW+RW, RW+SW, or RW+RW) that are separated by a contact discontinuity (Fig. 3.2e). There is no another general triple configuration for solution, for example, of SW+SW+SW or RW+RW+RW type. However, the left and right states in the Riemann problem flow can be selected so that the exact solution can be a single SW, CD, or RW, see Fig. 3.2a–d, or only twin configurations SW+SW or RW+RW, and so on. In the general case the contact discontinuity can separate two gases with arbitrary $\{v\}$ and $\{w\}$ and different EOSs.

The described solutions are self-similar with respect to the variable $\xi = x/t$. Since the initial data are automatically consistent with the dependence on ξ, we sought exact solutions of the Riemann problem in the form of $f(\xi)$. Other solutions $\tilde{f}(t,x)$ such that $\tilde{f}(t,x) \neq \tilde{f}(\xi)$ seem to be unknown by now. Moreover, if the solution is self-similar, it is unique.

For the two-term EOS the above exact solution as a function of ξ is unique. So far, for other, more complicated wide-range EOSs, there are no exact solutions available.

To determine the pressure P in the central zone, see Fig. 3.2e, only one equation suffices. To obtain this equation we subtract Eq. (3.3.18) (resp., Eq. (3.3.36)) from Eq. (3.3.17) (resp., Eq. (3.3.33)). Then, by eliminating U, we obtain the desired equation for P:

$$F(P) = f_{\rm I}(P, p_{\rm I}, \rho_{\rm I}) + f_{\rm II}(P, p_{\rm II}, \rho_{\rm II}) = u_{\rm I} - u_{\rm II}, \tag{3.3.37}$$

$$f_n = f_n(P, p_n, \rho_n) = \begin{cases} \dfrac{P - p_n}{\rho_n c_n \beta_n} & \text{if } P \geq p_n, \\ \dfrac{2c_n}{\gamma_n - 1}(\pi_n^{\chi} - 1) & \text{if } P < p_n, \end{cases} \qquad n = \text{I or II},$$

$$\pi_n = \frac{P + p_{0n}}{p_n + p_{0n}}, \quad \chi = \frac{\gamma_n - 1}{2\gamma_n}, \quad \beta_n = \sqrt{\frac{\gamma_n + 1}{2\gamma_n}\pi_n + \frac{\gamma_n - 1}{2\gamma_n}}, \quad c_n = \sqrt{\gamma_n \frac{p_n + p_{0n}}{\rho_n}}.$$

In these formulas the subscript $n = \text{I}, \text{II}$ refers to two different EOSs.

Straightforward calculation shows that

$$\frac{\partial f_n}{\partial P} = \begin{cases} \dfrac{(\gamma_n + 1)\pi_n + 3\gamma_n - 1}{4\rho_n c_n \gamma_n \beta_n{}^3} & \text{if } P \geq p_n, \\ \dfrac{c_n}{\gamma_n (P + p_{0n})}\pi_n^{\chi} & \text{if } P < p_n, \end{cases}$$

$$\frac{\partial^2 f_n}{\partial P^2} = \begin{cases} -\dfrac{(\gamma_n + 1)[(\gamma_n + 1)\pi_n + 7\gamma_n - 1]}{16\rho_n^2 c_n^3 \gamma_n \beta_n{}^5} & \text{if } P \geq p_n, \\ -\dfrac{(\gamma_n + 1)c_n}{2\gamma_n^2 (P + p_{0n})^2}\pi_n^{\chi} & \text{if } P < p_n. \end{cases}$$

Thus, $\partial F/\partial P > 0$ and $\partial^2 F/\partial P^2 < 0$. These properties of the derivatives indicate that $F(P)$ is monotone and convex upward, see Fig. 3.4. Therefore, Eq. (3.3.37) can be readily solved by Newton's or other iteration methods, see Ortega and Rheinboldt (1970). The positivity of the first derivative and the nonpositivity of the second derivative ensure the convergence of Newton's iterations. One can use the following iteration process:

$$P^{(m+1)} = P^{(m)} - \left[f_{\rm I}^{(m)} + f_{\rm II}^{(m)} - (u_{\rm I} - u_{\rm II})\right] \left[\left(\frac{\partial f_{\rm I}}{\partial P}\right)^{(m)} + \left(\frac{\partial f_{\rm II}}{\partial P}\right)^{(m)}\right]^{-1}. \tag{3.3.38}$$

Here $m = 0, 1, 2, \ldots$ is the number of the iteration and $P^{(0)}$ is the initial approximation. Since the function has a nonpositive second derivative, we can set $P^{(0)}$ equal to any number satisfying the inequality $F(P^{(0)}) - (u_{\rm I} - u_{\rm II}) \leq 0$. However, the most suitable starting point is the acoustic, or "sonic," approximation; then for all π_n and β_n occurring in f_n we have

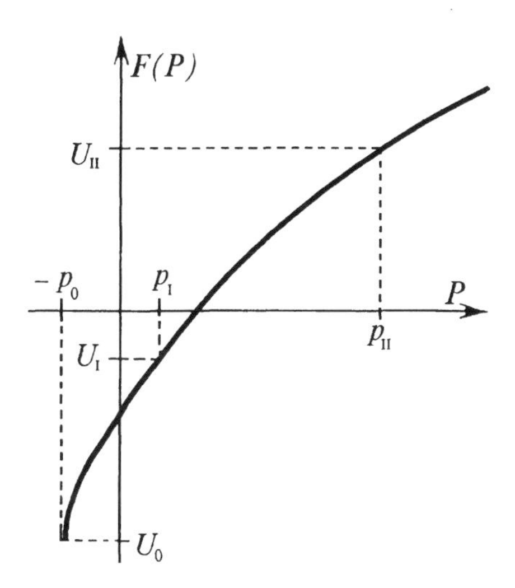

Figure 3.4 Function $F(P)$ for the case $p_I < p_{II}$.

$\pi_n = 1$ and $\beta_n = 1$, see Godunov et al. (1979). Then,

$$F(P^{(0)}) = \frac{P^{(0)} - p_I}{\rho_I c_I} + \frac{P^{(0)} - p_{II}}{\rho_{II} c_{II}} = u_I - u_{II},$$

$$P^{(0)} = \frac{p_I \rho_{II} c_{II} + p_{II} \rho_I c_I + (u_I - u_{II}) \rho_I c_I \rho_{II} c_{II}}{\rho_I c_I + \rho_{II} c_{II}}.$$

If $F(P^{(0)}) - (u_I - u_{II}) \leq 0$, then $P^{(0)}$ lies on the left of the exact solution, which ensures that Newton's method is monotonically convergent.

Note that for special cases of parameters of Eq. (3.3.37) an approximate formulas for solution P instead Newton's iterations can be used. In particular, in the case of large value of $u_I - u_{II}$, two shock waves configuration, and $P \gg p_{0n}$, the equation can be simplified and acquires the form

$$F(P) \approx \alpha_I \frac{P - p_I}{\sqrt{P}} + \alpha_{II} \frac{P - p_{II}}{\sqrt{P}} = \Delta u = u_I - u_{II}, \tag{3.3.39}$$

$$\alpha_I = \sqrt{\frac{2}{(\gamma_I + 1)\rho_I}}, \quad \alpha_{II} = \sqrt{\frac{2}{(\gamma_{II} + 1)\rho_{II}}}.$$

Solving Eq. (3.3.39), one can find that

$$\sqrt{P} = \frac{\Delta u + \sqrt{(\Delta u)^2 + 4(\alpha_I + \alpha_{II})(\alpha_I p_I + \alpha_{II} p_{II})}}{2(\alpha_I + \alpha_I)}.$$

Then the approximate formula for P is

$$P = \frac{\alpha_I p_I + \alpha_{II} p_{II}}{\alpha_I + \alpha_{II}} + \frac{(\Delta u)^2}{(\alpha_I + \alpha_{II})^2} + O(|\Delta u|^{-1}), \tag{3.3.40}$$

where term $O(|\Delta u|^{-1})$ is followed from asymptotic analysis (Charakhch'yan 2000a).

After the pressure P has been calculated we can determine U:

$$U = \tfrac{1}{2}(u_{\mathrm{I}} + u_{\mathrm{II}} + f_{\mathrm{II}} - f_{\mathrm{I}}).$$

This relation for U is obtained by summing Eq. (3.3.17) and Eq. (3.3.18) or (3.3.33) and (3.3.36) and eliminating P.

For the left shock wave (3.3.17) we obtain $|m_{\mathrm{I}}| = (u_{\mathrm{I}} - W_{\mathrm{I}})\rho_{\mathrm{I}}$, and hence

$$\begin{aligned}
m_{\mathrm{I}} &= \sqrt{\frac{\rho_{\mathrm{I}}}{2}[(P + p_{0\mathrm{I}})(\gamma_{\mathrm{I}} + 1) + (p_{\mathrm{I}} + p_{0\mathrm{I}})(\gamma_{\mathrm{I}} - 1)]},\\
W_{\mathrm{I}} &= u_{\mathrm{I}} - \frac{m_{\mathrm{I}}}{\rho_{\mathrm{I}}},\\
R_{\mathrm{I}} &= \rho_{\mathrm{I}}\frac{(P + p_{0\mathrm{I}})(\gamma_{\mathrm{I}} + 1) + (p_{\mathrm{I}} + p_{0\mathrm{I}})(\gamma_{\mathrm{I}} - 1)}{(P + p_{0\mathrm{I}})(\gamma_{\mathrm{I}} - 1) + (p_{\mathrm{I}} + p_{0\mathrm{I}})(\gamma_{\mathrm{I}} + 1)} = \rho_{\mathrm{I}}\frac{2m_{\mathrm{I}}^2/\rho_{\mathrm{I}}}{2m_{\mathrm{I}}^2/\rho_{\mathrm{I}} - 2(P - p_{\mathrm{I}})} =\\
&= \frac{\rho_{\mathrm{I}} m_{\mathrm{I}}}{m_{\mathrm{I}} - (u_{\mathrm{I}} - U)\rho_{\mathrm{I}}},\\
E_{\mathrm{I}} &= \varepsilon(R_{\mathrm{I}}, P),
\end{aligned}$$

where the values of P and U are known.

For the left rarefaction wave (3.3.33) we have

$$\begin{aligned}
c_{\mathrm{I}}^* &= c_{\mathrm{I}} + \frac{\gamma_{\mathrm{I}} - 1}{2}(u_{\mathrm{I}} - U) \qquad (w_{\mathrm{L}} = \mathrm{const}); \qquad (3.3.41)\\
R_{\mathrm{I}} &= \gamma_{\mathrm{I}}\frac{P + p_{0\mathrm{I}}}{(c_{\mathrm{I}}^*)^2} \qquad (S = \mathrm{const});\\
E_{\mathrm{I}} &= \varepsilon(R_{\mathrm{I}}, P).
\end{aligned}$$

Then the left rarefaction wave is bounded in the (t, x) plane by the rays $W_{\mathrm{I}} = u_{\mathrm{I}} - c_{\mathrm{I}}$ and $W_{\mathrm{I}}^* = U - c_{\mathrm{I}}^*$, see Fig. 3.2c. Let us seek the parameters inside the rarefaction wave for an arbitrary $\xi = \xi^*$: $W_{\mathrm{I}} \leq \xi^* \leq W_{\mathrm{I}}^*$. We denote the parameters for the ray $\xi = \xi^*$ by c^*, ρ^*, u^*, ε^*, and p^*. The following relations hold:

$$\begin{aligned}
u^* - c^* &= \xi^*,\\
u^* + \frac{2}{\gamma_{\mathrm{I}} - 1}c^* &= u_{\mathrm{I}} + \frac{2}{\gamma_{\mathrm{I}} - 1}c_{\mathrm{I}}. \qquad (3.3.42)
\end{aligned}$$

It follows that

$$c^* = \frac{2}{\gamma_{\mathrm{I}} + 1}c_{\mathrm{I}} + \frac{\gamma_{\mathrm{I}} - 1}{\gamma_{\mathrm{I}} + 1}(u_{\mathrm{I}} - \xi^*), \qquad (3.3.43)$$

$$u^* = \xi^* + c^*. \qquad (3.3.44)$$

Using the fact that $S = \mathrm{const}$, we obtain

$$\rho^* = \rho_{\mathrm{I}}\left(\frac{c^*}{c_{\mathrm{I}}}\right)^{\frac{2}{\gamma_{\mathrm{I}}-1}}, \qquad (3.3.45)$$

$$p^* + p_{0\mathrm{I}} = (p_{\mathrm{I}} + p_{0\mathrm{I}})\left(\frac{c^*}{c_{\mathrm{I}}}\right)^{\frac{2\gamma_{\mathrm{I}}}{\gamma_{\mathrm{I}}-1}}, \qquad (3.3.46)$$

$$\varepsilon^* = \varepsilon(\rho^*, p^*). \qquad (3.3.47)$$

For the right shock wave (3.3.18) we have $|m_{\mathrm{II}}| = -(u_{\mathrm{II}} - W_{\mathrm{II}})\rho_{\mathrm{II}}$, and hence,

$$
\begin{aligned}
m_{\mathrm{II}} &= \sqrt{\frac{\rho_{\mathrm{II}}}{2}[(P + p_{0\mathrm{II}})(\gamma_{\mathrm{II}} + 1) + (p_{\mathrm{II}} + p_{0\mathrm{II}})(\gamma_{\mathrm{II}} - 1)]},\\
W_{\mathrm{II}} &= u_{\mathrm{II}} + \frac{m_{\mathrm{II}}}{\rho_{\mathrm{II}}},\\
R_{\mathrm{II}} &= \frac{\rho_{\mathrm{II}} m_{\mathrm{II}}}{m_{\mathrm{II}} + (u_{\mathrm{II}} - U)\rho_{\mathrm{II}}},\\
E_{\mathrm{II}} &= \varepsilon(R_{\mathrm{II}}, P),
\end{aligned}
$$

where the values of P and U are known. For the right rarefaction wave (3.3.36) we obtain

$$
\begin{aligned}
c_{\mathrm{II}}^* &= c_{\mathrm{II}} - \frac{\gamma_{\mathrm{II}} - 1}{2}(u_{\mathrm{II}} - U) \qquad (w_{\mathrm{R}} = \mathrm{const});\\
R_{\mathrm{II}} &= \gamma_{\mathrm{II}} \frac{P + p_{0\mathrm{II}}}{(c_{\mathrm{II}}^*)^2} \qquad (S = \mathrm{const});\\
E_{\mathrm{II}} &= \varepsilon(R_{\mathrm{II}}, P).
\end{aligned}
\tag{3.3.48}
$$

Then the right rarefaction wave is bounded in the (t,x) plane by the rays $W_{\mathrm{II}} = u_{\mathrm{II}} + c_{\mathrm{II}}$ and $W_{\mathrm{II}}^* = U + c_{\mathrm{II}}^*$, see Fig. 3.2d. Let us seek the parameters inside the rarefaction wave for an arbitrary $\xi = \xi^*$: $W_{\mathrm{II}}^* \le \xi^* \le W_{\mathrm{II}}$. We denote the parameters for the ray $\xi = \xi^*$ by c^*, ρ^*, u^*, ε^*, and p^*. The following relations hold:

$$
\begin{aligned}
u^* + c^* &= \xi^*,\\
u^* - \frac{2}{\gamma_{\mathrm{II}} - 1}c^* &= u_{\mathrm{II}} - \frac{2}{\gamma_{\mathrm{II}} - 1}c_{\mathrm{II}}.
\end{aligned}
\tag{3.3.49}
$$

Whence,

$$
c^* = \frac{2}{\gamma_{\mathrm{II}} + 1}c_{\mathrm{II}} - \frac{\gamma_{\mathrm{II}} - 1}{\gamma_{\mathrm{II}} + 1}(u_{\mathrm{II}} - \xi^*), \tag{3.3.50}
$$

$$
u^* = \xi^* - c^*. \tag{3.3.51}
$$

Using the fact that $S = \mathrm{const}$, we obtain

$$
\rho^* = \rho_{\mathrm{II}} \left(\frac{c^*}{c_{\mathrm{II}}}\right)^{\frac{2}{\gamma_{\mathrm{II}}-1}}, \tag{3.3.52}
$$

$$
p^* + p_{0\mathrm{II}} = (p_{\mathrm{II}} + p_{0\mathrm{II}}) \left(\frac{c^*}{c_{\mathrm{II}}}\right)^{\frac{2\gamma_{\mathrm{II}}}{\gamma_{\mathrm{II}}-1}}, \tag{3.3.53}
$$

$$
\varepsilon^* = \varepsilon(\rho^*, p^*). \tag{3.3.54}
$$

All these results are summarized in the table on page 144. Further, we can determine the values of ρ, u, ε, p, v, and w for the ray $\xi = x/t$ in the plane (t,x). The case $\xi = 0$ corresponds to a stationary cell boundary and the case $\xi = D$ to a boundary moving with speed D, see Eqs. (3.2.7)–(3.2.10). The solution for an arbitrary ξ is $\mathbf{Y} = \mathbf{Y}(\xi) = [\rho(\xi), u(\xi), \varepsilon(\xi), p(\xi), v(\xi), w(\xi)]^{\mathrm{T}}$.

Flow type	Left position	Right position
Shock wave (SW)	If $U > \xi$ and $P \geq p_{\mathrm{I}}$, then $W_{\mathrm{I}} = u_{\mathrm{I}} - m_{\mathrm{I}}/\rho_{\mathrm{I}}$ is the SW velocity. If $\xi > W_{\mathrm{I}}$, then $\mathbf{Y} = [R_{\mathrm{I}}, U, E_{\mathrm{I}}, P, v_{\mathrm{I}}, w_{\mathrm{I}}]^{\mathrm{T}}$. If $\xi \leq W_{\mathrm{I}}$, then $\mathbf{Y} = [\rho_{\mathrm{I}}, u_{\mathrm{I}}, \varepsilon_{\mathrm{I}}, p_{\mathrm{I}}, v_{\mathrm{I}}, w_{\mathrm{I}}]^{\mathrm{T}}$.	If $U < \xi$ and $P \geq p_{\mathrm{II}}$, then $W_{\mathrm{II}} = u_{\mathrm{II}} + m_{\mathrm{II}}/\rho_{\mathrm{II}}$ is the SW velocity. If $\xi < W_{\mathrm{II}}$, then $\mathbf{Y} = [R_{\mathrm{II}}, U, E_{\mathrm{II}}, P, v_{\mathrm{II}}, w_{\mathrm{II}}]^{\mathrm{T}}$. If $\xi \geq W_{\mathrm{II}}$, then $\mathbf{Y} = [\rho_{\mathrm{II}}, u_{\mathrm{II}}, \varepsilon_{\mathrm{II}}, p_{\mathrm{II}}, v_{\mathrm{II}}, w_{\mathrm{II}}]^{\mathrm{T}}$.
Rarefaction wave (RW)	If $U > \xi$ and $P < p_{\mathrm{I}}$, then $W_{\mathrm{I}} = u_{\mathrm{I}} - c_{\mathrm{I}}$, $W_{\mathrm{I}}^* = U - c_{\mathrm{I}}^*$ are the RW boundaries, (3.3.41). If $\xi > W_{\mathrm{I}}^*$, then $\mathbf{Y} = [R_{\mathrm{I}}, U, E_{\mathrm{I}}, P, v_{\mathrm{I}}, w_{\mathrm{I}}]^{\mathrm{T}}$. If $\xi < W_{\mathrm{I}}$, then $\mathbf{Y} = [\rho_{\mathrm{I}}, u_{\mathrm{I}}, \varepsilon_{\mathrm{I}}, p_{\mathrm{I}}, v_{\mathrm{I}}, w_{\mathrm{I}}]^{\mathrm{T}}$. If $W_{\mathrm{I}} \leq \xi = \xi^* \leq W_{\mathrm{I}}^*$, then $\mathbf{Y} = [\rho^*, u^*, \varepsilon^*, p^*, v_{\mathrm{I}}, w_{\mathrm{I}}]^{\mathrm{T}}$, see formulas (3.3.43)–(3.3.47).	If $U < \xi$ and $P < p_{\mathrm{II}}$, then $W_{\mathrm{II}} = u_{\mathrm{II}} + c_{\mathrm{II}}$, $W_{\mathrm{II}}^* = U + c_{\mathrm{II}}^*$ are the RW boundaries, (3.3.48). If $\xi < W_{\mathrm{II}}^*$, then $\mathbf{Y} = [R_{\mathrm{II}}, U, E_{\mathrm{II}}, P, v_{\mathrm{II}}, w_{\mathrm{II}}]^{\mathrm{T}}$. If $\xi > W_{\mathrm{II}}$, then $\mathbf{Y} = [\rho_{\mathrm{II}}, u_{\mathrm{II}}, \varepsilon_{\mathrm{II}}, p_{\mathrm{II}}, v_{\mathrm{II}}, w_{\mathrm{II}}]^{\mathrm{T}}$. If $W_{\mathrm{II}}^* \leq \xi = \xi^* \leq W_{\mathrm{II}}$, then $\mathbf{Y} = [\rho^*, u^*, \varepsilon^*, p^*, v_{\mathrm{II}}, w_{\mathrm{II}}]^{\mathrm{T}}$, see formulas (3.3.50)–(3.3.54).

The exact solution of the Riemann problem provides continuity of the tangential velocity components $v(\xi)$ and $w(\xi)$ both in the shock wave, see Eqs. (3.3.13), and in the rarefaction wave, see Eqs. (3.3.27) and (3.3.30). The components $v(\xi)$ and $w(\xi)$ can be discontinuous at only a contact discontinuity. Thus, $v(\xi)$ and $w(\xi)$ for an arbitrary ray ξ are determined in accordance with the velocity $U(\xi)$ of the contact discontinuity, see the table. We put $v(\xi) = v_{\mathrm{I}}$ and $w(\xi) = w_{\mathrm{I}}$ if $U(\xi) > \xi$, otherwise $v(\xi) = v_{\mathrm{II}}$ and $w(\xi) = w_{\mathrm{II}}$.

Figure 3.2e presents the basic possible types of self-similar gas dynamic flows in the (t, x) plane. Here, two waves (shock and/or rarefaction) are separated by a contact discontinuity. All elementary solutions are separated by the regions with constant quantities. To solve the Riemann problem for the flows presented in Fig. 3.2e, it is sufficient to determine U and P in the central domain, the velocities of the discontinuities and rarefaction waves, and the flow parameters inside the rarefaction wave fan (or fans).

Note that the analysis of the function $F(P)$ permits one to determine the configuration of the exact solution *a priori*, without calculations (Godunov et al. 1979). Assume for definiteness that $p_{\mathrm{I}} < p_{\mathrm{II}}$. The behavior of the function $F(P)$ is illustrated in Fig. 3.4. The values of $F(P)$ at $P = p_{\mathrm{II}}$, $P = p_{\mathrm{I}}$, and $P = -p_0 = \max(-p_{0\mathrm{I}}, -p_{0\mathrm{II}})$ are given by

$$F(p_{\mathrm{II}}) = U_{\mathrm{II}} = \frac{p_{\mathrm{II}} - p_{\mathrm{I}}}{\rho_{\mathrm{I}} c_{\mathrm{I}} \beta_{\mathrm{I}}},$$

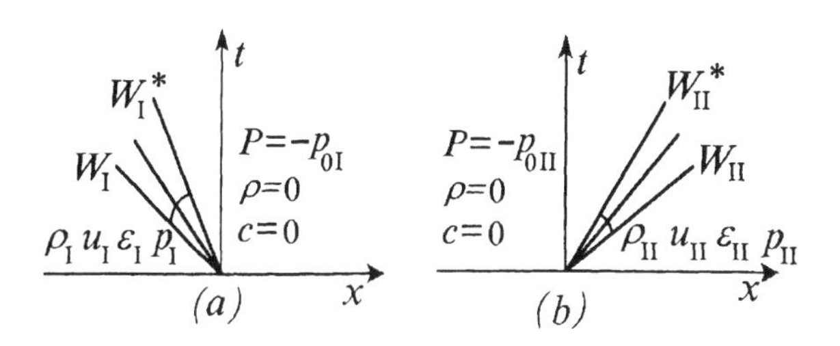

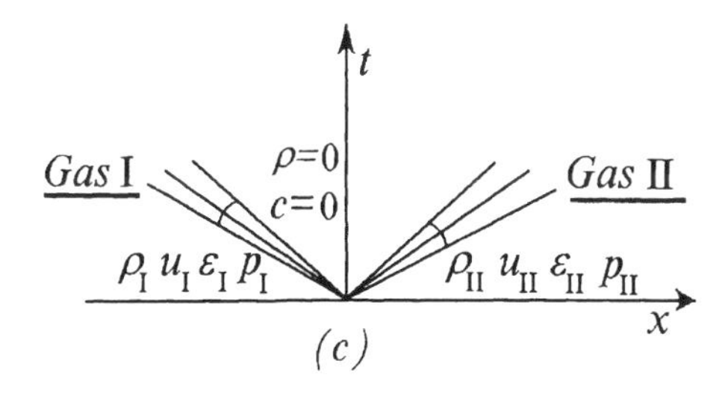

Figure 3.5 Flow configurations with a vacuum region.

$$\pi_I = \frac{p_{II} + p_{0I}}{p_I + p_{0I}}, \quad \beta_I = \sqrt{\frac{\gamma_I + 1}{2\gamma_I}\pi_I + \frac{\gamma_I - 1}{2\gamma_I}}, \quad c_I = \sqrt{\gamma_I \frac{p_I + p_{0I}}{\rho_I}};$$

$$F(p_I) = U_I = \frac{2c_{II}}{\gamma_{II} - 1}(\pi_{II}^{\chi} - 1),$$

$$\pi_{II} = \frac{p_I + p_{0II}}{p_{II} + p_{0II}}, \quad \chi = \frac{\gamma_{II} - 1}{2\gamma_{II}}, \quad c_{II} = \sqrt{\gamma_{II} \frac{p_{II} + p_{0II}}{\rho_{II}}};$$

$$F(-p_0) = U_0.$$

Depending on the value of $u_I - u_{II}$, the following cases are possible:

(i) If $u_I - u_{II} > U_{II}$, then $P > p_{II} > p_I$, i.e., the configuration of a flow with right and left shock waves is realized;

(ii) If $U_I < u_I - u_{II} < U_{II}$, then $p_I < P < p_{II}$, i.e., the configuration of a flow with a right rarefaction wave and a left shock wave is realized;

(iii) If $U_0 < u_I - u_{II} < U_I$, then $-p_0 < P < p_I$, i.e., the configuration of a flow with right and left rarefaction waves is realized;

(iv) If $u_I - u_{II} < U_0$, then a vacuum region with $\rho = 0$ and $c = 0$ occurs, see Fig. 3.5c.

Equation (3.3.37) has a unique solution if $u_I - u_{II} > U_0$. Let $u_I - u_{II} < U_0$. In this case, in general, a vacuum region ($\rho = 0$ and $c = 0$) is framed by two left and right rarefaction waves forms. Figure 3.5 illustrates special cases of a single rarefaction wave on the left of the vacuum region (Fig. 3.5a) and on the right of the vacuum region (Fig. 3.5b). A single wave configuration develops only if the density and the sound speed are zero on one side of the discontinuity at $t = 0$. Assuming $\rho_I = 0$ and $c_I = 0$, we obtain a single right rarefaction wave (Fig. 3.5b) that separates the domain where $u = u_{II}$, $\rho = \rho_{II}$, and $\varepsilon = \varepsilon_{II}$ from the zero-density domain in the plane (t,x). In this case,

$$W^*_{\mathrm{II}} = u_{\mathrm{II}} - \frac{2}{\gamma_{\mathrm{II}}-1}c_{\mathrm{II}}, \quad W_{\mathrm{II}} = u_{\mathrm{II}} + c_{\mathrm{II}}.$$

These expressions can be obtained from Eq. (3.3.49) by setting $c^* = 0$. On assuming further that $\rho_{\mathrm{II}} = 0$ and $c_{\mathrm{II}} = 0$, we obtain a single left rarefaction wave (Fig. 3.5a). In this case,

$$W^*_{\mathrm{I}} = u_{\mathrm{I}} + \frac{2}{\gamma_{\mathrm{I}}-1}c_{\mathrm{I}}, \quad W_{\mathrm{I}} = u_{\mathrm{I}} - c_{\mathrm{I}}.$$

These expressions can be obtained from Eq. (3.3.42) by setting $c^* = 0$. The general case (Fig. 3.5c) is characterized by the presence of both rarefaction waves. The flow parameters inside a rarefaction fan can be obtained from Eqs. (3.3.43)–(3.3.47) or Eqs. (3.3.50)–(3.3.54).

Note that there is a singularity in ε^* as $\rho^* \to 0$. It occurs for $p_0 \neq 0$ in the two-term EOS (3.1.5); for perfect gas with $p_0 = 0$ this singularity is absent. However, in all cases the values of the flux $\mathbf{F}^*$ for $\rho^* \to 0$ and $\rho^* = 0$ are bounded and equal to $\mathbf{F}^* = [0, -p_{0n}, 0, 0, 0]^{\mathrm{T}}$, where $n = \mathrm{I}$ or II.

For $u_{\mathrm{I}} - u_{\mathrm{II}} < U_0$ an exact solution always exists if $p_{0\mathrm{I}} = p_{0\mathrm{II}}$; in this case this solution contains the left and right rarefaction waves with a vacuum region between them (Fig. 3.5c). In the rare special case $-p_{00} = \min(-p_{0\mathrm{I}}, -p_{0\mathrm{II}}) < P < -p_0 = \max(-p_{0\mathrm{I}}, -p_{0\mathrm{II}})$ there is no exact solution, because the above values of the pressure are permissible for only one of the EOSs prescribed. Then one should use a regularization. In particular, the parameters in the vacuum region can be substituted by the parameters of a gas with small density, zero velocity, and initial background internal energy.

The described algorithm of the exact Riemann solver with a gas dynamic configuration analysis can be implemented in a simple numerical code with no more than 200 statements.

Equation for pressure for an arbitrary EOS

For an arbitrary EOS, $\varepsilon = \varepsilon(\rho, p)$, Eq. (3.3.37) transforms into the more general equation

$$F(P) = f_{\mathrm{I}}(P, p_{\mathrm{I}}, \rho_{\mathrm{I}}) + f_{\mathrm{II}}(P, p_{\mathrm{II}}, \rho_{\mathrm{II}}) = u_{\mathrm{I}} - u_{\mathrm{II}}, \tag{3.3.55}$$

where

$$f_n = \begin{cases} \sqrt{(P-p_n)\left(\dfrac{1}{\rho_n} - \dfrac{1}{\rho}\right)}, \quad \varepsilon - \varepsilon_n = \dfrac{P+p_n}{2}\left(\dfrac{1}{\rho_n} - \dfrac{1}{\rho}\right) & \text{if } P \geq p_n, \\ \displaystyle\int_{p_n}^{P} \frac{dp}{\rho c(\rho,p)}, \quad d\varepsilon(\rho,p) = -pd\left(\frac{1}{\rho}\right), \quad c = \sqrt{p_\rho + \frac{p p_\varepsilon}{\rho^2}} & \text{if } P < p_n. \end{cases}$$

Here $n = \mathrm{I}$ or II, and c is the speed of sound, see Eq. (3.1.24).

Equation (3.3.17) or (3.3.33) transforms into the more general equation

$$U - u_{\mathrm{I}} + f_{\mathrm{I}}(P, p_{\mathrm{I}}, \rho_{\mathrm{I}}) = 0, \tag{3.3.56}$$

and Eq. (3.3.18) or (3.3.36) into

$$U - u_{\mathrm{II}} - f_{\mathrm{II}}(P, p_{\mathrm{II}}, \rho_{\mathrm{II}}) = 0. \tag{3.3.57}$$

Then, by eliminating U, we obtain the desired equation (3.3.55) for P.

The function f for a shock wave ($P \geq p_n$) can be obtained from Eqs. (3.3.10), (3.3.14), (3.3.15), and (3.3.16) and for a rarefaction wave from the exact relations

$$du \pm \frac{1}{\rho c} dp = 0, \quad d\varepsilon = -pd\left(\frac{1}{\rho}\right). \tag{3.3.58}$$

These relations can be derived from the nonconservative gas dynamic system for the functions ρ, u, v, w, and ε:

$$\rho_t + u\rho_x + \rho u_x = 0, \quad u_t + uu_x + \frac{p_x}{\rho} = u_t + uu_x + \frac{p_\rho}{\rho}\rho_x + \frac{p_\varepsilon}{\rho}\varepsilon_x = 0,$$

$$v_t + uv_x = 0, \quad w_t + uw_x = 0, \quad \varepsilon_t + u\varepsilon_x + \frac{p}{\rho}u_x = 0; \quad p = p(\rho, \varepsilon).$$

The continuous flow in a rarefaction wave is described by self-similar form $\rho = \rho(\xi)$, $u = u(\xi)$, $v = v(\xi)$, $w = w(\xi)$, and $\varepsilon = \varepsilon(\xi)$, where $\xi = x/t$. In this case, instead of Eqs. (3.3.25)–(3.3.27), we have

$$u - c - \xi = 0, \tag{3.3.59}$$

$$d\varepsilon = -pd\left(\frac{1}{\rho}\right), \tag{3.3.60}$$

$$du + \frac{1}{\rho c} dp = 0, \tag{3.3.61}$$

$$v_\xi = 0, \quad w_\xi = 0 \quad \Longrightarrow \quad v = \text{const}, \quad w = \text{const}, \tag{3.3.62}$$

and instead of Eqs. (3.3.28)–(3.3.30) we obtain

$$u + c - \xi = 0, \tag{3.3.63}$$

$$d\varepsilon = -pd\left(\frac{1}{\rho}\right), \tag{3.3.64}$$

$$du - \frac{1}{\rho c} dp = 0, \tag{3.3.65}$$

$$v_\xi = 0, \quad w_\xi = 0 \quad \Longrightarrow \quad v = \text{const}, \quad w = \text{const}. \tag{3.3.66}$$

For a two-term EOS the systems of equations (3.3.59)–(3.3.62) and (3.3.63)–(3.3.66) are equivalent to systems (3.3.25)–(3.3.27) and (3.3.28)–(3.3.30), respectively. Integral of Eq. (3.3.61) (or (3.3.65)) is a left (or right) Riemann invariant.

The above equations can be used to solve the gas dynamic Riemann problems for specific EOSs, provided that the problem is well posed.

3.3.5 An arbitrary EOS. Previously we obtained exact formulas for the solution of the Riemann problem with the two-term EOS of (3.1.5),

$$Y(\rho, \varepsilon) = \varepsilon - \frac{p + \gamma p_0}{(\gamma - 1)\rho} + \frac{c_0^2}{\gamma - 1} = 0, \tag{3.3.67}$$

where ρ is the density, ε is the specific internal energy, and $p = p(\rho, \varepsilon)$ is the pressure. The quantities γ, p_0, and c_0 are some constants.

Below we describe an approach that permits one to use exact formulas for an arbitrary, physically relevant EOS. Let us approximate an arbitrary EOS locally by the two-term EOS (3.3.67). Then, three local values of γ, p_0, and c_0 can be calculated from the following equations:

$$Y = Y(\rho, \varepsilon) = 0, \quad \frac{\partial Y}{\partial \rho} = \frac{p + \gamma p_0}{(\gamma - 1)\rho^2} - \frac{p_\rho}{(\gamma - 1)\rho} = 0, \quad \frac{\partial Y}{\partial \varepsilon} = 1 - \frac{p_\varepsilon}{(\gamma - 1)\rho} = 0.$$

Here $p_\rho = (\partial p/\partial \rho)_\varepsilon$ and $p_\varepsilon = (\partial p/\partial \varepsilon)_\rho$. Then, for known fixed ρ and ε, we obtain the local approximate expressions

$$\gamma = 1 + \frac{p_\varepsilon}{\rho}, \quad p_0 = \frac{p_\rho \rho - p}{\gamma}, \quad c_0^2 = \frac{p + \gamma p_0}{\rho} - (\gamma - 1)\varepsilon. \tag{3.3.68}$$

Further, one can use the formulas of the exact solver in the following manner. Let us seek the values $\gamma_{\mathrm{I}}, p_{0\mathrm{I}}, c_{0\mathrm{I}}$ and $\gamma_{\mathrm{II}}, p_{0\mathrm{II}}, c_{0\mathrm{II}}$, which characterize the EOS on the left (subscript "I") and on the right (subscript "II") of the cell boundary. Such an approximation can be used for both one gas with a single EOS and two contacting gases with their own EOSs.

This approach requires the approximation of the derivatives p_ρ and p_ε. If the EOS has unusual properties, one may have to perform a regularization to avoid nonphysical results. For example, if the calculated pressure p (and/or energy ε) is less than the "cold" pressure p_{cold} (and/or cold energy $\varepsilon_{\mathrm{cold}}$), see (3.1.30) and (3.1.31), one must set $p = p_{\mathrm{cold}}$ (and/or $\varepsilon = \varepsilon_{\mathrm{cold}}$). If the calculated adiabatic index γ is less than unity, one must set $\gamma = 1 + \epsilon$, where ϵ is a small positive constant. If the numerical method requires too many corrections of this kind in the calculation, then it would be more reasonable to use another approximation of p_ρ and p_ε.

Only two-parameter EOSs are suitable and optimal for using the above two-term approximations, since one-parameter EOSs do not provide an approximation of the properties of two different media simultaneously. The two-term EOS (3.3.67) satisfies these requirements and has a wide area of application. The two-term approximation (3.3.68) was first proposed by A.V. Zabrodin (Godunov, Zabrodin, Pliner et al. 1968) and is known as "method B-71". This algorithm has been widely used for numerical modelling of physical and mechanical problems for more than 30 years; e.g., see the use of wide-range EOSs for metals in modelling by Charakhch'yan (1992, 1997a, 1997b), Charakhch'yan et al. (1999), and Charakhch'yan (2000b), or see also the recent two-term approximation in modelling of three-temperature gas dynamic flows (Zabrodin and Prokopov 1998; Prokopov 1999, 2000), etc.

In particular, for co-volume gases obeying the equation of state

$$\varepsilon = \frac{(1 - b\rho)p}{(\overline{\gamma} - 1)\rho}, \quad \overline{\gamma} = \text{const} > 1, \quad b = \text{const} \geq 0,$$

one can determine the approximation coefficients as

$$\gamma = \frac{\overline{\gamma} - b\rho}{1 - b\rho}, \quad p_0 = \frac{b\rho p}{\overline{\gamma} - b\rho}, \quad c_0^2 = \frac{bp}{1 - b\rho}.$$

The co-volume EOS (the Noble–Abel EOS) describes, in particular, gaseous products of power combustion. The exact solution of the gas dynamic Riemann problem for this EOS can be found in Toro (1997).

The one-temperature two-term approximation (3.3.68) can be generalized for the cases of two- and three-temperature gas dynamic equations, see Section 3.1.1, with taking into account the ionization process (Zabrodin and Prokopov 1998; Prokopov 1999, 2000). In this case the Riemann problem is solved exactly for the general EOS describing the behavior of ions, electrons, and/or thermal radiation, since they have the common density and velocity. This EOS is the sum of the corresponding two-term approximations (3.1.5) for the ion, electron, and thermal radiation EOSs. To obtain an exact solution of this Riemann problem by reduction to the solution of the P-equation (3.3.37), one should use additional physical relations: (i) $\{\varepsilon_2\} = 0$ and $\{p_3\} = 0$ across shocks, where ε_2 is the electron specific internal energy and p_3 is the thermal radiation pressure (Imshennik and Bobrova 1997); (ii) the conservation of the three (ion, electron, and thermal radiation) entropies in a rarefaction wave (see Zabrodin and Prokopov 1998). Another approach under investigation is the direct usage of the general EOS (Prokopov 2000),

$$p = (\gamma - 1)\rho\varepsilon + (\rho - \rho_0)c_0^2, \quad p = \sum_k p_k, \quad \varepsilon = \sum_k \varepsilon_k,$$

$$\gamma = \frac{1}{\varepsilon}\sum_k \gamma_k\varepsilon_k, \quad c_0^2 = \sum_k c_{0k}^2, \quad p_k = (\gamma_k - 1)\rho\varepsilon_k + (\rho - \rho_0)c_{0k}^2, \qquad k = 1, 2, 3.$$

After the Riemann problem for this EOS has been solved, one must determine the component pressures p_k and energies ε_k by using additional physical relations like those mentioned above.

Note that the two-term approximation (3.3.68) can be used without difficulties for a general EOS satisfying the condition of a normal gas. Let the EOS be given in the form $p = p(\eta, S)$ and $\varepsilon = \varepsilon(\eta, T)$, where $\eta = 1/\rho$ is the specific volume, S is the entropy, and T is the temperature. Then the relations

$$\frac{\partial p}{\partial \eta} < 0, \quad \frac{\partial^2 p}{\partial \eta^2} > 0, \quad \frac{\partial p}{\partial S} > 0, \quad \frac{\partial \varepsilon}{\partial T} > 0, \quad \lim_{\eta \to 0} p(\eta, S) \to \infty$$

must be satisfied (Weyl 1949). These relations define a so-called *normal* gas and guarantee that the self-similar solution of the Riemann problem exists, is unique, and is represented by the configuration shown in Fig. 3.2e (Rozhdestvenskii and Yanenko 1983). For other EOSs the solution of the Riemann problem can be nonunique, and by using the two-term approximation we can, in general, obtain some physically irrelevant solution.

There are other methods that take into account an arbitrary EOS. In particular, Naumova and Shmyglevskii (1978) created a freeware code for solving the gas dynamic Riemann problem for arbitrary EOSs prescribed in analytical form by the numerical solution of equation (3.3.55) for P. This code is based on the iteration methods for solving the corresponding rational equations, standard methods for numerical solution of the corresponding ordinary differential equations and calculation of the Riemann invariants, see Eq. (3.3.55). Results obtained with this code can be found in Zubov et al. (1980), Zubov et al. (1986), Krivtsov et al. (1992), and Zubov et al. (1993). The similar method based on the analytical form of certain EOSs can be found in Saurel, Larini, and Loraud (1994). They investigated the perfect gas, Van der Waals, and fifth-order virial EOSs. Introducing certain assumptions in the computations of rarefaction waves for particular EOSs, they suggested and also tested approximate algorithms of solution of the Riemann problem.

For arbitrary tabulated EOSs, an approach suggested by Charakhch'yan (2000a) may be recommended. In particular, let T_i and ρ_j, $i = 1, \ldots, n$ and $j = 1, \ldots, m$, be monotone increasing one-dimensional arrays of the temperature T and density ρ, so that $T_{i+1} > T_i$ and $\rho_{j+1} > \rho_j$. Let $p_{ij} = p(T_i, \rho_j)$ and $\varepsilon_{ij} = \varepsilon(T_i, \rho_j)$ be two-dimensional arrays of the pressure p and internal energy ε, where $\varepsilon_{i+1,j} > \varepsilon_{ij}$, and $p_{i+1,j} > p_{ij}$, $p_{i,j+1} > p_{ij}$. (The monotone properties of the arrays for p and ε permit one to rewrite them also in the form $\varepsilon = \varepsilon(\rho, p)$ or $T = T(\rho, p)$). Inside each region $T_i \leq T \leq T_{i+1}$, $\rho_j \leq \rho \leq \rho_{j+1}$, the values of p and ε can be approximated by bilinear functions $p(\widetilde{T}, \widetilde{\rho})$ and $\varepsilon(\widetilde{T}, \widetilde{\rho})$, respectively, where $\widetilde{T} = T - T_i$ and $\widetilde{\rho} = \rho - \rho_j$. In particular,

$$p(\widetilde{T}, \widetilde{\rho}) = p_0 + p_1\widetilde{T} + p_2\widetilde{\rho} + p_{12}\widetilde{T}\widetilde{\rho},$$

where p_0, p_1, p_2, and p_{12} are for $\rho_j \leq \rho \leq \rho_{j+1}$ and $T_i \leq T \leq T_{i+1}$ the constants defined by

$$p_0 = p_{ij}, \quad p_1 = \frac{p_{i+1,j} - p_{ij}}{T_{i+1} - T_i}, \quad p_2 = \frac{p_{i,j+1} - p_{ij}}{\rho_{j+1} - \rho_j},$$
$$p_{12} = \frac{p_{i+1,j+1} - p_{i+1,j} - p_{i,j+1} + p_{ij}}{(T_{i+1} - T_i)(\rho_{j+1} - \rho_j)}.$$

In the same manner one can obtain the coefficients ε_0, ε_1, ε_2, and ε_{12} in the bilinear approximation of ε:

$$\varepsilon(\widetilde{T}, \widetilde{\rho}) = \varepsilon_0 + \varepsilon_1\widetilde{T} + \varepsilon_2\widetilde{\rho} + \varepsilon_{12}\widetilde{T}\widetilde{\rho}.$$

Using p and ε in the form of bilinear functions of T and ρ permits us to solve analytically the ordinary differential equation in a rarefaction wave; see f_n in Eq. (3.3.55) for the case of $P < p_n$. In particular,

$$c(\widetilde{T}, \widetilde{\rho}) = \sqrt{p_2 + \widetilde{T}p_{12} + \frac{p_1 + \widetilde{\rho}p_{12}}{\varepsilon_1 + \widetilde{\rho}\varepsilon_{12}}\left(\frac{p(\widetilde{\rho}, \widetilde{T})}{(\rho_j + \widetilde{\rho})^2} - \varepsilon_2 - \widetilde{T}\varepsilon_{12}\right)}.$$

Then Eq. (3.3.55) must be transformed into an equation for P, which is convenient for the application of the Newton iteration method (3.3.38). This bilinear approach is fast and efficient if arrays of tabulated data are not too large. Otherwise, the calculation of the integral in a rarefaction wave, see (3.3.55), can be cumbersome. In these cases, a code similar to that developed by Naumova and Shmyglevskii (1978) for the exact solution of equation (3.3.55), the two-term approximation (3.3.68), or an approximate Riemann problem solver is more efficient.

There exist other methods of numerical solution of the Riemann problem for a complicated EOS. In particular, Fortov et al. (1996) used the following acceleration technique for the iteration method of solution of Eqs. (3.3.56)–(3.3.57). On the first step, an approximation of the unknown quantities U and P is calculated using the isentropicity assumption. This corresponds to the acoustic approximation that is suitable for small-amplitude initial jumps. In such cases, this approximation is quite accurate and can be used to determine the flux. For a strong jump, this solution is a good initial guess to start the iteration process.

Malama, Kestenboim and Hornung (2000), while solving Eqs. (3.3.56)–(3.3.57), used Newton's iteration method for shock waves and preliminary generated detailed tables for rarefaction waves. These tables represented isentropes in the form $\rho_{nm} = \rho(p_n, S_m)$, where m, $1 \le m \le 300$, is the isentrope number, and $1 \le n \le 400$. Note that the tables included data both for the left and right rarefaction waves, see Eq. (3.3.58). As far as a rarefaction wave is concerned, the initial values of ρ and p permit one to determine the isentrope corresponding to the exact solution of Eqs. (3.3.56)–(3.3.57) and further proceed in solving the Riemann problem. The numerical results obtained by this approach can be found in Hornung, Malama, and Thoma (1996).

3.4 Approximate Riemann problem solvers

This section describes approximate solutions for the one-dimensional gas dynamic Riemann problem. Here we present three basic approximate methods for solving the Riemann problem: the Courant–Isaacson–Rees (CIR), Roe, and Osher–Solomon methods. We present the formulas that are necessary for the implementation of approximate Riemann problem solvers in practice.

The exact Riemann problem solver. The exact solution consists of discontinuities and rarefaction waves separated by regions of constant quantities, see Fig. 3.2e. The exact solver is deduced for the two-term EOS (3.1.5). It admits the use of arbitrary EOSs by applying the local two-term approximation of (3.3.68).

The Courant–Isaacson–Rees (CIR) method. This method is based on the exact solution of the linearized gas dynamic system. The Riemann problem for this system always has a solution. The solution in this case is a linear combination of step-like travelling discontinuities separated by regions of constant quantities. It can be used for arbitrary EOSs.

Roe's method. This method is based on an approximate solution of the Riemann problem. This solution consists of only discontinuities and is exact for the linearized gas dynamic equations. Roe's solver preserves the Rankine–Hugoniot relations on shocks exactly. The solution can be constructed for EOSs more general than (3.1.5) with an arbitrary nonlinear term depending only on ρ.

Note that the exact solver permits one to simulate shock waves interaction of an arbitrary intensity with the CFL numbers close to unity, 0.8–1. In the case of intensive shock wave a pressure P is described by expression (3.3.40). Note that the first term in this expression can be considered as a result of linear interpolation for the pressure, and the second term can be interpreted as the quadratic artificial viscosity by von Neumann and Richtmyer (1950). Thus, in the region of strong shock wave interaction the Godunov scheme with exact solver has the viscous properties naturally similar to the quadratic artificial viscosity. The Godunov scheme based on CIR's and Roe's solvers always has a linear numerical viscosity. To simulate a strong shock wave one must occasionally reduce the CFL number from 1 to 0.2–0.6. This reduction is due to a shock wave intensity and EOS used; see Charakhch'yan (1979, 2000a). In particular, for pressure up to 1 Mbar behind the shock wave front there is no need to reduce the time increment. Otherwise, for pressure up to

30 Mbar the time increment of CIR's and Roe's methods is equal to 0.2–0.6 in comparison with Δt of the Godunov method with exact solver, see Charakhch'yan (2000a).

The Osher–Solomon method. This method is based on an approximate solution of the Riemann problem that is constructed of only Riemann waves.

3.4.1 The Courant–Isaacson–Rees method for an arbitrary EOS.

For Eq. (3.1.15), we can consider a one-dimensional finite-volume scheme of the form

$$\frac{\mathbf{U}_j^{k+1} - \mathbf{U}_j^k}{\Delta t} + \frac{\mathbf{F}_{j+1/2} - \mathbf{F}_{j-1/2}}{\Delta x} = \mathbf{0}, \tag{3.4.1}$$

$$\mathbf{F}_{m+1/2} = \tfrac{1}{2}(\mathbf{F}_m^k + \mathbf{F}_{m+1}^k) + \tfrac{1}{2}|A|_{m+1/2}^k(\mathbf{U}_m^k - \mathbf{U}_{m+1}^k), \quad m = j, j-1; \tag{3.4.2}$$

$$|A| = \Omega_{\mathrm{R}}|\Lambda|\Omega_{\mathrm{L}}, \tag{3.4.3}$$

see Eqs. (2.3.25)–(2.3.26). A nonconservative version of this scheme was first suggested by Courant, Isaacon, and Rees (1952). The above conservative version was first used by Kholodov (1978); see also Belotserkovskii, Kholodov, and Turchak (1986), and Magomedov and Kholodov (1988). The conservative and nonconservative CIR-type methods are successfully used for simulation of various multidimensional problems of gas dynamics; see, for example, Belotserkovskii et al. (1974a, 1974b), Kostrykin, Fomin, and Kholodov (1976), Belotserkovskii et al. (1978), Belotserkovskii, Kholodov, and Turchak (1986), Magomedov and Kholodov (1988), Andreev and Kholodov (1989), etc. There are more than two hundred works devoted to the application of the CIR method in gas dynamics simulations.

To implement scheme (3.4.1)–(3.4.3) one must calculate $|A|$ in accordance with Eq. (3.4.3), where $\Omega_{\mathrm{R}}, \Lambda$, and Ω_{L} are given by (3.1.21)–(3.1.23). To solve the equations on grids moving with velocity $D(x)$, see (3.2.7)–(3.2.10), we must modify the procedure of calculating $\mathbf{F}$ in accordance with Eq. (2.3.29). In doing so, we must take into account the variation of the cell volume in time, see (3.2.7)–(3.2.10), and modify the formulas for $\mathbf{F}$ as follows:

$$\begin{aligned}
\widetilde{\mathbf{F}}_{m+1/2} &= (\mathbf{F} - D\mathbf{U})_{m+1/2} \\
&= \tfrac{1}{2}(\mathbf{F}_m^k - D_{m+1/2}\mathbf{U}_m^k + \mathbf{F}_{m+1}^k - D_{m+1/2}\mathbf{U}_{m+1}^k) + \tfrac{1}{2}|A(D)|_{m+1/2}^k(\mathbf{U}_m^k - \mathbf{U}_{m+1}^k), \\
|A(D)| &= \Omega_{\mathrm{R}}[|\lambda_p - D|\,\delta_{pl}]\Omega_{\mathrm{L}}.
\end{aligned}$$

An implementation of formulas (3.1.21)–(3.1.23) requires that some expressions or approximations be used for p_ε and p_ρ, because these variables occur in Eqs. (3.1.22)–(3.1.23) explicitly. In the case where the function $p(\rho, \varepsilon)$ is known analytically and is sufficiently smooth, the implementation of the finite-volume scheme is relatively easy. In wide-range semi-empirical EOSs the pressure derivatives can be nonsmooth functions. In addition, these EOSs can often be given in tabulated form. This can result in a substantial loss of accuracy of calculations. Inadequate approximations of p_ε and p_ρ for multi-phase EOSs leads to numerical errors resulting in nonphysical values of thermodynamic variables. The numerical results thus obtained must be regularized. However, such corrections are undesirable if they are required frequently during the computation.

We can represent the formulas for $\mathbf{F}$ in (3.4.2)–(3.4.3) in a way allowing us to eliminate the explicit usage of p_ρ and p_ε, which makes their approximation unnecessary (Semenov

1997). Note that even though the expression (3.1.24) for the sound speed exists, it is often not used explicitly to calculate c. The sound speed c can be calculated for $j \pm 1/2$ directly from the wide-range equation of state without approximating p_ρ and p_ε. The determination of the sound speed in terms of ρ and ε is provided by various modern EOSs in the form of numerical codes. Moreover, in certain cases experimental values of c can be invoked. Note also that the following representation is used for the vector $\mathbf{Q} = |A|\Delta\mathbf{U}$:

$$\begin{aligned}
\mathbf{Q} &= |A|\Delta\mathbf{U} = \Omega_{\mathrm{R}}|\Lambda|\Omega_{\mathrm{L}}\Delta\mathbf{U} = \Omega_{\mathrm{R}}\,\mathrm{diag}[|u-c|, |u|, |u|, |u|, |u+c|]\Omega_{\mathrm{L}}\Delta\mathbf{U} \\
&= |u|\Delta\mathbf{U} + (|u-c|-|u|)\Omega_{\mathrm{R}}\,\mathrm{diag}[1,0,0,0,0]\Omega_{\mathrm{L}}\Delta\mathbf{U} \\
&\quad + (|u+c|-|u|)\Omega_{\mathrm{R}}\,\mathrm{diag}[0,0,0,0,1]\Omega_{\mathrm{L}}\Delta\mathbf{U}, \\
\Delta\mathbf{U} &= \mathbf{U}_m - \mathbf{U}_{m+1} = [\Delta\rho, \Delta(\rho u), \Delta(\rho v), \Delta(\rho w), \Delta e]^{\mathrm{T}},
\end{aligned}$$

or, equivalently,

$$\mathbf{Q} = |u|\begin{bmatrix}\Delta\rho \\ \Delta(\rho u) \\ \Delta(\rho v) \\ \Delta(\rho w) \\ \Delta e\end{bmatrix} + (|u-c|-|u|)(f+g)\begin{bmatrix}1 \\ u-c \\ v \\ w \\ h-uc\end{bmatrix} + (|u+c|-|u|)(f-g)\begin{bmatrix}1 \\ u+c \\ v \\ w \\ h+uc\end{bmatrix}, \tag{3.4.4}$$

where

$$f = \frac{p_\varepsilon}{2\rho c^2}[\theta\Delta\rho - u\Delta(\rho u) - v\Delta(\rho v) - w\Delta(\rho w) + \Delta e], \quad g = \frac{1}{2c}[u\Delta\rho - \Delta(\rho u)].$$

The derivatives of p_ρ and p_ε occur only in the formula for f. It can be verified that

$$\begin{aligned}
f &= \frac{p_\varepsilon}{2\rho c^2}[\theta\Delta\rho - u\Delta(\rho u) - v\Delta(\rho v) - w\Delta(\rho w) + \Delta e] = \frac{p_\varepsilon}{2\rho c^2}[(u^2+v^2+w^2 - \frac{e}{\rho} + \frac{\rho p_\rho}{p_\varepsilon})\Delta\rho \\
&\quad - u\Delta(\rho u) - v\Delta(\rho v) - w\Delta(\rho w) + \Delta(\tfrac{1}{2}\rho\,(u^2+v^2+w^2) + \rho\varepsilon)] \\
&= \frac{1}{2c^2}[p_\rho\Delta\rho + p_\varepsilon\Delta\varepsilon + O(\Delta^2)] = \frac{1}{2c^2}[\Delta p + O(\Delta^2)],
\end{aligned}$$

where Δ denotes the first finite differences. By the substitution of $\frac{1}{2}\Delta p/c^2$ for f, the derivatives of pressure can be eliminated from the final formulas (3.4.1)–(3.4.3). Then compact formulas for $\mathbf{Q} = [Q_1, Q_2, Q_3, Q_4, Q_5]^{\mathrm{T}}$ are the following:

$$\begin{aligned}
&Q_1 = |u|\Delta\rho + \beta_1, \quad Q_2 = |u|\Delta(\rho u) + \beta_1 u - \beta_2 c, \quad Q_3 = |u|\Delta(\rho v) + \beta_1 v, \\
&Q_4 = |u|\Delta(\rho w) + \beta_1 w, \quad Q_5 = |u|\Delta e + \beta_1 h - \beta_2 uc, \\
&\beta_1 = \alpha_1 + \alpha_2, \quad \beta_2 = \alpha_1 - \alpha_2, \\
&\alpha_1 = (|u-c|-|u|)(f+g), \quad \alpha_2 = (|u+c|-|u|)(f-g), \\
&f = \frac{\Delta p}{2c^2}, \quad g = \frac{u\Delta\rho - \Delta(\rho u)}{2c}.
\end{aligned} \tag{3.4.5}$$

This version of the scheme was used in numerical calculations by Vovchenko et al. (1994), Semenov (1997), and Batani et al. (1999a, 1999b). It proved to be reliable for any gradients

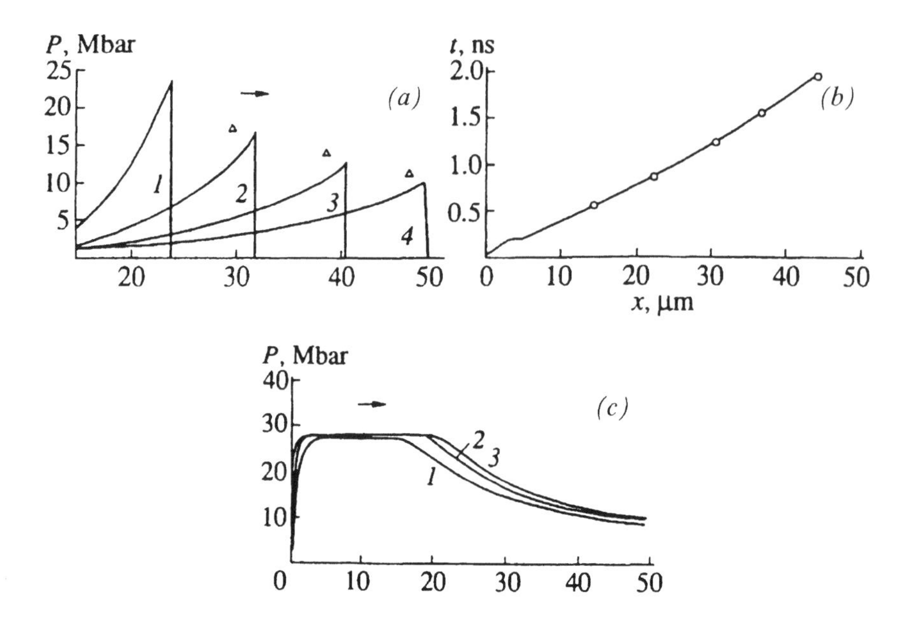

Figure 3.6 Shock-wave propagation in a plane aluminum target.

of the mesh functions. Algorithm (3.4.5) is applicable if $c \geq \epsilon > 0$. When an equation of state is given in analytical form, the current method reduces to the version of the method proposed for the one-dimensional case by Courant et al. (1952).

Further, we present some numerical results for shock wave propagation and shock-induced phenomena in real metals and plasma to demonstrate the key features of the techniques proposed above. More practical details can be found in Krasyuk and Semenov (1992), Vovchenko, Krasyuk, and Semenov (1992), Goncharov, Semenov, Serov, and Yanovskii (1992), Semenov (1997), and Batani et al. (1999a, 1999b).

The results discussed below were obtained in computations involving nonanalytical equations of state.

3.4.2 Computation of shock-induced phenomena by the CIR method.

Figure 3.6 shows the numerical results obtained in test simulations of the shock-wave propagation in a plane aluminum target using Lagrangian coordinates. The front side of a 50-μm-thick target (located to the left of the graph in Fig. 3.6a) was subjected to a rightward-propagating trapezoidal pressure pulse characterized by a total duration of 0.3 ns and rise and fall times of 0.03 and 0.01 ns, respectively. The initial pressure amplitude is equal to 28 Mbar. The computations were performed for a real, wide-range equation of state that was obtained by using both tabulated data and software packages by Bushman and Fortov (1983), Bushman, Lomonosov, and Fortov (1992), and Bushman, Kanel, Ni, and Fortov (1993). The results were compared with those obtained by applying the LASNEX code, see Trainor and Lee (1982). Figure 3.6a shows the shock-wave pressure profiles at

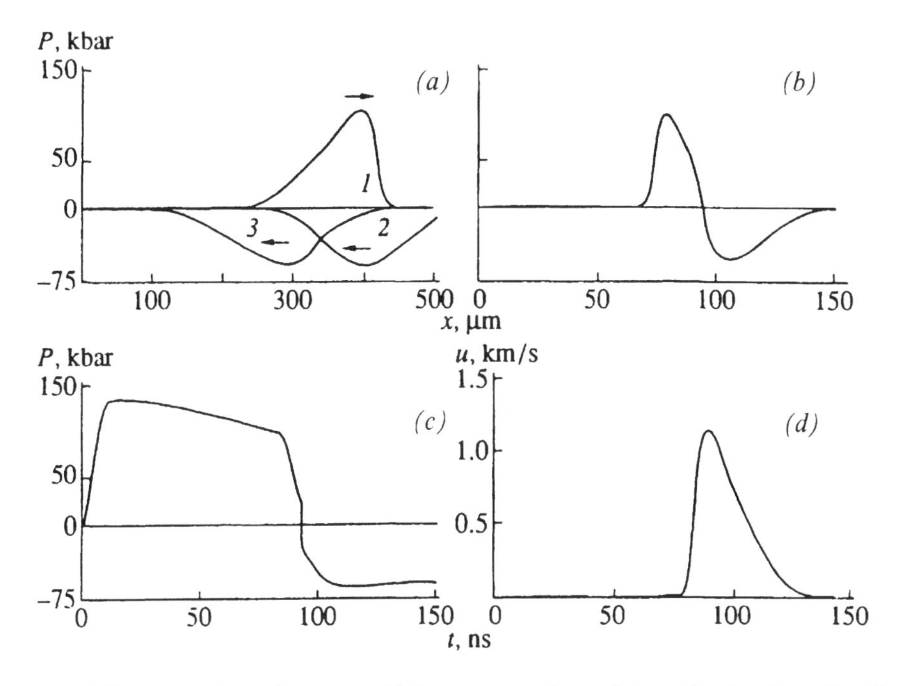

Figure 3.7 Propagation of a trapezoidal pressure pulse and its reflection from the free surface.

various moments after the starting moment of the pulse application; curves 1, 2, 3, and 4 correspond to 0.65, 0.9, 1.2, and 1.5 ns, respectively. Triangles represent the peak pressures that correspond to curves 2, 3, and 4 computed for a similar shock wave in Trainor and Lee (1982). Figure 3.6b illustrates the motion of a shock wave initiated by a 13 Mbar pressure pulse. The circles represent the results taken from Trainor and Lee (1982). In Figure 3.6c, the maximum pressure is shown versus the location of the corresponding peak to illustrate the convergence of the numerical results obtained with various mesh sizes Δx; curves 1, 2, and 3 correspond to $\Delta x = 0.5$, 0.25, and 0.125 μm, respectively. In the figure, the peak pressure is shown as a function of the current coordinate of the peak. Similar convergence checks were performed for all results to guarantee their reliability. The CFL number was set equal to 0.2 in all computations represented in Fig. 3.6. The calculations with large CFL numbers up to 0.8–0.9 occasionally contained nonphysical values of thermodynamic variables lying below the "cold" curves for the pressure and energy. The results obtained are in good agreement with those presented in Trainor and Lee (1982). Minor discrepancies, see in Fig. 3.6, can be attributed to the difference in the equations of state employed.

The computations were based on an up-to-date wide-range semi-empirical multiphase (solid–liquid–gas–plasma) equation with approximately 50 parameters, see Bushman, Lomonosov, and Fortov (1992) and Bushman et al. (1993), most of which are fundamental matter constants, whereas about 15 adjustable parameters could be chosen independently to represent the characteristic temperatures and densities at transition points (e.g., metal–dielectric) of phase diagrams. The computations employed either the tabulated equations

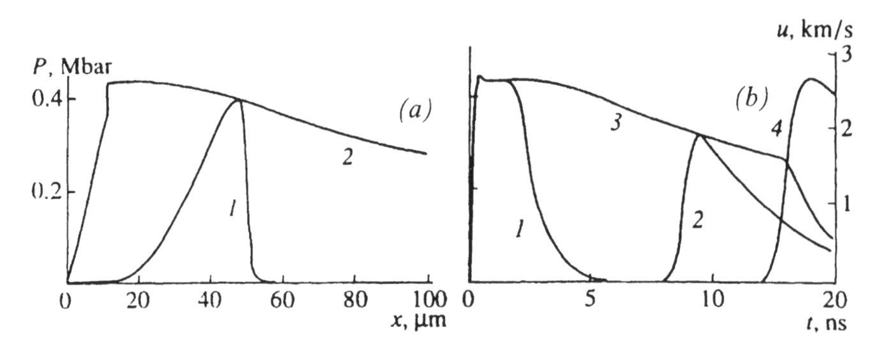

Figure 3.8 The collision of an aluminum target with a projectile.

of state, which were given as the tables for the sound speed and pressure versus density and internal energy, or software packages designed to calculate $p(\rho, \varepsilon)$ and $c(\rho, \varepsilon)$ directly to guarantee numerical accuracy. Typically, logarithmic-scale tables were generated, with a resolution of three to five data points per order of magnitude of the density and internal energy.

As mentioned above, when a version of the CIR method in which the thermodynamic derivatives p_ρ and p_ε are used in the explicit form was applied, the results obtained were corrected to avoid nonphysical values. In particular, if a value of the internal energy was found to lie below the "cold" curve, it was properly corrected. The modified algorithm of (3.4.5) is designed to eliminate numerical difficulties of this kind. Numerical results obtained for shock-induced phenomena are presented to demonstrate the reliability and efficiency of the proposed technique.

Figure 3.7 shows some results obtained for the propagation of a trapezoidal pressure pulse of amplitude 130 Kbar and duration 20 ns and its reflection from the free surface. The pulse was applied to a target from the left, see Fig. 3.7a. The target was made of aluminum and 500 μm thick; here Δx is equal to 1 μm. The solution of this problem was used to interpret the experimental results by Kilpio et al. (1992) and Vovchenko et al. (1992).

Figure 3.7a shows the pressure profiles at 70, 120, and 170 ns after the pulse hits the target (curves 1, 2, and 3, respectively). The arrows indicate the direction of the pulse propagation. We see that the initially compressing (positive) pressure pulse transforms into an expanding (negative) one after the reflection from the right free surface boundary. Figure 3.7b shows the graph of the pressure variation at the point located at a distance of 60 μm from the rare side of the target, where the spallation effect was observed, see Bushman et al. (1993), and Vovchenko et al. (1992). This highest expanding stress obtained numerically at the experimental location of spallation provides an estimate for the impact strength limit of the material. Figure 3.7c shows the highest absolute value of the pressure across the target as a function of time. Figure 3.7d shows the velocity of the free target surface as a function of time. Figure 3.7 illustrates the basic feature of shock-induced phenomena, i.e., the transformation of a shock-compressed flow into a highly rarefied one.

Figure 3.8 shows the results obtained for a somewhat different problem, in which the shock wave results from the collision of a 100-μm-thick aluminum target with a 10-μm-thick aluminum projectile with a velocity of 4 km/s; here the mesh size is equal to $\Delta x = 0.5$ μm.

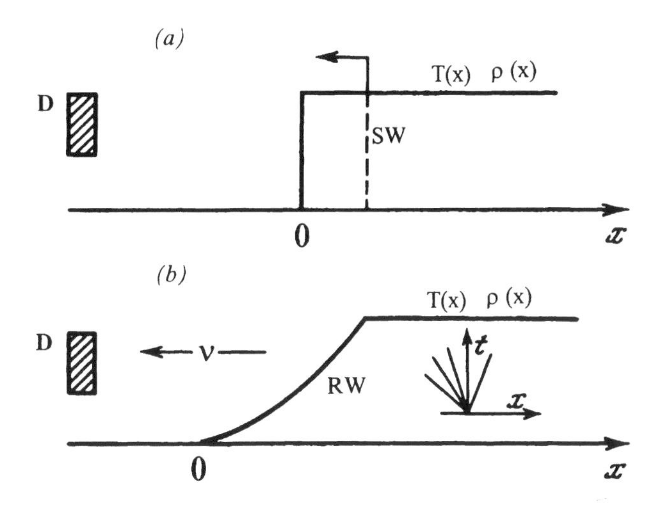

Figure 3.9 Scheme of luminescence detection.

Figure 3.8a shows the pressure profile in the shock wave at 4.96 ns (curve 1) and the peak pressure in the target (curve 2). Figure 3.8b shows the pressure versus time at the distances $x = 0$ and 80 μm from the front side of the target (curves 1 and 2, respectively), the peak pressure versus time (curve 3), and the target rare side velocity at the moment of the shock exit (curve 4). Analysis of this problem was also motivated by experiments. By solving it, one can find the projectile velocity from the measured times of the shock arrival at the free metal surface (Vovchenko et al. 1992).

In all computations, it was checked that thermodynamic variables had physically admissible values, i.e., $p \geq p_{\text{cold}}$, $\varepsilon \geq \varepsilon_{\text{cold}}$, and $c > 0$. This requirement was violated only in simulations of short pulse induced phenomena for which the CFL numbers were insufficiently small. By the reduction of its value from 1 to 0.2–0.6, numerical errors of this kind were completely eliminated. However, simulations of these problems based on the equation of state of an perfect gas could be performed with the CFL number close to unity.

Figures 3.9a and 3.9b show the scheme of physical experiments on heating measurements of a substance carried out in a study of strong laser-driven shock waves (Ng, Parfeniuk, and Da Silva 1985a, 1985b; Ng et al. 1989; Godwal et al. 1990; Kvitov et al. 1991; Vovchenko, Krasyuk, and Semenov 1992; Batani et al. 1999a, 1999b). The strong shock wave SW propagating in the metal (Fig. 3.9a–b) generates on the target–vacuum interface the luminescence of metal plasma in the rarefaction wave RW. The luminosity is measured in time by detector D for different light frequencies ν, in common for red and blue lights, and is used for measurements of the temperature and pressure of the material behind the shock wave front, for details see Batani et al. (1999a, 1999b). The application of dynamic methods of diagnostic for the investigation of thermodynamic properties of material in a wide range of parameters suggests stationarity and one-dimensionality of a hydrodynamic flow. The temperature measurements of shock-heated plasma in metals is a complicated problem, mainly due to screening of radiation that emerges from the hot regions at the target by cooled vapors of the material released by the rarefaction wave that follows the shock breakout at the target–vacuum interface, see Zel'dovich and Raizer (1967).

The measured time evolution of the temperature of the rear surface of the target is compared with that computed by the numerical code. For computing the time history of the temperature of the rear surface, the spatial profiles of the density ρ and temperature T at various times during the shock-unloading process were calculated (Fig. 3.9b).

First computations for aluminum, lead, and bismuth (Polishchuk, Semenov, Ternovoi, and Fortov 1991; Kvitov et al. 1991; Semenov 1995b) were based on exact rarefaction wave solution for wide-range EOSs. Further, to determine the shock-wave characteristics on the basis of experimental data, we use also the Godunov method based on the CIR scheme. The computational model employed hydrodynamic equations in the Lagrangian variables to which the wide-range semi-empirical multiphase EOS for aluminum was added (Busman, Lomonosov and Fortov 1992; Bushman et al. 1993). To compute the time history of the luminous radiation temperature of the material in the rarefaction wave, we applied the approach of Zel'dovich and Raizer (1967) based on the calculation of the spectral brightness of the emitting layer with known spatial density and temperature distributions shown schematically in Figure 3.9b.

Numerical simulation of this problem requires high accuracy in the calculation of gas dynamic parameters in a metal rarefaction wave. One can use special modification of the above numerical method. In fact, in a rarefaction wave we must use the equation

$$\frac{\partial \varepsilon}{\partial t} + p\frac{\partial \eta}{\partial t} = 0, \quad \eta = \frac{1}{\rho}, \tag{3.4.6}$$

for the calculation of the internal energy, see relation (3.3.60) or (3.3.64). Equation (3.4.6) follows also from the thermodynamic relation $T\,dS = d\varepsilon + p\,d\eta$, where T is the temperature and S the entropy. In the rarefaction wave, $dS = 0$, or $S = \text{const}$, and we obtain Eq. (3.4.6). In the calculations we use Eq. (3.4.6) in all space grid cells where $\partial u/\partial x \geq 0$ (u is the velocity). The necessity of applying Eq. (3.4.6) is due to the use of real EOSs. This algorithm provides a correct redistribution of the internal energy between its thermal and cold terms in the EOS (3.1.30)–(3.1.31) and can be used for an arbitrary Godunov-type scheme. Otherwise, the temperature values are considerably overestimated and the cases of $\lim_{\rho\to 0} T \neq 0$ and $\lim_{\rho\to 0} c \neq 0$ can be realized. Unfortunately, we can see such systematic errors in some publications dealing with calculations in rarefaction waves for wide-range EOSs. Note that for an ideal perfect gas this modification is not necessary.

3.4.3 The CIR-simulation of jet-like structures in laser plasma. This section describes the numerical results of two-dimensional modelling of physical phenomena that exist in unsteady laser plasma and are known as "jet-like structures" (JLS). Such specific density jets were detected experimentally in plasma produced in the interaction of a laser pulse with a solid surface, see Willi and Rumsby (1981), Willi, Rumsby, and Sartang (1981), Willi et al. (1982), Stamper et al. (1981), Bondarenko et al. (1981), Denus et al. (1983), Burges et al. (1985), Thiell and Meyer (1985), Xu et al. (1987, 1988, 1989), Goncharov, Serov, and Yanovskii (1987), Goncharov et al. (1992), Dhareswar et al. (1987), and Bol'shov et al. (1987). Our numerical results provide a qualitative understanding of the JLS generation and their dynamics and permit us to explain a number of experimental properties of JLS.

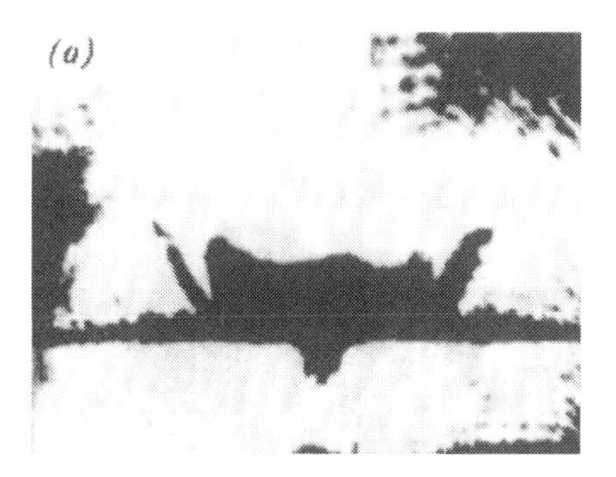

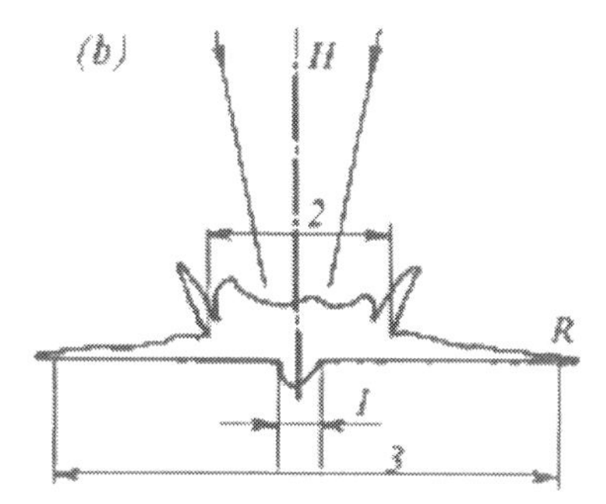

Figure 3.10 (a) Experimental shadowgram with jet-like structures; (b) the scheme of the experiment.

Figure 3.11 Experimental interferogram with jet-like structures.

In addition, numerical calculations permit us to verify the reliability of the Godunov-type methods for the case of large gradients of mesh functions.

There are a lot of experimental studies of ablative laser plasma produced by intensive laser pulse interaction with solid plane or spherical targets. In some experiments a strong jetting in plasma corona was detected, see experimental shadowgram with jets in Fig. 3.10a. A conventional scheme of experiments for the plane target is shown on Fig. 3.10b; H and R are the cylindrical coordinates, where R is the radial coordinate, and H the axial coordinate. The arrows at the top at the figure indicate the directions of laser radiation affecting the solid target. The numbers 1, 2, and 3 denote the regions of target ablation, jet generation, and cold plasma, respectively. The solid curve is the boundary of ablative plasma. This boundary was identified in accordance with some average characteristic value of the density. It was obtained from the experimental shadowgram shown in Fig. 3.10a. These shadowgrams were obtained in the General Physics Institute of the Russian Academy of Sciences by a UMI-35 laser installation; see for details Goncharov et al. (1992). In Fig. 3.10a, one can distinctly observe large jet-like protuberances. These structures are referred to as large-scale jet-like structures (JLS). Small-scale JLS are the rapid small density perturbations; they are not practically seen in the scale of this figure.

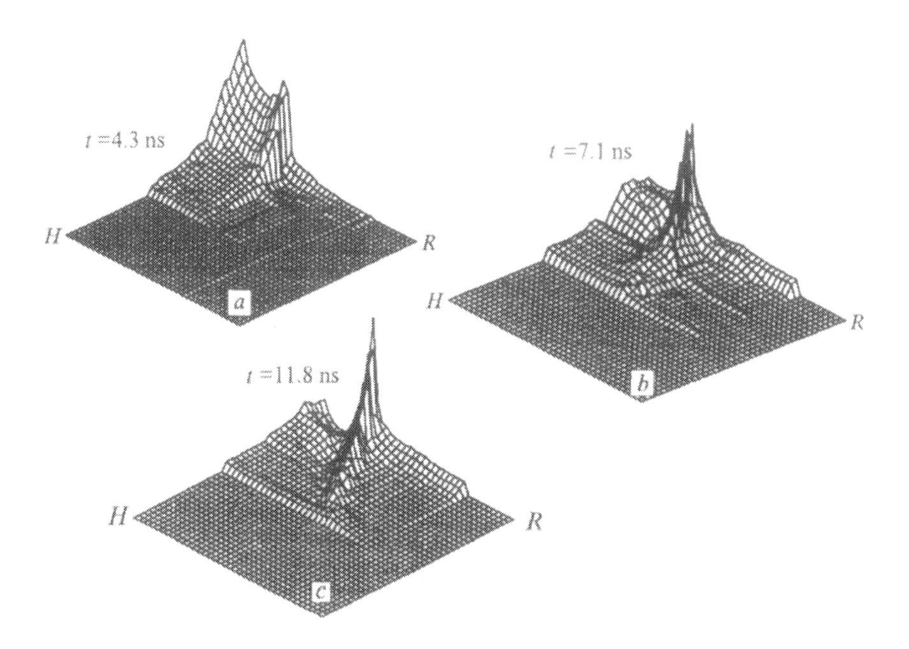

Figure 3.12 Plasma jets in experiment.

Figure 3.11 presents the interferogram of laser plasma near a solid wall, also obtained by the UMI-35 laser experiment. The image consists of many interferometric fringes (bands). These fringes approach each other and occasionally merge in the region of strong density gradients. Figure 3.11 shows these regions, which are actually jet-like structures. Both the shadowgram and the interferogram are quite typical. Some analogues are presented in a number of experimental data; for details see Goncharov et al. (1992).

The interferograms in Fig. 3.11 are used for the reconstruction of the plasma density distribution. This reconstruction was carried out by numerical solution of the Abelian integral equation by the Nestor–Olsen method (1960) with some modifications (Goncharov, Serov, and Yanovskii 1987). This modification is caused by difficulties in processing of experimental interferograms. They can have some regions where fringes intersect (Fig. 3.11) and therefore previous methods of processing cannot be used. Figures 3.12a–c show the reconstructed plasma density distributions at the fixed moments of time for the same experiment and demonstrate the JLS dynamics. Jet structures are represented here by the regions with abrupt gradients. Distributions are shown in cylindrical coordinates (R, H), where R is the radial coordinate, and H the axial coordinate. The origin is associated here with the center of the focal laser spot (Fig. 3.10b).

The JLS have the following properties. The jets change, in time, their length, thickness, spatial location, and inclination angle with respect to the free surface of the target. But the JLS boundaries at any time moments are easily seen, and in many cases the JLS are quite long-living in comparison with the laser pulse.

There are some theories of jetting in laser plasma. Jet formation is explained by

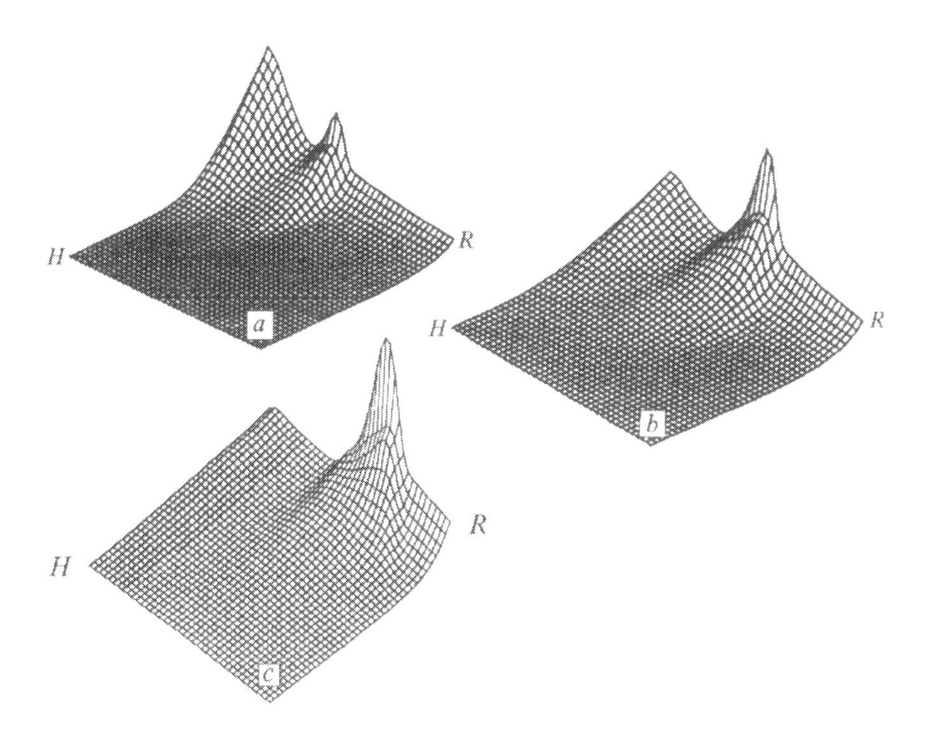

Figure 3.13 Jets in numerical simulation.

thermal and hydrodynamic instabilities produced in hot plasma or solid target. Among the thermal instabilities first noted were the electromagnetic instability (Tidman and Shanny 1974; Haines 1974; Ogasawara, Hirao, and Ohkubo 1980; Hirao and Ogasawara 1981), the Weibel instability (1959), the radiation cooling instability (Evans 1981), and others. These theories explain the dynamics of small-scale JLS fairly well. A theory of large-scale dynamics seems to be absent.

Goncharov et al. (1992) performed gas dynamic calculations of the JLS generation by solving the unsteady two-dimensional and two-temperature gas dynamic equations (3.1.26)–(3.1.29). The CIR method proved to give good resolution of strong shock interaction and strong gradients of the mesh function. The gas dynamic system of equations was chosen after analysis of the experimental data. Assume that there is a general physical mechanism of large-scale JLS generation and this mechanism has gas dynamic nature. This can explain why experimental data are the same in a number of quite different experiments. In these cases the influence of different instabilities is not commensurable but the data are analogous. The JLS live much longer than the laser pulse duration. Note that at later stages only gas dynamic processes are essential in plasma. Thus, the assumption about the gas dynamic character of large-scale JLS formation is quite reasonable.

The numerical results are presented in Figs. 3.13a–c for the same time moments as in the experiments (Figs. 3.12a–c). The jets are seen to form outside the laser spot and after

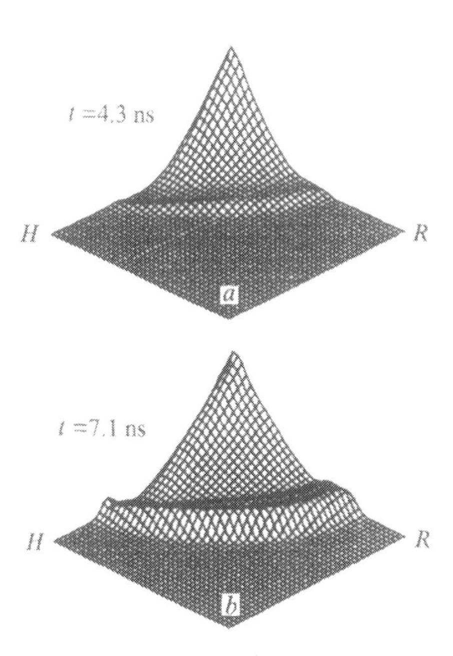

Figure 3.14 Spherical shock wave propagation.

the end of the laser pulse. One can clearly observe the qualitative and quantitative similarity of both the numerical and experimental density distributions. The results make it possible to reproduce the experimental data presented in Figs. 3.12a–c and to explain the dynamics of jets. The initial density distribution was chosen in the form

$$\rho(R,H) = \max\left[\rho_1 \exp\left(-\frac{R}{a_1} - \frac{H}{b_1}\right), \rho_2 \exp\left(-\frac{R}{a_2} - \frac{H}{b_2}\right)\right], \quad R \geq 0, \quad H \geq 0.$$

It is based on the experimental data and physical assumptions. The subscripts 1 and 2 correspond to cold (background) and hot (ablative) plasma, respectively. For $H = 0$ the impermeable wall condition was assumed. The quantities a and b are characteristic lengths of the density distributions along the R- and H-axes. The initial temperatures, T_1 and T_2 ($T_1 \ll T_2$), for cold and hot plasma, respectively, were constant. In some cases the background density distribution was also constant. Thus, to compare the numerical density profile with the experimental one, there were seven parameters to vary: a_1, b_1, a_2, b_2, T_1, T_2, and ρ_1/ρ_2. In numerical modelling we set $a_1 = 30\,\mu$m, $b_1 = 20\,\mu$m, $a_2 = 40\,\mu$m, $b_1 = 300\,\mu$m, $\rho_1/\rho_2 = 30$, and $T_1 = 400\,$eV and varied the temperature T_2. Note that the effect of two temperatures is not very significant and one can use Eqs. (3.1.1)–(3.1.3) as well.

The above values of parameters were determined by a series of calculations and by the analysis of experimental data. We also investigated the stability of numerical results to the variations of the parameters, the version of numerical method, and the space grid.

In fact, the large-scale jet-like structures result from the interaction between the extensive hot dense ablative plasma produced in the focal spot by the laser pulse and the background cold plasma. There are such parameter domains where JLS always exist. The two-dimensional effects, more exactly the nonuniformity of the density distribution in cold plasma, is essential in the JLS generation. Indeed, in the case of a constant density distribution in the cold plasma, only simple spherical shock wave structures (Korobeinikov 1991; Sedov 1993) were detected in the background two-temperature gas. The conical jet structures turned out to be absent, see Figs. 3.14a–b and Figs. 3.12a–b.

The jet inclination angle depends on T_1 and T_2. The larger T_1 and the smaller T_2, the larger this angle. Note also that our approach explains the dependence of the experimental angle on the atomic weight Z. Indeed, the characteristic lengths a and b depend on the substance of the target, or on Z, and therefore we can predict how the angle changes for various substances and for two-substance target (Goncharov et al. 1992).

Thus, the numerical results allow one to confirm the gas dynamic concept of the generation of large-scale jet-like structures in laser plasma under the action of a laser pulse. The results presented here, as well as those obtained in similar computations by Vovchenko et al. (1992, 1994), Krasyuk and Semenov (1992), and others, including those performed in the fixed Cartesian coordinate system and two-dimensional computations in moving coordinate systems, indicate that the proposed technique provides a reliable tool that can be recommended for computing strong transient interactions in systems with arbitrary equations of state (including those given in tabular form).

3.4.4 Roe's method.

Roe's method is based on an approximate solution of the Riemann problem for Eqs. (3.1.15)–(3.1.17). This method is similar to the CIR method (3.4.1)–(3.4.3) with a special choice of the matrix A, which we shall call $\mathcal{A}$. The matrix $\mathcal{A}$ is defined as follows:

$$\Delta\mathbf{F} = \mathcal{A}\Delta\mathbf{U}, \quad \Delta\mathbf{F} = \mathbf{F}_1 - \mathbf{F}_2, \quad \Delta\mathbf{U} = \mathbf{U}_1 - \mathbf{U}_2,$$
$$\mathbf{U}_1 = \mathbf{U}_m, \quad \mathbf{U}_2 = \mathbf{U}_{m+1}, \quad \mathbf{F}_1 = \mathbf{F}(\mathbf{U}_1), \quad \mathbf{F}_2 = \mathbf{F}(\mathbf{U}_2).$$

Since function $\mathbf{F}$ in (3.1.17) is a nonlinear function of $\mathbf{U}$, the choice the matrix $\mathcal{A}$ is not unique. Consider a nondegenerate transformation $\mathbf{U} = \mathbf{U}(\mathbf{Y})$. From the exact relations $\Delta\mathbf{F} = \mathcal{A}_F\Delta\mathbf{Y}$ and $\Delta\mathbf{U} = \mathcal{A}_U\Delta\mathbf{Y}$ it follows that $\mathcal{A} = \mathcal{A}_F\mathcal{A}_U^{-1}$. The exact analytical formulas for $\mathcal{A}_F$ and $\mathcal{A}_U$ can be written explicitly when $\mathbf{U}$ and $\mathbf{F}$ are either polynomials in the components of $\mathbf{Y}$ or rational functions of $\mathbf{Y}$. To find the entries of the matrix $\mathcal{A}_F$ and $\mathcal{A}_U$, we can use, in particular, the exact identities (2.3.66)–(2.3.69).

The main feature Roe's method can be stated as follows. Let the right and left quantities, $\mathbf{U}_1$ and $\mathbf{U}_2$, be related by the Rankine–Hugoniot conditions (3.3.4)–(3.3.8): $\Delta\mathbf{F} = W\Delta\mathbf{U}$, where W is the velocity of a discontinuity. Then

$$\Delta\mathbf{F} - W\Delta\mathbf{U} = (\mathcal{A}_F - W\mathcal{A}_U)\Delta\mathbf{Y} = (\mathcal{A} - WI)\mathcal{A}_U\Delta\mathbf{Y} = (\mathcal{A} - WI)\Delta\mathbf{U} = \mathbf{0},$$

where I is the 5×5 identity matrix. Thus, W is an eigenvalue of the matrix $\mathcal{A}$ and $\Delta\mathbf{U}$ is an eigenvector of $\mathcal{A}$. Roe's approach permits us to exactly satisfy the Rankine–Hugoniot relations on the discontinuity, see (3.3.4)–(3.3.8). Following Roe (1981) we choose

$$\mathbf{Y} = [R, U, V, W, H]^{\mathrm{T}} = [\sqrt{\rho}, \sqrt{\rho}u, \sqrt{\rho}v, \sqrt{\rho}w, \sqrt{\rho}h]^{\mathrm{T}}, \quad h = \frac{e+p}{\rho}. \tag{3.4.7}$$

In terms of $\mathbf{Y}$ we have

$$\mathbf{U} = [R^2, RU, RV, RW, HR - p]^{\mathrm{T}}, \tag{3.4.8}$$
$$\mathbf{F} = [RU, U^2 + p, UV, UW, UH]^{\mathrm{T}}. \tag{3.4.9}$$

Let us seek an appropriate function of the pressure $p = p(\rho, \varepsilon)$ in the general form

$$p(\rho, \varepsilon) = f(R) + a_1 H + a_2 RH + a_3 H^2 + a_4 \left(U^2 + V^2 + W^2\right), \tag{3.4.10}$$
$$a_1, a_2, a_3, a_4 = \text{const};$$

where $f(R)$ is a smooth function. We use here the fact that only linear and quadratic terms provide unique coefficients in a difference expansion, see Eqs. (2.3.66)–(2.3.69). Equation (3.4.10) defines the pressure p implicitly, since H depends on p. This equation must be solvable for p explicitly. Substituting the expressions R, U, V, W, and H into (3.4.10) yields

$$\begin{aligned}
p(\rho, \varepsilon) &= f(\sqrt{\rho}) + a_1 \frac{e + p}{\sqrt{\rho}} + a_2(e + p) + a_3 \frac{(e + p)^2}{\rho} + a_4 \rho \left(u^2 + v^2 + w^2\right) \\
&= f(\sqrt{\rho}) + a_1 \left[\sqrt{\rho}\,\varepsilon + \tfrac{1}{2}\sqrt{\rho}\left(u^2 + v^2 + w^2\right) + \frac{p}{\sqrt{\rho}}\right] \\
&\quad + a_2 \left[\rho\varepsilon + \tfrac{1}{2}\rho\left(u^2 + v^2 + w^2\right) + p\right] + a_3 \frac{(e + p)^2}{\rho} + a_4 \rho \left(u^2 + v^2 + w^2\right).
\end{aligned}$$

For $a_3 = 0$ we obtain a linear equation for p. The terms with velocities are eliminated by assuming $a_4 = -\frac{1}{2}a_2$ and $a_1 = 0$. Then the equation of state admitted by Roe's solver has the form

$$p(\rho, \varepsilon) = (\gamma - 1)\rho\varepsilon + g(\sqrt{\rho}), \tag{3.4.11}$$

where $g(R) = f(R)/(1 - a_2)$ is a smooth function and $\gamma = 1/(1 - a_2)$. The EOS (3.1.5) studied previously is a special case of (3.4.11). In terms of $\mathbf{Y}$, the quantities $\rho\varepsilon$, p, and e are expressed as

$$\begin{aligned}
\rho\varepsilon &= \frac{HR}{\gamma} - \frac{1}{2\gamma}\left(U^2 + V^2 + W^2\right) - \frac{g}{\gamma}, \\
p &= \frac{\gamma - 1}{\gamma} HR - \frac{\gamma - 1}{2\gamma}\left(U^2 + V^2 + W^2\right) + \frac{g}{\gamma}, \\
e &= \frac{HR}{\gamma} + \frac{\gamma - 1}{2\gamma}\left(U^2 + V^2 + W^2\right) - \frac{g}{\gamma}.
\end{aligned}$$

Then,

$$\mathbf{U} = \left[R^2, RU, RV, RW, \frac{HR}{\gamma} + \frac{\gamma - 1}{2\gamma}\left(U^2 + V^2 + W^2\right) - \frac{g}{\gamma}\right]^{\mathrm{T}},$$

$$\mathbf{F} = \left[RU,\ U^2 + \frac{\gamma - 1}{\gamma} HR - \frac{\gamma - 1}{2\gamma}\left(U^2 + V^2 + W^2\right) + \frac{g}{\gamma},\ UV, UW, UH\right]^{\mathrm{T}}.$$

Hence,

$$\Delta\mathbf{U} = \begin{bmatrix} 2\overline{R}\Delta R \\ \overline{U}\Delta R + \overline{R}\Delta U \\ \overline{V}\Delta R + \overline{R}\Delta V \\ \overline{W}\Delta R + \overline{R}\Delta W \\ \frac{1}{\gamma}(\overline{H} - \overline{g})\Delta R + \frac{\gamma-1}{\gamma}(\overline{U}\Delta U + \overline{V}\Delta V + \overline{W}\Delta W) + \frac{1}{\gamma}\overline{R}\Delta H \end{bmatrix}, \tag{3.4.12}$$

$$\Delta\mathbf{F} = \begin{bmatrix} \overline{U}\Delta R + \overline{R}\Delta U \\ 2\overline{U}\Delta U + (\frac{\gamma-1}{\gamma}\overline{H} + \frac{1}{\gamma}\overline{g})\Delta R - \frac{\gamma-1}{\gamma}(\overline{U}\Delta U + \overline{V}\Delta V + \overline{W}\Delta W) + \frac{\gamma-1}{\gamma}\overline{R}\Delta H \\ \overline{V}\Delta U + \overline{U}\Delta V \\ \overline{W}\Delta U + \overline{U}\Delta W \\ \overline{H}\Delta U + \overline{U}\Delta H \end{bmatrix}.$$

Here the bar over a quantity stands for the arithmetic average of the left and right values of this quantity, $\overline{a} = \frac{1}{2}(a_1 + a_2)$, where $a = R, U, V, W$, or H, see Eqs. (2.3.66)–(2.3.69),

$$\overline{g} = \frac{g(R_1) - g(R_2)}{R_1 - R_2} = \frac{g(\sqrt{\rho_1}) - g(\sqrt{\rho_2})}{\sqrt{\rho_1} - \sqrt{\rho_2}}.$$

To obtain more compact formulas, we divide all components of $\Delta\mathbf{U}$ and $\Delta\mathbf{F}$ by $\overline{R} = \frac{1}{2}(R_1 + R_2)$. Then the matrices $\mathcal{A}_U$ and $\mathcal{A}_F$, where $\Delta\mathbf{U} = \mathcal{A}_U\Delta\mathbf{Y}$ and $\Delta\mathbf{F} = \mathcal{A}_F\Delta\mathbf{Y}$, are expressed as

$$\mathcal{A}_U = \begin{bmatrix} 2 & 0 & 0 & 0 & 0 \\ \hat{u} & 1 & 0 & 0 & 0 \\ \hat{v} & 0 & 1 & 0 & 0 \\ \hat{w} & 0 & 0 & 1 & 0 \\ \frac{1}{\gamma}(\hat{h} - 2g_\rho) & \frac{\gamma-1}{\gamma}\hat{u} & \frac{\gamma-1}{\gamma}\hat{v} & \frac{\gamma-1}{\gamma}\hat{w} & \frac{1}{\gamma} \end{bmatrix}, \qquad \det\mathcal{A}_U = \frac{2}{\gamma};$$

$$\mathcal{A}_F = \begin{bmatrix} \hat{u} & 1 & 0 & 0 & 0 \\ \frac{\gamma-1}{\gamma}\hat{h} + \frac{2}{\gamma}g_\rho & \frac{\gamma+1}{\gamma}\hat{u} & \frac{1-\gamma}{\gamma}\hat{v} & \frac{1-\gamma}{\gamma}\hat{w} & \frac{\gamma-1}{\gamma} \\ 0 & \hat{v} & \hat{u} & 0 & 0 \\ 0 & \hat{w} & 0 & \hat{u} & 0 \\ 0 & \hat{h} & 0 & 0 & \hat{u} \end{bmatrix},$$

$$g_\rho = \frac{\overline{g}}{2\overline{R}} = \frac{g(R_1) - g(R_2)}{(R_1)^2 - (R_2)^2} = \frac{g(\sqrt{\rho_1}) - g(\sqrt{\rho_2})}{\rho_1 - \rho_2}.$$

Here $\hat{a}$ stands for Roe's average of $a = \{u, v, w, h\}$,

$$\hat{u} = \frac{\overline{U}}{\overline{R}} = \frac{\sqrt{\rho_1}\,u_1 + \sqrt{\rho_2}\,u_2}{\sqrt{\rho_1} + \sqrt{\rho_2}}, \qquad \hat{v} = \frac{\overline{V}}{\overline{R}} = \frac{\sqrt{\rho_1}\,v_1 + \sqrt{\rho_2}\,v_2}{\sqrt{\rho_1} + \sqrt{\rho_2}}, \tag{3.4.13}$$
$$\hat{w} = \frac{\overline{W}}{\overline{R}} = \frac{\sqrt{\rho_1}\,w_1 + \sqrt{\rho_2}\,w_2}{\sqrt{\rho_1} + \sqrt{\rho_2}}, \qquad \hat{h} = \frac{\overline{H}}{\overline{R}} = \frac{\sqrt{\rho_1}\,h_1 + \sqrt{\rho_2}\,h_2}{\sqrt{\rho_1} + \sqrt{\rho_2}}.$$

To establish an analytical formula for $\mathcal{A} = \mathcal{A}_F \mathcal{A}_U^{-1}$, we find a solution $\mathbf{x}^\mathrm{T}$ of the linear system

$$\mathbf{x}^\mathrm{T} \mathcal{A}_U = \mathbf{z}^\mathrm{T} \quad \Longrightarrow \quad \mathbf{x}^\mathrm{T} = \mathbf{z}^\mathrm{T} \mathcal{A}_U^{-1} : \tag{3.4.14}$$
$$x_1 = \tfrac{1}{2} z_1 - \tfrac{1}{2}\hat{u} z_2 - \tfrac{1}{2}\hat{v} z_3 - \tfrac{1}{2}\hat{w} z_4 + \tfrac{1}{2}(\gamma-1)(\hat{u}^2 + \hat{v}^2 + \hat{w}^2) z_5 - \tfrac{1}{2}(\hat{h} - g_\rho) z_5,$$
$$x_2 = z_2 - (\gamma-1)\hat{u} z_5, \quad x_3 = z_3 - (\gamma-1)\hat{v} z_5,$$
$$x_4 = z_4 - (\gamma-1)\hat{w} z_5, \quad x_5 = \gamma z_5.$$

Here $\mathbf{x}^\mathrm{T} = [x_1, x_2, x_3, x_4, x_5]$ and $\mathbf{z}^\mathrm{T} = [z_1, z_2, z_3, z_4, z_5]$. Thus, considering the nth row of the matrix $\mathcal{A}_F$ to be $\mathbf{z}^\mathrm{T}$, we calculate $\mathbf{x}^\mathrm{T}$, the nth row of the matrix $\mathcal{A} = \mathcal{A}_F \mathcal{A}_U^{-1}$, since $\mathcal{A}\mathcal{A}_U = \mathcal{A}_F$:

$$\mathcal{A} = \begin{bmatrix} 0 & 1 & 0 & 0 & 0 \\ -\hat{u}^2 + g_\rho + \frac{1}{2}(\gamma-1)q^2 & (3-\gamma)\hat{u} & (1-\gamma)\hat{v} & (1-\gamma)\hat{w} & \gamma-1 \\ -\hat{u}\hat{v} & \hat{v} & \hat{u} & 0 & 0 \\ -\hat{u}\hat{w} & \hat{w} & 0 & \hat{u} & 0 \\ -\hat{u}\hat{h} + \hat{u} g_\rho + \frac{1}{2}(\gamma-1)\hat{u} q^2 & \hat{h} - (\gamma-1)\hat{u}^2 & (1-\gamma)\hat{u}\hat{v} & (1-\gamma)\hat{u}\hat{w} & \gamma\hat{u} \end{bmatrix}, \tag{3.4.15}$$

$$q^2 = \hat{u}^2 + \hat{v}^2 + \hat{w}^2.$$

The matrices of (3.1.19) and (3.4.15) are identical for $u = \hat{u}$, $v = \hat{v}$, $w = \hat{w}$, $h = \hat{h}$, $b = \gamma - 1$, and $\theta = \frac{1}{2} q^2 + g_\rho/(\gamma - 1)$. Hence, from Eq. (3.1.25) we obtain that

$$\hat{c} = \sqrt{(\gamma - 1)\hat{h} - \tfrac{1}{2}(\gamma - 1) q^2 + g_\rho}. \tag{3.4.16}$$

Formula (3.4.16) defines the sound speed $\hat{c}$. Thus, from (3.1.21)–(3.1.23) we obtain the matrices required for the construction Roe method:

$$\hat{\Omega}_\mathrm{R} = \begin{bmatrix} 1 & 0 & 0 & 1 & 1 \\ \hat{u} - \hat{c} & 0 & 0 & \hat{u} & \hat{u} + \hat{c} \\ \hat{v} & 1 & 0 & \hat{v} & \hat{v} \\ \hat{w} & 0 & 1 & \hat{w} & \hat{w} \\ \hat{h} - \hat{u}\hat{c} & \hat{v} & \hat{w} & \hat{h} - \hat{c}^2/b & \hat{h} + \hat{u}\hat{c} \end{bmatrix}, \quad \det \hat{\Omega}_\mathrm{R} = \frac{2\hat{c}^3}{b}; \tag{3.4.17}$$

$$\hat{\Lambda} = \mathrm{diag}[\hat{u} - \hat{c},\ \hat{u},\ \hat{u},\ \hat{u},\ \hat{u} + \hat{c}], \tag{3.4.18}$$

$$\hat{\Omega}_\mathrm{L} = \frac{b}{2\hat{c}^2} \begin{bmatrix} \theta + \hat{u}\hat{c}/b & -\hat{u} - \hat{c}/b & -\hat{v} & -\hat{w} & 1 \\ -2\hat{v}\hat{c}^2/b & 0 & 2\hat{c}^2/b & 0 & 0 \\ -2\hat{w}\hat{c}^2/b & 0 & 0 & 2\hat{c}^2/b & 0 \\ 2\hat{h} - 2q^2 & 2\hat{u} & 2\hat{v} & 2\hat{w} & -2 \\ \theta - \hat{u}\hat{c}/b & -\hat{u} + \hat{c}/b & -\hat{v} & -\hat{w} & 1 \end{bmatrix},$$

$$b = \gamma - 1, \quad q^2 = \hat{u}^2 + \hat{v}^2 + \hat{w}^2, \quad \theta = \frac{1}{2} q^2 + \frac{g_\rho}{\gamma - 1}.$$

Note that for the model of an perfect gas (3.1.4) one can obtain

$$\hat{c}^2 = \frac{\sqrt{\rho_1}\,c_1^2+\sqrt{\rho_2}\,c_2^2}{\sqrt{\rho_1}+\sqrt{\rho_2}} + \frac{(\gamma-1)\sqrt{\rho_1\rho_2}}{2(\sqrt{\rho_1}+\sqrt{\rho_2})^2}[(u_1-u_2)^2+(v_1-v_2)^2+(w_1-w_2)^2] \geq 0,$$

where $c_1 = \sqrt{\gamma T_1}$ and $c_2 = \sqrt{\gamma T_2}$ are the left and right sound speeds, T_1 and T_2 the left and right temperatures in energy units. The function $\hat{c}^2$ is not natural, since it depends on velocities. Therefore the following regularization procedure can be recommended, see Yee (1989). One can use formula (3.4.16) for $\hat{c}$ only if $\min(c_1, c_2) \leq \hat{c} \leq \max(c_1, c_2)$. Otherwise, if $\hat{c} < \min(c_1, c_2)$ we put $\hat{c} = \min(c_1, c_2)$ and if $\hat{c} > \max(c_1, c_2)$ we put $\hat{c} = \max(c_1, c_2)$.

The finite-volume scheme for equations (3.1.15) has a familiar form:

$$\frac{\mathbf{U}_j^{k+1} - \mathbf{U}_j^k}{\Delta t} + \frac{\mathbf{F}_{j+1/2} - \mathbf{F}_{j-1/2}}{\Delta x} = \mathbf{0}, \tag{3.4.19}$$

$$\mathbf{F}_{m+1/2} = \tfrac{1}{2}(\mathbf{F}_m^k + \mathbf{F}_{m+1}^k) + \tfrac{1}{2}|\mathcal{A}|_{m+1/2}^k(\mathbf{U}_m^k - \mathbf{U}_{m+1}^k), \quad m = j, j-1; \tag{3.4.20}$$

$$|\mathcal{A}| = \hat{\Omega}_{\mathrm{R}}|\hat{\Lambda}|\hat{\Omega}_{\mathrm{L}}.$$

The values $\mathbf{U}_{m+1/2}$ can be found from the formula (2.3.70) for $\xi = 0$,

$$\mathbf{U}_{m+1/2} = \tfrac{1}{2}(\mathbf{U}_m^k + \mathbf{U}_{m+1}^k) + \tfrac{1}{2}\hat{S}_{m+1/2}^k(\mathbf{U}_m^k - \mathbf{U}_{m+1}^k), \quad m = j, j-1; \tag{3.4.21}$$

$$\hat{S} = \hat{\Omega}_{\mathrm{R}}[\operatorname{sgn}\hat{\lambda}_p\,\delta_{pl}]\hat{\Omega}_{\mathrm{L}}.$$

To solve equations on grids moving with velocity $D(x)$, see (3.2.7)–(3.2.10), we must modify the procedure of calculating $\mathbf{F}$ in accordance with Eq. (2.3.72). In doing so, we must take into account the variation of the cell volume in time, see (3.2.7)–(3.2.10), and modify the formulas for $\mathbf{F}$ as follows:

$$\begin{aligned}\widetilde{\mathbf{F}}_{m+1/2} &= (\mathbf{F} - D\mathbf{U})_{m+1/2} \\ &= \tfrac{1}{2}(\mathbf{F}_m^k + \mathbf{F}_{m+1}^k) - \tfrac{1}{2}D_{m+1/2}(\mathbf{U}_m^k + \mathbf{U}_{m+1}^k) + \tfrac{1}{2}|\mathcal{A}(D)|_{m+1/2}^k(\mathbf{U}_m^k - \mathbf{U}_{m+1}^k),\end{aligned} \tag{3.4.22}$$

$$|\mathcal{A}(D)| = \hat{\Omega}_{\mathrm{R}}[|\hat{\lambda}_p - D|\,\delta_{pl}]\hat{\Omega}_{\mathrm{L}}.$$

To solve the system of equations in curvilinear (cylindrical, spherical, and so on) coordinates x,

$$\frac{\partial s(x)\mathbf{U}}{\partial t} + \frac{\partial s(x)\mathbf{F}}{\partial x} = \mathbf{H}, \qquad \mathbf{H} = \left[0, p\frac{\partial s(x)}{\partial x}, 0, 0, 0\right]^{\mathrm{T}}, \tag{3.4.23}$$

we can use the modification of Roe's solver by Glaister (1988a). The substitution $\widetilde{\rho} = s(x)\rho$ transforms Eq. (3.4.23) into Eqs. (3.1.15)–(3.1.17), where we must assume $\rho = \widetilde{\rho}$. Then, one can use Roe's method (3.4.19)–(3.4.20) taking into account that $\widetilde{\rho}_m = s(x_m)\rho_m$ and $\widetilde{\rho}_{m+1} = s(x_{m+1})\rho_{m+1}$. In particular, the averaging (3.4.13) transforms into

$$\hat{u} = \frac{\sqrt{s_1\rho_1}\,u_1 + \sqrt{s_2\rho_2}\,u_2}{\sqrt{s_1\rho_1} + \sqrt{s_2\rho_2}}, \qquad \hat{v} = \frac{\sqrt{s_1\rho_1}\,v_1 + \sqrt{s_2\rho_2}\,v_2}{\sqrt{s_1\rho_1} + \sqrt{s_2\rho_2}},$$

$$\hat{w} = \frac{\sqrt{s_1\rho_1}\,w_1 + \sqrt{s_2\rho_2}\,w_2}{\sqrt{s_1\rho_1} + \sqrt{s_2\rho_2}}, \qquad \hat{h} = \frac{\sqrt{s_1\rho_1}\,h_1 + \sqrt{s_2\rho_2}\,h_2}{\sqrt{s_1\rho_1} + \sqrt{s_2\rho_2}},$$

$$s_1 = s(x_m), \quad s_2 = s(x_{m+1}), \quad f_1 = f_m, \quad f_2 = f_{m+1},$$

where $f = \rho, u, v, w,$ and h.

To construct the Godunov method only the formulas for $\mathbf{F}$ will suffice, see Eqs. (3.4.19)–(3.4.20). Formulas for $\mathbf{U}$ may be useful, for example, for approximating of the specific right-hand side of a system. At times one may need to calculate some arbitrary quantities $\mathbf{W}(\mathbf{U})$ at cell boundaries. In particular, pressure p at cell boundaries can be used for approximation of the right-hand side in (3.4.23). Let us derive the formulas for this case. Substituting $\mathbf{U} = \mathbf{U}(\mathbf{W})$ in system (3.1.18) we obtain the equations

$$\begin{aligned} &U_W \mathbf{W}_t + AU_W \mathbf{W}_x = 0, \quad U_W = \frac{\partial \mathbf{U}}{\partial \mathbf{W}}, \quad W_U = \frac{\partial \mathbf{W}}{\partial \mathbf{U}}, \\ &\mathbf{W}_t + W_U A U_W \mathbf{W}_x = 0. \end{aligned}$$

Using Eq. (3.4.21), we obtain

$$\begin{aligned} &\mathbf{W}_{m+1/2} = \tfrac{1}{2}(\mathbf{W}^k_m + \mathbf{W}^k_{m+1}) + \tfrac{1}{2}(W_U \hat{S} U_W)^k_{m+1/2}\,(\mathbf{W}^k_m - \mathbf{W}^k_{m+1}), \\ &\hat{S} = \hat{\Omega}_{\mathrm{R}}[\operatorname{sgn}\hat{\lambda}_p\,\delta_{pl}]\hat{\Omega}_{\mathrm{L}}. \end{aligned} \tag{3.4.24}$$

To preserve Roe's property we consider W_U such that

$$W_U \Delta\mathbf{U} = \Delta\mathbf{W}. \tag{3.4.25}$$

If a W_U is found which satisfies (3.4.25), then Eq. (3.4.24) becomes

$$\begin{aligned} &\mathbf{W}_{m+1/2} = \tfrac{1}{2}(\mathbf{W}^k_m + \mathbf{W}^k_{m+1}) + \tfrac{1}{2}(W_U \hat{S})^k_{m+1/2}\,(\mathbf{U}^k_m - \mathbf{U}^k_{m+1}), \\ &\hat{S} = \hat{\Omega}_{\mathrm{R}}[\operatorname{sgn}\hat{\lambda}_p\,\delta_{pl}]\hat{\Omega}_{\mathrm{L}}. \end{aligned} \tag{3.4.26}$$

From the relations $\Delta\mathbf{W} = \mathcal{A}_W \Delta\mathbf{Y}$ and $\Delta\mathbf{U} = \mathcal{A}_U \Delta\mathbf{Y}$, we conclude that $W_U = \mathcal{A}_W \mathcal{A}_U^{-1}$. Let

$$\begin{aligned} \mathbf{W} &= [g(\sqrt{\rho}), \rho u^2, \rho v^2, \rho w^2, p]^{\mathrm{T}} \\ &= \left[g(R), U^2, V^2, W^2, \frac{\gamma-1}{\gamma}HR - \frac{\gamma-1}{2\gamma}(U^2+V^2+W^2) + \frac{g}{\gamma}\right]^{\mathrm{T}}. \end{aligned}$$

Then

$$\Delta\mathbf{W} = \begin{bmatrix} \bar{g}\Delta R \\ 2\bar{U}\Delta U \\ 2\bar{V}\Delta V \\ 2\bar{W}\Delta W \\ (\frac{\gamma-1}{\gamma}\bar{H} + \frac{1}{\gamma}\bar{g})\Delta R - \frac{\gamma-1}{\gamma}(\bar{U}\Delta U + \bar{V}\Delta V + \bar{W}\Delta W) + \frac{\gamma-1}{\gamma}\bar{R}\Delta H \end{bmatrix},$$

$$\bar{g} = \frac{g(R_1) - g(R_2)}{R_1 - R_2} = \frac{g(\sqrt{\rho_1}) - g(\sqrt{\rho_2})}{\sqrt{\rho_1} - \sqrt{\rho_2}}.$$

Dividing all components of $\Delta\mathbf{W}$ and $\Delta\mathbf{Y}$ by $\bar{R} = \frac{1}{2}(R_1 + R_2)$ and using the equation $\Delta\mathbf{W} = \mathcal{A}_W \Delta\mathbf{Y}$, we obtain

$$\mathcal{A}_W = \begin{bmatrix} 2g_\rho & 0 & 0 & 0 & 0 \\ 0 & 2\hat{u} & 0 & 0 & 0 \\ 0 & 0 & 2\hat{v} & 0 & 0 \\ 0 & 0 & 0 & 2\hat{w} & 0 \\ \frac{\gamma-1}{\gamma}\hat{h} + \frac{2}{\gamma}g_\rho & \frac{1-\gamma}{\gamma}\hat{u} & \frac{1-\gamma}{\gamma}\hat{v} & \frac{1-\gamma}{\gamma}\hat{w} & \frac{\gamma-1}{\gamma} \end{bmatrix},$$

$$g_\rho = \frac{\overline{g}}{2\overline{R}} = \frac{g(R_1) - g(R_2)}{(R_1)^2 - (R_2)^2} = \frac{g(\sqrt{\rho_1}) - g(\sqrt{\rho_2})}{\rho_1 - \rho_2}.$$

Using the expressions of $\mathcal{A}_U$, and $\mathcal{A}_W$ and solution (3.4.14), we find $W_U = \mathcal{A}_W \mathcal{A}_U^{-1}$:

$$W_U = \begin{bmatrix} g_\rho & 0 & 0 & 0 & 0 \\ -\hat{u}^2 & 2\hat{u} & 0 & 0 & 0 \\ -\hat{v}^2 & 0 & 2\hat{v} & 0 & 0 \\ -\hat{w}^2 & 0 & 0 & 2\hat{w} & 0 \\ \frac{1}{2}(\gamma-1)q^2 + g_\rho & (1-\gamma)\hat{u} & (1-\gamma)\hat{v} & (1-\gamma)\hat{w} & \gamma-1 \end{bmatrix}.$$

Thus, in accordance with (3.4.26), we can determine p, ρu^2, ρv^2, and ρw^2 at the cell boundary. For $\mathbf{W} = \mathbf{F}$, Eq. (3.4.26) transforms into (3.4.20). For $\mathbf{W} = \mathbf{U}$, Eq. (3.4.26) transforms into (3.4.21) with $W_U = I$, where I is the 5×5 identity matrix. Formulas (3.4.26) are consistent with (3.4.20). In particular, it is easy to verify that

$$\langle \rho u^2 + p \rangle = \langle \rho u^2 \rangle + \langle p \rangle, \quad \langle f \rangle = f_{m+1/2}. \tag{3.4.27}$$

Here $\langle f \rangle$ denotes the boundary value of f. Relation (3.4.27) follows from the equality of the second row in the matrix $\mathcal{A}$ to the sum of second and fifth rows in the matrix W_U. Formulas (3.4.21) and (3.4.26) are consistent as well. In particular, it is easy to verify that

$$\langle e \rangle = \frac{\langle p \rangle}{\gamma - 1} + \tfrac{1}{2}\langle \rho u^2 \rangle + \tfrac{1}{2}\langle \rho v^2 \rangle + \tfrac{1}{2}\langle \rho w^2 \rangle - \frac{\langle g \rangle}{\gamma - 1}. \tag{3.4.28}$$

Equation (3.4.28) follows from the equality of the fifth row in the identity matrix I to the sum of the first through fifth rows in W_U with the corresponding coefficients $\pm 1/(\gamma - 1)$ and $\frac{1}{2}$.

3.4.5 Roe's Riemann problem solver for an arbitrary EOS.

The most general EOS that provides the uniqueness of the Roe formulas is provided by Eq. (3.4.11). For an arbitrary EOS the formulas are not unique, and this fact will be illustrated below.

Let $p = p(\rho, \varepsilon)$. From Eqs. (3.4.8)–(3.4.9) we obtain

$$\frac{\Delta \mathbf{F}}{\overline{R}} = \begin{bmatrix} \hat{u}\Delta R + \Delta U \\ 2\hat{u}\Delta U + \Delta p/\overline{R} \\ \hat{v}\Delta U + \hat{u}\Delta V \\ \hat{w}\Delta U + \hat{u}\Delta W \\ \hat{h}\Delta U + \hat{u}\Delta H \end{bmatrix}, \quad \frac{\Delta \mathbf{U}}{\overline{R}} = \begin{bmatrix} 2\Delta R \\ \hat{u}\Delta R + \Delta U \\ \hat{v}\Delta R + \Delta V \\ \hat{w}\Delta R + \Delta W \\ \Delta H + \hat{h}\Delta R - \Delta p/\overline{R} \end{bmatrix},$$

where $\overline{R} = \frac{1}{2}(R_1 + R_2)$. Let (3.1.19) be similar to a Roe matrix $\tilde{\mathcal{A}}$ if $u = \hat{u}$, $v = \hat{v}$, $w = \hat{w}$, $h = \hat{h}$, $q^2 = \hat{u}^2 + \hat{v}^2 + \hat{w}^2$, and $\theta = q^2 - h + c^2/b$, see Eq. (3.1.25). The coefficient b and the sound speed c will be determined later. Then, the equation

$$\frac{\Delta \mathbf{F}}{\overline{R}} = \tilde{\mathcal{A}} \frac{\Delta \mathbf{U}}{\overline{R}}$$

reduces one relation

$$\Delta p = \left[2c^2 + (q^2 - \hat{h})b\right] \frac{\overline{R}}{1+b} \Delta R + (\Delta H - \hat{u}\Delta U - \hat{v}\Delta V - \hat{w}\Delta W) \frac{b\overline{R}}{1+b}. \tag{3.4.29}$$

On the other hand,

$$\begin{aligned} &\rho = R^2, \quad \varepsilon = \frac{HR - p - \frac{1}{2}(U^2 + V^2 + W^2)}{R^2}, \\ &\Delta p = p_\rho \Delta\rho + p_\varepsilon \Delta\varepsilon, \end{aligned} \tag{3.4.30}$$

where the coefficients p_ρ and p_ε will be determined later. Then,

$$\begin{aligned} \Delta\rho &= 2\overline{R}\Delta R, \qquad (3.4.31) \\ \Delta\varepsilon &= (\overline{R}\Delta H + \overline{H}\Delta R - \Delta p - \overline{U}\Delta U - \overline{V}\Delta V - \overline{W}\Delta W)\overline{\left[\frac{1}{R^2}\right]} + \overline{\rho\varepsilon}\Delta\left[\frac{1}{R^2}\right] \\ &= (\Delta H + \hat{h}\Delta R - \hat{u}\Delta U - \hat{v}\Delta V - \hat{w}\Delta W)\overline{R}\overline{\left[\frac{1}{R^2}\right]} - \overline{\left[\frac{1}{R^2}\right]}\Delta p - \overline{\rho\varepsilon}\frac{2\overline{R}}{\hat{\rho}^2}\Delta R. \end{aligned}$$

Here $\overline{a} = \frac{1}{2}(a_1 + a_2)$ and $\hat{\rho} = \sqrt{\rho_1\rho_2}$. Substituting $\Delta\rho$ and $\Delta\varepsilon$ into (3.4.30) yields

$$\begin{aligned} \Delta p &= \left(2p_\rho + d - \overline{\rho\varepsilon}\frac{2p_\varepsilon}{\hat{\rho}^2}\right)\frac{\overline{R}\Delta R}{1+d} + (\Delta H - \hat{u}\Delta U - \hat{v}\Delta V - \hat{w}\Delta W)\frac{\overline{R}d}{1+d}, \qquad (3.4.32) \\ d &= \overline{\left[\frac{1}{R^2}\right]}p_\varepsilon = \frac{p_\varepsilon}{2}\left(\frac{1}{\rho_1} + \frac{1}{\rho_2}\right) = p_\varepsilon\frac{\overline{\rho}}{\hat{\rho}^2}. \end{aligned}$$

Comparing Eqs. (3.4.29) and (3.4.32), we establish that $b = d$ and

$$\begin{aligned} &c^2 = (\hat{h} - \hat{\varepsilon} - \tfrac{1}{2}q^2)p_\varepsilon\frac{\overline{\rho}}{\hat{\rho}^2} + p_\rho, \quad \hat{\varepsilon} = \frac{\overline{\rho\varepsilon}}{\overline{\rho}}, \\ \Longrightarrow \quad &c = \sqrt{p_\rho + \frac{\hat{p}p_\varepsilon\overline{\rho}}{\hat{\rho}^3}}, \quad \hat{p} = (\hat{h} - \hat{\varepsilon} - \tfrac{1}{2}q^2)\hat{\rho}. \end{aligned} \tag{3.4.33}$$

The quantities p_ε and p_ρ can be approximated as follows (Glaister 1988b):

$$p_\varepsilon = \begin{cases} \dfrac{1}{\Delta\varepsilon}\left(\dfrac{p(\rho_1,\varepsilon_1)+p(\rho_2,\varepsilon_1)}{2} - \dfrac{p(\rho_1,\varepsilon_2)+p(\rho_2,\varepsilon_2)}{2}\right) & \text{if } \Delta\varepsilon = \varepsilon_1 - \varepsilon_2 \neq 0, \\ \dfrac{1}{2}\left[\dfrac{\partial p(\rho_1,\varepsilon)}{\partial\varepsilon} + \dfrac{\partial p(\rho_2,\varepsilon)}{\partial\varepsilon}\right] & \text{if } \varepsilon_1 = \varepsilon_2 = \varepsilon, \end{cases} \tag{3.4.34}$$

$$p_\rho = \begin{cases} \dfrac{1}{\Delta\rho}\left(\dfrac{p(\rho_1,\varepsilon_1)+p(\rho_1,\varepsilon_2)}{2} - \dfrac{p(\rho_2,\varepsilon_1)+p(\rho_2,\varepsilon_2)}{2}\right) & \text{if } \Delta\rho = \rho_1 - \rho_2 \neq 0, \\ \dfrac{1}{2}\left[\dfrac{\partial p(\rho,\varepsilon_1)}{\partial\rho} + \dfrac{\partial p(\rho,\varepsilon_2)}{\partial\rho}\right] & \text{if } \rho_1 = \rho_2 = \rho. \end{cases} \tag{3.4.35}$$

The partial derivatives in Eqs. (3.4.34)–(3.4.35) should be calculated from the equation of state. The quantities p_ρ and p_ε can be approximated in a different way, for example,

$$\begin{aligned}
\text{(i)}\ \Delta p &= p(\rho_1,\varepsilon_1) - p(\rho_2,\varepsilon_2) = p(\rho_1,\varepsilon_1) - p(\rho_1,\varepsilon_2) + p(\rho_1,\varepsilon_2) - p(\rho_2,\varepsilon_2)\\
&\equiv \frac{p(\rho_1,\varepsilon_1) - p(\rho_1,\varepsilon_2)}{\varepsilon_1-\varepsilon_2}\Delta\varepsilon + \frac{p(\rho_1,\varepsilon_2) - p(\rho_2,\varepsilon_2)}{\rho_1-\rho_2}\Delta\rho,\\
p_\varepsilon^{(1)} &= \frac{p(\rho_1,\varepsilon_1) - p(\rho_1,\varepsilon_2)}{\varepsilon_1-\varepsilon_2}, \quad p_\rho^{(1)} = \frac{p(\rho_1,\varepsilon_2) - p(\rho_2,\varepsilon_2)}{\rho_1-\rho_2};
\end{aligned} \tag{3.4.36}$$

$$\begin{aligned}
\text{(ii)}\ \Delta p &= p(\rho_1,\varepsilon_1) - p(\rho_2,\varepsilon_2) = p(\rho_1,\varepsilon_1) - p(\rho_2,\varepsilon_1) + p(\rho_2,\varepsilon_1) - p(\rho_2,\varepsilon_2)\\
&\equiv \frac{p(\rho_2,\varepsilon_1) - p(\rho_2,\varepsilon_2)}{\varepsilon_1-\varepsilon_2}\Delta\varepsilon + \frac{p(\rho_1,\varepsilon_1) - p(\rho_2,\varepsilon_1)}{\rho_1-\rho_2}\Delta\rho,\\
p_\varepsilon^{(2)} &= \frac{p(\rho_2,\varepsilon_1) - p(\rho_2,\varepsilon_2)}{\varepsilon_1-\varepsilon_2}, \quad p_\rho^{(2)} = \frac{p(\rho_1,\varepsilon_1) - p(\rho_2,\varepsilon_1)}{\rho_1-\rho_2}.
\end{aligned} \tag{3.4.37}$$

A linear combination of (3.4.36) and (3.4.37) provides a new approximation of p_ρ and p_ε: $p_\varepsilon = \alpha p_\varepsilon^{(1)} + (1-\alpha)p_\varepsilon^{(2)}$ and $p_\rho = \beta p_\rho^{(1)} + (1-\beta)p_\rho^{(2)}$. For $\alpha = \beta = \frac{1}{2}$, we obtain (3.4.34)–(3.4.35).

Another difficulty of this approach is how to approximate p_ρ and p_ε adequately for the cases of $\Delta\rho \to 0$ and $\Delta\varepsilon \to 0$, see Kupriyanova, Mikhailov, and Chinilov (1991).

Some other versions of formulas for Roe's matrices for an arbitrary EOS can be found in Glaister (1988b) and Kupriyanova et al. (1991). The difference between all versions is of the order of $O(\Delta^2)$, where Δ are differences of gas dynamic variables. Unlike (3.4.29)–(3.4.33), Glaister (1988b) used a more sophisticated and complete analysis of eigenvector relations. In his results, $\hat{\varepsilon}$ coincided with the standard Roe average.

As far as the EOS (3.1.4) for a perfect gas is considered, one can see that the obtained Riemann problem solver provides formulas other than the initial Roe's Riemann problem solver (Roe 1981). Indeed, substituting the perfect gas EOS, we find from (3.4.34)–(3.4.35) that $p_\varepsilon = (\gamma-1)\overline{\rho}$ and $p_\rho = (\gamma-1)\overline{\varepsilon}$, where $\overline{\rho} = \frac{1}{2}(\rho_1+\rho_2)$ and $\overline{\varepsilon} = \frac{1}{2}(\varepsilon_1+\varepsilon_2)$. Thus, unlike the initial Roe's Riemann problem solver, the additional coefficients $\overline{\rho}$ and $\hat{\rho} = \sqrt{\rho_1\rho_2}$, and $\overline{\varepsilon}$ and $\hat{\varepsilon} = \overline{\rho\varepsilon}/\overline{\rho}$ appear in formulas for the matrix $\tilde{\mathcal{A}}$. These coefficients are the averages of the same variables but are defined by different formulas and, in general, $\overline{\rho} \neq \hat{\rho}$ and $\overline{\varepsilon} \neq \hat{\varepsilon}$.

This also pertains to Glaister's Riemann problem solver (Glaister 1988b). Thus, the construction of Roe's Riemann problem solvers for an arbitrary EOS is nonunique and encounters substantial difficulties.

3.4.6 Osher–Solomon numerical scheme.

Now we shall describe for completeness the implementation of the Osher finite-volume scheme for gas dynamic equations (Osher and Solomon 1982). Only the first order of accuracy numerical flux will be obtained. Extension to higher order can be made using standard TVD interpolation procedures. We also consider only a one-dimensional Euler gas dynamic system

$$\frac{\partial \mathbf{U}}{\partial t} + \frac{\partial \mathbf{F}}{\partial x} = \mathbf{0}, \tag{3.4.38}$$

where

$$\mathbf{U} = [\rho,\ \rho u,\ e]^{\mathrm{T}}, \quad \mathbf{F} = [\rho u,\ \rho u^2 + p,\ (e+p)u]^{\mathrm{T}}.$$

In order to determine the Osher numerical flux in the multidimensional case we must direct the coordinate line x along the normal to the computational cell boundary. In practice, system (3.4.38) is only considered as a supplementary one. Its solution is used to construct a global solution of the multidimensional system of the Euler system.

As usual, Eq. (3.4.38) must be supplemented by the caloric equation of state, and in the case of ideal perfect gas we obtain

$$e = \frac{p}{\gamma - 1} + \frac{\rho u^2}{2},$$

where γ is the adiabatic index.

In order to apply the Osher method (see Section 2.3.3) we need to know the Jacobian matrix

$$A = \frac{\partial \mathbf{F}}{\partial \mathbf{U}} = \begin{bmatrix} 0 & 1 & 0 \\ \frac{1}{2}(\gamma-3)u^2 & (3-\gamma)u & \gamma - 1 \\ (\gamma-1)u^3 - \gamma e u/\rho & \frac{3}{2}(1-\gamma)u^2 + \gamma e/\rho & \gamma u \end{bmatrix}. \tag{3.4.39}$$

The eigenvalues of this matrix obviously are

$$\lambda_1 = u, \quad \lambda_2 = u + c, \quad \lambda_3 = u - c,$$

where $c = \sqrt{\gamma p/\rho}$ is the adiabatic speed of sound.

Let us determine the numerical flux $\mathbf{F}_{j+1/2}$ using integration over the segments Γ_i^j of the path connecting $\mathbf{U}_j$ and $\mathbf{U}_{j+1}$ that correspond to negative eigenvalues of A, that is,

$$\mathbf{F}_{j+1/2} = \mathbf{F}(\mathbf{U}_j) + \int_{\mathbf{U}_j}^{\mathbf{U}_{j+1}} A^{-} d\mathbf{U}. \tag{3.4.40}$$

Each eigenvalue λ_i has a corresponding eigenvector $\mathbf{r}^i$. The integration path consists of the three subpaths Γ_i^j parallel to $\mathbf{r}^i$ in the $\mathbf{U}$-space. Each eigenvector has a corresponding simple wave that separates the constant states (see Section 1.4.1).

To perform the characteristic analysis, let us pass to the quasilinear form of (3.4.38),

$$\frac{\partial \mathbf{u}}{\partial t} + \tilde{A}\frac{\partial \mathbf{u}}{\partial x} = \mathbf{0}, \tag{3.4.41}$$

where the state vector $\mathbf{u}$, the coefficient matrix $\tilde{A}$, and its eigenvectors $\mathbf{r}$ and $\mathbf{l}$ are the following (compare with Section 1.3.1):

$$\mathbf{u} = [\rho,\ u,\ p]^{\mathrm{T}}, \quad \tilde{A} = \begin{bmatrix} u & \rho & 0 \\ 0 & u & 1/\rho \\ 0 & \rho c^2 & u \end{bmatrix},$$

$$\mathbf{l}^1 = \left[1,\ 0,\ -\frac{1}{c^2}\right], \quad \mathbf{l}^2 = \left[0,\ 1,\ \frac{1}{\rho c}\right], \quad \mathbf{l}^3 = \left[0,\ 1,\ -\frac{1}{\rho c}\right],$$

$$\mathbf{r}^1 = [1,\ 0,\ 0]^{\mathrm{T}}, \quad \mathbf{r}^2 = \left[\frac{\rho}{2c},\ \frac{1}{2},\ \frac{\rho c}{2}\right]^{\mathrm{T}}, \quad \mathbf{r}^3 = \left[-\frac{\rho}{2c},\ \frac{1}{2},\ -\frac{\rho c}{2}\right]^{\mathrm{T}}.$$

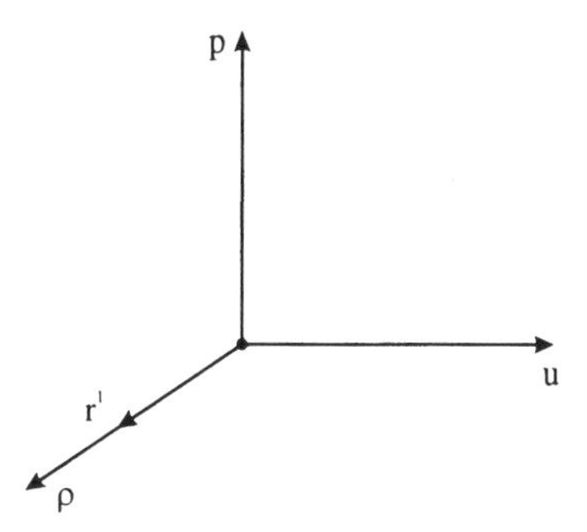

Figure 3.15 Phase space ρ–u–p.

Here $\mathbf{l}^i$ and $\mathbf{r}^i$ ($i = 1, 2, 3$) are the left and right eigenvectors corresponding to $\lambda_1 = u$, $\lambda_2 = u + c$, and $\lambda_3 = u - c$, respectively. The vector $\mathbf{r}^1$ in the $\mathbf{u}$-space is parallel to the ρ-axis (Fig. 3.15). This means that u and p remain constant along this segment of the path connecting $\mathbf{U}_j$ and $\mathbf{U}_{j+1}$.

It is obvious that

$$\frac{\partial \lambda_1}{\partial \mathbf{u}} \cdot \mathbf{r}^1 \equiv 0,$$

that is, this characteristic field is linearly degenerate. The variation of parameters in it corresponds to that on a contact discontinuity.

Consider now the variation of quantities in simple waves corresponding to the remaining genuinely nonlinear characteristic fields. Let us determine, first, the Riemann invariants w_i of the system. By definition,

$$dw_i = \mathbf{l}^i \cdot d\mathbf{u} = 0$$

along the corresponding characteristic path and, hence,

$$d\rho - \frac{1}{c^2} dp = 0 \quad \text{along} \quad \frac{dx}{dt} = u, \tag{3.4.42}$$

$$du + \frac{1}{\rho c} dp = 0 \quad \text{along} \quad \frac{dx}{dt} = u + c, \tag{3.4.43}$$

$$du - \frac{1}{\rho c} dp = 0 \quad \text{along} \quad \frac{dx}{dt} = u - c. \tag{3.4.44}$$

Since $p = p(\rho, S)$ and the squared speed of sound c^2, by definition, is equal to $(\partial p/\partial \rho)_S$, the relation (3.4.42) is equivalent to the conservation of the entropy function $S = p/\rho^\gamma$. This means that we can choose

$$w_1 = S. \tag{3.4.45}$$

By integration, from Eq. (3.4.43) we find that

$$u + \int \frac{dp}{\rho c} = \text{const}.$$

If isentropic flows are considered with $p = S\rho^\gamma$ and, hence, $c^2 = S\gamma\rho^{\gamma-1}$, we arrive at the relation

$$u + \int c(\rho) \frac{d\rho}{\rho} = \text{const},$$

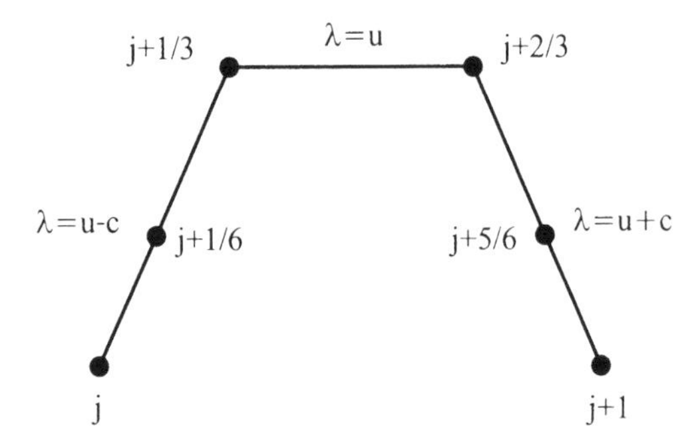

Figure 3.16 Path of the integration.

which results in

$$w_2 = u + \frac{2c}{\gamma - 1} = \text{const}\,. \tag{3.4.46}$$

Similarly from Eq. (3.4.44) we obtain

$$w_3 = u - \frac{2c}{\gamma - 1} = \text{const}\,. \tag{3.4.47}$$

Since we are looking for solutions in the form of a simple wave, that is, solutions in the form $\mathbf{u}(\xi(x,t))$, we have (see Section 1.4.1)

$$\frac{d\mathbf{U}^i}{d\xi} \parallel \mathbf{r}^i, \tag{3.4.48}$$

where $\mathbf{U}^i$ represents the simple wave solution corresponding to ith eigenvalue.

As $\mathbf{l}^k \cdot \mathbf{r}^i = 0$ for $k \neq i$, we obtain that in the simple wave corresponding to the ith eigenvalue all but the ith Riemann invariants are constant. Thus, w_1 and w_3 remain constant in the second simple wave and w_1 and w_2 do not change in the third one. This allows us to write out the following formulas relating $\mathbf{U}_j$ and $\mathbf{U}_{j+1}$ (see Fig. 3.16):

$$u_j + \frac{2c_j}{\gamma - 1} = u_{j+1/3} + \frac{2c_{j+1/3}}{\gamma - 1}, \quad \frac{p_j}{\rho_j^\gamma} = \frac{p_{j+1/3}}{\rho_{j+1/3}^\gamma}; \tag{3.4.49}$$

$$p_{j+1/3} = p_{j+2/3}, \quad u_{j+1/3} = u_{j+2/3}; \tag{3.4.50}$$

$$u_{j+1} - \frac{2c_{j+1}}{\gamma - 1} = u_{j+2/3} - \frac{2c_{j+2/3}}{\gamma - 1}, \quad \frac{p_{j+1}}{\rho_{j+1}^\gamma} = \frac{p_{j+2/3}}{\rho_{j+2/3}^\gamma}. \tag{3.4.51}$$

As noted in Section 2.3.3, the eigenvalues $u - c$ and $u + c$ can change their signs along the first and third segments, respectively. If we denote such points as $j + 1/6$ and $j + 5/6$, we can easily obtain the following supplementary relations:

$$c_{j+1/6} = u_{j+1/6} = \frac{\gamma - 1}{\gamma + 1}\left(u_j + \frac{2c_j}{\gamma - 1}\right), \quad \rho_{j+1/6} = \left(\frac{c_{j+1/6}^2}{\gamma S_j}\right)^{\frac{1}{\gamma - 1}}, \tag{3.4.52}$$

$$-c_{j+5/6} = u_{j+5/6} = \frac{\gamma - 1}{\gamma + 1}\left(u_{j+1} - \frac{2c_{j+1}}{\gamma - 1}\right), \quad \rho_{j+5/6} = \left(\frac{c_{j+5/6}^2}{\gamma S_{j+1}}\right)^{\frac{1}{\gamma - 1}}. \tag{3.4.53}$$

Relations (3.4.49)–(3.4.53) allow us to perform the integration along the path segments corresponding to negative eigenvalues as follows:

$$\begin{aligned}
\mathbf{F}_{j+1/2} &= \mathbf{F}(\mathbf{U}_j) + \int_{\mathbf{U}_j}^{\mathbf{U}_{j+1}} A^- \, d\mathbf{U} \qquad (3.4.54)\\
&= \mathbf{F}(\mathbf{U}_j)\\
&+ (\mathbf{F}(\mathbf{U}_{j+1/3}) - \mathbf{F}(\mathbf{U}_j)) \quad \text{if} \quad u_{j+1/3} - c_{j+1/3} < 0 \quad \text{and} \quad u_j - c_j < 0\\
&+ (\mathbf{F}(\mathbf{U}_{j+1/6}) - \mathbf{F}(\mathbf{U}_j)) \quad \text{if} \quad u_{j+1/3} - c_{j+1/3} \geq 0 \quad \text{and} \quad u_j - c_j < 0\\
&+ (\mathbf{F}(\mathbf{U}_{j+1/3}) - \mathbf{F}(\mathbf{U}_{j+1/6})) \quad \text{if} \quad u_{j+1/3} - c_{j+1/3} < 0 \quad \text{and} \quad u_j - c_j \geq 0\\
&+ (\mathbf{F}(\mathbf{U}_{j+2/3}) - \mathbf{F}(\mathbf{U}_{j+1/3})) \quad \text{if} \quad u_{j+1/3} = u_{j+2/3} < 0\\
&+ (\mathbf{F}(\mathbf{U}_{j+1}) - \mathbf{F}(\mathbf{U}_{j+2/3})) \quad \text{if} \quad u_{j+2/3} + c_{j+2/3} < 0 \quad \text{and} \quad u_{j+1} + c_{j+1} < 0\\
&+ (\mathbf{F}(\mathbf{U}_{j+1}) - \mathbf{F}(\mathbf{U}_{j+5/6})) \quad \text{if} \quad u_{j+2/3} + c_{j+2/3} \geq 0 \quad \text{and} \quad u_{j+1} + c_{j+1} < 0\\
&+ (\mathbf{F}(\mathbf{U}_{j+5/6}) - \mathbf{F}(\mathbf{U}_{j+2/3})) \quad \text{if} \quad u_{j+2/3} + c_{j+2/3} < 0 \quad \text{and} \quad u_{j+1} + c_{j+1} \geq 0
\end{aligned}$$

Note that the exact solution of the Riemann problem generally consists of the combination of shocks, rarefaction waves, and a contact discontinuity that separate the regions of constant quantities. In the Osher–Solomon scheme, shock waves are substituted by compression waves. Since sonic points are exactly separated along the integration path, this scheme turns out to be free from the entropy fix problems. It is also reported to give better resolution of contact discontinuities (Godlewski and Raviart 1996). The Osher–Solomon scheme is definitely more time-consuming than Roe's scheme, although its elements can be incorporated into numerical boundary condition implementation procedures (Sawada et al. 1986; Pogorelov 1993, 1995; Pogorelov and Semenov 1996b, 1997a). Extension of this method to systems more complicated than gas dynamic equations may cause substantial difficulties. For example, the relations in MHD rarefaction waves cannot be exactly integrated. As shown in Section 2.3.3, we actually do not need any exact or approximate simple-wave solutions of the system under consideration. Osher's scheme can be implemented if we only determine the zeros of all eigenvalues in the interval between $\mathbf{U}_j$ and $\mathbf{U}_{j+1}$. This approach is described in detail by Bell et al. (1989).

3.5 Shock-fitting methods

Shock-fitting methods provide an efficient tool for solving gas dynamic equations. If the structure of the flow is known beforehand and the number of discontinuities is not very large, such methods turn out to be extremely accurate (Lyubimov and Rusanov 1970). Shock-capturing methods can give the same level of precision only with the application of solution-adaptive grids (McRae and Lafli 1999; Zegeling 1999). Moreover, since in shock-fitting methods we approximate only derivatives of smooth functions, we can use finite-difference schemes of higher order of accuracy, taking less care about their monotonicity.

3.5.1 Discontinuities as boundaries of the computational region. There are several basic approaches to fit the discontinuity in the numerical modelling of gas dynamic flows. In one approach, the shock is regarded as a surface on which the unknown

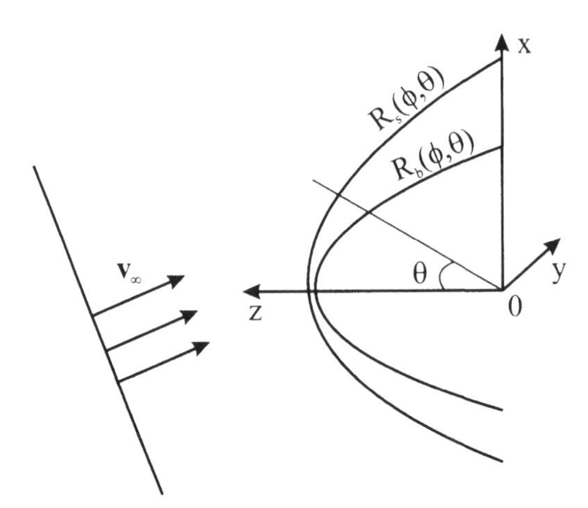

Figure 3.17 Supersonic blunt body problem.

functions are discontinuous. Thus, the calculated points that lie on the surface of discontinuity correspond to two sets of gas dynamic parameters satisfying the discontinuity relations (3.3.4)–(3.3.8). The values of the unknown functions at the new time layer, as well as the new position of the discontinuity, are determined by solving two coupled boundary-value problems, see, e.g., Roache (1976) and Magomedov and Kholodov (1988).

The simplest case occurs if there exists a global discontinuity (e.g., a shock) dividing the region of known parameters from that with the quantities to be found. The position of this shock can be unknown beforehand and must be found in the process of numerical solution of the problem. Fortunately, this case is frequent in one of the important aerodynamic problems, namely, in the problem on the supersonic gas flow near bodies. Since perturbations cannot propagate upstream in the supersonic flow, such a discontinuity, called a bow shock, divides the computational space into two parts. The flow parameters ahead of the bow shock are given and we need to find the correct shock position and the quantities behind it (see Fig. 3.17).

We shall illustrate the application of the shock-fitting method to solution of the problem of the hypersonic penetration of a blunt body into an air cloud possessing temperature and chemical inhomogeneities. Such a problem can be encountered in astrophysics (see Krebs and Hillebrandt 1983; Ikeuchi and Spitzer 1984), gas dynamic lasers (Gorbachev, Zhmakin, and Fursenko 1985), and high-speed aerodynamics (Champney, Chaussee, and Kutler 1982; Shugaev 1983). One can refer to both numerical and experimental studies of Zheleznyak, Mnatsakanyan, and Pervukhin (1986), Barkhudarov et al. (1987), Voynovich, Zhmakin, and Fursenko (1988). We consider the nonequilibrium air flow behind the bow shock originating near the blunt body that moves through the heated region in a detailed statement, taking into account a rather complete set of chemical reactions that occur in the shock layer. Air is treated as a mixture of ideal perfect gases.

Let the shapes of the body and shock be determined by the functions $R = R_b(\varphi, \vartheta)$ and $R = R_s(t, \varphi, \vartheta)$, respectively, where R is the distance to the coordinate system origin and φ and ϑ are the parameters. We assume that R, φ, and ϑ constitute a curvilinear coordinate system related to the Cartesian system by the formulas $x = x(R, \varphi, \vartheta)$, $y = y(R, \varphi, \vartheta)$, and $z = z(R, \varphi, \vartheta)$. Note that R_b is fixed, while the dependence of the function R_s on time is to be determined.

Statement of the problem

Let us normalize the physical variables $v^{(i)}$, p, and ρ occurring in (3.1.38) by $(p_\infty/\rho_\infty)^{1/2}$, p_∞, and ρ_∞, respectively. Characteristic length and time are L and $t^* = L/(p_\infty/\rho_\infty)^{1/2}$. The subscript "$\infty$" refers to the unperturbed quantities in front of the cloud.

In our calculation, the mixture consists of the following six species: (1) molecular oxygen O_2, (2) atomic oxygen O, (3) molecular nitrogen N_2, (4) atomic nitrogen N, (5) nitric oxide NO, and (6) argon Ar. The following 21 chemical reactions among the constituent species are taken into account:

$$\begin{aligned}
&O_2 + M \rightleftharpoons O + O + M,\\
&N_2 + M \rightleftharpoons N + N + M,\\
&NO + M \rightleftharpoons N + O + M,\\
&NO + O \rightleftharpoons O_2 + N,\\
&N_2 + O \rightleftharpoons NO + N,\\
&N_2 + O_2 \rightleftharpoons 2NO,
\end{aligned}$$

where M is any of the six species.

Using the law of the atomic element conservation and the condition $\sum_{l=1}^{6} c_l = 1$, we can eliminate c_1, c_3, and c_6 from Eq. (3.1.38):

$$\begin{aligned}
c_1 &= \frac{1}{2}\left(2\tilde{c}_O - \frac{c_2 M_1}{M_2} - \frac{c_5 M_1}{M_5}\right),\\
c_3 &= \frac{1}{2}\left(2\tilde{c}_N - \frac{c_4 M_3}{M_4} - \frac{c_5 M_3}{M_5}\right),\\
c_6 &= \tilde{c}_{Ar}
\end{aligned} \tag{3.5.1}$$

with $\tilde{c}_O \approx 0.2322$, $\tilde{c}_N \approx 0.7542$, $\tilde{c}_{Ar} \approx 0.0136$. These values are due to the usual fractions of the corresponding atoms in air.

To simplify the numerical calculation of the chemical production source terms, we can exclude the rate constants of direct reactions from Eq. (3.1.39) and introduce the equilibrium constants

$$K_r = \frac{k_r^+}{k_r^-}$$

instead of them. Note also that the equilibrium constants are usually available with substantially better precision than the rate constants.

The reverse reaction rates k_r^- and the equilibrium constants K_r are taken from Lin and

Teare (1963) with the corrections by Babenko et al. (1980). Their dimensionless form is

$$\mathcal{K}_{\bar{1}l} = k_{\bar{1}l}\frac{\rho_\infty^2 t^*}{M_l M_2}, \qquad \mathcal{K}_1 = K_1\frac{M_2^2}{\rho_\infty M_1},$$
$$\mathcal{K}_{\bar{2}l} = k_{\bar{2}l}\frac{\rho_\infty^2 t^*}{M_l M_4}, \qquad \mathcal{K}_2 = K_2\frac{M_4^2}{\rho_\infty M_3},$$
$$\mathcal{K}_{\bar{3}l} = k_{\bar{3}l}\frac{\rho_\infty^2 t^*}{M_1 M_4}, \qquad \mathcal{K}_3 = K_3\frac{M_2 M_4}{\rho_\infty M_5},$$
$$\mathcal{K}_{\bar{4}} = k_{\bar{4}}\frac{\rho_\infty t^* M_2}{M_1 M_4}, \qquad \mathcal{K}_4 = K_4\frac{M_1 M_4}{M_2 M_5},$$
$$\mathcal{K}_{\bar{5}} = k_{\bar{5}}\frac{\rho_\infty t^* M_2}{M_4 M_5}, \qquad \mathcal{K}_5 = K_5\frac{M_4 M_5}{M_2 M_3},$$
$$\mathcal{K}_{\bar{6}} = k_{\bar{6}}\frac{\rho_\infty t^*}{M_5}, \qquad \mathcal{K}_6 = K_6\frac{M_5^2}{M_1 M_3}.$$

The subscript l refers to reactions occurring with the participation of the lth catalytic component M. Thus, we have the following formulas for the source terms in the continuity equations for components:

$$\sigma_2 = \rho^2\left[2R_1\sum_{l=1}^{6}\mathcal{K}_{\bar{1}l}c_l + R_3\sum_{l=1}^{6}\mathcal{K}_{\bar{3}l}c_l - R_4\mathcal{K}_{\bar{4}} - R_5\mathcal{K}_{\bar{5}}\right],$$
$$\sigma_4 = \rho^2\left\{2R_2\sum_{l=1}^{6}\mathcal{K}_{\bar{2}l}c_l + \frac{M_4}{M_2}\left[R_3\sum_{l=1}^{6}\mathcal{K}_{\bar{3}l}c_l + R_4\mathcal{K}_{\bar{4}} + R_5\mathcal{K}_{\bar{5}},\right]\right\},$$
$$\sigma_5 = \rho^2\left\{\frac{M_5}{M_2}\left[-R_3\sum_{l=1}^{6}\mathcal{K}_{\bar{3}l}c_l - R_4\mathcal{K}_{\bar{4}} + R_5\mathcal{K}_{\bar{5}}\right] + 2R_6\mathcal{K}_{\bar{6}}\right\},$$
$$R_1 = \mathcal{K}_1 c_1 - \rho c_2^2, \quad R_2 = \mathcal{K}_2 c_3 - \rho c_4^2, \quad R_3 = \mathcal{K}_3 c_5 - \rho c_3 c_4,$$
$$R_4 = \mathcal{K}_4 c_2 c_5 - c_1 c_4, \quad R_5 = \mathcal{K}_5 c_2 c_3 - c_4 c_5, \quad R_6 = \mathcal{K}_6 c_1 c_3 - c_5^2.$$

The system of (3.1.38) and (3.5.1) must be supplemented by the thermal and caloric equations of state. The former one reads

$$T = \frac{pT_i\mu}{\rho}, \qquad T_i = \frac{p_\infty}{\rho_\infty R}, \qquad \mu = \left(\sum_{l=1}^{6}\frac{c_l}{M_l}\right)^{-1}, \tag{3.5.2}$$

where T is the temperature, μ is the mean molecular mass of the mixture, and R is the universal gas constant.

The specific internal energy of the mixture is

$$\varepsilon = \sum_{l=1}^{6} c_l\varepsilon_l.$$

In ideal gas, the energy is assumed to be equidistributed among the external modes of freedom, with $RT/2$ per each mode of a mole of gas. Thus

$$\varepsilon_l = c_{Vl}^{(0)}T + \varepsilon_l^{(v)} + h_{0l},$$

where the factor $c_{Vl}^{(0)}$ is independent of temperature and represents the specific heat at constant volume for the external (translational and rotational) modes of freedom, $\varepsilon_l^{(v)}$ is the specific energy of vibrational modes of freedom, and h_{0l} is specific formation energy of the lth component, respectively. It is natural to assume that the formation energies for molecular oxygen and nitrogen are zero. As for the remaining components, the experiments give $h_{02} = 1.537 \times 10^7\ \mathrm{m^2s^{-2}}$, $h_{04} = 3.364 \times 10^7\ \mathrm{m^2s^{-2}}$, and $h_{05} = 2.965 \times 10^6\ \mathrm{m^2s^{-2}}$. In addition,

$$c_{Vl}^{(0)} = \frac{1}{2} n_l \frac{R}{M_l},$$

where $n_l = 5$ for bi-atomic molecules and $n_l = 3$ for atoms.

Thus,

$$\varepsilon = c_V^{(0)} T + \varepsilon^{(v)} + h_0,$$

$$c_V^{(0)} = \sum_{l=1}^{6} c_{Vl}^{(0)}, \quad \varepsilon^{(v)} = \sum_{l=1}^{6} c_l \varepsilon_l^{(v)}, \quad h_0 = \sum_{l=1}^{6} c_l h_{0l}.$$

In this presentation we use a so-called ideal excitation of the vibrational mode of freedom for molecules (Lighthill 1957). In this model, the vibrational mode is assumed to be half-excited and we can write out the caloric equation of state in the form

$$\varepsilon = c_V T + h_0. \tag{3.5.3}$$

The specific heat is defined by the formula

$$c_V = R\left(\frac{3}{M_m} + \frac{1.5}{M_a}\right), \tag{3.5.4}$$

where

$$\frac{1}{M_m} = \frac{c_1}{M_1} + \frac{c_3}{M_3} + \frac{c_5}{M_5}, \quad \frac{1}{M_a} = \frac{c_2}{M_2} + \frac{c_4}{M_4} + \frac{c_6}{M_6}. \tag{3.5.5}$$

Combining Eqs. (3.5.2)–(3.5.5), we obtain the relationship similar to that in the case of perfect gas

$$(\varepsilon - h_0)\rho = \frac{p}{\gamma - 1}, \quad \gamma = \frac{4/M_m + 2.5/M_a}{3/M_m + 1.5/M_a}. \tag{3.5.6}$$

Otherwise, we can adopt the equilibrium excitation of the vibrational mode of freedom. In this case

$$\varepsilon_l^{(v)} = \frac{1}{M_l} \frac{R\theta_{\nu l}}{e^{\theta_{\nu l}/T} - 1},$$

where $\theta_{\nu l}$ are the characteristic temperatures. They are equal to 2230 K, 3340 K, and 2690 K for the molecules of oxygen, nitrogen, and nitric oxide, respectively. Thus, we can write out

$$\varepsilon_l = \int_0^T c_{Vl}\, dT + h_{0l},$$

where

$$c_{Vl} = c_{Vl}^{(0)} + \frac{d\varepsilon_l}{dT} = c_{Vl}^{(0)} + \frac{R}{M_l} e^{\theta_l} F^2(\theta_l)$$

and

$$F(\theta_l) = \frac{\theta_l}{e^{\theta_l} - 1}, \quad \theta_l = \frac{\theta_{vl}}{T}.$$

The caloric equation of state in this case acquires the form

$$\varepsilon - h_0 = R \sum_{l=1}^{6} \frac{c_l T}{M_l} \left[\frac{1}{2} n_l + F(\theta_l) \right]. \tag{3.5.7}$$

Initial and boundary conditions

We assume that the moving body has the shape of a spherically blunted cone with half-angle 20°. It moves at a supersonic speed with the Mach number $M_\infty = 15$ through the air with pressure $p_\infty = 24.12$ Pa and density $\rho_\infty = 3.32 \times 10^{-4}$ kg m^{-3}. The characteristic length is 1 m. We can easily predict that a bow shock will appear ahead of the moving body. We shall attach the coordinate system to the body. In this system we can introduce the angle of attack, that is, the angle between the velocity vector at infinity and the axis of the cone. Let this angle be equal to 20°.

The heated cloud is assumed to be a half-space with a plane front (see Fig. 3.17). Parameters inside the cloud vary only in the direction normal to its front. The free-stream velocity vector is perpendicular to the undisturbed boundary of the heated zone. At some initial moment of time the bow shock touches the boundary of the cloud. The temperature in the layer is defined by the formula

$$T = T_\infty \left[-18 \left(\frac{\zeta}{d} \right)^3 + 27 \left(\frac{\zeta}{d} \right)^2 + 1 \right],$$

where ζ is measured normally inside the cloud. The temperature $T_\infty = 253.7$ K corresponds to the quantities ρ_∞ and p_∞ at $\zeta = 0$. The temperature $T_\mathrm{f} = 2537$ K at $\zeta = d$ is adopted to be equal to the temperature of the core of the cloud. It is therefore assumed to be constant for $\zeta > d$. In the case analyzed, $d = 1$ m. The pressure both inside and outside the cloud is considered to be constant. The density and the concentrations of species are determined in the assumption of chemical equilibrium inside the cloud. Note that at $T = T_\mathrm{f}$ almost all oxygen is in the dissociated state, whereas nitrogen only starts dissociation. Since the velocity of the cloud is always supersonic, the body cannot affect its uniformity in the region ahead of the bow shock. The effective Mach number calculated via the parameters in the interior of the cloud is approximately equal to 4.3. The equilibrium parameters inside the heated layer are calculated by solving the system of equations $R_l = 0$ for known values of T and p by the method described by Pogorelov (1988b). This gives $\rho_\mathrm{f} = 0.275 \times 10^{-4}$ kg m^{-3} and

$$\begin{aligned}
c_{1\infty} &= 0.2322, & c_{1\mathrm{f}} &= 0.0061; \\
c_{2\infty} &= 0, & c_{2\mathrm{f}} &= 0.2234; \\
c_{3\infty} &= 0.7542, & c_{3\mathrm{f}} &= 0.7517; \\
c_{4\infty} &= 0, & c_{4\mathrm{f}} &= 0.471 \times 10^{-4}; \\
c_{5\infty} &= 0, & c_{5\mathrm{f}} &= 0.518 \times 10^{-2}; \\
c_{6\infty} &= 0.0136, & c_{6\mathrm{f}} &= 0.0136.
\end{aligned}$$

As initial parameters inside the shock layer around the body we adopt the steady-state solution of the governing system for the unperturbed motion of the body. We use the boundary condition of nonpenetration on the body surface. It is apparent that there exists a symmetry plane of the flow. We use the symmetry boundary conditions on this plane. The exit boundary is supersonic and we need no boundary condition on it. The bow shock is treated as a discontinuity surface.

For ideal gas the system of conservation relations on the gas dynamic shock is expressed by the formulas

$$\rho_2 \mathbf{u}_{2n} = \rho_1 \mathbf{u}_{1n},$$

$$\mathbf{v}_{2t} = \mathbf{v}_{1t}, \qquad \mathbf{u} = \mathbf{v} - \mathbf{s},$$

$$\rho_2 u_{2n}^2 + p_2 = \rho_1 u_{1n}^2 + p_1,$$

$$h_2 + \frac{\mathbf{u}^2}{2} = h_1 + \frac{\mathbf{u}_1^2}{2}.$$

Here v_n and $\mathbf{v}_t$ are the velocity vector projections on the unit outward normal to the shock wave surface and on the plane tangent to it, $\mathbf{s}$ is the shock velocity vector, and u_n is the normal component of the velocity in the frame attached to the shock. Note that $\mathbf{s}$ has only the normal component, since the shock propagates perpendicular to its surface. The enthalpy, by definition, is related to the internal energy as $h = \varepsilon + p/\rho$. We consider all these quantities at an arbitrarily fixed point on the discontinuity. The relation $\mathbf{v}_{t2} = \mathbf{v}_{t1}$ expresses the conservation of the tangential component of the velocity vector. It can also be written in the vector product form as

$$\mathbf{v}_2 \times \mathbf{n} = \mathbf{v}_1 \times \mathbf{n}.$$

The subscripts 1 and 2 correspond, respectively, to quantities in the inhomogeneous cloud ahead of the shock and to those in the compressed layer behind the shock. The above conservation relations can be rewritten as

$$h = \frac{\rho_1 h_1}{p_1} + \frac{1}{2}\left(1 + \frac{1}{\rho}\right)(p - 1), \tag{3.5.8}$$

$$\mathbf{U}_{1n}^2 = \frac{p - 1}{1 - 1/\rho}, \tag{3.5.9}$$

$$\rho|\mathbf{u}_{1n}| = |\mathbf{u}_{2n}|, \tag{3.5.10}$$

$$\mathbf{v}_{1t} = \mathbf{v}_{2t}, \tag{3.5.11}$$

$$h = \frac{h_2 \rho_1}{p_1}, \quad \rho = \frac{\rho_2}{\rho_1}, \quad p = \frac{p_2}{p_1}. \tag{3.5.12}$$

The bow shock is assumed to be thin, in the sense that the concentrations of components remain constant across the shock, that is, $c_{i2} = c_{i1}$. The applicability of this formula was investigated by Stulov (1969).

We shall rewrite the Rankine–Hugoniot relation (3.5.8) for internal energy as

$$\varepsilon = \frac{\rho_1 \varepsilon_1}{p_1} + \frac{1}{2}\left(1 - \frac{1}{\rho}\right)(p + 1). \tag{3.5.13}$$

Under the assumptions of the Lighthill model of ideal excitation of the vibrational mode of freedom in the shock layer and taking into account the equilibrium state of mixture in the cloud, we can easily obtain

$$\rho = \frac{(\gamma_2 + 1)p + \gamma_2 - 1}{(\gamma_2 - 1)(p + 2\beta + 1)}, \tag{3.5.14}$$

$$\beta = \frac{1}{T_1}\left\{\left(\frac{2.5}{M_m} + \frac{1.5}{M_a}\right)T_1 + \sum_{l=1,3,5} \frac{c_{l1}\theta_{\nu l}}{M_l[e^{\theta_{\nu l}/T_1} - 1]}\right\},$$

where γ_2 is defined by Eq. (3.5.6).

Let us consider the procedure of calculation the quantities behind the shock. Suppose that using some numerical scheme we managed to determine the value of pressure p behind the shock without approximation of the derivatives across the discontinuity. The values of the normal component of the velocity vector and the density behind the shock in terms of p can be obtained from relations (3.5.9) and (3.5.14). The velocity components normal and tangent to the shock surface can be found as

$$\mathbf{v}_{1n} = (\mathbf{v}_1 \cdot \mathbf{n}_s)\,\mathbf{n}_s, \quad \mathbf{v}_{1t} = \mathbf{v}_1 - \mathbf{v}_{1n}. \tag{3.5.15}$$

Here $\mathbf{n}_s$ is the unit outward vector normal to the shock surface

$$\mathbf{n}_s = \frac{1}{N}\left(\sqrt{g^{11}}\mathbf{e}^R - R'_{s\varphi}\sqrt{g^{22}}\mathbf{e}^\varphi - R'_{s\vartheta}\sqrt{g^{33}}\mathbf{e}^\vartheta\right), \tag{3.5.16}$$

where

$$N = \sqrt{g^{11} + g^{22}R'^2_{s\varphi} + g^{33}R'^2_{s\vartheta} - 2g^{12}R'_{s\varphi} - 2g^{13}R'_{s\vartheta} + 2g^{23}R'_{s\varphi}R'_{s\vartheta}}.$$

The vectors $\mathbf{e}^R$, $\mathbf{e}^\varphi$, and $\mathbf{e}^\vartheta$ in Eq. (3.5.16) are the unit contravariant basis vectors of our curvilinear coordinate system (R, φ, ϑ).

The projection of $\mathbf{v}_{1n}$ onto the normal can be found as

$$v_{1n} = \frac{1}{N}\left(\frac{-u_1}{\sqrt{g_{11}}} + \frac{v_1 R'_{s\varphi}}{\sqrt{g_{22}}} + \frac{w_1 R'_{s\vartheta}}{\sqrt{g_{33}}}\right). \tag{3.5.17}$$

Thus, we obtain

$$\mathbf{u}_{1n} = \mathbf{v}_{1n} - \mathbf{s} = (\mathbf{v}_1 \cdot \mathbf{n}_s)\mathbf{n}_s - s\mathbf{n}_s = [(\mathbf{v}_1 \cdot \mathbf{n}_s) - s]\,\mathbf{n}_s. \tag{3.5.18}$$

Note that s is the projection of $\mathbf{s}$ on the shock normal, that is, it has a sign. Hence,

$$s = \left(\frac{p - 1}{1 - 1/\rho}\right)^{1/2} + \frac{1}{N}\left(\frac{u_1}{\sqrt{g_{11}}} - \frac{v_1 R'_{s\varphi}}{\sqrt{g_{22}}} - \frac{w_1 R'_{s\vartheta}}{\sqrt{g_{33}}}\right). \tag{3.5.19}$$

The velocity vector behind the shock can also be decomposed into the vectors perpendicular and tangent to the shock surface and, since the tangent components remain continuous, we obtain

$$\mathbf{v}_2 = \mathbf{v}_{2n} + \mathbf{v}_{2t} = \mathbf{v}_{2n} + \mathbf{v}_{1t}. \tag{3.5.20}$$

From the mass conservation law in the attached coordinate system we can write out

$$\rho|\mathbf{u}_{2n}| = |\mathbf{u}_{1n}|, \qquad \mathbf{u}_{2n} = \mathbf{v}_{2n} - \mathbf{s}. \tag{3.5.21}$$

Hence,

$$|\mathbf{v}_{2n}| = \left(\frac{p-1}{1-1/\rho}\right)^{1/2}\left(\frac{1}{\rho}-1\right) + \frac{1}{N}\left(\frac{-u_1}{\sqrt{g_{11}}} + \frac{v_1 R'_{s\varphi}}{\sqrt{g_{22}}} + \frac{w_1 R'_{s\vartheta}}{\sqrt{g_{33}}}\right). \tag{3.5.22}$$

Now we can easily find $\mathbf{v}_{2n}$, $\mathbf{v}_{2t}$, and $\mathbf{v}_2$ as

$$\mathbf{v}_{2n} = -|\mathbf{v}_{2n}|\mathbf{n}_s, \quad \mathbf{v}_{2t} = \mathbf{v}_{1t} = \mathbf{v}_1 - \mathbf{v}_{1n}, \tag{3.5.23}$$

$$\mathbf{v}_2 = \mathbf{v}_{2n} - \mathbf{v}_{1n} + \mathbf{v}_1 = |\mathbf{u}_{1n}|\left(1 - \frac{1}{\rho}\right)\mathbf{n}_s + \mathbf{v}_\infty. \tag{3.5.24}$$

Let

$$b = \frac{|\mathbf{u}_{1n}|}{N}\left(1 - \frac{1}{\rho}\right),$$

then

$$\mathbf{v}_2 = \left(v_1^i + b\frac{\partial f}{\partial x^j} g^{ij}\sqrt{g_{ii}}\right)\mathbf{e}_i, \quad f = R - R_s(t, \varphi, \vartheta). \tag{3.5.25}$$

Now we can find the time derivative of the shock radius-vector as the component of the shock velocity in the R-direction,

$$\frac{\partial R_s}{\partial t} = s_R = \frac{s}{N}(g^{11} - g^{12}R'_{s\varphi} - g^{13}R'_{s\vartheta})\sqrt{g_{11}}. \tag{3.5.26}$$

Numerical results

It is important that in the considered case we can perform transformation of the independent variables in a way such that the shock layer between the body and the bow shock becomes rectangular, namely,

$$\tau = t, \quad \xi = \frac{R - R_b(\varphi, \vartheta)}{R_s(t, \varphi, \vartheta) - R_b(\varphi, \vartheta)}, \quad \phi = \varphi, \quad \theta = \vartheta. \tag{3.5.27}$$

Note that $\xi = 0$ and $\xi = 1$ correspond in this case to the body and bow shock, respectively. System (3.1.38) transforms now into

$$\frac{\partial \mathbf{U}}{\partial t} + \frac{\partial \mathbf{E}}{\partial x^1} + \frac{\partial \mathbf{F}}{\partial x^2} + \frac{\partial \mathbf{G}}{\partial x^3} = \mathbf{H} \tag{3.5.28}$$

with

$$\begin{aligned} &\mathbf{U} = \mathbf{U}', \quad \mathbf{E}' = f_1\mathbf{G}' + f_2\mathbf{E}' + f_3\mathbf{F}' + f_4\mathbf{U}', \quad \mathbf{F} = \mathbf{F}', \\ &\mathbf{G} = \mathbf{G}', \quad \mathbf{H} = \mathbf{H}' + f_{1\xi}\mathbf{G}' + f_{3\xi}\mathbf{F}' + f_{4\xi}\mathbf{U}', \\ &f_1 = \xi_\vartheta, \quad f_2 = \xi_R, \quad f_3 = \xi_\varphi, \quad f_4 = \xi_t. \end{aligned}$$

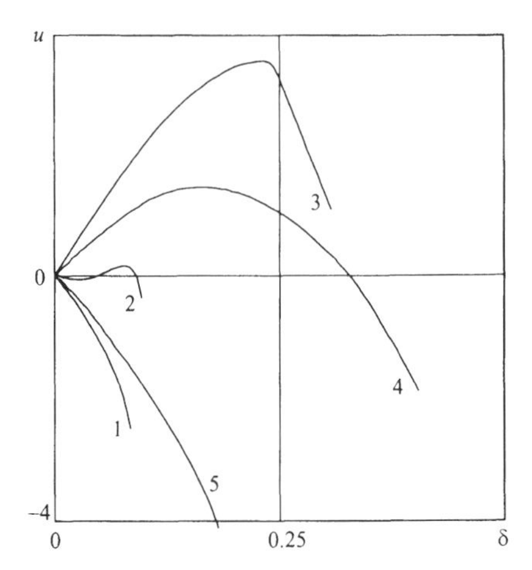

Figure 3.18 Distribution of the radial velocity component versus the distance to the body along the ray $\varphi = 0, \vartheta = \Delta\vartheta/2$ at $t = 0$ (1), 0.0125 (2), 0.05 (3), 0.075 (4), and 0.3 (5).

This simple coordinate transformation allows one to adjust the computational grid to the bow shock. No calculation is necessary in the outer region, which minimizes the computer memory and time resources. Since we fit the bow shock, it becomes possible to resolve the flow inside the shock layer with only a few computational cells. Namely, we use the mesh $R \times \phi \times \theta = 9 \times 13 \times 21$ with $\Delta R = (R_s - R_b)/8$, $\Delta\phi = \pi/8$, and $\Delta\theta = \pi/35$. We apply the second order of accuracy explicit MacCormack scheme (MacCormack 1969) to approximate the derivatives inside the shock layer. To avoid stiffness due to the exponential terms σ_l on the right-hand side of the governing system, we treat these source terms implicitly. This approach can also be used in the framework of shock-capturing methods based on the Godunov scheme.

If we introduce the grid functions $f_{l,m,n} = f[(l-1)\Delta\xi, (m-3)\Delta\phi, (n-2.5)\Delta\theta]$, the conservation relations (3.5.28) can be written out for each computational cell in the coordinate system (ξ, ϕ, θ) as a sequence of the following two time steps (I is the identity matrix):

Predictor step

$$\mathbf{U}^{(1)}_{l,m,n} = \mathbf{U}^k_{l,m,n} - \frac{\Delta t}{\Delta\xi}\left(\mathbf{E}^k_{l+1,m,n} - \mathbf{E}^k_{l,m,n}\right) - \frac{\Delta t}{\Delta\phi}\left(\mathbf{F}^k_{l,m+1,n} - \mathbf{F}^k_{l,m,n}\right) - \frac{\Delta t}{\Delta\theta}\left(\mathbf{G}^k_{l,m,n+1} - \mathbf{G}^k_{l,m,n}\right) + \Delta t\mathbf{H}^k_{l,m,n},$$

Corrector step

$$\left(I - \frac{\Delta t}{2}\frac{\partial \mathbf{H}^k}{\partial \mathbf{U}^k}\right)\left(\mathbf{U}^{k+1}_{l,m,n} - \mathbf{U}^k_{l,m,n}\right) = \frac{1}{2}\Big[\mathbf{U}^{(1)}_{l,m,n} + \mathbf{U}^k_{l,m,n} - \frac{\Delta t}{\Delta\xi}\left(\mathbf{E}^{(1)}_{l,m,n} - \mathbf{E}^{(1)}_{l-1,m,n}\right)$$

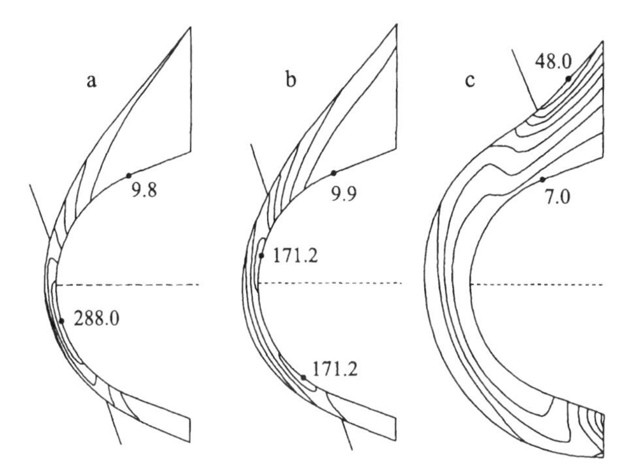

Figure 3.19 Isobar contours at $t = 0.0125$ (a), 0.025 (b), and 0.075 (c).

$$- \frac{\Delta t}{\Delta \phi} \left(\mathbf{F}^{(1)}_{l,m,n} - \mathbf{F}^{(1)}_{l,m-1,n} \right) - \frac{\Delta t}{\Delta \theta} \left(\mathbf{G}^{(1)}_{l,m,n} - \mathbf{G}^{(1)}_{l,m,n-1} \right) + \Delta t \mathbf{H}^{k}_{l,m,n} \Big] .$$

As the body penetrates into the region of smaller density, an intense motion of gas directed away from the body is observed in the shock layer. The radial velocity component u as a function of the bow shock stand-off distance δ_s in the vicinity of the critical streamline $\theta = \Delta\theta/2$ is shown in Fig. 3.18 at times $t = 0$ (1), 0.0125 (2), 0.05 (3), 0.075 (4), and 0.3 (5). The bow shock stand-off distance is initially very small, about 0.085, and the air flow occurs toward the body and around it. Later on there appear zones with velocities directed both to the body and away from it, until the relaxation occurs to the steady flow corresponding to the parameters inside the heated core of the cloud. The bow shock stand-off distance acquires the value of about 0.134 in this case. The charts of constant pressure in the symmetry plane of the flow at $t = 0.0125$, 0.025, and 0.075 are shown in Figs. 3.19a–c. The extremum points are marked. The boundary of the cloud is indicated by straight lines. The isobar distribution shown in Fig. 3.19a is familiar to that in the ordinary supersonic flow over blunt body at high angle of attack, with the maximum being located at the stagnation point and the minimum on the lee side. It is worth noting that soon after penetration (see Fig.3.19b), we get two maxima of the same value lying on the body surface symmetrically with respect to the stagnation point. As the body moves farther into the cloud, a high-pressure zone translocates to the region behind the bow shock that has not yet penetrated into the heated layer (Fig. 3.19c). In addition, an extended rarefaction zone is formed near the surface of the body. This process leads to a substantial redistribution of forces acting on the body. This can result in its stability loss. The shape of the bow shock also undergoes substantial changes, since penetration occurs at the fairly large angle of attack.

A rather long time is required for a new steady state to be established. This process is shown by the isobar charts in Figs. 3.20a–c. For $t > 0.3$ violent unsteady phenomena disappear and disturbances travel from the body to the bow shock and back until the latter reaches its position corresponding to $M_\infty = 4.3$.

The charts of constant temperatures at $t = 0.0375, 0.1$, and 0.3 are shown in Figs. 3.21a–

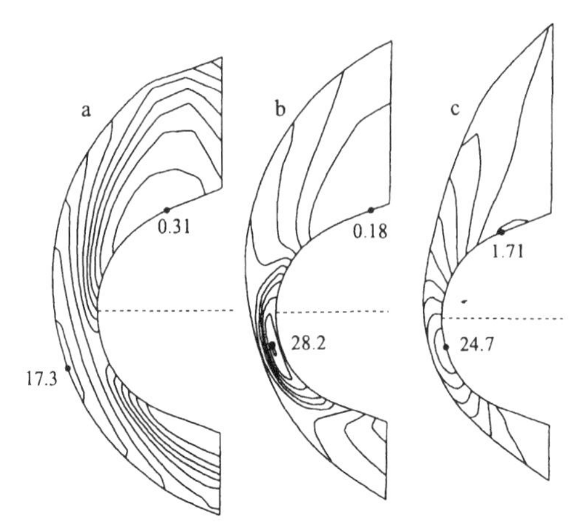

Figure 3.20 Isobar contours at $t = 0.1375$ (a), 0.1875 (b), and 0.3 (c).

c. One can see an abrupt temperature decrease in the thin layer behind the strong shock wave. It is caused by fast consumption of kinetic energy of the flow for dissociation of the mixture molecules. The concentration of molecular oxygen rapidly decreases and the temperature gradient becomes less steep as the body penetrates into the heated layer. The oxygen molecules are practically absent in the resulting steady-state flow.

3.5.2 Floating shock-fitting procedures.

In Section 2.9 we described the basic approach to shock fitting using the general form of the hyperbolic system. This method found the most intensive application in gas dynamics (Moretti 1963, 1974, 1979, 1987; Salas 1976). The basic idea of the floating shock-fitting approach lies in the application of the concept of characteristics incorporated in a finite-difference scheme. While solving the system numerically, the space derivatives are never approximated across discontinuities. Instead, the Rankine–Hugoniot relations are exactly satisfied on all of them. This allows us to avoid spurious oscillations intrinsic in higher-order finite-difference methods (see Section 2.7.1).

It is easy to notice that the code becomes more and more complicated if the number of discontinuities in the computational region increases. That is why, although fine solutions of very complicated three-dimensional problems were successfully obtained using shock-fitting technique, the shock-capturing approach has lately gained much greater popularity. On the other hand, the shock-capturing methods can attain the high quality of the floating shock-fitting methods only if we use numerical grids adapted to large gradients of solutions.

As an example the system in the form of (2.9.3) let us consider the one-dimensional gas dynamic system for the speed of sound c, the velocity u, and the entropy S. In this case (Di Giacinto and Valorani 1989),

$$\mathbf{u} = [c, u, S]^{\mathrm{T}}, \quad A = \begin{bmatrix} u & \beta c & 0 \\ c/\beta & u & -c^2 \\ 0 & 0 & u \end{bmatrix}, \quad \beta = \frac{\gamma - 1}{2}, \quad S = \log \frac{p}{\rho^{\gamma}}.$$

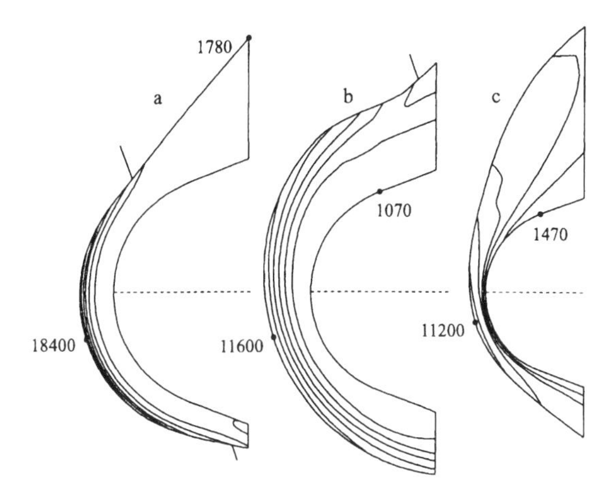

Figure 3.21 Isotherm contours at $t = 0.0375$ (a), 0.1 (b), and 0.3 (c).

The eigenvector matrix Ω_L and the eigenvalue ($\lambda_1 = u + c$, $\lambda_2 = u - c$, and $\lambda_3 = u$) matrix Λ of this system are

$$\Omega_L = \begin{bmatrix} 1/\beta & 1 & -c \\ 1/\beta & -1 & -c \\ 0 & 0 & 1 \end{bmatrix}, \quad \Lambda = \begin{bmatrix} u+c & 0 & 0 \\ 0 & u-c & 0 \\ 0 & 0 & u \end{bmatrix}.$$

By multiplying (2.9.3) by Ω_L, we obtain

$$\Omega_L \mathbf{u}_t + \Lambda \Omega_L \mathbf{u}_x = \mathbf{0}. \tag{3.5.29}$$

On introducing the set of new unknown variables (the characteristic variables)

$$w_1 = \frac{c}{\beta} + u, \quad w_2 = \frac{c}{\beta} - u, \quad w_3 = S, \tag{3.5.30}$$

we arrive at the system

$$w_{1t} + \lambda_1(w_{1x} - cw_{3x}) - cw_{3t} = 0, \tag{3.5.31}$$
$$w_{2t} + \lambda_2(w_{2x} - cw_{3x}) - cw_{3t} = 0, \tag{3.5.32}$$
$$w_{3t} + \lambda_3 w_{3x} = 0. \tag{3.5.33}$$

It is convenient to rewrite this system in terms of derivatives along the characteristic directions $dx/dt = \lambda_i$ ($i = 1, 2, 3$),

$$\left(\frac{\delta w_1}{\delta t}\right)_1 - c\left(\frac{\delta w_3}{\delta t}\right)_1 = 0, \tag{3.5.34}$$
$$\left(\frac{\delta w_2}{\delta t}\right)_2 - c\left(\frac{\delta w_3}{\delta t}\right)_2 = 0, \tag{3.5.35}$$
$$\left(\frac{\delta w_3}{\delta t}\right)_3 = 0. \tag{3.5.36}$$

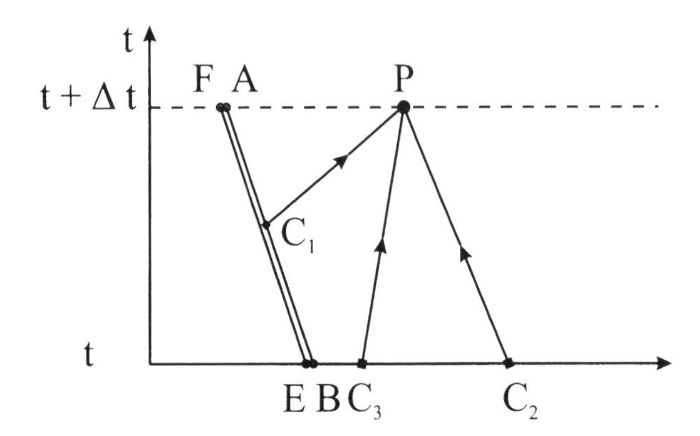

Figure 3.22 Approximation of derivatives near the shock.

The form of the system (2.9.3) implies that either no shocks are present in the computational region or, if they exist, the derivatives in Eqs. (3.5.34)–(3.5.36) must not be approximated across the shock. We need to know in this case the intersection points of the characteristics with discontinuities in the computational region. We can now apply the method of characteristics to obtain the solution at the point P at the $(n+1)$th time level (see Fig. 3.22). The quantities at $t = t^n$ are supposed to be known. We restrict ourselves to explicit numerical schemes in which all quantities at $t = t^{n+1}$ are expressed exclusively via the quantities at $t = t^n$.

Assuming that within a short interval between $t = t^n$ and $t = t^{n+1}$ the characteristics remain straight lines, we obtain

$$w_{1,P}^{n+1} - w_{1,C_1}^{n} = c_{C_1}\left(w_{3,P}^{n+1} - w_{3,C_1}^{n}\right), \tag{3.5.37}$$

$$w_{2,P}^{n+1} - w_{2,C_2}^{n} = c_{C_2}\left(w_{3,P}^{n+1} - w_{3,C_2}^{n}\right), \tag{3.5.38}$$

$$w_{3,P}^{n+1} - w_{3,C_3}^{n} = 0. \tag{3.5.39}$$

The points C_2 and C_3 can be easily found by taking into account that the slopes $dx/dt = \lambda_i$ of the corresponding characteristics are known on the line $t = t^n$ at the grid points and, hence, by the interpolation procedure they can be found at any intermediate point belonging to this line. Another subject is the characteristic line C_1 that intersects the shock represented in Fig. 3.22 by the two parallel lines corresponding to its sides. Let the side AB be the high-pressure side. This means that the velocity component normal to the shock is subsonic on this side. Otherwise, all three characteristics arrive at A from the right-hand side, thus allowing us to determine all quantities at A. After that we can determine the linear value distribution for λ_1 along AB. Note that the value of λ_1 at B can be found by interpolation along the line $t = t^n$. As a result, we can determine all the characteristic variables w_i at P.

From the high-pressure side only one compatibility relation is available, namely,

$$w_{j,A}^{n+1} = w_{j,C_j}^{n} + c_{C_j}(w_{3,A}^{n+1} - w_{3,C_j}^{n}), \tag{3.5.40}$$

where $j = 1$ for right-running shocks and $j = 2$ for left-running shocks (the points C_j are, of course, different for P and A).

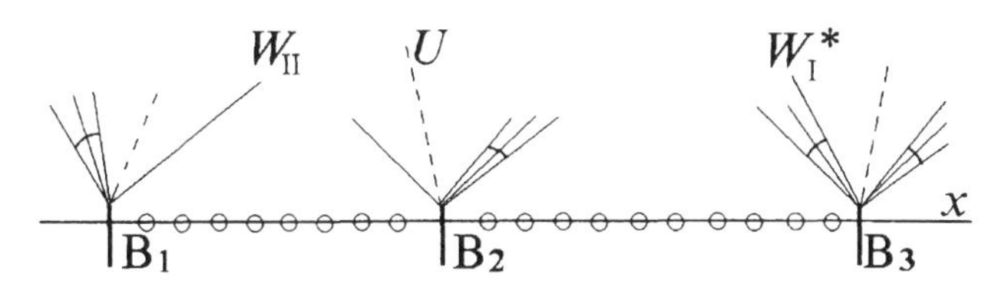

Figure 3.23 A one-dimensional example of the gas dynamic shock-fitting.

We need three additional relations to determine the lacking Riemann invariants and the shock velocity at the higher time level. They are provided by the Rankine–Hugoniot relations.

We used the shock velocity value, that is, the slope of the straight line AB basing on the data at $t = t^n$. The new shock velocity at $t = t^{n+1}$ is therefore updated using the compatibility relations along the characteristics. Treatment of the flow in the near vicinity of contact discontinuities or boundaries of the computational region can be done in a similar way (see, e.g., Di Giacinto and Valorani 1989).

If the first characteristic line does not intersect the shock profile, the previous step is not required and the quantities at the point P can be obtained directly from the compatibility relations (3.5.37)–(3.5.39). There still remains the problem of detecting new shocks in the course of calculations. This can be done by analyzing the behavior of characteristics.

Extension of the described method to multiple dimensions gave rise to a so-called λ-scheme (Moretti 1979). Convenient procedures of shock fitting were proposed by De Neef and Moretti (1980) and Moretti (1987).

3.5.3 Shock-fitting on moving grids.

As mentioned in Section 2.9.2, Godunov's method (1959) is ideologically different. In this approach the surface of discontinuity still consists of boundaries of the computational cells. The velocity at which the surface of discontinuity moves depends on the values of the gas dynamic variables in cells on either side of the discontinuity. To determine the flux across the surface of discontinuity, the boundary conditions, which depend on the type of the surface to be fitted, are used to calculate the values of the unknown functions. Unlike the traditional shock-fitting methods described in the previous subsections, Godunov's approach is naturally based on the exact Riemann problem solver. For example, suppose that in some one-dimensional gas dynamic problem the points B_1, B_2, and B_3 (Fig. 3.23) are associated, from of the flow analysis, with the right shock wave, contact discontinuity, and right boundary of the rarefaction wave, respectively; see the flow configurations in Fig. 3.2. Let the velocities of these points be W_{II}, U, and W_I^*, respectively. These velocities can be found in solving the gas dynamic Riemann problem at each time level. The velocities of the remaining cells can be determined, for instance, by linear interpolation of W_{II}, U, and W_I^*. Thus, we construct the moving grid associated with the specific character of the gas flow. After that, one can use the finite-volume scheme (3.2.7)–(3.2.10) for this moving grid.

The moving discontinuity-fitting grids for two- and three-dimensional cases can be constructed in a similar way. Figure 3.24 illustrates the two-dimensional discontinuity fitting for the particular problem of an axisymmetric blunt body flow. Let x be the axis of

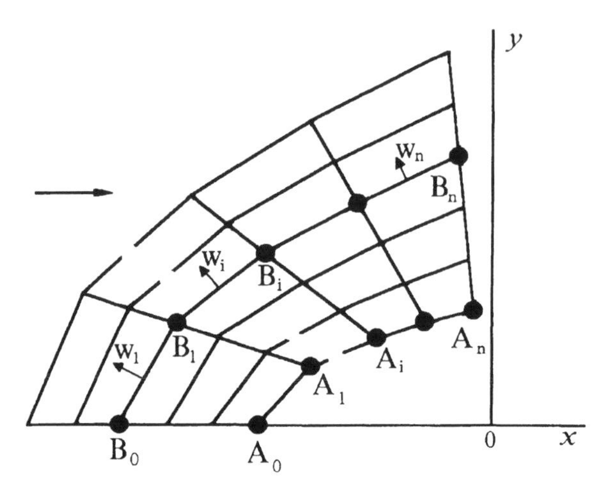

Figure 3.24 Construction of numerical grid.

symmetry. Then it suffices to study the flow in the half-plane. A body whose surface is approximated by a broken line $A_0A_1 \ldots A_n$ is located in a supersonic flow of an ideal perfect gas. A family of rays is drawn through the points A_i, $i = 0, 1, \ldots, n$. The points B_i are fixed on these rays and the segments A_iB_i are divided into equal parts. By connecting these points by straight lines, we divide the region of integration into computational cells. We will associate the surface $B_0B_1 \ldots B_n$, say, with the bow shock wave. Its position changes in the process of calculation. To determine the normal velocity directed along the outward normal at the center of each segment $B_{i-1}B_i$, we find the shock wave speed $\mathbf{W}_i$ by solving the Riemann problem. We assume that the velocities of the points B_i along fixed rays are equal to the weighted sum of the projections of $\mathbf{W}_i$ and $\mathbf{W}_{i+1}$ onto the ith ray. The velocities of the remaining points moving along the fixed ith rays can be determined, for instance, by linear interpolation of the velocities of the points B_i and A_i. Note that the velocity of A_i is here equal to zero.

This approach is close to the so-called shock-capturing method in which the calculation over the entire region, including the neighborhood of the discontinuity, is carried out by the same algorithm. Note that such an approach includes an additional algorithm of the configuration analysis to determine the fluxes across the surfaces of discontinuities being captured. An additional analysis is the main specific feature of the available methods of this type (Godunov, Zabrodin, and Prokopov 1961; Godunov et al. 1979; Kraiko, Makarov, and Tillyaeva 1980; Potapkin 1983). In these methods the determination of the velocity of discontinuities being captured is based essentially on the exact Riemann problem solver, since otherwise the shock wave would be smeared or would move across the grid.

3.5.4 Self-adjusting grids. On the basis of Roe's method here we describe so-called "self-adjusting" grid methods for solution of one- and two-dimensional gas dynamic equations on moving grids. These methods use special grids that can automatically detect and subsequently accurately track particular discontinuities arising in the flow. These discontinuities may not be initially present in the gas dynamic flow and may originate later. Unlike the methods by Godunov, Zabrodin, and Prokopov (1961) and Godunov et al. (1979), the self-adjusting shock-fitting is entirely within the framework of the shock-capturing approach and uses Roe's numerical method.

The construction of self-adjusting grids to fit discontinuities is based on the following typical property of the schemes of Godunov (1959) and Roe (1981). If the position of a contact discontinuity or a shock wave coincides with a cell boundary, the numerical solution given by these schemes is also an exact solution (contact discontinuity or a shock wave). This due to with the fact that both schemes guarantee the exact satisfaction of the jump conditions. Obviously, in the case of moving discontinuities, the numerical solution given these schemes will be the same as the exact solution if the cell boundaries move at the velocity of the discontinuity.

Harten and Hyman (1983) used this property to choose the velocity of cell boundaries in solving one-dimensional gas dynamic equations by Roe's method. Specifically, the velocity $W_{i+1/2}$ of the $(i+1/2)$th cell boundary was taken to be the velocity obtained as the weighted average of all eigenvalues of the hyperbolic system on this boundary, or of all velocities of discontinuities formed on the boundary. The weights were taken to be proportional to the "intensity" of the discontinuities. Specifically, for a one-dimensional hyperbolic system of n equations

$$\frac{\partial \mathbf{U}}{\partial t} + \frac{\partial \mathbf{F}(\mathbf{U})}{\partial x} = \mathbf{0}$$

the boundary velocity $W_{i+1/2}$ must be determined as

$$W_{i+1/2} = \left(\frac{\sum\limits_{k=1}^{n} \hat{\lambda}_k \, \delta w_k^2}{\sum\limits_{k=1}^{n} \delta w_k^2} \right)_{i+1/2}, \tag{3.5.41}$$

where the $\hat{\lambda}_k$ are the eigenvalues of Roe's matrix

$$\mathcal{A} = \Omega_{\mathrm{R}} \Lambda \Omega_{\mathrm{L}}, \quad \Delta\mathbf{F} = \mathcal{A} \Delta\mathbf{U}$$

and $\delta\mathbf{w} = [\delta w_1, \ldots, \delta w_n]^{\mathrm{T}}$ satisfies the relation

$$\mathbf{U}_{i+1} - \mathbf{U}_i = \Omega_{\mathrm{R}} \, \delta\mathbf{w} \quad \Longrightarrow \quad \delta\mathbf{w} = \Omega_{\mathrm{L}}(\mathbf{U}_{i+1} - \mathbf{U}_i).$$

For an ideal perfect gas we can use the expression for Ω_{R}, see Eqs. (3.4.17), to obtain the following formulas for $\delta\mathbf{w}$ in one-dimensional case with $\mathbf{U} = [\rho, \rho u, e]^{\mathrm{T}}$:

$$\begin{aligned}
\delta w_1 &= \frac{\gamma - 1}{2\hat{c}^2} \left[\left(\frac{1}{2}\hat{u}^2 + \frac{\hat{u}\hat{c}}{\gamma - 1} \right) \Delta\rho - \left(\hat{u} + \frac{\hat{c}}{\gamma - 1} \right) \Delta(\rho u) + \Delta e \right], \\
\delta w_2 &= \frac{\gamma - 1}{2\hat{c}^2} \left[\left(\frac{2\hat{c}^2}{\gamma - 1} - \hat{u}^2 \right) \Delta\rho + 2\hat{u}\Delta(\rho u) - 2\Delta e \right], \\
\delta w_3 &= \frac{\gamma - 1}{2\hat{c}^2} \left[\left(\frac{1}{2}\hat{u}^2 - \frac{\hat{u}\hat{c}}{\gamma - 1} \right) \Delta\rho - \left(\hat{u} - \frac{\hat{c}}{\gamma - 1} \right) \Delta(\rho u) + \Delta e \right],
\end{aligned} \tag{3.5.42}$$

where $\Delta f \equiv f_{i+1} - f_i$; the quantities $\hat{u}$ and $\hat{c}$ are defined in Eqs. (3.4.13) and (3.4.16). Additionally,

$$\hat{\lambda}_1 = \hat{u} - \hat{c}, \quad \hat{\lambda}_2 = \hat{u}, \quad \hat{\lambda}_3 = \hat{u} + \hat{c},$$

see Eq. (3.4.18). One can obtain more compact expression for (3.5.42) in the form

$$\delta w_1 = \tfrac{1}{2}(b_1 - b_2), \quad \delta w_2 = \Delta\rho - b_1, \quad \delta w_3 = \tfrac{1}{2}(b_1 + b_2),$$
$$b_1 = \frac{\gamma - 1}{\hat{c}^2}\left[\Delta e + \tfrac{1}{2}\hat{u}^2\Delta(\rho u) - \hat{u}\Delta(\rho u)\right], \quad b_2 = \frac{1}{\hat{c}}[\Delta(\rho u) - \hat{u}\Delta\rho].$$

For a single shock wave or a single contact discontinuity, the velocity W determined in this way is equal to the velocity of the discontinuity. The determination $W_{i+1/2}$ in accordance with that of (3.5.42) permits one to minimize the numerical diffusion of Roe's scheme. Whenever the cell boundaries are too close to each other or too far apart, the velocities W should be chosen (or corrected) so as to avoid this (Harten and Hyman 1983).

By means of a shock-capturing calculation, the algorithm by Harten and Hyman gives a very accurate solution to one-dimensional problems with practically no smearing of contact discontinuities or shock waves. This makes it competitive with shock-fitting methods. Furthermore, the algorithm can be used to detect the formation of discontinuities originating from smooth initial conditions.

However, the technique of determining the velocities of cell boundaries mentioned above is quite sophisticated and difficult for programming, since for each cell the velocities of the boundaries must be determined, see Eq. (3.5.42) and the check must be made that the boundaries are not too close or far apart. However, the number and the type of discontinuities that can form in one-dimensional problems are sometimes known *a priori*. Therefore, the algorithm by Harten and Hyman can be considerably simplified in these cases. It suffices to associate the velocity of only a few boundaries with the velocity of only several known discontinuities and then determine only these velocities by special way. The velocities of the remaining cells can be determined, for instance, by linear interpolation. The results of the numerical simulation by such a simplified algorithm will be demonstrated below.

The described simplification of the algorithm of Harten and Hyman (1983) in one dimension allows us to generalize it to the case of a larger number of space variables. A generalization of the proposed technique to the axisymmetric flows is described by Kamenetskii and Semenov (1993, 1994). The algorithm allows one to find a solution by the shock-capturing Roe method (see Section 3.4.4) of the second order of accuracy without any smearing of the bow shock wave.

The method by Kamenetskii and Semenov (1993, 1994) is based on a special procedure for determining the velocity of the cell boundaries and differs from the corresponding procedure in the method by Harten and Hyman (1983). We will describe this procedure first for the one-dimensional case. Let $\hat{u}$ and $\hat{c}$ be the flow velocity and the speed of sound obtained by Roe's averaging of the quantities to the right- and left-hand side of the discontinuity, see Eqs. (3.4.13) and (3.4.16). For an isolated contact discontinuity or a shock wave, Roe's Riemann problem solution is identical to the exact solution. In the case of a contact discontinuity its velocity is equal to $\hat{u}$, while for a shock wave it is equal to $\hat{u} \pm \hat{c}$, where the plus sign corresponds to a right, and the minus sign to a left shock wave, see Section 3.3. As mentioned above, we will associate with discontinuities only n fixed

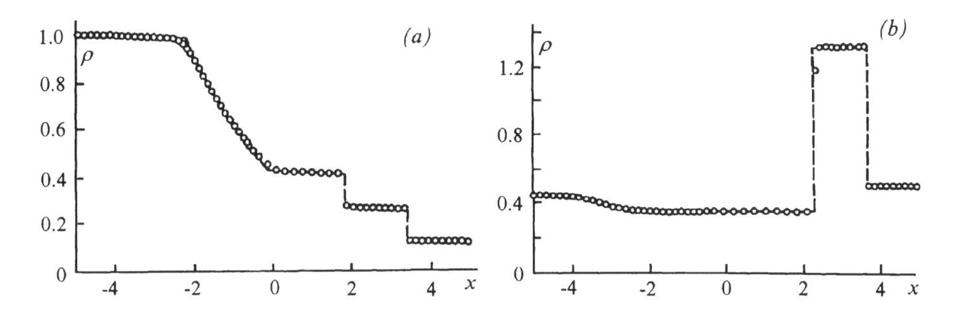

Figure 3.25 One-dimensional tests for the self-adjusting grid method.

cell boundaries, $n \ll N$, where N is the total number of cells; see Fig. 3.23 with $n = 3$ and $N = 21$. Let these be the right boundaries of cells with the indices B_i, $i = 1, \ldots, n$. With each of these boundaries we will associate a definite type of discontinuity. If it is a contact discontinuity, then the velocity of the boundary is equal to $\hat{u}$, and if it is a right or a left shock wave, then the velocity of the boundary is equal to $\hat{u} + \hat{c}$ or $\hat{u} - \hat{c}$, respectively. The velocity of the remaining cells will be determined, for instance, by linear interpolation of these velocities. Then, one can use Roe's method for a moving grid, see Section 3.2. Using a scheme of the second order of accuracy of the predictor–corrector type, one has to recalculate the grid velocity for both predictor and corrector stages.

In the one-dimensional case, we use the test problems due to Sod (1978) and Lax (Harten 1978; Harten and Hyman 1983). Both are the Riemann problems of a perfect gas. In the Sod problem, the quantities to the left and right of the discontinuity are given, respectively, by

$$\mathbf{U}_1 = [1, 0, 0, 0, 2.5]^{\mathrm{T}}, \quad \mathbf{U}_2 = [0.125, 0, 0, 0, 0.25]^{\mathrm{T}},$$

see (3.1.16). The exact solution consists of a left rarefaction wave, a contact discontinuity, and a right shock wave with the following parameters: $c_{\mathrm{I}} = -1.183$, $c_{\mathrm{I}}^* = -0.070$, $R_{\mathrm{I}} = 0.426$, $U = 0.928$, $E_{\mathrm{I}} = 0.758$, $R_{\mathrm{II}} = 0.266$, $E_{\mathrm{II}} = 0.758$, and $W_{\mathrm{II}} = 1.752$, see the notation in Fig. 3.2.

In the Lax problem the initial data are given as

$$\mathbf{U}_1 = [0.445, 0.311, 0, 0, 8.928]^{\mathrm{T}}, \quad \mathbf{U}_2 = [0.5, 0, 0, 0, 1.4275]^{\mathrm{T}}.$$

The exact solution consists also of a left rarefaction wave, a contact discontinuity, and a right shock wave and has the parameters $c_{\mathrm{I}} = -2.633$, $c_{\mathrm{I}}^* = -1.636$, $R_{\mathrm{I}} = 0.345$, $U = 1.529$, $E_{\mathrm{I}} = 6.166$, $R_{\mathrm{II}} = 1.304$, $E_{\mathrm{II}} = 6.166$, and $W_{\mathrm{II}} = 2.480$. The adiabatic index γ is equal to 1.4 in both tests. We will assume that the initial data are given on the interval $-5 < x < 5$, and the boundary of the initial discontinuity is at point $x = 0$. We divide the entire interval into 100 cells. We place 10 cells with $\Delta x = 0.05$ to the right and left of the initial discontinuity, and another 40 cells with $\Delta x = 0.1125$ to the right and left of those cells. Let us construct the moving grid. The first boundary corresponds to the 40th point, second to the 50th point, third to the 60th point, and fourth to the last, 100th point. We set the velocity of the first boundary equal to zero, of the second to $\hat{u}$, and the third to $\hat{u} + \hat{c}$. The velocity of the fourth boundary will be taken equal to that of the third, but all the cells lying to the left of the first boundary will be considered fixed.

Figures 3.25a–b show the values of the density obtained for the Sod and Lax problems, respectively. The size of the computational region varies with time. Only values that by the end of the calculation lie in the initial interval $-5 < x < 5$ are shown by circles in the figure. Both problems were solved with the CFL number equal to 0.9. The results for the Sod problem are shown after 60 time steps, and those for the Lax problem after 95. The dashed line is the exact solution. The results are in good agreement with theoretical values and no smearing of the shock wave is observed. The contact discontinuity is smeared only over two cells. Note that this smearing decreases with time and eventually the contact discontinuity coincides with a cell boundary.

The above one-dimensional scheme has a narrower range of application than the scheme of Harten and Hyman (1983) but is considerably simpler. The main limitation is due to the fact that the number of considered discontinuities is fixed *a priori*. Thus, in the analysis of problems with interaction of discontinuities, additional surfaces to be captured must be defined. Just as in the scheme of Harten and Hyman we allow the discontinuities to originate in the course of calculations. It is only necessary to associate a cell boundary with this discontinuity.

There are many multidimensional problems in which the flow is divided by the surfaces of discontinuity into a number of layers. Our algorithm can be extended to solve these problems. Consider the two-dimensional axisymmetric supersonic blunt body problem, see Fig. 3.24. In the two-dimensional case we can encounter certain difficulties in the attempt to adequately determine the velocities of the points B_i along the fixed rays to eliminate any movement of a discontinuity across the grid. Note that this is the main problem in constructing the algorithm for the self-adjusting grid capturing of discontinuities in the two- or three-dimensional cases. In particular, using the method of Godunov et al. (1961) resulted in the movement of captured discontinuities across the grid, so that they could not be fitted exactly. Note that in the method by Godunov et al. (1961) the velocity of the point B_i along a fixed ray is equal to the weighted sum of the projections of $\mathbf{W}_i$ and $\mathbf{W}_{i+1}$ on the ith ray, see Fig. 3.24.

To avoid these difficulties Kamenetskii and Semenov (1994) applied a procedure of limiting reconstruction (Semenov 1992), see also Section 2.7.3, for the determination of the velocities. This reconstruction was used to calculate more accurately the velocity distribution for the two-dimensional shock front $B_0B_1 \ldots B_n$, see Fig. 3.24.

Let $\mathbf{W}_i$ be the velocity of the boundary $B_{i-1}B_i$ in the direction of the outward normal. Assume that $|\mathbf{W}_i| = |\hat{u}_i \pm \hat{c}_i|$ for the shock wave, where the velocities $\hat{u}$ and $\hat{c}$ are determined from Roe's solution of the Riemann problem in the direction of the outward normal. The quantity $|\mathbf{W}_i|$ in this way determined is merely the average velocity. To calculate the velocity more accurately, one can reconstruct its spatial distribution along the segment $B_{i-1}B_i$. Let us choose the arc length s of the curve $B_0B_1 \ldots B_n$ as the independent coordinate. Then the velocity distribution $|\mathbf{W}(s)_i|$ in the direction of the outward normal for the segment $B_{i-1}B_i$ can be represented in the linear form

$$|\mathbf{W}(s)_i| = |\mathbf{W}_i| + (s - s_i^*)\alpha_i, \qquad s_{i-1} \leq s \leq s_i, \tag{3.5.43}$$

where $s = s_{i-1}$ corresponds to the point B_{i-1}, $s = s_i$ corresponds to B_i, and α_i is an unknown slope that can be obtained by the procedure of the minmod limiting reconstruction. Note that the center of the segment $B_{i-1}B_i$ corresponding to the quantity $s = s^*$ will not coincide, in general, with the middle of the segment, which we can calculate as $\frac{1}{2}(s_{i-1} + s_i)$. The difference of the velocities of the point B_i along the ith ray obtained by using the velocities on the left and right of the ith ray is given by

$$\Delta_i = \left[|\mathbf{W}_{i+1}| + (s_i - s^*_{i+1})\alpha_{i+1}\right] \beta_{i+1} - \left[|\mathbf{W}_i| + (s_i - s^*_i)\alpha_i\right] \beta_i. \tag{3.5.44}$$

The expressions in square brackets on the right-hand side of (3.5.44) have the coefficients β which are associated with the projection of the velocity normal to the corresponding ith ray direction. Try to find α_i such that the sum of boundary velocity jumps

$$J_\mathrm{b} = \sum_i |\Delta_i| \tag{3.5.45}$$

be minimal. Note that, in accordance with the approach of Godunov et al. (1961), one can determine the velocities of the points B_i along the ith ray with various versions of the weighted sum of the projections of $|\mathbf{W}_i|$ and $|\mathbf{W}_{i+1}|$ on the ith ray, for example,

$$\tfrac{1}{2}(|\mathbf{W}_{i+1}|\beta_{i+1} + |\mathbf{W}_i|\beta_i), \tag{3.5.46}$$

or

$$\frac{|\mathbf{W}_{i+1}|(s_i - s_{i-1})\beta_{i+1} + |\mathbf{W}_i|(s_{i+1} - s_i)\beta_i}{s_{i+1} - s_{i-1}},$$

and others. In particular, for the case of (3.5.46) the difference of the velocities (3.5.44) is given by

$$\Delta_i = |\mathbf{W}_{i+1}|\beta_{i+1} - |\mathbf{W}_i|\beta_i.$$

If Δ_i are sufficiently small, then the movement of a discontinuity across the self-adjusting grid is also small. Thus, we can interpret the minimization of sum (3.5.45) as the requirement that the grid does not move relative to the discontinuity.

The limiting reconstruction given by the minimization of (3.5.45) can be implemented to determine velocity distribution along a shock wave or a contact surface. The obtained distribution will preserve the continuity of the velocity distribution on the rays, that is $J_\mathrm{b} \to 0$. It was found from numerical calculations that such an approach permits one to adjust the grid to such an extent that the discontinuity is captured completely.

It is obvious that if the velocity distribution is sought in the form (3.5.43) and the velocity projections are only required to be equal on the rays, then it is sufficient to determine only one quantity among α_i, say α_1. A recurrence relation is easy to obtain for the other α's. We will assume that the quantities α_i obtained as a result of a limiting reconstruction are only preliminary values. The final quantities α^*_i are chosen so that the sum

$$S = \sum_{i=2}^{n-1} (\alpha^*_i - \alpha_i)^2$$

is minimal. Using the recurrence relations, we can represent S as a function of only α^*_1. It is obvious that if as a result of a limiting reconstruction the continuity of all velocity projections has already been obtained, then $\alpha^*_i = \alpha_i$.

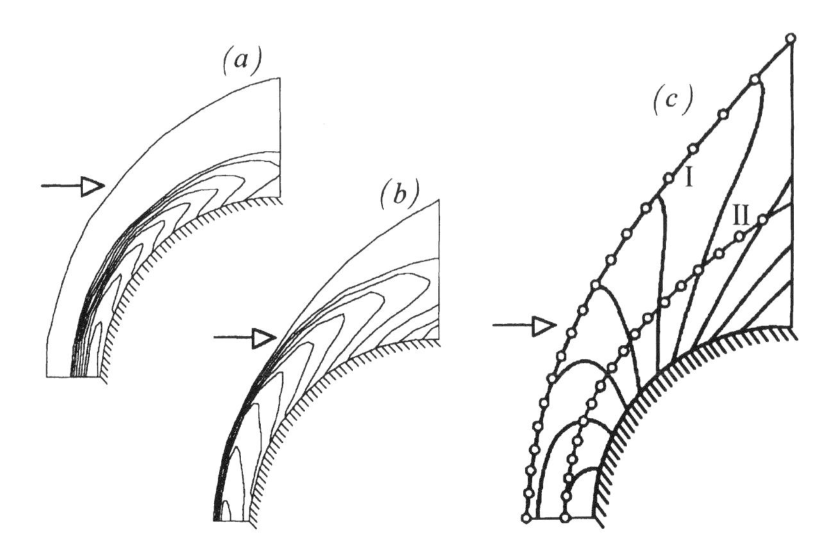

Figure 3.26 Two-dimensional simulation.

Note that this method for constructing a velocity distribution along a discontinuity depends on all points B_i. However, if the flow is supersonic, there cannot be any upstream effects. Generally speaking, the shock wave front might develop instabilities under these conditions. However, this was not observed in the calculations carried out in Kamenetskii and Semenov (1994).

Note that the described technique permits one to extend the method of Harten and Hyman (1983) to the axisymmetric case. This can be done by using the limiting reconstruction for determining the velocity distribution not for separate, as in above, but for all grid segments.

The application of the algorithm in the axisymmetric case is illustrated by the example of solving the problem of a supersonic blunt body flow about a spherical body (Fig. 3.24) with the Mach number M_0 equal to 2 and 6. Initially, the points B_i lie on a sphere that is concentric with the spherical body. The rays A_iB_i are drawn from the center of the body and the angles between them are equal. The constant parameters of the flow at infinity are used as the initial data. The calculations are carried out until the flow becomes steady. Isolines of the density and position of the capturing shock wave surface ($B_0B_1 \ldots B_n$ in Fig. 3.24) for intermediate moment of time for $M_0 = 2$ are shown in Fig. 3.26a–c. At the stage shown in Fig. 3.26a, the algorithm still operates like a shock-capturing calculation method, no shock having yet been captured. Figure 3.26b illustrates the application of the algorithm during the self-adjusting grid capture of the bow shock wave. The part of the shock closer to the axis of symmetry has already been captured. The remaining part of the discontinuity is still not fitted. And, finally, in Fig. 3.26c, we see a steady flow with the captured shock wave I. The isolines of the density are shown in the region between the sphere and the capturing shock wave surface ($B_0B_1 \ldots B_n$ in Fig. 3.24). Line II shows the position of this surface obtained for $M_0 = 6$. The circles show the position of the shock wave calculated by the shock-fitting method (Lyubimov and Rusanov 1970). It is apparent from the figure that the accuracy in determining the bow shock stand-off distance is very high. Thus, the self-adjusting grid method accurately tracks the position of the shock wave. There is also good agreement in the distribution of the gas dynamic parameters with respect to the flow field, with no more than 0.5% discrepancy with the tables by Lyubimov and Rusanov (1970). A mesh with an angular step of 5° was used, the segments of rays between the body and the isolated surface being divided into eight equal parts. Clearly, one can capture a larger number of discontinuities (Kamenetskii and Semenov 1994).

The methods based on the exact Riemann problem solver (Godunov et al. 1961, 1979; Kraiko et al. 1980, and Potapkin 1983) are not sensitive to the procedure of calculating the velocities for capturing discontinuities. The described self-adjusting grid algorithm based on Roe's Riemann problem solver is sensitive to velocity calculations. It is, however, a shock-capturing method. In this case the velocities of a moving grid are obtained by a separate procedure. To capture and track a discontinuity by a moving grid accurately the minmod limiting reconstruction can be used to determine the grid velocity distribution along the discontinuity surface.

The self-adjusting grid algorithms combine the advantages and simplicity of shock-capturing methods with the accuracy of shock-fitting methods and can be applied to increase the accuracy of shock-capturing calculations.

3.6 Stationary gas dynamics

In this section we describe the main specific features of the Godunov method for the hyperbolic system of equations that govern the stationary supersonic two-dimensional flows of an ideal perfect gas.

3.6.1 Systems of governing equations. The stationary equations of gas dynamics can be obtained if we assume that in Eqs. (3.1.1)–(3.1.3) the time derivatives are equal to zero. If the problem is two-dimensional, or $w \equiv 0$, we obtain

$$\operatorname{div}(\rho \mathbf{v}) = 0, \tag{3.6.1}$$

$$\operatorname{div}(\rho \mathbf{v}\mathbf{v} + p\hat{\mathbf{I}}) = \mathbf{0}, \tag{3.6.2}$$

$$\operatorname{div}[(e + p)\mathbf{v}] = 0. \tag{3.6.3}$$

Here $\rho = \rho(x, y)$ is the density, where x and y form a plane Cartesian coordinate system, $\mathbf{v} = \mathbf{v}(x, y) = [u, v]^{\mathrm{T}}$ is the flow velocity, $\hat{\mathbf{I}} = \operatorname{diag}[1, 1]$ is the 2×2 identity tensor, ε is the specific internal energy, $e = \rho\varepsilon + \frac{1}{2}\rho(u^2 + v^2)$ is the total energy per unit volume, and $p = p(\rho, \varepsilon)$ is the pressure. For simplicity we consider here the equation of state of an ideal perfect gas:

$$p = (\gamma - 1)\rho\varepsilon, \quad \varepsilon = \frac{p}{(\gamma - 1)\rho}, \tag{3.6.4}$$

where $\gamma > 1$ is an adiabatic index.

Equations (3.6.1)–(3.6.3) express the conservation of mass, momentum, and energy of the gas, respectively. In the expanded form, the system can be written as

$$\frac{\partial \rho u}{\partial x} + \frac{\partial \rho v}{\partial y} = 0, \tag{3.6.5}$$

$$\frac{\partial(\rho u^2 + p)}{\partial x} + \frac{\partial \rho u v}{\partial y} = 0, \tag{3.6.6}$$

$$\frac{\partial \rho v u}{\partial x} + \frac{\partial(\rho v^2 + p)}{\partial y} = 0, \tag{3.6.7}$$

$$\frac{\partial(e+p)u}{\partial x} + \frac{\partial(e+p)v}{\partial y} = 0. \tag{3.6.8}$$

If the flow is supersonic along one of the space coordinates, the system of stationary gas dynamic equations (3.6.5)–(3.6.8) is hyperbolic along this coordinate (see Section 1.3.2). For example, such flows are realized in air intakes, air ducts, gas-engine blowers, nozzles, and diffusers (Liepmann and Roshko 1957; Abramovich 1976). Note that stationary supersonic flows can be simulated by applying a time-dependent system and using a time relaxation method. Nevertheless, the use of stationary equations is more efficient and widely used if the system remains hyperbolic in the whole computational region, see, for example, Godunov et al. (1979).

We can represent Eqs. (3.6.5)–(3.6.8) in the familiar conservative form

$$\frac{\partial \mathbf{E}}{\partial x} + \frac{\partial \mathbf{F}}{\partial y} = \mathbf{0}, \tag{3.6.9}$$

$$\mathbf{E} = \left[\rho u, \rho u^2 + p, \rho u v, (e+p)u\right]^{\mathrm{T}}, \quad e = \rho\varepsilon + \tfrac{1}{2}\rho\,(u^2 + v^2); \tag{3.6.10}$$

$$\mathbf{F} = \left[\rho v, \rho u v, \rho v^2 + p, (e+p)v\right]^{\mathrm{T}}. \tag{3.6.11}$$

Equation (3.6.9) can be rewritten in the nonconservative form

$$\frac{\partial \mathbf{U}}{\partial x} + A\frac{\partial \mathbf{U}}{\partial y} = \mathbf{0}, \quad A = E_U^{-1} F_U, \tag{3.6.12}$$

$$E_U = \frac{\partial \mathbf{E}}{\partial \mathbf{U}} = \begin{bmatrix} U & R & 0 & 0 \\ \frac{\gamma-1}{\gamma}H & \frac{\gamma+1}{\gamma}U & \frac{1-\gamma}{\gamma}V & \frac{\gamma-1}{\gamma}R \\ 0 & V & U & 0 \\ 0 & H & 0 & U \end{bmatrix},$$

$$F_U = \frac{\partial \mathbf{F}}{\partial \mathbf{U}} = \begin{bmatrix} V & 0 & R & 0 \\ 0 & V & U & 0 \\ \frac{\gamma-1}{\gamma}H & \frac{1-\gamma}{\gamma}U & \frac{\gamma+1}{\gamma}V & \frac{\gamma-1}{\gamma}R \\ 0 & 0 & H & V \end{bmatrix},$$

$$\mathbf{U} = [R, U, V, H]^{\mathrm{T}} = [\sqrt{\rho}, \sqrt{\rho}\,u, \sqrt{\rho}\,v, \sqrt{\rho}\,h]^{\mathrm{T}}, \quad h = \frac{p+e}{\rho}, \tag{3.6.13}$$

where the function h is the total enthalpy. We consider here $\mathbf{U}$ to be a vector of Roe's variables, see (3.4.7). Of course, one can use other independent variables, in particular, $\mathbf{U} = [\rho, \rho u, \rho v, e]^{\mathrm{T}}$ or $\mathbf{U} = [\rho, u, v, p]^{\mathrm{T}}$, see Section 1.3.2, and so on.

The matrix A of Eq. (3.6.12) can be diagonalized as follows:

$$A = E_U^{-1} F_U = \Omega_R \Lambda \Omega_L, \tag{3.6.14}$$

where

$$\Omega_R = \begin{bmatrix} R_1(-r) & \sqrt{r^2+1} & -1 & R_1(r) \\ R_2(-r) & \dfrac{abc}{(\gamma-1)(b^2+1)} & 0 & R_2(r) \\ R_3(-r) & \dfrac{(b^2-1)ac}{2(\gamma-1)(b^2+1)} & 0 & R_3(r) \\ R_4(-r) & 0 & \dfrac{ac^2}{2(\gamma-1)} & R_4(r) \end{bmatrix}, \tag{3.6.15}$$

$$R_1(r) = (b^2+1)\sqrt{r^2+1}, \quad R_2(r) = 2(br+1)(r-b)c,$$

$$R_3(r) = [(br+1)^2 - (r-b)^2]c, \quad R_4(r) = \frac{(b^2+1)ac^2\sqrt{r^2+1}}{2(\gamma-1)},$$

$$a = (\gamma-1)r^2 + \gamma + 1,$$

$$r = \sqrt{\frac{u^2+v^2}{c^2} - 1}, \quad \frac{1}{2}\left(b - \frac{1}{b}\right) = \frac{v}{u}, \quad \Longrightarrow \quad b = \frac{v}{u} + \operatorname{sgn} u \times \sqrt{\frac{v^2}{u^2} + 1},$$

$$u = \frac{2bc}{b^2+1}\sqrt{r^2+1}, \quad v = \frac{(b^2-1)c}{b^2+1}\sqrt{r^2+1};$$

$$\det \Omega_R = -\frac{4a\gamma c^4}{(\gamma-1)^2}(b^2+1)^2 r\sqrt{r^2+1};$$

$$\Omega_L = \begin{bmatrix} L_1(-r) & L_2(-r) & L_3(-r) & L_4(-r) \\ \dfrac{(\gamma-1)(1-r^2)}{2\gamma\sqrt{r^2+1}} & \dfrac{2(\gamma-1)b}{(b^2+1)\gamma c} & \dfrac{(\gamma-1)(b^2-1)}{(b^2+1)\gamma c} & \dfrac{(\gamma-1)^2(1-r^2)}{\gamma ac^2\sqrt{r^2+1}} \\ -\dfrac{a}{4\gamma} & \dfrac{(\gamma-1)u}{2\gamma c^2} & \dfrac{(\gamma-1)v}{2\gamma c^2} & \dfrac{(\gamma-1)(4\gamma-a)}{2\gamma ac^2} \\ L_1(r) & L_2(r) & L_3(r) & L_4(r) \end{bmatrix},$$

$$L_1(r) = \frac{a}{8\gamma(b^2+1)\sqrt{r^2+1}}, \quad L_2(r) = \frac{2b(1-\gamma) + (1-b^2)\gamma/r}{4\gamma c(b^2+1)^2},$$

$$L_3(r) = \frac{(\gamma-1)(1-b^2) + 2b\gamma/r}{4\gamma c(b^2+1)^2}, \quad L_4(r) = \frac{\gamma-1}{4\gamma c^2(b^2+1)\sqrt{r^2+1}};$$

$$\Lambda = \operatorname{diag}[\lambda_-, \lambda_0, \lambda_0, \lambda_+], \tag{3.6.16}$$

$$\lambda_\pm = \frac{uv \pm c\sqrt{u^2+v^2-c^2}}{u^2-c^2}, \quad \lambda_0 = \frac{v}{u};$$

$$c = \sqrt{(\gamma-1)h - \tfrac{1}{2}(\gamma-1)(u^2+v^2)}. \tag{3.6.17}$$

Here the function c is the speed of sound and the variables b and r are introduced to simplify the presentation of the matrices. It is clear that (3.6.5)–(3.6.8) is hyperbolic if $u^2 + v^2 > c^2$. This does not resolve the question of whether the system is x-hyperbolic or y-hyperbolic; see Section 1.3.2. In the x-hyperbolic system $u^2 > c^2$, whereas for $v^2 > c^2$ the system is y-hyperbolic.

Note that it may be convenient to determine Ω_L in calculations by the direct numerical invertion of Ω_R.

The matrix $\widetilde{A} = F_U E_U^{-1}$ can also be diagonalized,

$$\widetilde{A} = F_U E_U^{-1} = \widetilde{\Omega}_R \Lambda \widetilde{\Omega}_L, \tag{3.6.18}$$

where

$$\widetilde{\Omega}_L = \begin{bmatrix} \widetilde{L}_1(-r) & \widetilde{L}_2(-r) & \widetilde{L}_3(-r) & \widetilde{L}_4(-r) \\ -c\sqrt{r^2+1} & \dfrac{2b}{b^2+1} & \dfrac{b^2-1}{b^2+1} & 0 \\ \dfrac{ac^2}{2(\gamma-1)} & 0 & 0 & -1 \\ \widetilde{L}_1(r) & \widetilde{L}_2(r) & \widetilde{L}_3(r) & \widetilde{L}_4(r) \end{bmatrix}, \tag{3.6.19}$$

$$\widetilde{L}_1(r) = \tfrac{1}{2}(b^2+1)ac^2\sqrt{r^2+1}, \quad \widetilde{L}_2(r) = (2br^2(1-\gamma) + (1-b^2)r - 2\gamma b)c,$$

$$\widetilde{L}_3(r) = (\gamma-1)(1-b^2)cr^2 + (1-b^2)\gamma c + 2bcr,$$

$$\widetilde{L}_4(r) = (\gamma-1)(b^2+1)\sqrt{r^2+1};$$

$$\det \widetilde{\Omega}_L = 2c^3(b^2+1)^2 r\sqrt{r^2+1};$$

$$\widetilde{\Omega}_R = \begin{bmatrix} \widetilde{R}_1(-r) & \dfrac{(\gamma-1)r^2+\gamma}{c\sqrt{r^2+1}} & \dfrac{\gamma-1}{c^2} & \widetilde{R}_1(r) \\ \widetilde{R}_2(-r) & \dfrac{2ab}{b^2+1} & \dfrac{(\gamma-1)u}{c^2} & \widetilde{R}_2(r) \\ \widetilde{R}_3(-r) & \dfrac{(b^2-1)a}{b^2+1} & \dfrac{(\gamma-1)v}{c^2} & \widetilde{R}_3(r) \\ \widetilde{R}_4(-r) & \dfrac{[(\gamma-1)r^2+\gamma]ac}{2(\gamma-1)\sqrt{r^2+1}} & \dfrac{(\gamma-1)(r^2+1)}{2} & \widetilde{R}_4(r) \end{bmatrix},$$

$$\widetilde{R}_1(r) = \frac{1}{2(b^2+1)c^2\sqrt{r^2+1}}, \quad \widetilde{R}_2(r) = \frac{2b + (1-b^2)/r}{2(b^2+1)^2 c},$$

$$\widetilde{R}_3(r) = \frac{b^2 - 1 + 2b/r}{2(b^2+1)^2 c}, \quad \widetilde{R}_4(r) = \frac{a}{4(\gamma-1)(b^2+1)\sqrt{r^2+1}}.$$

Note that the pth right vector of Ω_R satisfy

$$(F_U - \lambda_p E_U)\mathbf{r}^p = \mathbf{0},$$

and the qth left vector of $\tilde{\Omega}_{\mathrm{L}}$ satisfy

$$\tilde{\mathbf{l}}^q(F_U - \lambda_q E_U) = \mathbf{0}.$$

Hence,

$$\begin{aligned} &\tilde{\mathbf{l}}^q \cdot (F_U - \lambda_p E_U)\mathbf{r}^p = 0, \quad \tilde{\mathbf{l}}^q(F_U - \lambda_q E_U) \cdot \mathbf{r}^p = 0 \\ &\Longrightarrow \quad \tilde{\mathbf{l}}^q \cdot E_U \mathbf{r}^p = 0 \ \text{ if } \ \lambda_p \neq \lambda_q. \end{aligned}$$

Thus, $\tilde{\Omega}_{\mathrm{L}} E_U \Omega_{\mathrm{R}}$ is a diagonal matrix.

3.6.2 The Godunov method. The CIR and Roe's schemes. Let us introduce a discrete uniform space mesh on the y-axis with size Δy. Let $\mathbf{U}_i$ denote the mesh function values, where the integer subscript $i = 1, 2, \dots$ refers to the center of the ith space cell along the y-axis. The half-integer subscript $i \pm 1/2$ refers to the boundaries between the numerical cells with numbers i and $i \pm 1$. Assume that all mesh functions are constant inside each space cell. Let the integer superscript $k = 0, 1, \dots$ indicate the layer with respect to x and Δx be the increment along the x-axis. Then, for cell boundary $i + 1/2$ and for each x-step, we can solve the Riemann problem for Eqs. (3.6.5)–(3.6.8) with the following initial data: $\mathbf{U}_i^k = \text{const}$ for $y < y_{i+1/2}$ and $\mathbf{U}_{i+1}^k = \text{const}$ for $y > y_{i+1/2}$. Let $\mathbf{U}_{i+1/2}$ be a solution of this problem. In the same manner, we can calculate $\mathbf{U}_{i-1/2}$ for boundary $i - 1/2$. Then, the Godunov scheme has the form

$$\frac{\mathbf{E}_i^{k+1} - \mathbf{E}_i^k}{\Delta x} + \frac{\mathbf{F}_{i+1/2} - \mathbf{F}_{i-1/2}}{\Delta y} = \mathbf{0}, \quad \mathbf{F}_{i\pm 1/2} = \mathbf{F}(\mathbf{U}_{i\pm 1/2}). \tag{3.6.20}$$

The spectral analysis (Richtmyer and Morton 1967) of the linearized equations (3.6.20) leads to the CFL stability condition

$$C = \max |\lambda_\pm| \frac{\Delta x}{\Delta y} \leq 1. \tag{3.6.21}$$

Scheme (3.6.20) is described by the system of nonlinear equations

$$(\rho u)_i^{k+1} = K_i, \quad (\rho u^2 + p)_i^{k+1} = L_i, \quad (\rho uv)_i^{k+1} = M_i, \quad [(e + p)u]_i^{k+1} = N_i. \tag{3.6.22}$$

We can rewrite Eq. (3.6.22) in terms of Roe's variables

$$\mathbf{Y} = [R, U, V, H]^{\mathrm{T}} = [\sqrt{\rho}, \sqrt{\rho}\, u, \sqrt{\rho}\, v, \sqrt{\rho}\, h]^{\mathrm{T}}, \quad h = \frac{e + p}{\rho},$$

where

$$p = \frac{\gamma - 1}{\gamma} HR - \frac{\gamma - 1}{2\gamma}(U^2 + V^2), \quad e = \frac{1}{\gamma} HR + \frac{\gamma - 1}{2\gamma}(U^2 + V^2).$$

Thus, Eq. (3.6.22) becomes

$$\begin{aligned} &(RU)_i^{k+1} = K_i, \quad (U^2 + p)_i^{k+1} = \left[U^2 + \frac{\gamma - 1}{\gamma} HR - \frac{\gamma - 1}{2\gamma}(U^2 + V^2)\right]_i^{k+1} = L_i, \\ &(UV)_i^{k+1} = M_i, \quad (HU)_i^{k+1} = N_i. \end{aligned}$$

Using the relations

$$R_i^{k+1} = \frac{K_i}{U}, \quad V_i^{k+1} = \frac{M_i}{U}, \quad H_i^{k+1} = \frac{N_i}{U}, \quad U = U_i^{k+1},$$

we obtain the following equation for U:

$$U^2 + \frac{\gamma - 1}{\gamma}\frac{K_i N_i}{U^2} - \frac{\gamma - 1}{2\gamma}\left(U^2 + \frac{M_i^2}{U^2}\right) = L_i,$$

or

$$U^4 - \frac{2\gamma L_i}{\gamma + 1}U^2 + \frac{\gamma - 1}{\gamma + 1}(2K_i N_i - M_i) = 0.$$

Hence,

$$U = U_i^{k+1} = \operatorname{sgn} U_i^k \times \sqrt{\frac{\gamma L_i}{\gamma + 1} - \sqrt{\frac{\gamma^2 L_i^2}{(\gamma + 1)^2} - \frac{\gamma - 1}{\gamma + 1}(2K_i N_i - M_i^2)}}.$$

After that ρ, u, v, and ε can be calculated.

The above scheme is of the first order of accuracy. A higher-order scheme can be constructed in a manner similar to that for the time-dependent system of equations, see Section 3.2.

The Courant–Isaacson–Rees (CIR) and Roe's schemes

Consider first the CIR and Roe versions of the Godunov method based on approximate solution of the Riemann problem. A one-dimensional finite-volume scheme of the CIR type, for example, (2.3.25)–(2.3.26) for Eq. (3.6.9) can be written as follows:

$$\frac{\mathbf{E}_i^{k+1} - \mathbf{E}_i^k}{\Delta x} + \frac{\mathbf{F}_{i+1/2} - \mathbf{F}_{i-1/2}}{\Delta y} = \mathbf{0}, \tag{3.6.23}$$

$$\begin{aligned} &\mathbf{F}_{m+1/2} = \tfrac{1}{2}(\mathbf{F}_m^k + \mathbf{F}_{m+1}^k) + \tfrac{1}{2}(E_U|A|E_U^{-1})_{m+1/2}^k(\mathbf{E}_m^k - \mathbf{E}_{m+1}^k), \quad m = i, i{-}1; \\ &\text{or} \ \ \mathbf{F}_{m+1/2} = \tfrac{1}{2}(\mathbf{F}_m^k + \mathbf{F}_{m+1}^k) + \tfrac{1}{2}(E_U|A|)_{m+1/2}^k(\mathbf{U}_m^k - \mathbf{U}_{m+1}^k), \\ &|A| = \Omega_R|\Lambda|\Omega_L. \end{aligned} \tag{3.6.24}$$

To implement (3.6.23)–(3.6.24), one must calculate $|A|$ in accordance with Eq. (3.6.24), where Ω_L, Λ, and Ω_R are defined in (3.6.15)–(3.6.16). Instead of A, one can use the matrix $\tilde{A}$ of (3.6.18). In this case, the finite-volume scheme (3.6.23) involves the formulas more simple in calculations

$$\begin{aligned} &\mathbf{F}_{m+1/2} = \tfrac{1}{2}(\mathbf{F}_m^k + \mathbf{F}_{m+1}^k) + \tfrac{1}{2}|\tilde{A}|_{m+1/2}^k(\mathbf{E}_m^k - \mathbf{E}_{m+1}^k), \quad m = i, i{-}1; \\ &\text{or} \ \ \mathbf{F}_{m+1/2} = \tfrac{1}{2}(\mathbf{F}_m^k + \mathbf{F}_{m+1}^k) + \tfrac{1}{2}(|\tilde{A}|E_U)_{m+1/2}^k(\mathbf{U}_m^k - \mathbf{U}_{m+1}^k), \\ &|\tilde{A}| = \tilde{\Omega}_R|\Lambda|\tilde{\Omega}_L. \end{aligned}$$

Roe's method is similar to the method of (3.6.23)–(3.6.24), with the matrix A (or $\widetilde{A}$) denoted $\mathcal{A}$ (or $\widetilde{\mathcal{A}}$) and defined as

$$\mathcal{A} = \mathcal{E}_U^{-1}\mathcal{F}_U, \quad \widetilde{\mathcal{A}} = \mathcal{F}_U\mathcal{E}_U^{-1}, \quad \Delta\mathbf{F} = \mathcal{F}_U\Delta\mathbf{U}, \quad \Delta\mathbf{E} = \mathcal{E}_U\Delta\mathbf{U},$$
$$\Delta\mathbf{U} = \mathbf{U}_{m+1} - \mathbf{U}_m, \quad \Delta\mathbf{E} = \mathbf{E}(\mathbf{U}_{m+1}) - \mathbf{E}(\mathbf{U}_m), \quad \Delta\mathbf{F} = \mathbf{F}(\mathbf{U}_{m+1}) - \mathbf{F}(\mathbf{U}_m),$$

where $\mathcal{E}_U$ and $\mathcal{F}_U$ are matrices that provide the exact relation between corresponding finite differences. This choice of the matrix $\mathcal{A}$ permits one to preserve exactly the discontinuity relations, see Section 3.4.4. Since $\mathbf{U}$ is a vector of Roe's variables, the matrix $\mathcal{A}$ coincides with (3.6.14), where Roe's averages $\hat{u}$, $\hat{v}$, and $\hat{c}$, see Eqs. (3.4.13) and (3.4.16), must be substituted for u, v, and c, respectively (Semenov, unpublished).

To use the Godunov method based on the exact solution, one must use the exact solution of the corresponding Riemann problem. Below we describe the basic stages of the construction of this solution for stationary gas dynamic equations.

3.6.3 Exact solution of the Riemann problem.

Let us construct an exact solution of the Riemann problem for gas dynamic equations for ideal perfect gas. In constructing the solution, we follow the logical scheme of the stationary gas dynamic presentation in Godunov et al. (1979).

Elementary solution 1: Shock wave. The first elementary solution of the gas dynamic equations is a stationary discontinuity (shock wave or tangential discontinuity). For obtaining the discontinuity relations let us integrate Eq. (3.6.9) over x and y and consider its integral form

$$\oint_L (\mathbf{E}\,dy - \mathbf{F}\,dx) = \mathbf{0}, \tag{3.6.25}$$

where L is the boundary of a region in the (x, y) plane. Let us seek a discontinuous solution of Eqs. (3.6.5)–(3.6.8) as a rectilinear wave of the form $f(x, y) = f(\eta) \equiv f(y - \theta x)$ where $\theta = \text{const}$ is the wave slope. Consider Eq. (3.6.25) in orthogonal coordinates (η, τ) associated with discontinuity, where $\eta = y - \theta x$ and $\tau = \theta y + x$. The coordinate η is normal and coordinate τ is tangential with respect to the discontinuity. Using transformation

$$y = \frac{\eta + \theta\tau}{1 + \theta^2}, \quad x = \frac{-\theta\eta + \tau}{1 + \theta^2},$$

we can rewrite Eq. (3.6.25) as

$$\frac{1}{1 + \theta^2}\oint_L [(\theta\mathbf{E} - \mathbf{F})\,d\tau + (\mathbf{E} + \theta\mathbf{F})\,d\eta] = \mathbf{0}. \tag{3.6.26}$$

Let us integrate (3.6.26) for the rectangular region $\tau_0 - \delta\tau \le \tau \le \tau_0 + \delta\tau$ and $\eta_0 - \delta\eta \le \eta \le \eta_0 + \delta\eta$, where $\eta = \eta_0$ corresponds to the discontinuity. We can find

$$(\theta\mathbf{E} - \mathbf{F})_1 2\delta\tau - (\theta\mathbf{E} - \mathbf{F})_2 2\delta\tau = \mathbf{0}$$
$$\Longrightarrow \quad \theta\{\mathbf{E}\} - \{\mathbf{F}\} = \mathbf{0}, \tag{3.6.27}$$

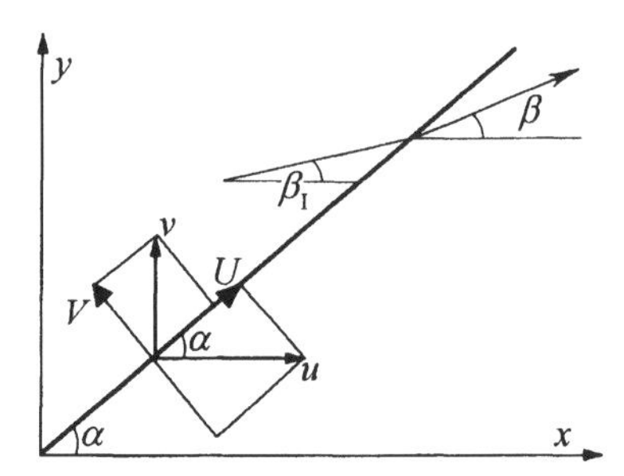

Figure 3.27 Coordinate system associated with the discontinuity.

where $\{q\} \equiv q_1 - q_2$, and indices 1 and 2 denote the variables on the left- and right-hand side of the discontinuity.

Consider the Riemann problem with the following initial data: $\mathbf{U}_1 = [\rho_1, u_1, v_1, \varepsilon_1, p_1]^T$, for $y > 0$, and $\mathbf{U}_2 = [\rho_2, u_2, v_2, \varepsilon_2, p_2]^T$ for $y < 0$. Both $\mathbf{U}_1$ and $\mathbf{U}_2$ satisfy Eqs. (3.6.5)–(3.6.8). Assume that this initial discontinuity has the slope θ and is a solution of the Riemann problem. Then we can rewrite Eq. (3.6.27) in the form

$$\theta\{\rho u\} - \{\rho v\} = 0, \tag{3.6.28}$$
$$\theta\{\rho u^2 + p\} - \{\rho uv\} = 0, \tag{3.6.29}$$
$$\theta\{\rho uv\} - \{\rho v^2 + p\} = 0, \tag{3.6.30}$$
$$\theta\{(e + p)u\} - \{(e + p)v\} = 0. \tag{3.6.31}$$

These equations relate θ with $\mathbf{U}_1$ and $\mathbf{U}_2$. If these conditions are satisfied, then the discontinuity is a solution of Eqs. (3.6.5)–(3.6.8). Let us introduce new variables V and U associated with the discontinuity:

$$V = v\cos\alpha - u\sin\alpha, \tag{3.6.32}$$
$$U = v\sin\alpha + u\cos\alpha, \tag{3.6.33}$$

where $\tan\alpha = \theta$, see Fig. 3.27. Since $\{\theta\} = 0$, one can obtain from Eqs. (3.6.28)–(3.6.31) the more compact relations

$$\{\rho V\} = 0, \tag{3.6.34}$$
$$\{\rho V^2 + p\} = 0, \tag{3.6.35}$$
$$\{\tfrac{1}{2}\rho V^3 + pV + \rho\varepsilon V\} = 0, \tag{3.6.36}$$
$$\{\rho UV\} = 0 \quad \Longrightarrow \quad \{U\} = 0 \;\text{ if }\; \rho V = \rho_1 V_1 = \rho_2 V_2 \neq 0. \tag{3.6.37}$$

The case $\rho V = 0$ corresponding a tangential discontinuity will be considered separately. Note that the form of these relations is equivalent to that of the relations for nonstationary equations, see Eqs. (3.3.9)–(3.3.12). Equations (3.6.34)–(3.6.36) permit us to find compact formulas for ρ_2, V_2, U_2, and ε_2 as functions of p_2, p_1, ρ_1, and θ, see Godunov et al. (1979).

From (3.6.34) we have

$$V_2 = \frac{\rho_1 V_1}{\rho_2}.$$

Using this relation and Eq. (3.6.35), we obtain

$$V_1 = \pm\sqrt{\frac{p_2 - p_1}{\rho_2 - \rho_1}\frac{\rho_2}{\rho_1}}.$$

Substituting V_1 and V_2 into Eq. (3.6.36) yields

$$\varepsilon_2 - \varepsilon_1 = \frac{p_2 + p_1}{2}\left(\frac{1}{\rho_1} - \frac{1}{\rho_2}\right). \tag{3.6.38}$$

Using the EOS of (3.6.4) to eliminate ε_1 and ε_2 from Eq. (3.6.38), we arrive at the Rankine–Hugoniot relation

$$\rho_2 = \rho_1\frac{(\gamma+1)p_2 + (\gamma-1)p_1}{(\gamma-1)p_2 + (\gamma+1)p_1}, \quad \text{or} \quad \frac{\rho_2 - \rho_1}{\rho_1} = \frac{2(p_2 - p_1)}{(\gamma-1)p_2 + (\gamma+1)p_1}. \tag{3.6.39}$$

Thus, we obtain V_1 as a function of p_2, p_1, and ρ_1,

$$V_1^2 = \frac{(\gamma+1)p_2 + (\gamma-1)p_1}{2\rho_1}.$$

Let $m = \rho V$ denote the mass flux across the discontinuity. From Eq. (3.6.34)–(3.6.35) it follows that $\{m\} = 0$ and $\{mV + p\} = 0$. Then,

$$m_1 = \rho_1 V_1 = \pm\sqrt{\frac{\rho_1}{2}\left[(\gamma+1)p_2 + (\gamma-1)p_1\right]}. \tag{3.6.40}$$

Since only compression shocks can exist, it is clear that V_1 must be directed to the high-pressure side of the shock. This allows one to determine the sign in (3.6.40).

Upper shock wave

We say that the shock wave is *upper* if the gas flow through it is directed along the y-direction from top to bottom (Fig. 3.28a), i.e., $m = \rho V < 0$. Denote the gas parameters above the shock wave by ρ_{I}, u_{I}, v_{I}, ε_{I}, and p_{I} and those below by ρ, u, v, ε, and p, see Fig. 3.28a. Using the relations $\{U\} = 0$ and $\{m\} = 0$, we obtain the following equations for u and ρ, where we assume $p \geq p_{\mathrm{I}}$:

$$v_{\mathrm{I}}\sin\alpha + u_{\mathrm{I}}\cos\alpha = v\sin\alpha + u\cos\alpha, \tag{3.6.41}$$

$$(v_{\mathrm{I}}\cos\alpha - u_{\mathrm{I}}\sin\alpha)\rho_{\mathrm{I}} = (v\cos\alpha - u\sin\alpha)\rho. \tag{3.6.42}$$

Introducing the notation

$$\tan\beta_{\mathrm{I}} = \frac{v_{\mathrm{I}}}{u_{\mathrm{I}}}, \quad \tan\beta = \frac{v}{u},$$

see Fig. 3.27, we rewrite Eqs. (3.6.41)–(3.6.42) as

$$(\tan\beta_{\mathrm{I}}\tan\alpha + 1)u_{\mathrm{I}} = (\tan\beta\tan\alpha + 1)u, \tag{3.6.43}$$

$$(\tan\beta_{\mathrm{I}} - \tan\alpha)\rho_{\mathrm{I}}u_{\mathrm{I}} = (\tan\beta - \tan\alpha)\rho u. \tag{3.6.44}$$

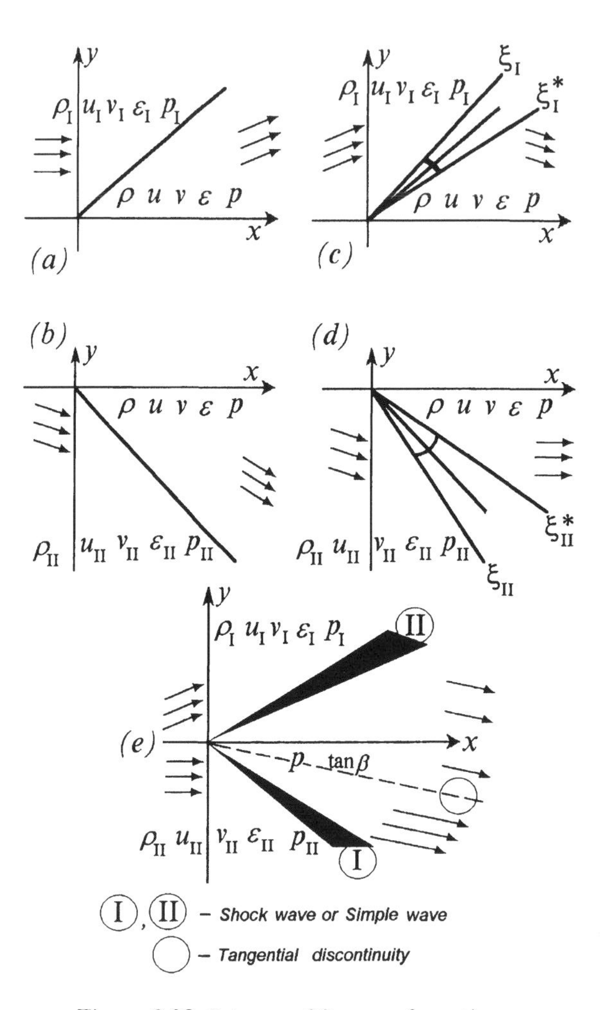

Figure 3.28 Scheme of flow configurations.

Dividing Eq. (3.6.44) by Eq. (3.6.43) and using the identity

$$\tan(\mu \pm \nu) = \frac{\tan\mu \pm \tan\nu}{1 \mp \tan\mu \tan\nu},$$

we arrive at the relation

$$\tan(\alpha - \beta) = \frac{\rho_{\rm I}}{\rho} \tan(\alpha - \beta_{\rm I}), \tag{3.6.45}$$

where $\alpha > \beta$, since $V < 0$, and $\beta > \beta_{\rm I}$, since $p > p_{\rm I}$ and $\rho > \rho_{\rm I}$; see Eq. (3.6.39). Let us transform relation (3.6.32) for $V = V_{\rm I}$ in the following way:

$$V_{\rm I} = v_{\rm I} \cos\alpha - u_{\rm I} \sin\alpha = -q_{\rm I} \sin(\alpha - \beta_{\rm I}) < 0, \quad q_{\rm I} = \sqrt{u_{\rm I}^2 + v_{\rm I}^2}.$$

Thus,

$$\sin(\alpha - \beta_{\rm I}) = \frac{|V_{\rm I}|}{q_{\rm I}} = \frac{|m_{\rm I}|}{\rho_{\rm I} q_{\rm I}} = \frac{1}{q_{\rm I}} \sqrt{\frac{(\gamma+1)p + (\gamma-1)p_{\rm I}}{2\rho_{\rm I}}},$$

or

$$\tan(\alpha - \beta_{\rm I}) = \left(\frac{1}{\sin^2(\alpha - \beta_{\rm I})} - 1 \right)^{-1/2} = \left(\frac{2\rho_{\rm I} q_{\rm I}^2}{(\gamma+1)p + (\gamma-1)p_{\rm I}} - 1 \right)^{-1/2}. \tag{3.6.46}$$

Using Eqs. (3.6.45)–(3.6.46), one can find $\tan(\beta - \beta_{\rm I})$ as a function of $\rho - \rho_{\rm I}$:

$$\tan(\beta - \beta_{\rm I}) = \frac{\tan(\alpha - \beta_{\rm I}) - \tan(\alpha - \beta)}{1 + \tan(\alpha - \beta_{\rm I}) \tan(\alpha - \beta)} = \frac{(\rho - \rho_{\rm I}) \tan(\alpha - \beta_{\rm I})}{\rho + \rho_{\rm I} \tan^2(\alpha - \beta_{\rm I})}.$$

Using (3.6.39) and (3.6.46) yields

$$\tan(\beta - \beta_{\rm I}) = \frac{\tan\beta - \tan\beta_{\rm I}}{1 + \tan\beta \tan\beta_{\rm I}} = \frac{p - p_{\rm I}}{p_{\rm I} - p + \rho_{\rm I} q_{\rm I}^2} \sqrt{\frac{2\gamma p_{\rm I} {\rm M}_{\rm I}^2}{(\gamma+1)p + (\gamma-1)p_{\rm I}} - 1}, \tag{3.6.47}$$

$${\rm M}_{\rm I} = \frac{q_{\rm I}}{c_{\rm I}}, \quad q_{\rm I} = \sqrt{u_{\rm I}^2 + v_{\rm I}^2}, \quad c_{\rm I} = \sqrt{\frac{\gamma p_{\rm I}}{\rho_{\rm I}}}.$$

Hence,

$$\tan\beta = L_{\rm I} \tan\beta_{\rm I} + (p - p_{\rm I}) N_{\rm I}, \tag{3.6.48}$$

$$L_{\rm I} = \frac{1}{1 + (p_{\rm I} - p) J_{\rm I} \tan\beta_{\rm I}}, \quad N_{\rm I} = L_{\rm I} J_{\rm I},$$

$$J_{\rm I} = \left(p_{\rm I} - p + \rho_{\rm I} q_{\rm I}^2\right)^{-1} \left[\frac{2\gamma p_{\rm I} {\rm M}_{\rm I}^2}{(\gamma+1)p + (\gamma-1)p_{\rm I}} - 1 \right]^{1/2}$$

To determine $\tan\alpha$, one can use the relation

$$\tan\alpha = \frac{\tan(\alpha - \beta_{\rm I}) + \tan\beta_{\rm I}}{1 - \tan(\alpha - \beta_{\rm I}) \tan\beta_{\rm I}}, \tag{3.6.49}$$

which depends on $\tan(\alpha - \beta_{\rm I})$, see Eq. (3.6.46), and $\tan\beta_{\rm I}$. Using relation (3.6.41) and $v = u \tan\beta$, one can determine u and v.

Lower shock wave

We say that the shock wave is *lower* if the gas flow through it is directed from bottom to top and $m = \rho V > 0$. Denote the gas parameters below the shock wave by $\rho_{\rm II}$, $u_{\rm II}$, $v_{\rm II}$, $\varepsilon_{\rm II}$, and $p_{\rm II}$ and those above the shock wave by ρ, u, v, ε, and p, see Fig.3.28b. Using the relations $\{U\} = 0$ and $\{m\} = 0$, we obtain, just as in Eqs. (3.6.41)–(3.6.47), the following equation for u and ρ, where we assume $p \geq p_{\rm II}$:

$$\tan(\beta_{\rm II} - \beta) = \frac{p - p_{\rm II}}{p_{\rm II} - p + \rho_{\rm II} q_{\rm II}^2} \sqrt{\frac{2\gamma p_{\rm II} {\rm M}_{\rm II}^2}{(\gamma+1)p + (\gamma-1)p_{\rm II}} - 1}, \tag{3.6.50}$$

$${\rm M}_{\rm II} = \frac{q_{\rm II}}{c_{\rm II}}, \quad q_{\rm II} = \sqrt{u_{\rm II}^2 + v_{\rm II}^2}, \quad c_{\rm II} = \sqrt{\frac{\gamma p_{\rm II}}{\rho_{\rm II}}};$$

hence,

$$\tan\beta = L_{\rm II} \tan\beta_{\rm II} - (p - p_{\rm II}) N_{\rm II}, \tag{3.6.51}$$

$$L_{\rm II} = \frac{1}{1 + (p - p_{\rm II}) J_{\rm II} \tan\beta_{\rm II}}, \qquad N_{\rm II} = L_{\rm II} J_{\rm II},$$

$$J_{\rm II} = \left(p_{\rm II} - p + \rho_{\rm II} q_{\rm II}^2\right)^{-1} \left[\frac{2\gamma p_{\rm II} {\rm M}_{\rm II}^2}{(\gamma+1)p + (\gamma-1)p_{\rm II}} - 1\right]^{1/2}.$$

To determine $\tan\alpha$, one can use the relation

$$\tan\alpha = \frac{\tan\beta_{\rm II} - \tan(\beta_{\rm II} - \alpha)}{1 + \tan\beta_{\rm II} \tan(\beta_{\rm II} - \alpha)}, \tag{3.6.52}$$

which depends on $\tan\beta_{\rm II}$ and $\tan(\alpha - \beta_{\rm II})$, where

$$\tan(\beta_{\rm II} - \alpha) = \left(\frac{2\rho_{\rm II} q_{\rm II}^2}{(\gamma+1)p + (\gamma-1)p_{\rm II}} - 1\right)^{-1/2},$$

where $\beta_{\rm II} > \alpha$, since $V > 0$. Further, by using relation (3.6.41) and $v = u\tan\beta$, one can determine u and v.

A stationary shock wave is a self-similar solution in $\eta = y - \theta x$, $\theta = \text{const}$. It is also self-similar in $\xi = y/x$. In fact, a discontinuity in the (x, y) coordinates is a straight line defined by the equation $\xi = \theta = \text{const}$.

Elementary solution 2: Tangential discontinuity. This elementary solution of gas dynamic equations is a special case of the discontinuity (3.6.34)–(3.6.37) with zero mass flow across the discontinuity: $m = \rho V = \rho_1 V_1 = \rho_2 V_2 = 0$. Thus,

$$V = V_1 = V_2 = 0, \quad \{p\} = 0,$$

see Eq. (3.6.35).

Moreover at the top (bottom) of the discontinuity the parameters of the gas flow are constant. The top variables are $p = p_1$, $u = u_1$, and $v = v_1$ and the bottom values are to $p = p_2$, $u = u_2$, and $v = v_2$. Then from the relations $\{p\} = 0$ and $V = 0$ one obtains

$$p_1 = p_2 = p, \quad \tan\alpha = \tan\beta_1 = \frac{v_1}{u_1} = \tan\beta_2 = \frac{v_2}{u_2}. \tag{3.6.53}$$

Elementary solution 3: Simple wave. Another solution of the stationary gas dynamic equations is a simple wave. This solution is continuous. Since we seek a continuous solution, we can consider the gas dynamic system (3.6.5)–(3.6.8) in the nonconservative form

$$u\rho_x + v\rho_y + (u_x + v_y)\rho = 0, \tag{3.6.54}$$

$$uu_x + vu_y + \frac{p_x}{\rho} = 0, \quad uv_x + vv_y + \frac{p_y}{\rho} = 0, \tag{3.6.55}$$

$$up_x + vp_y + (u_x + v_y)\gamma p = 0. \tag{3.6.56}$$

Let us seek a continuous solution in the self-similar form $f(x,y) = f(\xi) \equiv f(y/x)$. Substituting $\rho = \rho(\xi)$, $u = u(\xi)$, $v = v(\xi)$, and $p = p(\xi)$ into Eqs. (3.6.54)–(3.6.56) yields

$$(-u\xi + v)\rho_\xi - \xi\rho u_\xi + \rho v_\xi = 0, \tag{3.6.57}$$

$$(-u\xi + v)u_\xi - \xi\frac{p_\xi}{\rho} = 0, \quad (-u\xi + v)v_\xi + \frac{p_\xi}{\rho} = 0, \tag{3.6.58}$$

$$-\gamma p\xi u_\xi + \gamma p v_\xi + (-u\xi + v)p_\xi = 0. \tag{3.6.59}$$

System (3.6.57)–(3.6.59) can have a nonzero solution for ρ_ξ, u_ξ, v_ξ, and p_ξ only if the determinant of the 4×4 coefficient matrix system is equal to zero. Thus, we obtain the following three different conditions for a nontrivial solution to exist:

$$-u\xi + v = 0, \tag{3.6.60}$$

$$-u\xi + v = +c\sqrt{1+\xi^2}, \tag{3.6.61}$$

$$-u\xi + v = -c\sqrt{1+\xi^2}, \quad c = \sqrt{\frac{\gamma p}{\rho}}. \tag{3.6.62}$$

Using the notation $\xi = \tan\alpha$ and $\tan\beta = v/u$, see Fig. 3.27, one can transform the Eqs. (3.6.61)–(3.6.62) into the relation

$$\sin(\alpha - \beta) = \mp\frac{c\cos\beta}{u} = \mp\frac{c}{q} = \mp\frac{1}{\mathrm{M}}, \quad \mathrm{M} = \frac{q}{c}, \quad q = \sqrt{u^2 + v^2}. \tag{3.6.63}$$

Consider Eqs. (3.6.57)–(3.6.59) for the three cases of (3.6.60)–(3.6.62); denote these cases by I, II, and III, respectively.

Case I:

$$-u\xi + v = 0, \quad p_\xi = 0, \quad -\xi u_\xi + v_\xi = 0.$$

Here, the exact, however, trivial solution is given by $p = \text{const}$, $u = 0$, and $v = 0$.

Case II:

$$-u\xi + v = c\sqrt{1+\xi^2},$$
$$(-u\xi + v)u_\xi - \xi\frac{p_\xi}{\rho} = 0, \quad (-u\xi + v)v_\xi + \frac{p_\xi}{\rho} = 0,$$
$$(-u\xi + v)\rho_\xi - \xi\rho u_\xi + \rho v_\xi = 0,$$

and, hence,

$$-u\xi + v = c\sqrt{1+\xi^2}, \tag{3.6.64}$$
$$u_\xi = \frac{\xi p_\xi}{\rho c\sqrt{1+\xi^2}}, \quad v_\xi = -\frac{p_\xi}{\rho c\sqrt{1+\xi^2}}, \tag{3.6.65}$$
$$\gamma\frac{\rho_\xi}{\rho} = \frac{p_\xi}{p} \quad \Longrightarrow \quad S = \frac{p}{\rho^\gamma} = \text{const}, \tag{3.6.66}$$

where S is the entropy function. Then, using formulas (3.6.65) and relation (3.6.63), we find that

$$\frac{\partial \tan\beta}{\partial \xi} = \frac{\partial}{\partial \xi}\left(\frac{v}{u}\right) = \frac{uv_\xi - vu_\xi}{u^2} = -\frac{(u+v\xi)}{\rho c u^2\sqrt{1+\xi^2}}\frac{\partial p}{\partial \xi} = -\frac{\cos(\alpha-\beta)}{\rho c u\cos\beta}\frac{\partial p}{\partial \xi}$$
$$= -\frac{\cos(\alpha-\beta)(1+\tan^2\beta)\cos\beta}{\rho c u}\frac{\partial p}{\partial \xi} = -\frac{(1+\tan^2\beta)\sqrt{\mathrm{M}^2-1}}{\rho q^2}\frac{\partial p}{\partial \xi} = -R(\beta,p)\frac{\partial p}{\partial \xi}$$
$$\Longrightarrow \quad \frac{\partial \tan\beta}{\partial \xi} = -R(\beta,p)\frac{\partial p}{\partial \xi}, \quad R(\beta,p) = (1+\tan^2\beta)\frac{\sqrt{\mathrm{M}^2-1}}{\rho q^2}. \tag{3.6.67}$$

Using formulas (3.6.65) and taking into account relations (3.6.64) and (3.6.66), we can verify that

$$\frac{\partial h}{\partial \xi} = \frac{\partial}{\partial \xi}\left(\frac{e+p}{\rho}\right) = 0.$$

Thus, in the simple wave the following relation holds:

$$h = \frac{e+p}{\rho} = \text{const}. \tag{3.6.68}$$

Case III:

$$-u\xi + v = -c\sqrt{1+\xi^2},$$
$$(-u\xi + v)u_\xi - \xi\frac{p_\xi}{\rho} = 0, \quad (-u\xi + v)v_\xi + \frac{p_\xi}{\rho} = 0,$$
$$(-u\xi + v)\rho_\xi - \xi\rho u_\xi + \rho v_\xi = 0.$$

Thus, by analogy with case II, we have

$$\frac{\partial \tan\beta}{\partial \xi} = \frac{(1+\tan^2\beta)\sqrt{\mathrm{M}^2-1}}{\rho q^2}\frac{\partial p}{\partial \xi} = R(\beta,p)\frac{\partial p}{\partial \xi}, \tag{3.6.69}$$

It is not difficult to show that, in this case, relations (3.6.66) and (3.6.68) hold as well.

Upper simple wave

We say that the simple wave is *upper* if the gas flow through it is directed along the y-direction from top to bottom, i.e., $m = \rho V < 0$. In this case, we deal with the relations

$$\begin{aligned}
-u\xi + v &= -c\sqrt{1+\xi^2},\\
\frac{\partial \tan\beta}{\partial \xi} &= R(\beta,p)\frac{\partial p}{\partial \xi},\\
S = \frac{p}{\rho^\gamma} &= \text{const}, \quad h = \frac{e+p}{\rho} = \text{const}.
\end{aligned} \tag{3.6.70}$$

We have $m = \rho V = (v - u\xi)\rho\cos\alpha = -\rho c < 0$. Denote the flow parameters above the simple wave by ρ_{I}, u_{I}, v_{I}, ε_{I} and those below by ρ, u, v, and ε, see Fig. 3.28c. Integrating Eq. (3.6.70) across the simple wave with respect to ξ yields the following relation for $\tan\beta$ and p:

$$\tan\beta = \tan\beta_{\mathrm{I}} + (p - p_{\mathrm{I}})r_{\mathrm{I}}, \tag{3.6.71}$$

$$\begin{aligned}
&r_{\mathrm{I}} = r_{\mathrm{I}}(\beta,p) = \frac{1}{p-p_{\mathrm{I}}}\int_{\xi_{\mathrm{I}}}^{\xi_{\mathrm{I}}^*} R(\beta,p)\frac{\partial p}{\partial \xi}\,d\xi, \quad R(\beta,p) = (1+\tan^2\beta)\frac{\sqrt{\mathrm{M}^2-1}}{\rho q^2},\\
&r_{\mathrm{I}}(\beta,p) \simeq \tfrac{1}{2}[R(\beta_{\mathrm{I}},p_{\mathrm{I}}) + R(\beta,p)],
\end{aligned} \tag{3.6.72}$$

where the sign $\simeq$ stands for the approximation of the second order of accuracy with respect to ξ. Here we assume that the simple wave is bounded in the (x, y) plane by the rays $\xi = \xi_{\mathrm{I}}$ and $\xi = \xi_{\mathrm{I}}^*$, see Fig. 3.28c. Note that other approximating relations can also be used instead of (3.6.72). To increase the accuracy of the approximation for strong simple waves, a finer mesh in the y-direction can be used.

Lower simple wave

We say that the simple wave is *lower* if the gas flow through it is directed from bottom to top, i.e., $m = \rho V > 0$. Here we deal with the relations

$$\begin{aligned}
-u\xi + v &= c\sqrt{1+\xi^2},\\
\frac{\partial \tan\beta}{\partial \xi} &= -R(\beta,p)\frac{\partial p}{\partial \xi},\\
S = \frac{p}{\rho^\gamma} &= \text{const}, \quad h = \frac{e+p}{\rho} = \text{const}.
\end{aligned} \tag{3.6.73}$$

We have $m = \rho V = (v - u\xi)\rho\cos\alpha = \rho c > 0$. Denote the flow parameters below the simple wave by ρ_{II}, u_{II}, v_{II}, and $\varepsilon_{\mathrm{II}}$ and those above by ρ, u, v, and ε, see Fig. 3.28d. Integrating Eq. (3.6.73) across the simple wave with respect to ξ yields the following relation for $\tan\beta$ and p:

$$\tan\beta = \tan\beta_{\mathrm{II}} - (p - p_{\mathrm{II}})r_{\mathrm{II}}, \tag{3.6.74}$$

$$\begin{aligned}
&r_{\mathrm{II}} = r_{\mathrm{II}}(\beta,p) = \frac{1}{p-p_{\mathrm{II}}}\int_{\xi_{\mathrm{II}}}^{\xi_{\mathrm{II}}^*} R(\beta,p)\frac{\partial p}{\partial \xi}\,d\xi, \quad R(\beta,p) = (1+\tan^2\beta)\frac{\sqrt{\mathrm{M}^2-1}}{\rho q^2},\\
&r_{\mathrm{II}}(\beta,p) \simeq \tfrac{1}{2}[R(\beta_{\mathrm{II}},p_{\mathrm{II}}) + R(\beta,p)].
\end{aligned} \tag{3.6.75}$$

Here we assume that the simple wave is bounded in the (x, y) plane by the rays $\xi = \xi_{\text{II}}^*$ and $\xi = \xi_{\text{II}}$, see Fig. 3.28d.

3.6.4 General exact solution.

The general solution of the Riemann problem consists of two shock and/or simple waves with a tangential discontinuity that are separated by flow regions with constant parameters, see Fig. 3.28e. Special cases of this general solution are a single shock wave or a simple wave, see Fig. 3.28a–d.

It is noteworthy that p and $\tan\beta$ in the central zone for the case shown in Fig. 3.28e can be determined from two coupled equations. Considering Eq. (3.6.51) (resp., Eq. (3.6.74)), Eq. (3.6.48) (resp., Eq. (3.6.71)), and the relation at the tangential discontinuity (3.6.53), we arrive at the following system of equations for p and $\tan\beta$:

$$\tan\beta = \psi_{\text{II}}(p)\tan\beta_{\text{II}} - (p - p_{\text{II}})\varphi_{\text{II}}(\beta, p), \tag{3.6.76}$$

$$\tan\beta = \psi_{\text{I}}(p)\tan\beta_{\text{I}} + (p - p_{\text{I}})\varphi_{\text{I}}(\beta, p), \tag{3.6.77}$$

where

$$\psi_n(p) = \begin{cases} \dfrac{1}{1 - (p - p_n)J_n \tan\beta_n} & \text{if } n = \text{I and } p \geq p_{\text{I}}, \\ \dfrac{1}{1 + (p - p_n)J_n \tan\beta_n} & \text{if } n = \text{II and } p \geq p_{\text{II}}, \\ 1 & \text{if } p < p_n, \end{cases}$$

$$\varphi_n(\beta, p) = \begin{cases} \psi_n J_n & \text{if } p \geq p_n, \\ \frac{1}{2}[R(\beta_n, p_n) + R(\beta, p)] & \text{if } p < p_n, \end{cases}$$

$n = \text{I}$ or II, and

$$J_n = \left(p_n - p + \rho_n q_n^2\right)^{-1}\left[\frac{2\gamma p_n \text{M}_n^2}{(\gamma+1)p + (\gamma-1)p_n} - 1\right]^{1/2},$$

$$R(\beta, p) = (1 + \tan^2\beta)\frac{\sqrt{\text{M}^2 - 1}}{\rho q^2}, \quad \text{M} = \frac{q}{c}, \quad q = \sqrt{u^2 + v^2}, \quad c = \sqrt{\frac{\gamma p}{\rho}}.$$

One can solve Eqs. (3.6.76)–(3.6.77) by an iteration method (Godunov et al. 1979). In particular, the first approximation can be written as

$$p^{(1)} = \frac{\varphi_{\text{I}} p_{\text{I}} + \varphi_{\text{II}} p_{\text{II}} + \psi_{\text{II}}\tan\beta_{\text{II}} - \psi_{\text{I}}\tan\beta_{\text{I}}}{\varphi_{\text{I}} + \varphi_{\text{II}}},$$

$$\tan\beta^{(1)} = \psi_{\text{I}}\tan\beta_{\text{I}} + (p^{(1)} - p_{\text{I}})\varphi_{\text{I}} = \psi_{\text{II}}\tan\beta_{\text{II}} - (p^{(1)} - p_{\text{II}})\varphi_{\text{II}},$$

$$\psi_n = \psi_n(p_n), \quad \varphi_n = \varphi_n(\beta_n, p_n), \quad n = \text{I, II}.$$

Then, for the kth iteration we have ($k = 1, 2, \ldots$),

$$p^{(k+1)} = \left[\frac{\varphi_{\text{I}} p_{\text{I}} + \varphi_{\text{II}} p_{\text{II}} + \psi_{\text{II}}\tan\beta_{\text{II}} - \psi_{\text{I}}\tan\beta_{\text{I}}}{\varphi_{\text{I}} + \varphi_{\text{II}}}\right]^{(k)},$$

$$\tan\beta^{(k+1)} = \psi_{\text{I}}^{(k)}\tan\beta_{\text{I}} + (p^{(k+1)} - p_{\text{I}})\varphi_{\text{I}}^{(k)} = \psi_{\text{II}}^{(k)}\tan\beta_{\text{II}} - (p^{(k+1)} - p_{\text{II}})\varphi_{\text{II}}^{(k)},$$

$$\psi_n^{(k)} = \psi_n(p^{(k)}), \quad \varphi_n^{(k)} = \varphi_n(\beta^{(k)}, p^{(k)}), \quad n = \text{I, II}.$$

To calculate ρ, u, v, and ε in the rarefaction fan, one should use the formulas

$$h = \frac{e+p}{\rho} = \text{const}, \quad S = \frac{p}{\rho^\gamma} = \text{const}, \quad v = u \tan\beta, \quad u = \frac{q}{\sqrt{1+\tan^2\beta}}. \tag{3.6.78}$$

Just as in the case of the time-dependent system of equations, see Section 3.3, after solving Eqs. (3.6.76)–(3.6.77), one must analyze the flow configuration and determine the solution $\rho(\xi)$, $u(\xi)$, $v(\xi)$, and $p(\xi)$ for the desired ray ξ, in particular, for the stationary boundary $\xi = 0$. To determine the flow parameters, one can use relations (3.6.78) for the configurations with simple waves, and relations (3.6.49) and/or (3.6.52) for shock waves. In the three-dimensional case the tangential component of the velocity, w, in the exact solution is determined as follows: we put $w(\xi) = w_{\mathrm{I}}$ if $\xi > \tan\beta$, otherwise $w(\xi) = w_{\mathrm{II}}$.

It should be noted that, to accelerate the convergence of iterations, one can perform a change of variables in Eqs. (3.6.76)–(3.6.77), or in Eqs. (3.6.47), (3.6.50), (3.6.70), and (3.6.73), respectively. For example, the change of variables

$$\mathcal{P} = p^{\frac{\gamma-1}{2\gamma}}$$

can be helpful, see Godunov et al. (1979). One can organize another iteration process as will be described for stationary shallow water equations in Section 4.6. In that case φ_n is a continuous function for $\beta = \beta_n$ and $p = p_n$.

3.7 Solar wind – interstellar medium interaction

In this section we present the results of numerical solution of the gas dynamic problem modelling the interaction between the solar wind and the local interstellar medium. The solution is obtained on the basis of the MUSCL TVD numerical scheme applying Roe's approximate Riemann problem solver to determine fluxes at the cell interfaces. This problem is characterized by the presence of several shocks and tangential discontinuities, and its astrophysical importance represents a challenging task for computational fluid dynamics. We shall also use this problem later to illustrate the application of high-resolution numerical methods to MHD equations. Describing the numerical method, we shall present the details of application of the far-field boundary conditions outlined in Section 2.8.

3.7.1 Physical formulation of the problem. The interpretation of measurements performed by space vehicles requires the development of theoretical models for the stellar wind, including the solar wind (SW), and the local interstellar medium (LISM) interaction. Of great interest are nonstationary problems connected with the variable stellar activity. Application of the continuum equations to the SW–LISM interaction is systematically discussed by Baranov and Krasnobaev (1977). The first qualitative model for the interaction of the solar wind and the local interstellar medium was proposed by Parker (1961). He assumed the interstellar medium to be a subsonic stream with the Mach number $M_\infty \ll 1$. Generally speaking the above assumption is not correct, since the velocity of the interstellar medium is $V_\infty \sim 20\ \mathrm{km\,s^{-1}}$ and the scattering experiments for the solar radiation show that the temperature T_∞ of charged particles constituting the LISM is about 10^4 K. Taking into account the fact that the number density of charged particles is usually considered as $n_\infty \sim 0.1\ \mathrm{cm}^{-3}$, the LISM flow can be supposed more likely supersonic than subsonic. The supersonic model of the interaction was first proposed by Baranov, Krasnobaev, and Kulikovskii (1971). The important feature of that approach lies in application of the continuum (Euler gas dynamic) equations only to charged particles of the both counteracting winds. Although the presence of turbulent pulsations in plasma is supposed to be insignificant for the mean flow structure, their influence is realized by a remarkable change of the transport coefficients due to the possibility of scattering of charged particles on electromagnetic plasma fluctuations. This results in a substantial decrease of their mean free path compared with that calculated on the basis of the Coulomb collisions. The solar wind plasma consists mainly of electrons and protons with the number density $n_e \sim 10\ \mathrm{cm}^{-3}$ and velocity $V_e \sim 400 - 500\ \mathrm{km\,s}^{-1}$ and is also considered supersonic. The subscript e corresponds here to the quantities measured at $1\ \mathrm{AU} = 1.5 \times 10^{11}$ m, that is, at the Earth distance from the Sun. Thus, we can consider this problem, from the gas dynamic viewpoint, as interaction of the supersonic, spherically symmetric (or sometimes asymmetric, see Pauls and Zank 1996) source flow of the SW with the uniform supersonic LISM flow. This assumption gave rise to a so-called two-shock model. The flow pattern in this case is essentially a combination of a supersonic jet and a blunt-body flow, which are perfectly well studied and described in classical gas dynamics. Different flow regimes and shock-wave flow patterns for the SW–LISM interaction were discussed by Wallis and Dryer (1976).

Baranov, Lebedev, and Ruderman (1979) and Zaitsev and Radvogin (1990) considered the axisymmetric problem of the interaction on the basis of the shock-fitting approach, although owing to limitations of the applied numerical method they calculated only the upwind part of the flow. Sawada et al. (1986) and Matsuda et al. (1989) analyzed the problem of the stellar wind interaction with the interstellar medium in the closed region surrounding the star. Their numerical results confirmed the scheme of Wallis and Dryer (1976), though a bullet shape of the internal shock was obtained for a broader range of parameters. The general schematic picture is shown in Fig. 3.29. Here TS is the inner shock terminating the solar wind within the tangential discontinuity called heliopause (HP), BS is the bow shock, or the outer shock. TS at a certain point may turn to form a Mach disk (MD). At this triple point a reflected shock (RS) and a slip line (SL) originate. The shape of the termination shock is stipulated by its Mach-type reflection from the z-axis.

We believe it is worth noting at this point that some important effects governing the SW–LISM interaction are omitted in our presentation, since they are not directly related to the subject of this monograph. The main effect omitted is the resonance charge exchange between the neutral and charged particles constituting the wind (see Blum and Fahr 1969; Wallis 1971, 1975; Baranov et al. 1979). The neutral particle motion cannot be described in the framework of the continuum approach, as their mean free-path is much larger than the characteristic length of the problem. For this reason, Baranov, Ermakov, and Lebedev (1981) suggested an approximate method to account for their influence. Later, Baranov and Malama (1993) developed a self-consistent model that takes into account the charge exchange processes. The Monte Carlo method was used to calculate the trajectories of neutral particles. The charge exchange process effectively decreases the Mach number of the LISM flow. This justifies the development of the subsonic interaction models (Steinolfson, Pizzo, and Holzer 1994; Khabibrakhmanov and Summers 1996; Zank et al. 1996). In our

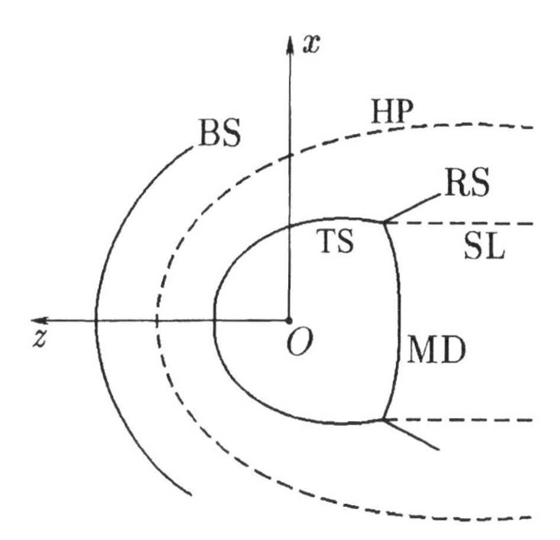

Figure 3.29 Schematic picture of the supersonic SW–LISM interaction.

further gas dynamic and MHD calculations we shall adopt the continuum approach.

Matsuda et al. (1989) discussed instabilities of the tangential discontinuity dividing the SW and the LISM flow. These instabilities originate in its lateral region and near the stagnation point of the flow. Unstable solutions were obtained in a parameter range that was not exactly suitable for the solar wind and local interstellar medium flow and therefore were disputed by Steinolfson et al. (1994). Recently, Liewer, Karmesin, and Brackbill (1996) presented results on the hydrodynamic instability of the heliopause driven by the plasma–neutral charge exchange processes. Belov and Myasnikov (1999) have recently studied the stability of the contact surface dividing the counteracting winds. They showed that this surface is only convectively unstable. Thus, the amplitude of the HP oscillations will not grow in time and will tend to zero at the forward stagnation point of the winds. Thus, such instability can develop if we use very fine numerical grids.

As shown in Fig. 3.29 and widely accepted, the solar wind changes its speed from supersonic to subsonic through the termination shock. A number of reasons can cause temporal asymmetries of the termination shock and, therefore, of the interaction pattern as a whole. Among them are short-time perturbations of the solar wind (Pogorelov 1993, 1997) and its 11-year periodicity (Pogorelov 1995, 2000). Owing to these reasons the heliospheric shock will move in response to the variation of the upstream solar wind conditions. Baranov (1990) showed that the Struchal number of the nonstationary gas dynamic problem is negligibly small only for the region size $L \ll 10^{13}$ m. The characteristic size of the heliopause is about 10^{13} m. For this reason, the interaction problem cannot be considered, strictly speaking, as a stationary problem. Worth mentioning is also unsteadiness caused by the nonuniformity of the LISM (Pogorelov 2000), although we shall disregard it in the further presentation.

The system of governing equations (1.3.1) for the axisymmetric flow of ideal perfect gas in the Cartesian coordinate system x, z shown in Fig. 3.29 acquires the form

$$\frac{\partial \mathbf{U}}{\partial t} + \frac{\partial \mathbf{E}}{\partial x} + \frac{\partial \mathbf{G}}{\partial z} + \mathbf{H} = \mathbf{0}, \tag{3.7.1}$$

where

$$\mathbf{U} = \begin{bmatrix} \rho \\ \rho u \\ \rho w \\ e \end{bmatrix}, \quad \mathbf{E} = \begin{bmatrix} \rho u \\ \rho u^2 + p \\ \rho uw \\ (e+p)u \end{bmatrix}, \quad \mathbf{G} = \begin{bmatrix} \rho w \\ \rho uw \\ \rho w^2 + p \\ (e+p)w \end{bmatrix}, \quad \mathbf{H} = \frac{\rho u}{x} \begin{bmatrix} 1 \\ u \\ w \\ (e+p)/\rho \end{bmatrix}.$$

This system is valid in the half-plane $x0z$ and can be obtained from (1.3.1) in the assumption of cylindrical symmetry. Its other form is

$$\frac{\partial x\mathbf{U}}{\partial t} + \frac{\partial x\mathbf{E}}{\partial x} + \frac{\partial x\mathbf{G}}{\partial z} + \tilde{\mathbf{H}} = \mathbf{0}, \tag{3.7.2}$$

where

$$\tilde{\mathbf{H}} = [0, -p, 0, 0]^{\mathrm{T}}.$$

Though both presentations of governing equations are mathematically equivalent, for numerical reasons it is often more convenient to use the former one, since it is supposed to give more stable results in the vicinity of the geometrical singularity $x = 0$.

In these equations $e = p/(\gamma - 1) + \rho(u^2 + w^2)/2$. The quantities of density, pressure, and velocity are normalized by ρ_∞, $\rho_\infty V_\infty^2$, and V_∞, respectively, where the subscript ∞ corresponds to the quantities in the uniform LISM flow. The time and linear dimensions are respectively normalized by L/V_∞ and L. Here L is equal to 1 AU.

As we consider only the motion of charged particles and the main charged component of the winds is represented by the ionized hydrogen atoms H^+, by the number density we imply the H^+ number density. In this calculation the LISM number density, its velocity and the speed of sound are $n_\infty = 0.1\,\mathrm{cm}^{-3}$, $V_\infty = 20\,\mathrm{km/s}$, and $c_\infty = 10\,\mathrm{km/s}$, respectively. The parameters of the solar wind vary substantially within the 11-year period of the solar activity. The concentration of charged particles $n_e = 1.56\,\mathrm{cm}^{-3}$ and the radial SW velocity $V_e = 400\,\mathrm{km/s}$ are chosen for the minimum of the solar activity, while $n_e = 8\,\mathrm{cm}^{-3}$ and $V_e = 500\,\mathrm{km/s}$ correspond to its maximum (see Brandt 1970). The dimensionless parameters of the problem are the LISM and SW Mach numbers M_∞ and M_e and the ratios of their dynamic pressures $K = \rho_e V_e^2/\rho_\infty V_\infty^2$ and stagnation temperatures $\chi = T_{0e}/T_{0\infty}$. The stationary supersonic spherically symmetric source flow, if we take into account its isentropicity and the equation of mass conservation $\rho U R^2 = \mathrm{const}$ (here U is the radial velocity component and R is the distance from the point under consideration to the Sun), with a very large accuracy can be represented by the following formulas:

$$\rho = \frac{\rho_e}{R^2}, \quad U = U_e = \mathrm{const}, \quad p = \frac{p_e}{R^{2\gamma}}. \tag{3.7.3}$$

It follows from Eq. (3.7.3) that the speed of sound and the Mach number vary as

$$c = \frac{c_e}{R^{\gamma-1}}, \quad M = M_e R^{\gamma-1}. \tag{3.7.4}$$

Taking into account that $\gamma = 5/3$ for a fully ionized plasma, we have $M = M_e R^{2/3}$. This means that the SW Mach number will be very large at typical TS distances from the Sun and, therefore, the results will only slightly depend on M_e itself.

For the chosen values of physical quantities, the dimensionless parameters take the values $K = 6250$, $\chi = 256$, $M_\infty = 2$, and $M_e = 5$ at the minimum and $K = 50000$, $\chi = 400$, $M_\infty = 2$, and $M_e = 5$ at the maximum of the solar activity. These values are chosen as the basic points for the sine function approximating the time dependence of K and χ within 11 years. The dimensionless time unit is 86.8 days. The time step is chosen to be $\Delta t \approx 0.4$ days. The specific heat ratios are supposed to be $5/3$ both for LISM and SW. The steady-state quantity distribution corresponding to the time-convergent solution in the minimum of the solar activity is chosen as initial data. To obtain this stationary solution we apply the numerical method described below. The same numerical method is later used to solve the time-dependent problem.

The Euler equations of gas dynamics in the form (3.7.1) are solved numerically using the MUSCL TVD scheme with the second order of accuracy in space and time. Initial conditions can be fairly arbitrary, though they must satisfy the boundary conditions that consist of the fixed SW parameters on the inner boundary (spherical for convenience) and the uniform LISM parameters at infinity.

Let us introduce a polar mesh in the xOz half-plane ($x > 0$)

$$R_l = R_{\min} + (l-1)\Delta R,\ l = 1, 2, \ldots, L; \quad \Delta R = (R_{\max} - R_{\min})/(L-1);$$
$$\theta_n = (n - 2.5)\Delta\theta,\ n = 1, 2, \ldots, N; \quad \Delta\theta = \pi/(N-4);$$

with the center at the source position. This mesh is the result of the cut of the general spherical mesh by the above-mentioned half-plane; the angle θ is counted off the z-axis. The system of (3.7.1) written out for each cell at some fixed time instant acquires the form

$$R_l \Delta R \Delta\theta \frac{\partial \mathbf{U}_{l,n}}{\partial t} + (R_{l+1/2}\bar{\mathbf{E}}_{l+1/2,n} + R_{l-1/2}\bar{\mathbf{E}}_{l-1/2,n})\Delta\theta + (\bar{\mathbf{E}}_{l,n+1/2} + \bar{\mathbf{E}}_{l,n-1/2})\Delta R + R_l \Delta R \Delta\theta\, \mathbf{H}_{l,n} = \mathbf{0}. \quad (3.7.5)$$

Here $\bar{\mathbf{E}}$ is the flux normal to the boundary defined as

$$\bar{\mathbf{E}} = n_1 \mathbf{E} + n_2 \mathbf{G}, \quad (3.7.6)$$

where $\mathbf{n} = (n_1, n_2)$ is a unit outward vector normal to the cell surface. It is clear that for the polar grid system we have $(n_1)_{l,n} = \sin\theta_n$ and $(n_2)_{l,n} = \cos\theta_n$ for the cell surfaces perpendicular to the radial coordinate lines and $(n_1)_{l,n+1/2} = \cos\theta_{n+1/2}$ and $(n_2)_{l,n+1/2} = -\sin\theta_{n+1/2}$ for the surfaces perpendicular to the angular coordinate lines. The general procedure described in Section 2.7 requires the determination of quantities on the left- and right-hand sides of the cell interface using proper interpolation of quantities available at the cell centers. This procedure must include certain slope limiting, which allows us to satisfy the TVD property. We assume a linear distribution of parameters in the radial and angular directions. The results presented in this section were obtained with the use of the smooth slope limiter (van Albada et al. 1982):

$$\bar{\mathbf{E}}_{l+1/2,n} = \bar{\mathbf{E}}(\mathbf{U}^{\mathrm{L}}, \mathbf{U}^{\mathrm{R}}),\ \mathbf{U}^{\mathrm{L}} = \mathbf{U}_{l,n} + \frac{\delta\mathbf{U}_{l,n}}{2},\ \mathbf{U}^{\mathrm{R}} = \mathbf{U}_{l+1,n} - \frac{\delta\mathbf{U}_{l+1,n}}{2}, \quad (3.7.7)$$

$$\delta\mathbf{U}_{l,n} = (\Omega_{\mathrm{R}})_{l,n}\delta\mathbf{w}_{l,n}, \quad (3.7.8)$$

$$\delta\mathbf{w}_{l,n} = \frac{(\delta\mathbf{w}^2_{l-1/2,n} + \epsilon)\delta\mathbf{w}_{l+1/2,n} + (\delta\mathbf{w}^2_{l+1/2,n} + \epsilon)\delta\mathbf{w}_{l-1/2,n}}{\delta\mathbf{w}^2_{l+1/2,n} + \delta\mathbf{w}^2_{l-1/2,n} + 2\epsilon}, \quad (3.7.9)$$

$$\delta\mathbf{w}_{l+1/2,n} = (\Omega_{\mathrm{L}})_{l,n}(\mathbf{U}_{l+1,n} - \mathbf{U}_{l,n}),\ \delta\mathbf{w}_{l-1/2,n} = (\Omega_{\mathrm{L}})_{l,n}(\mathbf{U}_{l,n} - \mathbf{U}_{l-1,n}). \quad (3.7.10)$$

Here ϵ is a small positive constant used to avoid the division by zero. Note that the matrices Ω_L and Ω_R are the 4×4 matrices constructed of the left and the right eigenvectors of the Jacobian matrix $\partial\bar{\mathbf{E}}/\partial\mathbf{U}$, respectively. The numerical flux inherent in the first formula of Eq. (3.7.7) is determined using Roe's approximate Riemann problem solver. Fluxes through another pair of interfaces are obtained similarly. It is worth noting that we applied here the limiting procedure to the increments of characteristic variables. Though this approach is more time-consuming than, for example, limiting the increments of the primitive variables, it proved to give the best quality of results for a large variety of gas dynamic problems. The advance of the solution in time is implemented using the second order of accuracy predictor–corrector method

$$\mathbf{U}_{l,n}^{(1)} = \mathbf{U}_{l,n}^{k} + \frac{\Delta t}{2}\frac{\partial \mathbf{U}_{l,n}^{k}}{\partial t}, \quad \mathbf{U}_{l,n}^{k+1} = \mathbf{U}_{l,n}^{k} + \Delta t\frac{\partial \mathbf{U}_{l,n}^{(1)}}{\partial t} \tag{3.7.11}$$

for $t = k\Delta t,\ k = 1, 2, \ldots$ and Δt is defined by the CFL condition.

Calculation is performed in the ring region with the inner and outer circle radii $R_{\min} = 14$ and $R_{\max} = 700$, respectively. Since the radial velocity is supersonic, at the inner circle all the parameters are specified as functions of time. To obtain the initial conditions corresponding to the solar activity minimum, we must fix corresponding steady-state quantities at this boundary. The points on the outer boundary can be either entrance or exit flow points. At the entrance points we can safely fix the quantities corresponding to the uniform supersonic LISM flow. The exit flow velocity can be both supersonic and subsonic. In the former case all quantities can be extrapolated downwind. In the latter case, however, such an extrapolation will violate the characteristic properties of the Euler system. This case seems to be a good candidate for application of nonreflecting boundary conditions described in Section 2.8. This will require, however, a special consideration, which will be given in the next subsection. We must only add, as far as the boundary conditions are concerned, that there is no necessity, of course, to solve the problem in the whole circular region, since we can impose a symmetry boundary conditions on the symmetry axis z.

3.7.2 Nonreflecting boundary conditions. In Section 2.8 we described various approaches to numerical nonreflecting boundary conditions. The main aim of establishing such conditions is to find an appropriate way to pass the knowledge of physical conditions at infinity to finite distances or, at least, to minimize reflections from numerical boundaries into the computational region. Far-field boundary conditions are frequently encountered in astrophysical applications and determine to a great extent the efficiency of the numerical algorithm. We are mainly interested in this section in the subsonic exit boundaries. The statement of similar conditions at the entrance segments of the boundary is analogous. Remaining in the framework of the local approach, we can follow the line of the previously described boundary conditions in which a layer of artificial boundary cells is introduced outside the boundary. This layer is filled by the "infinity" gas. Numerical flux through the boundary can be found in this case using the solution of the one-dimensional (linear axis is perpendicular to the cell surface) Riemann problem of the disintegration of an arbitrary jump between the quantities at infinity and those in the cell adjacent to the boundary. The peculiarity of our problem is that the LISM flow is always supersonic at infinity. This means that if the exit flow is subsonic there must be a sonic point transition somewhere between the outer boundary and infinity. The application of the Riemann-problem solution fails in this case, since it contains a shock wave propagating inside the computational region. This shock eventually deteriorates the solution.

Pogorelov and Semenov (1996b) described an approach based on the two nonreflecting conditions: an extrapolation condition for a supersonic exit that provides a characteristically consistent approximation of equations on the boundary and a rarefaction wave condition. The idea of applying the relations in the rarefaction wave for implementation of the far-field boundary conditions is closely related to the artificial locating of the sonic point on the exit boundary. If the flow is supersonic at infinity, such a procedure gives reasonable results and allows us to perform calculations in the cases for which other known approaches fail to give acceptable quality of solution. Another interpretation of this method is the following. Assume that parameters inside the chosen computational region fully define the flow behavior outside the boundary. In the case of subsonic exit the only possible elementary Riemann problem configuration for the above system is a rarefaction wave whose fan covers the boundary. In this case, if the value of the self-similar variable is known, we can locally continue the internal flow field to the boundary. That is why the additional condition is provided by the assumption that the flow velocity attains the sonic value there.

Consider the hyperbolic system for the vector $\mathbf{U}$ of unknown variables in the vicinity of the boundary (the right one for definiteness) in the form

$$\frac{\partial \mathbf{U}}{\partial t} + A\frac{\partial \mathbf{U}}{\partial x} = \mathbf{0}, \tag{3.7.12}$$

where $\mathbf{U} = \mathbf{U}(x,t)$, and x is the variable in the direction normal to the boundary Γ, t is time, and $A(\mathbf{U})$ is the coefficient matrix with a complete set of eigenvectors and only real eigenvalues. We seek the solution in the form of a simple wave $\mathbf{U} = \mathbf{U}(x,t) = \mathbf{U}(\xi)$, where $\xi = x/t$. By substituting this representation into Eq. (3.7.12), we obtain

$$(A - \lambda I)\mathbf{U}_\xi = \mathbf{0}, \quad \lambda = \xi, \tag{3.7.13}$$

where I is the identity matrix. Owing to Eq. (3.7.13), the vector $\mathbf{U}_\xi$ is the eigenvector of A for the eigenvalue $\lambda = \xi$. This means that we need to solve the following system of ordinary differential equations supplemented by the nondifferential relation:

$$\mathbf{U}_\xi = d(\mathbf{U},\lambda)\,\mathbf{r}(\mathbf{U},\lambda)\,, \quad \lambda(\mathbf{U}) = \xi\,, \tag{3.7.14}$$

where $\mathbf{r}$ is the right eigenvector (the vector-column) of A defined up to a scalar multiplier d. The eigenvalue in the simple wave varies like $\lambda(\mathbf{U}) = \xi$. This condition completes the system for determining $\mathbf{U}$ and d. While implementing this boundary condition, we must integrate Eq. (3.7.14) over ξ from $\xi_0 = \lambda(\mathbf{U}_0)$, where $\mathbf{U}_0$ represents the subsonic quantities inside the region, to $\xi = \xi_\Gamma = 0$, that is, to the sonic point. It is interesting to look at the described approach from the viewpoint of the Osher numerical scheme. One can see that the proposed procedure implies integration in the phase space starting from the state in the center of the cell adjacent to the boundary (or from the state on the inner side of the boundary surface) to the first sonic point on the path connecting the inner state with the state at infinity. Other possible sonic points and, hence, other possible segments of integration are disregarded.

Consider this approach in the application to pure gas dynamics. Let us choose the form $\mathbf{U} = [\rho, u, v, w, c]^{\mathrm{T}}$ of the unknown vector in Eq. (3.7.12), where ρ is the density, u is the velocity vector component normal to Γ, v and w are its tangential components, and c is the speed of sound. The smallest eigenvalue in this case is $\lambda = u - c$ and the related eigenvector is

$$\mathbf{r} = \left[1,\ -\frac{c}{\rho},\ 0,\ 0,\ \frac{(\gamma-1)c}{2\rho}\right]^{\mathrm{T}}, \tag{3.7.15}$$

where γ is the adiabatic index. System (3.7.14) in this case acquires the form

$$\begin{aligned} &\rho_\xi = d, \quad u_\xi = -\frac{cd}{\rho}, \quad v_\xi = 0, \quad w_\xi = 0, \\ &c_\xi = \frac{(\gamma-1)cd}{2\rho}, \quad u - c = \xi. \end{aligned} \tag{3.7.16}$$

This system can be exactly integrated, as its invariants are

$$\left(\frac{p}{\rho^\gamma}\right)_\xi = 0, \quad \left(u + \frac{2c}{\gamma-1}\right)_\xi = 0, \quad v_\xi = 0, \quad w_\xi = 0. \tag{3.7.17}$$

Thus, we obtain

$$\begin{aligned} &c_\Gamma = \frac{\gamma-1}{\gamma+1}\left(u_0 + \frac{2}{\gamma-1}c_0\right), \quad u_\Gamma = c_\Gamma, \quad v_\Gamma = v_0, \\ &w_\Gamma = w_0, \quad \rho_\Gamma = \rho_0\left(\frac{c_\Gamma}{c_0}\right)^{\frac{2}{\gamma-1}}. \end{aligned} \tag{3.7.18}$$

The subscript 0 indicates the values belonging to the inner region. On the discreet mesh this means that they are taken from the center (or from the left-hand side) of the cell adjacent to the boundary At the supersonic exit, for $(u/c)_0 \geq 1$, Eq. (3.7.18) must be supplemented by the conditior

$$\mathbf{U}_\Gamma = \mathbf{U}_0.$$

These two groups of conditions are mutually consistent and coincide for $u_0 = c_0$. Note that in this case we need no explicit expression for d, which can be easily found from the second and fifth equations in the system (3.7.16). Besides the exact derivation based on the relations in the rarefaction wave, we give here the approximate relations, keeping in mind such systems for which no exact expressions can be written out. For this purpose, we first exclude d by substituting the first equation from (3.7.16) into the other ones. Then we obtain

$$\begin{aligned} &u_\xi = -\frac{c\rho_\xi}{\rho}, \quad v_\xi = 0, \quad w_\xi = 0, \\ &c_\xi = \frac{(\gamma-1)c\rho_\xi}{2\rho}, \quad u - c = \xi. \end{aligned} \tag{3.7.19}$$

Now approximating Eqs. (3.7.19) by finite differences we arrive at the following relations:

$$c_\Gamma = \frac{\gamma-1}{\gamma+1}\left(u_0 + \frac{2}{\gamma-1}c_0\right), \quad u_\Gamma = c_\Gamma, \quad v_\Gamma = v_0,$$

$$w_\Gamma = w_0, \quad \rho_\Gamma = \left[1 + \frac{2}{\gamma - 1}\left(\frac{c_\Gamma}{c_0} - 1\right)\right]. \tag{3.7.20}$$

Solution (3.7.20) in this approximation differs from (3.7.18) only in the entropy invariant and represents its linearization.

Since relations (3.7.18) and (3.7.20) are based on the artificial shift of the sonic point to the outer boundary, the resulting numerical conditions are only approximate. To determine the real influence on the inner flow quantities, one must experiment with the location of the outer boundary. The compromise is between its location "close" to infinity and better resolution in the smaller computational region. The application of the method described in this section turned out to give stable results in contrast to the attempts of the straightforward application of the well-known nonreflecting boundary conditions (Thompson 1987).

3.7.3 Numerical results.

The contours of constant pressure (below the symmetry axis) and density logarithms corresponding to the initial data are presented in Fig. 3.30a. The results are shown in the polar region with the inner and outer circle radii equal to 14 and 560 AU, respectively. All peculiarities are seen of the shock wave structure shown in Fig. 3.29 (the flow is from the left to the right).

Let us look at the initial stage of the development of the flow. The increase of the parameter K in time leads to the inner shock motion from the Sun. A compression wave intersects the inner shock resulting in the origin of an additional time-dependent shock wave propagating toward the heliopause (see Fig. 3.30b). This wave penetrates through it in the upwind side and moves toward the bow shock. Since the density of the interstellar medium on the outer side of the heliopause is larger than that of the solar wind on its inner side, the speed of the shock propagation becomes smaller. Its intensity decreases as well due to the radial damping. The inner shock shape substantially changes in the course of time (see Figs. 3.30c,d), since the perturbations from the Sun reach its surface at different times. The decreasing part of periodic perturbations causes the opposite motion of the inner shock to the Sun. This leads to the origin of additional flow division surfaces, reverse flow zones, and vortices of variable size in the wake region. In Fig. 3.30e we can see the next nonstationary shock wave following the first one. We can also see the increase of the bow shock stand-off distance. A sequence of newly developed jumps emerges later in time (see Fig. 3.30f).

It is interesting to follow the flow pattern variation within an 11-year cycle sufficiently remote from the initial data. This variation is presented in Fig. 3.31a–d. We see that a definite 11-year periodicity is established in the shape of the inner shock wave. It is quite clear that this shape differs very much from the initial one and hardly ever coincides with any steady-state solution from the chosen range of the solar wind variation. The presence of consecutively moving shock waves is also a new feature of the nonstationary flow.

It is worth noting that the calculations performed on the basis of the newly available observational data (see Baranov and Zaitsev 1998) show that the variation of the dynamic pressure ration is not so large as accepted in the presented model solution. As a result the motion of the bow shock as well as the amplitude of the inner shock oscillations are not so large, though their behavior remains qualitatively the same. On the other hand, Pogorelov (1995) has already attracted attention to the fact that numerical modelling of the LISM

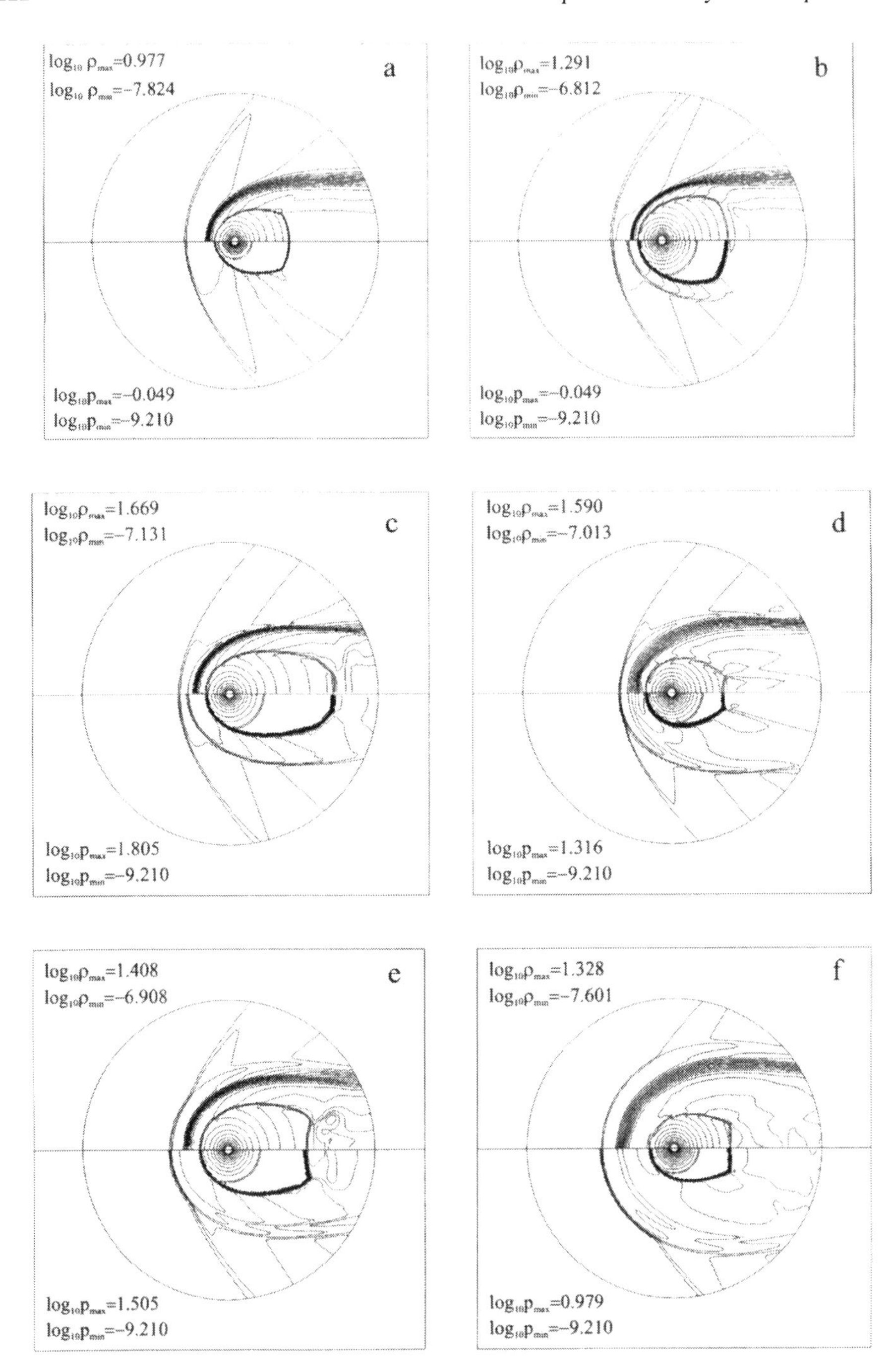

Figure 3.30 Pressure and density isolines, $t = 0$ (a), 20 (b), 36 (c), 60 (d), 84 (e), 148 (f).

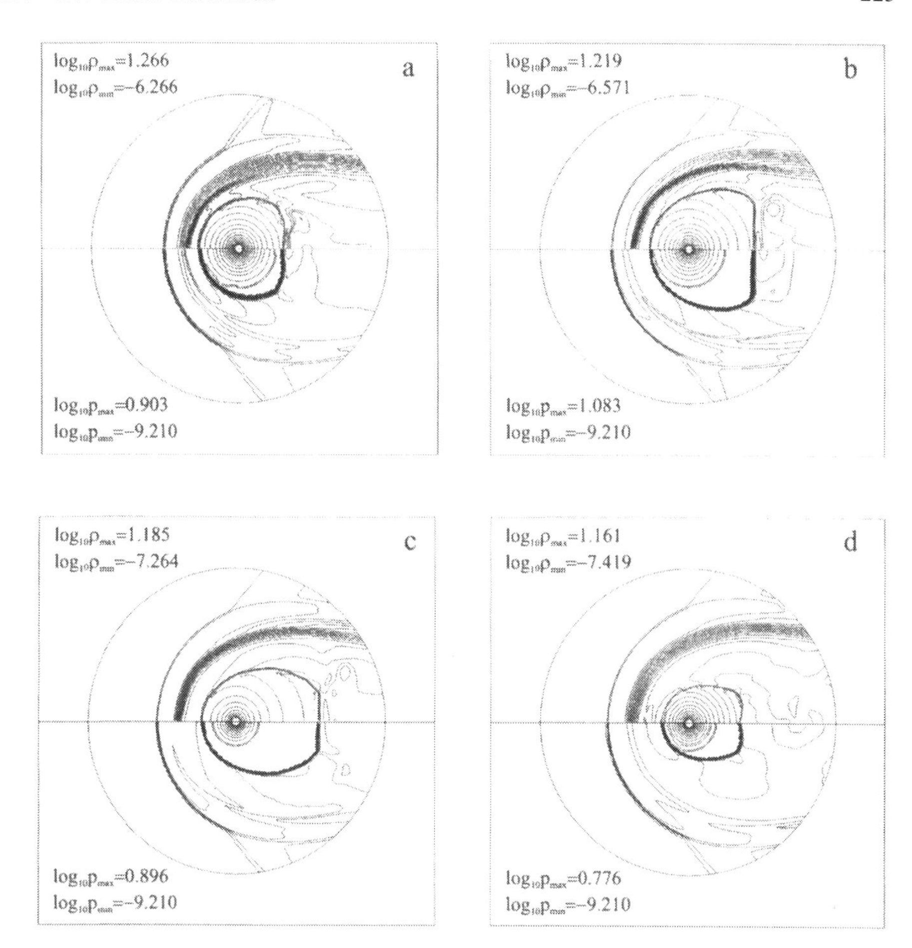

Figure 3.31 Pressure and density isolines, $t = 160$ (a), 172 (b), 184 (c), 192 (d).

interaction with the periodic solar wind cannot be adequately performed in the axisymmetric statement of the problem, since the solar wind is not spherically symmetric, especially around its activity minimum. The measurements obtained by the Ulysses spacecraft (Neugebauer 1999) made it possible to perform a preliminary analysis based on the three-dimensional model (Tanaka and Washimi 1999). In that paper, the solar wind spatial nonuniformity was taken into account. Namely, the SW speed increases from 400 km s^{-1} to 800 km s^{-1} if we go from the ecliptic plane to the pole. This increase is rather steep and occurs at the helioaltitude angle varying within the solar cycle (from 30° to 80°). Besides, the effect was considered of the interplanetary magnetic field polarity change at the solar maximum.

Here we illustrated the application of the high-resolution, Godunov-type numerical scheme based on Roe's gas dynamic Riemann problem solver and van Albada's slope limiter to the nonstationary axisymmetric problem. Combined with an effective outer boundary

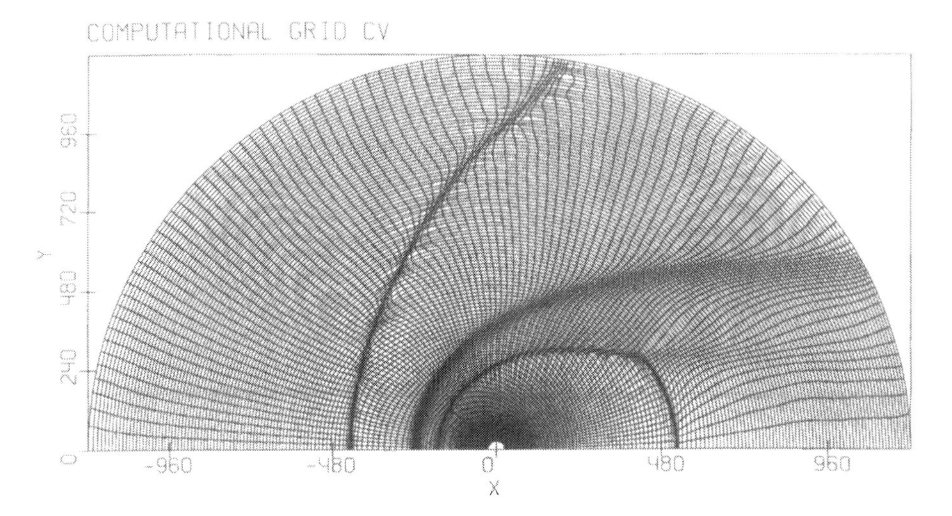

Figure 3.32 Solution-adapted grid.

conditions it allows one to attain a reasonable accuracy. Further improvement of the code can be made by introducing a moving grid adjusting to discontinuities. In Fig. 3.32, we show the example of such grid generated for the SW–LISM interaction problem (Ivanov, Kryukov, and Pogorelov 1999).

3.7.4 A note on Godunov-type methods for relativistic hydrodynamics.

The problem considered in this section was solved on the basis of gas dynamic equation in the Newtonian approximation. Characteristic velocities in this case are much smaller than the speed of light. On the other hand, a great variety of astrophysical problems are characterized by relativistic velocities and the effect of the special theory of relativity must be taken into account. Here we mention just a few references that contributed to the development of the Godunov-type methods for relativistic hydrodynamics. Martí and Müller (1994) obtained an exact solution of the Riemann problem in special relativistic hydrodynamics (SRHD). In 1996, on the basis of this solution the same authors developed a SRHD version of the PPM (piecewise parabolic method). The SRHD version of Roe's method was suggested by Eulderink and Mellema (1995). Worth mentioning also are the papers by Marquina et al. (1992), Schneider et al. (1993), Falle and Komissarov (1996), and Dolezal and Wong (1995). The last authors extended the ENO method to SRHD. Since this list is, of course, not exhaustive, interested readers can refer to the review by Martí and Müller (1999).

Chapter 4
Shallow Water Equations

In this chapter we describe Godunov-type explicit shock-capturing methods designed to compute fluid flows governed by the one- and two-dimensional shallow water equations. These methods are based on the solution of the corresponding one-dimensional Riemann problem. We present exact and approximate solvers for the shallow water equations. Among them are exact Riemann problem solver, and the approximate Courant–Isaacson–Rees, Roe, and Osher–Solomon solvers. All these solvers permit one to perform calculations involving an arbitrary bottom relief. Special attention is paid to the exact Riemann problem solver for the shallow water equations. It allows one to calculate flows with shoaling and dry bottom regions by the explicit shock-capturing method without applying any data-fitting techniques.

The Godunov-type methods are suitable for computing flows characterized by moderate and high values of the Froude number, which can involve strong gradients or even jumps in the values of flow variables. The Froude number in the shallow water equations plays a role similar to that of the Mach number in gas dynamics (Rozhdestvenskii and Yanenko 1983). The Froude number is defined as $\mathrm{Fr} = |\mathbf{v}|/c$, where $\mathbf{v}$ is the flow velocity and $c = \sqrt{gh}$ is the speed of the propagation of small disturbances (h is the water depth, and g is the vertical gravitational acceleration). The explicit methods to be described in this chapter can be applied even to flows with low Froude numbers, although their computational efficiency will degrade in this case because of the more restrictive CFL condition for the time step. The developed numerical algorithms can adequately predict the propagation dynamics of strong hydraulic jumps—analogues of gas dynamic shock waves—and preserve the monotonicity of profiles of grid variables in the vicinity of them.

4.1 System of governing equations

The time-dependent system of two-dimensional shallow water equations (the Saint-Venant equations) that describe planar fluid flow have the form

$$\frac{\partial h}{\partial t} + \operatorname{div}(h\mathbf{v}) = 0, \tag{4.1.1}$$

$$\frac{\partial h\mathbf{v}}{\partial t} + \operatorname{div}(h\mathbf{v}\mathbf{v}) + gh\operatorname{div}(\varsigma\hat{\mathbf{I}}) = \mathbf{f}. \tag{4.1.2}$$

Here $h = h(t,x,y)$ is the water depth, see Fig. 4.1, t is time, and (x,y) are the Cartesian coordinates on the horizontal plane; $\mathbf{v} = \mathbf{v}(t,x,y) = [u,v]^{\mathrm{T}}$ is the velocity averaged over the vertical coordinate; g is the free-fall acceleration; $b = b(x,y)$ is the bottom relief

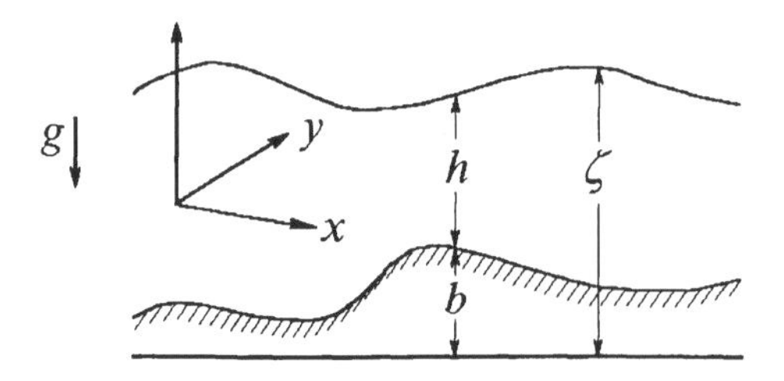

Figure 4.1 Notations for shallow water equations.

measured relative to a horizontal level; the quantity $\zeta = h + b$ is called the free-surface level; $\hat{\mathbf{I}} = \text{diag}[1, 1]$ is the 2×2 identity tensor; and $\mathbf{f} = [f_1, f_2]^{\mathrm{T}}$ represents external forces, e.g., the Coriolis or a friction force. In particular, the friction force can be equal to $\mathbf{f} = \frac{1}{2}\lambda\mathbf{v}|\mathbf{v}|$, where the hydraulic friction coefficient is equal to $\lambda = 2gn^2h^{-1/3}$, with n being the Manning's roughness coefficient (Voltsinger and Pyaskovskii 1977).

Equations (4.1.1) and (4.1.2) express the conservation of the water mass and momentum, respectively. In the expanded form, system (4.1.1)–(4.1.2) can be represented as

$$\frac{\partial h}{\partial t} + \frac{\partial hu}{\partial x} + \frac{\partial hv}{\partial y} = 0, \tag{4.1.3}$$

$$\frac{\partial hu}{\partial t} + \frac{\partial (hu^2 + \frac{1}{2}gh^2)}{\partial x} + \frac{\partial huv}{\partial y} = f_1 - gh\frac{\partial b}{\partial x}, \tag{4.1.4}$$

$$\frac{\partial hv}{\partial t} + \frac{\partial hvu}{\partial x} + \frac{\partial (hv^2 + \frac{1}{2}gh^2)}{\partial y} = f_2 - gh\frac{\partial b}{\partial y}. \tag{4.1.5}$$

Water flows described by the shallow water equations are rather common. They include the propagation of breakers and tidal bores in rivers (Stoker 1948), the propagation of tsunamis (Marchuk, Chubarov, and Shokin 1983), flows in tailraces of hydroelectric power stations, flows in culverts and chutes (Stoker 1957), global atmosphere motion in weather prediction (Spotz, Taylor, and Swarztrauber 1998; Göttelmann 1999), etc. The shallow water equations are derived by averaging the time-dependent three-dimensional Euler or Navier–Stokes equations over the vertical coordinate. Their derivation and validation can be found in Friedrichs (1948), Stoker (1957), Ovsiannikov (1985), and Makarenko (1985).

Smooth solutions of the shallow water type equations were first obtained by Riemann (1860). The quasilinear hyperbolic system (4.1.1)–(4.1.2) admits discontinuous solutions, see Section 1.4. Discontinuities can arise from initially smooth data (Riemann 1860). The complete general exact solution of the shallow water equations was first obtained by Stoker (1948, 1957).

System (4.1.3)–(4.1.5) in the one-dimensional case acquires the form

$$\frac{\partial h}{\partial t} + \frac{\partial hu}{\partial x} = 0, \tag{4.1.6}$$

$$\frac{\partial hu}{\partial t} + \frac{\partial (hu^2 + \frac{1}{2}gh^2)}{\partial x} = f_1 - gh\frac{\partial b}{\partial x}, \tag{4.1.7}$$

$$\frac{\partial hv}{\partial t} + \frac{\partial hvu}{\partial x} = f_2. \tag{4.1.8}$$

In the vector notion Eqs. (4.1.6)–(4.1.8) read

$$\frac{\partial \mathbf{U}}{\partial t}+\frac{\partial \mathbf{F}}{\partial x}=\mathbf{B}, \tag{4.1.9}$$

where

$$\mathbf{U}=[h,\ hu,\ hv]^{\mathrm{T}}, \tag{4.1.10}$$
$$\mathbf{F}=[hu,\ hu^2+\tfrac{1}{2}gh^2,\ huv]^{\mathrm{T}}, \tag{4.1.11}$$
$$\mathbf{B}=D\mathbf{U}+\mathbf{d},\quad D=\begin{bmatrix}0 & 0 & 0\\ -gb_x & 0 & 0\\ 0 & 0 & 0\end{bmatrix},\quad b_x=\frac{\partial b}{\partial x},$$
$$\mathbf{d}=[0, f_1, f_2]^{\mathrm{T}}. \tag{4.1.12}$$

This system can be represented in the nonconservative form

$$\frac{\partial \mathbf{U}}{\partial t}+A\frac{\partial \mathbf{U}}{\partial x}=\mathbf{B}, \tag{4.1.13}$$

where the matrix $A=\partial\mathbf{F}/\partial\mathbf{U}$ is defined as

$$A=\begin{bmatrix}0 & 1 & 0\\ gh-u^2 & 2u & 0\\ -uv & v & u\end{bmatrix}. \tag{4.1.14}$$

System (4.1.6)–(4.1.8) is hyperbolic, since its coefficient matrix A has only real eigenvalues and a complete set of eigenvectors. Thus, the matrix A can be diagonalized,

$$A=\Omega_{\mathrm{R}}\Lambda\Omega_{\mathrm{L}}.$$

One can find that

$$\Omega_{\mathrm{R}}=\begin{bmatrix}1 & 0 & 1\\ u-c & 0 & u+c\\ v & -1 & v\end{bmatrix},\quad \Omega_{\mathrm{L}}=\frac{1}{2c}\begin{bmatrix}c+u & -1 & 0\\ 2vc & 0 & -2c\\ c-u & 1 & 0\end{bmatrix}, \tag{4.1.15}$$

$$\Lambda=\mathrm{diag}[u-c,\ u,\ u+c], \tag{4.1.16}$$

$$c=\sqrt{gh}. \tag{4.1.17}$$

The function c is the speed of the propagation of small disturbances in shallow water or a celerity. Note that $\Omega_{\mathrm{R}}\Omega_{\mathrm{L}}=I$ and $\det\Omega_{\mathrm{R}}=2c$. The matrix A can be rewritten in the form

$$A=\begin{bmatrix}0 & 1 & 0\\ c^2-u^2 & 2u & 0\\ -uv & v & u\end{bmatrix}. \tag{4.1.18}$$

When applied to specific problems, Eqs. (4.1.1)–(4.1.2) must be supplemented with appropriate initial and boundary conditions. At $t=0$, $\mathbf{v}(0,x,y)=\mathbf{v}_0(x,y)$ and $h(0,x,y)=$

$h_0(x, y)$ are prescribed. On the boundaries of the computational domain, boundary conditions are set; for example, conditions in terms of fluxes, impermeable wall conditions, etc., are used. The well-posedness of boundary conditions is discussed in Section 2.8, see also Baklanovskaya, Pal'tsev, and Chechel (1979).

From the time-dependent two-dimensional shallow water equations (4.1.1)–(4.1.2), one can obtain the hyperbolic system of equations, which governs the stationary two-dimensional supercritical shallow water flows. Godunov-type methods for this system of equations will be described in Section 4.6.

4.2 The Godunov method for shallow water equations

The shallow water equations can be solved by adapting numerical methods originally developed for solving gas dynamics equations. This approach is quite natural, since the shallow water equations are similar to the equations of motion of a barotropic gas with the polytropic exponent equal to two. For example, the local Lax–Friedrichs scheme developed by Rusanov (1961) was modified in Lyatkher and Militeev (1978) to analyze a whirling flow in an expanding channel. The scheme by MacCormack (1969) was used to compute the flow in channels with curved walls (Pandolfi 1975). By analogy with gas dynamics, Alalykin et al. (1970) constructed an implicit jump-fitting finite-difference scheme for solving the one-dimensional shallow water equations without taking into account variable bottom relief. The concept of invariant scheme, which was developed for the gas dynamic equations by Shokin (1973) and Shokin and Yanenko (1985), was applied to solve the shallow water equations by Fedotova (1978). Papa (1984) solved numerically the shallow water equations by a modified Courant–Isaacson–Rees method. The Roe gas dynamic method was adopted for shallow water equations (Glaister 1991, 1995) and used to simulate various shallow water flows, for example, calculations on unstructured triangular grids (Garsia-Navarro, Hubbard, and Priestley 1995; Hubbard and Baines 1997), flows in channels with irregular geometry (Vázquez-Cendón 1999), etc.

The integral form of Eqs. (4.1.1)–(4.1.2) is

$$\frac{d}{dt}\left(\iint_G h\,dG\right) + \oint_L hv_n\,dL = 0, \tag{4.2.1}$$

$$\frac{d}{dt}\left(\iint_G h\mathbf{v}\,dG\right) + \oint_L h\mathbf{v}v_n\,dL + \tfrac{1}{2}g\oint_L h^2\,\mathbf{n}\,dL + g\iint_G h\nabla b\,dG = \iint_G \mathbf{f}\,dG. \tag{4.2.2}$$

Here, G is a domain in the (x, y) plane, $dG = dx\,dy$ is the area element, L is the boundary of G, dL is the arc element, $v_n = \mathbf{v}\cdot\mathbf{n}$, where $\mathbf{n}$ is the outward normal to dL, $\nabla = [\partial/\partial x,\ \partial/\partial y]^{\mathrm{T}}$, and $\mathbf{p}\cdot\mathbf{q}$ stands for the scalar product of the vectors $\mathbf{p}$ and $\mathbf{q}$.

To construct the Godunov finite-volume scheme for equations (4.2.1)–(4.2.2) written in integral form, we discretize the computational domain by constructing a grid of arbitrary convex polygons having areas G_i, $i = 1, 2, \ldots$, with $m = m(i)$ sides having lengths L_j and outward normals $\mathbf{n}_j$, where $j = 1, 2, \ldots, m(i)$. On each polygon, the integral equations are approximated as follows:

$$G_i\frac{h_i^{k+1} - h_i^k}{\Delta t} + \sum_{j=1}^{m(i)}(HV_nL)_j = 0, \tag{4.2.3}$$

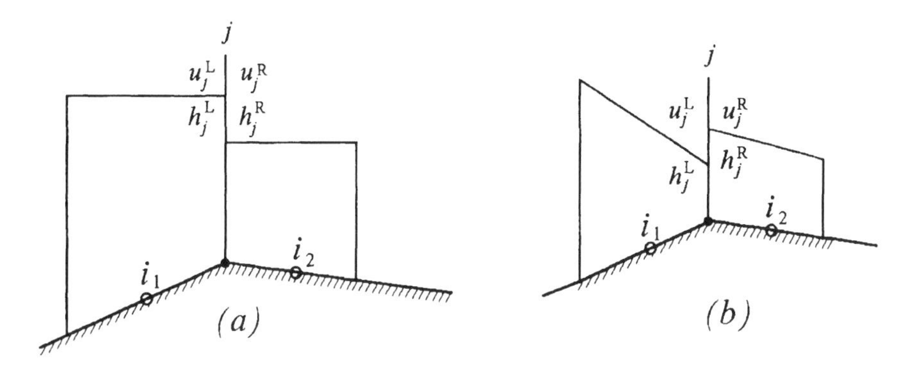

Figure 4.2 Selection of function slopes.

$$G_i \frac{(h\mathbf{v})_i^{k+1} - (h\mathbf{v})_i^k}{\Delta t} + \sum_{j=1}^{m(i)} (H\mathbf{V}V_n L)_j + \tfrac{1}{2} g \sum_{j=1}^{m(i)} [H_j^2 - (h_j^\alpha)^2](\mathbf{n}L)_j = G_i \mathbf{f}_i^\mu, \tag{4.2.4}$$

$$\mathbf{f}_i^\mu = \mu \mathbf{f}_i^{k+1} + (1-\mu)\mathbf{f}_i^k, \quad h_j^\alpha = \alpha h_j^{k+1} + (1-\alpha) h_j^k,$$

where $V_n = \mathbf{V} \cdot \mathbf{n}$. The integer subscript i in (4.2.3) and (4.2.4) denotes the values of grid variables calculated at the center of mass of the ith polygon, and the subscript j denotes their values on the jth side of the cell. The integer superscript k denotes the number of the time layer, and Δt is the time increment. The corresponding capital letters denote the depth H and velocity $\mathbf{V}$ on the cell boundaries. Their values are calculated by solving the Riemann problem for the shallow water equations. The coefficient α specifies the time approximation of the term accounting for the bottom relief. In addition, the coefficient μ specifies the time approximation of $\mathbf{f}$ on the right-hand side of the discrete equations. In particular, we approximated the friction expressed as $\mathbf{f} = \frac{1}{2}\lambda \mathbf{v}|\mathbf{v}|$ with $\lambda = 2gn^2h^{-1/3}$ by

$$\mathbf{f} \simeq \mathbf{f}_i^\mu = [(h\mathbf{v})_i^{k+1}\mu + (h\mathbf{v})_i^k(1-\mu)]\, gn^2 \frac{|\mathbf{v}_i^k|}{(h_i^{k+1})^{4/3}}. \tag{4.2.5}$$

The idea of the Godunov method can be explained by analyzing the following one-dimensional example. Consider two adjacent cells, with numbers i_1 and i_2, separated by the jth boundary, see Fig. 4.2a. Treating the velocity and the free-surface level, $\zeta = h + b$, inside each cell as piecewise constant functions, and the bottom relief as a piecewise-linear continuous function in G, we calculate the values of the depths, h_j^L and h_j^R, and velocities, u_j^L and u_j^R, on the left- and right-hand sides of the jth boundary. Then, we use these values to solve the Riemann problem and find the depth H and velocity U in the direction normal to the boundary. Note that the calculation of H_j and U_j can be performed without any limitation if even either h_j^R or h_j^L is initially equal to zero (i.e., if the cell is incompletely filled by water), since a solution to the Riemann problem exists even in this case.

In the two-dimensional problem, the vector $\mathbf{V} = [U, V]^T$ is represented as the sum of the components that are normal (U) and tangential (V) to the cell boundary. The flow depths, h_j^L and h_j^R, are calculated at the middle points of the corresponding sides of the polygon. When required, the free-surface levels could be determined at polygon's vertices.

In Eq. (4.2.2) the bottom relief is approximated as

$$g \iint_G h\nabla b \, dG \simeq -\tfrac{1}{2} g \sum_{j=1}^{m(i)} h_j^2 \mathbf{n}_j L_j. \tag{4.2.6}$$

Let us fix the initial values of h and take into account that $b = \zeta - h$, where $\zeta = \text{const}$. Using the Gauss theorem, we obtain

$$\begin{aligned} g \iint_G h\nabla b \, dG &= g \iint_G h\nabla(\zeta - h) \, dG = -g \iint_G h\nabla h \, dG = -g \iint_G \tfrac{1}{2}\nabla h^2 \, dG \\ &= -\tfrac{1}{2} g \oint_L h^2 \mathbf{n} \, dL \simeq -\tfrac{1}{2} g \sum_{j=1}^{m(i)} h_j^2 \mathbf{n}_j L_j. \end{aligned}$$

One important property of this approximation is that the equilibrium solutions, $\mathbf{v} = 0$ and $\zeta = \text{const}$, to the shallow water equations are also solutions to the finite-volume equations (4.2.3) and (4.2.4) (Semenov 1983; Belikov and Semenov 1985).

The numerical scheme is stable on a uniform Cartesian grid if

$$\max(|C_x| + |C_y|) \le \max |C_x| + \max |C_y| \le 1, \tag{4.2.7}$$

where C_x and C_y are the CFL numbers that correspond to the x-axis and y-axis, respectively,

$$C_x = (|u| + c)\frac{\Delta t}{\Delta x}, \quad C_y = (|v| + c)\frac{\Delta t}{\Delta y},$$

where Δx and Δy are the mesh sizes in the corresponding directions. Condition (4.2.7) was generalized for an arbitrary mesh of convex polygons in the following way (Belikov and Semenov 1997c):

$$\max_i \frac{1}{2G_i} \left[\sum_{j=1}^{m(i)} (|V_n| - V_n + c)_j L_j \right] \Delta t \le 1, \qquad V_n = \mathbf{V} \cdot \mathbf{n}.$$

For the uniform Cartesian grid and constant flow velocity this criterion transforms into $\max(|C_x| + |C_y|) \le 1$.

The scheme described above is first-order accurate in time and space. The accuracy of schemes of this type can be improved by applying some well-known algorithms (see Sections 2.5 and 2.7). Higher-order accuracy may be required for the following reasons: (i) if the grid is highly nonuniform in space, the first-order scheme may be unacceptable as an approximation (the error may be of order unity); (ii) the first-order schemes are characterized by considerable numerical viscosity, which may significantly flatten the profiles of grid variables in the domains of smooth flow.

The order of accuracy with respect to the space coordinates can be increased by applying the interpolation algorithm. Let u and h to be linear functions within each cell, see Fig. 4.2b, and calculate the initial values of h^{L}, h^{R}, u^{L}, and u^{R} at the midpoints of the polygon sides. We construct linear distributions of the velocity and free-surface level inside the cell by applying a reconstruction procedure described in Section 2.7.

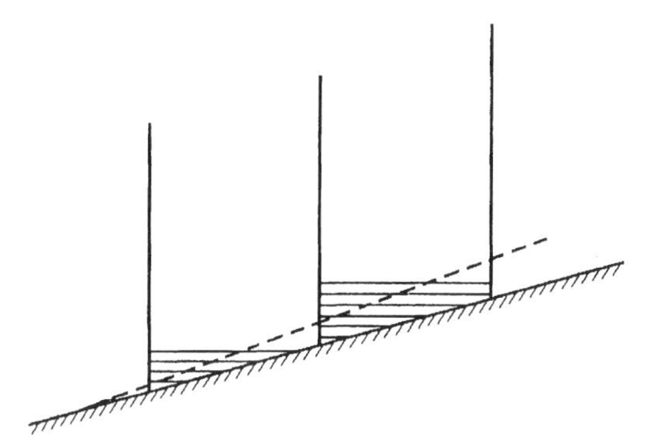

Figure 4.3 The piecewise constant and linear approximations of the free water surface.

As applied to the shallow water equations, a piecewise-linear approximation of the free surface of water is more natural than a piecewise constant one when a shallow water flow over a steep bottom is considered. In this case, the solution computed with a first-order accurate scheme was found to contain nonphysical discontinuities and pulsating shoal regions, see Fig. 4.3, and no relaxation to a steady limit was observed. A steady-state solution can be obtained, for example, by applying the second-order numerical method developed by Belikov and Semenov (1997b, 1997c). The order of accuracy in time can be increased by using standard predictor–corrector methods, see Section 2.5.

For the construction of Godunov-type methods for arbitrary moving, jump-fitting, and solution-adaptive grids, one can use the methods that were described for gas dynamic equations, see Sections 2.9 and 3.5.

4.3 Exact solution of the Riemann problem

Let us construct an exact solution of the Riemann problem for the shallow water equations in accordance with Stoker (1948, 1957). The importance of this exact solution stems from the fact that it has universal character and permits one to calculate flows over an arbitrary bottom relief, as well as flows with shoaling and dry bottom without any special jump-fitting techniques. In constructing the solution, we follow the logical scheme of Section 3.3 dealing with gas dynamics.

First we shall construct the basic elementary solutions. Then an exact solution will be constructed using the superposition principle.

4.3.1 Elementary solution 1: Hydraulic jump.

The first elementary solution of the shallow water equations is a moving discontinuity, or a hydraulic jump. For obtaining the discontinuity relations let us integrate Eq. (4.1.9) over t and x and consider the integral form

$$\oint_{\Gamma} (\mathbf{U}\, dx - \mathbf{F}\, dt) = \iint_{\Theta} \mathbf{B}\, dt\, dx, \tag{4.3.1}$$

where Γ is the boundary of a region Θ in the (t, x) plane. Let us seek discontinuous solution of Eq. (4.1.9) in the form of a travelling wave $f(t,x) = f(\eta) \equiv f(x - Wt)$, where $W = \text{const}$ is the wave velocity. Consider Eq. (4.3.1) in orthogonal coordinates (η, τ) associated with discontinuity: $\eta = x - Wt$ and $\tau = Wx + t$. The coordinate η is normal and coordinate τ is tangential with respect to the discontinuity. Using transformation

$$x = \frac{\eta + W\tau}{1 + W^2}, \quad t = \frac{-W\eta + \tau}{1 + W^2}$$

we can rewrite Eq. (4.3.1) as

$$\frac{1}{1 + W^2} \oint_L [(W\mathbf{U} - \mathbf{F})\, d\tau + (\mathbf{U} + W\mathbf{F})\, d\eta] = \frac{1}{1 + W^2} \iint_G \mathbf{B}\, d\tau\, d\eta. \tag{4.3.2}$$

Let us integrate (4.3.2) for the rectangular region $\tau_0 - \delta\tau \le \tau \le \tau_0 + \delta\tau$ and $\eta_0 - \delta\eta \le \eta \le \eta_0 + \delta\eta$, where $\eta = \eta_0$ corresponds to the discontinuity. We can find

$$(W\mathbf{U} - \mathbf{F})_1 2\delta\tau - (W\mathbf{U} - \mathbf{F})_2 2\delta\tau = \int_{\tau_0 - \delta\tau}^{\tau_0 + \delta\tau} \int_{\eta_0 - \delta\eta}^{\eta_0 + \delta\eta} \mathbf{B}\, d\tau\, d\eta.$$

For $\delta\eta \to 0$ we obtain

$$W\{\mathbf{U}\} - \{\mathbf{F}\} = \mathbf{0}, \tag{4.3.3}$$

where $\{q\} \equiv q_1 - q_2$, and indices 1 and 2 denote the variables on the left- and right-hand side of the discontinuity.

Consider the Riemann problem for shallow water equations with the following initial data: $\mathbf{U}_1 = [h_1, u_1, v_1]^{\mathrm{T}}$ for $x < 0$ and $\mathbf{U}_2 = [h_2, u_2, v_2]^{\mathrm{T}}$ for $x > 0$. Both $\mathbf{U}_1$ and $\mathbf{U}_2$ satisfy Eq. (4.3.3). Let this initial discontinuity moving with the velocity W be a solution. Then we find from Eq. (4.3.3) the desired formulas relating $\mathbf{U}_1$, $\mathbf{U}_2$, and W:

$$W\{h\} - \{hu\} = 0, \tag{4.3.4}$$
$$W\{hu\} - \{hu^2 + \tfrac{1}{2}gh^2\} = 0, \tag{4.3.5}$$
$$W\{hv\} - \{huv\} = 0. \tag{4.3.6}$$

If these conditions are satisfied, then the moving discontinuity is a solution of Eq. (4.1.9) with $b = \text{const}$. Let us introduce a new variable $V = u - W$ that is the velocity component normal to the discontinuity in coordinate system attached to it. Since $\{W\} = 0$, we obtain more compact relations:

$$\{hV\} = 0, \tag{4.3.7}$$
$$\{hV^2 + \tfrac{1}{2}gh^2\} = 0, \tag{4.3.8}$$
$$\{hvV\} = 0. \tag{4.3.9}$$

Suppose that the mass flux $m = hV = h_1 V_1 = h_2 V_2$ through the discontinuity is not equal to zero. In this case we have a hydraulic jump for which

$$\{v\} = v_1 - v_2 = 0. \tag{4.3.10}$$

The case $m = hV = 0$ that corresponds to a tangential discontinuity will be considered separately.

Equations (4.3.7)–(4.3.8) permit us to find compact formulas for u_2 and h_2 as functions of h_1, u_1, and W.

From (4.3.7) we have

$$V_2 = \frac{h_1 V_1}{h_2}.$$

Using this relation and Eq. (4.3.8), we arrive at the following formula for V_1:

$$V_1 = \pm\sqrt{\tfrac{1}{2}g(h_1 + h_2)\frac{h_2}{h_1}}.$$

Thus, we obtained V_1 as a function of h_2 and h_1. From Eqs. (4.3.7)–(4.3.8) it follows that $\{m\} = 0$ and $\{mV + \frac{1}{2}gh^2\} = 0$. Then,

$$m_1 = h_1 V_1 = \pm\sqrt{\tfrac{1}{2}g(h_1 + h_2)h_1 h_2}.$$

Let us decide which sign must be chosen in the above equation for m_1. The rule can be stated as follows: there can only exist a hydraulic jump with increasing h, and hence the depth h behind the hydraulic jump must be greater than that ahead of it. This condition was first formulated by Rayleigh (see Stoker 1957), and is used to select physically relevant solutions by analogy with the principle of the entropy (or pressure) increase across shock waves in gas dynamics (Rozhdestvenskii and Yanenko 1983). The Rayleigh condition is used to choose the correct sign of the square root to provide the existence a hydraulic jump with only increasing h.

Let us describe all configurations of a hydraulic jump.

Left hydraulic jump

We say that the hydraulic jump is *left* if the fluid flow is directed from left to right along the x-direction (Fig. 4.4a), i.e., $m = (u - W)h > 0$. Denote the flow parameters to the left of the jump by h_I and u_I and those to the right of the jump by h and u, see Fig. 4.4a. Using the relation $\{mV + \frac{1}{2}gh^2\} = 0$, we obtain the following equations for h and u $(h \geq h_\text{I})$, respectively:

$$u - u_\text{I} + \tfrac{1}{2}g\frac{h^2 - h_\text{I}^2}{|m_\text{I}|} = 0, \tag{4.3.11}$$

$$m_\text{I} = \sqrt{\tfrac{1}{2}g(h_\text{I} + h)h_\text{I} h}.$$

Right hydraulic jump

We say that the hydraulic jump is *right* if the fluid flow is directed from right to left, i.e., $m = (u - W)h < 0$. Denote the flow parameters to the right of the jump by h_II and

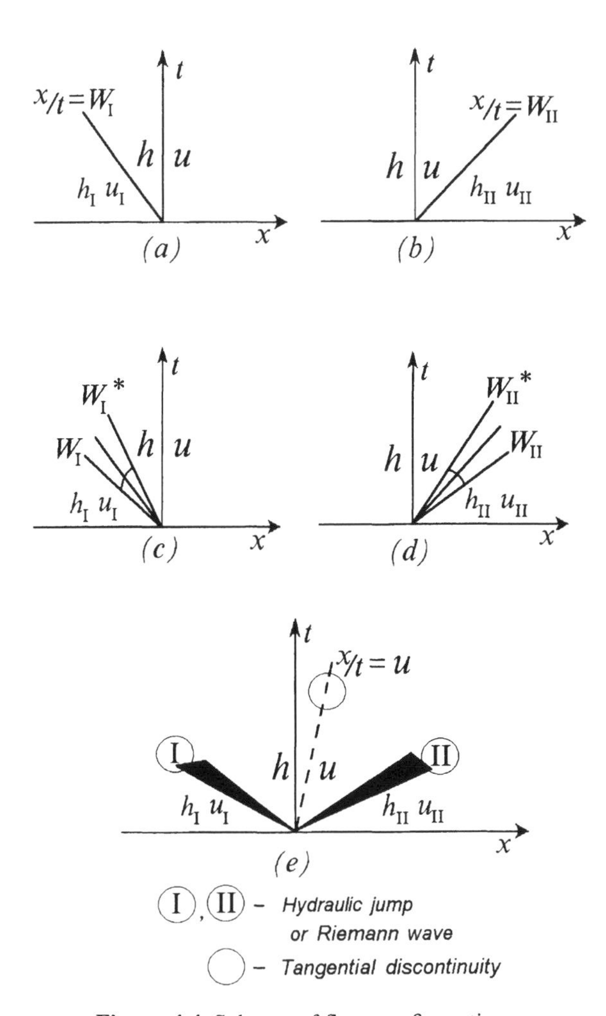

Figure 4.4 Scheme of flow configurations.

u_{II} and those to the left of the jump by h and u, see Fig. 4.4b. Applying the relation $\{mV + \frac{1}{2}gh^2\} = 0$, we obtain the following equation for h and u $(h \geq h_{\text{II}})$, respectively:

$$u - u_{\text{II}} - \tfrac{1}{2}g\frac{h^2 - h_{\text{II}}^2}{|m_{\text{II}}|} = 0, \tag{4.3.12}$$

$$m_{\text{II}} = \sqrt{\tfrac{1}{2}g(h_{\text{II}} + h)h_{\text{II}}h}.$$

A hydraulic jump is a self-similar solution with respect to the variable $\eta = x - Wt$, $W = \text{const}$. This solution is also self-similar with respect to the variable $\xi = x/t$. In fact, a discontinuity in the (t, x)-coordinates is a straight line defined by the equation $\xi = W = \text{const}$.

4.3.2 Elementary solution 2: Tangential discontinuity.

Another elementary solution of the shallow water equations is the so-called tangential discontinuity which is a special case of a discontinuity moving with a velocity $W = u = \text{const}$ and having zero mass flux across it, i.e., $m = hV = h_1V_1 = h_2V_2 = 0$. Denote the variables on the left- and right-hand side of the tangential discontinuity by h_1, u_1, and v_1 and h_2, u_2, and v_2, respectively. Then from formulas (4.3.7)–(4.3.8) we obtain $u_1 = u_2 = u$ and $h_1 = h_2 = h$, and from relation (4.3.9) we obtain that v_1 and v_2 can have an arbitrary jump.

4.3.3 Elementary solution 3: Riemann wave.

There also exist continuous elementary solutions of the shallow water equations. They can be sought for system (4.1.6)–(4.1.8) written in the nonconservative quasilinear form in assumption that the bottom is horizontal, $b = \text{const}$, and $\mathbf{f} = \mathbf{0}$:

$$h_t + uh_x + hu_x = 0, \tag{4.3.13}$$

$$u_t + uu_x + gh_x = 0, \tag{4.3.14}$$

$$v_t + uv_x = 0. \tag{4.3.15}$$

Let us seek a continuous solution in the self-similar form $f(x,t) = f(\xi) \equiv f(x/t)$. Substituting $h = h(\xi)$, $u = u(\xi)$, and $v = v(\xi)$ into Eqs. (4.3.13)–(4.3.15) we obtain

$$(u - \xi)h_\xi + hu_\xi = 0, \tag{4.3.16}$$

$$gh_\xi + (u - \xi)u_\xi = 0, \tag{4.3.17}$$

$$(u - \xi)v_\xi = 0. \tag{4.3.18}$$

System (4.3.16)–(4.3.18) admits a nonzero solution h_ξ, u_ξ, and v_ξ only if the determinant of the system is equal to zero. Thus, we have $u - \xi = 0$ and $u - \xi = \pm\sqrt{gh}$. Consider three corresponding cases, I, II, and III, for (4.3.16)–(4.3.18).

Case I:

$$u - \xi = 0, \quad h = 0, \quad h_\xi = 0.$$

The exact solution of the system is $u = \xi$ and $h = 0$. The function v is arbitrary. This formal solution has no physical meaning.

Case II:

$$\begin{aligned} &u - c - \xi = 0, \quad c = \sqrt{gh}, \\ &gh_\xi + cu_\xi = 0, \\ &v_\xi = 0 \quad \Longrightarrow \quad v = \text{const}. \end{aligned} \tag{4.3.19}$$

Here, the second equation is equivalent to $(u + 2c)_\xi = 0$, or $w_\mathrm{L} = u + 2c = \text{const}$, where w_L is the left Riemann invariant.

Case III:

$$\begin{aligned} &u + c - \xi = 0, \quad c = \sqrt{gh}, \\ &gh_\xi - cu_\xi = 0, \\ &v_\xi = 0 \quad \Longrightarrow \quad v = \text{const}. \end{aligned} \tag{4.3.20}$$

It follows that $w_\mathrm{R} = u - 2c = \text{const}$, where w_R is the right Riemann invariant.

Left Riemann wave

We say that the Riemann wave is *left* if the fluid flow through it is directed along the x-direction from left to right (Fig. 4.4c), i.e., $m = (u - \xi)h > 0$. It means that we have the relations

$$\begin{aligned} &u - c = \xi, \\ &\{w_\mathrm{L}\} = \{u + 2c\} = 0, \end{aligned} \tag{4.3.21}$$

and, hence, $m = (u - \xi)h = hc > 0$. Denote the flow parameters to the left of the Riemann wave by h_I and u_I and those to the right of it by h and u, see Fig. 4.4c. Using Eq. (4.3.21), we obtain the following equation for h and u:

$$u - u_\mathrm{I} + 2\left(\sqrt{gh} - \sqrt{gh_\mathrm{I}}\right) = 0. \tag{4.3.22}$$

Right Riemann wave

We say that the Riemann wave is *right* if the fluid flow through it is directed from right to left, i.e., $m = (u - \xi)h < 0$. It indicates that we deal with the relations

$$\begin{aligned} &u + c = \xi, \\ &\{w_\mathrm{R}\} = \{u - 2c\} = 0, \end{aligned} \tag{4.3.23}$$

and, hence, $m = (u - \xi)h = -hc < 0$. Denote the flow parameters to the right of the Riemann wave by h_II and u_II and those to the left of it by h and u, see Fig. 4.4d. Equation (4.3.23) leads to the following relation for h and u:

$$u - u_\mathrm{II} - 2\left(\sqrt{gh} - \sqrt{gh_\mathrm{II}}\right) = 0. \tag{4.3.24}$$

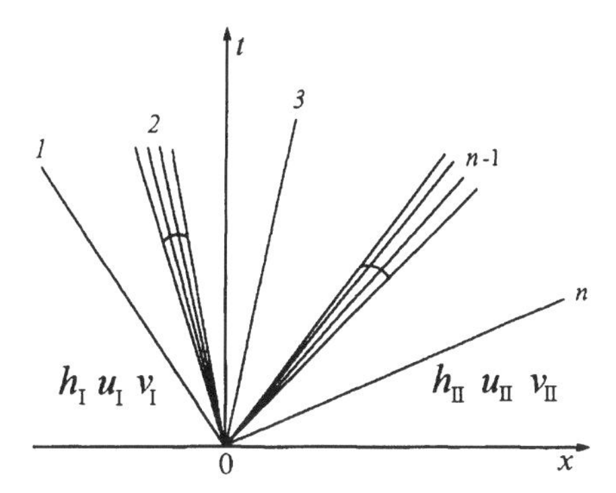

Figure 4.5 Construction of the general solution for shallow water equations.

4.3.4 General exact solution. The general exact solution of the shallow water Riemann problem with horizontal bottom consists of the elementary solutions separated by regions with constant quantities and can be represented on the plane (t, x) in the way shown in Fig. 4.5. The solution consists of, in total, n hydraulic jumps (HJ), tangential discontinuities (TD), and Riemann waves (RW). The RW is represented by a bundle of straight lines, or a rarefaction fan. The elementary solutions are separated by regions with constant quantities, since a constant flow is also an exact solution of the equations. The n basic solutions provide $3n$ relations for the constant flow parameters. These $3n$ relations must relate $3(n-1)$ unknowns h, u, and v located in $(n-1)$ inner regions with constant quantities, and n unknown velocities of the HJ, TD, and right boundaries of the RW. The number of unknowns must be equal to the number of relations. Thus, we have $3(n-1)+n=3n$ and $n=3$. We conclude, by analogy with gas dynamics, that the general solution is a combination of two hydraulic jumps and/or Riemann waves separated by a tangential discontinuity (Fig. 4.4e). Note that the left and right states in the Riemann problem flow can be selected so that the exact solution can be a single HJ, TD, or RW, see Fig. 4.4a–d, or only twin configurations HJ+HJ or RW+RW, and so on. In particular, Stoker (1948, 1957) investigated the shallow water flows with $v \equiv 0$. In this case there are no tangential discontinuities.

The above exact solutions are self-similar solutions with respect to the variable $\xi = x/t$. We seek an exact solutions of the Riemann problem in the form $f(\xi)$ since the initial step-like data for the Riemann problem are automatically consistent with this dependence on ξ. Alternative solution $\tilde{f}(t, x)$, such that $\tilde{f}(t, x) \neq \tilde{f}(\xi)$, is unknown. Moreover, if the solution is self-similar, it is unique.

It is noteworthy that h in the central zone for the case shown in Fig. 4.4e can be determined from only one equation. Subtracting Eq. (4.3.12) (resp., Eq. (4.3.24)) from Eq. (4.3.11) (resp., Eq. (4.3.22)) and eliminating u, we arrive at the following equation for h:

$$\mathcal{F}(h) = f_{\mathrm{I}}(h, h_{\mathrm{I}}) + f_{\mathrm{II}}(h, h_{\mathrm{II}}) = u_{\mathrm{I}} - u_{\mathrm{II}}, \tag{4.3.25}$$

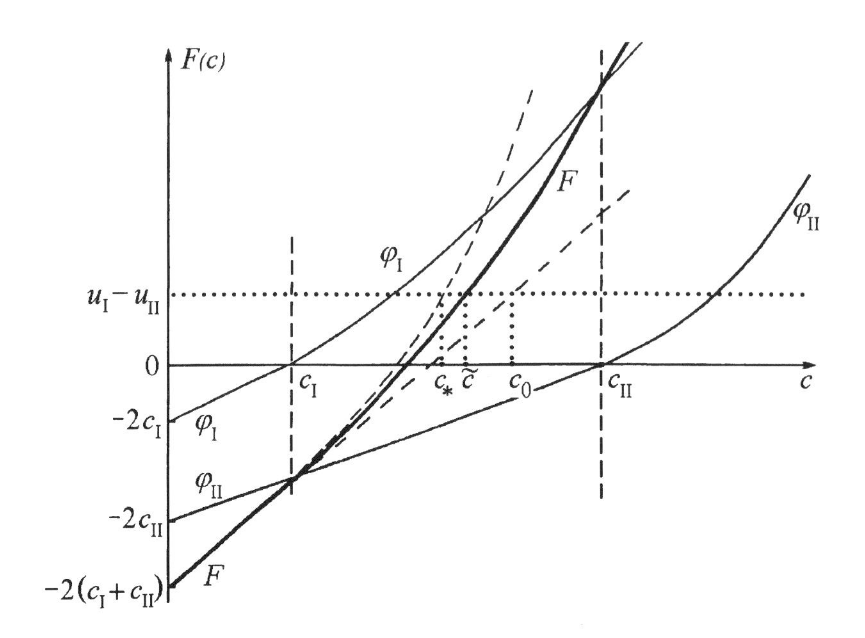

Figure 4.6 The graph of $F(c)$ for the case $c_I < c_{II}$.

$$f_n = f_n(h, h_n) = \begin{cases} \frac{1}{2}g\dfrac{h^2 - h_n^2}{\sqrt{\frac{1}{2}g(h + h_n)h_n h}} & \text{if } h \geq h_n, \\ 2(\sqrt{gh} - \sqrt{gh_n}\,) & \text{if } h < h_n, \end{cases} \qquad n = \text{I} \text{ or } \text{II}.$$

We can rewrite (4.3.25) in a more compact form by introducing $c = \sqrt{gh}$, $c_I = \sqrt{gh_I}$, and $c_{II} = \sqrt{gh_{II}}$ (Belikov and Semenov 1985, 1997b):

$$F(c) = \varphi_I(c, c_I) + \varphi_{II}(c, c_{II}) = u_I - u_{II}, \tag{4.3.26}$$

$$\varphi_n = \varphi_n(c, c_n) = \begin{cases} (c - c_n)(\sigma_n + 1)\sqrt{\frac{1}{2}(1 + \sigma_n^{-2})} & \text{if } \sigma_n = c/c_n \geq 1, \\ 2(c - c_n) & \text{if } \sigma_n < 1, \end{cases}$$

$$n = \text{I} \text{ or } n = \text{II}; \quad c_n = \sqrt{gh_n}.$$

Equation (4.3.26) holds for $c_n > 0$ (n = I or II). If $c_I = 0$ or $c_{II} = 0$, then there exists a single Riemann wave, and either relation (4.3.22) or (4.3.24) holds. Thus, the Riemann problem can be reduced to solving the single equation (4.3.26) for c. We denote its solution by $c = \tilde{c}$. Figure 4.6 illustrates the behavior of the function $F(c)$.

Straightforward calculation shows that $\partial F/\partial c > 0$ and $\partial^2 F/\partial c^2 \geq 0$. Specifically,

$$\frac{\partial \varphi_n}{\partial c} = \begin{cases} \dfrac{2\sigma_n^2 + 1 + \sigma_n^{-2}}{\sqrt{2(1 + \sigma_n^2)}} & \text{if } \sigma_n = c/c_n \geq 1, \\ 2 & \text{if } \sigma_n < 1, \end{cases}$$

$$\frac{\partial^2 \varphi_n}{\partial c^2} = \begin{cases} \dfrac{\sqrt{2}(\sigma_n - \sigma_n^{-1})(\sigma_n^2 + \sigma_n^{-2} + \frac{5}{2})}{(1+\sigma_n^2)^{3/2} c_n} & \text{if } \sigma_n = c/c_n \geq 1, \\ 0 & \text{if } \sigma_n < 1. \end{cases}$$

These properties of the derivatives indicate that $F(c)$ is monotone and convex downward. Equation (4.3.26), therefore, can be readily solved by Newton's or other iteration methods, see Ortega and Rheinboldt (1970). The positivity of the first and second derivatives ensures the convergence of the Newton iterations. Specifically, we can use the iterative process

$$c^{(m+1)} = c^{(m)} - \left[\varphi_{\rm I}^{(m)} + \varphi_{\rm II}^{(m)} - (u_{\rm I} - u_{\rm II})\right] \left[\left(\frac{\partial \varphi_{\rm I}}{\partial c}\right)^{(m)} + \left(\frac{\partial \varphi_{\rm II}}{\partial c}\right)^{(m)}\right]^{-1}$$

Here $m = 0, 1, 2, \ldots$ is the number of the iteration and c^0 is the initial approximation. Since the function has a positive second derivative, we can set c^0 equal to any positive number satisfying $F(c^0) - (u_{\rm I} - u_{\rm II}) \geq 0$. However, the most suitable starting point is the linear acoustic approximation:

$$c_0 = \tfrac{1}{4}(u_{\rm I} - u_{\rm II}) + \tfrac{1}{2}(c_{\rm I} + c_{\rm II}), \tag{4.3.27}$$

where c_0 is the exact solution to (4.3.26) if $c < c_{\rm I}$ and $c < c_{\rm II}$. Otherwise (see Fig. 4.6), c_0 lies to the right of the exact solution, which ensures that Newton's method is monotonically convergent.

To speed up the computational procedure, in practical realization we use the fact that $1 < q = 1 + \sigma_n^{-2} < 2$ for $\sigma_n > 1$ and calculate $\sqrt{q}$ as given by the best uniformly approximating trinomial, $\sqrt{q} \approx q_0 = 0.443451 + 0.629462q - 0.072268q^2$, which is characterized by a maximum relative error of 0.00066. If necessary, we can refine the result by performing one step of Newton's method: $\sqrt{q} \approx q_1 = \frac{1}{2}(q_0 + q/q_0)$. Thus, we reduce the maximum error to 3×10^{-7}, which is more than sufficient to compute any solution to the Riemann problem.

On calculating the speed c, we can find h and u, specifically,

$$h = \frac{c^2}{g}, \qquad u = \tfrac{1}{2}[u_{\rm I} + u_{\rm II} + \varphi_{\rm II}(c, c_{\rm II}) - \varphi_{\rm I}(c, c_{\rm I})].$$

The above equation for u is obtained by summing Eq. (4.3.11) (resp., (4.3.22)) and Eq. (4.3.12) (resp., (4.3.24)).

For the left hydraulic jump (4.3.11) we obtain $|m_{\rm I}| = (u_{\rm I} - W_{\rm I})h_{\rm I}$; hence,

$$m_{\rm I} = \sqrt{\tfrac{1}{2} g(h_{\rm I} + h) h_{\rm I} h},$$

$$W_{\rm I} = u_{\rm I} - \frac{m_{\rm I}}{h_{\rm I}} = u_{\rm I} - \sqrt{\tfrac{1}{2} g(h_{\rm I} + h)\frac{h}{h_{\rm I}}} = u_{\rm I} - c\sqrt{\tfrac{1}{2}(1 + \sigma_{\rm I}^2)}.$$

For the left Riemann wave (4.3.22) we have

$$c_{\rm I}^* = c_{\rm I} + \tfrac{1}{2}(u_{\rm I} - u) \quad (w_{\rm L} = \text{const}).$$

Then the left Riemann wave is bounded in the (t, x) plane by the rays $W_{\mathrm{I}} = u_{\mathrm{I}} - c_{\mathrm{I}}$ and $W_{\mathrm{I}}^* = u - c_{\mathrm{I}}^*$, see Fig. 4.4c. Denote the parameters for the ray ξ^*, $W_{\mathrm{I}} \leq \xi^* \leq W_{\mathrm{I}}^*$, by h^* and u^*, i.e., $h^* = h(\xi^*)$ and $u^* = u(\xi^*)$. The following relations are satisfied:

$$\begin{aligned} u^* - c^* &= \xi^*, \\ u^* + 2c^* &= u_{\mathrm{I}} + 2c_{\mathrm{I}}. \end{aligned} \tag{4.3.28}$$

Thus,

$$c^* = \tfrac{2}{3}c_{\mathrm{I}} + \tfrac{1}{3}(u_{\mathrm{I}} - \xi^*), \tag{4.3.29}$$

$$h^* = \frac{(c^*)^2}{g}, \tag{4.3.30}$$

$$u^* = \xi^* + c^*. \tag{4.3.31}$$

For the right hydraulic jump (4.3.12) we obtain $|m_{\mathrm{II}}| = -(u_{\mathrm{II}} - W_{\mathrm{II}})h_{\mathrm{II}}$, and hence,

$$\begin{aligned} m_{\mathrm{II}} &= \sqrt{\tfrac{1}{2}g(h_{\mathrm{II}} + h)h_{\mathrm{II}}h}, \\ W_{\mathrm{II}} &= u_{\mathrm{II}} + \frac{m_{\mathrm{II}}}{h_{\mathrm{II}}} = u_{\mathrm{II}} + c\sqrt{\tfrac{1}{2}(1 + \sigma_{\mathrm{II}}^2)}. \end{aligned}$$

For the right Riemann wave (4.3.24) we have

$$c_{\mathrm{II}}^* = c_{\mathrm{II}} - \tfrac{1}{2}(u_{\mathrm{II}} - u) \quad (w_{\mathrm{R}} = \text{const}).$$

Then the right Riemann wave is bounded in the (t, x) plane by the rays $W_{\mathrm{II}} = u_{\mathrm{II}} + c_{\mathrm{II}}$ and $W_{\mathrm{II}}^* = u + c_{\mathrm{II}}^*$, see Fig. 4.4d. We denote the parameters for the ray ξ^*, $W_{\mathrm{II}}^* \leq \xi^* \leq W_{\mathrm{II}}$, by h^* and u^*, i.e., $h^* = h(\xi^*)$ and $u^* = u(\xi^*)$. The following relations hold:

$$\begin{aligned} u^* + c^* &= \xi^*, \\ u^* - 2c^* &= u_{\mathrm{II}} - 2c_{\mathrm{II}}. \end{aligned} \tag{4.3.32}$$

Thus,

$$c^* = \tfrac{2}{3}c_{\mathrm{II}} + \tfrac{1}{3}(\xi^* - u_{\mathrm{II}}), \tag{4.3.33}$$

$$h^* = \frac{(c^*)^2}{g}, \tag{4.3.34}$$

$$u^* = \xi^* - c^*. \tag{4.3.35}$$

All these results are summarized in the table on page 241 (see Belikov and Semenov 1985, 1997b). Next, we can determine the values of h, u, and v, which lie on an arbitrary ray $\xi = x/t$ in the plane (t, x), where $-\infty < \xi < +\infty$. The case $\xi = 0$ corresponds to a stationary cell boundary. We write the solution for an arbitrary ξ as $H = H(\xi)$, $U = U(\xi)$, and $V = V(\xi)$.

The exact solution of the Riemann problem provides a continuity of the tangential velocity component $V = V(\xi)$ both in a hydraulic jump, see Eq. (4.3.10), and in a Riemann wave, see Eqs. (4.3.19) and (4.3.20). The component $V(\xi)$ can be discontinuous at only a tangential discontinuity. Thus, $V(\xi)$ for an arbitrary ray ξ are determined in accordance

Flow type	Left position	Right position
Hydraulic jump	If $u > \xi$ and $c \geq c_{\mathrm{I}}$, then $W_{\mathrm{I}}=u_{\mathrm{I}}-c\sqrt{\frac{1}{2}(1+\sigma_{\mathrm{I}}^2)}$ and $\sigma_{\mathrm{I}}=c/c_{\mathrm{I}}$, W_{I} is the hydraulic jump velocity. If $\xi > W_{\mathrm{I}}$, then $H=h, U=u$, and $V=v_{\mathrm{I}}$. If $\xi \leq W_{\mathrm{I}}$, then $H=h_{\mathrm{I}}, U=u_{\mathrm{I}}$, and $V=v_{\mathrm{I}}$.	If $u < \xi$ and $c \geq c_{\mathrm{II}}$, then $W_{\mathrm{II}}=u_{\mathrm{II}}+c\sqrt{\frac{1}{2}(1+\sigma_{\mathrm{II}}^2)}$ and $\sigma_{\mathrm{II}}=c/c_{\mathrm{II}}$, W_{II} is the hydraulic jump velocity. If $\xi < W_{\mathrm{II}}$, then $H=h, U=u$, and $V=v_{\mathrm{II}}$. If $\xi \geq W_{\mathrm{II}}$, then $H=h_{\mathrm{II}}, U=u_{\mathrm{II}}$, and $V=v_{\mathrm{II}}$.
Riemann wave (RW)	If $u > \xi$ and $c < c_{\mathrm{I}}$, then $W_{\mathrm{I}}=u_{\mathrm{I}}-c_{\mathrm{I}}$ and $W_{\mathrm{I}}^*=u-c_{\mathrm{I}}^*$ are the RW boundaries. If $\xi > W_{\mathrm{I}}^*$, then $H=h, U=u$, and $V=v_{\mathrm{I}}$. If $\xi < W_{\mathrm{I}}$, then $H=h_{\mathrm{I}}, U=u_{\mathrm{I}}$, and $V=v_{\mathrm{I}}$. If $W_{\mathrm{I}} \leq \xi = \xi^* \leq W_{\mathrm{I}}^*$, then $H=h^*, U=u^*$, and $V=v_{\mathrm{I}}$, see formulas (4.3.29)–(4.3.31).	If $u < \xi$ and $c < c_{\mathrm{II}}$, then $W_{\mathrm{II}}=u_{\mathrm{II}}+c_{\mathrm{II}}$ and $W_{\mathrm{II}}^*=u+c_{\mathrm{II}}^*$ are the RW boundaries. If $\xi < W_{\mathrm{II}}^*$, then $H=h, U=u$, and $V=v_{\mathrm{II}}$. If $\xi > W_{\mathrm{II}}$, then $H=h_{\mathrm{II}}, U=u_{\mathrm{II}}$, and $V=v_{\mathrm{II}}$. If $W_{\mathrm{II}}^* \leq \xi = \xi^* \leq W_{\mathrm{II}}$, then $H=h^*, U=u^*$, and $V=v_{\mathrm{II}}$, see formulas (4.3.33)–(4.3.35).

with the velocity u of the tangential discontinuity, see the table. We put $V(\xi)=v_{\mathrm{I}}$ if $u \geq \xi$, otherwise $V(\xi)=v_{\mathrm{II}}$.

Figure 4.4e presents the basic possible types of a self-similar flow in the (t,x) plane. Here, the hydraulic jumps or Riemann waves are separated by a tangential discontinuity. All elementary solutions are separated by the regions with constant quantities. To solve the Riemann problem for the flows presented in Fig. 4.4e, it suffices to determine u and h in the central domain, the propagation velocities of the jumps, tangential discontinuity and Riemann waves, and the flow characteristics inside the Riemann wave fan.

Note that an investigation of the function $F(c)$ permits one to determine the configuration of an exact solution without calculations (Belikov and Semenov 1997b, 1997c). For definiteness we assume $c_{\mathrm{I}} < c_{\mathrm{II}}$. The behavior of $\varphi_{\mathrm{I}}(c)$, $\varphi_{\mathrm{II}}(c)$ and their sum, $F(c)$, for this case are shown in Fig. 4.6. The values of $F(c)$ at $c=c_{\mathrm{II}}$, $c=c_{\mathrm{I}}$, and $c=0$ are

$$F(c_{\mathrm{II}}) = U_{\mathrm{II}} = \frac{1}{\sqrt{2}}(c_{\mathrm{II}}-c_{\mathrm{I}})\left(\frac{c_{\mathrm{II}}}{c_{\mathrm{I}}}+1\right)\sqrt{1+\left(\frac{c_{\mathrm{I}}}{c_{\mathrm{II}}}\right)^2},$$
$$F(c_{\mathrm{I}}) = U_{\mathrm{I}} = 2(c_{\mathrm{I}}-c_{\mathrm{II}}), \quad F(0) = U_0 = -2(c_{\mathrm{I}}+c_{\mathrm{II}}).$$

Depending on the value of $u_{\mathrm{I}}-u_{\mathrm{II}}$, the following cases are possible:

(i) If $u_{\mathrm{I}}-u_{\mathrm{II}} > U_{\mathrm{II}}$, then $c > c_{\mathrm{II}} > c_{\mathrm{I}}$, i.e., the configuration of a flow with right and left hydraulic jumps is realized;

(ii) If $U_{\mathrm{I}} < u_{\mathrm{I}}-u_{\mathrm{II}} < U_{\mathrm{II}}$, then $c_{\mathrm{I}} < c < c_{\mathrm{II}}$, i.e., the configuration with a right Riemann wave and a left hydraulic jump is realized;

(iii) If $U_0 < u_{\mathrm{I}}-u_{\mathrm{II}} < U_{\mathrm{I}}$, then $0 < c < c_{\mathrm{I}}$, i.e., the configuration with right and left Riemann waves is realized;

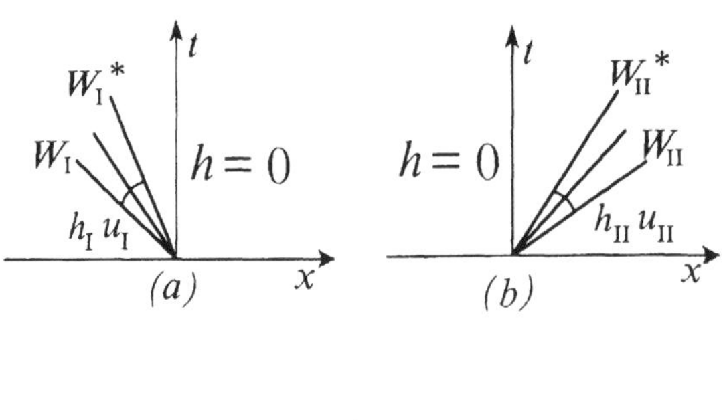

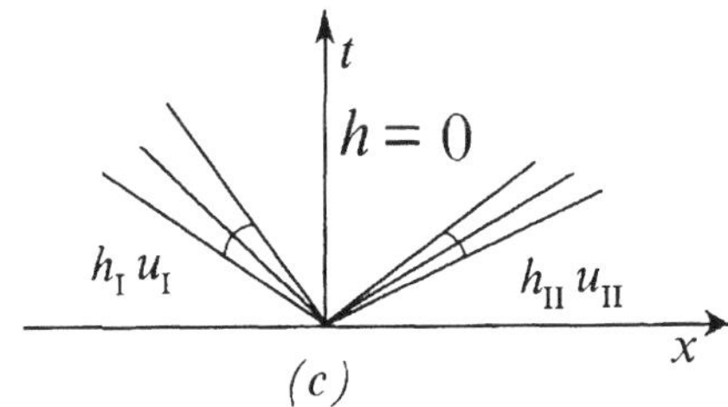

Figure 4.7 Configurations for flow with shoaling.

(iv) If $u_{\mathrm{I}} - u_{\mathrm{II}} < U_0$, then a dry region with $h = 0$ and $c = 0$ occurs. The dry region is framed by either twin Riemann waves or a single (rightward or leftward propagating) Riemann wave, see Fig. 4.7.

Equation (4.3.26) has a unique solution if $u_{\mathrm{I}} - u_{\mathrm{II}} > U_0$. Consider the cases where Eq. (4.3.26) does not have a solution, which takes place if $u_{\mathrm{I}} - u_{\mathrm{II}} < U_0 = -2(c_{\mathrm{I}} + c_{\mathrm{II}})$. In the general case a domain of dry bottom ($h = 0$) is framed by two left and right Riemann wave forms, see Fig. 4.7c. Figures 4.7a–b illustrate special cases of a single Riemann wave on the left and on the right of the dry-bottom domains. The configuration of a single right Riemann wave (Fig. 4.7b) develops if the depth and velocity are zero on the left of the discontinuity at $t = 0$. Assuming $h_{\mathrm{I}} = h = 0$ and $u_{\mathrm{I}} = u = 0$, we obtain a single right Riemann wave that separates the domain where $u = u_{\mathrm{II}}$ and $h = h_{\mathrm{II}}$ from the zero-depth domain in the plane (t, x). In this case, $W_{\mathrm{II}}^* = u_{\mathrm{II}} - 2c_{\mathrm{II}}$ and $W_{\mathrm{II}} = u_{\mathrm{II}} + c_{\mathrm{II}}$. These expressions can be obtained from Eq. (4.3.32) by setting $c^* = 0$. On assuming further that $h_{\mathrm{II}} = h = 0$ and $u_{\mathrm{II}} = u = 0$, we obtain a single left Riemann wave, see Fig. 4.7a. In this case, $W_{\mathrm{I}}^* = u_{\mathrm{I}} + 2c_{\mathrm{I}}$ and $W_{\mathrm{I}} = u_{\mathrm{I}} - c_{\mathrm{I}}$. These expressions can be obtained from Eq. (4.3.28) by setting $c^* = 0$. The general case (Fig. 4.7c) is the combination of the above left and right Riemann waves with a shoal region. The flow parameters inside a left or right Riemann wave can be obtained from Eqs. (4.3.29)–(4.3.31) or Eqs. (4.3.33)–(4.3.35).

The generalized Riemann problem

So far, we have considered the exact solution of Eq. (4.1.9) for a horizontal bottom $b(x) = \text{const}$ and $\mathbf{f} = \mathbf{0}$. For piecewise-linear initial data $\mathbf{U}_0(x)$ for $\mathbf{U}$ and continuous piecewise-linear bottom $b(x)$ (Fig. 4.2b) we can solve the generalized Riemann problem, see Section 2.4. In this case the vector $\mathbf{U}_0$ and the function $b(x)$ have the following form:

$$\mathbf{U}_0(x) = \begin{cases} \mathbf{Q}_{+0} + \mathbf{Q}_{+1}x, & x > 0, \\ \mathbf{Q}_{-0} + \mathbf{Q}_{-1}x, & x < 0, \end{cases} \qquad b(x) = \begin{cases} b_0 + B_+x, & x \geq 0, \\ b_0 + B_-x, & x < 0, \end{cases} \tag{4.3.36}$$

where $\mathbf{Q}_{+0}$, $\mathbf{Q}_{+1}$, $\mathbf{Q}_{-0}$, and $\mathbf{Q}_{-1}$ are constant vectors and b_0, B_-, and B_+ are constants.

We shall seek a solution of Eq. (4.1.9) in the form of a series,

$$\mathbf{U}(t,x) \equiv \mathbf{U}(\xi,\theta) = \sum_{m=0}^{+\infty} \mathbf{V}_m(\xi)\theta^m = \mathbf{V}_0(\xi) + \mathbf{V}_1(\xi)\theta + O(\theta^2), \tag{4.3.37}$$

$$\xi = x/t, \quad \theta = \sqrt{x^2+t^2}; \qquad t = \theta/\sqrt{1+\xi^2}, \quad x = \theta\xi/\sqrt{1+\xi^2}.$$

Equation (4.3.37) is in agreement with the initial value of (4.3.36) if we have

$$\lim_{t\to 0} \mathbf{V}_0(\xi) = \begin{cases} \mathbf{Q}_{+0}, & x > 0, \\ \mathbf{Q}_{-0}, & x < 0; \end{cases} \qquad \lim_{t\to 0} \mathbf{V}_1(\xi) = \begin{cases} \mathbf{Q}_{+1}, & x > 0, \\ \mathbf{Q}_{-1}, & x < 0. \end{cases}$$

Substituting $\mathbf{U}(\xi,\theta)$ from (4.3.37) into Eq. (4.1.9) with $\mathbf{d} = 0$, we obtain

$$\left(-\xi\frac{\partial \mathbf{U}}{\partial \xi} + A\frac{\partial \mathbf{U}}{\partial \xi}\right)\sqrt{1+\xi^2} + \left(\frac{\partial \mathbf{U}}{\partial \theta} + \xi A\frac{\partial \mathbf{U}}{\partial \theta}\right)\frac{\theta}{\sqrt{1+\xi^2}} = \theta D\mathbf{U}. \tag{4.3.38}$$

Thus, the zeroth approximation term $\mathbf{V}_0$ can be found from the equation

$$-\xi\frac{\partial \mathbf{V}_0}{\partial \xi} + A(\mathbf{V}_0)\frac{\partial \mathbf{V}_0}{\partial \xi} = (A_0 - \xi I)\frac{\partial \mathbf{V}_0}{\partial \xi} = \mathbf{0}, \tag{4.3.39}$$

where I is the 3×3 identity matrix and $A_0 = A(\mathbf{V}_0)$, see (4.1.14). The zeroth approximation term $\mathbf{V}_0$ of Eq. (4.3.39) is an exact solution of the Riemann problem with horizontal bottom in terms of the self-similar variable ξ and does not depend on the bottom relief. Thus, to construct the Godunov method of the first order of accuracy we can use only this exact solution and take into account the bottom relief additionally, in particular, by (4.2.6).

To construct the Godunov method of the second order of accuracy we can use higher-order time-approximation procedures, e.g., Runge–Kutta predictor–corrector methods, see Section 2.5.

Also, one may approximate the solution by two terms of series (4.3.37). The formulas of $\mathbf{V}_1$ in gas dynamic equations were derived by Men'shov (1990, 1991). Such an approach permits us to determine exact formulas for boundary fluxes that may be used for the construction of the Godunov method of the second order of accuracy. However, the analytical formulas for the generalized Riemann problem are rather complicated. Note that the first approximation term $\mathbf{V}_1$ can be obtained from Eq. (4.3.38) as follows. Substituting series (4.3.37) into Eq. (4.3.38) and taking into account the terms with θ only, we obtain the following equation for $\mathbf{V}_1$:

$$-\frac{\partial \mathbf{V}_1}{\partial \xi}\xi\sqrt{1+\xi^2} + A_0\frac{\partial \mathbf{V}_1}{\partial \xi}\sqrt{1+\xi^2} + \left(\frac{\partial A}{\partial \mathbf{U}}\right)_0 : \left(\mathbf{V}_1\frac{\partial \mathbf{V}_0}{\partial \xi}\right)\sqrt{1+\xi^2}$$
$$+\mathbf{V}_1\frac{1}{\sqrt{1+\xi^2}} + A_0\mathbf{V}_1\frac{\xi}{\sqrt{1+\xi^2}} = D\mathbf{V}_0,$$

where the colon stands for the contraction of tensors and/or matrices over two indices. The notation $\mathbf{v}_1 = \mathbf{V}_1\sqrt{1+\xi^2}$ permits us to represent the equation in a more compact form,

$$(A_0 - \xi I)\frac{\partial \mathbf{v}_1}{\partial \xi} + \mathbf{v}_1 + \left(\frac{\partial A}{\partial \mathbf{U}}\right)_0 : \left(\mathbf{v}_1 \frac{\partial \mathbf{V}_0}{\partial \xi}\right) = D\mathbf{V}_0. \qquad (4.3.40)$$

Equation (4.3.40) is linear in $\mathbf{v}_1$, and if $\mathbf{V}_0$ is known, we can always determine $\mathbf{V}_1$ from the solution of Eq. (4.3.40).

In particular, let the initial data be equal to $\mathbf{Q}_0 + \mathbf{Q}_1 x$, $b(x) = b_0 + Bx$, $-\infty < x < +\infty$, where $\mathbf{Q}_0$ and $\mathbf{Q}_1$ are constant vectors, and b_0 and B are constants. Then we obtain $\mathbf{V}_0(\xi) = \mathbf{Q}_0$ and $\partial\mathbf{V}_0/\partial\xi = 0$. Hence from Eq. (4.3.40) it follows that $\mathbf{v}_1 = \mathbf{Q}_1\xi - A_0\mathbf{Q}_1 + D\mathbf{Q}_0$, or $\mathbf{V}_1 = (\mathbf{Q}_1\xi - A\mathbf{Q}_1 + D\mathbf{Q}_0)/\sqrt{1+\xi^2}$. Thus, with $O(\theta^2)$ error we have

$$\mathbf{U}(t,x) = \mathbf{Q}_0 + (\mathbf{Q}_1\xi - A\mathbf{Q}_1 + D\mathbf{Q}_0)\frac{\theta}{\sqrt{1+\xi^2}} = \mathbf{Q}_0 + \mathbf{Q}_1 x - A_0\mathbf{Q}_1 t + D\mathbf{Q}_0 t.$$

Linear approximation estimates

Consider now the error of the linear approximation (4.3.27) for c_0 as compared with the exact solution $c = \tilde{c}$ (Belikov and Semenov 1988a, 1997b, 1997c). Without loss of generality, we can assume that $c_{\mathrm{I}} < c_{\mathrm{II}}$. As noted above, c_0 is the exact solution to (4.3.26) in the domain $c \le c_{\mathrm{I}}$. Consider the domain $c_{\mathrm{I}} < c \le c_{\mathrm{II}}$. The estimate

$$\varphi_{\mathrm{I}}(c, c_{\mathrm{I}}) = \frac{c^2 - c_{\mathrm{I}}^2}{\sqrt{2}c_{\mathrm{I}}}\sqrt{1 + \frac{c_{\mathrm{I}}^2}{c^2}} < \frac{c^2 - c_{\mathrm{I}}^2}{c_{\mathrm{I}}}$$

results in

$$F_*(c) = \frac{c^2 - c_{\mathrm{I}}^2}{c_{\mathrm{I}}} + 2(c - c_{\mathrm{II}}) > F(c).$$

Consequently, the solution c_* to the equation $F_*(c) = u_{\mathrm{I}} - u_{\mathrm{II}}$ is given by $c_* = 2\sqrt{c_{\mathrm{I}}c_0} - c_{\mathrm{I}}$. This solution lies to the left of the exact solution, see Fig. 4.6. Setting $c_0 = (1+\alpha)c_{\mathrm{I}}$, where $0 < \alpha < 1$, we obtain an estimate for the relative error δ_{I}:

$$\delta_{\mathrm{I}} = \frac{c_0 - \tilde{c}}{\tilde{c}} < \frac{c_0 - c_*}{c_{\mathrm{I}}} = 2 + \alpha - 2\sqrt{1+\alpha} = \tfrac{1}{4}\alpha^2 - \tfrac{1}{8}\alpha^3 + O(\alpha^4). \quad (4.3.41)$$

Consider now the domain where $c > c_{\mathrm{II}}$; we have

$$\varphi_{\mathrm{I}}(c, c_{\mathrm{I}}) < \frac{c^2 - c_{\mathrm{I}}^2}{c_{\mathrm{I}}}, \quad \varphi_{\mathrm{II}}(c, c_{\mathrm{II}}) < \frac{c^2 - c_{\mathrm{II}}^2}{c_{\mathrm{II}}}.$$

Therefore, the solution

$$c_{**} = \sqrt{c_{\mathrm{I}}c_{\mathrm{II}}\left(1 + \frac{u_{\mathrm{I}} - u_{\mathrm{II}}}{c_{\mathrm{I}} + c_{\mathrm{II}}}\right)}$$

to the equation

$$F_{**}(c) = \frac{c^2 - c_{\mathrm{I}}^2}{c_{\mathrm{I}}} + \frac{c^2 - c_{\mathrm{II}}^2}{c_{\mathrm{II}}} = u_{\mathrm{I}} - u_{\mathrm{II}}$$

also lies to the left of the exact solution $c = \tilde{c}$. Suppose $c_{II} = (1+\beta)c_I$, where $1 > \beta > \alpha > 0$, since $c_{II} > c_0 > c_I$. From the relation

$$c_0 = (1+\alpha)c_I = \tfrac{1}{4}(u_I - u_{II}) + \tfrac{1}{2}(c_I + c_{II}) = \tfrac{1}{4}(u_I - u_{II}) + (1 + \tfrac{1}{2}\beta)c_I$$

we obtain $u_I - u_{II} = (2\alpha - \beta)2c_I$. Then,

$$\begin{aligned} \delta_{II} &= \frac{c_0 - \tilde{c}}{\tilde{c}} < \frac{c_0 - c_{**}}{c_I} = 1 + \alpha - \sqrt{(1+\beta)\left(1 + \frac{2\alpha - \beta}{1 + \frac{1}{2}\beta}\right)} \\ &= \tfrac{1}{4}\alpha^2 + \tfrac{1}{4}(\alpha - \beta)^2 - \tfrac{1}{2}(\alpha - \beta)\alpha^2 - \tfrac{1}{8}\beta^3 + O([\alpha + \beta]^4). \end{aligned} \tag{4.3.42}$$

If $\alpha = \beta$ or $c_0 = c_{II}$, the estimated δ_I and δ_{II} are equal; this ensures the continuity of the error for $c > c_I$. It can be readily shown that the maximum errors estimated by (4.3.41) and (4.3.42) are obtained at $\alpha = 0$ or $c_I = c_0$ and expressed as

$$\delta_{max} = \max(\delta_I, \delta_{II}) = \tfrac{1}{4}\beta^2 - \tfrac{1}{8}\beta^3 + O(\beta^4) < \tfrac{1}{4}\beta^2.$$

Thus, the maximum relative error of the linear approximation c_0 does not exceed $\frac{1}{4}\beta^2$; therefore, the error estimate for $h = c^2/g$ does not exceed $\frac{1}{2}\beta^2$. By similar considerations, it can be shown that the error of the linear approximation for the velocity

$$u_0 = \tfrac{1}{2}(u_I + u_{II}) + c_I - c_{II}$$

is, for $\alpha = 0$,

$$\delta_u = \frac{u_0 - u}{c_I} = \tfrac{3}{8}\beta^3 - \tfrac{3}{32}\beta^4 + O(\beta^5).$$

The formulas presented above can be used to make an *a priori* estimation of the error of a solution to (4.3.26) by comparing the values of c_0 and c_I. For example, if $\alpha = (c_0 - c_I)/c_I = 0.1$ and $u_I \approx u_{II}$, then we use (4.3.27) to obtain $c_{II} = 1.2c_I$, $\beta = 0.2$, and $h_{II} = 1.44h_I$. Therefore, when the ratio of the depths across the discontinuity is less than 1:1.44, we can use the linear approximation to calculate the values of c with an error within 1%, since $\frac{1}{4}\beta^2 \leq 0.01$.

Practical application of the exact Riemann solver described here has shown that an error of 1% is acceptable in almost any problem. In only a few problems analyzed by this method, the values on the cell boundaries that correspond to strong jumps must be further refined by Newton's iteration method. Thus, by using the value given by (4.3.27) as the initial approximation, one can substantially reduce the number of iteration cycles involved in the most frequently executed steps of the Riemann solver and improve the efficiency of the method as a whole.

4.4 Results of numerical analysis

In this section we present numerical results obtained for some one- and two-dimensional problems to illustrate the main features of the Godunov method based on the exact Riemann solver. Some of these results were reported in more detail by Belikov and Semenov (1985, 1988a, 1988b, 1997b, 1997c, 1998).

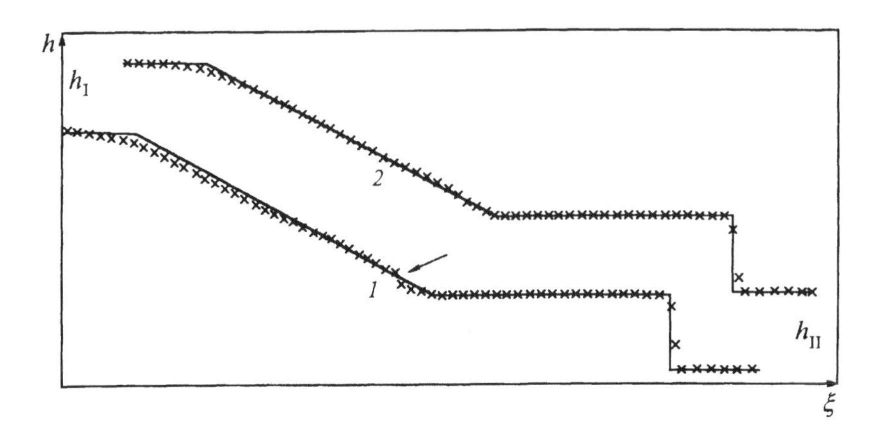

Figure 4.8 Disintegration of $u_I = u_{II} = 0$ and $h_I{:}h_{II} = 10{:}1$.

Figure 4.8 shows the results computed for the one-dimensional Riemann problem with the initial depth ratio $h_I{:}h_{II} = 10{:}1$ and zero initial velocity; the bottom was assumed to be horizontal. This problem was solved analytically by Stoker (1948, 1957) and is widely used for testing the performance of numerical algorithms. Here, the distribution of h calculated on a 100-cell domain is presented as a function of the similarity variable $\xi = x/(t\sqrt{gh_I})$. The solid lines represent analytical solutions, the crosses at curve 1 represent results computed with a first-order accurate scheme, and those at curve 2 represent results computed with a second-order scheme with the minmod limiter. By using the second-order scheme, we eliminated the jump obtained with the first-order scheme at the point of initial discontinuity, the so-called start-off error, indicated by the arrow in Fig. 4.8. In the calculations of the depth with respect to the initial level $h_I = 10$ by the second-order scheme, the error did not exceed 1%. Both schemes permitted us to obtain the monotone profiles of the solution and smear the hydraulic jump over two or three cells only.

Figure 4.9 shows the results of a one-dimensional test that turned out to be rather difficult for numerical simulation. The test is formulated as follows. At the initial moment, the computational domain $0 \leq x \leq L$ is divided into 100 cells. In two or four cells that are adjacent to its left boundary, the depth is set equal to the constant value $h = h_0 = 100$ m; in the remaining cells, $h = 0$. The initial velocity is zero, and the bottom is horizontal. The flow is assumed to be symmetric about $x = 0$. The physical process described by this problem is interpreted as the evolution of a narrow planar column of water under the action of the gravity force in terms of the shallow water theory. This solution is an analogue of the solution to the problem of gas dynamic point source explosion by Sedov (1946) and Taylor (1950); see also Korobeinikov (1991) and Sedov (1993). Note that this test is rather formal, because physically meaningful results cannot be guaranteed when the shallow water approximation is used without taking into account the effects of hydrodynamic instability and three-dimensional nature of the real flow. However, this test is used here to check the numerical algorithm described above.

Figure 4.9 illustrates the dynamics of the disintegration of a water column in the coor-

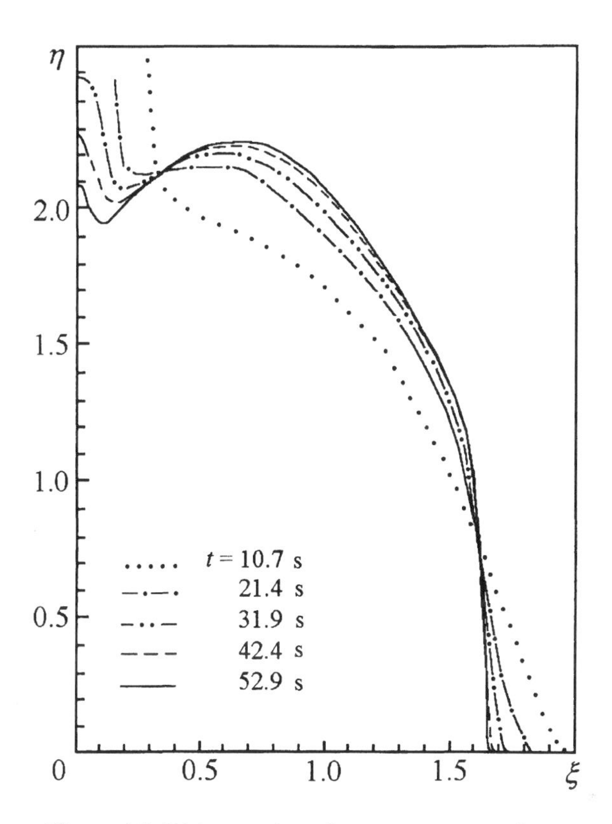

Figure 4.9 Disintegration of a narrow water column.

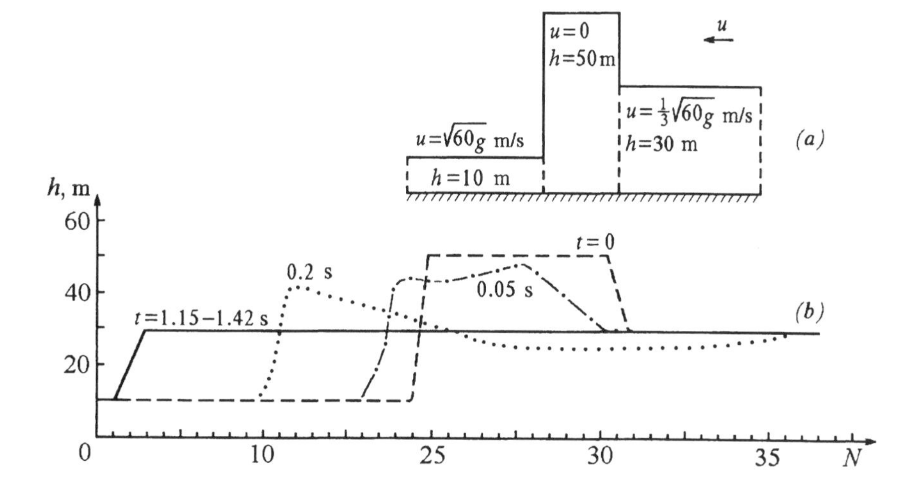

Figure 4.10 Modelling of a steady jump.

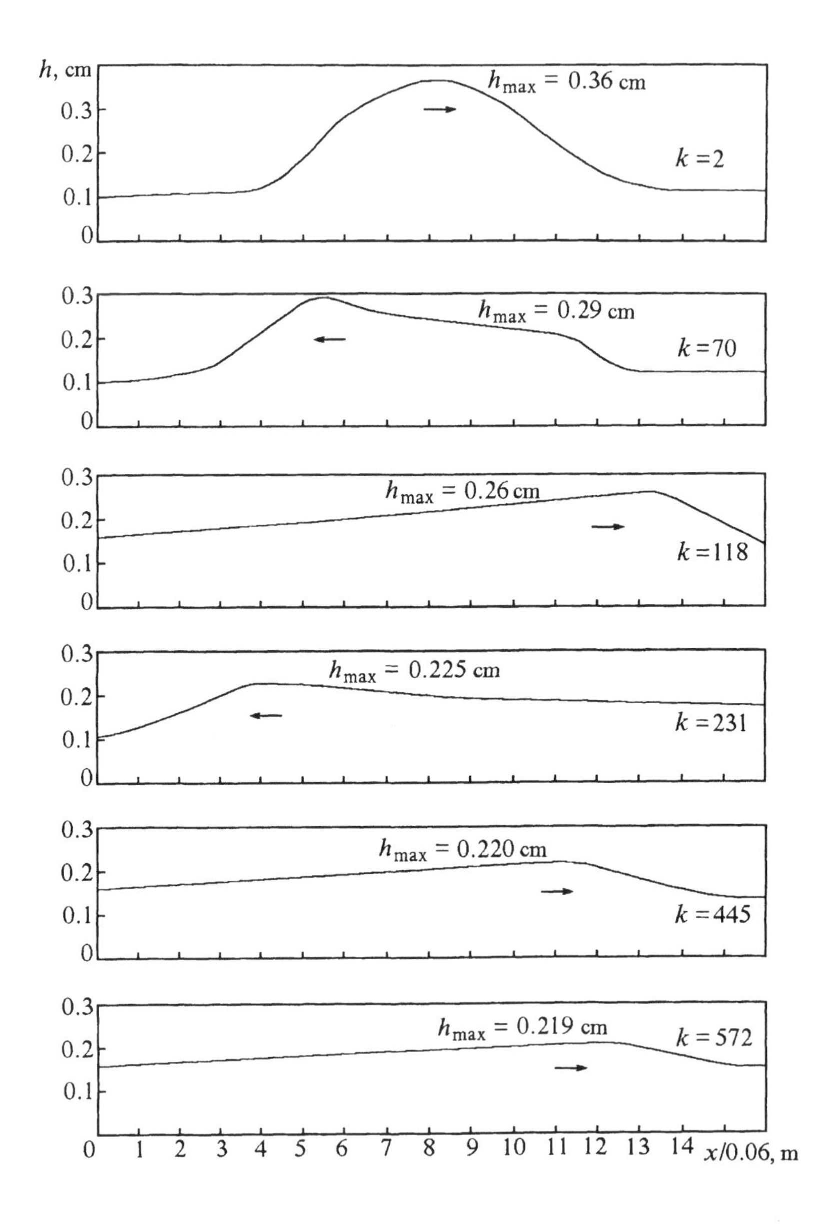

Figure 4.11 Modelling of soliton propagation.

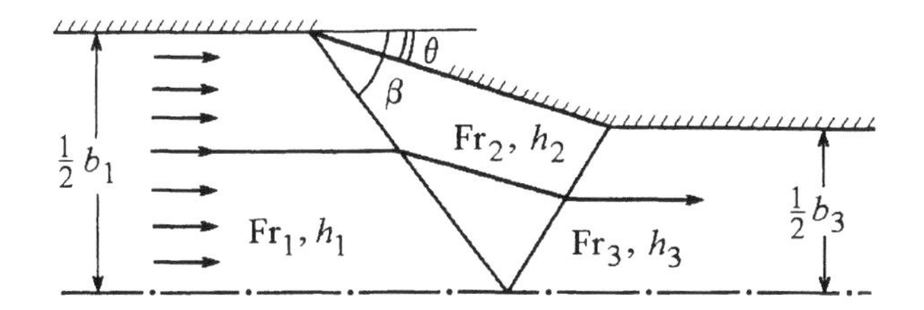

Figure 4.12 Scheme of flow into a narrowing channel.

dinates $\eta = th(t,x)/h_0$ and $\xi = x/(t\sqrt{gh_0})$, where t is time measured in seconds. Note that with the time elapsed, the curves tend to an asymptotic self-similar solution. As ξ increases, the velocity linearly increases and then drops to zero at the propagating jump front. The results shown in Fig. 4.9 were obtained with the first-order scheme. The results computed with higher order schemes showed small but unremovable oscillations.

Figure 4.10b shows the results of computation of a steady hydraulic jump, which is important as a test for various numerical schemes; see Ivanov, Koretskii, and Kurochkina (1980a, 1980b, 1980c). Here, we demonstrate sequential stages of the time variation of h as the steady-state regime of a hydraulic jump is approached.

The initial conditions, see Fig. 4.10a, were chosen so as to guarantee the development of a steadily propagating jump. The numerical jump was smeared over two cells.

To study the dispersion properties of the scheme, see Ivanov et al. (1980a, 1980b, 1980c), we investigate numerically a soliton propagation for large times. A soliton, see Fig. 4.11, propagates in a finite region (70 cells) with boundary conditions of solid wall, where k is the time level number and $\Delta x = 0.06$. The initial soliton transforms with time into a wave of triangular form with the front smeared over three cells. The initial length of the soliton was 10 cells.

We also performed computations for some two-dimensional problems. One of them is schematically illustrated in Fig. 4.12. Here, we consider supercritical water flow with Froude number $\mathrm{Fr} = |u|/\sqrt{gh} = |u|/c > 1$ flowing from a rectangular flume, through a contracting symmetric chute, into a narrowing channel whose walls converge at an angle θ. The hydraulic jumps that make an angle β with the walls of the contraction are reflected from the symmetry line of the flow, then from the walls, etc., forming a pattern consisting of elevations and depressions of the free-surface level. The results presented in Fig. 4.13 were obtained for $\theta = 10°$, $\mathrm{Fr}_1 = 3$, $h_1 = 100$ mm, $b_1 = 1$ m, and $b_3 = 0.511$ m (with zero bottom slope) on progressively finer grids having 7×26 (a), 14×52 (b), and 28×104 (c) cells. The numerical results, which were calculated with the first-order scheme along the downstream coordinate and with the second-order scheme along the transverse coordinate, converge to the exact solution shown by dashed lines, which describes a system of steady hydraulic jumps, see Section 4.6. Here, solid curves represent the calculated isopleths of the free-surface level; the numbers indicates the level measured in millimeters. As the grid is made finer, the localization of hydraulic jumps becomes more distinct.

Figure 4.14 compares the calculated and measured free surface levels h shown by solid curves and crosses, respectively. We see that the numerical results are in good agreement with the experiment. In this problem, we used slightly different values of chute parameters

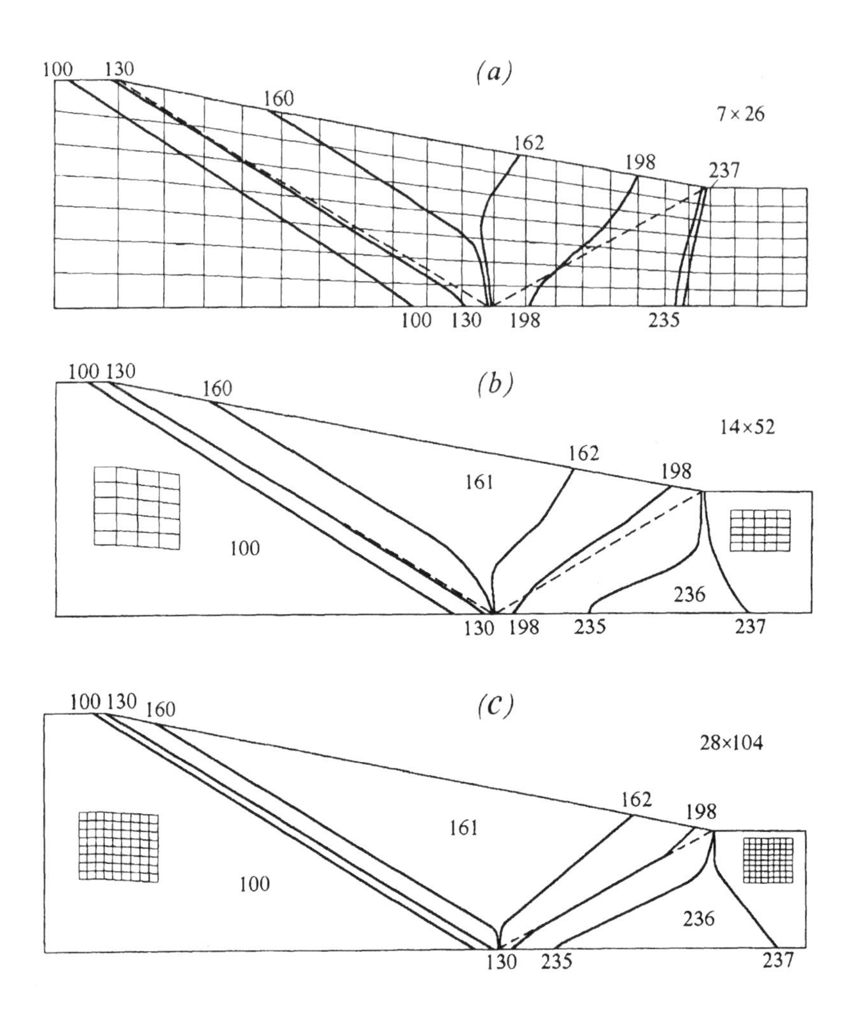

Figure 4.13 Convergence of numerical results for finer grids.

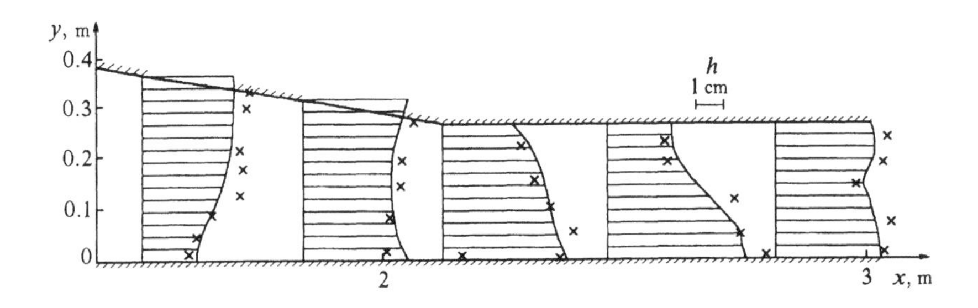

Figure 4.14 Comparison of experimental and numerical data.

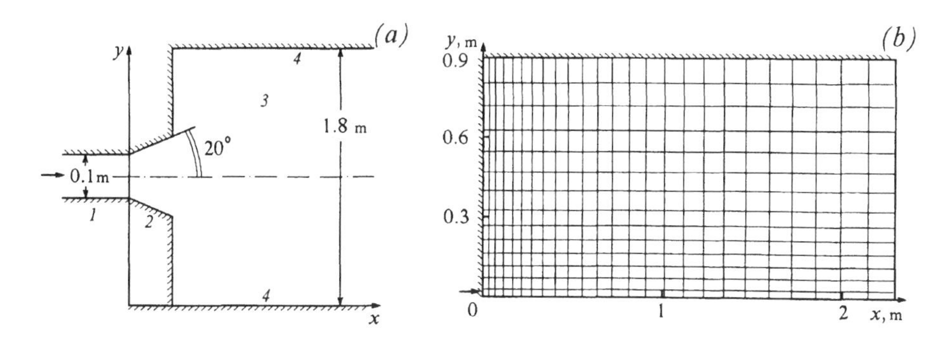

Figure 4.15 (a) Scheme of pipe with outlet: 1, pipe, 2, outlet, 3, bottom (concrete), and 4, walls (concrete); (b) fragment of the space grid for calculations.

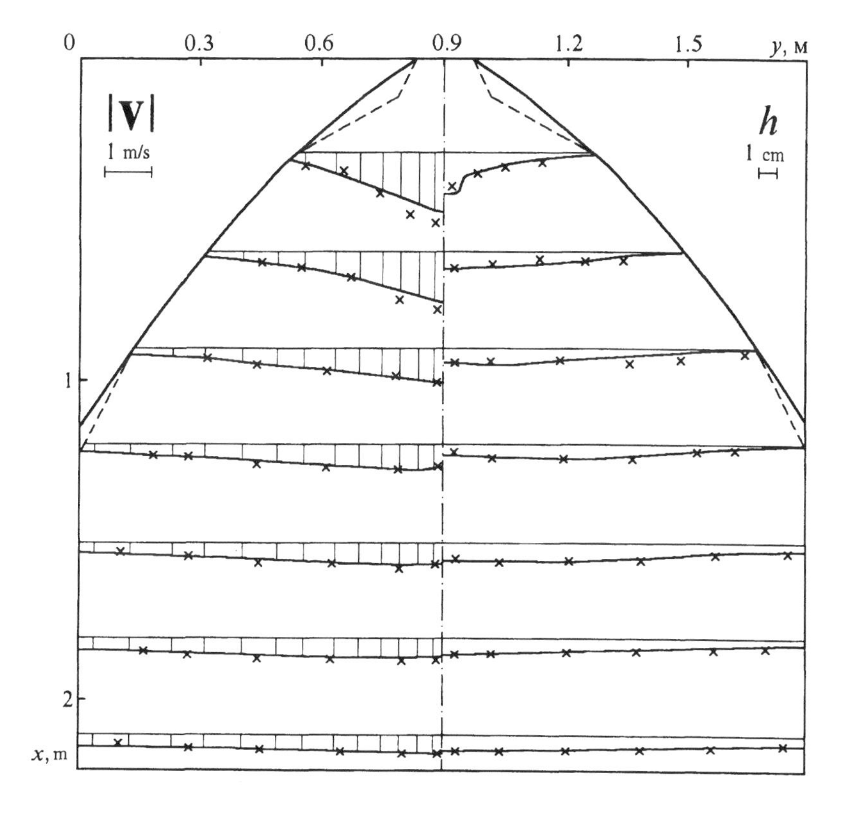

Figure 4.16 Flow in pipe with outlet.

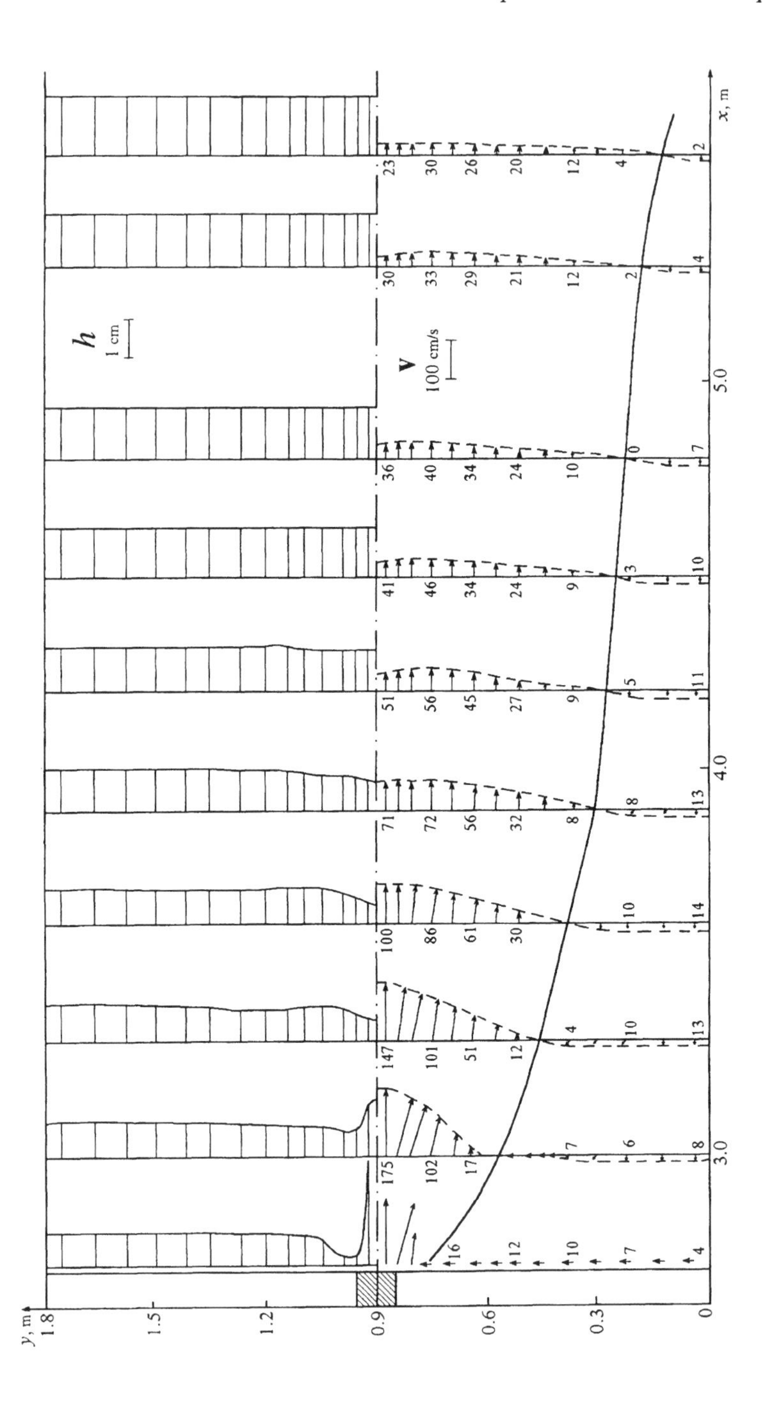

Figure 4.17 Calculations of whirlpool flow.

taken from the experiments by Volchenkov et al. (1985); the bottom slope angle was set equal to 0.1, the roughness factor was assumed to be $n = 0.014$ (for concrete), $\mathrm{Fr}_1 = 2.37$, $h_1 = 30$ mm, and $u_1 = 1.3$ m/s. The geometry of the chute was characterized by $\theta = 10^\circ$, $b_1 = 0.52$ m, and $b_3 = 0.26$ m. Friction was calculated by formula (4.2.5), with $\frac{1}{2} < \mu < 1$.

Figures 4.15 and 4.16 illustrate the numerical dynamics of steady spreading of water flowing out of a pipe 0.1 m in diameter, with a flare angle of 20°, see Fig. 4.15a, through an outlet section onto a concrete plane having a slope of 0.05 and characterized by the roughness factor $n = 0.014$. The problem is stated in accordance with the experiments conducted in Volchenkov et al. (1985). Taking into account the symmetry of the problem, we performed computations in a 3.3 m long and 0.9 m wide rectangular domain with a second-order scheme and the space mesh shown in Fig. 4.15b; the numerical slopes were corrected once every ten time layers. Figure 4.16 compares the experimental velocity profiles (measured in meters per second) and flow depths (shown by crosses) with those calculated for the bottom slope equal to 0.05. We see that a good agreement is obtained. In Fig. 4.16, solid and dashed curves represent the calculated boundaries of spreading water and the boundaries of spreading water observed in the experiment, respectively. An outlet section was not modelled in these computations. It is apparent that the influence of the outlet section is localized in the close vicinity of the outlet pipe. It should be noted here that the flow was computed using a jump-capturing method, which was advantageous in that we did not need special procedures to determine the boundaries of spreading water. Figure 4.17 presents results of numerical modelling for smaller slope 0.001. In this flow, shoaling domains are absent and a whirlpool flow arises.

Figure 4.18 shows the calculated shapes of long solitary waves, with amplitude and velocity proportional to $\cos^2(\lambda x - l)$, $|\lambda x - l| \leq \frac{1}{2}\pi$, approaching a shore at a slope of 0.3, see Fig. 4.18a. In Fig. 4.18b, the ordinate is the elevation or depression of the water level relative to the equilibrium level: $\Delta h_- = |h_{\min} - h_\infty|$ and $\Delta h_+ = |h_{\max} - h_\infty|$. The abscissa is the relative length l/λ. Solid lines represent the results calculated with the first-order scheme for various values of l and λ; crosses denote the experimental data taken from Mishuev, Sladkevich, and Silchenko (1984) and Lyatkher et al. (1986). For $l/\lambda > 0.67$, the calculated curves of $\Delta h_\pm / h_\infty$ lie below the experimental points. This can be explained by the fact that the shallow water equations are inapplicable in this case. For small l/λ, the calculated results are in good agreement with experiments.

Figures 4.19 and 4.20 present numerical modelling of tsunami-type flows. Initial conditions and data are shown in the inserts; a shaded water volume has an initial velocity 5 m/s.

Figure 4.21 presents initial data and typical numerical results of modelling of flow–underwater barrier interaction within the framework of the shallow water theory. One can see that a hydraulic jump travelling upstream is generated. It is interesting to compare these results with numerical modelling of such problems within the framework of the two-dimensional Navier–Stokes equations with free water surface by Kon'shin (1985) and Belotserkovskii, Gushchin, and Kon'shin (1987). Within some acceptable variations the hydraulic jump velocity W is equal to 0.28 m/s and its height Δh is equal to 2.30 m (V. N. Kon'shin, private communication). In shallow water modelling we obtain $W = 0.6$ m/s and $\Delta h = 2.35$ m. The difference in velocities can be explained by the two-dimensional water rotatory motion that exists in viscous flow near the hydraulic jump.

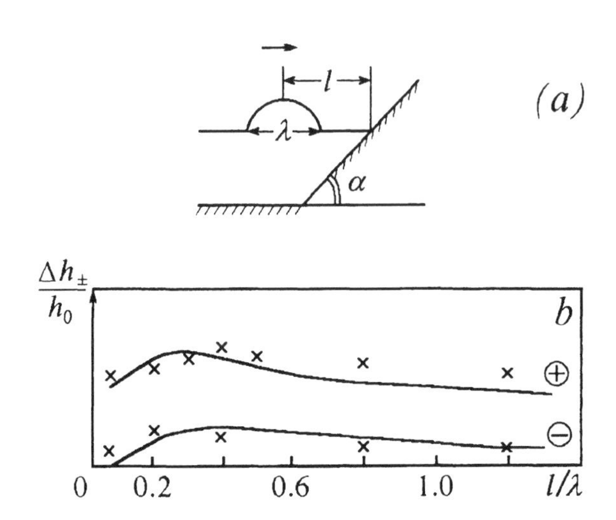

Figure 4.18 (a) Shape of solitary wave; (b) elevation (depression) of the water level.

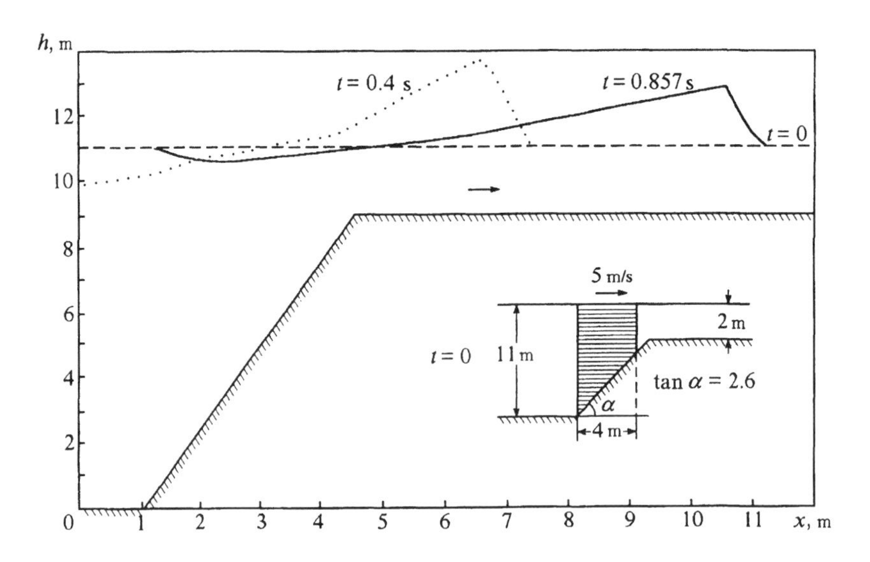

Figure 4.19 Tsunami-type flow.

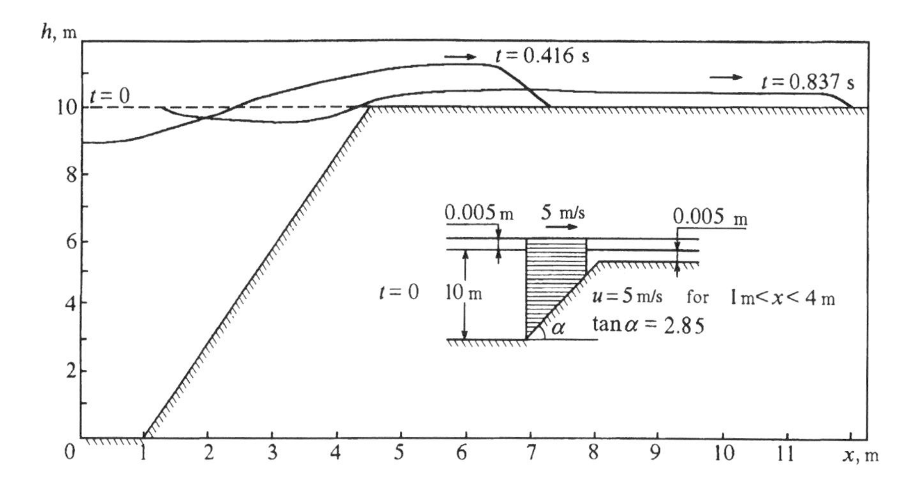

Figure 4.20 Tsunami-type flow with shoaling bottom.

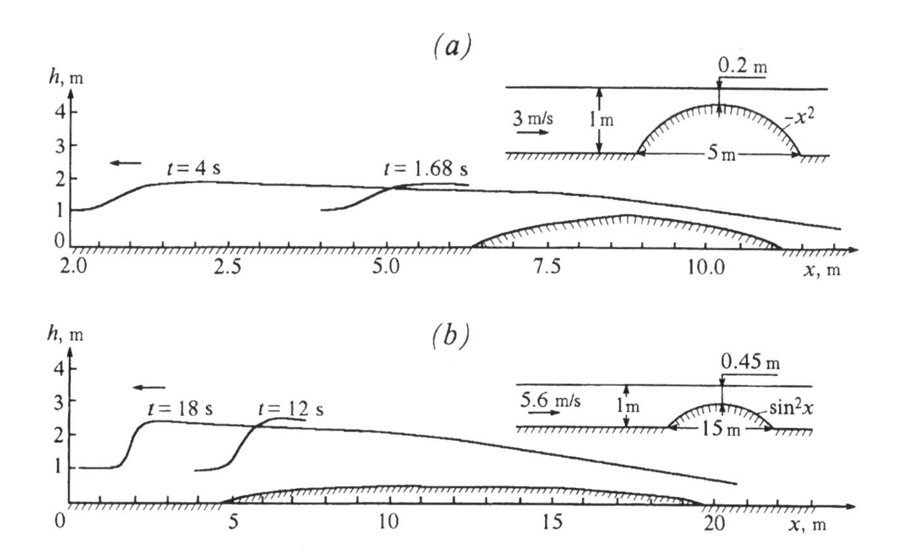

Figure 4.21 Modelling of flow–underwater barrier interaction.

The Godunov method is particularly suitable for computing flows characterized by moderate and high values of the Froude number, which can involve gradients or even jumps in the values of flow variables. Here, the role played by the Froude number is similar to that of the Mach number in gas dynamics. The methods developed here can also be applied to flows with low Froude numbers, $\mathrm{Fr} = |u|/c \ll 1$, but their computational efficiency would be poor because of the more restrictive CFL condition for the time step. Therefore, implicit finite-difference schemes, such as those described in Alalykin et al. (1970), should be applied to compute these flows. Another important issue here is that other models can be used instead of the shallow water equations; in particular, equations of the diffusion wave type may serve as a better model for the flow under consideration (Makhanov and Semenov 1994, 1995, 1996, 1998). The diffusion wave model, see Voltsinger and Pyaskovskii (1977), describes shallow and slow water flows by the space parabolic equation

$$\frac{\partial \zeta}{\partial t} = \mathrm{div}(D\nabla\zeta) + R, \tag{4.4.1}$$

where $D \geq 0$ and the quantity $\zeta = h + b$ is the free-surface level, see Fig. 4.1. This model is derived under the assumption that the inertia force is negligible small and the term $h\mathbf{v}$ is independent of time. Substituting $\partial h\mathbf{v}/\partial t = \mathrm{div}(h\mathbf{v}\mathbf{v}) = 0$ into Eq. (4.1.2) and assuming the linear friction law $\mathbf{f} = \lambda\mathbf{v}$, where λ is the hydraulic friction coefficient, we obtain

$$\mathbf{v} = \frac{gh}{\lambda}\nabla\zeta.$$

Substituting the above into Eq. (4.1.1), we obtain Eq. (4.4.1), where $D = gh^2/\lambda$. For the quadratic friction law $\mathbf{f} = \frac{1}{2}\lambda\mathbf{v}|\mathbf{v}|$ the coefficient D is expressed as

$$D = \sqrt{\frac{2gh^3}{\lambda|\nabla\zeta|}}.$$

The algorithm developed in this chapter can adequately predict the propagation dynamics of strong hydraulic jumps—analogues of gas dynamic shock waves—and simulate the monotone profiles of grid variables in the vicinity of a hydraulic jump. The algorithm can also be used to calculate the values of flow variables in computational cells incompletely filled by water so as to simulate shoaling and dry-bottom regions without treating them separately.

Here, we focus on the following two features that distinguish our approach from those known previously. We have developed a fast and efficient numerical algorithm of the exact solution of the general Riemann problem and obtained estimates for the rate of convergence of the iteration process. We have developed and tested the algorithms designed to allow for smooth bottom relief of arbitrary shape (including shoal regions) to be incorporated as a single-valued function of coordinates into Godunov-type front-capturing schemes. Here, one difficulty is that the shallow water equations cannot be solved exactly when the bottom is not horizontal, see the generalized Riemann problem (Section 4.3, page 242). Another difficulty is associated with the construction of a shock-capturing algorithm that allows for shoaling and dry bottom. By applying the exact formulas derived for the Riemann problem instead of the formulas used in other Riemann solvers (the Courant–Isaacson–Rees and Roe algorithms), which involve division by the square root of the depth, see Section 4.5, these difficulties can be overcome in a straightforward manner without invoking any regularization procedures. Thus, the exact Riemann problem solver is more universal as compared with approximate solvers to be described below.

4.5 Approximate Riemann problem solvers

This section describes the Godunov-type methods based on approximate solutions of the one-dimensional shallow water Riemann problem. We describe here the basic approximate Riemann problem solvers: the Courant–Isaacson–Rees (CIR), Roe, and Osher–Solomon solvers.

4.5.1 The CIR method. Let us consider a one-dimensional numerical scheme of the CIR type, for example, (2.3.25)–(2.3.26) for Eq. (4.1.9):

$$\frac{\mathbf{U}_i^{k+1}-\mathbf{U}_i^k}{\Delta t}+\frac{\mathbf{F}_{i+1/2}-\mathbf{F}_{i-1/2}}{\Delta x}=\mathbf{B}_i^k, \tag{4.5.1}$$

$$\mathbf{F}_{m+1/2}=\tfrac{1}{2}(\mathbf{F}_m^k+\mathbf{F}_{m+1}^k)+\tfrac{1}{2}|A|_{m+1/2}^k(\mathbf{U}_m^k-\mathbf{U}_{m+1}^k),\quad m=i,i-1; \tag{4.5.2}$$

$$|A|=\Omega_{\mathrm{R}}[|\lambda_p|\,\delta_{pl}]\Omega_{\mathrm{L}}. \tag{4.5.3}$$

Here Δx is the mesh size and Δt is the time increment. The integer subscripts $i=1,2,\ldots$ refer to centers of mesh cells, and half-integer ones to boundaries. The integer superscript is the index indicating the time level. To implement scheme defined by (4.5.1)–(4.5.3), one must calculate $|A|$ in accordance with Eq.(4.5.3), where Ω_{R}, Λ, and Ω_{L} are defined in (4.1.15)–(4.1.16). The following convenient representation can be used for the calculation of the vector $\mathbf{Q}=|A|\Delta\mathbf{U}$:

$$\begin{aligned}\mathbf{Q}&=|A|\Delta\mathbf{U}=\Omega_{\mathrm{R}}|\Lambda|\Omega_{\mathrm{L}}\Delta\mathbf{U}=\Omega_{\mathrm{R}}\,\mathrm{diag}[|u-c|,|u|,|u+c|]\Omega_{\mathrm{L}}\Delta\mathbf{U}\\&=|u|\Delta\mathbf{U}+(|u-c|-|u|)\Omega_{\mathrm{R}}\,\mathrm{diag}[1,0,0]\Omega_{\mathrm{L}}\Delta\mathbf{U}+(|u+c|-|u|)\Omega_{\mathrm{R}}\,\mathrm{diag}[0,0,1]\Omega_{\mathrm{L}}\Delta\mathbf{U},\end{aligned}$$

where

$$\Delta\mathbf{U}=\mathbf{U}_m-\mathbf{U}_{m+1}=[\Delta h,\Delta(hu),\Delta(hv)]^{\mathrm{T}},$$

or, equivalently,

$$\mathbf{Q}=|u|\begin{bmatrix}\Delta h\\ \Delta(hu)\\ \Delta(hv)\end{bmatrix}+(|u-c|-|u|)(p+q)\begin{bmatrix}1\\ u-c\\ v\end{bmatrix}+(|u+c|-|u|)(p-q)\begin{bmatrix}1\\ u+c\\ v\end{bmatrix}, \tag{4.5.4}$$

$$p=\frac{\Delta h}{2},\quad q=\frac{u\Delta h-\Delta(hu)}{2c}.$$

The final compact formulas to calculate $\mathbf{Q}=[Q_1,Q_2,Q_3]^{\mathrm{T}}$ are

$$\begin{aligned}&Q_1=|u|\Delta h+\beta_1,\quad Q_2=|u|\Delta(hu)+\beta_1 u-\beta_2 c,\quad Q_3=|u|\Delta(hv)+\beta_1 v, \\&\beta_1=\alpha_1+\alpha_2,\quad \beta_2=\alpha_1-\alpha_2,\\&\alpha_1=(|u-c|-|u|)(p+q),\quad \alpha_2=(|u+c|-|u|)(p-q),\\&p=\frac{\Delta h}{2},\quad q=\frac{u\Delta h-\Delta(hu)}{2c}.\end{aligned} \tag{4.5.5}$$

To use Eqs. (4.5.5) for $c \to 0$ one can transform the quantities $\alpha_1 \sim 1/c$ and $\alpha_2 \sim 1/c$ in the following way:

$$\alpha_1 = (|u-c|-|u|)(p+q) = -(p+q)c\,\mathrm{sgn}\,u = -\tfrac{1}{2}[c\Delta h - \Delta(hu) + u\Delta h],$$
$$\alpha_2 = (|u+c|-|u|)(p-q) = (p-q)c\,\mathrm{sgn}\,u = \tfrac{1}{2}[c\Delta h + \Delta(hu) - u\Delta h].$$

For the two-dimensional case, one can calculate the flux $\mathbf{F}$ for jth cell boundary in accordance with (4.5.2) and (4.5.5) and use the approximations (4.2.3)–(4.2.4):

$$G_i \frac{\mathbf{U}_i^{k+1} - \mathbf{U}_i^k}{\Delta t} + \sum_{j=1}^{m(i)} L_j \mathbf{F}_j = \mathbf{B}_i,$$

$$\mathbf{B}_i = [B_1, B_2, B_3]_i^{\mathrm{T}} = \left[0, \tfrac{1}{2} g \sum_{j=1}^{m(i)} (h_j^{\alpha})^2 \mathbf{n}_j L_j + G_i \mathbf{f}_i^{\mu}\right]^{\mathrm{T}}.$$

As to the approximation of the bottom and friction terms, see Eqs. (4.2.6) and (4.2.5), and various approaches, for example, by Garsia-Navarro et al. (1995), Hubbard and Baines (1997), and Vázquez-Cendón (1999).

4.5.2 Roe's method.

The Roe method is based on an approximate solution of the Riemann problem for Eqs. (4.1.6)–(4.1.8). This method is similar to the CIR method with a special choice of the matrix A; we shall call it $\mathcal{A}$ (see Sections 2.3.2 and 3.4.4). The matrix $\mathcal{A}$ is defined as follows:

$$\Delta\mathbf{F} = \mathcal{A}\Delta\mathbf{U}, \quad \Delta\mathbf{F} = \mathbf{F}(\mathbf{U}_1) - \mathbf{F}(\mathbf{U}_2), \quad \Delta\mathbf{U} = \mathbf{U}_1 - \mathbf{U}_2,$$
$$\mathcal{A} = \mathcal{A}(\mathbf{U}_1, \mathbf{U}_2), \quad \mathcal{A}(\mathbf{U}, \mathbf{U}) = A(\mathbf{U}),$$
$$\mathbf{U}_1 = \mathbf{U}_m, \quad \mathbf{U}_2 = \mathbf{U}_{m+1}.$$

Consider a nondegenerate nonlinear transformation $\mathbf{U} = \mathbf{U}(\mathbf{Y})$. Then, from the exact relations $\Delta\mathbf{F} = \mathcal{A}_F\Delta\mathbf{Y}$ and $\Delta\mathbf{U} = \mathcal{A}_U\Delta\mathbf{Y}$ it follows that $\mathcal{A} = \mathcal{A}_F\mathcal{A}_U^{-1}$. Since the function $\mathbf{F}$ given by (4.1.11) is a nonlinear function of $\mathbf{U}$, the obtained matrix $\mathcal{A}$ is not unique, see example in Section 2.3.2. Exact analytical formulas for $\mathcal{A}_F$ and $\mathcal{A}_U$ can be written explicitly for the cases where $\mathbf{U}$ and $\mathbf{F}$ are polynomials in $\mathbf{Y}$ or rational functions of $\mathbf{Y}$. To derive the entries of $\mathcal{A}_F$ and $\mathcal{A}_U$, we can use the exact identities of the type (2.3.66)–(2.3.69).

Consider the main feature of the Roe method. Let the left and right quantities $\mathbf{U}_1$ and $\mathbf{U}_2$ be related by Eqs. (4.3.4)–(4.3.6): $\Delta\mathbf{F} = W\Delta\mathbf{U}$, where W is the discontinuity velocity. Then W is an eigenvalue of the matrix $\mathcal{A}$. We have

$$\Delta\mathbf{F} - W\Delta\mathbf{U} = (\mathcal{A}_F - W\mathcal{A}_U)\Delta\mathbf{Y} = (\mathcal{A} - WI)\mathcal{A}_U\Delta\mathbf{Y} = (\mathcal{A} - WI)\Delta\mathbf{U} = \mathbf{0},$$

where I is the 3×3 identity matrix. To obtain nontrivial solutions we must put $\det(\mathcal{A} - WI) = 0$, and hence $\Delta\mathbf{U}$ is an eigenvector of $\mathcal{A}$. Thus, with the Roe approach, the discontinuity relations are satisfied exactly.

Let us choose the parametric vector $\mathbf{Y}$ by analogy with the gas dynamic Roe variables (see Section 3.4.4):

$$\mathbf{Y} = [R, U, V]^{\mathrm{T}} = [\sqrt{h}, \sqrt{h}\,u, \sqrt{h}\,v]^{\mathrm{T}}. \tag{4.5.6}$$

In terms of these variables $\mathbf{U}$ and $\mathbf{F}$ of (4.1.10)–(4.1.11) can be rewritten as

$$\mathbf{U} = [R^2, RU, RV]^{\mathrm{T}}, \quad \mathbf{F} = [RU, U^2 + \tfrac{1}{2}gR^4, UV]^{\mathrm{T}}. \tag{4.5.7}$$

Thus,

$$\Delta\mathbf{U} = \begin{bmatrix} 2\overline{R}\Delta R \\ \overline{U}\Delta R + \overline{R}\Delta U \\ \overline{V}\Delta R + \overline{R}\Delta V \end{bmatrix}, \quad \Delta\mathbf{F} = \begin{bmatrix} \overline{U}\Delta R + \overline{R}\Delta U \\ 2\overline{U}\Delta U + 2g\overline{R^2}\,\overline{R}\Delta R \\ \overline{V}\Delta U + \overline{U}\Delta V \end{bmatrix},$$

where $\overline{f} = \frac{1}{2}(f_1 + f_2)$ denotes the arithmetic average of the function f, where f is R, R^2, U, or V. To obtain more compact formulas, we divide all entries of $\Delta\mathbf{U}$ and $\Delta\mathbf{F}$ by $\overline{R}$. Then the matrices $\mathcal{A}_U$ and $\mathcal{A}_F$ acquire the form

$$\mathcal{A}_U = \begin{bmatrix} 2 & 0 & 0 \\ \hat{u} & 1 & 0 \\ \hat{v} & 0 & 1 \end{bmatrix}, \quad \mathcal{A}_F = \begin{bmatrix} \hat{u} & 1 & 0 \\ 2g\hat{h} & 2\hat{u} & 0 \\ 0 & \hat{u} & \hat{v} \end{bmatrix}. \tag{4.5.8}$$

The hat over a symbol stands for the Roe average, specifically,

$$\hat{u} = \frac{\overline{U}}{\overline{R}} = \frac{\sqrt{h_1}u_1 + \sqrt{h_2}u_2}{\sqrt{h_1} + \sqrt{h_2}}, \quad \hat{v} = \frac{\overline{V}}{\overline{R}} = \frac{\sqrt{h_1}v_1 + \sqrt{h_2}v_2}{\sqrt{h_1} + \sqrt{h_2}},$$
$$\hat{h} = \overline{h} = \tfrac{1}{2}(h_1 + h_2). \tag{4.5.9}$$

To obtain an analytical formula for $\mathcal{A} = \mathcal{A}_F\mathcal{A}_U^{-1}$, we find a solution $\mathbf{x}^{\mathrm{T}}$ of the linear system of equations

$$\mathbf{x}^{\mathrm{T}}\mathcal{A}_U = \mathbf{z}^{\mathrm{T}} \quad \Longrightarrow \quad \mathbf{x}^{\mathrm{T}} = \mathbf{z}^{\mathrm{T}}\mathcal{A}_U^{-1}; \tag{4.5.10}$$
$$x_1 = \tfrac{1}{2}z_1 - \tfrac{1}{2}\hat{u}z_2 - \tfrac{1}{2}\hat{v}z_3,$$
$$x_2 = z_2, \quad x_3 = z_3,$$

where $\mathbf{x}^{\mathrm{T}} = [x_1, x_2, x_3]$ and $\mathbf{z}^{\mathrm{T}} = [z_1, z_2, z_3]$. Thus, considering the nth row of the matrix $\mathcal{A}_F$ ($n = 1, 2, 3$) to be $\mathbf{z}^{\mathrm{T}}$, we calculate $\mathbf{x}^{\mathrm{T}}$ which is the nth row of the matrix $\mathcal{A} = \mathcal{A}_F\mathcal{A}_U^{-1}$ to obtain

$$\mathcal{A} = \begin{bmatrix} 0 & 1 & 0 \\ g\hat{h} - \hat{u}^2 & 2\hat{u} & 0 \\ -\hat{u}\hat{v} & \hat{v} & \hat{u} \end{bmatrix}. \tag{4.5.11}$$

The matrices of (4.1.18) and (4.5.11) are identical, provided that $u = \hat{u}$, $v = \hat{v}$, and

$$\hat{c} = \sqrt{g\hat{h}}. \tag{4.5.12}$$

Formula (4.5.12) defines the average celerity $\hat{c}$. Then, in accordance with (4.1.15)–(4.1.16), we obtain the matrices necessary for the implementation of the Roe method:

$$\hat{\Omega}_{\mathrm{R}} = \begin{bmatrix} 1 & 0 & 1 \\ \hat{u} - \hat{c} & 0 & \hat{u} + \hat{c} \\ \hat{v} & -1 & \hat{v} \end{bmatrix}, \quad \hat{\Omega}_{\mathrm{L}} = \frac{1}{2\hat{c}} \begin{bmatrix} \hat{c} + \hat{u} & -1 & 0 \\ 2\hat{v}\hat{c} & 0 & -2\hat{c} \\ \hat{c} - \hat{u} & 1 & 0 \end{bmatrix},$$
$$\hat{\Lambda} = \mathrm{diag}[\hat{u} - \hat{c}, \hat{u}, \hat{u} + \hat{c}].$$

Thus, for Eq. (4.1.9) we have the following finite-volume scheme similar to (4.5.1)–(4.5.3) with specially chosen matrices $\hat{\Omega}_{\rm R}$, $\hat{\Lambda}$, and $\hat{\Omega}_{\rm L}$:

$$\frac{\mathbf{U}_i^{k+1}-\mathbf{U}_i^k}{\Delta t}+\frac{\mathbf{F}_{i+1/2}-\mathbf{F}_{i-1/2}}{\Delta x}=\mathbf{B}_i^k, \tag{4.5.13}$$

$$\mathbf{F}_{m+1/2}=\tfrac{1}{2}(\mathbf{F}_m^k+\mathbf{F}_{m+1}^k)+\tfrac{1}{2}|\mathcal{A}|_{m+1/2}^k(\mathbf{U}_m^k-\mathbf{U}_{m+1}^k),\quad m=i,i-1; \tag{4.5.14}$$

$$|\mathcal{A}|=\hat{\Omega}_{\rm R}|\hat{\Lambda}|\hat{\Omega}_{\rm L},\quad |\hat{\Lambda}|=[|\hat{\lambda}_p|\delta_{pl}].$$

For practical implementation of the formulas for $\mathbf{F}$, one can use Eq. (4.5.5).

To construct the Godunov-type method, we need here only formulas for $\mathbf{F}$, see Eq. (4.5.14). For $\mathbf{U}_{m+1/2}$ we have (2.3.70) for $\xi=0$,

$$\mathbf{U}_{m+1/2}=\tfrac{1}{2}(\mathbf{U}_m^k+\mathbf{U}_{m+1}^k)+\tfrac{1}{2}\hat{S}_{m+1/2}^k(\mathbf{U}_m^k-\mathbf{U}_{m+1}^k),\quad m=i,i-1; \tag{4.5.15}$$

$$S=\hat{\Omega}_{\rm R}[\operatorname{sgn}\hat{\lambda}_p\,\delta_{pl}]\hat{\Omega}_{\rm L}.$$

Formulas for $\mathbf{U}$ may be useful, for example, for approximation of the right-hand terms, calculation of function slopes in the procedure of reconstruction, and in moving grid calculations; see (2.3.72) and (3.4.22). Occasionally, one may need to calculate boundary values $\mathbf{W}_{m+1/2}$ of some function $\mathbf{W}(\mathbf{U})$. Let us derive formulas for this case. Substituting $\mathbf{U}=\mathbf{U}(\mathbf{W})$ in system (4.1.13), we obtain the equations

$$U_W\mathbf{W}_t+AU_W\mathbf{W}_x=\mathbf{B},\quad U_W=\frac{\partial\mathbf{U}}{\partial\mathbf{W}},\quad W_U=\frac{\partial\mathbf{W}}{\partial\mathbf{U}},$$

$$\mathbf{W}_t+W_UAU_W\mathbf{W}_x=W_U\mathbf{B}.$$

Applying Eq. (4.5.15), we arrive at the formulas

$$\mathbf{W}_{m+1/2}=\tfrac{1}{2}(\mathbf{W}_m^k+\mathbf{W}_{m+1}^k)+\tfrac{1}{2}(W_U\hat{S}U_W)_{m+1/2}^k\,(\mathbf{W}_m^k-\mathbf{W}_{m+1}^k), \tag{4.5.16}$$

$$\hat{S}=\hat{\Omega}_{\rm R}[\operatorname{sgn}\hat{\lambda}_p\,\delta_{pl}]\hat{\Omega}_{\rm L}.$$

To retain the property of the Roe method, we choose a W_U such that

$$W_U\Delta\mathbf{U}=\Delta\mathbf{W},\quad W_U=\frac{\Delta\mathbf{W}}{\Delta\mathbf{U}}. \tag{4.5.17}$$

If we find a U_W with property (4.5.17), then Eq. (4.5.16) becomes

$$\mathbf{W}_{m+1/2}=\tfrac{1}{2}(\mathbf{W}_m^k+\mathbf{W}_{m+1}^k)+\tfrac{1}{2}(W_U\hat{S})_{m+1/2}^k\,(\mathbf{U}_m^k-\mathbf{U}_{m+1}^k), \tag{4.5.18}$$

$$\hat{S}=\hat{\Omega}_{\rm R}[\operatorname{sgn}\hat{\lambda}_p\,\delta_{pl}]\hat{\Omega}_{\rm L}.$$

Using the relations $\Delta\mathbf{W}=\mathcal{A}_W\Delta\mathbf{Y}$ and $\Delta\mathbf{U}=\mathcal{A}_U\Delta\mathbf{Y}$, we find that $W_U=\mathcal{A}_W\mathcal{A}_U^{-1}$. For example, let us choose $\mathbf{W}$ as follows:

$$\mathbf{W}=[h,hu^2,\tfrac{1}{2}gh^2]^{\rm T}=[R^2,U^2,\tfrac{1}{2}gR^4]^{\rm T}.$$

Then,

$$\Delta\mathbf{W}=\begin{bmatrix}2\overline{R}\Delta R\\ 2\overline{U}\Delta U\\ 2g\overline{R}\,\overline{R^2}\Delta R\end{bmatrix}.$$

Dividing all components of $\Delta\mathbf{W}$ by $\overline{R}$, we obtain

$$\mathcal{A}_W = \begin{bmatrix} 2 & 0 & 0 \\ 0 & 2\hat{u} & 0 \\ 2\hat{c}^2 & 0 & 0 \end{bmatrix}. \tag{4.5.19}$$

Using formulas (4.5.8) and (4.5.19) for $\mathcal{A}_U$ and $\mathcal{A}_W$ and solution (4.5.10), we can find $W_U = \mathcal{A}_W \mathcal{A}_U^{-1}$:

$$W_U = \begin{bmatrix} 1 & 0 & 0 \\ -\hat{u}^2 & 2\hat{u} & 0 \\ \hat{c}^2 & 0 & 0 \end{bmatrix}.$$

Thus, in accordance with Eq. (4.5.18), one can determine hu^2 and $\frac{1}{2}gh^2$ at the cell boundary. For $\mathbf{W} = \mathbf{F}$, Eq. (4.5.18) transforms into Eq. (4.5.14). For $\mathbf{W} = \mathbf{U}$, Eq. (4.5.18) transforms into Eq. (4.5.15) with $W_U = I$, where I is the 3×3 identity matrix. Formulas (4.5.14) and (4.5.18) are consistent. Indeed, it is easy to verify that

$$\langle hu^2 + \tfrac{1}{2}gh^2 \rangle = \langle hu^2 \rangle + \langle \tfrac{1}{2}gh^2 \rangle, \quad \langle f \rangle \equiv f_{m+1/2}. \tag{4.5.20}$$

In particular, relation (4.5.20) follows from the fact that the second row of the matrix $\mathcal{A}$ is equal to the sum of the second and third rows of W_U.

To construct a modification of the Roe method for an arbitrary curvilinear moving system of coordinates, one can use, in particular, methods described earlier for gas dynamic equations in Section 3.2.

4.5.3 The Osher–Solomon solver. The Osher–Solomon method, described in Section 3.4.6 for gas dynamic equations, can be easily extended to the shallow water equations. We need to construct the Riemann invariants for the system (4.1.1)–(4.1.2). This can be done if we recall that the Riemann invariant w_i corresponding to the eigenvalue λ_i must satisfy the relation

$$\frac{\partial w_i}{\partial \mathbf{U}} \cdot \mathbf{r}^i = 0.$$

It is easy to check, using formulas (4.1.15) for the right eigenvectors, that the Riemann invariants corresponding to $\lambda_1 = u - c$ are

$$v = \text{const}, \qquad u + 2c = u + 2\sqrt{gh} = \text{const}\,.$$

Similarly, for the eigenvalue $\lambda_2 = u$ the invariants are

$$h = \text{const}, \qquad u = \text{const}\,.$$

This subpath of the integral curve connecting $\mathbf{U}_j$ and $\mathbf{U}_{j+1}$ corresponds to the tangential discontinuity dividing the fluids with the same speeds u and free surface levels h possessing an arbitrary jump in the tangential velocity component v.

For the subpath corresponding to $\lambda_3 = u + c$, we have

$$v = \text{const}, \qquad u - 2c = u - 2\sqrt{gh} = \text{const}\,.$$

As usual for the Osher–Solomon scheme, we introduce the points $\mathbf{U}_{j+1/3}$ and $\mathbf{U}_{j+2/3}$ corresponding to the end points of the first and second subpaths (see Fig. 3.16). The quantities at these points are related by the formulas

$$u_j + 2c_j = u_{j+1/3} + 2c_{j+1/3}, \qquad v_{j+1/3} = v_j,$$
$$c_{j+2/3} = c_{j+1/3}, \qquad u_{j+2/3} = u_{j+1/3},$$
$$u_{j+1} - 2c_{j+1} = u_{j+2/3} - 2c_{j+2/3}, \qquad v_{j+1} = v_{j+2/3}.$$

Hence,

$$c_{j+1/3} = \frac{1}{4}\left[u_j - u_{j+1} + 2(c_j + c_{j+1})\right], \tag{4.5.21}$$
$$u_{j+1/3} = \frac{1}{2}(u_j + u_{j+1}) + c_j - c_{j+1}, \tag{4.5.22}$$
$$h_{j+2/3} = h_{j+1/3}, \tag{4.5.23}$$
$$u_{j+2/3} = u_{j+1/3}. \tag{4.5.24}$$

It is clear that the eigenvalues λ_1 and λ_3 can change their signs along the respective subpaths. If we denote these points as $\mathbf{U}_{j+1/6}$ and $\mathbf{U}_{j+5/6}$, respectively, it is obvious that

$$v_{j+1/6} = v_j, \tag{4.5.25}$$
$$u_j + 2c_j = u_{j+1/6} + 2c_{j+1/6},$$
$$u_{j+1/6} = c_{j+1/6}, \tag{4.5.26}$$

thus giving

$$c_{j+1/6} = \frac{u_j + 2c_j}{3}. \tag{4.5.27}$$

Similarly, we have

$$v_{j+5/6} = v_{j+1}, \tag{4.5.28}$$
$$a_{j+5/6} = \frac{2c_{j+1} - u_{j+1}}{3}, \tag{4.5.29}$$
$$u_{j+5/6} = -c_{j+5/6}. \tag{4.5.30}$$

On noting that $c = \sqrt{gh}$, we conclude that formulas (4.5.21)–(4.5.30) allow us to calculate the Osher–Solomon flux using the formula (3.4.54).

4.6 Stationary shallow water equations

In this section we describe the main specific features of the Godunov method for the system of equations that governs the stationary two-dimensional supercritical shallow water flows (Semenov, unpublished).

4.6.1 System of governing equations.

The stationary shallow water equations can be obtained if we assume that in Eqs. (4.1.1)–(4.1.2) the time derivatives are equal to zero; thus, we obtain

$$\operatorname{div}(h\mathbf{v}) = 0, \tag{4.6.1}$$

$$\operatorname{div}(h\mathbf{v}\mathbf{v}) + gh\operatorname{div}(\zeta\hat{\mathbf{I}}) = \mathbf{f}. \tag{4.6.2}$$

Equations (4.6.1) and (4.6.2) express the conservation of the water mass and momentum, respectively. In expanded form, system (4.6.1)–(4.6.2) can be represented as

$$\frac{\partial hu}{\partial x} + \frac{\partial hv}{\partial y} = 0, \tag{4.6.3}$$

$$\frac{\partial(hu^2 + \frac{1}{2}gh^2)}{\partial x} + \frac{\partial huv}{\partial y} = f_1 - gh\frac{\partial b}{\partial x}, \tag{4.6.4}$$

$$\frac{\partial hvu}{\partial x} + \frac{\partial(hv^2 + \frac{1}{2}gh^2)}{\partial y} = f_2 - gh\frac{\partial b}{\partial y}. \tag{4.6.5}$$

Water flows governed by the stationary shallow water equations include supercritical flows with the Froude number $\mathrm{Fr}_x = |u|/c > 1$ and/or $\mathrm{Fr}_y = |v|/c > 1$, where $c = \sqrt{gh}$. Such flows are realized in tailraces of hydroelectric power stations, flows in culverts and chutes (Stoker 1957), etc., see the stationary flows in Figs. 4.12–4.16. Note that stationary supercritical flows can be simulated by solving Eqs. (4.1.1)–(4.1.2) by a time relaxation method. However, the use of Eqs. (4.6.1)–(4.6.2) can turn out to be more fast and efficient.

In the vector notation Eqs. (4.6.3)–(4.6.5) read

$$\frac{\partial \mathbf{E}}{\partial x} + \frac{\partial \mathbf{F}}{\partial y} = \mathbf{B}, \tag{4.6.6}$$

where

$$\mathbf{E} = [hu,\ hu^2 + \tfrac{1}{2}gh^2,\ huv]^{\mathrm{T}}, \tag{4.6.7}$$

$$\mathbf{F} = [hv,\ huv,\ hv^2 + \tfrac{1}{2}gh^2]^{\mathrm{T}}, \tag{4.6.8}$$

$$\mathbf{B} = [0, f_1 - ghb_x, f_2 - ghb_y]^{\mathrm{T}}, \quad b_x = \frac{\partial b}{\partial x}, \quad b_y = \frac{\partial b}{\partial y}. \tag{4.6.9}$$

This equation can be represented in the nonconservative form

$$\frac{\partial \mathbf{U}}{\partial x} + A\frac{\partial \mathbf{U}}{\partial y} = \tilde{\mathbf{B}}, \quad A = E_U^{-1}F_U, \quad \tilde{\mathbf{B}} = E_U^{-1}\mathbf{B}, \tag{4.6.10}$$

$$E_U = \frac{\partial \mathbf{E}}{\partial \mathbf{U}} = \begin{bmatrix} U & R & 0 \\ 2gR^3 & 2U & 0 \\ 0 & V & U \end{bmatrix}, \quad F_U = \frac{\partial \mathbf{F}}{\partial \mathbf{U}} = \begin{bmatrix} V & 0 & R \\ 0 & V & U \\ 2gR^3 & 0 & 2V \end{bmatrix},$$

$$\mathbf{U} = [R, U, V]^{\mathrm{T}} = [\sqrt{h}, \sqrt{h}\,u, \sqrt{h}\,v]^{\mathrm{T}}. \tag{4.6.11}$$

Here, we consider $\mathbf{U}$ to be a vector of Roe variables, see (4.5.6). Of course, one can use other independent variables, in particular, $\mathbf{U} = [h, hu, hv]^{\mathrm{T}}$ or $\mathbf{U} = [h, u, v]^{\mathrm{T}}$.

The matrix A of Eq. (4.6.10) can be diagonalized as follows:

$$A = E_U^{-1} F_U = \Omega_R \Lambda \Omega_L, \tag{4.6.12}$$

where

$$\Omega_R = \begin{bmatrix} u\lambda_1 - v & 0 & u\lambda_2 - v \\ (u^2 - 2c^2)\lambda_1 - uv & u & (u^2 - 2c^2)\lambda_2 - uv \\ 2c^2 - v^2 + uv\lambda_1 & v & 2c^2 - v^2 + uv\lambda_2 \end{bmatrix}, \tag{4.6.13}$$

$$\det \Omega_R = -2(u^2 + v^2)c^2 \Delta\lambda;$$

$$\Omega_L = \frac{1}{\det \Omega_R} \begin{bmatrix} (u + v\lambda_2)2c^2 & (u\lambda_2 - v)v & (u\lambda_2 - v)(-u) \\ (u^2 + v^2 - 2c^2)2c^2\Delta\lambda & -2uc^2\Delta\lambda & -2vc^2\Delta\lambda \\ (u + v\lambda_1)(-2c^2) & (u\lambda_1 - v)(-v) & (u\lambda_1 - v)u \end{bmatrix},$$

$$\Lambda = \text{diag}[\lambda_1, \lambda_0, \lambda_2], \tag{4.6.14}$$

$$\lambda_{2,1} = \lambda_\pm = \frac{uv \pm c\sqrt{u^2 + v^2 - c^2}}{u^2 - c^2}, \quad \lambda_0 = \frac{v}{u}, \quad \Delta\lambda = \lambda_2 - \lambda_1, \quad c = \sqrt{gh}.$$

The function c is the speed of propagation of small disturbances in shallow water flow. It is clear that (4.6.3)–(4.6.5) is hyperbolic if $u^2 + v^2 > c^2$. This does not resolve the question of whether the system is x-hyperbolic or y-hyperbolic; see Section 1.3.2. In the x-hyperbolic system $u^2 > c^2$, whereas for $v^2 > c^2$ the system is y-hyperbolic.

The matrix $\tilde{A} = F_U E_U^{-1}$ can also be diagonalized,

$$\tilde{A} = F_U E_U^{-1} = \tilde{\Omega}_R \Lambda \tilde{\Omega}_L, \tag{4.6.15}$$

where

$$\tilde{\Omega}_L = \begin{bmatrix} 2(u\lambda_1 - v) & -\lambda_1 & 1 \\ u^2 + v^2 & -u & -v \\ 2(u\lambda_2 - v) & -\lambda_2 & 1 \end{bmatrix}, \qquad \det \tilde{\Omega}_L = (u^2 + v^2)\Delta\lambda; \tag{4.6.16}$$

$$\tilde{\Omega}_R = \frac{1}{\det \tilde{\Omega}_L} \begin{bmatrix} -(u + v\lambda_2) & -\Delta\lambda & u + v\lambda_1 \\ -(u^2 - v^2 + 2uv\lambda_2) & -2u\Delta\lambda & u^2 - v^2 + 2uv\lambda_1 \\ -(2uv + (v^2 - u^2)\lambda_2) & -2v\Delta\lambda & 2uv + (v^2 - u^2)\lambda_1 \end{bmatrix}.$$

Note that the pth right vector of Ω_R satisfy

$$(F_U - \lambda_p E_U)\mathbf{r}^p = \mathbf{0},$$

and the qth left vector of $\tilde{\Omega}_L$ satisfy

$$\tilde{\mathbf{l}}^q (F_U - \lambda_q E_U) = \mathbf{0}.$$

Hence,

$$\tilde{\mathbf{l}}^q \cdot (F_U - \lambda_p E_U)\mathbf{r}^p = 0, \quad \tilde{\mathbf{l}}^q (F_U - \lambda_q E_U) \cdot \mathbf{r}^p = 0$$
$$\implies \quad \tilde{\mathbf{l}}^q \cdot E_U \mathbf{r}^p = 0 \quad \text{if} \quad \lambda_p \neq \lambda_q.$$

Thus, $\tilde{\Omega}_L E_U \Omega_R$ is a diagonal matrix.

4.6.2 The Godunov method. The CIR and Roe's schemes. Let us introduce a discrete uniform space mesh along the y-axis with size Δy. Let $\mathbf{U}_i$ denote the mesh function values, where the integer subscript $i = 1, 2, \ldots$ refers to the center of the ith space cell along the y-axis. The half-integer subscript $i \pm 1/2$ refers to the boundaries between the numerical cells with numbers i and $i \pm 1$. Assume that all mesh functions are constant inside each space cell. Let the integer superscript $k = 0, 1, \ldots$ indicate the layer with respect to x and Δx be the space increment along x-axis. Then, for cell boundary $i + 1/2$ and each x-step, we can solve the Riemann problem for Eqs. (4.6.1)–(4.6.2) with the following initial data: $\mathbf{U}_i^k = \text{const}$ for $y < y_{i+1/2}$ and $\mathbf{U}_{i+1}^k = \text{const}$ for $y > y_{i+1/2}$. Let $\mathbf{U}_{i+1/2}$ be a solution of this problem. In the same manner we can calculate $\mathbf{U}_{i-1/2}$ for boundary $i - 1/2$. Then, the Godunov scheme for the case in question has the form

$$\frac{\mathbf{E}_i^{k+1} - \mathbf{E}_i^k}{\Delta x} + \frac{\mathbf{F}_{i+1/2} - \mathbf{F}_{i-1/2}}{\Delta y} = \mathbf{B}_i, \quad \mathbf{F}_{i\pm 1/2} = \mathbf{F}(\mathbf{U}_{i\pm 1/2}), \tag{4.6.17}$$

where $\mathbf{B}_i$ approximates the bottom relief term, see Eq. (4.2.6).

The spectral analysis (Richtmyer and Morton 1967) of the linearized equations (4.6.17) leads to the stability CFL condition

$$\max |\lambda_\pm| \frac{\Delta x}{\Delta y} \le 1.$$

The finite-volume scheme expressed by (4.6.17) is described by the system of nonlinear equations

$$(hu)_i^{k+1} = Q_i, \quad (hu^2 + \tfrac{1}{2}gh^2)_i^{k+1} = P_i, \quad (huv)_i^{k+1} = R_i. \tag{4.6.18}$$

One can solve this system for h, u, and v to obtain

$$v_i^{k+1} = \frac{R_i}{Q_i}, \quad u_i^{k+1} = \frac{Q_i}{h}, \quad h = h_i^{k+1},$$

$$h^3 + 3ph + 2q = 0, \qquad p = -\frac{2P_i}{3g} < 0, \quad q = \frac{Q_i^2}{g} > 0, \quad q^2 + p^3 < 0.$$

Using formulas for the roots of a cubic equation (Korn and Korn 1968; Bronshtein and Semendyayev 1973), one can select the corresponding physically relevant solution for h,

$$h = 2r\cos\left(\frac{\pi}{3} + \frac{\psi}{3}\right), \quad \cos\psi = \frac{q}{r^3}, \quad r = \sqrt{|p|},$$

and evaluate u_i^{k+1} and v_i^{k+1}.

The above scheme is of the first order of accuracy. A higher-order scheme can be constructed in the same manner as that for a time-dependent system of equations, see Section 4.2.

The Courant–Isaacson–Rees (CIR) and Roe's schemes

Consider first the CIR and Roe versions of the Godunov method based on approximate solution of the Riemann problem. A one-dimensional scheme of the CIR type, for example, (2.3.25)–(2.3.26) for Eq. (4.6.6), can be represented as follows:

$$\frac{\mathbf{E}_i^{k+1}-\mathbf{E}_i^k}{\Delta x}+\frac{\mathbf{F}_{i+1/2}-\mathbf{F}_{i-1/2}}{\Delta y}=\mathbf{B}_i^k, \tag{4.6.19}$$

$$\begin{aligned}
&\mathbf{F}_{m+1/2}=\tfrac{1}{2}(\mathbf{F}_m^k+\mathbf{F}_{m+1}^k)+\tfrac{1}{2}(E_U|A|E_U^{-1})_{m+1/2}^k(\mathbf{E}_m^k-\mathbf{E}_{m+1}^k), \quad m=i,i-1;\\
&\text{or}\ \ \mathbf{F}_{m+1/2}=\tfrac{1}{2}(\mathbf{F}_m^k+\mathbf{F}_{m+1}^k)+\tfrac{1}{2}(E_U|A|)_{m+1/2}^k(\mathbf{U}_m^k-\mathbf{U}_{m+1}^k),\\
&|A|=\Omega_{\mathrm{R}}[|\lambda_p|\delta_{pl}]\Omega_{\mathrm{L}}.
\end{aligned} \tag{4.6.20}$$

To implement the scheme expressed by (4.6.19)–(4.6.20), one must calculate $|A|$ in accordance with Eq. (4.6.20), where Ω_{R}, Ω_{L}, and Λ are defined in (4.6.13)–(4.6.14). Instead of the matrix A, one can use the matrix $\tilde{A}$ of (4.6.15). Then, the finite-volume scheme defined by (4.6.19) involves the formulas

$$\begin{aligned}
&\mathbf{F}_{m+1/2}=\tfrac{1}{2}(\mathbf{F}_m^k+\mathbf{F}_{m+1}^k)+\tfrac{1}{2}|\tilde{A}|_{m+1/2}^k(\mathbf{E}_m^k-\mathbf{E}_{m+1}^k), \quad m=i,i-1;\\
&\text{or}\ \ \mathbf{F}_{m+1/2}=\tfrac{1}{2}(\mathbf{F}_m^k+\mathbf{F}_{m+1}^k)+\tfrac{1}{2}(|\tilde{A}|E_U)_{m+1/2}^k(\mathbf{U}_m^k-\mathbf{U}_{m+1}^k),\\
&|\tilde{A}|=\tilde{\Omega}_{\mathrm{R}}[|\lambda_p|\delta_{pl}]\tilde{\Omega}_{\mathrm{L}}.
\end{aligned}$$

The Roe method is similar to the method of (4.6.19)–(4.6.20); here the matrix A (or $\tilde{A}$) is denoted by $\mathcal{A}$ (or $\tilde{\mathcal{A}}$) and defined as

$$\mathcal{A}=\mathcal{E}_U^{-1}\mathcal{F}_U, \quad \tilde{\mathcal{A}}=\mathcal{F}_U\mathcal{E}_U^{-1}, \quad \Delta\mathbf{F}=\mathcal{F}_U\Delta\mathbf{U}, \quad \Delta\mathbf{E}=\mathcal{E}_U\Delta\mathbf{U},$$

$$\mathcal{F}_U=\mathcal{F}_U(\mathbf{U}_{m+1},\mathbf{U}_m)=\mathcal{F}_U[\tfrac{1}{2}(\mathbf{U}_{m+1}+\mathbf{U}_m)], \tag{4.6.21}$$

$$\mathcal{E}_U=\mathcal{E}_U(\mathbf{U}_{m+1},\mathbf{U}_m)=\mathcal{E}_U[\tfrac{1}{2}(\mathbf{U}_{m+1}+\mathbf{U}_m)], \tag{4.6.22}$$

$$\Delta\mathbf{U}=\mathbf{U}_{m+1}-\mathbf{U}_m, \quad \Delta\mathbf{E}=\mathbf{E}(\mathbf{U}_{m+1})-\mathbf{E}(\mathbf{U}_m), \quad \Delta\mathbf{F}=\mathbf{F}(\mathbf{U}_{m+1})-\mathbf{F}(\mathbf{U}_m),$$

where $\mathcal{E}_U$ and $\mathcal{F}_U$ are matrices that provide the exact relation between the corresponding finite differences. The matrices $\mathcal{F}_U$ and $\mathcal{E}_U$ satisfy relation (4.6.21) and (4.6.22), since $\mathbf{U}$ is a vector of Roe variables. This choice of the matrix $\mathcal{A}$ permits one to preserve exactly the hydraulic jump relations, see Section 4.3. Note that the matrix $\mathcal{A}$ coincides with the matrix of (4.6.12), where the Roe averages $\hat{u}$, $\hat{v}$, and $\hat{c}$, see Eqs. (4.5.9), must be substituted for u, v, and c, respectively.

To take advantage of the Godunov method based on the exact solution, one must use the exact solution of the corresponding Riemann problem. Below we describe the basic stages of the construction of this solution for stationary supercritical shallow water equations.

4.6.3 Exact solution of the Riemann problem. In constructing the exact solution, we follow the logical scheme of Section 3.6 dealing with stationary supersonic gas dynamics. First we shall construct the basic elementary solutions. After that an exact solution will be described using the superposition principle.

Elementary solution 1: Hydraulic jump. The first elementary solution of the shallow water equations is a stationary discontinuity (hydraulic jump or tangential discontinuity). For obtaining of the discontinuity relations let us integrate Eq. (4.6.6) over x and y and consider their integral form

$$\oint_L (\mathbf{E}\,dy - \mathbf{F}\,dx) = \iint_G \mathbf{B}\,dx\,dy, \tag{4.6.23}$$

where L is the boundary of a region G in the (x,y) plane. Let us seek a discontinuous solution of Eqs. (4.6.3)–(4.6.5) as a rectilinear wave of the form $f(x,y) = f(\eta) \equiv f(y-\theta x)$, where $\theta = \text{const}$ is the wave slope. Consider Eqs. (4.6.23) in orthogonal coordinates (η,τ) associated with discontinuity: $\eta = y - \theta x$ and $\tau = \theta y + x$. The coordinate η is normal and coordinate τ is tangential with respect to the discontinuity. Using transformation

$$y = \frac{\eta + \theta\tau}{1+\theta^2}, \qquad x = \frac{-\theta\eta + \tau}{1+\theta^2}$$

we can rewrite Eq. (4.6.23) as

$$\frac{1}{1+\theta^2}\oint_L [(\theta\mathbf{E} - \mathbf{F})\,d\tau + (\mathbf{E} + \theta\mathbf{F})\,d\eta] = \frac{1}{1+\theta^2}\iint_G \mathbf{B}\,d\tau\,d\eta. \tag{4.6.24}$$

Let us integrate (4.6.24) for the rectangular region $\tau_0 - \delta\tau \le \tau \le \tau_0 + \delta\tau$ and $\eta_0 - \delta\eta \le \eta \le \eta_0 + \delta\eta$, where $\eta = \eta_0$ corresponds to the discontinuity. We can find

$$(\theta\mathbf{E} - \mathbf{F})_1 2\delta\tau - (\theta\mathbf{E} - \mathbf{F})_2 2\delta\tau = \int_{\tau_0-\delta\tau}^{\tau_0+\delta\tau}\int_{\eta_0-\delta\eta}^{\eta_0+\delta\eta} \mathbf{B}\,d\tau\,d\eta.$$

For $\delta\eta \to 0$ we obtain

$$\theta\{\mathbf{E}\} - \{\mathbf{F}\} = \mathbf{0}, \tag{4.6.25}$$

where $\{q\} \equiv q_1 - q_2$, and indices 1 and 2 denote the variables on the left- and right-hand side of the discontinuity.

Consider the Riemann problem with the following initial data: $\mathbf{U}_1 = [h_1, u_1, v_1]^{\mathrm{T}}$ for $y > 0$ and $\mathbf{U}_2 = [h_2, u_2, v_2]^{\mathrm{T}}$ for $y < 0$. Both $\mathbf{U}_1$ and $\mathbf{U}_2$ satisfy Eqs. (4.6.3)–(4.6.5) with $\mathbf{f} = 0$ and $b = \text{const}$. Suppose that this initial discontinuity has the slope θ and is a solution of the Riemann problem. Then we can rewrite Eq. (4.6.25) in the form

$$\theta\{hu\} - \{hv\} = 0, \tag{4.6.26}$$
$$\theta\{hu^2 + \tfrac{1}{2}gh^2\} - \{huv\} = 0, \tag{4.6.27}$$
$$\theta\{huv\} - \{hv^2 + \tfrac{1}{2}gh^2\} = 0, \tag{4.6.28}$$

which relates θ to $\mathbf{U}_1$ and $\mathbf{U}_2$. If these conditions are satisfied, then the discontinuity in question is a solution of Eqs. (4.6.3)–(4.6.5) with $\mathbf{f} = 0$ and $b = \text{const}$.

Let us introduce new variables V and U associated with the jump:

$$V = v\cos\alpha - u\sin\alpha, \tag{4.6.29}$$
$$U = v\sin\alpha + u\cos\alpha, \tag{4.6.30}$$

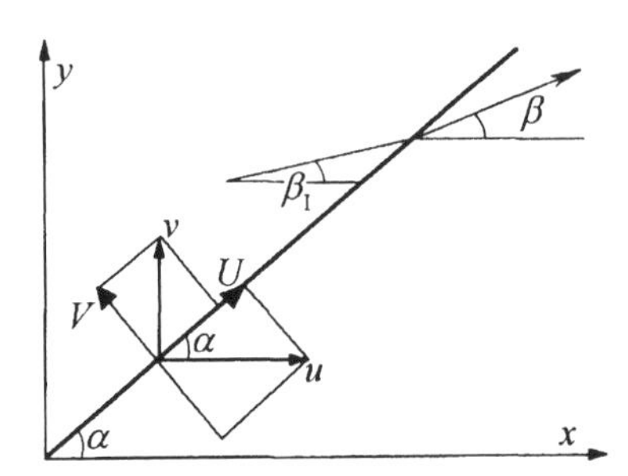

Figure 4.22 Coordinate system associated with the jump.

where $\tan\alpha = \theta$, see Fig. 4.22. Since $\{\theta\} = 0$, from Eqs. (4.6.26)–(4.6.28) one can obtain the compact relations

$$\{hV\} = 0, \tag{4.6.31}$$

$$\{hV^2 + \tfrac{1}{2}gh^2\} = 0, \tag{4.6.32}$$

$$\{hVU\} = 0 \quad \Longrightarrow \quad \{U\} = 0 \ \text{ if } \ hV = h_1V_1 = h_2V_2 \neq 0. \tag{4.6.33}$$

The case $hV = 0$ corresponding a tangential discontinuity will be considered separately. Note that the above relations have a form equivalent to that of the relations for time-dependent equations, see Eqs. (4.3.7)–(4.3.9). Equations (4.6.31)–(4.6.32) permit one to find the formulas for V_2 and h_2 as functions of h_1 and V_1.

From (4.6.31) we have

$$V_2 = \frac{h_1V_1}{h_2}.$$

Using this relation and Eq. (4.6.32), we arrive at the following formula for V_1:

$$V_1 = \pm\sqrt{\tfrac{1}{2}g(h_1 + h_2)\frac{h_2}{h_1}}.$$

Thus, we have obtained V_1 as a function of h_2 and h_1. Let $m = hV$ denote the mass flux across the discontinuity. From Eq. (4.6.31)–(4.6.32) it follows that $\{m\} = 0$ and $\{mV + \frac{1}{2}gh^2\} = 0$. Then,

$$m_1 = h_1V_1 = \pm\sqrt{\tfrac{1}{2}g(h_1 + h_2)h_1h_2}. \tag{4.6.34}$$

Since only hydraulic jump with increasing h can exist, it is clear that V_1 must be directed to the high elevation side of the jump. This allows one to determine the sign in (4.6.34).

Upper hydraulic jump

We say that the hydraulic jump is *upper* if the fluid flow through it is directed along the y-direction from top to bottom (Fig. 4.23a), i.e., $m = hV < 0$. Denote the flow parameters above the jump by h_1, u_1, and v_1 and those below the jump by h, u, and v, see Fig. 4.23a.

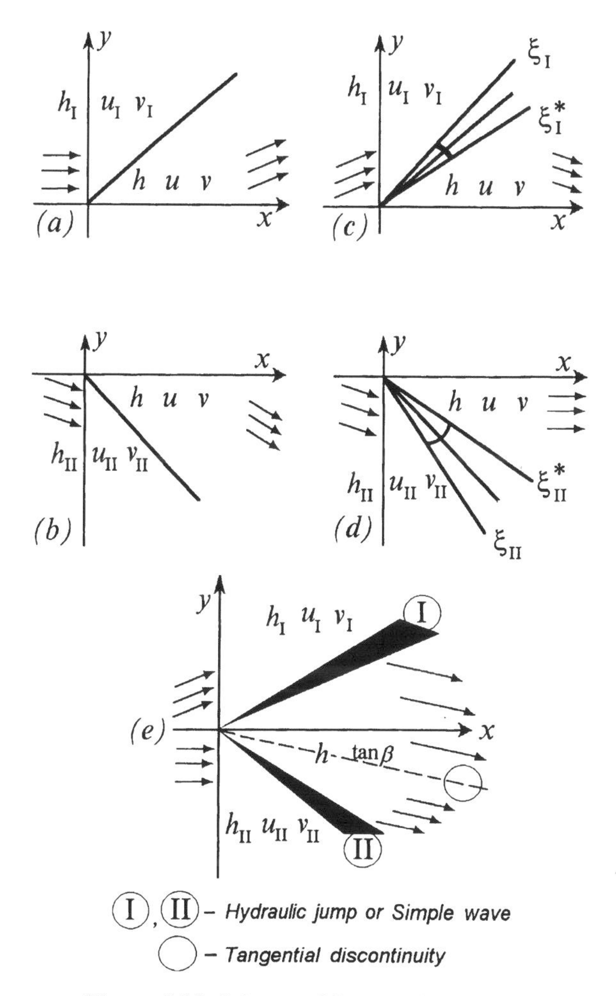

Figure 4.23 Scheme of flow configurations.

Using the relations $\{U\} = 0$ and $\{m\} = 0$, we obtain the following equations for h, u, and v, where $h > h_{\rm I}$, or $c > c_{\rm I}$, respectively:

$$v_{\rm I} \sin\alpha + u_{\rm I} \cos\alpha = v \sin\alpha + u \cos\alpha, \tag{4.6.35}$$

$$(v_{\rm I} \cos\alpha - u_{\rm I} \sin\alpha) h_{\rm I} = (v \cos\alpha - u \sin\alpha) h. \tag{4.6.36}$$

Introducing the notation

$$\tan\beta_{\rm I} = \frac{v_{\rm I}}{u_{\rm I}}, \quad \tan\beta = \frac{v}{u},$$

see Fig. 4.22, we rewrite Eqs. (4.6.35)–(4.6.36) as

$$(\tan\beta_{\rm I} \tan\alpha + 1) u_{\rm I} = (\tan\beta \tan\alpha + 1) u, \tag{4.6.37}$$

$$(\tan\beta_{\rm I} - \tan\alpha) h_{\rm I} u_{\rm I} = (\tan\beta - \tan\alpha) h u. \tag{4.6.38}$$

Dividing Eq. (4.6.38) by Eq. (4.6.37) and using the identity

$$\tan(\mu \pm \nu) = \frac{\tan\mu \pm \tan\nu}{1 \mp \tan\mu \tan\nu},$$

we arrive at the relation

$$\tan(\alpha - \beta) = \frac{h_{\rm I}}{h} \tan(\alpha - \beta_{\rm I}), \tag{4.6.39}$$

where $\beta > \beta_{\rm I}$, since $h > h_{\rm I}$, and $\alpha > \beta_{\rm I}$, since $V < 0$. Let us transform Eq. (4.6.29) for $V = V_{\rm I}$ in the following way:

$$V_{\rm I} = v_{\rm I} \cos\alpha - u_{\rm I} \sin\alpha = -q_{\rm I} \sin(\alpha - \beta_{\rm I}) < 0, \quad q_{\rm I} = \sqrt{u_{\rm I}^2 + v_{\rm I}^2}.$$

Thus,

$$\sin(\alpha - \beta_{\rm I}) = \frac{|V_{\rm I}|}{q_{\rm I}} = \frac{|m_{\rm I}|}{h_{\rm I} q_{\rm I}} = \frac{1}{q_{\rm I}} \sqrt{\tfrac{1}{2} g (h + h_{\rm I}) \frac{h}{h_{\rm I}}},$$

or

$$\tan(\alpha - \beta_{\rm I}) = \left(\frac{1}{\sin^2(\alpha - \beta_{\rm I})} - 1 \right)^{-1/2} = \left(\frac{2 h_{\rm I} q_{\rm I}^2}{(h + h_{\rm I}) g h} - 1 \right)^{-1/2}. \tag{4.6.40}$$

Using Eqs. (4.6.39)–(4.6.40), one can find $\tan(\beta - \beta_{\rm I})$ as a function of $h - h_{\rm I}$:

$$\tan(\beta - \beta_{\rm I}) = \frac{\tan(\alpha - \beta_{\rm I}) - \tan(\alpha - \beta)}{1 + \tan(\alpha - \beta_{\rm I}) \tan(\alpha - \beta)} = \frac{(h - h_{\rm I}) \tan(\alpha - \beta_{\rm I})}{h + h_{\rm I} \tan^2(\alpha - \beta_{\rm I})}.$$

Finally,

$$\tan(\beta - \beta_{\rm I}) = \frac{h - h_{\rm I}}{\left(\dfrac{2 h_{\rm I}^2 {\rm Fr}_{\rm I}^2}{(h + h_{\rm I})} + h_{\rm I} - h \right)} \sqrt{\frac{2 h_{\rm I}^2 {\rm Fr}_{\rm I}^2}{(h + h_{\rm I}) h} - 1}, \tag{4.6.41}$$

$${\rm Fr}_{\rm I} = \frac{q_{\rm I}}{c_{\rm I}}, \quad q_{\rm I} = \sqrt{u_{\rm I}^2 + v_{\rm I}^2}, \quad c_{\rm I} = \sqrt{g h_{\rm I}}.$$

Hence,

$$\tan\beta - \tan\beta_{\mathrm{I}} = (c - c_{\mathrm{I}})(1 + \tan\beta\tan\beta_{\mathrm{I}})J_{\mathrm{I}}, \tag{4.6.42}$$

$$J_{\mathrm{I}} = (c + c_{\mathrm{I}})\left(\frac{2gh_{\mathrm{I}}^2\mathrm{Fr}_{\mathrm{I}}^2}{(h + h_{\mathrm{I}})} + gh_{\mathrm{I}} - gh\right)^{-1}\sqrt{\frac{2h_{\mathrm{I}}^2\mathrm{Fr}_{\mathrm{I}}^2}{(h + h_{\mathrm{I}})h} - 1}.$$

Note that Eq. (4.6.42) can be solved also for $\tan\beta$; then,

$$\tan\beta = L_{\mathrm{I}}\tan\beta_{\mathrm{I}} + (c - c_{\mathrm{I}})N_{\mathrm{I}},$$

$$L_{\mathrm{I}} = \frac{1}{1 + (c_{\mathrm{I}} - c)J_{\mathrm{I}}\tan\beta_{\mathrm{I}}}, \quad N_{\mathrm{I}} = L_{\mathrm{I}}J_{\mathrm{I}}.$$

This approach was earlier developed for stationary gas dynamics in Section 3.6. It provides another equation to determine an exact solution and can be also used in shallow water equations.

To determine $\tan\alpha$ for known h, one can use the relation

$$\tan\alpha = \frac{\tan(\alpha - \beta_{\mathrm{I}}) + \tan\beta_{\mathrm{I}}}{1 - \tan(\alpha - \beta_{\mathrm{I}})\tan\beta_{\mathrm{I}}}, \tag{4.6.43}$$

which depends on $\tan(\alpha - \beta_{\mathrm{I}})$, see Eq. (4.6.40), and $\tan\beta_{\mathrm{I}}$. Using Eq. (4.6.35) and the relation $v = u\tan\beta$, one can determine u and v.

Lower hydraulic jump

We say that the hydraulic jump is *lower* if the fluid flow through it is directed from bottom to top, i.e., $m = hV > 0$. Denote the flow parameters below the jump by h_{II}, u_{II}, and v_{II} and those above the jump by h, u, and v, see Fig. 4.23b. Applying the relations $\{U\} = 0$ and $\{m\} = 0$, we obtain, just as in Eqs. (4.6.35)–(4.6.40), the following equation for β and h, where $h > h_{\mathrm{II}}$, or $c > c_{\mathrm{II}}$, respectively:

$$\tan(\beta_{\mathrm{II}} - \beta) = \frac{h - h_{\mathrm{II}}}{\left(\dfrac{2h_{\mathrm{II}}^2\mathrm{Fr}_{\mathrm{II}}^2}{(h + h_{\mathrm{II}})} + h_{\mathrm{II}} - h\right)}\sqrt{\frac{2h_{\mathrm{II}}^2\mathrm{Fr}_{\mathrm{II}}^2}{(h + h_{\mathrm{II}})h} - 1}, \tag{4.6.44}$$

$$\mathrm{Fr}_{\mathrm{II}} = \frac{q_{\mathrm{II}}}{c_{\mathrm{II}}}, \quad q_{\mathrm{II}} = \sqrt{u_{\mathrm{II}}^2 + v_{\mathrm{II}}^2}, \quad c_{\mathrm{II}} = \sqrt{gh_{\mathrm{II}}};$$

hence,

$$\tan\beta_{\mathrm{II}} - \tan\beta = (c - c_{\mathrm{II}})(1 + \tan\beta\tan\beta_{\mathrm{II}})J_{\mathrm{II}}, \tag{4.6.45}$$

$$J_{\mathrm{II}} = (c + c_{\mathrm{II}})\left(\frac{2gh_{\mathrm{II}}^2\mathrm{Fr}_{\mathrm{II}}^2}{(h + h_{\mathrm{II}})} + gh_{\mathrm{II}} - gh\right)^{-1}\sqrt{\frac{2h_{\mathrm{II}}^2\mathrm{Fr}_{\mathrm{II}}^2}{(h + h_{\mathrm{II}})h} - 1}.$$

To determine $\tan\alpha$, we express it in terms of $\tan\beta_{\mathrm{II}}$ and $\tan(\beta_{\mathrm{II}} - \alpha)$ by using the relation

$$\tan\alpha = \frac{\tan\beta_{\mathrm{II}} - \tan(\beta_{\mathrm{II}} - \alpha)}{1 + \tan\beta_{\mathrm{II}}\tan(\beta_{\mathrm{II}} - \alpha)}, \quad \tan(\beta_{\mathrm{II}} - \alpha) = \left(\frac{2h_{\mathrm{II}}q_{\mathrm{II}}^2}{(h + h_{\mathrm{II}})gh} - 1\right)^{-1/2}, \tag{4.6.46}$$

where $\beta_{II} > \alpha$, since $V > 0$. Further, by using Eq. (4.6.35) and the relation $v = u \tan\beta$, one can determine u and v.

A stationary hydraulic jump is a self-similar solution in $\eta = y - \theta x$, $\theta =$ const. This solution is also self-similar in $\xi = y/x$. In fact, a discontinuity in the (x, y)-coordinates is a straight line defined by the equation $\xi = \theta =$ const.

Elementary solution 2: Tangential discontinuity. This elementary solution of shallow water equations is a special case of discontinuity with zero water flow across the discontinuity: $m = hV = h_1 V_1 = h_2 V_2 = 0$. Thus

$$V = V_1 = V_2 = 0, \quad \{h\} = 0,$$

see Eq. (4.6.32).

The top variables are $h = h_1$, $u = u_1$, and $v = v_1$ and the bottom values are to $h = h_2$, $u = u_2$, and $v = v_2$. Then from relations $\{h\} = 0$ and $V = 0$, one obtains

$$h_1 = h_2 = h, \quad \tan\alpha = \tan\beta_1 = \frac{v_1}{u_1} = \tan\beta_2 = \frac{v_2}{u_2}. \tag{4.6.47}$$

Elementary solution 3: Simple wave. Another solution of the stationary shallow water equations is a simple wave. This solution is continuous. Since we seek a continuous solution, we can take advantage of the shallow water system (4.6.3)–(4.6.5) in the nonconservative form for horizontal botton relief, $b =$ const, and $\mathbf{f} = \mathbf{0}$:

$$uh_x + vh_y + (u_x + v_y)h = 0, \tag{4.6.48}$$

$$uu_x + vu_y + gh_x = 0, \tag{4.6.49}$$

$$uv_x + vv_y + gh_y = 0. \tag{4.6.50}$$

Let us seek a continuous solution in the self-similar form $f(x, y) = f(\xi) \equiv f(y/x)$. Substituting $h = h(\xi)$, $u = u(\xi)$, and $v = v(\xi)$ into Eqs. (4.6.48)–(4.6.50) yields

$$(-u\xi + v)h_\xi - \xi h u_\xi + h v_\xi = 0, \tag{4.6.51}$$

$$-g\xi h_\xi + (-u\xi + v)u_\xi = 0, \tag{4.6.52}$$

$$gh_\xi + (-u\xi + v)v_\xi = 0. \tag{4.6.53}$$

System (4.6.51)–(4.6.53) can have a nonzero solution for h_ξ, u_ξ, and v_ξ only if the determinant of the 3×3 system matrix is zero. Thus, we obtain the following three necessary conditions for a nontrivial solution to exist:

$$-u\xi + v = 0, \tag{4.6.54}$$

$$-u\xi + v = +c\sqrt{1 + \xi^2}, \tag{4.6.55}$$

$$-u\xi + v = -c\sqrt{1 + \xi^2}, \quad c = \sqrt{gh}. \tag{4.6.56}$$

Using the notation $\xi = \tan\alpha$ and $\tan\beta = v/u$, see Fig. 4.22, one can transform Eqs. (4.6.55)–(4.6.56) into the relation

$$\sin(\alpha - \beta) = \mp\frac{c\cos\beta}{u} = \mp\frac{c}{q} = \mp\frac{1}{\mathrm{Fr}}, \quad \mathrm{Fr} = \frac{q}{c}, \quad q = \sqrt{u^2 + v^2}. \tag{4.6.57}$$

Consider Eqs. (4.6.51)–(4.6.53) for the three cases of (4.6.54)–(4.6.56); we refer to these cases as I, II, and III, respectively.

Case I:

$$-u\xi + v = 0, \quad h_\xi = 0, \quad -\xi u_\xi + v_\xi = 0.$$

Here, the solution is given by h = const, $u = 0$, and $v = 0$. This solution is physically trivial.

Case II:

$$\begin{aligned} &-u\xi + v = c\sqrt{1+\xi^2}, \\ &-g\xi h_\xi + cu_\xi\sqrt{1+\xi^2} = 0, \quad gh_\xi + cv_\xi\sqrt{1+\xi^2} = 0. \end{aligned}$$

Thus,

$$u_\xi = \frac{g\xi h_\xi}{c\sqrt{1+\xi^2}}, \qquad v_\xi = -\frac{gh_\xi}{c\sqrt{1+\xi^2}}. \tag{4.6.58}$$

Then, using these formulas and relation (4.6.57), we find that

$$\begin{aligned} \frac{\partial \tan\beta}{\partial\xi} &= \frac{\partial}{\partial\xi}\left(\frac{v}{u}\right) = \frac{uv_\xi - vu_\xi}{u^2} = -\frac{(u+v\xi)g}{cu^2\sqrt{1+\xi^2}}\frac{\partial h}{\partial\xi} = -\frac{g\cos(\alpha-\beta)}{cu\cos\beta}\frac{\partial h}{\partial\xi} \\ &= -\frac{g\cos(\alpha-\beta)(1+\tan^2\beta)\cos\beta}{cu}\frac{\partial h}{\partial\xi} = -\frac{(1+\tan^2\beta)g\sqrt{\mathrm{Fr}^2-1}}{q^2}\frac{\partial h}{\partial\xi} = -R(\beta,c)\frac{\partial c}{\partial\xi} \\ \Longrightarrow \quad & \frac{\partial\tan\beta}{\partial\xi} = -R(\beta,c)\frac{\partial c}{\partial\xi}, \quad R(\beta,c) = 2(1+\tan^2\beta)\frac{\sqrt{q^2-c^2}}{q^2}. \end{aligned} \tag{4.6.59}$$

Using formulas (4.6.58) we also obtain

$$\frac{\partial}{\partial\xi}\left(\tfrac{1}{2}q^2\right) = uu_\xi + vv_\xi = -gh_\xi = -\frac{\partial}{\partial\xi}(gh).$$

Thus, in the simple wave the following relation holds:

$$\frac{\partial}{\partial\xi}\left(\tfrac{1}{2}q^2 + gh\right) = 0 \quad \Longrightarrow \quad \tfrac{1}{2}q^2(\xi) + gh(\xi) = \text{const}. \tag{4.6.60}$$

Case III:

$$\begin{aligned} &-u\xi + v = -c\sqrt{1+\xi^2}, \\ &-g\xi h_\xi - cu_\xi\sqrt{1+\xi^2} = 0, \quad gh_\xi - cv_\xi\sqrt{1+\xi^2} = 0. \end{aligned}$$

Thus, by analogy with case II, we have

$$\frac{\partial\tan\beta}{\partial\xi} = \frac{(1+\tan^2\beta)g\sqrt{\mathrm{Fr}^2-1}}{q^2}\frac{\partial h}{\partial\xi} = R(\beta,c)\frac{\partial c}{\partial\xi}. \tag{4.6.61}$$

It is not difficult to show that, in this case, relation (4.6.60) holds as well.

Upper simple wave

We say that the simple wave is *upper* if the fluid flow through it is directed along the y-direction from top to bottom, i.e., $m = hV < 0$. In this case we deal with relations (4.6.56), (4.6.61), and (4.6.60):

$$\begin{aligned} &-u\xi + v = -c\sqrt{1+\xi^2}, \\ &\frac{\partial \tan\beta}{\partial \xi} = R(\beta, c)\frac{\partial c}{\partial \xi}, \\ &\tfrac{1}{2}q^2 + gh = \text{const}. \end{aligned} \tag{4.6.62}$$

We have $m = hV = (v - u\xi)h\cos\alpha = -hc < 0$. Denote the flow parameters above the simple wave by h_{I}, u_{I}, and v_{I} and those below by h, u, and v, see Fig. 4.23c. Integrating Eq. (4.6.62) across the simple wave with respect to ξ yields the following relation for $\tan\beta$ and c:

$$\tan\beta - \tan\beta_{\text{I}} = (c - c_{\text{I}})r_{\text{I}}, \tag{4.6.63}$$

$$r_{\text{I}} = r_{\text{I}}(\beta, c) = \frac{1}{c - c_{\text{I}}}\int_{\xi_{\text{I}}}^{\xi_{\text{I}}^*} R(\beta, c)\frac{\partial c}{\partial \xi}\, d\xi, \quad R(\beta, c) = 2(1 + \tan^2\beta)\frac{\sqrt{q^2 - c^2}}{q^2},$$
$$r_{\text{I}}(\beta, c) \simeq \tfrac{1}{2}[R(\beta_{\text{I}}, c_{\text{I}}) + R(\beta, c)], \tag{4.6.64}$$

where the sign $\simeq$ denotes the approximation of the second order of accuracy with respect to ξ. Here we assume that the simple wave is bounded in the (x, y) plane by the rays $\xi = \xi_{\text{I}}$ and $\xi = \xi_{\text{I}}^*$, see Fig. 4.23c. Note that another approximating relation can be also be used instead of (4.6.64). To increase the accuracy of approximation for strong simple waves, a finer mesh in the y-direction should be used.

Lower simple wave

We say that the simple wave is *lower* if the fluid flow through it is directed from bottom to top, i.e., $m = hV > 0$. Here we deal with relations (4.6.55), (4.6.59), and (4.6.60):

$$\begin{aligned} &-u\xi + v = c\sqrt{1+\xi^2}, \\ &\frac{\partial \tan\beta}{\partial \xi} = -R(\beta, c)\frac{\partial c}{\partial \xi}, \\ &\tfrac{1}{2}q^2 + gh = \text{const}. \end{aligned} \tag{4.6.65}$$

We have $m = hV = (v - u\xi)h\cos\alpha = hc > 0$. Denote the flow parameters below the simple wave by h_{II}, u_{II}, and v_{II}, and those above by h, u, and v, see Fig. 4.23d. Integrating Eq. (4.6.65) across the simple wave with respect to ξ yields the following relation for $\tan\beta$ and c:

$$\tan\beta_{\text{II}} - \tan\beta = (c - c_{\text{II}})r_{\text{II}}, \tag{4.6.66}$$

$$r_{\text{II}} = r_{\text{II}}(\beta, c) = \frac{1}{c - c_{\text{II}}}\int_{\xi_{\text{II}}}^{\xi_{\text{II}}^*} R(\beta, c)\frac{\partial c}{\partial \xi}\, d\xi, \quad R(\beta, c) = 2(1 + \tan^2\beta)\frac{\sqrt{q^2 - c^2}}{q^2},$$
$$r_{\text{II}}(\beta, c) \simeq \tfrac{1}{2}[R(\beta_{\text{II}}, c_{\text{II}}) + R(\beta, c)]. \tag{4.6.67}$$

Here we assume that the simple wave is bounded in the (x, y) plane by the rays $\xi = \xi_{\mathrm{II}}^*$ and $\xi = \xi_{\mathrm{II}}$, see Fig. 4.23d.

4.6.4 General exact solution. The general solution of the Riemann problem consists of two hydraulic jumps and/or simple waves with a tangential discontinuity that are separated by flow regions with constant parameters, see Fig. 4.23e. Special cases of this general solution are a single hydraulic jump or a simple wave, see Fig. 4.23a–d.

Below we show how the general exact solution can be constructed on the basis of the elementary solutions described above.

It is noteworthy that c and $\tan\beta$ in the central zone for the case shown in Fig. 4.23e can be determined from two coupled equations. Considering Eq. (4.6.45) (resp., Eq. (4.6.66)) and Eq. (4.6.42) (resp., Eq. (4.6.63)), and the relation at the tangential discontinuity (4.6.47), we arrive at the following system of equations for c and $\tan\beta$:

$$\tan\beta_{\mathrm{II}} - \tan\beta = (c - c_{\mathrm{II}})\varphi_{\mathrm{II}}(\beta, c), \tag{4.6.68}$$

$$\tan\beta - \tan\beta_{\mathrm{I}} = (c - c_{\mathrm{I}})\varphi_{\mathrm{I}}(\beta, c), \tag{4.6.69}$$

where

$$\varphi_n = \begin{cases} (1+\tan\beta\tan\beta_n)(c+c_n)\left(\dfrac{2h_n^2\mathrm{Fr}_n^2}{(h+h_n)h} - 1\right)^{1/2}\left(\dfrac{2gh_n^2\mathrm{Fr}_n^2}{(h+h_n)} + gh_n - gh\right)^{-1} & \text{if } c > c_n, \\ \frac{1}{2}[R(\beta_n, c_n) + R(\beta, c)], \quad R(\beta, c) = 2(1+\tan^2\beta)\dfrac{\sqrt{q^2-c^2}}{q^2} & \text{if } c \le c_n. \end{cases}$$

Here $n = \mathrm{I}$ or II, $\mathrm{Fr} = q/c$, $q = \sqrt{u^2+v^2}$, and $c = \sqrt{gh}$. One can solve Eqs. (4.6.68)–(4.6.69) by an iteration method. For example, the first approximation can be written as

$$c^{(1)} = \frac{\varphi_{\mathrm{I}}c_{\mathrm{I}} + \varphi_{\mathrm{II}}c_{\mathrm{II}} + \tan\beta_{\mathrm{II}} - \tan\beta_{\mathrm{I}}}{\varphi_{\mathrm{I}} + \varphi_{\mathrm{II}}},$$
$$\tan\beta^{(1)} = \tan\beta_{\mathrm{I}} + (c^{(1)} - c_{\mathrm{I}})\varphi_{\mathrm{I}} = \tan\beta_{\mathrm{II}} - (c^{(1)} - c_{\mathrm{II}})\varphi_{\mathrm{II}},$$
$$\varphi_n = \varphi_n(\beta_n, c_n), \quad n = \mathrm{I}, \mathrm{II}.$$

Then, for the kth iteration we have $(k = 1, 2, \ldots)$

$$c^{(k+1)} = \left[\frac{\varphi_{\mathrm{I}}c_{\mathrm{I}} + \varphi_{\mathrm{II}}c_{\mathrm{II}} + \tan\beta_{\mathrm{II}} - \tan\beta_{\mathrm{I}}}{\varphi_{\mathrm{I}} + \varphi_{\mathrm{II}}}\right]^{(k)},$$

$$\tan\beta^{(k+1)} = \tan\beta_{\mathrm{I}} + (c^{(k+1)} - c_{\mathrm{I}})\varphi_{\mathrm{I}}^{(k)} = \tan\beta_{\mathrm{II}} - (c^{(k+1)} - c_{\mathrm{II}})\varphi_{\mathrm{II}}^{(k)},$$

$$\varphi_n^{(k)} = \varphi_n(\beta^{(k)}, c^{(k)}), \quad n = \mathrm{I}, \mathrm{II}.$$

To evaluate h, q, u, and v in the rarefaction fan, one should use the formulas

$$h = \frac{c^2}{g}, \quad v = u\tan\beta, \quad u = \frac{q}{\sqrt{1+\tan^2\beta}}, \tag{4.6.70}$$

and relation (4.6.60).

In many cases, the iterative process converges after two to eight iterations and provides an accuracy of 1% for c. To determine the exact solution, other iteration methods can also be used, e.g., see Section 3.6.

Just as in the case of a time-dependent system of equations, see Section 4.3, after solving Eqs. (4.6.68)–(4.6.69), one must analyze the flow configuration and determine the solution $h(\xi)$, $u(\xi)$, $v(\xi)$ for the desired ray ξ, in particular, for the stationary boundary $\xi = 0$. To determine the flow parameters, one can use relations (4.6.70) and (4.6.60) for the configurations with simple waves, and relations (4.6.43) and/or (4.6.46) for hydraulic jumps.

Chapter 5 Magnetohydrodynamic Equations

In Section 1.3.3 we presented the quasilinear system of MHD equations for ideal and infinitely conducting plasma. It was shown to be hyperbolic, and a nondegenerate system of eigenvectors was written out. As mentioned earlier, solutions to hyperbolic systems, in general, must contain discontinuities. This, of course, is also valid for MHD systems and therefore necessitates the development of numerical methods for their solution in the conservation-law form. This system must include the integral laws of mass, momentum, and energy conservation with the action of electromagnetic forces taken into account. It must also be supplemented by the Maxwell equations describing behavior of the electromagnetic field. The latter subsystem under certain conditions can also be written in the conservation-law form. For completeness, in Section 5.1 we are going to describe the assumptions that are usually adopted when obtaining the classical hyperbolic MHD system. In Sections 5.2 and 5.3 we outline the possible types of discontinuities intrinsic in solutions of the MHD equations and indicate those that satisfy the evolutionary property. In Section 5.4 we present various approaches to obtaining approximate solutions to the MHD Riemann problem in the one-dimensional statement. This section is devoted to numerical methods for MHD equations based, in particular, on the Riemann problem solvers. We also present the results of numerous numerical tests. In Section 5.5 we describe some peculiarities of numerical solution of MHD equations by shock-capturing methods. They are related to the fact that numerical viscosity and conductivity can never be avoided in the finite difference and finite volume methods. This sometimes results in the origin of nonevolutionary discontinuities or combinations of discontinuities. In Section 5.6 we discuss the possibility of extension of the Roe-type procedure to MHD problems involving strong background magnetic field. Section 5.7 contains some considerations on the numerical implementation of the magnetic field divergence-free condition (the absence of magnetic charge). And, finally, in Section 5.8 we present the application of numerical methods to the two- and three-dimensional problem of the solar wind interaction with the magnetized interstellar medium. We discuss in this chapter only nonrelativistic MHD. A robust Godunov-type scheme for relativistic MHD has recently been suggested by Komissarov (1999).

5.1 MHD system in the conservation-law form

Consider, first, a subsystem consisting of the equations of mass, momentum, and energy conservation in the presence of electromagnetic forces. Note that the general form of such

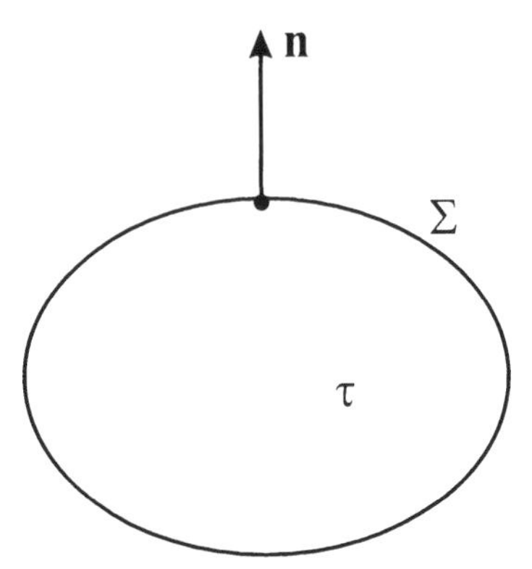

Figure 5.1 Integration volume.

equations for dynamics of continuous media can written in the form

$$\frac{d}{dt}\int_\tau \rho\, d\tau = 0, \tag{5.1.1}$$

$$\frac{d}{dt}\int_\tau \rho\mathbf{v}\, d\tau = \oint_\Sigma \mathbf{p}_n\, d\Sigma + \int_\tau \mathbf{f}\, d\tau, \tag{5.1.2}$$

$$\frac{d}{dt}\int_\tau \rho\left(\varepsilon + \frac{\mathbf{v}^2}{2}\right) d\tau = -\int_\Sigma q_n\, d\Sigma + \int_\Sigma \mathbf{p}_n \cdot \mathbf{v}\, d\Sigma + \int_\tau A\, d\tau. \tag{5.1.3}$$

This system is based on the Lagrangian treatment of continuous media in which we follow the time-variation of quantities in an initially fixed individual volume. The derivative d/dt in this case is a so-called substantive derivative that is equal to the time-derivative of the corresponding integrals taken over individual volumes τ surrounded by the surface Σ with the unit outward vector **n** (see Fig. 5.1).

In Eqs. (5.1.1)–(5.1.3), $\mathbf{v}$ is velocity, ρ is density, and ε is specific internal energy. The physical meaning of the system reads that the time-rate of variation of the quantity of mass (5.1.1), momentum (5.1.2), and energy (5.1.3) consists of volumetric contributions and surface effects. From this viewpoint, $\mathbf{p}_n$ is the surface force per unit area, $\mathbf{f}$ is the volumetric force density, q_n is the heat flux through Σ, and A is the energy loss/supply due to local volumetric sources and sinks (e. g., gravitation or radiation). As far as electromagnetic field is concerned, its action on charged particles is manifested via the Lorentz force, which in nonrelativistic case $|\mathbf{v}| \ll c$ (c is the speed of light) is given by the formula

$$\mathbf{f} = \rho_e \mathbf{E} + \frac{1}{c}(\mathbf{j} \times \mathbf{B}), \tag{5.1.4}$$

where ρ_e and $\mathbf{j}$ are the charge and the electric current density and $\mathbf{E}$ and $\mathbf{B}$ are the electric and magnetic field strengths, respectively. This formula under adopted assumptions is a Galilean invariant. This means that it remains valid in any coordinate system moving with velocity $\mathbf{v}$. Recall that the electromagnetic field in this coordinate system is transformed by the formulas (Kulikovskii and Lyubimov 1965)

$$\mathbf{E}' = \mathbf{E} + \frac{1}{c}(\mathbf{v} \times \mathbf{B}), \tag{5.1.5}$$

$$\mathbf{B}' = \mathbf{B} - \frac{1}{c}(\mathbf{v} \times \mathbf{E}). \tag{5.1.6}$$

It is important that the quantity $\mathbf{j}$ in (5.1.4) is the current density in a fixed coordinate system and can be represented as

$$\mathbf{j} = \mathbf{j}' + \rho_e \mathbf{v},$$

where $\mathbf{j}'$ is a genuine conductivity current, whereas the second term accounts for currents connected with the motion of the medium as a whole. The total energy supply provided by the electromagnetic force to the unit volume per unit time is given by the formula

$$A = \mathbf{j} \cdot \mathbf{E}. \tag{5.1.7}$$

The heat release J (the Joule heat) due to conductivity currents is the difference between the total energy A and the work of the Lorentz force

$$J = A - \mathbf{f} \cdot \mathbf{v} = \mathbf{j}' \cdot \mathbf{E}'.$$

Let us denote the stress tensor as

$$\hat{\mathbf{P}} = [p_{ij}], \qquad p_{ij} = p_{ji},$$

where p_{ij} is the stress acting in the direction $\mathbf{e}_i$ on the surface perpendicular to the unit basis vector $\mathbf{e}_j$ of a coordinate system. It can also be represented in the vector form $\mathbf{p}_j = p_{ij}\mathbf{e}_i$. Thus, the total force per unit area acting on the surface Σ with the normal vector $\mathbf{n}$ is

$$\mathbf{p}_n = \mathbf{p}_j n_j, \tag{5.1.8}$$

where n_j is the projection of $\mathbf{n}$ onto $\mathbf{e}_j$. Summation is adopted here over repeated indices.

The surface integrals on the right-hand side of Eqs. (5.1.1)–(5.1.3) can be transformed into the volume integrals using the Gauss formula

$$\oint_\Sigma \mathbf{p}_n \, d\Sigma = \int_\tau \operatorname{div} \hat{\mathbf{P}} \, d\tau \tag{5.1.9}$$

and

$$\oint_\Sigma \mathbf{p}_n \cdot \mathbf{v} \, d\Sigma = \int_\tau \operatorname{div} \left(\hat{\mathbf{P}} \cdot \mathbf{v}\right) d\tau. \tag{5.1.10}$$

If we consider only flows of ideal gas without viscosity and heat conductivity, then

$$\hat{\mathbf{P}} = -p\,\hat{\mathbf{I}}, \tag{5.1.11}$$

where $\hat{\mathbf{I}}$ is the identity tensor. Using conventional formulas that establish the relationship between the Lagrangian time-derivative d/dt and the Eulerian partial time-derivative $\partial/\partial t$ (see any textbook on dynamics of continuous media, e.g., Landau and Lifshitz 1984), Eqs. (5.1.1)–(5.1.3) acquire the form

$$\int_\tau \left[\frac{\partial \rho}{\partial t} + \operatorname{div}(\rho \mathbf{v})\right] d\tau = 0, \tag{5.1.12}$$

$$\int_\tau \left[\frac{\partial \rho \mathbf{v}}{\partial t} + \operatorname{div}\left(\rho \mathbf{v}\mathbf{v} + p\hat{\mathbf{I}}\right) - \mathbf{f}\right] d\tau = \mathbf{0}, \tag{5.1.13}$$

$$\int_\tau \left[\frac{\partial e}{\partial t} + \operatorname{div}\left[(e + p)\mathbf{v}\right] - \mathbf{j} \cdot \mathbf{E}\right] d\tau = 0. \tag{5.1.14}$$

Here we introduced the total energy (excluding electromagnetic contribution, of course) per unit volume

$$e = \rho\left(\varepsilon + \frac{\mathbf{v}^2}{2}\right).$$

This system must be accompanied by the Maxwell equations (see Landau and Lifshitz 1984)

$$\operatorname{curl}\mathbf{B} = \frac{4\pi}{c}\mathbf{j} + \frac{1}{c}\frac{\partial \mathbf{E}}{\partial t}, \tag{5.1.15}$$

$$\operatorname{curl}\mathbf{E} = -\frac{1}{c}\frac{\partial \mathbf{B}}{\partial t}, \tag{5.1.16}$$

$$\operatorname{div}\mathbf{E} = 4\pi\rho_e, \tag{5.1.17}$$

$$\operatorname{div}\mathbf{B} = 0, \tag{5.1.18}$$

where ρ_e is the charge density.

We do not make difference here between the magnetic field strength $\mathbf{H}$ and the induction $\mathbf{B}$, since in the chosen Gauss system of units

$$\mathbf{B} = \mu\mathbf{H}$$

and magnetic permittivity μ is supposed to be equal to 1 for fully ionized gaseous plasma.

By substituting the quantities of ρ_e and $\mathbf{j}$ from the Maxwell equations into the expression for the Lorentz force, it can be rewritten as

$$\mathbf{f} = \rho_e\mathbf{E} + \frac{1}{c}(\mathbf{j}\times\mathbf{B}) = -\frac{\partial \mathbf{g}}{\partial t} + \operatorname{div}\hat{\mathbf{T}}, \tag{5.1.19}$$

where

$$\mathbf{g} = \frac{1}{4\pi c}(\mathbf{E}\times\mathbf{B}); \quad \hat{\mathbf{T}} = [T_{ij}],$$

$$T_{ij} = \frac{1}{4\pi}(B_iB_j + E_iE_j) - \frac{1}{8\pi}(B^2 + E^2)\delta_{ij}.$$

Similarly,

$$\mathbf{j}\cdot\mathbf{E} = -\operatorname{div}\mathbf{s} - \frac{\partial w}{\partial t}, \tag{5.1.20}$$

where

$$\mathbf{s} = \frac{c}{4\pi}(\mathbf{E}\times\mathbf{B}), \qquad w = \frac{1}{8\pi}(E^2 + B^2).$$

The vector $\mathbf{g}$ has the meaning of the electromagnetic momentum per unit volume, $\hat{\mathbf{T}}$ is the electromagnetic stress tensor, w is the electromagnetic energy per unit volume, and $\mathbf{s}$ is the Pointing vector, that is, the vector of electromagnetic energy flux per unit area.

Thus, if we take into account the fact that the integration in Eqs. (5.1.12)–(5.1.14) is performed over an arbitrary volume τ, we can write out the governing system in the conservation-law form

$$\frac{\partial \rho}{\partial t} + \operatorname{div}(\rho\mathbf{v}) = 0, \tag{5.1.21}$$

$$\frac{\partial \mathbf{G}}{\partial t} + \operatorname{div} \hat{\mathbf{\Pi}} = \mathbf{0}, \tag{5.1.22}$$

$$\frac{\partial W}{\partial t} + \operatorname{div} \mathbf{S} = 0, \tag{5.1.23}$$

where

$$\mathbf{G} = \rho \mathbf{v} + \mathbf{g}, \qquad \hat{\mathbf{\Pi}} = [\pi_{ij}] = [\rho v_i v_j + p\delta_{ij} - T_{ij}],$$
$$W = e + \frac{E^2 + B^2}{8\pi}, \qquad \mathbf{S} = \mathbf{s} + (e + p)\,\mathbf{v},$$

can be interpreted as the total momentum per unit volume, the total momentum flux tensor per unit area, the total energy per unit volume, and the total energy flux per unit area.

The above system can be rewritten back in the integral form, which is more suitable for numerical solution,

$$\frac{\partial}{\partial t}\int_{\tau} \rho\, d\tau + \oint_{\Sigma} \rho v_n\, d\Sigma = 0, \tag{5.1.24}$$

$$\frac{\partial}{\partial t}\int_{\tau} \mathbf{G}\, d\tau + \oint_{\Sigma} \hat{\mathbf{\Pi}} \cdot \mathbf{n}\, d\Sigma = \mathbf{0}, \tag{5.1.25}$$

$$\frac{\partial}{\partial t}\int_{\tau} W\, d\tau + \oint_{\Sigma} S_n\, d\Sigma = 0, \tag{5.1.26}$$

where

$$\hat{\mathbf{\Pi}} \cdot \mathbf{n} = \Pi_{ij} n_i \mathbf{e}_j = \left(\rho v_i v_j + p\delta_{ij} - T_{ij}\right) n_i \mathbf{e}_j,$$
$$S_n = (e + p)\, v_n + \frac{c}{4\pi}(\mathbf{E} \times \mathbf{B}) \cdot \mathbf{n}.$$

To resolve the system of (5.1.24)–(5.1.26) supplemented by the Maxwell equations (5.1.15)–(5.1.18), we must add the caloric equation of state

$$\varepsilon = \varepsilon(p, \rho) \tag{5.1.27}$$

and also the formula (Ohm's law) for the current density $\mathbf{j}$. In the simplified form of the classical MHD this formula is

$$\mathbf{j} = \sigma\left(\mathbf{E} + \frac{1}{c}\mathbf{v} \times \mathbf{B}\right) + \rho_e \mathbf{v}, \tag{5.1.28}$$

where the coefficient σ is called a conductivity coefficient. This formula can be considered as an experimental fact. The first term in the sum is called a conductivity current and $\rho_e \mathbf{v}$ is called a convective current, since the latter is connected with the motion of the ionized gas as a whole in the fixed coordinate system.

Leaving aside the details that can be found in any textbook on MHD (e.g. Kulikovskii and Lyubimov 1965), we assume that the conductivity is so large that

$$\frac{1}{\sigma t_*} \ll 1, \qquad \frac{v_*}{\sigma L} \ll 1, \tag{5.1.29}$$

where t_*, v_*, and L are the characteristic time, velocity, and length, respectively. It is easy to see in this case that the terms $\partial \mathbf{E}/\partial t$ and $4\pi\rho_e \mathbf{v}$ in the equation

$$\operatorname{curl}\mathbf{B} - \frac{1}{c}\frac{\partial \mathbf{E}}{\partial t} = \frac{4\pi}{c}\left[\sigma\left(\mathbf{E} + \frac{1}{c}\mathbf{v}\times\mathbf{B}\right) + \rho_e\mathbf{v}\right]$$

following from Eqs. (5.1.15) and (5.1.28) can be neglected in comparison with $4\pi\sigma\mathbf{E}$.

Thus, under assumptions (5.1.29), the electric field strength can be excluded from the system of (5.1.24)–(5.1.26) and (5.1.15)–(5.1.18),

$$\mathbf{E} = \frac{1}{c}\left[\nu_m \operatorname{curl}\mathbf{B} - \mathbf{v}\times\mathbf{B}\right], \qquad \nu_m = \frac{c^2}{4\pi\sigma}. \tag{5.1.30}$$

The coefficient ν_m has the same physical dimension as the kinematic viscosity and is therefore called a *magnetic viscosity*. Thus, from Eqs. (5.1.16) and (5.1.30) we obtain

$$\frac{\partial \mathbf{B}}{\partial t} = \operatorname{curl}(\mathbf{v}\times\mathbf{B}) - \operatorname{curl}(\nu_m \operatorname{curl}\mathbf{B}). \tag{5.1.31}$$

Comparing the magnitude of the terms on the right-hand side of Eq. (5.1.31), we find that if

$$Re_m = \frac{v_* L}{\nu_m} \equiv \frac{4\pi\sigma v_* L}{c^2} \gg 1, \tag{5.1.32}$$

then the second term can be neglected, and we obtain

$$\frac{\partial \mathbf{B}}{\partial t} = \operatorname{curl}(\mathbf{v}\times\mathbf{B}). \tag{5.1.33}$$

By analogy with hydrodynamics Re_m is called the *magnetic Reynolds number*. Equation (5.1.33) is called the induction, or the Faraday, equation. It is easy to check component by component that

$$\operatorname{curl}(\mathbf{v}\times\mathbf{B}) = \operatorname{div}(\mathbf{vB} - \mathbf{Bv}),$$

that is, the set of the Maxwell equations (5.1.15)–(5.1.18) reduces to two equations for the magnetic field strength

$$\frac{\partial \mathbf{B}}{\partial t} = \operatorname{div}(\mathbf{vB} - \mathbf{Bv}), \tag{5.1.34}$$

$$\operatorname{div}\mathbf{B} = 0. \tag{5.1.35}$$

Note that if $\sigma = \text{const}$, then

$$\operatorname{curl}(\nu_m \operatorname{curl}\mathbf{B}) = -\nu_m \Delta\mathbf{B}$$

and equation (5.1.31) acquires the form

$$\frac{\partial \mathbf{B}}{\partial t} = \operatorname{curl}(\mathbf{v}\times\mathbf{B}) + \nu_m \Delta\mathbf{B}. \tag{5.1.36}$$

In nonrelativistic cases with

$$\frac{v_*^2}{c^2} \ll 1 \tag{5.1.37}$$

direct estimation under assumptions (5.1.29) shows that the displacement currents $(1/4\pi)\partial\mathbf{E}/\partial t$ and the convective currents $\rho_e\mathbf{v}$ are small compared with the total current $c \times \operatorname{curl}\mathbf{B}/4\pi$ and, hence,

$$\mathbf{j} = \sigma\left(\mathbf{E} + \frac{\mathbf{v}}{c} \times \mathbf{B}\right). \tag{5.1.38}$$

Thus, equation (5.1.15) acquires the form

$$\operatorname{curl}\mathbf{B} = \frac{4\pi}{c}\mathbf{j}, \tag{5.1.39}$$

that is, the current density $\mathbf{j}$ and the electric field strength $\mathbf{E}$ can be found using Eqs. (5.1.39) and (5.1.30), respectively, once the MHD system is solved.

It is easy to see that under assumptions (5.1.29), (5.1.32), and (5.1.37) the electric force $\rho_e\mathbf{E}$ and the electric field energy $E^2/8\pi$ can be neglected as compared with the magnetic force $\mathbf{j} \times \mathbf{B}/c$ and the magnetic energy $B^2/8\pi$, respectively. Hence, we can adopt

$$w = \frac{B^2}{8\pi}, \qquad \mathbf{f} = \frac{1}{c}\,(\mathbf{j} \times \mathbf{B}).$$

Note also that under assumptions (5.1.29) and (5.1.37) it also turns out that $\mathbf{B}$ and $\mathbf{j}$ remain unchanged in any inertial coordinate system. In fact, the Lorentz transform gives formula (5.1.6) for the magnetic field in the coordinate system moving with velocity $\mathbf{v}$. The invariance of $\mathbf{B}$ follows from the estimate for the magnitude of $\mathbf{E}$ on the basis of Eq. (5.1.30),

$$E \leq \max\left\{\frac{cB_*}{\sigma L}, \frac{V_*}{c}B_*\right\},$$

where L and the quantities with the subscript $*$ are the characteristic length and the characteristic magnitudes of the corresponding vectors. Then,

$$\left|\frac{\mathbf{v}}{c} \times \mathbf{E}\right| \leq B_* \max\left\{\frac{V_*}{\sigma L}, \frac{V_*^2}{c^2}\right\}.$$

Thus, we arrive at the final form of the MHD system in the conservation-law form,

$$\frac{\partial \rho}{\partial t} + \operatorname{div}\rho\mathbf{v} = 0, \tag{5.1.40}$$

$$\frac{\partial \rho\mathbf{v}}{\partial t} + \operatorname{div}\left[\rho\mathbf{v}\mathbf{v} + p_0\hat{\mathbf{I}} - \frac{\mathbf{B}\,\mathbf{B}}{4\pi}\right] = \mathbf{0}, \tag{5.1.41}$$

$$\frac{\partial e}{\partial t} + \operatorname{div}\left[(e + p_0)\,\mathbf{v} - \frac{\mathbf{B}\,(\mathbf{v}\cdot\mathbf{B})}{4\pi}\right] = 0. \tag{5.1.42}$$

Here we took into account the formula

$$(\mathbf{v} \times \mathbf{B}) \times \mathbf{B} = -\mathbf{v}\,B^2 + \mathbf{B}\,(\mathbf{v}\cdot\mathbf{B}),$$

redefined the total energy per unit volume e by adding the energy of magnetic field

$$e \equiv W = e_{\text{old}} + \frac{B^2}{8\pi},$$

and introduced the total pressure

$$p_0 = p + \frac{B^2}{8\pi}.$$

As previously noted in Section 1.3.3, Eq. (5.1.35) is mathematically excessive, since by applying the divergence operation to the Faraday equation and taking into account that the vector analysis formula $\operatorname{div}\operatorname{curl}\mathbf{a} \equiv 0$, we obtain

$$\frac{\partial}{\partial t}(\operatorname{div}\mathbf{B}) = 0.$$

This means that if $\operatorname{div}\mathbf{B} = 0$ at $t = 0$, it will always remain zero, that is, magnetic charge cannot originate later on. Though this is true in the mathematical sense, some numerical magnetic charge, in fact, can be accumulated if we solve the MHD system in the conservation-law form.

In Section 5.4 we shall consider the application of numerical methods to the system of (5.1.40)–(5.1.42) and (5.1.34). This system, as can be shown by direct differentiation, is equivalent to the system (1.3.24) and is hyperbolic. Solution of the system written in the integral form allows us to widen the class of solutions by including discontinuous functions.

Written for the components of the state vector $\mathbf{U}$, the conservative system of ideal MHD acquires the form

$$\frac{\partial \mathbf{U}}{\partial t} + \frac{\partial \mathbf{E}}{\partial x} + \frac{\partial \mathbf{F}}{\partial y} + \frac{\partial \mathbf{G}}{\partial z} = \mathbf{0}, \tag{5.1.43}$$

where

$$\mathbf{U} = \begin{bmatrix} \rho \\ \rho u \\ \rho v \\ \rho w \\ e \\ B_x \\ B_y \\ B_z \end{bmatrix}, \quad \mathbf{E} = \begin{bmatrix} \rho u \\ \rho u^2 + p_0 - \dfrac{B_x^2}{4\pi} \\ \rho u v - \dfrac{B_x B_y}{4\pi} \\ \rho u w - \dfrac{B_x B_z}{4\pi} \\ (e + p_0)u - \dfrac{B_x}{4\pi}(\mathbf{v}\cdot\mathbf{B}) \\ 0 \\ u B_y - v B_x \\ u B_z - w B_x \end{bmatrix}, \quad \mathbf{F} = \begin{bmatrix} \rho v \\ \rho u v - \dfrac{B_x B_y}{4\pi} \\ \rho v^2 + p_0 - \dfrac{B_y^2}{4\pi} \\ \rho u w - \dfrac{B_y B_z}{4\pi} \\ (e + p_0)v - \dfrac{B_y}{4\pi}(\mathbf{v}\cdot\mathbf{B}) \\ v B_x - u B_y \\ 0 \\ v B_z - w B_y \end{bmatrix},$$

$$\mathbf{G} = \begin{bmatrix} \rho w \\ \rho u w - \dfrac{B_x B_z}{4\pi} \\ \rho v w - \dfrac{B_y B_z}{4\pi} \\ \rho w^2 + p_0 - \dfrac{B_z^2}{4\pi} \\ (e + p_0)w - \dfrac{B_z}{4\pi}(\mathbf{v}\cdot\mathbf{B}) \\ w B_x - u B_z \\ w B_y - v B_z \\ 0 \end{bmatrix}.$$

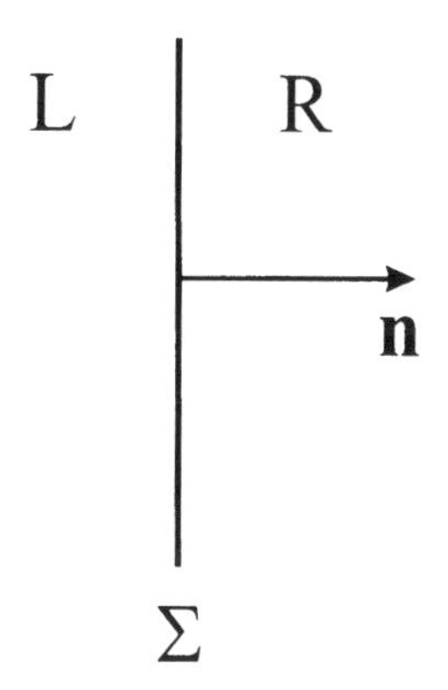

Figure 5.2 Discontinuity surface.

5.2 Classification of MHD discontinuities

In this section we classify possible types of discontinuities intrinsic in MHD flows and indicate those of them that satisfy the evolutionary property. Since this subject is perfectly well described in various monographs on MHD, we shall only discuss the facts necessary for the further discussion. It is obvious from Eqs. (5.1.24)–(5.1.26) that the following relations must be satisfied in the coordinate system attached to the discontinuity surface Σ:

$$\{\rho v_n\} = 0, \quad \{\hat{\mathbf{\Pi}} \cdot \mathbf{n}\} = 0, \quad \{S_n\} = 0, \tag{5.2.1}$$

where $\{f\} = f^{\mathrm{R}} - f^{\mathrm{L}}$ is the jump of the function f at the point belonging to Σ (see Fig. 5.2). These relations express the conservation of the mass, momentum, and energy fluxes through Σ.

The Faraday equation (5.1.34) gives

$$\{(\mathbf{vB} - \mathbf{Bv}) \cdot \mathbf{n}\} = 0. \tag{5.2.2}$$

Here $\mathbf{n}$ is the normal vector of unit length. Equations (5.2.1)–(5.2.2) can be projected onto the normal and tangential directions giving

$$\{\rho v_n\} = 0, \tag{5.2.3}$$

$$\left\{p + \rho v_n^2 + \frac{1}{8\pi}(\mathbf{B}_t^2 - B_n^2)\right\} = 0, \tag{5.2.4}$$

$$\left\{\rho v_n \mathbf{v}_t - \frac{B_n \mathbf{B}_t}{4\pi}\right\} = \mathbf{0}, \tag{5.2.5}$$

$$\left\{(e + p_0)v_n - \frac{B_n(\mathbf{v} \cdot \mathbf{B})}{4\pi}\right\} = 0, \tag{5.2.6}$$

$$\{B_n \mathbf{v}_t - v_n \mathbf{B}_t\} = \mathbf{0}, \tag{5.2.7}$$

$$\{B_n\} = 0. \tag{5.2.8}$$

If we denote the mass flux as $m = \rho v_n$, the system (5.2.3)–(5.2.8) acquires the form

$$\{m\} = 0, \tag{5.2.9}$$

$$\{p\} + m^2 \left\{ \frac{1}{\rho} \right\} + \frac{1}{8\pi} \{\mathbf{B}_t^2\} = 0, \tag{5.2.10}$$

$$m\{\mathbf{v}_t\} - \frac{B_n}{4\pi} \{\mathbf{B}_t\} = \mathbf{0}, \tag{5.2.11}$$

$$m \left\{ h + \frac{m^2}{2\rho^2} + \frac{\mathbf{v}_t^2}{2} + \frac{\mathbf{B}_t^2}{4\pi\rho} \right\} = \frac{B_n}{4\pi} \{\mathbf{B}_t \cdot \mathbf{v}_t\}, \tag{5.2.12}$$

$$B_n\{\mathbf{v}_t\} = m \left\{ \frac{\mathbf{B}_t}{\rho} \right\}. \tag{5.2.13}$$

Here we introduced the specific enthalpy $h = \varepsilon + p/\rho$.

MHD discontinuities are usually subdivided into four types: contact surfaces, tangential discontinuities, rotational, or Alfvén, discontinuities, and shocks. The first two occur if there is no flux through the discontinuity surface, that is, $m \equiv 0$. This means that $v_n^{\mathrm{R}} = v_n^{\mathrm{L}} = 0$ and the gas moves along the both sides of the discontinuity. In this case, taking into account that $B_n^{\mathrm{R}} = B_n^{\mathrm{R}} = \text{const}$, we obtain

$$\left\{ p + \frac{\mathbf{B}_t^2}{8\pi} \right\} = \{p_0\} = 0, \quad -\frac{B_n}{4\pi}\{\mathbf{B}_t\} = \mathbf{0}, \quad \frac{B_n}{4\pi}\{\mathbf{B}_t \cdot \mathbf{v}_t\} = 0, \quad B_n\{\mathbf{v}_t\} = \mathbf{0}.$$

If $B_n \neq 0$, then

$$\{\mathbf{B}_t\} = \mathbf{0}, \quad \{\mathbf{v}_t\} = \mathbf{0}, \quad \{\mathbf{B}_t \cdot \mathbf{v}_t\} = 0, \quad \{p\} = 0, \tag{5.2.14}$$

which means that velocity, magnetic field, and pressure are continuous, while density has an arbitrary jump. This is a contact discontinuity.

If $B_n \equiv 0$, then only the total pressure p_0 remains continuous, while the other quantities have an arbitrary jump. This is a tangential discontinuity.

If $m \neq 0$, then the two choices are important. If $\{\rho\} = 0$, then the normal component of the velocity vector is continuous and from Eqs. (5.2.11) and (5.2.13) we obtain

$$m = B_n \sqrt{\frac{\rho}{4\pi}}, \qquad v_n = \frac{B_n}{\sqrt{4\pi\rho}}, \tag{5.2.15}$$

that is, this discontinuity moves with the Alfvén velocity with respect to the gas ahead of it and

$$\{\mathbf{v}_t\} = \frac{\{\mathbf{B}_t\}}{4\pi\rho}.$$

It easy to check that all thermodynamic quantities and the magnitudes of **B** and **v** are continuous across this discontinuity: these vectors undergo only rotation. For this reason such discontinuities are called rotational.

If $m \neq 0$ and $\{\rho\} \neq 0$, we have MHD shocks. Note that an arbitrary constant vector can be added to $\mathbf{v}_t$ in Eqs. (5.2.11)–(5.2.13). This means that the coordinate system attached to the shock is fixed in space only by the shock velocity in its normal direction. Thus, the shock relations remain unchanged in any coordinate system moving with a constant velocity along the shock surface. This allows us, for example, to choose a coordinate system in which $\mathbf{v}^{\mathrm{L}}$

lies in the plane defined by the vectors $\mathbf{B}^{\mathrm{L}}$ and $\mathbf{n}$. Comparing Eqs. (5.2.11) and (5.2.13), we see that for $B_n \neq 0$,

$$\mathbf{B}_t^{\mathrm{L}} - \mathbf{B}_t^{\mathrm{R}} \parallel \mathbf{v}_t^{\mathrm{L}} - \mathbf{v}_t^{\mathrm{R}}, \quad \left(\frac{\mathbf{B}_t}{\rho}\right)^{\mathrm{L}} - \left(\frac{\mathbf{B}_t}{\rho}\right)^{\mathrm{R}} \parallel \mathbf{v}_t^{\mathrm{L}} - \mathbf{v}_t^{\mathrm{R}}.$$

As $\rho^{\mathrm{L}} \neq \rho^{\mathrm{R}}$, then $\mathbf{B}_t^{\mathrm{L}} \parallel \mathbf{B}_t^{\mathrm{R}}$ and, hence, the vectors $\mathbf{v}^{\mathrm{L}}$, $\mathbf{B}^{\mathrm{L}}$, $\mathbf{B}^{\mathrm{R}}$, and $\mathbf{n}$ lie in the same plane. As $\mathbf{v}_t^{\mathrm{R}} - \mathbf{v}_t^{\mathrm{L}}$ also belongs to this plane, then the same is valid for the vector $\mathbf{v}^{\mathrm{R}}$ itself. The result remains unchanged for $B_n \equiv 0$, since in this case

$$\left(\frac{\mathbf{B}_t}{\rho}\right)^{\mathrm{L}} = \left(\frac{\mathbf{B}_t}{\rho}\right)^{\mathrm{R}} \tag{5.2.16}$$

and, hence, $\mathbf{B}_t^{\mathrm{L}} \parallel \mathbf{B}_t^{\mathrm{R}}$. The described feature means that MHD shocks are plane by their nature.

Worth interest for future discussion are MHD shocks in the following two special cases: (i) $B_n \equiv 0$ and (ii) $\mathbf{B}_t^{\mathrm{L}} = \mathbf{B}_t^{\mathrm{R}} = \mathbf{0}$. It is easy to see that in the former case one-dimensional motions in MHD do not differ from pure hydrodynamic motions if we substitute thermal pressure by the total pressure p_0. Note that in this case $B/\rho = \text{const}$ at shocks. These shocks are called perpendicular in the sense that magnetic field ahead of and behind them is perpendicular to the normal vector $\mathbf{n}$.

In the case (ii) the shocks are called parallel. The shocks that are neither parallel nor perpendicular are called oblique. Let the right-hand-side quantities belong to the flow ahead of the shock. If $\mathbf{B}_t^{\mathrm{R}} = \mathbf{0}$ and $\mathbf{B}_t^{\mathrm{L}} \neq \mathbf{0}$, the shock is called a switch-on shock. On switch-off shocks, on the contrary, $\mathbf{B}_t^{\mathrm{R}} \neq \mathbf{0}$ and $\mathbf{B}_t^{\mathrm{L}} = \mathbf{0}$. Earlier we mentioned that the tangential component of the velocity ahead of the shock can be rotated in the shock plane by the choice a moving coordinate system. If $B_n \neq 0$, we can choose the coordinate system with the velocity

$$\mathbf{v}_t^{\mathrm{L}} - \frac{v_n}{B_n}\mathbf{B}_t^{\mathrm{L}} = \mathbf{v}_t^{\mathrm{R}} - \frac{v_n}{B_n}\mathbf{B}_t^{\mathrm{R}},$$

thus making $\mathbf{v}^{\mathrm{L}} \parallel \mathbf{B}^{\mathrm{L}}$ and $\mathbf{v}^{\mathrm{R}} \parallel \mathbf{B}^{\mathrm{R}}$.

Equations (5.2.11) and (5.2.13) give us

$$\left\{\frac{\mathbf{B}_t}{\rho}\right\} - \frac{B_n^2}{4\pi m^2}\{\mathbf{B}_t\} = \mathbf{0},$$

or

$$\left(\frac{1}{\rho^{\mathrm{L}}} - \frac{B_n^2}{4\pi m^2}\right)\mathbf{B}_t^{\mathrm{L}} = \left(\frac{1}{\rho^{\mathrm{R}}} - \frac{B_n^2}{4\pi m^2}\right)\mathbf{B}_t^{\mathrm{R}}. \tag{5.2.17}$$

For switch-off shocks $\mathbf{B}_t^{\mathrm{L}} = \mathbf{0}$ and, hence,

$$m^2 = \frac{B_n^2}{4\pi\rho^{\mathrm{R}}} \quad \text{or} \quad v_n^{\mathrm{R}} = v_{\mathrm{A}n}^{\mathrm{R}}.$$

Combined with Eq. (5.2.13) this gives

$$\mathbf{v}^{\mathrm{R}} = \mathbf{v}_{\mathrm{A}}^{\mathrm{R}}. \tag{5.2.18}$$

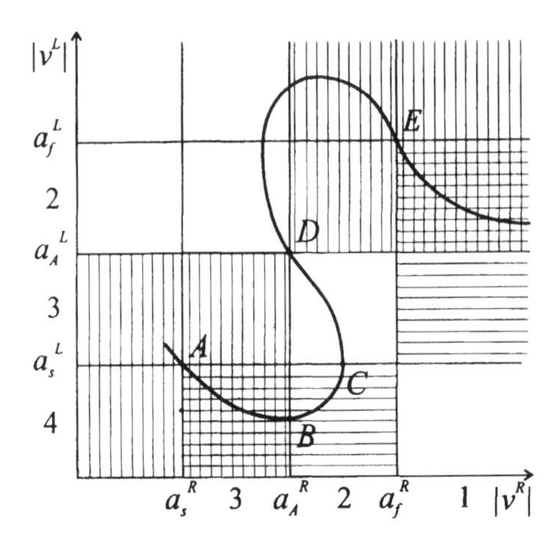

Figure 5.3 MHD Hugoniot curve.

Similarly, for switch-on shocks

$$\mathbf{v}^{\mathrm{L}} = \mathbf{v}_{\mathrm{A}}^{\mathrm{L}}. \tag{5.2.19}$$

Note that the relations (5.2.18) or (5.2.19) are only valid in coordinate systems ensuring that $\mathbf{v}^{\mathrm{R}} \parallel \mathbf{B}^{\mathrm{R}}$ or $\mathbf{v}^{\mathrm{L}} \parallel \mathbf{B}^{\mathrm{L}}$, respectively.

Combining the conservation equations on the shock, we can obtain the shock adiabatic relation similar to the Rankine–Hugoniot relation in pure gas dynamics (see Landau and Lifshitz 1984),

$$\{\varepsilon\} + \bar{p}\left\{\frac{1}{\rho}\right\} + \frac{1}{16\pi}\left\{\frac{1}{\rho}\right\}\{\mathbf{B}_t\}^2 = 0, \tag{5.2.20}$$

where $\bar{p} = \frac{1}{2}(p^{\mathrm{L}} + p^{\mathrm{R}})$.

5.3 Evolutionary MHD shocks

5.3.1 Evolutionary diagram. Suppose the half-space (see Fig. 5.2) to the right of the shock surface is occupied by the unperturbed gas ahead of the shock and the normal component v_n^{R} of its velocity with respect to the shock is negative. The half-space L is therefore occupied by the gas that crossed the shock. The shock surface is the boundary between the continuous regions. As shown in Section 1.4.4, the number of relations at this boundary must be equal to $n+1$, where n is the number of outgoing characteristics. In this case the shock is evolutionary, that is, the problem of its interaction with small perturbations is well-posed. This shock can exist without disintegration into other types of shocks.

As noted previously, we can choose a coordinate system moving at a constant speed in the shock plane in a way such that the velocity, magnetic field, and shock normal vectors lie in the same plane. It is very well known (Akhiezer, Lyubarskii, and Polovin 1958) that the MHD shock relations linearized with respect to small perturbations of the states ahead of and behind the shock (this also results in some small perturbation of the shock velocity) split into two groups. The first one relates only those perturbations of the vector components that lie in the above-mentioned plane. Only out-of-plane perturbations occur in the second group. That is why the evolutionary property must be checked separately for perturbations tangent and normal to the shock surface.

It is also widely known, see, e.g., Landau and Lifshitz (1984), that in magnetohydrodinamics only fast and slow shocks are evolutionary (this problem was first solved by Akhiezer et al. 1958). The relations

$$-v_n^{\mathrm{R}} > a_{\mathrm{f}}^{\mathrm{R}}, \qquad a_{\mathrm{A}}^{\mathrm{L}} < -v_n^{\mathrm{L}} < a_{\mathrm{f}}^{\mathrm{L}} \tag{5.3.1}$$

define fast shocks and the relations

$$a_{\mathrm{s}}^{\mathrm{R}} < -v_n^{\mathrm{R}} < a_{\mathrm{A}}^{\mathrm{R}}, \qquad -v_n^{\mathrm{L}} < a_{\mathrm{s}}^{\mathrm{L}} \tag{5.3.2}$$

define slow shocks. The magnitudes of the slow velocity a_{s}, the fast velocity a_{f}, and the Alfvén velocity a_{A} are given by the formulas

$$a_{\mathrm{A}} = \frac{|B_n|}{\sqrt{4\pi\rho}}, \quad a_{f,s} = \frac{1}{2}\left[\sqrt{c_e^2 + \frac{\mathbf{B}^2}{4\pi\rho} + \frac{|B_n|c_e}{\sqrt{\pi\rho}}} \pm \sqrt{c_e^2 + \frac{\mathbf{B}^2}{4\pi\rho} - \frac{|B_n|c_e}{\sqrt{\pi\rho}}}\right], \tag{5.3.3}$$

where c_e is the acoustic speed of sound.

The diagram showing the evolutionary regions is presented in Fig. 5.3. The regions corresponding to shocks evolutionary only with respect to transversal perturbations or perturbations in magnetosonic quantities are shown in this figure by vertical and horizontal crossing, respectively.

It is easy to check, after converting Eq. (5.2.17) into

$$\left(\frac{B_n^2}{4\pi m} - v_n^{\mathrm{L}}\right)\mathbf{B}_t^{\mathrm{L}} = \left(\frac{B_n^2}{4\pi m} - v_n^{\mathrm{R}}\right)\mathbf{B}_t^{\mathrm{R}},$$

or

$$\left(\frac{a_{\mathrm{A}}^2 - v_n^2}{v_n}\right)^{\mathrm{L}}\mathbf{B}_t^{\mathrm{L}} = \left(\frac{a_{\mathrm{A}}^2 - v_n^2}{v_n}\right)^{\mathrm{R}}\mathbf{B}_t^{\mathrm{R}}, \tag{5.3.4}$$

that the tangential components of the magnetic field are not only collinear on both sides of the shock, but have also the same direction.

For slow shocks

$$|v_n| < \frac{a_{\mathrm{A}}^2}{|v_n|}$$

on both their sides. Since $|v_n^{\mathrm{L}}| < |v_n^{\mathrm{R}}|$, from Eq. (5.3.4) it follows that $|\mathbf{B}_t^{\mathrm{L}}| < |\mathbf{B}_t^{\mathrm{R}}|$. This means that the magnitude of the tangential component of the magnetic field decreases across slow MHD shocks. Similarly we can obtain that it increases across fast MHD shocks.

It is sometimes convenient to represent Fig. 5.3 in terms of the shock velocity W. In this case we must rewrite Eqs. (5.3.1)–(5.3.2) in the fixed coordinate system. Then we obtain

$$\lambda_{\mathrm{s}}^{\mathrm{R}} < W < \lambda_{\mathrm{A}}^{\mathrm{R}}, \qquad \lambda_e^{\mathrm{L}} < W < \lambda_{\mathrm{s}}^{\mathrm{L}} \tag{5.3.5}$$

for slow shocks and

$$W > \lambda_{\mathrm{f}}^{\mathrm{R}}, \qquad \lambda_{\mathrm{A}}^{\mathrm{L}} < W < \lambda_{\mathrm{f}}^{\mathrm{L}} \tag{5.3.6}$$

for fast shocks.

Here we introduced the characteristic velocities

$$\lambda_e = \tilde{v}_n, \quad \lambda_s = \tilde{v}_n + a_s, \quad \lambda_A = \tilde{v}_n + a_A, \quad \lambda_f = \tilde{v}_n + a_f,$$

with

$$\tilde{v}_n = v_n + W$$

being the flow velocity in the fixed coordinate system. The eigenvalues are written in the increasing order. The left inequality in the second formula of Eq. (5.3.5) is not a direct consequence of Eq. (5.3.1), but is the result of our choice of the direction of the normal **n** for which the shock speed cannot be smaller than the velocity ahead of the shock.

5.3.2 Convenient relations on MHD shocks. Let us write out convenient formulas relating the quantities on the evolutionary MHD shocks. Suppose that we chose the coordinate system in which the vectors **B**, **v**, and **n** lie in the same plane on both sides of the shock. Let us introduce two dimensionless parameters,

$$\eta = \frac{B_t^{\mathrm{L}}}{B_t^{\mathrm{R}}} - 1, \quad \xi = -h_1 \frac{\{v_n\}}{\{v_t\}}, \qquad \left(h_1 = \frac{B_t^{\mathrm{R}}}{B_n}\right). \tag{5.3.7}$$

Due to the choice of the coordinate system, knowing the properties of the tangential components of the magnetic field and velocity vectors across the shock, we can omit the vector notions for these components.

It can be shown (Barmin 1962) that the jumps of all quantities across the shock can be expressed in terms of the introduced parameters as

$$\begin{gathered}
\frac{\rho^{\mathrm{L}}}{\rho^{\mathrm{R}}} = \frac{1 + (1+\eta)\xi}{1+\xi}, \quad p^{\mathrm{L}} = p^{\mathrm{R}} + \eta \frac{2\xi B_n^2 - (2+\eta)(B_t^{\mathrm{R}})^2}{8\pi}, \\
\tilde{v}_n^{\mathrm{L}} = \tilde{v}_n^{\mathrm{R}} - \frac{\eta \xi B_n}{\sqrt{4\pi\rho^{\mathrm{R}}[1 + (1+\eta)\xi]}}, \\
\tilde{v}_t^{\mathrm{L}} = \tilde{v}_t^{\mathrm{R}} + \frac{\eta B_t^{\mathrm{R}}}{\sqrt{4\pi\rho^{\mathrm{R}}[1 + (1+\eta)\xi]}}, \\
v_n^{\mathrm{R}} = B_n \sqrt{\frac{1 + (1+\eta)\xi}{4\pi\rho^{\mathrm{R}}}}, \quad v_n^{\mathrm{L}} = B_n \sqrt{\frac{1+\xi}{4\pi\rho^{\mathrm{L}}}}, \quad B_t^{\mathrm{L}} = (1+\eta) B_t^{\mathrm{R}}.
\end{gathered} \tag{5.3.8}$$

The tilde sign refers in these equations to the velocity components in the fixed coordinate system. Thus, the exact relations $W = \tilde{v}_n - v_n$ for the shock velocity W can be obtained. These relations are very convenient if we need to check the precision of numerically obtained shock speeds and quantity profiles. In specific perfect gas calculations with

$$\varepsilon = \frac{p}{(\gamma - 1)\rho}$$

one can notice that

$$\eta = 2\frac{\xi^2 - (h_1^2 + \gamma \bar{p}^{\mathrm{R}} - 1)\xi - h_1^2}{(\gamma - 1)\xi^2 + (2-\gamma)h_1^2\xi + h_1^2}, \quad \bar{p} = \frac{4\pi p^{\mathrm{R}}}{B_n^2}. \tag{5.3.9}$$

5.3.3 Evolutionarity of perpendicular, parallel, and singular shocks.

Let us consider the evolutionary property for the limiting cases of perpendicular, parallel, and singular shocks.

If the shock is perpendicular, then $B_n^{\mathrm{R}} = B_n^{\mathrm{L}} = 0$ and, hence, $a_{\mathrm{A}n}^{\mathrm{R}} = a_{\mathrm{A}n}^{\mathrm{L}} = 0$. Thus,

$$a_{\mathrm{s}} = 0, \quad a_{\mathrm{f}} = \sqrt{c_e^2 + \frac{\mathbf{B}_t^2}{4\pi\rho}}.$$

In this case slow shocks are absent and one-dimensional MHD motions coincide with one-dimensional motions in pure gas dynamics for which all compression shocks are evolutionary.

For parallel shocks $\mathbf{B}_t^{\mathrm{R}} = \mathbf{B}_t^{\mathrm{L}} = 0$. As mentioned earlier in Section 5.2, we can always choose a coordinate system moving in the shock plane for which $\mathbf{v}_t^{\mathrm{R}} = \mathbf{v}_t^{\mathrm{L}} = 0$ as well. This means that

$$\begin{aligned} &a_{\mathrm{s}} = c_e, \quad a_{\mathrm{f}} = a_{\mathrm{A}} \qquad \text{for} \quad a_{\mathrm{A}} > c_e, \\ &a_{\mathrm{s}} = a_{\mathrm{A}}, \quad a_{\mathrm{f}} = c_e \qquad \text{for} \quad c_e > a_{\mathrm{A}}. \end{aligned}$$

The presence of these two possibilities affects the evolutionary conditions for parallel shocks.

Let $a_{\mathrm{A}}^{\mathrm{R}} < c_e^{\mathrm{R}}$. Since the density increases across the shock, a_{A} and c_e are decreasing and increasing functions of density, respectively. Thus, $a_{\mathrm{A}}^{\mathrm{L}}$ remains larger than c_e^{L} and, thus,

$$a_{\mathrm{s}}^{\mathrm{L}} = a_{\mathrm{A}}^{\mathrm{L}}, \quad a_{\mathrm{f}}^{\mathrm{L}} = c_e^{\mathrm{L}}.$$

The coincidence of $a_{\mathrm{s}}^{\mathrm{R}}$ and $a_{\mathrm{A}}^{\mathrm{R}}$ means that the evolutionary condition (5.3.2) cannot be satisfied for slow shocks. As for the fast shocks, the evolutionary condition (5.3.1) transforms into

$$|v^{\mathrm{R}}| > c_e^{\mathrm{R}}, \quad a_{\mathrm{A}}^{\mathrm{L}} < |v^{\mathrm{L}}| < c_e^{\mathrm{L}}.$$

This condition is valid for all shock intensities $\rho^{\mathrm{L}}/\rho^{\mathrm{R}}$. Note that fast shocks remain super-Alfvénic in this case.

The situation is more complicated if $a_{\mathrm{A}}^{\mathrm{R}} > c_e^{\mathrm{R}}$. It is obvious in this case that for small shock intensities $a_{\mathrm{A}}^{\mathrm{L}}$ will remain larger than c_e^{L}, while for larger intensities, alternatively, $a_{\mathrm{A}}^{\mathrm{L}} < c_e^{\mathrm{L}}$. Let us determine the range of v^{R} for which the evolutionary property is satisfied. The shock relations (5.2.9), (5.2.10), and (5.2.12) for parallel shocks acquire the form coincident with those in pure gas dynamics of an ideal perfect gas with the adiabatic index γ

$$\begin{aligned} &\rho^{\mathrm{L}} v^{\mathrm{L}} = \rho^{\mathrm{R}} v^{\mathrm{R}}, \\ &p^{\mathrm{L}} + \rho^{\mathrm{L}}(v^{\mathrm{L}})^2 = p^{\mathrm{R}} + \rho^{\mathrm{R}}(v^{\mathrm{R}})^2, \\ &\frac{\gamma}{\gamma - 1}\frac{p^{\mathrm{L}}}{\rho^{\mathrm{L}}} + \frac{(v^{\mathrm{L}})^2}{2} = \frac{\gamma}{\gamma - 1}\frac{p^{\mathrm{R}}}{\rho^{\mathrm{R}}} + \frac{(v^{\mathrm{R}})^2}{2}. \end{aligned}$$

It is easy to obtain from these relations that

$$v^{\mathrm{L}} = \frac{(\gamma - 1)(v^{\mathrm{R}})^2 + 2(c_e^{\mathrm{R}})^2}{(\gamma + 1)v^{\mathrm{R}}}, \tag{5.3.10}$$

where $c_e = \sqrt{\gamma p/\rho}$.

Similarly, we can write out the relations that define the dependence of the acoustic speed of sound and of the Alfvén velocity behind the shock on the velocity ahead of it,

$$c_e^{\mathrm{L}} = \frac{(\gamma-1)\sqrt{\left((v^{\mathrm{R}})^2 + \frac{2}{\gamma-1}(c_e^{\mathrm{R}})^2\right)\left(\frac{2\gamma}{\gamma-1}(v^{\mathrm{R}})^2 - (c_e^{\mathrm{R}})^2\right)}}{(\gamma+1)|v^{\mathrm{R}}|}, \tag{5.3.11}$$

$$a_{\mathrm{A}}^{\mathrm{L}} = \frac{a_{\mathrm{A}}^{\mathrm{R}}}{|v^{\mathrm{R}}|}\sqrt{\frac{(\gamma-1)(v^{\mathrm{R}})^2 + 2(c_e^{\mathrm{R}})^2}{\gamma+1}}. \tag{5.3.12}$$

The functions c_e^{L} and $a_{\mathrm{A}}^{\mathrm{L}}$ are increasing and decreasing functions of $|v^{\mathrm{R}}|$, respectively. The point at which $c_e^{\mathrm{L}} = a_{\mathrm{A}}^{\mathrm{L}}$ corresponds to

$$|v^{\mathrm{R}}| = v_* = \sqrt{\frac{\gamma+1}{2\gamma}(a_{\mathrm{A}}^{\mathrm{R}})^2 + \frac{\gamma-1}{2\gamma}(c_e^{\mathrm{R}})^2} < a_{\mathrm{A}}^{\mathrm{R}}.$$

At $|v^{\mathrm{R}}| = c_e^{\mathrm{R}}$,

$$|v^{\mathrm{L}}| = c_e^{\mathrm{L}} = c_e^{\mathrm{R}}, \quad a_{\mathrm{A}}^{\mathrm{L}} = a_{\mathrm{A}}^{\mathrm{R}} > c_e^{\mathrm{R}} = c_e^{\mathrm{L}}.$$

The function $|v^{\mathrm{L}}|$ is a decreasing function of $|v^{\mathrm{R}}|$ until $|v^{\mathrm{R}}| = c_e^{\mathrm{R}}\sqrt{2/(\gamma-1)}$ and starts to increase further on. It is easy to see that, at $|v^{\mathrm{R}}| = a_{\mathrm{A}}^{\mathrm{R}}$,

$$\frac{|v^{\mathrm{L}}|}{a_{\mathrm{A}}^{\mathrm{L}}} = \frac{1}{a_{\mathrm{A}}^{\mathrm{R}}}\sqrt{\frac{(\gamma-1)(a_{\mathrm{A}}^{\mathrm{R}})^2 + 2(c_e^{\mathrm{R}})^2}{\gamma+1}} < 1.$$

The point at which $|v^{\mathrm{L}}| = a_{\mathrm{A}}^{\mathrm{L}}$ is attained at

$$|v^{\mathrm{R}}| = v_{**} = \sqrt{\frac{\gamma+1}{\gamma-1}(a_{\mathrm{A}}^{\mathrm{R}})^2 - \frac{2}{\gamma-1}(c_e^{\mathrm{R}})^2}.$$

Let $|v^{\mathrm{R}}| \in [c_e^{\mathrm{R}}, v_*^{\mathrm{R}}]$, then $|v^{\mathrm{L}}| \le c_e^{\mathrm{L}} = a_s^{\mathrm{L}}$ and we have an evolutionary slow shock.

Let $|v^{\mathrm{R}}| \in [v_*^{\mathrm{R}}, a_{\mathrm{A}}^{\mathrm{R}}]$, then $|v^{\mathrm{L}}| \le a_{\mathrm{A}}^{\mathrm{L}} = a_s^{\mathrm{L}}$ and we still have an evolutionary slow shock. Thus, parallel slow shocks are evolutionary for $a_{\mathrm{A}}^{\mathrm{R}} > c_e^{\mathrm{R}}$.

For $|v^{\mathrm{R}}| \in [a_{\mathrm{A}}^{\mathrm{R}}, v_{**}]$, $|v^{\mathrm{L}}| \le a_{\mathrm{A}}^{\mathrm{L}}$ and this fast shock is nonevolutionary (note that it is trans-Alfvénic).

For $|v^{\mathrm{R}}| \in [v_{**}, \infty)$, we have $a_{\mathrm{A}}^{\mathrm{L}} \le |v^{\mathrm{L}}| < c_e^{\mathrm{L}}$. Thus, fast parallel shocks are evolutionary only for $|v^{\mathrm{R}}| > v_{**}$.

Let us consider the evolutionary property for singular shocks and choose for definiteness a fast switch-on shock. As shown in Section 5.2, $|v^{\mathrm{L}}|$ is exactly equal to the Alfvén velocity $a_{\mathrm{A}}^{\mathrm{L}}$ in this case. The corresponding characteristic wave cannot be considered as outgoing, which makes questionable its evolutionarity. On the other hand, singular shocks are always present in the solution of the piston problem (Barmin and Gogosov 1960). As tangential components of the magnetic field are usually perturbed in realistic problems, the singular shock can be treated as the limit of a nonsingular shock for $|\mathbf{B}_t^{\mathrm{R}}| \to 0$.

For a switch-on shock, Eq. (5.3.10) must be substituted by

$$v_n^{\mathrm{L}} = \frac{(a_{\mathrm{A}}^{\mathrm{R}})^2}{v_n^{\mathrm{R}}}. \tag{5.3.13}$$

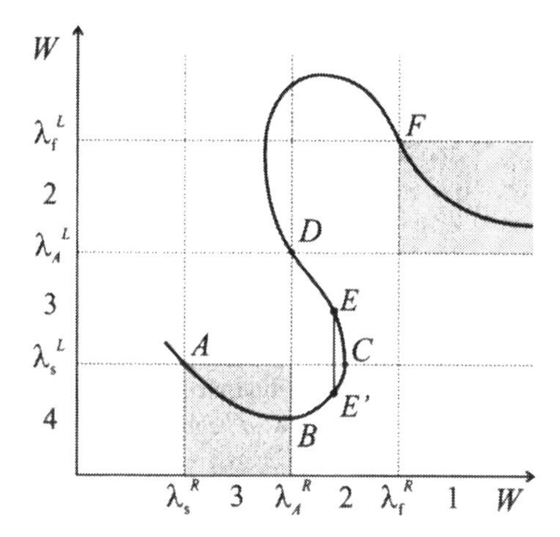

Figure 5.4 MHD Hugoniot curve.

Polovin and Demutskii (1960) showed that if a small tangential disturbance is added ahead of the shock to the magnetic field vector initially parallel to the shock normal, there will appear two possible configurations of oblique shocks close to the initial switch-on shock. One of them corresponds to the nonevolutionary and another one to the evolutionary oblique shock. Thus, in a small vicinity of a switch-on shock we have an evolutionary branch of the shock, and it can be considered evolutionary as a limit of the evolutionary oblique shock. It is not evolutionary, however, for all values of v_n^{R}.

In fact, manipulation with the shock relations (5.2.9)–(5.2.13) gives us the following expression for the tangential component of the magnetic field behind the shock:

$$\frac{(\mathbf{B}_t^{\mathrm{L}})^2}{8\pi\rho} = \frac{\gamma-1}{2(a_{\mathrm{A}}^{\mathrm{R}})^2}\left[(v_n^{\mathrm{R}})^2 - (a_{\mathrm{A}}^{\mathrm{R}})^2\right]\left[\frac{\gamma+1}{\gamma-1}(a_{\mathrm{A}}^{\mathrm{R}})^2 - \frac{2}{\gamma-1}(c_e^{\mathrm{R}})^2 - (v_n^{\mathrm{R}})^2\right].$$

For fast shocks $|v_n^{\mathrm{R}}| > a_{\mathrm{A}}^{\mathrm{R}}$. Thus, the second multiplier in brackets must be positive, that is, for the fast switch-on shock to be evolutionary, we must demand

$$a_{\mathrm{A}}^{\mathrm{R}} < |v_n^{\mathrm{R}}| < v_{**} = \sqrt{\frac{\gamma+1}{\gamma-1}(a_{\mathrm{A}}^{\mathrm{R}})^2 - \frac{2}{\gamma-1}(c_e^{\mathrm{R}})^2}. \tag{5.3.14}$$

This means that fast switch-on shocks are evolutionary exactly in the same interval within which fast parallel shocks are nonevolutionary. Thus, fast shocks with the magnetic field parallel to the shock normal can exist for all intensities of the shock. If the condition (5.3.14) is satisfied, then we shall have a singular switch-on shock. Otherwise, for $v_n^{\mathrm{R}} > v_{**}$ the shock will be parallel (sometimes it is also called hydrodynamic, since the shock relations defining it coincide with those for purely hydrodynamic shocks).

The relations (5.3.5) and (5.3.6) coincide with the general conditions of the shock evolutionarity, also called as the Lax conditions, presented in Section 1.4.4.

5.3.4 Jouget points.

The Hugoniot curve can be considered as a curve in the functional space with the initial point at $\mathbf{U}^{\mathrm{R}}$. If we know all seven components of the state vector ahead of the shock and one component of the vector $\mathbf{U}^{\mathrm{L}}$, than we have enough relations to determine the shock speed W and the other components of the state vector behind the shock. In this sense the Hugoniot curve is the function of only one parameter. This means that v_n^{L}, as well as the shock velocity W, can be represented as

$$W = W(\sigma), \qquad v_n^{\mathrm{L}} = v_n^{\mathrm{L}}(\sigma), \tag{5.3.15}$$

where σ is a parameter on the Hugoniot curve, say, the arc length calculated from the initial point $\mathbf{U}^{\mathrm{R}}$ in the functional space. The Hugoniot curve in the $(\lambda^{\mathrm{R}}, \lambda^{\mathrm{L}})$ space is shown schematically in Fig. 5.4. In such diagrams, $W(\sigma)$ can be plotted in the real scale along the horizontal axis and schematically along the vertical axis, by retaining only the inequalities between the $W(\sigma)$ and $\lambda_i(\sigma)$ and by preserving the continuity of the shock velocity. As mentioned earlier, the evolutionary segments of the Hugoniot curve lie inside the dashed rectangles. The vicinity of the points $\lambda_i^{\mathrm{R}} = \lambda_i^{\mathrm{L}}$ obviously correspond to weak shocks. As shown for this case by Lax (1957) (see Section 1.4.4), the Hugoniot curve and the integral curve of the Riemann wave with the simple characteristic velocity λ have the second-order tangent point at the initial point $\sigma = 0$ and

$$W = \frac{1}{2}\left(\lambda^{\mathrm{R}} + \lambda^{\mathrm{L}}\right).$$

Since we have two types of evolutionary shocks, the curve $W(\sigma)$ passes through the points $\lambda_{\mathrm{s}}^{\mathrm{R}} = \lambda_{\mathrm{s}}^{\mathrm{L}}$ and $\lambda_{\mathrm{f}}^{\mathrm{R}} = \lambda_{\mathrm{f}}^{\mathrm{L}}$.

Another important feature of the Hugoniot curve behavior concerns so-called Jouget points. These are the points where the shock speed W coincides with one of the characteristic velocities behind the shock. In the vicinity of these points we have additional information (Hanyga 1976; Kulikovskii 1979). Namely, the relations

$$\frac{dW}{d\sigma} = 0 \qquad \text{and} \qquad W = \lambda^{\mathrm{L}} \tag{5.3.16}$$

can be satisfied only simultaneously (see Section 7.3).

There is an important consequence of such behavior of the shock curve. Consider the Jouget point C in Fig. 5.4 and note that the segment BC of the curve corresponds to slow MHD shocks nonevolutionary only with respect to transversal perturbations, whereas the segment CD is also nonevolutionary with respect to perturbations in magnetosonic quantities. Since $dW/d\sigma = 0$ at C, alongside the initial discontinuity we can consider another small-amplitude shock EE' moving with the same velocity $W = \lambda_{\mathrm{s}}^{\mathrm{L}}$. The shock EE' is an evolutionary slow shock as well as the primary shock at C, since its speed is larger than $\lambda_{\mathrm{s}}^{\mathrm{L}}$ ahead of and smaller behind the shock. To understand the solution behavior on the shock EE', we can refer to the point A of the Hugoniot curve whose vicinity corresponds to weak slow MHD shocks. As shown in Section 1.4.3, the weak shock and the Riemann wave have the second-order tangent point at A. This Riemann wave has convergent characteristics if we move in the direction of increasing λ^{R}, that is, into the evolutionary rectangle. Conversely, we have a rarefaction fan if we move in the direction of decreasing λ^{R}, that is outside of the evolutionary rectangle. Thus, the integral curve of an expanding Riemann wave at the point C is a continuation of the evolutionary (with respect to perturbations in magnetosonic quantities) Hugoniot curve with the common tangent at the Jouget point corresponding to the slow shock and we can construct a solution at the point C that consists of a slow shock and a rarefaction fan moving together at the same speed $W = \lambda_s^L$. Note that this compound wave lies in the part of the Hugoniot curve that is nonevolutionary in three dimensions. We shall return to this subject later when discussing the origin of nonevolutionary solutions in MHD calculations.

5.4 High-resolution numerical schemes for MHD equations

TVD upwind and symmetric differencing schemes have recently become a very efficient tool for solving complex multi-shocked gas dynamic flows. This is due to their robustness for strong shock wave calculations. The extension of these schemes to the equations of ideal magnetohydrodynamics is not straightforward. The exact solution to the MHD Riemann problem (Gogosov 1961) is too multivariant to be used in regular calculations. For this reason various approximate solutions are expected to be applied. The first candidates among them are the Roe and the Osher methods described in the previous chapters. In these methods numerical flux is constructed on the basis of the solution to the one-dimensional system of governing equations at the interface between computational cells. The one-dimensional system of MHD equations has the form

$$\frac{\partial \mathbf{U}}{\partial t} + \frac{\partial \mathbf{F}}{\partial x} = \mathbf{0}, \tag{5.4.1}$$

where $\mathbf{U} = [\rho,\ \rho u,\ \rho v,\ \rho w,\ e,\ B_y,\ B_z]^{\mathrm{T}}$ and

$$\mathbf{F}(\mathbf{U}) = \begin{bmatrix} \rho u \\ \rho u^2 + p_0 - B_x^2/4\pi \\ \rho u v - B_x B_y/4\pi \\ \rho u w - B_x B_z/4\pi \\ (e + p_0)u - (uB_x + vB_y + wB_z)B_x/4\pi \\ uB_y - vB_x \\ uB_z - wB_x \end{bmatrix}.$$

In these formulas $e = p/(\gamma - 1) + \rho(u^2 + v^2 + w^2)/2 + (B_x^2 + B_y^2 + B_z^2)/8\pi$ is the total energy per unit volume, $p_0 = p + (B_x^2 + B_y^2 + B_z^2)/8\pi$ is the total pressure, p and ρ are pressure and density, $\mathbf{v} = [u, v, w]^{\mathrm{T}}$ is the velocity vector, $\mathbf{B} = [B_x, B_y, B_z]^{\mathrm{T}}$ is the magnetic field vector, and γ is the adiabatic index. We assume that all functions depend only on time t and on the linear coordinate x directed perpendicular to the computational cell boundary. Equation (5.4.1) can be approximated on the basis of a finite-volume method as

$$\mathbf{U}_i^{k+1} = \mathbf{U}_i^k - \frac{\Delta t}{\Delta x}[\bar{\mathbf{F}}_{i+1/2} - \bar{\mathbf{F}}_{i-1/2}]. \tag{5.4.2}$$

Our aim is to construct the solution to Eq. (5.4.1) for $t > 0$ assuming a piecewise-constant initial distribution of $\mathbf{U}$: $\mathbf{U} = \mathbf{U}^{\mathrm{L}}$ for $x < 0$ and $\mathbf{U} = \mathbf{U}^{\mathrm{R}}$ for $x > 0$. It is clear that owing to the divergence-free condition, $B_x = B_x^{\mathrm{L}} = B_x^{\mathrm{R}} \equiv \text{const}$.

5.4.1 The Osher-type method. We previously formulated the Osher method for an arbitrary strictly hyperbolic system of conservation laws. The system of gas dynamic equations is not strictly hyperbolic, since the eigenvalue corresponding to the entropy characteristic wave has the multiplicity 3. This degeneracy, however, is independent of the solution to the problem and can be easily overcome, as was done in the Osher–Solomon scheme, by using exact relations for the Riemann invariants of the Euler system. This allows us to avoid tracking the degeneracy points along the path connecting $\mathbf{U}_i$ and $\mathbf{U}_{i+1}$ in the phase space.

The situation is completely different in the MHD case. If the magnetic field vector component normal to the direction of motion (x-direction in Eq. (5.4.1)) is equal to zero, the slow magnetosonic velocity and the Alfvén velocity are zero, and the corresponding characteristic waves become degenerate with the entropy wave. If $B_y^2 + B_z^2 = 0$,

$$a_{f,s} = \frac{1}{2}(|c_e + a_{\mathrm{A}}| \pm |c_e - a_{\mathrm{A}}|).$$

Thus,

- if $a_{\mathrm{A}} > c_e$, then $a_{\mathrm{f}} = a_{\mathrm{A}}$ and $a_{\mathrm{s}} = c_e$, and the fast magnetosonic wave is degenerate with the Alfvén wave;
- if $a_{\mathrm{A}} < c_e$, then $a_{\mathrm{f}} = c_e$ and $a_{\mathrm{s}} = a_{\mathrm{A}}$, and the slow magnetosonic wave is degenerate with the Alfvén wave;
- if $a_{\mathrm{A}} = c_e$, then all these waves are degenerate.

The main problem is that this degeneracy is solution-dependent and can manifest itself in the vicinity of an isolated singular point. The general way out from this situation suggested in the extension of the Osher scheme to the general hyperbolic system made by Bell et al. (1989) lies in tracing the degeneracy points, since the eigenvectors corresponding to degenerate waves are expected to be ill-defined in the vicinity of these points. In the MHD case, however, the system of nondegenerate eigenvectors always exists. It was written out in Section 1.3.4. On the other hand, MHD waves that are originally genuinely nonlinear (e.g., slow and fast magnetosonic waves) can become linearly degenerate if their eigenvalues coincide with those of already linearly degenerate waves (the entropy and Alfvén waves). If this happens along the path of integration connecting $\mathbf{U}_i$ and $\mathbf{U}_{i+1}$, the Osher formulation fails (see Section 2.3.3) and the entropy condition may not be satisfied. We imply that $\mathbf{U}^{\mathrm{L}} = \mathbf{U}_i$ and $\mathbf{U}^{\mathrm{R}} = \mathbf{U}_{i+1}$ are the vectors of conservative variables in adjacent computational cells. The superscripts L and R refer to the left and right cell positions with respect to their interface boundary, respectively.

In the Godunov-type methods, we construct the solution to Eq. (5.4.1) using the flux $\bar{\mathbf{F}}(\mathbf{U}^{\mathrm{L}}, \mathbf{U}^{\mathrm{R}})$ calculated along the ray $x/t = 0$ in the local Riemann problem for this equation with initial conditions defined by the quantities in adjacent cells. In fact, we do not need the entire solution of this Riemann problem. Though this solution exists for MHD, it may not exist or be nonunique for other hyperbolic systems. This concerns not only nonstrictly hyperbolic systems mentioned by Zachary and Colella (1992). The solution can be nonunique (though entropy-consistent) even for a single conservation law with a nonconvex flux function (Kulikovskii 1984; see also Chapter 7). The important point is that, instead of obtaining the solution as a whole, it suffices to only develop the numerical flux as an approximation to the Riemann problem solution. It is questionable, however, that

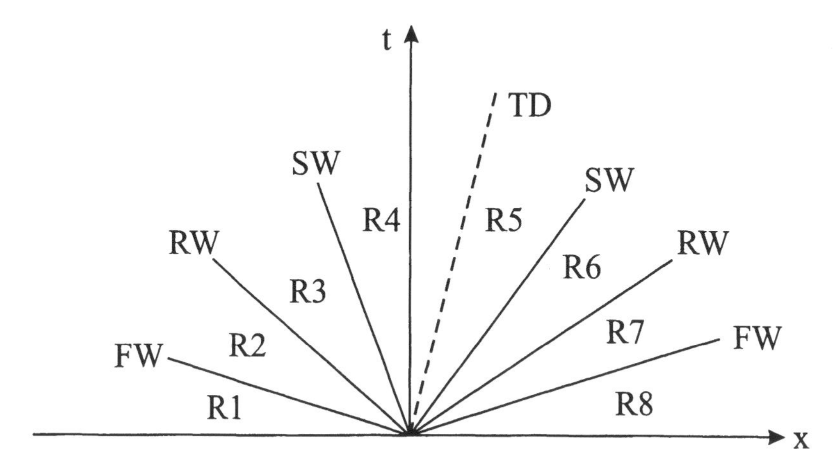

Figure 5.5 General scheme of decomposition of the arbitrary discontinuity (SW and FW can be either compression or rarefaction shocks).

such formulas can be applied if the solution of the Riemann problem is nonunique.

The MHD numerical flux obtained on the basis of the version of the Osher scheme developed by Bell et al. (1989) was suggested by Zachary and Colella (1992) and further developed and tested for multishocked problems by Zachary, Malagoli, and Colella (1994). The tests showed a good resolution of shocks, while it turned out to introduce small perturbations of rarefaction waves. These perturbations do not affect the solution substantially. The Osher scheme automatically adds some numerical viscosity at transonic points. As pointed out in Section 2.3.3, this viscosity must be increased in regions containing both transonic points and points of the eigenvalue degeneracy. This is done by switching from the Osher scheme to the Lax–Friedrichs-type scheme in approximation of these waves.

The addition of viscosity can be expected to help in selection of a physically admissible solution in this case, since the higher-order small-scale MHD model accounting for finite conductivity, molecular viscosity, and heat conduction contains only the second-order derivatives. We shall show later in Chapter 7 that this is not true in cases for which a smaller-scale model involves third-order derivatives.

5.4.2 Piecewise-parabolic method. Dai and Woodward (1994b) suggested the extension of the piecewise parabolic method (PPM) to MHD problems. PPM was initially developed for gas dynamic flows by Colella and Woodward (1984). As mentioned earlier, the exact solution to the MHD Riemann problem is too complicated to be used in regular calculations. For this reason, PPM uses an approximate, though nonlinear, solver. Simplification is attained by approximating rarefaction waves by rarefaction shocks obeying the jump relations and involving decreases in entropy. This allows one to organize the iteration process in order to find the wave speeds and parameter distributions in the smooth regions $R1$–$R8$ (see Fig. 5.5). The interfaces between these regions are represented by fast (FW), Alfvén, or rotational (RW), and slow waves (SW) propagating on the both sides of a tangential discontinuity (TD).

It is clear that the iteration process strongly depends on the initial guess for the transverse components B_{y2}, B_{y4}, and B_{y7} of the magnetic field in regions $R2$, $R4$, and $R7$ and the orientation ψ of the transverse part of the magnetic field in regions $R3$, $R4$, $R5$, and $R6$ adopted in the method. Here $\tan\psi = B_z/B_y$. This guess in PPM is made on the basis of the preliminary solution of the linearized Riemann problem that possesses invariants along corresponding characteristic curves.

To increase the space resolution, cubic polynomials are used to interpolate either the increments of the Riemann or primitive physical variables in order to determine the quantities at the interfaces of computational zones. These quantities are used as initial conditions for the corresponding Riemann problem.

Since the described method is iterative, difficulties can exist with the iteration convergence. The PPM developers suggest remaining in this case in the framework of the solution of the linear problem (Dai and Woodward 1994a). The strong point of the method is evidently in its ability of exact reproduction of shock relations and in the permanent presence of the Alfvén discontinuity in the possible configuration shown in Fig. 5.5. Note that it must exist even if the Riemann problem is coplanar. This is very important, taking into account the fact that MHD equations are three-dimensional by nature and the restriction of the variety of admissible solutions only to coplanar ones may result in the origin of nonevolutionary shocks. This subject will be discussed in Section 5.5.

5.4.3 Roe's characteristic decomposition method.

The numerical flux (5.4.2) in Roe's approach is constructed by the simple wave decomposition of the flux calculated on the basis of the frozen Jacobian matrix $J = \partial\mathbf{F}/\partial\mathbf{U}$. After linearization of the initial system, the flux can be written out as

$$\bar{\mathbf{F}}_{i+1/2} = \frac{1}{2}\left[\mathbf{F}_i + \mathbf{F}_{i+1} - (\Omega_{\mathrm{R}}|\Lambda|\Omega_{\mathrm{L}}\delta\mathbf{U})_{i+1/2}\right].$$

The peculiarity of Roe's approach to nonlinear systems lies in such specific freezing of J, which allows us to satisfy exactly the conservation relations on discontinuities. This procedure in the application to gas dynamic equations was described in Section 3.4.4. It cannot, however, be extended straightforwardly to MHD equations. Brio and Wu (1988) found out that an original Roe-type method is realizable only for the special case with the specific heat ratio $\gamma = 2$. They suggested using a simple arithmetic averaging of nonconservative variables to find the flux $\bar{\mathbf{F}}_{i+1/2}$. Thus, this scheme can be attributed to the Courant–Isaacson–Rees type. As a result, a second-order upwind scheme was constructed that demonstrated several advantages in comparison with the Lax–Friedrichs, Lax–Wendroff, and flux-corrected transport (DeVore 1991) schemes. As will be shown later by direct mathematical manipulations, the reason of such property of the MHD equations is in the fact that there is no single averaging procedure to find a frozen Jacobian matrix of the system. Another linearization approaches were used by Hanawa, Nakajima, and Kobuta (1994), Pogorelov et al. (1995), Aslan (1996a, 1996b), Cargo and Gallice (1995, 1997), and Balsara (1998a, 1998b). In these papers the frozen Jacobian matrix is not a function of a single averaged set of variables like in pure gas dynamics, but depends in a complicated way on the variables on the right- and left-hand sides of the computational cell surface. Pogorelov and Semenov (1996a, 1997b, 1997c) showed that this procedure is nonunique.

A multiparametric family of linearized MHD approximate Riemann problem solutions was presented that assured an exact satisfaction of the conservation relations on discontinuities. A proper choice of parameters is necessary to avoid physically inconsistent solutions.

We shall follow the general approach to create a Roe-type solver. It is applicable to any kind of hyperbolic systems. Let us first find the exact expression for the matrix A satisfying the relation $\Delta\mathbf{F} = A\Delta\mathbf{U}$, where $\Delta\mathbf{U} = \mathbf{U}^{\mathrm{L}} - \mathbf{U}^{\mathrm{R}}$ and $\Delta\mathbf{F} = \mathbf{F}^{\mathrm{L}} - \mathbf{F}^{\mathrm{R}}$. The function $\mathbf{F}$ being nonlinear, the expression for A is determined nonuniquely. In fact, if we choose some nondegenerate substitution $\mathbf{s} = \mathbf{s}(\mathbf{U})$, then from the equalities $\Delta\mathbf{F} = A_F\Delta\mathbf{s}$ and $\Delta\mathbf{U} = A_U\Delta\mathbf{s}$ it follows that $A = A_F A_U^{-1}$. The exact analytic expressions for A_F and A_U can be written out explicitly if $\mathbf{U}$ and $\mathbf{F}$ are linear functions of $\mathbf{s}$ or polynomials with respect to its components. We use here the equivalent transforms of the type $\Delta(BC) = \frac{1}{2}(B^{\mathrm{L}}+B^{\mathrm{R}})\Delta C + \frac{1}{2}(C^{\mathrm{L}}+C^{\mathrm{R}})\Delta B$. The structure and simplicity of A depends on the choice of $\mathbf{s}$. The matrix A can be considered as an approximation to the Jacobian matrix $J = \partial\mathbf{F}/\partial\mathbf{U}$ and must preserve the main hyperbolic properties of J. It must be representable in the form $A = \Omega_{\mathrm{R}}\Lambda\Omega_{\mathrm{L}}$, where Ω_{L} and Ω_{R} are nondegenerate matrices of its left and right eigenvectors, respectively; $\Omega_{\mathrm{R}}\Omega_{\mathrm{L}} = I$ (I is the identity matrix); $\Lambda = [\lambda_i\delta_{ij}]$ is the diagonal matrix composed of the real eigenvalues of A (δ_{ij} is the Kronecker delta).

Then the sought self-similar solution to Eq. (5.4.2) for $t > 0$ acquires the form

$$\mathbf{U}(\xi) = \frac{1}{2}(\mathbf{U}^{\mathrm{L}} + \mathbf{U}^{\mathrm{R}} + \Omega_{\mathrm{R}}S(\xi)\Omega_{\mathrm{L}}\Delta\mathbf{U}), \tag{5.4.3}$$

$$\mathbf{F}(\xi) = \frac{1}{2}(\mathbf{F}^{\mathrm{L}} + \mathbf{F}^{\mathrm{R}} + \Omega_{\mathrm{R}}|\Lambda(\xi)|\Omega_{\mathrm{L}}\Delta\mathbf{U}), \tag{5.4.4}$$

where $\xi = x/t$, $S(\xi) = [\operatorname{sgn}(\lambda_i-\xi)\times\delta_{ij}]$, and $|\Lambda(\xi)| = [\lambda_i \operatorname{sgn}(\lambda_i-\xi)\times\delta_{ij}]$. Equations (5.4.3) and (5.4.4) define piecewise-constant functions $\mathbf{U}(\xi)$ and $\mathbf{F}(\xi)$ that connect the right and left states of the initial distribution via system of jumps. If the Hugoniot-type condition is valid, $\Delta\mathbf{F} = \lambda\Delta\mathbf{U}$, where λ is the jump velocity, then λ is one of the eigenvalues of A, since $\Delta\mathbf{F} - \lambda\Delta\mathbf{U} = (A_F - \lambda A_U)\Delta\mathbf{s} = (A - \lambda I)A_U\Delta\mathbf{s} = (A - \lambda I)\Delta\mathbf{U} = \mathbf{0}$. Thus, $\det(A - \lambda I) = 0$, $\Delta\mathbf{U} \neq 0$ is the eigenvector of A, and relations (5.4.3)–(5.4.4) describe the jump exactly.

The solution of this kind was first constructed for equations of ideal gas dynamics by Roe (1981). Brio and Wu (1988) obtained such a solution for the MHD system in the case of $\gamma = 2$. The approximate solution for an arbitrary adiabatic index was probably first proposed by Hanawa et al. (1994). Here we present a regular procedure (Pogorelov et al. 1995; Pogorelov and Semenov 1996a, 1997b, 1997c) for obtaining the extension of Roe's solution for the linearized MHD equations and show that it is not unique. As a result, the solution of Hanawa et al. (1994) turns out to be a particular case of the multiparametric family of approximate solutions to the MHD Riemann problem.

We choose the vector $\mathbf{s}$ as a generalization of that for pure gas dynamics

$$\mathbf{s} = \begin{bmatrix} R \\ U \\ V \\ W \\ \mathcal{I} \\ Y \\ Z \end{bmatrix} = \begin{bmatrix} \sqrt{\rho} \\ \sqrt{\rho}\,u \\ \sqrt{\rho}\,v \\ \sqrt{\rho}\,w \\ \sqrt{\rho}\,H \\ B_y/\sqrt{\rho} \\ B_z/\sqrt{\rho} \end{bmatrix},$$

where $H = (e + p_0)/\rho$ is the total enthalpy.

Thus,

$$\mathbf{U} = \left[R^2,\ RU,\ RV,\ RW,\ U_5,\ RY,\ RZ\right]^{\mathrm{T}},$$

where

$$U_5 = \frac{1}{\gamma}R\mathcal{I} + \frac{\gamma-1}{2\gamma}(U^2+V^2+W^2) - \frac{2-\gamma}{8\pi\gamma}[B_x^2 + (Y^2+Z^2)R^2].$$

The vector $\mathbf{F}$ in terms of the components of the vector $\mathbf{s}$ acquires the form

$$\mathbf{F} = \begin{bmatrix} RU \\ F_2 \\ UV - \dfrac{RYB_x}{4\pi} \\ UW - \dfrac{RZB_x}{4\pi} \\ U\mathcal{I} - \dfrac{(UB_x/R + VY + WZ)B_x}{4\pi} \\ UY - \dfrac{VB_x}{R} \\ UZ - \dfrac{WB_x}{R} \end{bmatrix},$$

where

$$F_2 = U^2 + \frac{\gamma-1}{\gamma}R\mathcal{I} - \frac{\gamma-1}{2\gamma}(U^2+V^2+W^2) + \frac{2-\gamma}{8\pi\gamma}[B_x^2 + (Y^2+Z^2)R^2] - \frac{1}{4\pi}B_x^2.$$

Using the new expressions for $\mathbf{U}$ and $\mathbf{F}$, we can find the matrices A_U, A_F, and $A_\lambda = A_F - \lambda A_U$. We present only the expression for A_λ with omitted multiplier $\overline{R}$:

$$A_\lambda = \begin{bmatrix} u-2\lambda & 1 & 0 & 0 & 0 & 0 & 0 \\ A_{21} & \frac{\gamma+1}{\gamma}u-\lambda & \frac{1-\gamma}{\gamma}v & \frac{1-\gamma}{\gamma}w & \frac{\gamma-1}{\gamma} & q_y & q_z \\ A_{31} & v & u-\lambda & 0 & 0 & -\frac{B_x}{4\pi} & 0 \\ A_{41} & w & 0 & u-\lambda & 0 & 0 & -\frac{B_x}{4\pi} \\ A_{51} & A_{52} & A_{53} & A_{54} & u-\frac{\lambda}{\gamma} & A_{56} & A_{57} \\ A_{61} & b_y & -\frac{B_x}{\rho} & 0 & 0 & u-\lambda & 0 \\ A_{71} & b_z & 0 & -\frac{B_x}{\rho} & 0 & 0 & u-\lambda \end{bmatrix},$$

where

$$A_{21} = \frac{\gamma-1}{\gamma}\mathcal{H} - \lambda u + q, \quad A_{31} = -\frac{B_x b_y}{4\pi} - \lambda v, \quad A_{41} = -\frac{B_x b_z}{4\pi} - \lambda w,$$

$$A_{51} = \frac{uB_x^2}{4\pi\rho} + \lambda\left(q - \frac{\mathcal{H}}{\gamma}\right), \quad A_{52} = \mathcal{H} - \frac{B_x^2}{4\pi\rho} + \lambda u\frac{1-\gamma}{\gamma},$$

$$A_{53} = -\frac{B_x b_y}{4\pi} + \lambda v\frac{1-\gamma}{\gamma}, \quad A_{54} = -\frac{B_x b_z}{4\pi} + \lambda w\frac{1-\gamma}{\gamma}, \quad A_{56} = -\frac{vB_x}{4\pi} + \lambda q_y,$$

$$A_{57} = -\frac{wB_x}{4\pi} + \lambda q_z, \quad A_{61} = \frac{vB_x}{\rho} - \lambda b_y, \quad A_{71} = \frac{wB_x}{\rho} - \lambda b_z.$$

The following notions are adopted in the above relations:

$$\rho = \sqrt{\rho^{\mathrm{L}}\rho^{\mathrm{R}}}, \quad u = \frac{\overline{Ru}}{\overline{R}}, \quad v = \frac{\overline{Rv}}{\overline{R}}, \quad w = \frac{\overline{Rw}}{\overline{R}},$$
$$\mathcal{H} = \frac{\overline{RH}}{\overline{R}}, \quad b_y = \frac{\overline{B_y/R}}{\overline{R}}, \quad b_z = \frac{\overline{B_z/R}}{\overline{R}},$$

where $\overline{f}$ means arithmetic averaging. Besides,

$$q = \frac{2-\gamma}{4\pi\gamma}\left(\overline{Y}^2 + \overline{Z}^2 + \frac{\theta_1}{4}(\Delta Y)^2 + \frac{\theta_2}{4}(\Delta Z)^2 + \frac{\eta_1}{4\overline{R}}\overline{Y}\Delta Y\Delta R + \frac{\eta_2}{4\overline{R}}\overline{Z}\Delta Z\Delta R\right),$$
$$q_y = \frac{2-\gamma}{4\pi\gamma}\left(\overline{Y}\,\overline{R} + \frac{1-\theta_1}{4}\Delta Y\Delta R + \frac{1-\eta_1}{4\overline{R}}(\Delta R)^2\overline{Y}\right),$$
$$q_z = \frac{2-\gamma}{4\pi\gamma}\left(\overline{Z}\,\overline{R} + \frac{1-\theta_2}{4}\Delta Z\Delta R + \frac{1-\eta_2}{4\overline{R}}(\Delta R)^2\overline{Z}\right),$$

where θ_1, θ_2, η_1, and η_2 are arbitrary parameters. Their origin is caused by the presence of the terms containing the factors $\Delta R\Delta Y$ and $\Delta R\Delta Z$ in the expressions for $\Delta\mathbf{F}$. These factors can both be attributed to the terms proportional to ΔR and ΔY or ΔZ. This results in additional parametrization of entries in the matrices A_F and A_U. It is not difficult to find that

$$\det A_\lambda = \frac{2K}{\gamma}\left(K^2 - \frac{B_x^2}{4\pi\rho}\right)\left\{(K^2 - c_e^2 - \alpha) \times \left(K^2 - \frac{B_x^2}{4\pi\rho}\right) - K^2\left[\frac{\rho}{4\pi}(b_y^2 + b_z^2) + \beta\right]\right\},$$

where

$$K = u - \lambda, \quad c_e^2 = (\gamma - 1)\left[\mathcal{H} - \frac{u^2 + v^2 + w^2}{2} - \frac{B_x^2}{4\pi\rho} - \frac{(b_y^2 + b_z^2)\rho}{4\pi}\right],$$
$$\alpha = \frac{\gamma(\delta - \delta_y b_y - \delta_z b_z)}{2} \equiv \frac{\gamma(q - q_y b_y - q_z b_z)}{2}, \quad \beta = \gamma(b_y\delta_y + b_z\delta_z),$$
$$\delta = q - \frac{2-\gamma}{4\pi\gamma}\left(b_y^2 + b_z^2\right)\rho, \quad \delta_y = q_y - \frac{2-\gamma}{4\pi\gamma}b_y\rho, \quad \delta_z = q_z - \frac{2-\gamma}{4\pi\gamma}b_z\rho.$$

Here c_e is the expression for the averaged acoustic speed of sound. As is well known (Yee 1989), special care must be taken that c_e does not fall from the interval determined by c_e^{L} and c_e^{R}. If this occurs on the left- or right-hand side of this interval, we put $c_e = \min(c_e^{\mathrm{L}}, c_e^{\mathrm{R}})$ or $c_e = \max(c_e^{\mathrm{L}}, c_e^{\mathrm{R}})$, respectively.

The expression for α can be rewritten in the form

$$\alpha = \frac{2-\gamma}{32\pi}\left[\theta_1(\Delta Y)^2 + (\theta_1 + \eta_1 - 1)\frac{\overline{Y}\Delta Y\Delta R}{\overline{R}} + (\eta_1 - 1)\left(\frac{\overline{Y}\Delta R}{\overline{R}}\right)^2 + \theta_2(\Delta Z)^2\right.$$
$$\left. + (\theta_2 + \eta_2 - 1)\frac{\overline{Z}\Delta Z\Delta R}{\overline{R}} + (\eta_2 - 1)\left(\frac{\overline{Z}\Delta R}{\overline{R}}\right)^2\right].$$

The eigenvalues of A are equal to u, $u \pm a_{\mathrm{A}}$, where $a_{\mathrm{A}} = |B_x|/\sqrt{4\pi\rho}$, and to the four roots of the biquadratic equation

$$K^4 - 2pK^2 + Q = 0, \tag{5.4.5}$$

where

$$2p = c_e^2 + \alpha + a_A^2 + \frac{(b_y^2 + b_z^2)\rho}{4\pi} + \beta, \quad Q = (c_e^2 + \alpha)a_A^2.$$

If $c_e^2 + \alpha \geq 0$ and $(b_y^2 + b_z^2)\rho/4\pi + \beta \geq 0$, the roots of this equation are real and the diagonal matrix composed of the eigenvalues acquires the form

$$\Lambda = \text{diag}[u + a_f,\ u + a_A,\ u + a_s,\ u,\ u - a_s,\ u - a_A,\ u - a_f].$$

The quantities a_f and a_s are the largest and the smallest wave propagation velocities (fast and slow magnetosonic waves), and a_A corresponds to Alfvén waves. The eigenvalue $\lambda = u$ corresponds to entropy waves. Our approach assures strict ordering of the eigenvalues. This guarantees the absence of their additional nonphysical degeneration, which is not originally inherent in J. Note that other choices of the parameter vector $\mathbf{s}$ can break this property. Besides, such a degeneration occurs if q_y and q_z are not proportional to $\overline{Y}$ and $\overline{Z}$, respectively. This leads to the most simple admissible choice of θ: $\theta_1 = \theta_2 = 1$. In the MHD case, in contrast to pure gas dynamics, it is not possible to construct the matrix A depending on a single average vector. The exception occurring for $\gamma = 2$ considered by Brio and Wu (1988) is obvious, since in this case $q = q_y = q_z \equiv 0$.

It is easy to check that

$$\beta = \frac{(2-\gamma)(\Delta R)^2}{16\pi}\left[(2-\eta_1)b_y^2 + (2-\eta_2)b_z^2\right],$$

thus giving $\beta \equiv 0$ and $\alpha \geq 0$ for $\eta_1 = \eta_2 = 2$ (compare with Balsara 1998a). We shall accept this simple choice later on. In addition, we can introduce a new notation

$$\tilde{c}_e^2 = c_e^2 + \alpha.$$

Let us determine Ω_R and Ω_L. It is convenient to introduce the matrix Ω_r instead, for which $\Omega_R = A_U\Omega_r$. This matrix consists of the seven columns $\mathbf{r}^i$. For the eigenvalues $\lambda = u + sa$, where $s = \pm 1$ and $a = a_f$, a_s, or 0, the corresponding vector-columns $\mathbf{r} = \mathbf{r}(s, a)$ are the following:

$$\mathbf{r}(s,a) = [1,\ u + 2sa,\ v - sb_yM,\ w - sb_zM,\ r_5,\ b_yN,\ b_zN]^{\mathrm{T}}, \tag{5.4.6}$$

where

$$r_5 = -\mathcal{H} + u^2 + v^2 + w^2 + 2sau - (vb_y + wb_z)sM + \frac{[2a^2 - q - (q_yb_y + q_zb_z)N]\gamma}{(\gamma - 1)}, \tag{5.4.7}$$

$$M = M(a) = \frac{aB_x}{2\pi(a^2 - a_A^2)}, \quad N = N(a) = \frac{a^2 + a_A^2}{a^2 - a_A^2}. \tag{5.4.8}$$

For the eigenvalues $\lambda = u + sa_{\mathrm{A}}$ ($s = \pm 1$) the corresponding vectors are

$$\mathbf{r}(s) = \begin{bmatrix} 0 \\ 0 \\ b_z^* \\ -b_y^* \\ vb_z^* - wb_y^* \\ -sb_z^*\sqrt{\frac{4\pi}{\rho}}\,\mathrm{sgn}\,B_x \\ sb_y^*\sqrt{\frac{4\pi}{\rho}}\,\mathrm{sgn}\,B_x \end{bmatrix}, \tag{5.4.9}$$

where $b_y^* = b_y/|\mathbf{b}|$ and $b_z^* = b_z/|\mathbf{b}|$ ($\mathbf{b} = [b_y, b_z]$).

When using the above formulas for $|\mathbf{b}| \to 0$, the indeterminacies of the type $0/0$ must be resolved. This can be done, e.g., by the substitution $b_y = |\mathbf{b}| \sin\varphi$ and $b_z = |\mathbf{b}| \cos\varphi$. It is clear that the choice of the phase angle φ is not important. In our calculations, for $|\mathbf{b}| = 0$, we assume (see Brio and Wu 1988)

$$b_y^* = b_z^* = \frac{1}{\sqrt{2}}.$$

The matrix Ω_{L} can be found similarly by introducing $\Omega_{\mathrm{l}} = D\,\Omega_{\mathrm{L}}$, where D is a diagonal matrix specified by the equality $A_U\Omega_{\mathrm{r}}D^{-1}\Omega_{\mathrm{l}} = I$. It consists of seven rows. For the eigenvalues $\lambda = u + sa$, where $s = \pm 1$ and $a = a_{\mathrm{f}}$, a_{s}, or 0, the corresponding vector-rows $\mathbf{l} = \mathbf{l}(s, a)$ are the following:

$$\begin{aligned} l_1 &= \frac{c_e^2 - a^2 + sau}{\gamma - 1} - \frac{u^2 + v^2 + w^2}{2} - 2s\pi(vr_y + wr_z)M + (b_y r_y + b_z r_z)L, \\ l_2 &= u - \frac{sa}{\gamma - 1}, \quad l_3 = v + 2\pi Msr_y, \quad l_4 = w + 2\pi Msr_z, \\ l_5 &= -1, \quad l_6 = -r_y L + \frac{\rho b_y}{4\pi}, \quad l_7 = -r_z L + \frac{\rho b_z}{4\pi}, \end{aligned} \tag{5.4.10}$$

where

$$r_y = \frac{b_y}{4\pi} + \frac{\gamma q_y}{\rho(\gamma - 1)}, \quad r_z = \frac{b_z}{4\pi} + \frac{\gamma q_z}{\rho(\gamma - 1)}, \quad L(a) = \frac{[1 + N(a)]\rho}{2}.$$

For $\lambda = u + sa_{\mathrm{A}}$ ($s = \pm 1$) we obtain

$$\begin{aligned} &l_1 = wb_y^* - vb_z^*, \quad l_2 = l_5 = 0, \quad l_3 = b_z^*, \quad l_4 = -b_y^*, \\ &l_6 = -sb_z^*\sqrt{\frac{\rho}{4\pi}}\,\mathrm{sgn}\,B_x, \quad l_7 = sb_y^*\sqrt{\frac{\rho}{4\pi}}\,\mathrm{sgn}\,B_x. \end{aligned} \tag{5.4.11}$$

The matrix D has the form

$$D = \mathrm{diag}[d(a_{\mathrm{f}}),\ 2,\ d(a_{\mathrm{s}}),\ -d(0),\ d(a_{\mathrm{s}}),\ 2,\ d(a_{\mathrm{f}})],$$

where

$$d(a) = -\frac{2}{\gamma - 1}\left[a^2 + \tilde{c}_e^2 + \frac{(1 + N)N}{2} \times \frac{(b_y^2 + b_z^2)\rho}{4\pi}\right].$$

In practice, if $B_x \to 0$ or $|\mathbf{b}| \to 0$, the indeterminacy of the type $0/0$ must be resolved in the above relations for $M = M(a)$ and $N = N(a)$ at $a = a_s$ and $a = a_f$. Using the biquadratic equation for the roots, we obtain

$$a_s^2 a_f^2 = a_A^2 \tilde{c}_e^2, \tag{5.4.12}$$
$$(a_f^2 - \tilde{c}_e^2)(a_f^2 - a_A^2) = |\mathbf{b}|^2 \varepsilon a_f^2, \tag{5.4.13}$$
$$(a_s^2 - \tilde{c}_e^2)(a_s^2 - a_A^2) = |\mathbf{b}|^2 \varepsilon a_s^2, \tag{5.4.14}$$
$$(a_f^2 - \tilde{c}_e^2)(\tilde{c}_e^2 - a_s^2) = |\mathbf{b}|^2 \varepsilon \tilde{c}_e^2, \tag{5.4.15}$$

where $\varepsilon = \rho/4\pi$.

It is not difficult to find that

$$N(a_s) = -\frac{(\tilde{c}_e^2 + a_f^2)(\tilde{c}_e^2 - a_s^2)}{\varepsilon \tilde{c}_e^2 |\mathbf{b}|^2}, \tag{5.4.16}$$

$$M(a_s) = -a_f \sqrt{\frac{\rho \tilde{c}_e^2}{\pi}} \frac{(\tilde{c}_e^2 - a_s^2)\operatorname{sgn} B_x}{\varepsilon \tilde{c}_e^2 |\mathbf{b}|^2}, \tag{5.4.17}$$

$$N(a_f) = \frac{(\tilde{c}_e^2 + a_s^2)(a_f^2 - \tilde{c}_e^2)}{\varepsilon \tilde{c}_e^2 |\mathbf{b}|^2}, \tag{5.4.18}$$

$$M(a_f) = a_s \sqrt{\frac{\rho \tilde{c}_e^2}{\pi}} \frac{(a_f^2 - \tilde{c}_e^2)\operatorname{sgn} B_x}{\varepsilon \tilde{c}_e^2 |\mathbf{b}|^2}. \tag{5.4.19}$$

It is apparent that for $|\mathbf{b}| = 0$, either a_f or a_s are equal to $\tilde{c}_e$. To avoid this indeterminacy in Eqs. (5.4.16)–(5.4.19), we multiply $N(a_s)$ and $M(a_s)$ by α_s and $N(a_f)$ and $M(a_f)$ by α_f, where

$$\alpha_s = \sqrt{\frac{a_f^2 - \tilde{c}_e^2}{a_f^2 - a_s^2}}, \qquad \alpha_f = \sqrt{\frac{\tilde{c}_e^2 - a_s^2}{a_f^2 - a_s^2}},$$

and transform the result using Eq. (5.4.15).

Note that

$$\bar{N}(a_s) = \alpha_s N(a_s)|\mathbf{b}| = -\frac{\alpha_f (a_f^2 + \tilde{c}_e^2)}{\tilde{c}_e \sqrt{\varepsilon}},$$

$$\bar{M}(a_s) = \alpha_s M(a_s)|\mathbf{b}| = -\sqrt{\frac{\rho \tilde{c}_e^2}{\pi}} \frac{\alpha_f a_f \operatorname{sgn} B_x}{\tilde{c}_e \sqrt{\varepsilon}},$$

$$\bar{N}(a_f) = \alpha_f N(a_f)|\mathbf{b}| = \frac{\alpha_s (a_s^2 + \tilde{c}_e^2)}{\tilde{c}_e \sqrt{\varepsilon}},$$

$$\bar{M}(a_f) = \alpha_f M(a_f)|\mathbf{b}| = \sqrt{\frac{\rho \tilde{c}_e^2}{\pi}} \frac{\alpha_s a_s \operatorname{sgn} B_x}{\tilde{c}_e \sqrt{\varepsilon}},$$

always have nonzero denominators.

To resolve the indeterminacy, eigenvectors $\mathbf{r}(\pm 1,\ a_f)$ and $\mathbf{l}(\pm 1,\ a_f)$ are multiplied by α_f and the corresponding $d(a_f)$ by α_f^2. Then, the substitution is made similar to that used for the regularization of the Alfvén eigenvectors and we obtain

$$\mathbf{r}(s, a_f) = \left[\alpha_f,\ (u + 2sa_f)\alpha_f,\ \alpha_f v - s b_y^* \bar{M}(a_f),\ \alpha_f w - s b_z^* \bar{M}(a_f),\ \bar{r}_5,\ b_y^* \bar{N}(a_f),\ b_z^* \bar{N}(a_f)\right]^{\mathrm{T}}$$

with

$$\bar{r}_5 = -\alpha_f(\mathcal{H} + u^2 + v^2 + w^2 + 2sa_f u) - (vb_y^* + wb_z^*)s\bar{M}(a_f)$$
$$+ \frac{\gamma}{\gamma - 1}\Big[(2a_f^2 - q)\alpha_f - (q_y b_y^* + q_z b_z^*)\bar{N}(a_f)\Big].$$

Similarly we obtain the formulas for the components of the eigenvectors $\mathbf{l}(s, a_f)$,

$$l_1 = \alpha_f\left(\frac{c_e^2 - a_f^2 + sa_f u}{\gamma - 1} - \frac{u^2 + v^2 + w^2}{2}\right) - 2s\pi(vr_y^* + wr_z^*)\bar{M}(a_f) + (b_y r_y^* + b_z r_z^*)\bar{L}(a_f),$$
$$l_2 = \alpha_f\left(u - \frac{sa_f}{\gamma - 1}\right), \quad l_3 = \alpha_f v + 2\pi s r_y^* \bar{M}(a_f), \quad l_4 = \alpha_f w + 2\pi s r_z^* \bar{M}(a_f),$$
$$l_5 = -\alpha_f, \quad l_6 = -r_y^* \bar{L}(a_f) + \frac{\alpha_f \rho b_y}{4\pi}, \quad l_7 = -r_z^* \bar{L}(a_f) + \frac{\alpha_f \rho b_z}{4\pi},$$

where

$$r_y^* = \frac{r_y}{|\mathbf{b}|}, \quad r_z^* = \frac{r_z}{|\mathbf{b}|}, \quad \bar{L}(a) = \frac{\rho[\alpha_f|\mathbf{b}| + \bar{N}(a)]}{2}.$$

Note also that in this case

$$d(a_f) = -\frac{2}{\gamma - 1}\left\{\alpha_f^2(a_f^2 + \tilde{c}_e^2) + \frac{[\alpha_f|\mathbf{b}| + \bar{N}(a_f)]\bar{N}(a_f)\rho}{8\pi}\right\}.$$

The formulas for $\mathbf{l}(s, a_s)$ and $\mathbf{r}(s, a_s)$ are the same, with the only exception that a_f and α_f must be substituted by a_s and α_s, respectively. It is easy to notice that no degeneration of the eigenvectors can occur, since α_s and α_f can never vanish simultaneously unless $|\mathbf{b}| = 0$ and $a_A = \tilde{c}_e$. In this case we suggest to put $B_x = B_x(1 + \epsilon)$, where ϵ is a small constant.

No nonphysical degeneration of the eigenvalue set occurs for $\gamma = 2$, since in this case $\alpha \equiv 0$ and $\beta \equiv 0$ and the characteristic equation has only real roots. The choice $\theta_1 = \theta_2 = 1$ and $\eta_1 = \eta_2 = 2$ for an arbitrary γ (Hanawa et al. 1994) results in $\alpha \geq 0$ and $\beta \equiv 0$, thus giving only real roots of the characteristic equation for any admissible right and left values. The family of approximate solutions to the MHD Riemann problem presented here generalizes the approximate quasi-linearized and linearized solutions to this problem (see Koldoba et al. 1992) and preserves the Hugoniot-type relations on the jumps. The Roe-type solver was also extended for isothermal MHD equations by Nakajima and Hanawa (1996) (see also Balsara 1998a; Kim et al. 1999). Cargo and Gallice (1997) constructed Roe's method for the Lagrangian form of the ideal MHD system.

5.4.4 Numerical tests with the Roe-type scheme. As described in Section 2.7, we can increase the spatial accuracy in Godunov's schemes by replacing piecewise constant initial data of the Riemann problem with piecewise linear or piecewise polynomial initial data. By using proper reconstruction techniques, we can increase the order of accuracy in obtaining the fluxes in Eq. (5.4.2). In this case the superscripts R and L must be attributed to quantities on the right- and the left-hand side of the computational cell.

Let us consider the one-dimensional case in which all variables depend only on time t and the space coordinate x. The MHD system for this case is presented by Eq. (5.4.1).

Owing to the condition $\operatorname{div}\mathbf{B} = (B_x)_x = 0$ and the Maxwell equation for B_x, we have $B_x \equiv \text{const}$. For this reason the equation for B_x is omitted in the above system. We start with the aim to solve the ideal MHD Riemann problem with the initial data consisting of two constant states L and R lying to the left and to the right from the centerline of the computation region.

Let us introduce the mesh functions:

$$\mathbf{U}_i^k = \mathbf{U}(t^k, x_i);\ \mathbf{F}_i^k = \mathbf{F}(\mathbf{U}_i^k),\ t^k = k\Delta t;\ x_i = (i-1)\Delta x;\ k = 0, 1, \ldots;\ i = 1, \ldots, I$$

with the space increment Δx and time increment Δt defined by the CFL condition.

In order to obtain the second order of accuracy both in space and in time we use the predictor–corrector approach. In the predictor step we apply the first order of accuracy Roe's scheme

"Predictor"

$$\mathbf{U}_i^{k+1/2} = \mathbf{U}_i^k - \left(\mathbf{F}_{i+1/2}^k - \mathbf{F}_{i-1/2}^k\right)\frac{\Delta t}{2\Delta x}, \tag{5.4.20}$$

where $\mathbf{F}_{i+1/2}^k$ is the first-order Roe's flux

$$\mathbf{F}_{i+1/2}^k = \frac{1}{2}[(\mathbf{F}_{i+1}^k + \mathbf{F}_i^k - (\Omega_\mathrm{R}|\Lambda|\Omega_\mathrm{L})_{i+1/2}^k(\mathbf{U}_{i+1}^k - \mathbf{U}_i^k)] \tag{5.4.21}$$

with Ω_R, Ω_L, and $|\Lambda|$ calculated by the formulas from Section 5.4.3. After that we perform interpolation of functions known in the cell centers to find their values $\mathbf{U}_{i+1/2}^\mathrm{L}$ and $\mathbf{U}_{i+1/2}^\mathrm{R}$ on the left- and right-hand sides of each cell and pass to the corrector stage

"Corrector"

$$\mathbf{U}_i^{k+1} = \mathbf{U}_i^k - \left(\bar{\mathbf{F}}_{i+1/2}^{k+1/2} - \bar{\mathbf{F}}_{i-1/2}^{k+1/2}\right)\frac{\Delta t}{\Delta x} \tag{5.4.22}$$

$$\bar{\mathbf{F}}_{i+1/2}^{k+1/2} = \frac{1}{2}\left[\mathbf{F}\left(\mathbf{U}_{i+1/2}^\mathrm{R}\right) + \mathbf{F}\left(\mathbf{U}_{i+1/2}^\mathrm{L}\right) - (\Omega_\mathrm{R}|\Lambda|\Omega_\mathrm{L})_{i+1/2}^{k+1/2}(\mathbf{U}_{i+1/2}^\mathrm{R} - \mathbf{U}_{i+1/2}^\mathrm{L})\right]. \tag{5.4.23}$$

In what follows we present the results of numerical test on the basis of various reconstruction procedures. Several of them can be written out as a two-parametric family (van Leer 1979):

$$\mathbf{u}_{i+1/2}^\mathrm{R} = \mathbf{u}_i^{k+1/2} - \tfrac{1}{4}[(1-\eta)\tilde{\Delta}_{i+3/2} + (1+\eta)\tilde{\tilde{\Delta}}_{i+1/2}] \tag{5.4.24}$$

$$\mathbf{u}_{i+1/2}^\mathrm{L} = \mathbf{u}_i^{k+1/2} + \tfrac{1}{4}[(1-\eta)\tilde{\tilde{\Delta}}_{i-1/2} + (1+\eta)\tilde{\Delta}_{i+1/2}] \tag{5.4.25}$$

$$\tilde{\Delta}_{i+1/2} = \operatorname{minmod}(\Delta_{i+1/2}, \omega\Delta_{i-1/2}) \tag{5.4.26}$$

$$\tilde{\tilde{\Delta}}_{i+1/2} = \operatorname{minmod}(\Delta_{i+1/2}, \omega\Delta_{i+3/2}) \tag{5.4.27}$$

with $\Delta_{i+1/2} = \mathbf{u}_{i+1}^{k+1/2} - \mathbf{u}_i^{k+1/2}$.

We chose here interpolation of the primitive variables $\mathbf{u}$, which is certainly the less time-consuming approach. The space approximation is defined by the value of η, namely, $\eta = -1$ corresponds to a fully upwind scheme, $\eta = 1$ to a three-point central difference scheme, and $\eta = 1/3$ to a third-order upwind-biased scheme. The parameter ω is responsible for the artificial compression which allows us to improve the resolution of contact and Alfvén discontinuities.

We shall also test the following procedure:

$$\mathbf{u}^{\mathrm{R}}_{i+1/2} = \mathbf{u}^{k+1/2}_{i+1} - \tfrac{1}{2}\Delta^{\mathrm{A}}_{i+1}, \tag{5.4.28}$$

$$\mathbf{u}^{\mathrm{L}}_{i+1/2} = \mathbf{u}^{k+1/2}_{i} + \tfrac{1}{2}\Delta^{\mathrm{A}}_{i}, \tag{5.4.29}$$

where $\Delta^{\mathrm{A}}_{\mathrm{L}}$ is a smooth limiter function due to van Albada et al. (1982)

$$\Delta^{\mathrm{A}}_{i} = \frac{\Delta_{i-1/2}\left[\left(\Delta_{i+1/2}\right)^2 + \epsilon\right] + \Delta_{i+1/2}\left[\left(\Delta_{i-1/2}\right)^2 + \epsilon\right]}{\left(\Delta_{i+1/2}\right)^2 + \left(\Delta_{i-1/2}\right)^2 + 2\epsilon} \tag{5.4.30}$$

with ϵ being a small constant in the range 10^{-7} to 10^{-5} used to avoid division by zero.

Instead of Δ^{A}, in Eqs. (5.4.28)–(5.4.29) we can use the limiter proposed by Colella and Woodward (1984),

$$\Delta^{\mathrm{CW}}_{i} = \mathrm{minmod}\left(2\Delta_{i+1/2}, 2\Delta_{i-1/2}, 0.5\left(\Delta_{i+1/2} + \Delta_{i-1/2}\right)\right), \tag{5.4.31}$$

which corresponds to a piecewise-parabolic parameter distribution inside computational cells or

$$\Delta^{\mathrm{Sbee}}_{i} = S \times \max\left[0, \min(2|\Delta_{i+1/2}|, S\,\Delta_{i-1/2}), \min(|\Delta_{i+1/2}|, 2S\,\Delta_{i-1/2})\right], \tag{5.4.32}$$

where $S = \mathrm{sgn}\,\Delta_{i+1/2}$. The last formula represents a so-called superbee limiter (Roe 1985). It is known to be extremely effective in sharpening the contact discontinuity profiles in pure gas dynamics. We shall show later that it is also very useful for sharp resolution of rotational discontinuities.

As a test problem we choose the MHD Riemann problem with the following initial conditions: $[\rho, p, u, v, w, B_y, B_z] = [0.18405, 0.3541, 3.8964, 0.5361, 2.4866, 2.394, 1.197]$ for $x < 0.5$ and $[0.1, 0.1, -5.5, 0, 0, 2, 1]$ for $x > 0.5$ with $B_x \equiv 4$. The results will be presented for $t = 0.15$ (400 cells are taken between 0 and 1). The solution of this problem is characterized by presence of all types of discontinuities intrinsic to MHD. This problem was also chosen as one of the tests by Dai and Woodward (1994a).

At first, we put $\omega = 1$, thus switching the limiting function to the least compressive minmod function. We choose $\eta = 1/3$, though for the chosen mesh and the value of the parameter ω other choices of η give essentially the same results. In Fig. 5.6 the plots of all primitive variables are shown. As could be expected, resolution of fast MHD shocks is very good. They are smeared only over three computational cells. Slow MHD shocks occupy four cells. As known from pure gas dynamics, resolution of contact discontinuities is generally not so good if the minmod slope limiter is used (see Hirsch 1990). In MHD, besides the contact discontinuity, Alfvén, or rotational, discontinuities are also rather smeared over the computational grid. Such behavior of rotational discontinuities is expectable, since they are known to have no stationary viscous structure (Landau and Lifshitz 1984). The problem of the dissipative and dispersive structures of a discontinuity plays an important role in the resolution of the question of its admissibility. In fact, the ideal plasma approximation is just an idealization made to emphasize that dissipative and dispersive effects are negligible in comparison with convective processes. The width of a discontinuity in this case is supposed

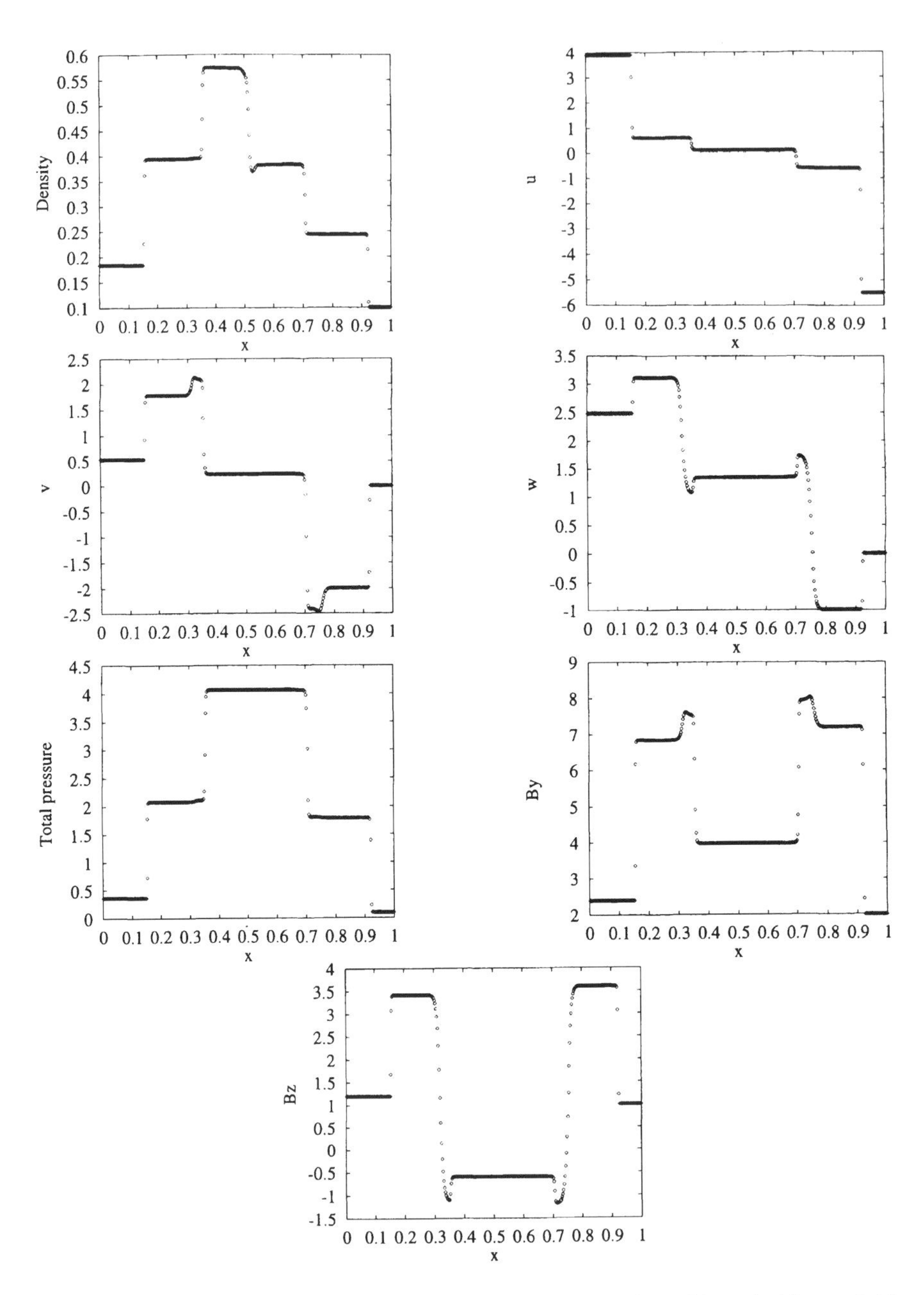

Figure 5.6 Solution of the test problem. van Leer's reconstruction of the primitive variables with $\omega = 1$ and $\eta = 1/3$.

to be zero. Flows of real media are always continuous. From this viewpoint, the criterion of the shock admissibility can be formulated as existence of a viscous profile in the solution to the higher-order differential equations that take into account the processes neglected in the ideal statement. As far as MHD flows are concerned, these processes include viscosity, heat transfer, the Joule heat release, finite conductivity, etc. Admissible shocks are only those that originate from viscous profiles in the limit of vanishing dissipation.

The physical reason explaining the absence of a stationary structure of rotational discontinuities is quite simple. Kinetic, internal, and magnetic energies are the same on both sides of them (see Section 5.2). Thus, there is no source of energy covering the dissipation connected with rotation of the velocity and magnetic field lines. On the other hand, rotational discontinuities are inevitable, for example, in the general solution of the MHD piston problem (Barmin and Gogosov 1960). Such discontinuities must therefore have nonstationary structure, that is, their width D_A caused by dissipation will increase with time according to the following relation:

$$D_A \sim \sqrt{(\nu + \nu_m)\, t},$$

where ν and ν_m are molecular and magnetic viscosities, respectively. Note, on the other hand, that numerical viscosity and dispersion are always present in calculations of ideal flows. This means that in order to obtain sharply resolved contact and rotational discontinuities in accordance with truly small dissipative processes we must use some artificial sharpening of them. One of the possible procedures lies in choosing the values of the parameter ω in Eqs. (5.4.26)–(5.4.27) greater than 1. In Figs. 5.7, 5.8, and 5.9, we show the solution for $\omega = 2$ and $\eta = -1$, $1/3$, and 1, respectively. It is apparent that application of the upwind interpolation results in this case in high-frequency oscillations, especially near fast MHD shocks. There also exist noticeable oscillations in the regions of constant distributions. Application of the third-order upwind-biased scheme ($\eta = 1/3$) gives much better results, though oscillations do not disappear. The best results are obtained in the case with $\eta = 1$, that is, symmetric interpolation works much better for MHD equations than in pure gas dynamics. On the other hand, the improvement in the resolution of rotational discontinuity, though apparent, is smaller in the last case than that for $\eta = 1/3$.

It is worth mentioning that we can apply different types of reconstruction for different quantities. In the previous cases we used the same formulas for density, thermal pressure, velocity, and magnetic field components. Experience shows that the overall performance of interpolation can be improved if the simplest minmod reconstruction is used for pressure and for the normal component of the velocity vector. The amplitude of spurious oscillations is smaller in this case, though they still exist, especially for the upwind interpolation.

We would also like to emphasize that application of the arithmetic averaging to find parameters on the cell interfaces gives essentially the same result for the one-dimensional problem under consideration and for the chosen space discretization.

In Fig. 5.10 and 5.11 we present the numerical solution with the application of van Albada's and Colella and Woodward's limiters for all variables except p and u, for which the minmod reconstruction is used. These approaches give increasingly better resolution of the contact and rotational discontinuities, though they are also not free from spurious oscillations. Application of the former of these two limiters results also in noticeable distortion of the smooth regions.

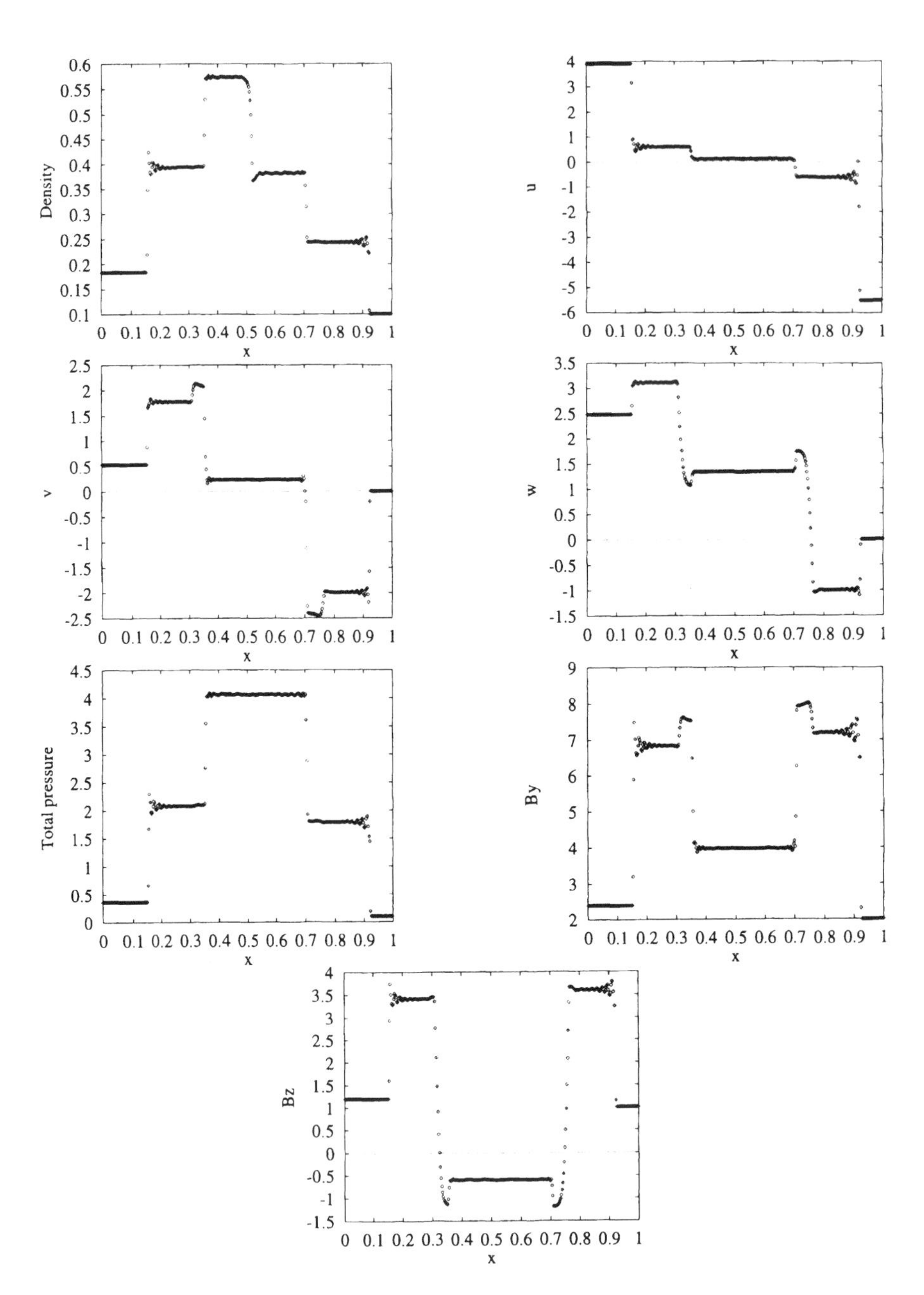

Figure 5.7 Solution of the test problem. van Leer's reconstruction of the primitive variables with $\omega = 2$ and $\eta = -1$.

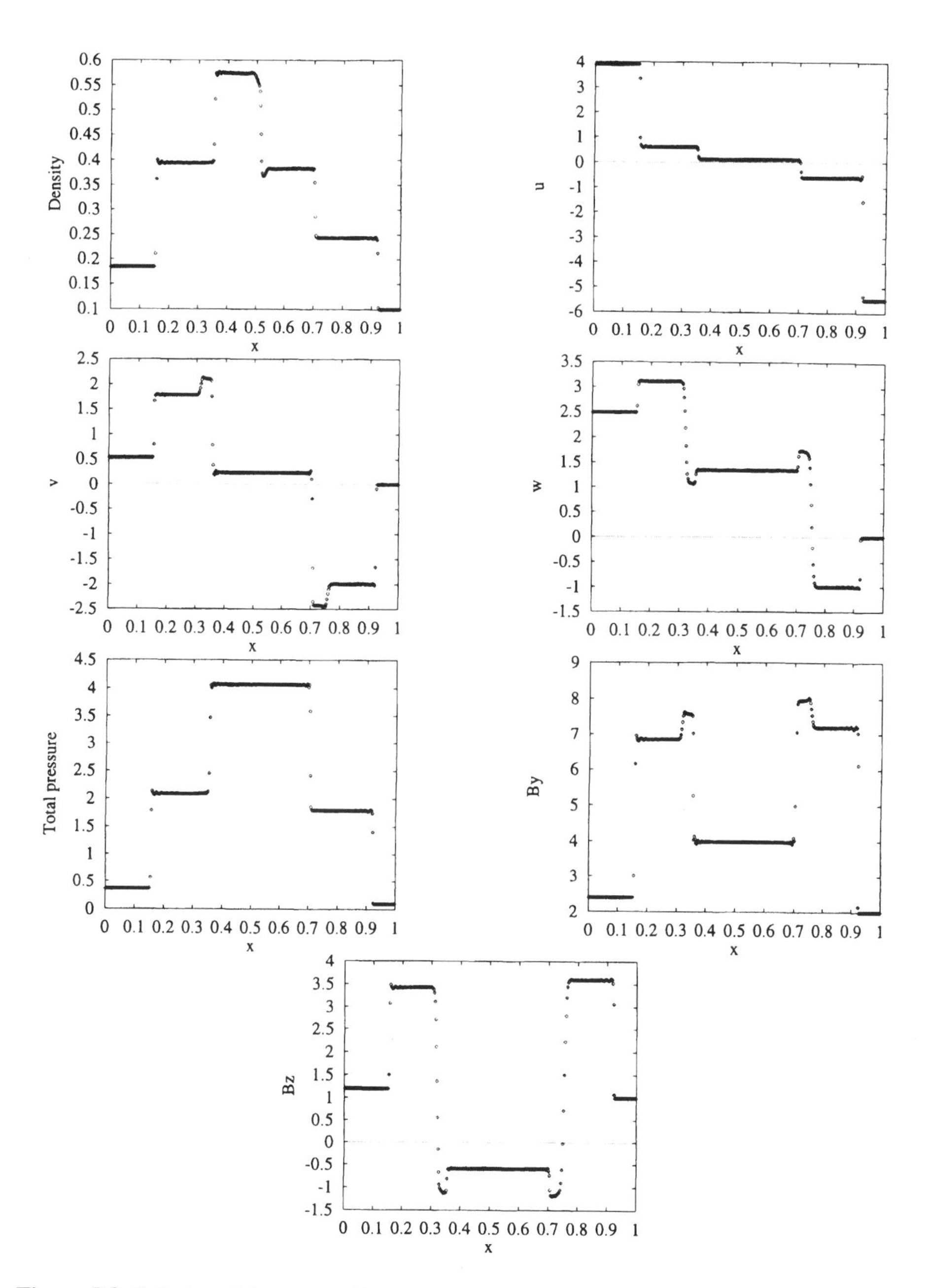

Figure 5.8 Solution of the test problem. van Leer's reconstruction of the primitive variables with $\omega = 2$ and $\eta = 1/3$.

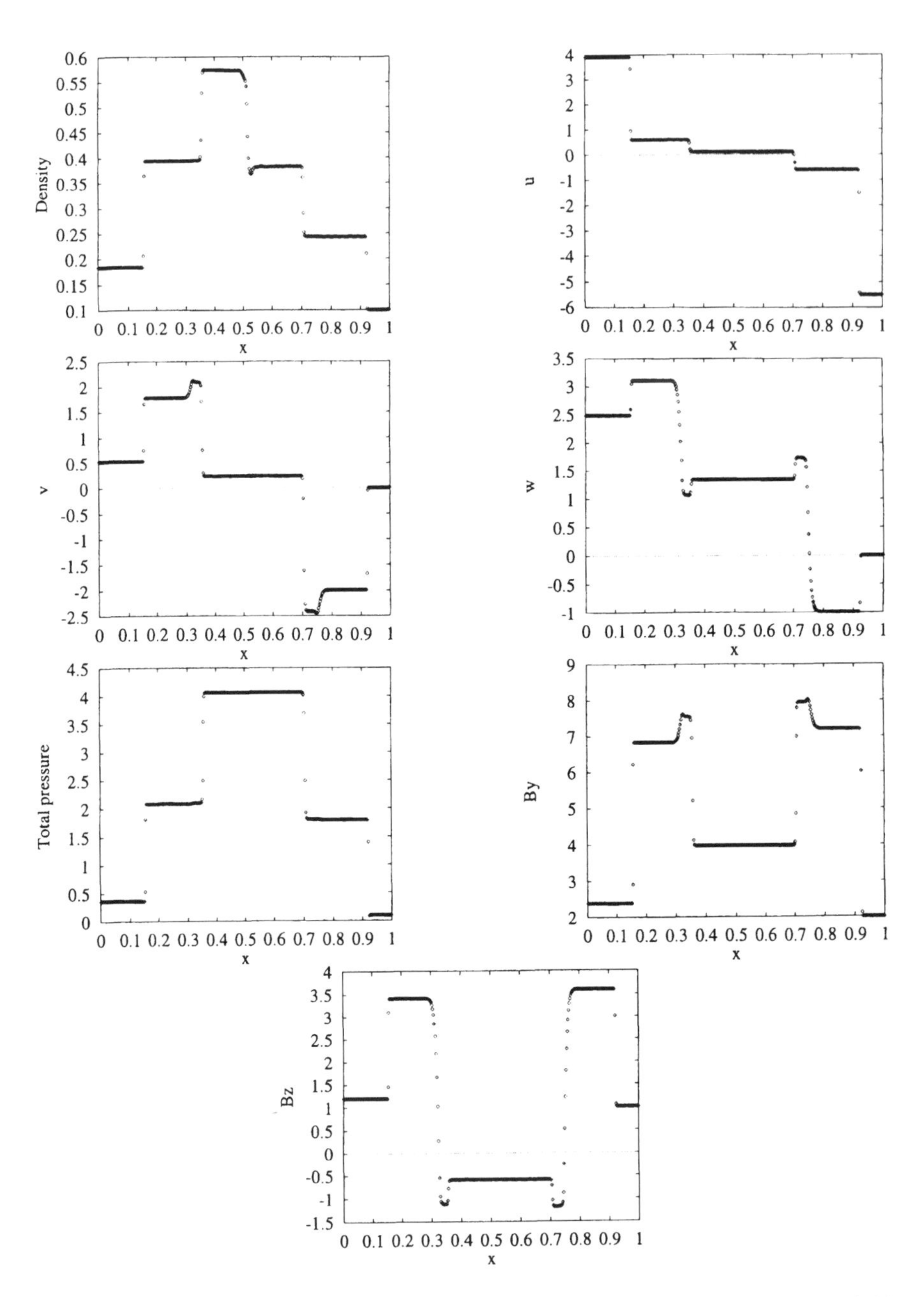

Figure 5.9 Solution of the test problem. van Leer's reconstruction of the primitive variables with $\omega = 2$ and $\eta = 1$.

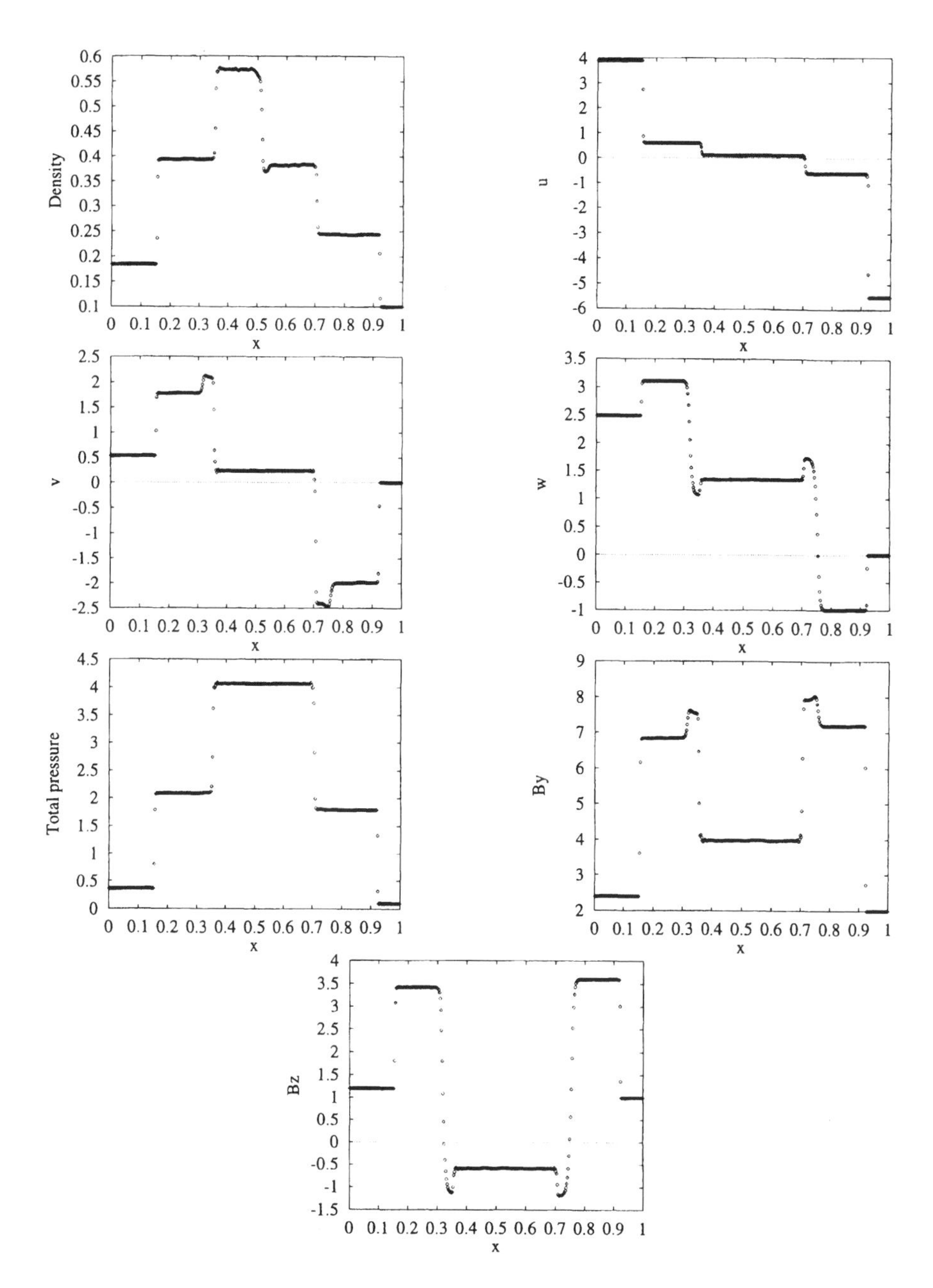

Figure 5.10 Solution of the test problem. van Albada's reconstruction of the primitive variables.

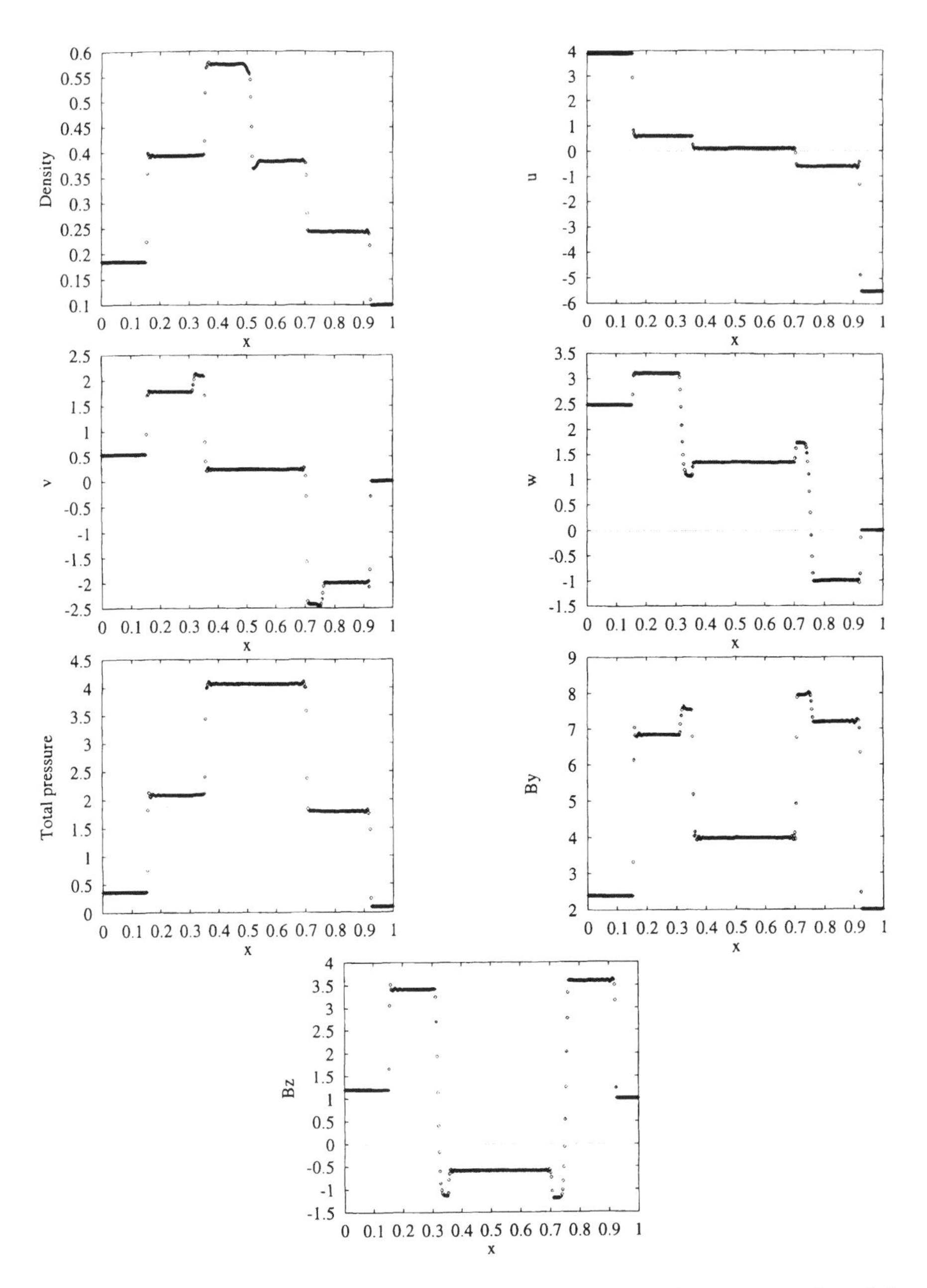

Figure 5.11 Solution of the test problem. Colella and Woodward's reconstruction of the primitive variables.

Analysis of the presented results shows that only symmetric interpolation of primitive variables preserves the monotonicity of the numerical scheme. The presence of oscillations in all other cases makes them hardly reliable for multidimensional problems for which the choice of functions to be interpolated is even more complicated. The oscillations can be damped, in principle, by an artificial dissipation (Dai and Woodward 1994a), though this seems to be a step back in the methods of obtaining high-resolution monotone discontinuity profiles. It is worth noting that the amplitude of oscillations can be somewhat decreased if we adopt the central, second order of accuracy in space, approximation of the flux derivative in the predictor step (5.4.23)–(5.4.22) (Barmin, Kulikovskii, and Pogorelov 1996; Kulikovskii, Pogorelov, and Semenov 1998)

To determine more consistent approaches applicable to multidimensional problems, we are urged to apply interpolation procedures with limiting the increments of characteristic variables. Besides, they showed the best quality for purely gas dynamic flows. On the other hand, application of characteristic variables in the framework of the variable interpolation adopted in the MUSCL approach is very costly in MHD. In this case, the flux interpolation approach can be preferable.

Let us introduce positive and negative flux differences by relations

$$\mathbf{F}^k_{i+1/2} - \mathbf{F}_i = \delta\mathbf{F}^-_{i+1/2}, \qquad \mathbf{F}^k_{i-1/2} - \mathbf{F}_i = -\delta\mathbf{F}^+_{i-1/2}, \tag{5.4.33}$$

where $\mathbf{F}^k_{i+1/2}$ is the first-order numerical flux defined by Eq. (5.4.21). As usual, we use the following decomposition into positive and negative fluxes:

$$\delta\mathbf{F}^-_{i+1/2} = \Omega_{\mathrm{R}}\Lambda^-\Omega_{\mathrm{L}}\,\delta\mathbf{U}_{i+1/2}, \qquad \delta\mathbf{F}^+_{i+1/2} = \Omega_{\mathrm{R}}\Lambda^+\Omega_{\mathrm{L}}\,\delta\mathbf{U}_{i+1/2}, \tag{5.4.34}$$

where diagonal matrices Λ^+ and Λ^- are constructed of positive and negative eigenvalues of the Jacobian matrix $\partial\mathbf{F}/\partial\mathbf{U}$, namely,

$$\Lambda^+ = \mathrm{diag}\left[\frac{|\lambda_j| + \lambda_j}{2}\right], \qquad \Lambda^- = \mathrm{diag}\left[\frac{|\lambda_j| - \lambda_j}{2}\right].$$

The eigenvalues and eigenvectors in Eq. (5.4.34) are evaluated via solution of the corresponding Riemann problem using the parameters at the cell centers as initial conditions.

Extrapolating the fluxes using the formulas (5.4.24)–(5.4.27) applied earlier to the primitive variables, we can write out a second-order numerical flux as

$$\mathbf{F}^{(2)}_{i+1/2} = \mathbf{F}^{k+1/2}_{i+1/2} + \frac{1}{4}\left[(1-\eta)\widetilde{\widetilde{\delta\mathbf{F}}}{}^+_{i-1/2} + (1+\eta)\widetilde{\delta\mathbf{F}}{}^+_{i+1/2}\right]$$

$$-\frac{1}{4}\left[(1-\eta)\widetilde{\delta\mathbf{F}}{}^-_{i+3/2} + (1+\eta)\widetilde{\widetilde{\delta\mathbf{F}}}{}^-_{i+1/2}\right], \tag{5.4.35}$$

$$\widetilde{\delta\mathbf{F}}_{i+1/2} = \mathrm{minmod}(\delta\mathbf{F}_{i+1/2}, \omega\delta\mathbf{F}_{i-1/2}), \tag{5.4.36}$$

$$\widetilde{\widetilde{\delta\mathbf{F}}}_{i+1/2} = \mathrm{minmod}(\delta\mathbf{F}_{i+1/2}, \omega\delta\mathbf{F}_{i+3/2}). \tag{5.4.37}$$

As discussed above, MHD problems turn out to be rather sensitive to the interpolation procedure. The choice of the basis for the flux difference splitting is also very important.

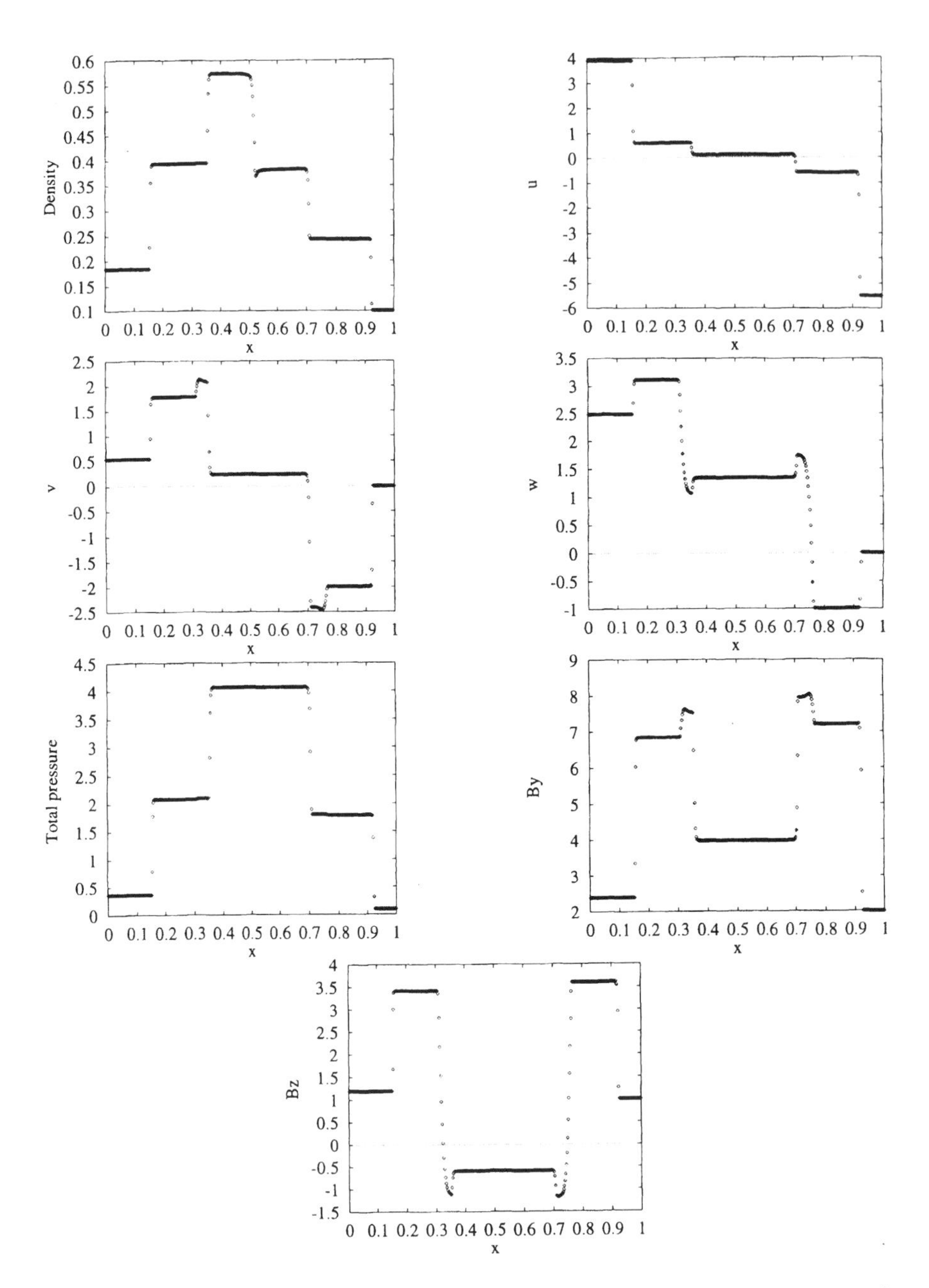

Figure 5.12 Solution of the test problem: non-MUSCL approach. van Leer's reconstruction with $\omega = 2$ and $\eta = 1$.

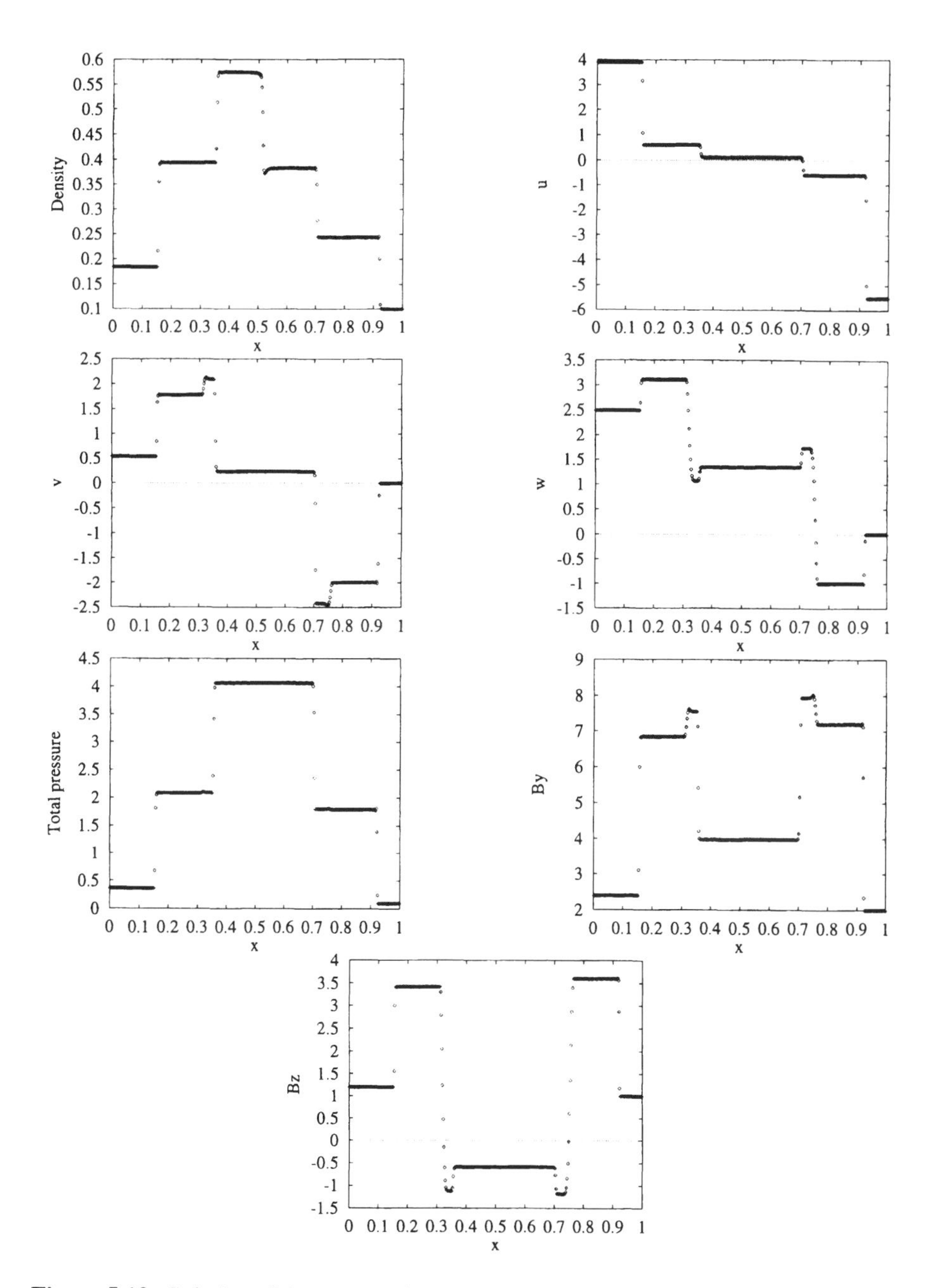

Figure 5.13 Solution of the test problem: non-MUSCL approach. van Albada's limiter.

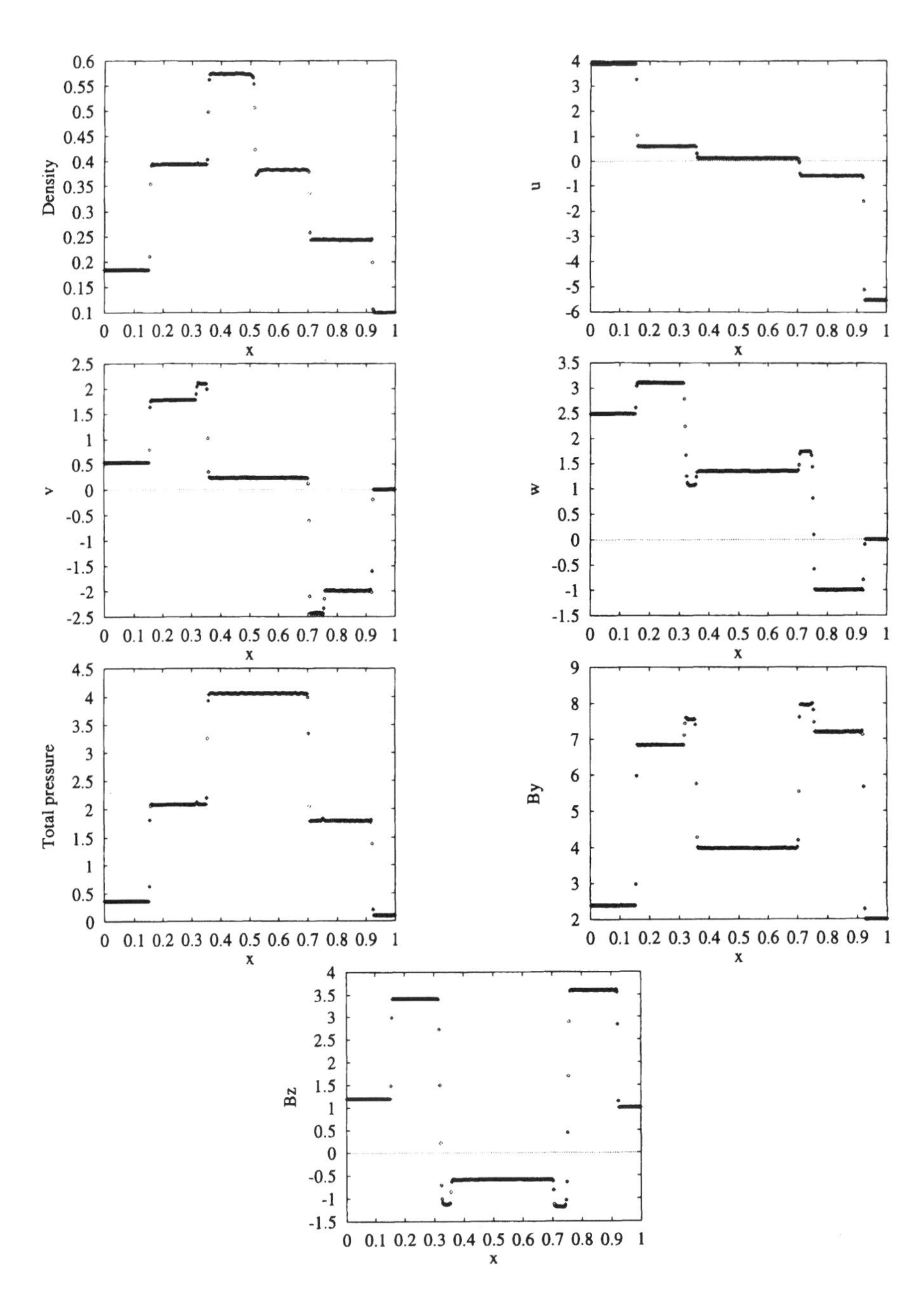

Figure 5.14 Solution of the test problem: non-MUSCL approach. Superbee limiter.

Fukuda and Hanawa (1999) have recently shown that there is a specific choice of the decomposition basis that is especially successful in the conservation of the travelling Alfvén wave profile. Namely, if we want to calculate the numerical flux at the cell interface $i+1/2$, then all the flux differences $\delta\mathbf{F}^-$ and $\delta\mathbf{F}^+$ occurring in Eq. (5.4.35) must be decomposed in the coordinate basis fixed at this interface. Taking this into account, we easily obtain

$$\mathbf{F}^{(2)}_{i+1/2} = \mathbf{F}^{k+1/2}_{i+1/2} + \frac{1}{4}(\Omega_{\mathrm{R}}\Lambda^+)^{k+1/2}_{i+1/2}\,[(1-\eta)\tilde{\tilde{\alpha}}_{i-1/2} + (1+\eta)\tilde{\alpha}_{i+1/2}]$$
$$-\frac{1}{4}(\Omega_{\mathrm{R}}\Lambda^-)^{k+1/2}_{i+1/2}\,[(1-\eta)\tilde{\alpha}_{i+3/2} + (1+\eta)\tilde{\tilde{\alpha}}_{i+1/2}], \tag{5.4.38}$$

where

$$\alpha_{i+1/2} = (\Omega_{\mathrm{L}})^{k+1/2}_{i+1/2}\,(\mathbf{U}^{k+1/2}_{i+1} - \mathbf{U}^{k+1/2}_{i}),$$
$$\tilde{\alpha}_{i+1/2} = \mathrm{minmod}(\alpha_{i+1/2}, \omega\alpha_{i-1/2}),$$
$$\tilde{\tilde{\alpha}}_{i+1/2} = \mathrm{minmod}(\alpha_{i+1/2}, \omega\alpha_{i+3/2})$$

and $\mathbf{F}^{k+1/2}_{i+1/2}$ are obtained from the solution of the Riemann problem at the time level $k+1/2$. Thus, the predictor step (5.4.20)–(5.4.21) must be followed by the corrector step

$$\mathbf{U}^{k+1}_i = \mathbf{U}^k_i - \left(\mathbf{F}^{(2)}_{i+1/2} - \mathbf{F}^{(2)}_{i-1/2}\right)\frac{\Delta t}{\Delta x}. \tag{5.4.39}$$

Various other types of the flux reconstruction formulas can be obtained using the limiters from Eqs. (5.4.28)–(5.4.32).

Besides its excellent properties in the Alfvén wave profile conservation, the presented flux-reconstruction approach is very economical from the viewpoint of computational efficiency. Let us consider its performance for the chosen Riemann problem. In Fig. 5.12 we present the solution obtained using $\eta = 1$ and $\omega = 2$. The choice of $\eta = -1$ gives virtually the same result in this case. This speaks in favor of applying the flux-reconstruction scheme with interpolation of characteristic variables, since the numerical scheme becomes more robust. If we compare our results with those in Fig. 5.6, we see that the resolution of shocks is almost the same. On the other hand, resolution of the contact and rotational discontinuities, though not sufficient, is still better than in the case of interpolation of the primitive variables. If we apply a more compressive limiter, e.g., van Albada's (Fig. 5.13) or the superbee (Fig. 5.14) limiter, resolution can be substantially improved. The superbee limiter, for example, gives remarkably thin contact and rotational discontinuities (four to five computational cells). One can admit, however, a definite presence of wavy distributions in the regions of constant quantities. Another advantage, however, of interpolating the characteristic variables is that we can consciously apply more compressive limiters to the characteristic fields responsible for the entropy and Alfvén wave behavior and regular minmod limiters to fast and slow magnetosonic fields. This allows us to eliminate unwanted oscillations (see Fig. 5.15).

Thus, there exists a family of solutions to the linearized MHD Riemann problem that can be effectively incorporated into various TVD algorithms designed to obtain discontinuous solutions to the MHD system. The shock relations in this case are satisfied exactly. On the other hand, MHD equations, in contrast to pure gas dynamic equations, are substantially

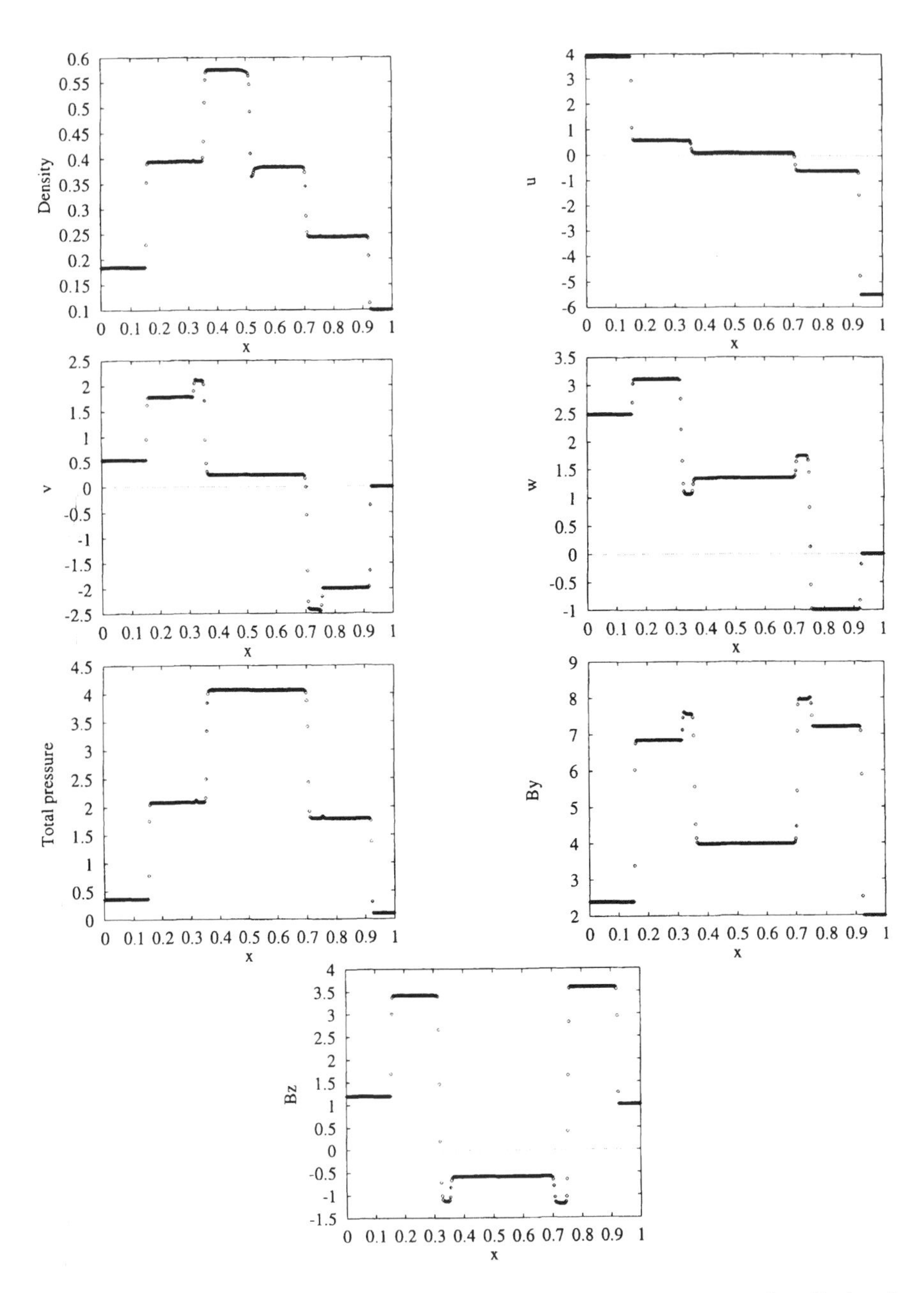

Figure 5.15 Solution of the test problem: non-MUSCL approach. Superbee limiter for selected characteristic fields.

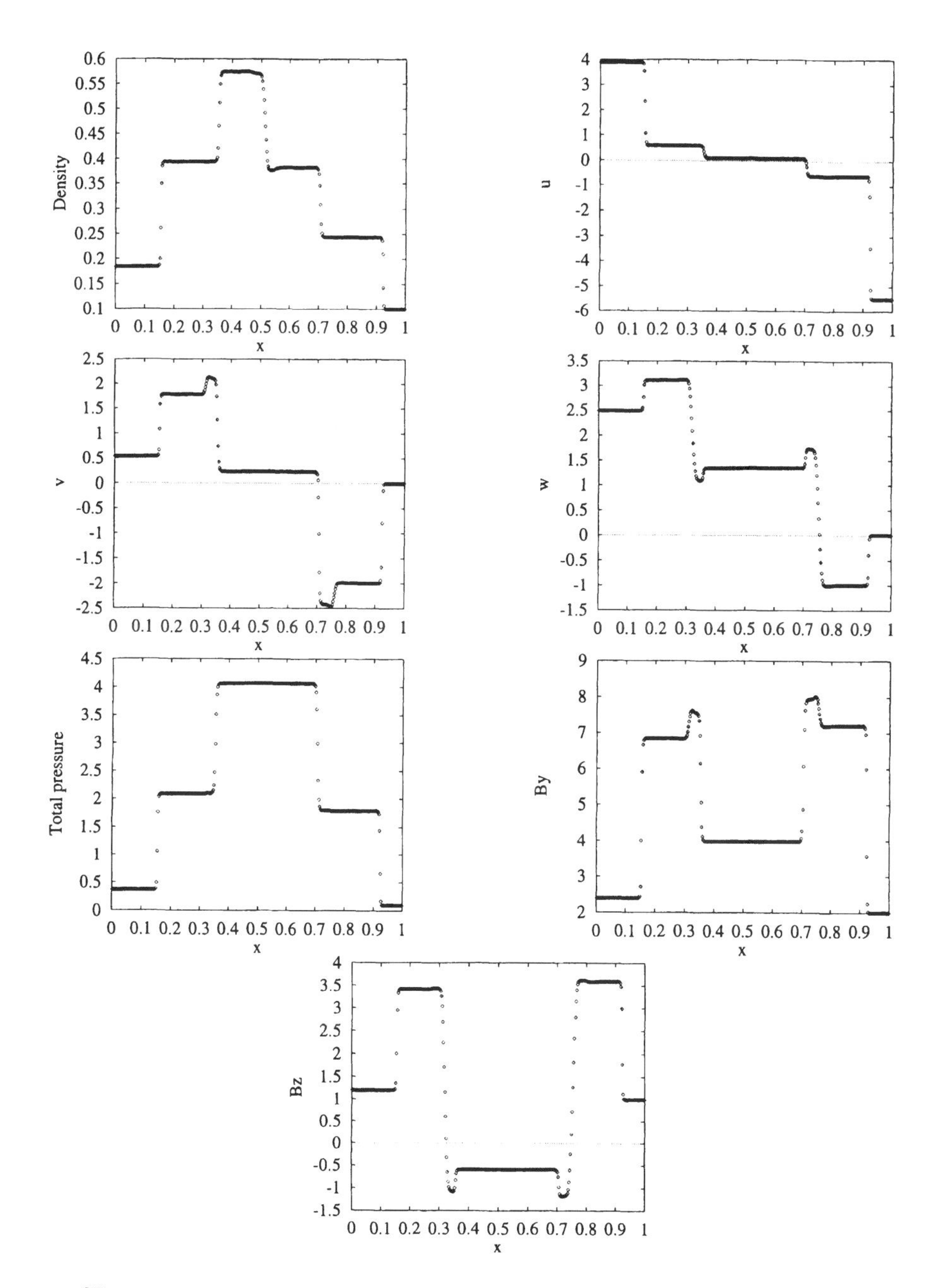

Figure 5.16 Solution of the test problem. TVD Lax–Friedrichs-type scheme.

three-dimensional. If we apply the solution of the one-dimensional Riemann problem to find the numerical flux at the cell interface in the multidimensional problem, the controversy originates in treating the divergence-free (absence of magnetic charge) condition. In fact, in the one-dimensional treatment we assume that the component of the magnetic field normal to the cell interface is constant. This is not true, however, for multidimensional problems. This disagreement must be eliminated by special magnetic charge sweep procedures that will be described in the separate section.

Due to the complexity of formulas, the described approach is much more cumbersome than Roe's linearization method in pure gas dynamics. This concerns, even to a larger extent, the MHD PPM or Osher-type methods. Taking into account the above remark, some simplified approaches are welcome that should (i) satisfy the TVD property and (ii) be enough economical and robust. Barmin et al. (1996) suggested applying a second order of accuracy in time and space TVD extension of the Lax–Friedrichs scheme that gives a substantial simplification of the finite-volume numerical algorithm comparing with the schemes which use precise characteristic splitting of the Jacobian matrices. The results obtained by this scheme were compared with those by Brio and Wu (1988), Dai and Woodward (1994a), and the Roe-type method of Pogorelov et al. (1995) and Pogorelov and Semenov (1996a, 1997c, 1997d) and a good agreement was observed. In this scheme we substitute the diagonal eigenvalue matrix in Eq. (5.4.2) by the diagonal matrix $\hat{\mathbf{R}}_{i+1/2}$ with the spectral radius r (the maximum of eigenvalue magnitudes) of the Jacobian matrix $J = \partial \mathbf{F}/\partial \mathbf{U}$ on its diagonal

$$r = |V_n| + a_{\mathrm{f}}, \quad a_{\mathrm{f}}^2 = \frac{1}{2}\left((a^*)^2 + \sqrt{(a^*)^4 - 4c_e^2 b_n^2}\right),$$
$$b_n = B_n/\sqrt{4\pi\rho}, \quad (a^*)^2 = (\gamma p + B^2/4\pi)/\rho, \quad c_e^2 = \gamma p/\rho,$$

where V_n, B_n, and B are the normal velocity and magnetic field components and the magnitude of magnetic field. In this case we avoid the calculation of the eigenvectors and obtain the numerical flux in the form

$$\mathbf{F}_{i+1/2} = \frac{1}{2}\left[\mathbf{F}\left(\mathbf{U}_{i+1/2}^{\mathrm{R}}\right) + \mathbf{F}\left(\mathbf{U}_{i+1/2}^{\mathrm{L}}\right) + \mathbf{\Phi}_{i+1/2}\right], \tag{5.4.40}$$
$$\mathbf{\Phi}_{i+1/2} = -\hat{\mathbf{R}}_{i+1/2}\left(\mathbf{U}_{i+1/2}^{\mathrm{R}} - \mathbf{U}_{i+1/2}^{\mathrm{L}}\right). \tag{5.4.41}$$

We can obtain a higher-order approximation by using proper parameter reconstruction to find quantities on the computational cell surfaces. Having the second order of accuracy, the proposed scheme is much less dissipative than the original Lax–Friedrichs method and provides incomparably better shock resolution. Being much simpler than the scheme based on Roe's linearization method, it still gives numerical results with a reasonable accuracy. For illustration, we present in Fig. 5.16 the solution of our test Riemann problem obtained using the Lax–Friedrichs-type method and the reconstruction procedures based on Eqs. (5.4.24)–(5.4.27) with $\eta = 1/3$ and $\omega = 2$ for all primitive variables except p and u for which $\omega = 1$ is used. We see that this simple method allows us to obtain the shocks smeared over five cells compared with three cells when using the Roe-type method. Besides, other discontinuities are also rather "thick," which is not surprising if we take into account very rough treatment of characteristic fields by this method. The advantages of the Lax–Friedrichs-type approach are in its robustness and small computer time consumption. Also, this scheme automatically satisfies the entropy inequality and no special modifications of the scheme are necessary if we look for steady-state solutions (see Section 2.10). The TVD Lax–Friedrichs scheme was first suggested and studied both in nonstaggered and staggered versions by Nessyahu and Tadmor (1988, 1990). It is an extension to the second order of accuracy of the local Lax–Friedrichs scheme by Rusanov (1961). We shall return to discussion of the latter method in the section dealing with multidimensional MHD problems. Several numerical tests of the TVD scheme applications to multidimensional MHD problems are presented by Tóth & Odstrčil (1996).

5.4.5 Modified MHD system. Let us try to construct Roe's solver for modified 8×8 MHD equations (eight equations for eight unknowns) that represent a direct mathematical consequence (Powell 1994) of the quasilinear system (1.3.20)–(1.3.23) transformed to the conservative form. The extension of the procedure described in Section 5.4.3 is not straightforward. The reason for this can be illustrated if we consider the initial differential MHD system. The conservation form of MHD equations has the form (5.1.43). The formal nonconservative form of these equations is

$$\frac{\partial \mathbf{U}}{\partial t} + \tilde{A}\frac{\partial \mathbf{U}}{\partial x} + \tilde{B}\frac{\partial \mathbf{U}}{\partial y} + \tilde{C}\frac{\partial \mathbf{U}}{\partial z} = \mathbf{0}, \tag{5.4.42}$$

$$\tilde{A} = \frac{\partial \mathbf{E}}{\partial \mathbf{U}}, \quad \tilde{B} = \frac{\partial \mathbf{F}}{\partial \mathbf{U}}, \quad \tilde{C} = \frac{\partial \mathbf{G}}{\partial \mathbf{U}}.$$

It is easy to see, however, that all matrices $\tilde{A}$, $\tilde{B}$, and $\tilde{C}$ have one row of zero entries, and, hence, they have nonphysical zero eigenvalue. To obtain nonsingular matrices we must transform Eq. (5.4.42) in the following way. In Eq. (5.4.42) let us select all terms proportional to $\operatorname{div}\mathbf{B}$ as follows:

$$\frac{\partial \mathbf{U}}{\partial t} + A\frac{\partial \mathbf{U}}{\partial x} + B\frac{\partial \mathbf{U}}{\partial y} + C\frac{\partial \mathbf{U}}{\partial z} + \operatorname{div}(\mathbf{B})\,\mathbf{f} = \mathbf{0}, \tag{5.4.43}$$

$$\mathbf{f} = \left[0, -\frac{B_x}{4\pi}, -\frac{B_y}{4\pi}, -\frac{B_z}{4\pi}, -\frac{\mathbf{v}\cdot\mathbf{B}}{4\pi}, -u, -v, -w\right]^{\mathrm{T}},$$

$$A\frac{\partial \mathbf{U}}{\partial x} = \tilde{A}\frac{\partial \mathbf{U}}{\partial x} - \mathbf{f}\frac{\partial B_x}{\partial x}, \quad B\frac{\partial \mathbf{U}}{\partial y} = \tilde{B}\frac{\partial \mathbf{U}}{\partial y} - \mathbf{f}\frac{\partial B_y}{\partial y}, \quad C\frac{\partial \mathbf{U}}{\partial z} = \tilde{C}\frac{\partial \mathbf{U}}{\partial z} - \mathbf{f}\frac{\partial B_z}{\partial z}.$$

If $\operatorname{div}\mathbf{B}$ is identically equal to zero, we obtain the system of equations

$$\frac{\partial \mathbf{U}}{\partial t} + A\frac{\partial \mathbf{U}}{\partial x} + B\frac{\partial \mathbf{U}}{\partial y} + C\frac{\partial \mathbf{U}}{\partial z} = \mathbf{0} \tag{5.4.44}$$

with nonsingular matrices A, B, and C.

While constructing one-dimensional Roe's solver we will use an analogous procedure. We select and then omit from $\Delta\mathbf{E}$ the differences that correspond to the approximation all the terms proportional to $\operatorname{div}\mathbf{B}$. Instead of the formal approximation of matrices $\tilde{A}$, $\tilde{B}$, and $\tilde{C}$, in the Roe solver we shall approximate the physically relevant matrices A, B, and C.

Similarly to the case of the 7×7 MHD solver, we choose the vector **s** as

$$\mathbf{s} = \begin{bmatrix} R \\ U \\ V \\ W \\ \mathcal{I} \\ X \\ Y \\ Z \end{bmatrix} = \begin{bmatrix} \sqrt{\rho} \\ \sqrt{\rho}u \\ \sqrt{\rho}v \\ \sqrt{\rho}w \\ \sqrt{\rho}H \\ B_x/\sqrt{\rho} \\ B_y/\sqrt{\rho} \\ B_z/\sqrt{\rho} \end{bmatrix},$$

where $H = (e + p_0)/\rho$ is the total enthalpy.

Thus,

$$\mathbf{U} = \left[R^2,\ RU,\ RV,\ RW,\ U_5,\ RX,\ RY,\ RZ\right]^{\mathrm{T}},$$

where

$$U_5 = \frac{1}{\gamma}R\mathcal{I} + \frac{\gamma-1}{2\gamma}(U^2+V^2+W^2) - \frac{2-\gamma}{8\pi\gamma}(X^2+Y^2+Z^2)R^2.$$

The vector **E** in terms of the components of the vector **s** acquires the form

$$\mathbf{E} = \begin{bmatrix} RU \\ E_2 \\ UV - \frac{1}{4\pi}XYR^2 \\ UW - \frac{1}{4\pi}XZR^2 \\ U\mathcal{I} - \frac{1}{4\pi}(UX+VY+WZ)XR \\ 0 \\ UY - VX \\ UZ - WX \end{bmatrix},$$

where

$$E_2 = U^2 + \frac{\gamma-1}{\gamma}R\mathcal{I} - \frac{\gamma-1}{2\gamma}(U^2+V^2+W^2) + \frac{2-\gamma}{8\pi\gamma}(X^2+Y^2+Z^2)R^2 - \frac{1}{4\pi}X^2R^2.$$

In the procedure of constructing Roe's solver, from equalities $\Delta\mathbf{E} = A_E\Delta\mathbf{s}$ and $\Delta\mathbf{U} = A_U\Delta\mathbf{s}$ we must determine the matrix $\mathcal{A}$ satisfying $\Delta\mathbf{E} = \mathcal{A}\Delta\mathbf{U}$, whence $\mathcal{A} = A_E A_U^{-1}$. Consider $\Delta\mathbf{U}$ first. We have

$$\frac{1}{\overline{R}}\Delta U_1 = 2\Delta R, \quad \frac{1}{\overline{R}}\Delta U_2 = u\Delta R + \Delta U, \quad \frac{1}{\overline{R}}\Delta U_3 = v\Delta R + \Delta V, \quad \frac{1}{\overline{R}}\Delta U_4 = w\Delta R + \Delta W,$$

$$\frac{1}{\overline{R}}\Delta U_5 = \frac{1}{\gamma}\mathcal{H}\Delta R + \frac{1}{\gamma}\Delta H + \frac{\gamma-1}{\gamma}(u\Delta U + v\Delta V + w\Delta W) - q\Delta R - q_x\Delta X - q_y\Delta Y - q_z\Delta Z,$$

$$\frac{1}{\overline{R}}\Delta U_6 = b_x\Delta R + \Delta X, \quad \frac{1}{\overline{R}}\Delta U_7 = b_y\Delta R + \Delta Y, \quad \frac{1}{\overline{R}}\Delta U_8 = b_z\Delta R + \Delta Z,$$

where

$$u = \frac{\overline{Ru}}{\overline{R}}, \quad v = \frac{\overline{Rv}}{\overline{R}}, \quad w = \frac{\overline{Rw}}{\overline{R}}, \quad \mathcal{H} = \frac{\overline{RH}}{\overline{R}},$$
$$b_x = \frac{\overline{B_x/R}}{\overline{R}}, \quad b_y = \frac{\overline{B_y/R}}{\overline{R}}, \quad b_z = \frac{\overline{B_z/R}}{\overline{R}},$$

and $\overline{f}$ means arithmetic averaging. Besides,

$$q = \frac{2-\gamma}{4\pi\gamma}\Big(\overline{X}^2 + \overline{Y}^2 + \overline{Z}^2 + \frac{\theta_1}{4}(\Delta X)^2 + \frac{\theta_2}{4}(\Delta Y)^2 + \frac{\theta_3}{4}(\Delta Z)^2$$
$$+ \frac{\eta_1}{4\overline{R}}\overline{X}\Delta X\Delta R + \frac{\eta_2}{4\overline{R}}\overline{Y}\Delta Y\Delta R + \frac{\eta_3}{4\overline{R}}\overline{Z}\Delta Z\Delta R\Big),$$
$$q_x = \frac{2-\gamma}{4\pi\gamma}\left(\overline{X}\,\overline{R} + \frac{1-\theta_1}{4}\Delta X\Delta R + \frac{1-\eta_1}{4\overline{R}}(\Delta R)^2\overline{X}\right),$$
$$q_y = \frac{2-\gamma}{4\pi\gamma}\left(\overline{Y}\,\overline{R} + \frac{1-\theta_2}{4}\Delta Y\Delta R + \frac{1-\eta_2}{4\overline{R}}(\Delta R)^2\overline{Y}\right),$$
$$q_z = \frac{2-\gamma}{4\pi\gamma}\left(\overline{Z}\,\overline{R} + \frac{1-\theta_3}{4}\Delta Z\Delta R + \frac{1-\eta_3}{4\overline{R}}(\Delta R)^2\overline{Z}\right),$$

where θ_1, θ_2, θ_3, η_1, η_2, and η_3 are some parameters. Their origin is caused by the presence of the terms containing the factors $\Delta R\Delta X$, $\Delta R\Delta Y$, and $\Delta R\Delta Z$ in the expressions for $\Delta\mathbf{E}$. These factors can be attributed both to the terms proportional to ΔR and ΔX, ΔY, or ΔZ. This results in additional parametrization of entries in the matrices A_E and A_U. From investigation of the 7×7 MHD solver it follows that the preferable values of these parameters are $\theta_1 = \theta_2 = \theta_3 = 1$ and $\eta_1 = \eta_2 = \eta_3 = 2$.

The matrix A_U has the form

$$A_U = \overline{R}\begin{bmatrix} 2 & 0 & 0 & 0 & 0 & 0 & 0 & 0 \\ u & 1 & 0 & 0 & 0 & 0 & 0 & 0 \\ v & 0 & 1 & 0 & 0 & 0 & 0 & 0 \\ w & 0 & 0 & 1 & 0 & 0 & 0 & 0 \\ \frac{1}{\gamma}\mathcal{H}-q & \frac{\gamma-1}{\gamma}u & \frac{\gamma-1}{\gamma}v & \frac{\gamma-1}{\gamma}w & \frac{1}{\gamma} & -q_x & -q_y & -q_z \\ b_x & 0 & 0 & 0 & 0 & 1 & 0 & 0 \\ b_y & 0 & 0 & 0 & 0 & 0 & 1 & 0 \\ b_z & 0 & 0 & 0 & 0 & 0 & 0 & 1 \end{bmatrix}.$$

To construct Roe's solver for the 8×8 MHD equations we must preliminary transform $\Delta\mathbf{E} = [\mathcal{E}_1, \dots, \mathcal{E}_8]^{\mathrm{T}}$ in the following way:

$$\mathcal{E}_1 = \Delta(RU),$$
$$\mathcal{E}_2 = \Delta\left(U^2 + \frac{\gamma-1}{\gamma}R\mathcal{I} - \frac{\gamma-1}{2\gamma}(U^2+V^2+W^2) + \frac{2-\gamma}{8\pi\gamma}(X^2+Y^2+Z^2)R^2\right)$$
$$- \frac{1}{4\pi}\overline{XR}\,\Delta(XR) - \left[\frac{1}{4\pi}\overline{XR}\,\Delta(XR)\right]_-,$$
$$\mathcal{E}_3 = \Delta(UV) - \frac{1}{4\pi}\overline{XR}\,\Delta(YR) - \left[\frac{1}{4\pi}\overline{YR}\,\Delta(XR)\right]_-,$$

$$\mathcal{E}_4 = \Delta(UW) - \frac{1}{4\pi}\overline{XR}\,\Delta(ZR) - \left[\frac{1}{4\pi}\overline{ZR}\,\Delta(XR)\right]_-,$$

$$\mathcal{E}_5 = \Delta(UH) - \frac{1}{4\pi}\overline{XR}[\Delta(UX)+\Delta(UY)+\Delta(UZ)] - \left[\frac{1}{4\pi}(\overline{UX}+\overline{UY}+\overline{UZ})\Delta(XR)\right]_-,$$

$$\mathcal{E}_6 = \left[\overline{\left(\frac{U}{R}\right)}\Delta(XR)\right]_+,$$

$$\mathcal{E}_7 = \Delta(UY) - \overline{XR}\,\Delta\left(\frac{V}{R}\right) - \left[\overline{\left(\frac{V}{R}\right)}\Delta(XR)\right]_-,$$

$$\mathcal{E}_8 = \Delta(UZ) - \overline{XR}\,\Delta\left(\frac{W}{R}\right) - \left[\overline{\left(\frac{W}{R}\right)}\Delta(XR)\right]_-.$$

Here, in accordance with the transformation made in Eqs. (5.4.43)–(5.4.44), the omitted terms are marked by the minus subscript, $[\ ]_-$, and a new term originated in $\mathcal{E}_6$ is marked by the plus subscript, $[\ ]_+$. Note that to obtain in the Roe solver an approximation consistent with the differential case, we should postulate

$$\overline{\left(\frac{U}{R}\right)} \equiv u, \tag{5.4.45}$$

since, in general,

$$u = \frac{\overline{Ru}}{\overline{R}} \neq \overline{\left(\frac{U}{R}\right)} = \bar{u}.$$

It is a single assumption in the particular approach discussed.

After such modification of the expression for $\Delta\mathbf{E}$, the matrix A_E acquires the form

$$A_E = \overline{R}\begin{bmatrix} u & 1 & 0 & 0 & 0 & 0 & 0 & 0 \\ \frac{\gamma-1}{\gamma}\mathcal{H}+q-\frac{B_x b_x}{4\pi} & \frac{\gamma+1}{\gamma}u & \frac{1-\gamma}{\gamma}v & \frac{1-\gamma}{\gamma}w & \frac{\gamma-1}{\gamma} & q_x-\frac{B_x}{4\pi} & q_y & q_z \\ -\frac{B_x b_y}{4\pi} & v & u & 0 & 0 & 0 & -\frac{B_x}{4\pi} & 0 \\ -\frac{B_x b_z}{4\pi} & w & 0 & u & 0 & 0 & 0 & -\frac{B_x}{4\pi} \\ 0 & \mathcal{H}-\frac{B_x b_x}{4\pi} & -\frac{B_x b_y}{4\pi} & -\frac{B_x b_z}{4\pi} & u & -\frac{B_x}{4\pi}u & -\frac{B_x}{4\pi}v & -\frac{B_x}{4\pi}w \\ u b_x & 0 & 0 & 0 & 0 & u & 0 & 0 \\ \frac{B_x}{\rho}v & b_y & -\frac{B_x}{\rho} & 0 & 0 & 0 & u & 0 \\ \frac{B_x}{\rho}w & b_z & 0 & -\frac{B_x}{\rho} & 0 & 0 & 0 & u \end{bmatrix},$$

where

$$\rho = \sqrt{\rho^{\mathrm{L}}\rho^{\mathrm{R}}}, \quad B_x = \overline{B_x}.$$

Then, we can find the eigenvalues λ from the equation

$$\det\,(A_E - \lambda A_U) = 0,$$

or

$$K^2(K^2 - a_{\mathrm{A}}^2)\left\{(K^2 - c_e^2 - \alpha) \times (K^2 - a_{\mathrm{A}}^2) - K^2\left[\frac{\rho}{4\pi}(b_y^2 + b_z^2) + \beta\right]\right\} = 0,$$

where

$$K = u - \lambda, \quad a_{\mathrm{A}} = \frac{|B_x|}{\sqrt{4\pi\rho}},$$

$$c_e^2 = (\gamma - 1)\Big[\mathcal{H} - \tfrac{1}{2}(u^2 + v^2 + w^2) - \frac{1}{4\pi}B_x b_x - \frac{\rho}{4\pi}(b_y^2 + b_z^2)\Big],$$

$$\alpha = \frac{\gamma}{2}(q - q_x b_x - q_y b_y - q_z b_z), \quad \beta = \gamma(b_y\delta_y + b_z\delta_z),$$

$$\delta_y = q_y - \frac{2-\gamma}{4\pi\gamma}b_y\rho, \quad \delta_z = q_z - \frac{2-\gamma}{4\pi\gamma}b_z\rho.$$

Here c_e is the expression for the averaged acoustic speed of sound. Note that this expression has a drawback, since it contains the terms proportional to

$$(B_x^{\mathrm{R}} - B_x^{\mathrm{L}})\left(\frac{B_x^{\mathrm{R}}}{\sqrt{\rho^{\mathrm{R}}}} - \frac{B_x^{\mathrm{L}}}{\sqrt{\rho^{\mathrm{L}}}}\right),$$

which is not, in general, positive. However, we can take special care in order that c_e^2 does not fall from the interval determined by $(c_e^{\mathrm{L}})^2$ and $(c_e^{\mathrm{R}})^2$ (Yee 1989). If this occurs on the left- or right-hand side of this interval, we put $c_e^2 = \min\left((c_e^{\mathrm{L}})^2, (c_e^{\mathrm{R}})^2\right)$ or $c_e^2 = \max\left((c_e^{\mathrm{L}})^2, (c_e^{\mathrm{R}})^2\right)$, respectively.

The expression for α can be rewritten in the form

$$\alpha = \frac{2-\gamma}{32\pi}\Big[\theta_1(\Delta X)^2 + (\theta_1 + \eta_1 - 1)\frac{\overline{X}\Delta X\Delta R}{\overline{R}} + (\eta_1 - 1)\Big(\frac{\overline{X}\Delta R}{\overline{R}}\Big)^2$$
$$+ \theta_2(\Delta Y)^2 + (\theta_2 + \eta_2 - 1)\frac{\overline{Y}\Delta Y\Delta R}{\overline{R}} + (\eta_2 - 1)\Big(\frac{\overline{Y}\Delta R}{\overline{R}}\Big)^2$$
$$+ \theta_3(\Delta Z)^2 + (\theta_3 + \eta_3 - 1)\frac{\overline{Z}\Delta Z\Delta R}{\overline{R}} + (\eta_3 - 1)\Big(\frac{\overline{Z}\Delta R}{\overline{R}}\Big)^2\Big].$$

The eigenvalues of A are equal to u, $u \pm a_{\mathrm{A}}$ and to the four roots of the biquadratic equation

$$K^4 - 2pK^2 + Q = 0, \tag{5.4.46}$$

where

$$2p = c_e^2 + \alpha + a_{\mathrm{A}}^2 + \frac{\rho}{4\pi}(b_y^2 + b_z^2) + \beta, \quad Q = (c_e^2 + \alpha)a_{\mathrm{A}}^2.$$

If $c_e^2 + \alpha \geq 0$ and $(b_y^2 + b_z^2)\rho/4\pi + \beta \geq 0$, the roots of this equation are real and the diagonal matrix composed of the eigenvalues acquires the form

$$\Lambda = \mathrm{diag}[u + a_{\mathrm{f}},\ u + a_{\mathrm{A}},\ u + a_{\mathrm{s}},\ u,\ u,\ u - a_{\mathrm{s}},\ u - a_{\mathrm{A}},\ u - a_{\mathrm{f}}].$$

The quantities a_{f} and a_{s} are the largest and the smallest wave propagation velocities (fast and slow magnetosonic waves) and a_{A} corresponds to Alfvén waves. The two eigenvalues $\lambda = u$ correspond to entropy and magnetic charge waves, respectively. In this case approach assures strict ordering of the eigenvalues. It guarantees the absence of their additional nonphysical degeneration, which is not originally inherent in matrix A. Similarly to the 7×7 solver, we put $\theta_1 = \theta_2 = \theta_3 = 1$. It is easy to check that

$$\beta = \frac{(2-\gamma)(\Delta R)^2}{16\pi}\left[(2-\eta_2)b_y^2 + (2-\eta_3)b_z^2\right],$$

thus giving $\beta \equiv 0$ and $\alpha \geq 0$ for $\eta_1 = \eta_2 = \eta_3 = 2$, since for such values of η and θ,

$$\alpha = \frac{2-\gamma}{32\pi\overline{R}^2}[(\Delta \mathrm{B_x})^2 + (\Delta \mathrm{B_y})^2 + (\Delta \mathrm{B_z})^2].$$

The eigenvectors $\mathbf{r}$ and $\mathbf{l}$ can be determined similarly to the case of truly the one-dimensional 7×7 MHD solver.

Powell et al. (1995) suggested another 8×8 solver that turned out to be not very much suitable for regular calculations, since even the Alfvén velocity a_{A} could be imaginary. Gallice (1996) constructed yet another solver of this type. In the expression for $\mathcal{A}$ obtained for the system of seven equations, Gallice formally added one row and column with a nonzero entry u. The obtained solver is truly free of degeneracies, although it can hardly be attributed to the Roe type. This happens only for $B_x \equiv$ const. Otherwise, we have a CIR-type solver.

Although we believe that our approach based on the only assumption (5.4.45) provides a natural and physically relevant algorithm, one can easily see that the extension of Roe's solver to the modified system is nonunique. Even if special care is taken in order that the eigenvalues are real, the resulting scheme cannot preserve the exact satisfaction of the Hugoniot relations on multidimensional discontinuities. This seems to be the reason Powell et al. (1999) used in numerical simulations a CIR-type solver involving only simple arithmetic averaging of the variables. This approach coincides with that of Brio and Wu (1988). To implement this method, it obviously suffices to know the formulas for eigenvalues and eigenvectors written out in Section 1.3.4.

5.5 Shock-capturing approach and nonevolutionary solutions in MHD

5.5.1 Preliminary remarks. The important subject of discussion is related to the fact that certain initial- and boundary-value problems can be solved nonuniquely using different shocks or different combinations of shocks, whereas physically one would expect only unique solutions. The situation in MHD differs from that in pure gas dynamics of perfect gas, where all entropy-increasing solutions are evolutionary and physically admissible. The term "evolutionary" means that the necessary conditions of well-posedness for the linearized problem of the shock interaction with small disturbances are satisfied. The term "admissible" implies that the shock has a dissipative structure. Such structures can occur if we use a higher-order physical model including the effects of molecular viscosity, heat conduction, and plasma resistivity. In MHD, the condition of entropy increase is necessary, but not sufficient for the shock to be admissible. Only slow and fast MHD shocks were found to be evolutionary, while intermediate (or improper slow) shocks were not and therefore must

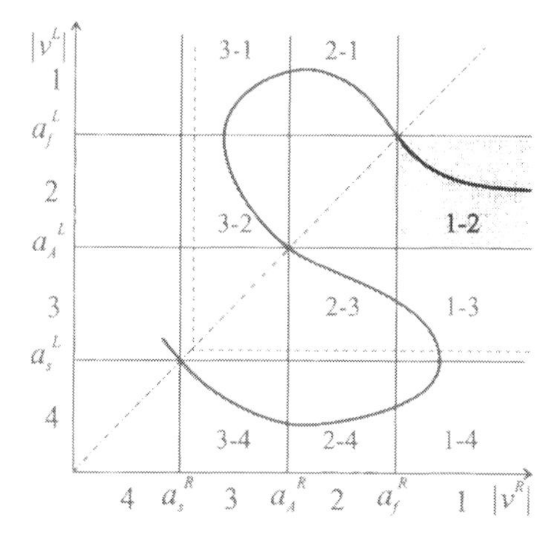

Figure 5.17 MHD Hugoniot curve.

be excluded in ideal MHD (Section 5.3). All MHD shocks are plane-polarized, that is, the magnetic field and the shock normal vectors lie in the same plane both ahead and behind the shock. In contrast to evolutionary shocks, the tangential component of the magnetic field behind a nonevolutionary shock acquires the direction opposite to that which it had ahead of the shock.

There is a simple rule (see the Hugoniot curve in Fig. 5.3) to determine whether the shock is evolutionary. Fast MHD shocks are always super-Alfvénic, that is, the magnitude of the velocity component normal to the shock is larger than the Alfvén velocity $a_{\mathrm{A}} = |B_n|/2\sqrt{\pi\rho}$ both ahead of the shock and behind it. Slow shocks are sub-Alfvénic. Nonevolutionary shocks are trans-Alfvénic. In other words, we call the shock super-Alfvénic, sub-Alfvénic, or trans-Alfvénic if the following relations between the Alfvén numbers $A = |v_n|/a_{\mathrm{A}}$ ahead of and behind the shock are respectively satisfied: $A^{\mathrm{R}} > A^{\mathrm{L}} > 1$, $A^{\mathrm{L}} < A^{\mathrm{R}} < 1$, or $A^{\mathrm{R}} > 1 > A^{\mathrm{L}}$. Thus, the velocities of Alfvén waves that move in the same direction with respect to the medium have opposite signs with respect to the shock on its different sides.

The Hugoniot shock curve mapped onto the evolutionary diagram for the most general case has the form shown in Fig. 5.17 (compare it with Fig. 5.3).

All shocks can be classified by the location of their states ahead of and behind the shock with respect to the intervals marked by the numbers 1 through 4. Germain (1959) showed that only the transition $i \to j$ with $i < j$ satisfy the condition of nondecreasing entropy. Such shocks are represented in Fig. 5.17 by rectangles lying below the bisectrix of the coordinate quadrant. Evolutionary fast and slow shocks are represented by the transitions $1 \to 2$ and $3 \to 4$, respectively. The corresponding rectangles are shaded. In accordance with Section 5.3, the intermediate shocks $2 \to 3$, $1 \to 3$, $2 \to 4$, and $1 \to 4$ are nonevolutionary with respect to Alfvén disturbances. These shocks are nonevolutionary since the number of incoming Alfvén waves is greater than the number of outgoing ones. In this case the solution of the linearized problem on the shock interaction with small Alfvén disturbances has no solution. The shocks $2 \to 3$ and $1 \to 4$ are also nonevolutionary with respect to disturbances in magnetosonic quantities, although the former ones satisfy the evolutionary conditions for the total variety of disturbances. The linearized problem on the interaction of these shocks with small magnetosonic disturbances has no solution for the shock $1 \to 4$. For the shock $2 \to 3$ the solution is nonunique.

The study of the nonevolutionary shock behavior under the action of disturbances is generally performed from the following two different viewpoints. One of them treats these shocks as discontinuities in ideal plasma with negligible electrical resistance. Evolutionarity of plane, oblique MHD shocks to small normal (depending only on time and linear coordinate normal to the shock) disturbances was performed by Akhiezer et al. (1959) and Syrovatskii (1959) (see the discussion of the evolutionary property in Section 5.3). We call the shock oblique if it is not parallel or perpendicular. Those authors found out that all intermediate, or trans-Alfvénic, shocks cannot emit enough small-amplitude plane waves to adjust themselves to small normal disturbances. It was also determined that fast oblique and slow oblique shocks are evolutionary with respect to a general small disturbance but intermediate shocks are not. Akhiezer et al. (1958) also showed that parallel shocks are unstable only if they are trans-Alfvénic, null switch-on or null switch-off. As shown earlier in Section 5.2, $A^{R} > A^{L} = 1$ in switch-on shocks and $A^{L} < A^{R} = 1$ in switch-off shocks. The evolutionary conditions for singular (switch-on and switch-off) shocks are discussed in Section 5.3.3. The limits of switch-on and switch-off shocks for which, respectively, $\mathbf{B}_t^L$ or $\mathbf{B}_t^R$ vanish are called null switch-on and null switch-off shocks. The described nonevolutionary property of certain shocks resulted in the conclusion on their physical impossibility. This viewpoint is also supported by the fact that if we specify the states on the both sides of the nonevolutionary shock as initial conditions for the Riemann problem, there always can be found a solution of this problem involving only evolutionary shocks (Gogosov 1961). The important feature of nonevolutionary shock waves is that they can be represented as a combination of discontinuities and rarefaction waves, whereas evolutionary shocks cannot (Lyubarskii and Polovin 1960; Polovin and Cherkasova 1962). Another argument against existence of nonevolutionary shocks is that they are isolated solutions to the Riemann problem that do not have neighboring solutions corresponding to small deviations in the boundary conditions (Kantrovitz and Petschek 1966).

Another approach treats MHD shocks as thin layers in nonideal plasma with dissipative coefficients (including conductivity) being responsible for the so-called shock structure—the distribution of parameters inside these layers.

Note that dissipative effects are still considered small in this statement compared with convective processes and do not play any substantial role outside of the shock layer. Shocks in the ideal sense can be considered admissible if they are limit solutions of the higher-order dissipative equations under vanishing dissipation. It turns out that in the presence of different dissipative mechanisms both evolutionary and nonevolutionary shocks can have a stationary structure represented by an invariable travelling wave (Germain 1959). This approach was applied by Todd (1964, 1965, 1966) to consideration of transient processes occurring in nonevolutionary shocks. He showed that intermediate, trans-Alfvénic, shocks are unstable to normal perturbations in transverse and/or magnetosonic quantities.

The evolution of switch-on and switch-off shocks, which are the limiting cases of fast and slow shocks, was investigated both analytically and numerically by Todd (1966) and Chu and Taussig (1967).

Rotational, or Alfvén, discontinuities always have nonstationary structure. It can be shown, however, that in the general case their width increases in time as $\sqrt{t}$ (Landau and Lifshitz 1984). On the other hand, if we consider nonstationary solutions at large times when the characteristic linear scales are of the order of t, rotational waves are relatively narrow and still can be considered as discontinuities.

The structure of nonevolutionary MHD shocks is generically nonplanar (Todd 1965). Let us write out the equation for the out-of-plane z component of the magnetic field in the one-dimensional case (see Eq. (5.1.36)):

$$\frac{\partial B_z}{\partial t} = -\frac{\partial}{\partial x}\left(B_z v_x - B_x v_z - \nu_m \frac{\partial B_z}{\partial x}\right). \tag{5.5.1}$$

If the boundary conditions outside of the shock structure satisfy the Hugoniot conditions, then the shock is planar at $x = \pm\infty$. Thus, integrating Eq. (5.5.1) over the shock structure we obtain

$$\frac{\partial I_z}{\partial t} = 0, \quad I_z = \int_{-\infty}^{\infty} B_z \, dx. \tag{5.5.2}$$

Similarly, from the momentum equation we can obtain

$$\frac{\partial}{\partial t}\int_{-\infty}^{\infty} \rho v_z \, dx = 0. \tag{5.5.3}$$

For plane evolutionary shocks the integral is identically equal to zero. If nonevolutionary shocks have nonplanar structure, the integrals (5.5.2) and (5.5.3) are constants (Todd 1964). If we fix the value of the integral characterizing the shock structure (Todd 1964, 1965, 1966; Kennel, Blandford, and Wu 1990), the nonplanar structure will also be fixed.

Wu (1990) and Hada (1994) suggested that amplitudes of purely dissipative unsteady linear waves must be included in the solution of the evolutionary problem for MHD shocks. It is obvious, however, that purely dissipative waves die out within the length of the order of the shock thickness. Thus, their amplitudes do not occur in the boundary conditions and they must be disregarded in the solution of the evolutionary problem (see also the discussion in Markovskii 1998a, 1998b).

Myong and Roe (1997), by developing the admissibility condition, showed that intermediate shocks are necessary to ensure that the planar Riemann problem is well-posed. They showed that in the general three-dimensional case the solution of the Riemann problem is not necessarily unique. In agreement with Kennel et al. (1990), it was shown that this solution depends not only on the right and left states, but also on the associated internal structure of the shock (on the integral I_z).

Important results concerning nonstationary processes occurring inside the structure of nonevolutionary shocks have lately been obtained by Wu (1990), Kennel et al. (1990), and Wu and Kennel (1992a, 1992b, 1993). By means of numerical experiments it has been shown that the structures of nonevolutionary shocks are fairly stable with respect to splitting under the action of Alfvén perturbations. They can absorb these disturbances changing either their internal structure, if it is described by a free parameter, or intensity. A nonevolutionary shock can also split into an evolutionary shock and a nonevolutionary shock of another type. On absorbing sufficiently large (in the sense of the integral over time or a space coordinate) Alfvén disturbances, such a secondary nonevolutionary shock approaches by its parameters to an Alfvén discontinuity with rotation of the magnetic field vector by

180°. The amount of Alfvén disturbances necessary for this to occur is related to the value of dissipative coefficients determining the width of the structure and, consequently, the ability of a nonevolutionary shock wave to absorb the incoming disturbances. In the limit of vanishing dissipation, infinitesimal Alfvén disturbances suffice to make the nonevolutionary shock indistinguishable from a rotational discontinuity. In this sense the above-mentioned results concerning nonevolutionary shocks do not contradict the classical results related to ideal dissipation-free MHD. The exception can make such cases where, as the dissipative coefficients tend to zero, more and more stiff restrictions are imposed on the smallness of incoming Alfvén disturbances.

Below we consider several topics related to the behavior of nonevolutionary shocks and the results of the numerical experiments performed on the basis of MHD equations simplified in the case of small angles of propagation relative to the upstream magnetic field (Rogister 1971; Cohen and Kulsrud 1974; Kennel et al. 1990). The choice of these equations makes the interpretation of the results of the investigation of nonevolutionary shocks more descriptive.

5.5.2 Simplified MHD equations and related discontinuities.

Let us consider the propagation of one-dimensional MHD waves with all the quantities depending only on x and t. If the angle between the magnetic field vector and the x-axis is small, then in the absence of dissipation the velocities of two characteristic families turn out to be close. One of these families corresponds to rotational waves and the other corresponds to fast, if $a_A > c_e$, or slow, if $a_A < c_e$, magnetosonic waves. Following Kennel et al. (1990) and Wu and Kennel (1993), we shall consider only the first of the indicated cases.

It turns out that for small-amplitude waves one can obtain a system of equations for the disturbances related to the above-mentioned characteristic families. They have the form

$$\frac{\partial B_i}{\partial t} + \frac{\partial}{\partial x}\left[(c_0 + \alpha B_\tau^2)B_i\right] = 0, \quad i = y, z, \tag{5.5.4}$$

where $B_\tau^2 = B_y^2 + B_z^2$ and c_0 and α are constants. If $a_A > c_e$, then $\alpha > 0$. System (5.5.4) describes plane-polarized magnetosonic waves with the invariable direction of $\mathbf{B}_\tau$ that correspond to the characteristic velocity equal to $c_0 + 3\alpha B_\tau^2$ and rotational (Alfvén) waves with the characteristic velocity $c_0 + \alpha B_\tau^2$. The system has a divergent form and gives rise to the following relations on discontinuities:

$$\{\alpha B_\tau^2 B_i\} + (c_0 - W)\{B_i\} = 0. \tag{5.5.5}$$

These relations admit discontinuities of two types, that is, rotational discontinuities and shock waves,

$$\{B_\tau\} = 0, \quad W' = \alpha B_\tau^2; \tag{5.5.6}$$

$$B_z^{\mathrm{R}} = B_z^{\mathrm{L}}, \quad W' = \frac{\{\alpha B_y^3\}}{\{B_y\}}, \tag{5.5.7}$$

where $W' = W - c_0$ and a proper orientation of axes y and z is chosen.

Consider the evolutionary properties of the shocks. In Figs. 5.18a,b we show the plot of αB_y^3. According to (5.5.6), the velocity of a rotational discontinuity and that of Alfvén

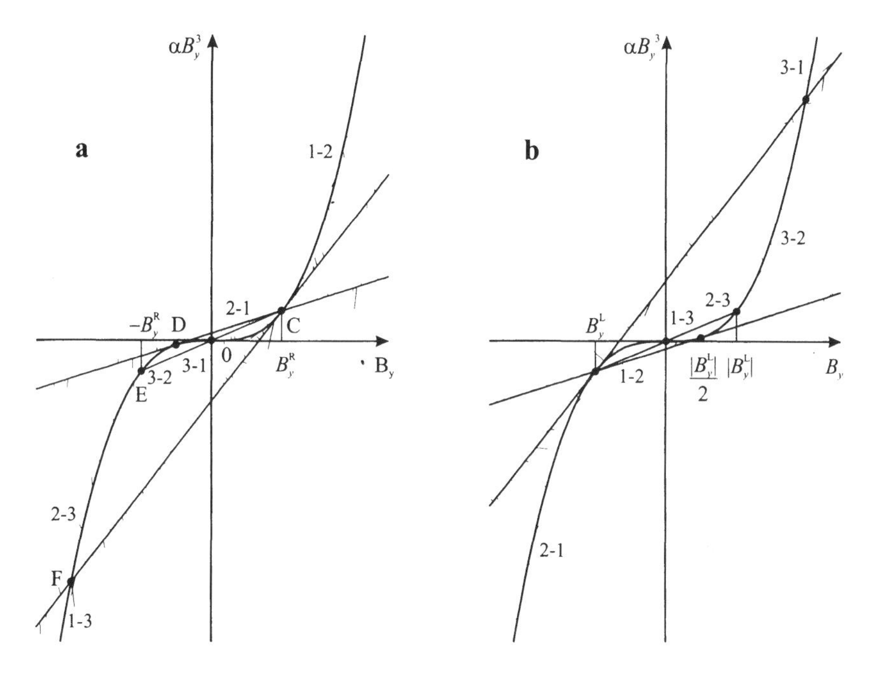

Figure 5.18 Shock transitions for the simplified MHD system.

waves are characterized by the slope of the secant drawn through the origin and the point of the plot with the fixed B_y. The velocity of small magnetosonic disturbances is determined by the slope of the tangent to the plot at the point under consideration. According to (5.5.7), the velocity W' of the shock is equal to the slope of the secant drawn through the plot points with abscissae B_y^R and B_y^L. This fairly easily allows us to indicate the types of the states ahead of and behind various shocks. This is done in Fig. 5.18a for fixed B_y^R and in Fig 5.18b for fixed B_y^L.

The comparison of the discontinuities shown in Fig. 5.18a with Fig. 5.17 shows that those parts of the Hugoniot curve are absent in the former figure that correspond to gas velocities with respect to the shock that are smaller or comparable with the slow magnetosonic velocity a_s. Besides, the case shown in Fig. 5.18a corresponds to the part of Fig. 5.17 lying above and to the right of the dotted lines in the latter figure. Recall that only the shocks with increasing the state number agree with the condition of nondecreasing entropy.

5.5.3 Shock structure in solutions of the simplified system.

If dissipative processes are taken into account, Eq. (5.5.4) acquires the Cohen–Kulsrud–Burgers (CKB) form (Kennel et al. 1990)

$$\frac{\partial B_i}{\partial t} + \frac{\partial}{\partial x}\left[(c_0 + \alpha B_\tau^2)B_i\right] = \nu\frac{\partial^2 B_i}{\partial x^2}, \quad i = y, z, \tag{5.5.8}$$

where ν is a certain combination of dissipative coefficients regarded as a constant quantity. Consider a solution depending on $\xi = -x + Wt$. In addition, let us assume that $\xi = -\infty$

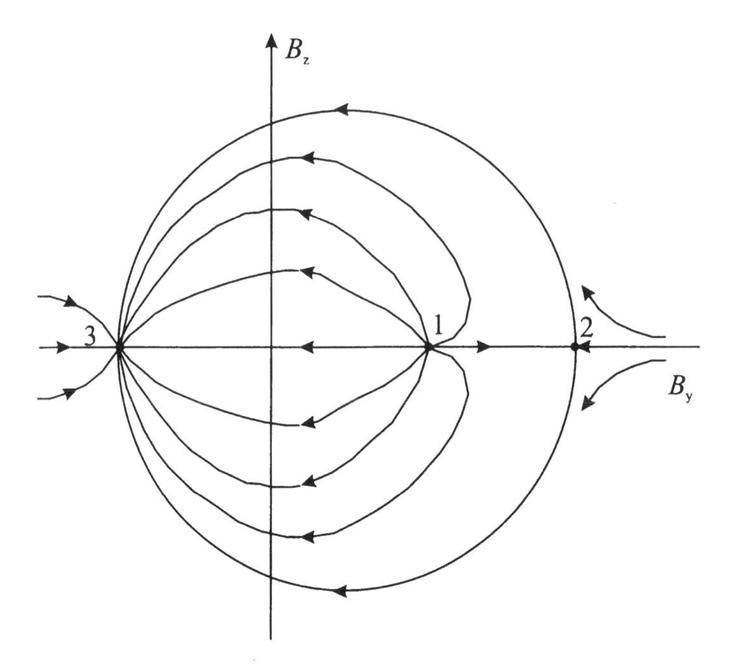

Figure 5.19 Integral curves of the simplified MHD system.

and $\xi = \infty$ correspond to the states ahead of and behind the shock, respectively. Taking into account the fact that all derivatives are zero for $\mathbf{B} = \mathbf{B}^{\mathrm{R}}$, from (5.5.8) we obtain

$$\nu \frac{dB_i}{d\xi} = W'(B_i - B_i^{\mathrm{R}}) - \alpha B_i \left[B_\tau^2 - (B_\tau^{\mathrm{R}})^2\right]. \tag{5.5.9}$$

Let $B_y^{\mathrm{R}} > 0$ and $B_z^{\mathrm{R}} = 0$. Then the integral curves of (5.5.9) have the form shown in Fig. 5.19. For a fixed state 3, each of the states 1 or 2 can be chosen as a state ahead of the shock front. As seen from Fig. 5.18b,

$$B_{y(1)}^{\mathrm{R}} \leq \frac{|B_{y(3)}^{\mathrm{L}}|}{2}, \qquad \frac{|B_{y(3)}^{\mathrm{L}}|}{2} \leq B_{y(2)} \leq |B_{y(3)}^{\mathrm{L}}|.$$

Relationships between $B_{y(1)}$, $B_{y(2)}$, and $B_{y(3)}$ depend on the velocity $W' = W - c_0$. The embraced numbers in the subscripts denote the state types.

The shock $1 \to 2$ is a fast evolutionary shock. The nonevolutionary shock $2 \to 3$ has a structure represented by one of the two symmetric integral curves. A continuous variety of structures can be assigned to the nonevolutionary shock $1 \to 3$. As mentioned above, it is convenient to choose the integral $I_z = \int_{-\infty}^{\infty} B_z d\xi$ as such a parameter. Then it is obvious from Fig. 5.19 that for the shock $1 \to 3$ we have

$$|I_z| \leq I_z^*,$$

where I_z^* is the value of I for the shock $2 \to 3$ corresponding to the same velocity W'.

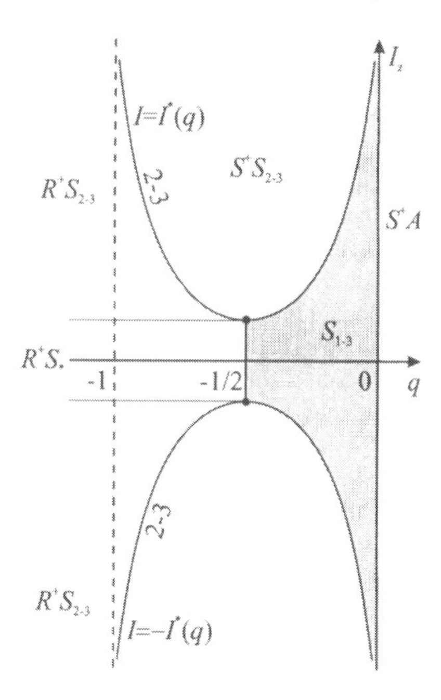

Figure 5.20 Dependence of I_z on q.

Instead of W', we can choose $q = B_y^{\mathrm{R}}/B_y^{\mathrm{L}}$ as a parameter characterizing the shock. This quantity, as seen from Fig. 5.18b, varies between 0 and -1. For $q = -0.5$, the shock velocity is minimal. The values $q < -0.5$ and $q > -0.5$ correspond to the $2 \to 3$ and $1 \to 3$ shocks, respectively. The dependence of I_z on q obtained numerically (Kennel et al. 1990) for $B_y^{\mathrm{L}} = \text{const}$ is qualitatively shown in Fig. 5.20.

The shadowed part of the figure corresponds to admissible values of I_z for $1 \to 3$ shocks. Tending of the curves to infinity at $q \to 0$ and $q \to -1$ is stipulated by the fact that the $2 \to 3$ shock, as seen from Fig. 5.18b, tends to a switch-on shock and a rotational discontinuity in these cases. In addition, the width of the structure and, hence, the value of I_z^* tend to zero as functions of q.

5.5.4 Nonstationary processes in the structure of nonevolutionary shock waves. From Eq. (5.5.8) written for B_z we obtain the conservation law

$$\frac{d}{dt}\int_{x_1}^{x_2} B_z dx + Q_z|_{x=x_2} - Q_z|_{x=x_1} = 0, \tag{5.5.10}$$

where x_1 and x_2 are constants and

$$Q_z = (c_0 + \alpha B_\tau^2)B_z - \nu\frac{\partial B_z}{\partial x} \tag{5.5.11}$$

is the flux of B_z through $x = \text{const}$. If B_z vanishes sufficiently fast as $x \to \pm\infty$, we obtain Eq. (5.5.2) and

$$I_z = \int_{-\infty}^{\infty} B_z dx = \text{const}. \tag{5.5.12}$$

If we consider large-scale phenomena, the main contribution to the flux Q_z is given by the convection (the first term in Eq. (5.5.11)) related to the transfer of B_z at the Alfvén wave velocity. For intermediate waves, which are trans-Alfvénic, this means that possible fluxes of B_z are directed toward the structure of the $2 \to 3$ and $1 \to 3$ waves under consideration. The signs of the relative velocities of Alfvén waves is seen from Figs. 5.18b and 5.17. The change in I_z occurs due to incoming Alfvén disturbances. If no disturbances of this kind approach the shock structure, the integral of B_z over the structure of a nonevolutionary shock remains constant.

As mentioned earlier, various numerical experiments exhibited high stability of the nonevolutionary shock structure. We shall outline some of these results.

Outside of a segment of the x-axis some constant vectors $\mathbf{B}^{0\mathrm{L}}$ and $\mathbf{B}^{0\mathrm{R}}$ directed along or opposite the y-axis were specified with the assumption that $\mathbf{B}^{0\mathrm{R}} > 0$. On the segment itself, some distributions of $B_y^0(x)$ and $B_z^0(x)$ were prescribed (see the details in Kennel et al. 1990 and Wu and Kennel 1993). If $B_y^{0\mathrm{L}} > 0$ and $B_z^0(x) \equiv 0$, then in accordance with the classical results, depending on the relation between $B_y^{0\mathrm{R}}$ and $B_y^{0\mathrm{L}}$, either a fast shock and a Riemann wave occurred in the solution. If $B_z^0 \neq 0$, then exactly the same fast wave is formed accompanied by a rotational wave accumulating the component B_z. These waves move at different velocities and the rotational wave expands in time due to dissipative processes.

If $B_y^{0\mathrm{L}} < 0$, the originating wave structures depend not only on $q^0 = B_y^{0\mathrm{R}}/B_y^{0\mathrm{L}}$ but also on I_z^0 representing the integral of B_z^0 over x. For definiteness, let $I_z^0 > 0$. If q^0 and I_z^0 belong to the plot of $I_z^*(q)$ or to the shaded region of Fig. 5.20, then with increasing time the solution reaches the asymptotics consisting of one nonevolutionary shock. If a point with coordinates q^0, I_z^0 does not belong to the indicated variety and $|I_z^0| > I_z^*(-0.5)$, the $2 \to 3$ intermediate shock is formed that corresponds to the given $I_z = I_z^0$. Besides, a fast shock or Riemann wave goes forward with B_y behind it acquiring such a value of B_y^{R} that is required by the curve $I_z^*(q)$ for the $2 \to 3$ shock at $I_z = I_z^0$.

If $q^0 < -0.5$ and $|I_z^0| < I_z^*(-0.5)$, a fast Riemann wave goes forward with $B_y = |B_y^{0\mathrm{L}}|/2$ behind it. The nonevolutionary shock with $q = -0.5$ that forms in this case has the velocity equal to the fast magnetosonic velocity in the state ahead of this wave (see Fig. 5.18b). This state can be denoted as $1 = 2$, while the state behind it is 3. This wave, similarly to the $1 \to 3$ waves has a variety of structures with an arbitrary I_z from the interval $|I_z| < I_z^*(-0.5)$. In the considered problem the wave of the described type forms with the structure corresponding to the initial value $I_z = I_z^0$. For large t the variety of solutions described above is shown in the plane (I_z, q) (see Fig. 5.20). We used the following designations: S^+ and R^+ are fast shock and Riemann waves; S_{2-3} and S_{1-3} are intermediate waves; S_* is the intermediate wave $1 = 2 \to 3$ corresponding to $q = -0.5$.

Thus, if the $1 \to 3$ intermediate wave is formed, its parameters are determined by I_z^0 and q_0 belonging to the shadowed part of Fig. 5.20. If the $2 \to 3$ or $1 = 2 \to 3$ waves are formed, they are determined only by I_z^0. The necessary change in B_y that occurs in the fast wave is determined by the curve representing these waves in Fig. 5.20 (the curves $I_z = \pm I_z^*(q)$ and the segment of the straight line $q = -0.5$).

If we consider a similar problem in which the angle between the vectors $\mathbf{B}^{0\mathrm{R}}$ and $\mathbf{B}^{0\mathrm{L}}$ differs from 180°, the solution, as shown by Wu and Kennel (1993), preserves many features of the solutions described above. This especially clear if $\mathbf{B}^{0\mathrm{R}}$ and $\mathbf{B}^{0\mathrm{L}}$ form an angle close to 180°. In this case the integral I_z over the structure of nonevolutionary shocks will be time-dependent. If this variation is not very fast, the structure of a nonevolutionary shock adjusts itself to the current value of I_z. In this case the evolution of the solution is determined by the variation of I_z.

Consider, for example, the case in which the state (q^0, I_z^0) determined by the initial data belongs to the shadowed part of Fig. 5.20. This means that initially a $1 \to 3$ shock is formed. As I_z increases, the point representing this shock moves parallel to the I_z-axis of the (q, I_z)-plane. In Fig. 5.19 this case is characterized by the transition from the lower to higher integral curves connecting the points 1 and 3. This process lasts until the $1 \to 3$ integral curve splits into the $1 \to 2$ and $2 \to 3$ curves. This implies the emission of the fast wave $1 \to 2$ and the transfer of the point of the (q, I_z)-plane representing the nonevolutionary shock to the curve $I_z^*(q)$ corresponding to the $2 \to 3$ waves and lying in the region with $-1 < q < -0.5$. Further increase in I_z leads to the motion of the point representing the nonevolutionary wave $2 \to 3$ upward along the curve $I_z^*(q)$. In this case q approaches -1 and the discontinuity becomes close to a rotational one. Its velocity also tends to the velocity of a rotational discontinuity.

If the directions of $\mathbf{B}_\tau^{0L}$ and $\mathbf{B}_\tau^{0R}$ constitute an angle differing from 180° by 10 or more degrees, then, although all possible shocks have co-linear $\mathbf{B}_\tau^{L}$ and $\mathbf{B}_\tau^{R}$, a wave expanding in time is formed in which all rotation of $\mathbf{B}_\tau$ is located. There is no rotation of $\mathbf{B}_\tau$ outside this wave. The value of I_z increases in time, and this wave finally becomes a rotational wave. The width of this nonstationary, nonevolutionary wave increases as $\sqrt{t}$. At the initial stages of its evolution, this wave is close to nonevolutionary waves $1 \to 3$ or $2 \to 3$ and its width is substantially larger than the width of the Alfvén discontinuity if they originated simultaneously. More detailed description of the solution behavior for various initial data is given by Wu and Kennel (1993).

5.5.5 Numerical experiments based on the full set of MHD equations.

Recently, the properties of trans-Alfvénic shocks have been studied in numerical simulations by Markovskii and Skorokhodov (2000a, 2000b). In those papers the behavior of nonevolutionary shocks was considered in the framework of full MHD equations including the terms accounting for dissipative processes. The case with $a_A < c_e$ was discussed. As compared with the simplified equations previously considered in Section 5.5.2, the role of the $1 \to 3$ wave was played by the $2 \to 4$ wave, while the $2 \to 3$ wave preserved its properties. The numerical experiments performed in the above-mentioned papers dealt with the response of the stationary structures of $2 \to 3$ and $2 \to 4$ waves to Alfvén disturbances coming to them from the forward side. If the simplified system had been written out for the case $a_A < c_e$, as a result of the interaction we would have obtained slow MHD waves (in contrast with fast ones occurring if $a_A > c_e$) propagating in the same direction with a nonevolutionary shock wave. However, since the full MHD system was solved, in addition to the mentioned slow wave all other waves formed which could be formed in that case. Their amplitudes turned out to be substantially smaller than that of the slow wave.

The changes occurring with nonevolutionary shock waves under the action of Alfvén disturbances of invariable sign qualitatively correspond to those describe in the previous subsection on the basis of the simplified system. In particular, $2 \to 3$ waves without delay responded to incoming disturbances by the change in their parameters (the point representing this wave moved along the curve $2 \to 3$ in Fig. 5.20) and the radiation of waves of other types. The $2 \to 4$ wave started to generate outgoing waves only after the time interval necessary for I_z to acquire a critical value. This also qualitatively agrees with the description of this phenomenon on the basis of the simplified system.

Markovskii and Skorohodov (2000a) also performed a numerical modelling of the $2 \to 3$ wave interaction with the incoming periodic Alfvén wave. If the value of I_z does not belong to the interval $[-I^*(-\frac{1}{2}), I^*(\frac{1}{2})]$, the result can be qualitatively described as the motion of the point representing this nonevolutionary wave along the curve $2-3$ in Fig. 5.20. If in the process of oscillations the values of I_z fall in the mentioned interval, the nonevolutionary wave acquires the type $2 \to 3 = 4$ (similar to the type $1 = 2 \to 3$ described in the previous subsection) and the variation in I_z occurs with invariable $q = -1/2$.

The numerical experiment showed that all changes of the nonevolutionary wave were reversible even in those cases for which the oscillations resulted in very large absolute values of I_z and the $2 \to 3$ wave became practically indistinguishable from a rotational discontinuity. Alongside the periodic variations of I_z, in agreement with Fig. 5.20, a periodic variation of q took place. In the numerical experiment this resulted in the radiation of a periodic slow wave together with other waves with substantially smaller amplitude.

Another series on numerical experiments (Markovskii and Skorohodov 2000b) relates to the response of nonevolutionary waves with an abrupt change in dissipative coefficients of the medium. It is worth mentioning that, strictly speaking, instead of I_z we must plot $\nu^{-1}I_z$ along the vertical axis in Fig. 5.20. Here ν is some generalized dissipative coefficient determining the width of the shock wave structure. The results of the before-mentioned numerical experiments with the full system of MHD equations can be qualitatively interpreted as follows. Under the variation in ν, the point representing a nonevolutionary wave on the $(q,\ \nu^{-1}I_z)$-plane changes its second coordinate. As a result, the disintegration of the initial wave can take place in accordance with the scenario described in the previous subsection.

In connection with the above results, it is necessary to make a few comments concerning stability of intermediate shock waves. These waves are fairly stable with respect to Alfvén disturbances if the latter are estimated by the value of the integral I_z supplied by the incoming wave into the structure of the intermediate shock (if $1 \to 3$ or $2 \to 4$ waves are considered, it is assumed that the value of $|I_z|$ is not equal to a maximum permissible value). In this case, if the variation in I_z is sufficiently small, the change in parameters of the intermediate shock will also be small. This ensures the smallness of the emitted magnetosonic wave. At the same time, an intermediate wave is unstable if we characterize the incoming Alfvén waves only by their amplitudes (as it is done in the classical case), regardless of the variation in I_z. The smallness of incoming Alfvén waves by itself does not guarantee the smallness of the variation in parameters of the intermediate wave and emitted magnetosonic wave.

In fact, trans-Alfvénic shocks can exist in nature, although only as transient time-dependent structures. They can be observed in the interplanetary plasma where $Re_m \gg 1$, although not so frequently as evolutionary shocks, in agreement with the observations of Chao et al. (1993).

In the approximation of ideal MHD, nonevolutionary shocks are not simply unstable. Their disintegration into evolutionary jumps occurs under infinitesimal perturbation within infinitesimal time. On the other hand, numerical viscosity (including the presence of finite conductivity) and numerical dispersion of a numerical scheme make such intermediate structures existent for a certain time interval before disintegration. It is very important to realize that this interval has nothing to do with the real interval of existence of intermediate waves in non-ideal plasma and is grid- and numerical scheme-dependent. If numerical viscosity and/or resistivity are substantial, intermediate structures can exist for a long time depending on the amount of dissipation.

In conclusion, let us once again emphasize that all the described effects are related to the presence of dissipative effects ensuring the existence of the shock structures with nonzero width. On the other hand, of great importance is the absence of a stationary structure in rotational discontinuities and the possibility of an infinite growth of the width in nonevolutionary shock structures accompanied by an appropriate variation of their parameters. That is why, even in the problems with high Reynolds numbers we must realize that, in fact, nonevolutionary shocks are initially formed that differ from rotational discontinuities. It is especially important to have this in mind if we use shock-capturing numerical methods in which the shock width does not correspond to that in reality.

5.5.6 Numerical disintegration of a compound wave. In this section we shall consider, as a numerical example, the ideal MHD Riemann problem with the initial data consisting of two constant states lying to the right and to the left of the centerline of the computational domain. If the problem is solved as a strictly coplanar one, a compound wave (see Section 5.3.4) can originate (Brio and Wu 1988). For large magnetic Reynolds numbers, such a wave suffers disintegration under the action of Alfvén disturbances and is not realizable in physical problems. As mentioned earlier, the peculiarity of MHD is that there exist discontinuities that are nonevolutionary only with respect to Alfvén (rotational) disturbances. That is why, if a strictly coplanar problem is considered (velocity and magnetic field vectors lie in the same plane with the shock normal and the system of MHD equations includes only two vector components) the construction of the solution is possible both with evolutionary and nonevolutionary shock waves. The solution in this case is nonunique. Depending on the applied method, a nonevolutionary solution in the form of a compound wave can be realized in numerical calculations, since in the phase space it lies closer to initial conditions. If the full set of three-dimensional MHD equations is solved and a small transversal disturbance is added to the magnetic field vector, at sufficiently large times this compound wave results in a rotational discontinuity and a slow shock. This means that the compound wave is unstable against normal perturbations in transverse quantities and is nonevolutionary in three dimensions (Barmin et al. 1996).

The necessity of three-dimensional consideration of MHD Riemann problem was also noted by Dai and Woodward (1994a). As noted in Section 5.4.2, the numerical method developed by these authors directly includes rotational discontinuities as components of the approximate solution to the MHD Riemann problem, even if the problem is plane polarized. This allows them to use only evolutionary elementary solutions in the construction of the global solution to the multidimensional MHD problem. In what follows we investigate the process of the flow reconstruction leading, finally, to the disintegration of the nonevolutionary solution.

We shall return now to the shock adiabatic diagram shown in Fig. 5.4. Let the vector components on the both sides of the shock lie in the same plane (xy). The domains corresponding to the evolutionary shocks are shaded. The interval AB ($3 \to 4$ shocks) corresponds to slow shock waves, with $B_\tau^{\mathrm{L}} = 0$ at the point B (a switch-off shock). The other shaded rectangle corresponds to fast shock waves. The interval BCD of the Hugoniot curve corresponds to nonevolutionary shock waves. The shocks from the parts of the Hugoniot curve that lie above the bisectrix passing through A, D, and E cannot have structure because they correspond to rarefaction shocks and are not entropy-consistent. The interval BCD of the Hugoniot curve is nonevolutionary with respect to rotational (out-of-plane) disturbances. The structure of the shocks belonging the portion BC ($2 \to 4$ shocks) of the shock curve is nonunique. The interval BC corresponds to solutions nonevolutionary with respect to rotational disturbances, while it is stable with respect to magnetosonic (xy) ones. As indicated in Section 5.4.1, the consequence is that, if a coplanar flow is considered (z-components of the magnetic field and of the momentum are omitted), a compound wave can exist with the junction point at C.

Two relations (5.2.5) and (5.2.7) on the discontinuity used to determine the amplitudes of rotational disturbances describe the conservation of fluxes for the z-component of the momentum and the y-component of the electric field in the coordinate system attached to the jump. Due to Faraday's law, this means that the flux of the z-component of the magnetic field is also continuous along the axis x. It is easy to check that we have only one outgoing Alfvén wave on the shocks belonging to the interval BC. Thus, only one dependent variable exists for these shock waves in the linearized problem of the interaction with rotational disturbances. For this reason, the system of two independent equations cannot be met. As noted earlier, ρw and B_z will be accumulated in the layer δx containing a nonevolutionary jump, i.e., a nonstationary process is to be expected. This process serves to fix a unique solution in the form of a dissipative shock structure. It is not close to a quasi-stationary one in the case of the ideal MHD model, since the linearized problem has no solution. A self-similar solution that occurs as a final result of this process will consist of the evolutionary slow shock and a rotational discontinuity. Since they leave the Jouget point C, these waves will propagate at different speeds and will never meet again in the nonevolutionary jump after such disintegration. The compound wave disintegration occurs instantaneously in ideal MHD. We would want to observe this process under the application of the shock-capturing numerical method. It can be expected that numerical dissipation will slow down the process of disintegration.

To perform this study, we choose the same ideal MHD Riemann problem as that considered by Brio and Wu (1988). This is a Cauchy problem with the initial data consisting of constant states $\mathbf{u}^{\mathrm{L}}$ and $\mathbf{u}^{\mathrm{R}}$. Let

$$\rho^{\mathrm{L}} = 1, \quad u^{\mathrm{L}} = 0, \quad v^{\mathrm{L}} = 0, \quad p^{\mathrm{L}} = 1, \quad B_y^{\mathrm{L}} = 1,$$
$$\rho^{\mathrm{R}} = 0.125, \quad u^{\mathrm{R}} = 0, \quad v^{\mathrm{R}} = 0, \quad p^{\mathrm{R}} = 0.1, \quad B_y^{\mathrm{R}} = -1.$$

In addition, $B_x \equiv 0.75$ and $\gamma = 2$. Note that the magnetic field component B_y changes its sign, or rotates to the angle π across the initial jump.

Brio and Wu (1988) tested different numerical schemes for this problem in the purely coplanar statement. All of them gave qualitatively the same picture of the flow, but Roe's scheme produced much sharper discontinuities with the absence of spurious oscillations. Note that the case with $\gamma = 2$ allows us to construct a genuine Roe-type scheme for MHD flows.

It is worth emphasizing that Alfvén discontinuities cannot originate in the numerical solution of this problem even if we formally use three space coordinates. There is simply no

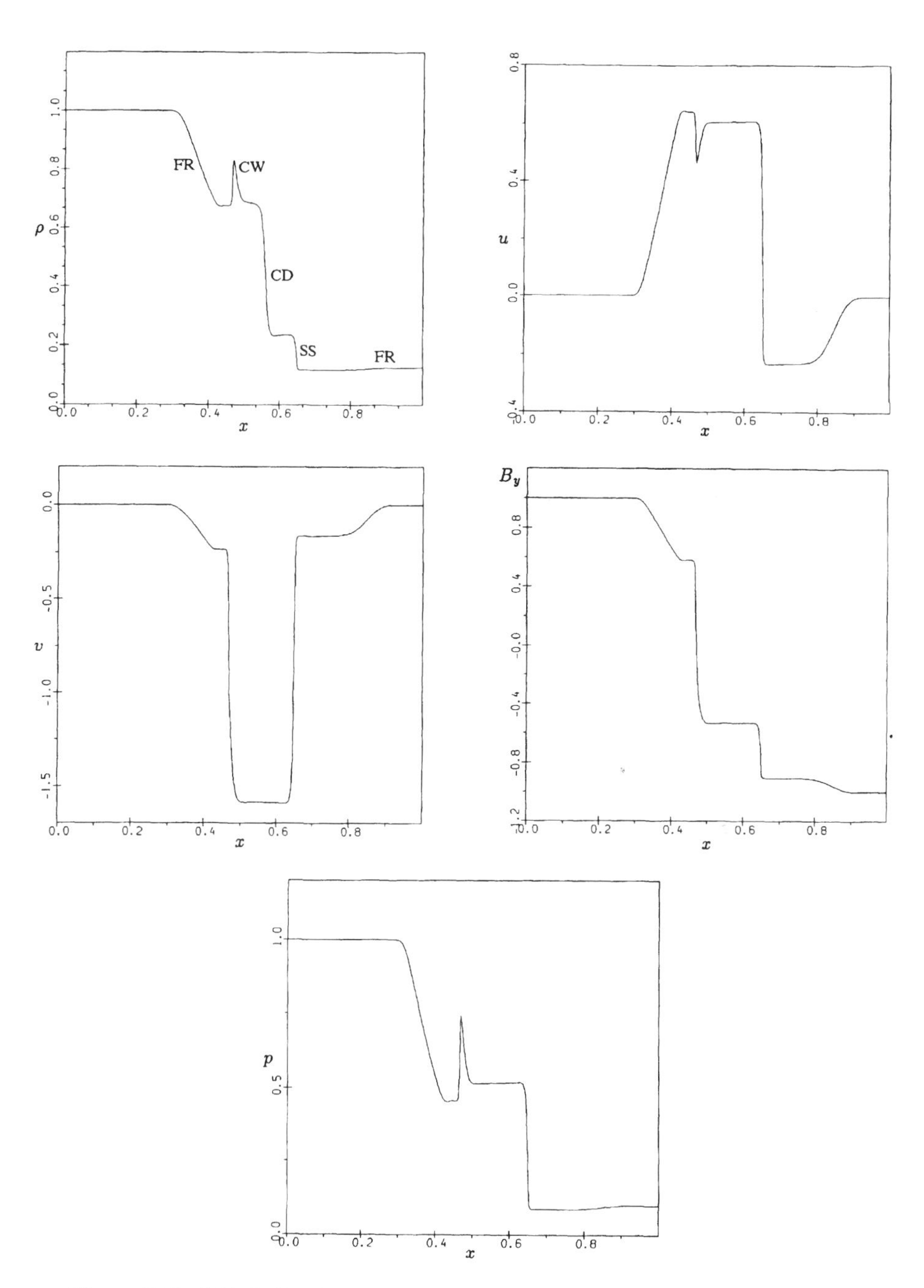

Figure 5.21 Self-similar distributions of dependent variables (coplanar problem).

mechanism to change the value of w and B_z. As a result, since evolutionary shocks cannot change the sign of the transverse component of the magnetic field, this appears to be done by a nonevolutionary compound wave. Figure 5.21 shows the self-similar distributions of the density, the velocity components u and v, the magnetic field component B_y, and the thermal pressure obtained for the coplanar problem by the Lax–Friedrichs-type numerical scheme applied to MHD by Barmin et al. (1996).

To attain the second order of accuracy in time we use a predictor–corrector procedure. For simplicity, in contrast to the approach described in Section 5.4.4, the central difference approximation of space derivatives is adopted for the predictor step.

"Predictor"

$$u_i^{k+1/2} = u_i^k - \left(f_{i+1}^k - f_{i-1}^k\right) \frac{\Delta t}{4\Delta x}. \tag{5.5.13}$$

This approximation itself is known to be unstable. That is why the corrector step must be added. For this purpose we adopt a linear distribution of quantities in the computational cells and use the slope limiting procedure to determine them on the cell boundaries. In this calculation, the above-mentioned procedure is applied to the primitive flow variables. We use for this purpose the interpolation formulas (5.4.24)–(5.4.27) with $\eta = 1/3$ and $\omega = 2$ for all variables except for the thermal pressure and the normal component u of the velocity vector. This choice seems to be preferable for the TVD Lax–Friedrichs-type scheme. Interpolation of characteristic variables can be considered as too time consuming for this scheme. Roe's method will be more suitable if this type of interpolation is used.

After the reconstruction procedure, the modified Lax–Friedrichs step follows:

"Corrector"

$$\mathbf{u}_i^{k+1} = \mathbf{u}_i^k - \left(\tilde{f}_{i+1/2}^{k+1/2} - \tilde{f}_{i-1/2}^{k+1/2}\right) \frac{\Delta t}{\Delta x}, \tag{5.5.14}$$

$$\tilde{f}_{i+1/2} = \frac{1}{2}\left[\mathbf{f}\left(\mathbf{u}_{i+1/2}^{\mathrm{R}}\right) + \mathbf{f}\left(\mathbf{u}_{i+1/2}^{\mathrm{L}}\right) - \hat{\mathbf{R}}_{i+1/2}\left(\mathbf{u}_{i+1/2}^{\mathrm{R}} - \mathbf{u}_{i+1/2}^{\mathrm{L}}\right)\right], \tag{5.5.15}$$

where $\hat{R}_{i+1/2}$ is a diagonal matrix with the same elements on its diagonal equal to the spectral radius r (the maximum of eigenvalue magnitudes) of the Jacobian matrix $\partial\mathbf{f}/\partial\mathbf{u}$.

The number of grid points is 800 with $\Delta x = 1$ and $\Delta t = 0.2$ (CFL number ~ 0.8). This solution is shown after 400 time steps. Initial discontinuity is located in the middle of the computational interval. One can see fast rarefaction waves FR, a contact discontinuity CD, a slow shock wave SS, and a slow compound wave CW. The results are in a good agreement with those obtained by the method of Brio and Wu (1988). The second order of accuracy in time and space, high-resolution Lax–Friedrichs-type scheme (5.5.13)–(5.5.15) gives meanwhile a substantial simplification of the numerical algorithm, since we need not calculate the eigenvectors of the Jacobian matrices.

Brio and Wu (1988) adduced a number of arguments to justify the existence of the before-mentioned compound wave. In fact, such a wave will disintegrate as nonevolutionary one under the action of transverse disturbances. Disturbances of this type cannot manifest themselves in the Riemann problem considered as a strictly coplanar one. In nature, of course, they can fairly easily exist unless the flow is artificially restricted in one of the tangential directions.

If we consider a three-dimensional problem with parameters depending only on t and x and give a uniform disturbance $B_z = 0.1$, just to make the proper equations work, the solution

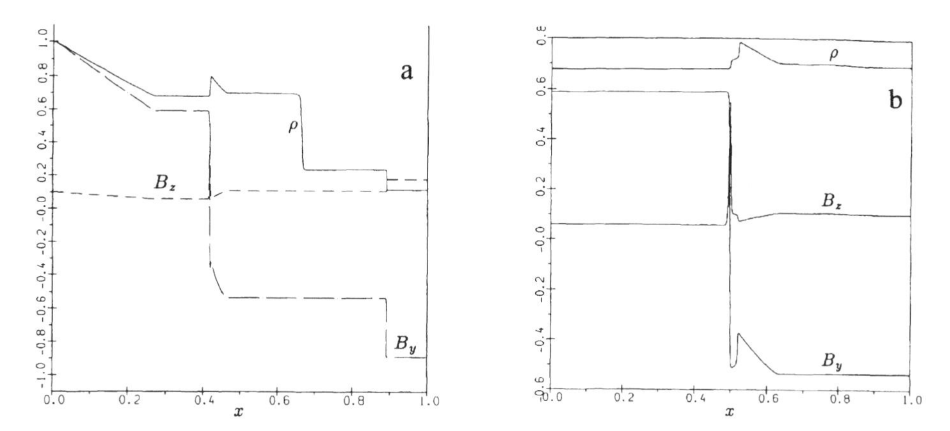

Figure 5.22 Quantity distributions after 8000 (a) and 16,000 (b) time steps (three-dimensional problem).

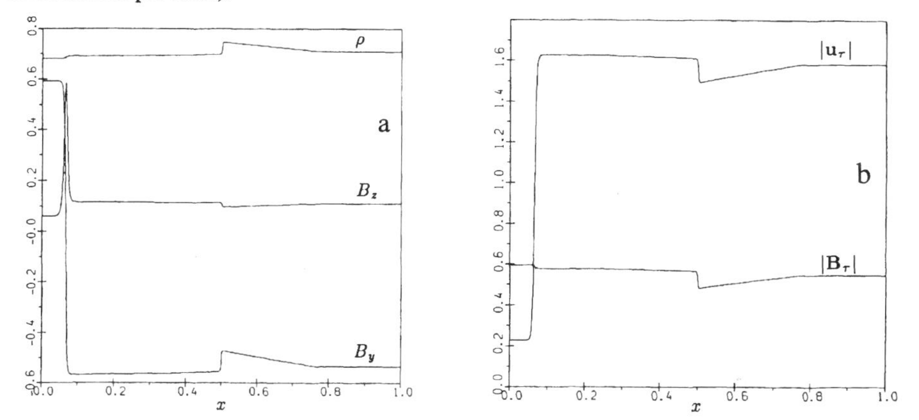

Figure 5.23 Quantity distributions after 58,000 time steps (three-dimensional problem).

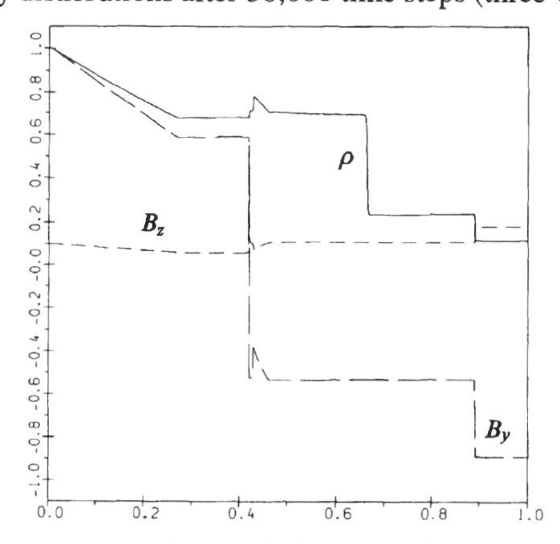

Figure 5.24 The same shock transitions as in Figure 5.22a for smaller numerical viscosity.

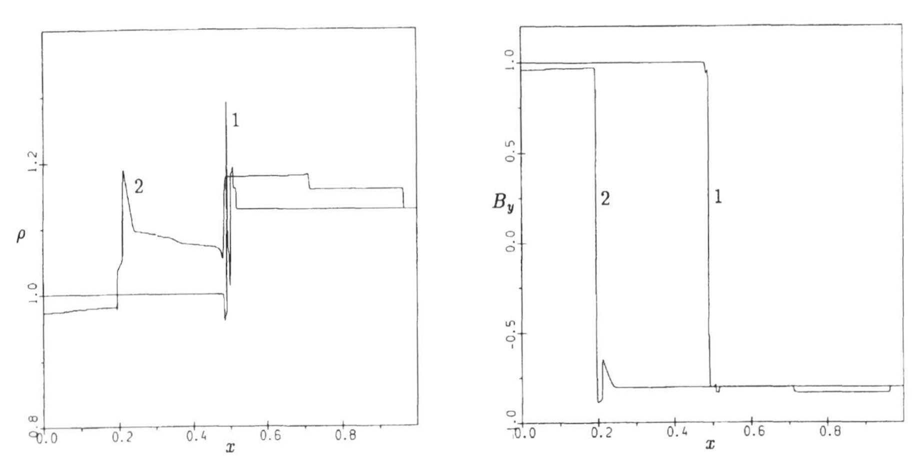

Figure 5.25 Nonevolutionary shock decomposition. Density (a) and B_y (b) distributions after 200 (1) and 6000 (2) time steps.

substantially changes and another configuration is realized. The component B_z gradually increases and a time-dependent shock-like structure finally splits from the compound wave. In the infinite time limit this structure will transform into the rotational discontinuity. The slow shock remaining after this disintegration is no longer a Jouget jump and it starts interaction with the rarefaction wave in accordance with the path AB of the Hugoniot curve (Fig. 5.4). The intensity of this shock decreases until the rarefaction wave vanishes.

The quantity variation in time is presented in Figs. 5.22 and 5.23, where the distributions are presented of the density, B_y and B_z components of the magnetic field in the vicinity of the zone in question. The figures are placed in the floating computational window with the number of points equal to 6000 for Fig. 5.22a and 4000 for Figs. 5.22b and 5.23. The initial mesh size is $\Delta x = 2$ and CFL number is 0.8. After the first 8000 steps the results were interpolated into the grid with $\Delta x = 1$ and 4000 mesh point window was placed around the compound wave. On the plots, the window size is normalized. To look inside the parameter distribution in the structure close to a rotational discontinuity, the values of the tangential components of the velocity vector $\mathbf{v}$ and magnetic field $\mathbf{B}$ are presented in Fig. 5.23b.

We mentioned earlier that stability of the dissipative structure corresponding in ideal MHD to the intermediate shock depends on the magnitude of the perturbation and on its duration. For this reason, if we consider numerical solutions to the ideal MHD equations, stability of the compound wave depends on the mesh size, that is, on the introduced amount of numerical viscosity. This property is illustrated in Fig. 5.24, where the distributions of ρ, B_y, and B_z are shown for the considered Riemann problem exactly at the same moment of time and in the same computational region as in Fig. 5.22a, but for the number of cells equal to 12,000 and $\Delta x = 1$. It is apparent that the compound wave in this case has already been disintegrated. Thus, the solution turns out to be grid-dependent. This must be taken into account by those dealing with ideal MHD problems, especially if an adaptive mesh refinement is used.

Consider another illustration of the described instabilities. Let us choose the following set of right and left parameters:

$$\rho^{\mathrm{L}} = 1, \quad u^{\mathrm{L}} = 0, \quad v^{\mathrm{L}} = 0, \quad p^{\mathrm{L}} = 1, \quad B_y^{\mathrm{L}} = 1, \quad B_z^{\mathrm{L}} = 0.1,$$
$$\rho^{\mathrm{R}} = 1.12895, \quad u^{\mathrm{R}} = -0.117236, \quad v^{\mathrm{R}} = -1.753716,$$
$$p^{\mathrm{R}} = 1.12895, \quad B_y^{\mathrm{R}} = -0.8, \quad B_z^{\mathrm{R}} = 0.1.$$

In addition, $B_x \equiv 1$ and $\gamma = 5/3$. If the value of B_z is not taken into account, these boundary conditions correspond to a nonevolutionary rarefaction shock. In Fig. 5.25, the distributions of density and B_y are shown after 200 (1) and 6000 (2) time steps. The number of grid points is 8000, $\Delta x = 0.5$, and CFL $= 0.8$. Immediate disintegration of the rarefaction shock is evident, whereas the resulting structure contains an intermediate shock unstable to rotational disturbances. It is apparent that after about 6000 time steps another nonevolutionary jump splits from the initially formed compound wave. This nonevolutionary shock finally transforms into a rotational discontinuity.

The results presented show that, in order to obtain truly evolutionary solutions, one should be very careful in reducing the dimension of MHD problems. This should also be taken into account in the application of shock-capturing methods for ideal MHD flows. A self-similar evolutionary solution to MHD equations will be obtained automatically in the infinite time limit if the full set of them is used and rotations of the magnetic field vector are admitted. Development of numerical schemes that might facilitate the disintegration of nonevolutionary shocks is discussed for the model problem in the paper by Myong and Roe (1998).

5.6 Strong background magnetic field

In this section we shall consider a modification to the MHD system in the conservation-law form that allows us to perform high-resolution calculations in the presence of a strong background magnetic field. This is important if we deal with astrophysical problems. The potential magnetic field of a white dwarf or a neutron star is so strong that it can lead to certain difficulties in *numerical* simulation due to round-off errors accumulated if we operate with large numbers. The approach based on the elimination of the background dipole magnetic field can be easily implemented in the Lax–Friedrichs, the CIR, or Roe's schemes. It was first applied by Tanaka (1994) in the framework of the CIR-type scheme for MHD equations (Brio and Wu 1988). It turns out that Roe's scheme requires only a slight modification to become applicable to the modified system to be derived below.

If we assume that the background field $\mathbf{B}_0$ is potential, then

$$\operatorname{curl} \mathbf{B}_0 = \mathbf{0} \tag{5.6.1}$$

and, hence,

$$\operatorname{curl} \mathbf{B}_0 \times \mathbf{B}_0 = \mathbf{0}. \tag{5.6.2}$$

Relation (5.6.2) can be transformed into

$$\frac{1}{2}\nabla B_0^2 - (\mathbf{B}_0 \cdot \nabla)\,\mathbf{B}_0 = \mathbf{0}. \tag{5.6.3}$$

This vector relation should be subtracted from the momentum equation (5.1.41). The resulting equation will be written out below when we present the final form of the modified MHD system.

The continuity equation (5.1.40) is not affected by $\mathbf{B}_0$. The Faraday equation (5.1.33) cannot be simplified on its right-hand side. To transform the energy equation, let us introduce the notation

$$\mathbf{B}_1 = \mathbf{B} - \mathbf{B}_0, \tag{5.6.4}$$

$$e_1 = e - \frac{\mathbf{B}_1 \cdot \mathbf{B}_0}{4\pi} - \frac{B_0^2}{8\pi} = \frac{p}{\gamma - 1} + \frac{\rho \mathbf{v}^2}{2} + \frac{\mathbf{B}_1^2}{8\pi}. \tag{5.6.5}$$

In this notation we obtain

$$\frac{\partial e_1}{\partial t} + \frac{\mathbf{B}_0}{4\pi}\frac{\partial \mathbf{B}_1}{\partial t} + \operatorname{div}\left[\left(e_1 + p + \frac{\mathbf{B}_1^2}{8\pi}\right)\mathbf{v} + \left(\frac{\mathbf{B}_1 \cdot \mathbf{B}_0}{2\pi} + \frac{B_0^2}{4\pi}\right)\mathbf{v} \right.$$
$$\left. - \frac{\mathbf{B}_1(\mathbf{B}_1 \cdot \mathbf{v})}{4\pi} - \frac{\mathbf{B}_1(\mathbf{B}_0 \cdot \mathbf{v})}{4\pi} - \frac{\mathbf{B}_0(\mathbf{B}_1 \cdot \mathbf{v})}{4\pi} - \frac{\mathbf{B}_0(\mathbf{B}_0 \cdot \mathbf{v})}{4\pi}\right] = 0. \tag{5.6.6}$$

From the Faraday equation (5.1.33) we have

$$\frac{\mathbf{B}_0}{4\pi}\frac{\partial \mathbf{B}_1}{\partial t} = \frac{\mathbf{B}_0}{4\pi}\operatorname{curl}(\mathbf{v} \times \mathbf{B}) = \frac{\mathbf{B}_0}{4\pi}\left[\operatorname{curl}(\mathbf{v} \times \mathbf{B}_0) + \operatorname{curl}(\mathbf{v} \times \mathbf{B}_1)\right]. \tag{5.6.7}$$

Using the vector analysis formula

$$\operatorname{div}(\mathbf{a} \times \mathbf{b}) = \mathbf{b} \cdot \operatorname{curl}\mathbf{a} - \mathbf{a} \cdot \operatorname{curl}\mathbf{b},$$

we can write out

$$\mathbf{B}_0 \cdot \operatorname{curl}(\mathbf{v} \times \mathbf{B}_0) - (\mathbf{v} \times \mathbf{B}_0) \cdot \operatorname{curl}\mathbf{B}_0 = \operatorname{div}[(\mathbf{v} \times \mathbf{B}_0) \times \mathbf{B}_0]. \tag{5.6.8}$$

Note that the second term on the left-hand side of Eq. (5.6.8) is identically zero. Similarly,

$$\mathbf{B}_0 \cdot \operatorname{curl}(\mathbf{v} \times \mathbf{B}_1) = \operatorname{div}[(\mathbf{v} \times \mathbf{B}_1) \times \mathbf{B}_0]. \tag{5.6.9}$$

Using the formula

$$\mathbf{a} \times (\mathbf{b} \times \mathbf{c}) = \mathbf{b}(\mathbf{a} \cdot \mathbf{c}) - \mathbf{c}(\mathbf{a} \cdot \mathbf{b}),$$

we obtain

$$(\mathbf{v} \times \mathbf{B}_0) \times \mathbf{B}_0 = \mathbf{B}_0(\mathbf{v} \cdot \mathbf{B}_0) - \mathbf{v}B_0^2, \tag{5.6.10}$$

$$(\mathbf{v} \times \mathbf{B}_1) \times \mathbf{B}_0 = \mathbf{B}_1(\mathbf{v} \cdot \mathbf{B}_0) - \mathbf{v}(\mathbf{B}_0 \cdot \mathbf{B}_1). \tag{5.6.11}$$

Combining Eqs. (5.6.6)–(5.6.11), we arrive at

$$\frac{\partial e_1}{\partial t} + \operatorname{div}\left[\left(e_1 + p + \frac{B_1^2}{8\pi} + \frac{\mathbf{B}_1 \cdot \mathbf{B}_0}{4\pi}\right)\mathbf{v} - \frac{\mathbf{B}(\mathbf{v} \cdot \mathbf{B}_1)}{4\pi}\right] = 0. \tag{5.6.12}$$

This form of the energy equation is rather convenient, since it is well known that this equation is frequently responsible for the accuracy loss if we resolve e for pressure.

Thus, the final form of the transformed system is

$$\frac{\partial \tilde{\mathbf{U}}}{\partial t} + \frac{\partial \tilde{\mathbf{E}}}{\partial x} + \frac{\partial \tilde{\mathbf{F}}}{\partial y} + \frac{\partial \tilde{\mathbf{G}}}{\partial z} = \mathbf{0}, \tag{5.6.13}$$

where

$$\tilde{\mathbf{U}} = \left[\rho, \rho u, \rho v, \rho w, e_1, B_{1x}, B_{1y}, B_{1z}\right]^{\mathrm{T}},$$

$$\tilde{\mathbf{E}} = \begin{bmatrix} \rho u \\ \rho u^2 + p + \dfrac{\mathbf{B}^2}{8\pi} - \dfrac{\mathbf{B}_0^2}{8\pi} - \dfrac{B_x^2}{4\pi} + \dfrac{B_{0x}^2}{4\pi} \\ \rho uv - \dfrac{B_x B_y}{4\pi} + \dfrac{B_{0x}B_{0y}}{4\pi} \\ \rho uw - \dfrac{B_x B_z}{4\pi} + \dfrac{B_{0x}B_{0z}}{4\pi} \\ \left(e_1 + p + \dfrac{\mathbf{B}_1^2}{8\pi} + \dfrac{\mathbf{B}_1 \cdot \mathbf{B}_0}{4\pi}\right)u - \dfrac{B_x}{4\pi}(\mathbf{v} \cdot \mathbf{B}_1) \\ 0 \\ uB_y - vB_x \\ uB_z - wB_x \end{bmatrix},$$

$$\tilde{\mathbf{F}} = \begin{bmatrix} \rho v \\ \rho uv - \dfrac{B_x B_y}{4\pi} + \dfrac{B_{0x}B_{0y}}{4\pi} \\ \rho v^2 + p + \dfrac{\mathbf{B}^2}{4\pi} - \dfrac{\mathbf{B}_0^2}{4\pi} - \dfrac{B_y^2}{4\pi} + \dfrac{B_{0y}^2}{4\pi} \\ \rho uw - \dfrac{B_y B_z}{4\pi} + \dfrac{B_{0y}B_{0z}}{4\pi} \\ \left(e_1 + p + \dfrac{\mathbf{B}_1^2}{8\pi} + \dfrac{\mathbf{B}_1 \cdot \mathbf{B}_0}{4\pi}\right)v - \dfrac{B_y}{4\pi}(\mathbf{v} \cdot \mathbf{B}_1) \\ vB_x - uB_y \\ 0 \\ vB_z - wB_y \end{bmatrix},$$

$$\tilde{\mathbf{G}} = \begin{bmatrix} \rho w \\ \rho uw - \dfrac{B_x B_z}{4\pi} + \dfrac{B_{0x}B_{0z}}{4\pi} \\ \rho vw - \dfrac{B_y B_z}{4\pi} + \dfrac{B_{0y}B_{0z}}{4\pi} \\ \rho w^2 + p + \dfrac{\mathbf{B}^2}{8\pi} - \dfrac{\mathbf{B}_0^2}{8\pi} - \dfrac{B_z^2}{4\pi} + \dfrac{B_{0z}^2}{4\pi} \\ \left(e_1 + p + \dfrac{\mathbf{B}_1^2}{8\pi} + \dfrac{\mathbf{B}_1 \cdot \mathbf{B}_0}{4\pi}\right)w - \dfrac{B_z}{4\pi}(\mathbf{v} \cdot \mathbf{B}_1) \\ wB_x - uB_z \\ wB_y - vB_z \\ 0 \end{bmatrix}.$$

In order to construct the Roe-type method for the new system, it is necessary to represent the increment of $\Delta\tilde{\mathbf{E}}$ in terms of $\Delta\mathbf{E}$ in the initial form of the MHD system (5.1.43). If we want to remain in the framework of the 7×7 one-dimensional system, the sixth equation for B_{1x} must be omitted. It is easy to check that

$$\Delta\tilde{\mathbf{E}} = R_E \Delta\mathbf{E},$$

where

$$R_E = \begin{bmatrix} 1&0&0&0&0&0&0 \\ 0&1&0&0&0&0&0 \\ 0&0&1&0&0&0&0 \\ 0&0&0&1&0&0&0 \\ 0&0&0&0&1&-\frac{B_{0y}}{4\pi}&-\frac{B_{0z}}{4\pi} \\ 0&0&0&0&0&1&0 \\ 0&0&0&0&0&0&1 \end{bmatrix}.$$

Similarly,

$$\Delta\mathbf{U} = R_U^{-1} \Delta\tilde{\mathbf{U}},$$

where

$$R_U^{-1} = \begin{bmatrix} 1&0&0&0&0&0&0 \\ 0&1&0&0&0&0&0 \\ 0&0&1&0&0&0&0 \\ 0&0&0&1&0&0&0 \\ 0&0&0&0&1&\frac{B_{0y}}{4\pi}&\frac{B_{0z}}{4\pi} \\ 0&0&0&0&0&1&0 \\ 0&0&0&0&0&0&1 \end{bmatrix}.$$

Thus,

$$\Delta\tilde{\mathbf{E}} = R_E A_{\text{Roe}} \Delta\mathbf{U} = R_E A_{\text{Roe}} R_U^{-1} \Delta\tilde{\mathbf{U}} = R_E \Omega_{\text{R}} \Lambda \Omega_{\text{L}} R_U^{-1} \Delta\tilde{\mathbf{U}},$$

where A_{Roe}, Ω_{R}, and Ω_{L} are the matrices obtained in Section 5.4.3.

If we denote

$$\tilde{\Omega}_{\text{R}} = R_E \Omega_{\text{R}}, \quad \tilde{\Omega}_{\text{L}} = \Omega_{\text{L}} R_U^{-1},$$

we obtain the required relation

$$\Delta\tilde{\mathbf{E}} = \tilde{\Omega}_{\text{R}} \Lambda \tilde{\Omega}_{\text{L}} \Delta\tilde{\mathbf{U}},$$

which allows us to construct the Roe-type solver for the modified system if we know it for the initial system. It is apparent that the eigenvalues occurring in both solvers coincide. Besides,

$$\tilde{\Omega}_{\text{R}} \tilde{\Omega}_{\text{L}} = R_E \Omega_{\text{R}} \Omega_{\text{L}} R_U^{-1} = R_E R_U^{-1} = I.$$

Thus, normalization of the eigenvector system is also preserved.

5.7 Elimination of numerical magnetic charge

5.7.1 Preliminary remarks. As mentioned earlier in Section 1.3.4, the condition of the magnetic charge absence (or the divergence-free condition for magnetic field) is automatically satisfied in a strict mathematical sense if we start with some magnetic charge-free initial conditions. This is not so, numerically, in the multidimensional case owing to the approximation errors due to discretization, especially in the cases for which we evaluate derivatives of discontinuous functions.

Another obvious reason of the magnetic charge accumulation is the application of the one-dimensional Riemann problem solvers to solution of a multidimensional MHD system. This introduces additional, sometimes very large, errors, and an additional mechanism is necessary to eliminate the extraneous magnetic charge. This requirement, in the sense of the Gauss law in the form of an integral over the surface S of the whole computational volume τ, reads

$$\oint_S \mathbf{B} \cdot \mathbf{n}\, dS = 0. \tag{5.7.1}$$

Note that, since the Lorentz force is equal to

$$\mathbf{f} = \frac{1}{4\pi} \operatorname{div} \left(\mathbf{BB} - \frac{\mathbf{B}^2}{2} \hat{\mathbf{I}} \right),$$

for $\operatorname{div} \mathbf{B} \neq 0$, this force acquires a spurious component parallel to the magnetic field vector

$$f_B = \mathbf{f} \cdot \mathbf{B} = \frac{1}{4\pi} (\mathbf{B} \cdot \mathbf{B}) \operatorname{div} \mathbf{B}, \tag{5.7.2}$$

which can destroy the correct physical behavior of the flow (see Brackbill and Barnes 1980).

Another important consequence of the magnetic charge accumulation is the difficulty in implementation of the multidimensional operator splitting in which a fully multidimensional step is achieved by combining a series of one-dimensional steps in alternating directions. The condition $\operatorname{div} \mathbf{B} = 0$ introduces cross-coupling between these spatial directions (Zachary et al. 1994). The latter problem can be resolved if we approximate the momentum equation (1.3.21) in the nonconservative form. This fact was first pointed out by Brackbill and Barnes (1980). Note that the component f_B of the magnetic force does not occur in the momentum equation (1.3.21) written out in the form of Newton's law. Thus, if we solve the mass, energy, and Faraday induction equations in the conservation-law form and approximate momentum from the nonconservative vector equation, the spurious magnetic charge cannot affect the flow. We must admit, however, that application of the modified system is not suitable for some numerical schemes and, more importantly, can lead to nonphysical results, since the conservation of momentum is disregarded.

5.7.2 Application of the vector potential. There exist several approaches in managing the divergence-free problem. The most straightforward one is to rewrite the induction equation in terms of the vector potential $\mathbf{A}$. This potential is defined by the formula

$$\mathbf{B} = \operatorname{curl} \mathbf{A}. \tag{5.7.3}$$

Combining Eq. (5.7.3) with Eq. (5.1.33), we obtain

$$\frac{\partial \mathbf{A}}{\partial t} = \mathbf{v} \times \mathbf{B}. \tag{5.7.4}$$

Note that the vector potential **A** is defined up to addition of the gradient of an arbitrary function Φ.

The equation for the vector potential becomes especially simple if we solve an axisymmetric problem with the toroidal component of the magnetic field equal to zero. It is apparent that the vector potential **A** has only one component in this case. In fact, in the cylindrical coordinate system z, ϕ, r we have

$$B_r = -\frac{\partial A_\phi}{\partial z}, \qquad B_z = \frac{1}{r}\frac{\partial r A_\phi}{\partial r}.$$

Thus, Eq. (5.7.4) reduces to

$$\frac{\partial A_\phi}{\partial t} + v_z \frac{\partial A_\phi}{\partial z} + \frac{v_r}{r}\frac{\partial r A_\phi}{\partial r} = 0.$$

Although this approach exactly ensures the divergence-free condition, since

$$\operatorname{div}\operatorname{curl}\mathbf{A} \equiv 0,$$

we must approximate the second derivatives of the vector potential in the Lorentz force term of the momentum equation. This makes impossible the construction of the numerical scheme based only on the application of standard, characteristically consistent, high-resolution approximations providing nonoscillatory behavior of functions in the vicinity of discontinuities. As indicated by Evans and Hawley (1988), it will not be surprising if certain high-resolution schemes designed for hydrodynamics produce anomalous spikes in second derivatives (see also Ostapenko 1998). This results in the current $\mathbf{j} = c/4\pi \operatorname{curl}\operatorname{curl}\mathbf{A}$ reversals, which can be a major problem in attempts to use the vector potential in MHD simulations.

5.7.3 The use of an artificial scalar potential.

Another method incorporates an artificial scalar potential φ to clean off the magnetic charge from the computational region (see Brackbill and Barnes 1980). It is easy to notice that if, after several time steps, we obtained such a magnetic field that $\operatorname{div}\mathbf{B} \neq 0$, then we can introduce φ satisfying the Poisson equation

$$\Delta\varphi = -\operatorname{div}\mathbf{B} \tag{5.7.5}$$

and after that modify the magnetic field vector **B** using a simple formula

$$\mathbf{B}' = \mathbf{B} + \nabla\varphi.$$

It is apparent that

$$\operatorname{div}\mathbf{B}' = \operatorname{div}\mathbf{B} + \operatorname{div}\nabla\varphi \equiv 0.$$

The Poisson equation (5.7.5) can be solved numerically after discretization, for example in a two-dimensional Cartesian coordinate system x and y,

$$\frac{\varphi_{l+2,m} - 2\varphi_{l,m} + \varphi_{l-2,m}}{4(\Delta x)^2} + \frac{\varphi_{l,m+2} - 2\varphi_{l,m} + \varphi_{l,m-2}}{4(\Delta y)^2} = -\left[\frac{B_{x,l+1,m} - B_{x,l-1,m}}{2\Delta x}\right] - \left[\frac{B_{y,l,m+1} - B_{y,l,m-1}}{2\Delta y}\right]. \tag{5.7.6}$$

If the field structure is periodic on some space, we can use the fast Fourier transform, which provides an exact solution to $\varphi_{l,m}$. Otherwise, the method of straightforward integration of the discretized Poisson equation can be used (see Tanaka 1994). In the latter case numerical solution can require up to 30% of the total CPU time (Ryu, Jones, and Frank 1995). Note also that the boundary conditions for Eq. (5.7.6) can be rather complicated, like nonreflecting conditions, and that it is usually necessary to approximate nonsmooth functions φ and $\mathbf{B}$. Such approximation can give rise to spurious oscillations of unknown functions in the vicinity of discontinuities. Regardless of these drawbacks, this method of artificial scalar potential is frequently used in MHD calculations and has proved to be rather effective for modelling of a series of astrophysical problems (Tanaka 1994; Nakajima and Hanawa 1996; Ryu et al. 1998)

5.7.4 Application of the modified MHD system. An ideologically different approach was described by Powell et al. (1999) (see also Aslan 1993; Powell 1994; Powell et al. 1995; Aslan 1999). The main idea of this approach is to construct a method that can be easily incorporated in existing high-resolution numerical methods of the Godunov type, in particular those based on Roe's approximate Riemann solver. To understand the possibilities of this approach, it is necessary to determine the conservative form of the MHD equations resulting from its quasilinear form (1.3.20)–1.3.23) in which the condition $\operatorname{div}\mathbf{B} = 0$ is disregarded. It easy to see that we arrive in this case at the following nonhomogeneous system:

$$\frac{\partial}{\partial t}\begin{bmatrix}\rho\\ \rho\mathbf{v}\\ e\\ \mathbf{B}\end{bmatrix} + \operatorname{div}\begin{bmatrix}\rho\mathbf{v}\\ \rho\mathbf{v}\mathbf{v} + p_0\hat{\mathbf{I}} - \dfrac{\mathbf{B}\mathbf{B}}{4\pi}\\ (e + p_0)\,\mathbf{v} - \dfrac{\mathbf{B}(\mathbf{v}\cdot\mathbf{B})}{4\pi}\\ \mathbf{v}\mathbf{B} - \mathbf{B}\mathbf{v}\end{bmatrix} = -\begin{bmatrix}0\\ \dfrac{\mathbf{B}}{4\pi}\\ \dfrac{\mathbf{v}\cdot\mathbf{B}}{4\pi}\\ \mathbf{v}\end{bmatrix} \times \operatorname{div}\mathbf{B}. \tag{5.7.7}$$

It is worth noting that the source term is proportional to $\operatorname{div}\mathbf{B}$ and that at the partial differential equation level only the terms that are equal to zero are added to the standard conservation form of the MHD equations (5.1.40)–(5.1.42). The resulting system (i) has a nonstrictly conservative form that is favorable for the elimination of magnetic charge (Brackbill and Barnes 1980) and (ii) has on its left-hand side the form suitable for approximation by the CIR and Roe-type methods. As noted earlier, the Jacobian matrix acquires in this case the 8×8 form (5.4.43) and we can solve a one-dimensional Riemann problem with a discontinuous normal component of the magnetic field vector. This can fairly easily occur if we apply a one-dimensional Riemann solver to multidimensional problems.

Let us find the difference in the magnetic charge $\operatorname{div}\mathbf{B}$ evolution represented by the two forms of the governing equations. In the classical situation we obviously have

$$\frac{\partial}{\partial t}(\operatorname{div}\mathbf{B}) = 0$$

and the magnetic charge remains absent if it does not exist initially.

If we apply the divergence operator to the Faraday equation in system (5.7.7), we obtain

$$\frac{\partial}{\partial t}(\operatorname{div}\mathbf{B}) + \operatorname{div}(\mathbf{v}\operatorname{div}\mathbf{B}) = 0,$$

that is, the addition of the source term transforms the condition of the magnetic charge absence to a convection equation for this charge. The approach implies that the magnetic charge will hopefully be convected away from the computational region. This elegant construction seems to work very well for problems with open boundaries (Linde et al. 1998, Pogorelov and Matsuda 1998a) but not for others. It is also questionable if it can be applied in the vicinity of stagnation points or, in general, to time-dependent problems. The latter remark is caused by the fact that the method based on the extended MHD system satisfies the divergence-free condition only in the infinite time limit, when a steady-state solution is established.

5.7.5 Application of staggered grids. In the above subsections we outlined the difficulties arising if the vector potential is introduced or if we eliminate magnetic charge using the artificial scalar potential approach. These difficulties are mainly caused by the necessity to approximate second-order derivatives of the functions $\mathbf{A}$ and φ that are discontinuous in the computational region. In principle, magnetic charge is absent in this case only within the accuracy of the chosen numerical scheme. Besides, when solving the Poisson equation, one needs global information that is not favorable in existing parallel codes. Moreover, the clean-up procedure can destroy the properties of the Godunov-like schemes (Dai and Woodward 1998).

Let the magnetic field and other hydrodynamic properties exactly satisfy the conservation laws before the magnetic charge elimination. After the nonconservative clean-up procedure, the conservation laws for the energy and for the magnetic field vector are satisfied only to the order of accuracy of the scheme and truncation errors may be $O(1)$. This kind of $O(1)$ truncation errors can be easily detected when a strong MHD shock, propagating obliquely through a computational region, is simulated using a nonconservative scheme.

Another important observation which can be made is that the calculation of the magnetic charge (or $\operatorname{div}\mathbf{B}$) inside a computational cell in the finite-volume-consistent manner requires magnetic field values rather on the cell surfaces than at the cell centers. This can encourage the application of staggered grids in which magnetic field is evaluated at the cell interfaces, while all other dependent functions are determined at the cell centers. The idea itself seems to be very fruitful, though there exists a drawback caused by the fact that in order to perform a time integration step in the cell center we also need the value of magnetic field at it. This can be done only by an interpolation procedure, which can diminish the accuracy of calculations if we admit that magnetic field can be discontinuous.

Let us illustrate the staggered grid approach (Dai and Woodward 1998) on the uniform Cartesian mesh x_l, y_m, and z_n with the space steps Δx, Δy, and Δz, respectively. The points at the cell interfaces will then have half-integer indices. Generalization of the approach to other types of regular meshes is straightforward.

Consider the system of (5.1.43) written in the conservation-law form. A general finite volume scheme for obtaining the unknown vector $\mathbf{U}$ at the higher time level $k+1$ using the values at the lower level k is the following:

$$\mathbf{U}^{k+1}_{l,m,n} = \mathbf{U}^{k}_{l,m,n} - \frac{\Delta t}{\Delta x}(\bar{\mathbf{E}}_{l+1/2,m,n} - \bar{\mathbf{E}}_{l-1/2,m,n}) - \frac{\Delta t}{\Delta y}(\bar{\mathbf{E}}_{l,m+1/2,n} - \bar{\mathbf{E}}_{l,m-1/2,n})$$
$$-\frac{\Delta t}{\Delta z}(\bar{\mathbf{E}}_{l,m,n+1/2} - \bar{\mathbf{E}}_{l,m,n-1/2}), \tag{5.7.8}$$

where $\mathbf{U}^{k+1}_{l,m,n}$ is the cell averaged value of $\mathbf{U}$ at $t^{k+1} = (k+1)\Delta t$, while $\bar{\mathbf{E}}$ represents the averaged fluxes at different cell interfaces, depending on the choice of indices l, m, and n. They can be written out as

$$\mathbf{U}_{l,m,n} = \frac{1}{\Delta x \Delta y \Delta z}\int_{z_{n-1/2}}^{z_{n+1/2}}\int_{y_{m-1/2}}^{y_{m+1/2}}\int_{x_{l-1/2}}^{x_{l+1/2}} \mathbf{U}(t,x,y,z)\,dx\,dy\,dz, \tag{5.7.9}$$

$$\bar{\mathbf{E}}_{l+1/2,m,n} = \frac{1}{\Delta t \Delta y \Delta z}\int_{t^k}^{t^{k+1}}\int_{y_{m-1/2}}^{y_{m+1/2}}\int_{z_{n-1/2}}^{z_{n+1/2}} \mathbf{E}(\mathbf{U}(t,x_{l+1/2},y,z))\,dy\,dz\,dt, \tag{5.7.10}$$

$$\bar{\mathbf{E}}_{l,m+1/2,n} = \frac{1}{\Delta t \Delta x \Delta z}\int_{t^k}^{t^{k+1}}\int_{x_{l-1/2}}^{x_{l+1/2}}\int_{z_{n-1/2}}^{z_{n+1/2}} \mathbf{F}(\mathbf{U}(t,x,y_{m+1/2},z))\,dx\,dz\,dt, \tag{5.7.11}$$

$$\bar{\mathbf{E}}_{l,m,n+1/2} = \frac{1}{\Delta t \Delta x \Delta y}\int_{t^k}^{t^{k+1}}\int_{x_{l-1/2}}^{x_{l+1/2}}\int_{y_{m-1/2}}^{y_{m+1/2}} \mathbf{G}(\mathbf{U}(t,x,y,z_{n+1/2}))\,dx\,dy\,dt. \tag{5.7.12}$$

The time-averaged fluxes can be determined, for example, by using Roe's approximate MHD Riemann problem solver. The procedure becomes especially simple if in Eqs. (5.7.10)–(5.7.12) we calculate the integrals over time with the first order of accuracy using the quantities at the lower time level.

In accordance with the idea outlined above, we can introduce the surface-averaged magnetic field vector components perpendicular to the corresponding cell interface, that is,

$$b_{x,l,m,n} = \frac{1}{\Delta y\,\Delta z}\int_{y_{m-1/2}}^{y_{m+1/2}}\int_{z_{n-1/2}}^{z_{n+1/2}} B_x(x_{l+1/2},y,z)\,dy\,dz, \tag{5.7.13}$$

$$b_{y,l,m,n} = \frac{1}{\Delta x\,\Delta z}\int_{x_{l-1/2}}^{x_{l+1/2}}\int_{z_{n-1/2}}^{z_{n+1/2}} B_y(x,y_{m+1/2},z)\,dx\,dz, \tag{5.7.14}$$

$$b_{z,l,m,n} = \frac{1}{\Delta x\,\Delta y}\int_{x_{l-1/2}}^{x_{l+1/2}}\int_{y_{m-1/2}}^{y_{m+1/2}} B_z(x,y,z_{n+1/2})\,dx\,dy. \tag{5.7.15}$$

Note the x-, y-, and z-components (5.7.13)–(5.7.15) of the vector $\mathbf{b}$ are defined on the different surfaces of the cell. The divergence-free condition can easily be written out in terms of the vector $\mathbf{b}$ as

$$\oint_{S_{l,m,n}} \mathbf{B}\cdot d\mathbf{S} = \Delta y\,\Delta z\,(b_{x,l,m,n} - b_{x,l-1,m,n})$$
$$+\,\Delta x\,\Delta z\,(b_{y,l,m,n} - b_{y,l,m-1,n}) + \Delta x\,\Delta y\,(b_{z,l,m,n} - b_{z,l,m,n-1}) = 0, \tag{5.7.16}$$

where $S_{l,m,n}$ is the computational cell surface and $d\mathbf{S}$ is the vector of magnitude equal to the area of the elementary surface and oriented along its normal.

In the two-dimensional statement with derivatives with respect to z equal to zero, $b_{z,l,m,n+1/2} = b_{z,l,m,n-1/2}$ and we obtain a two-dimensional variant of Eq. (5.7.16) in the form

$$\Delta y\,(b_{x,l,m} - b_{x,l-1,m}) + \Delta x\,(b_{y,l,m} - b_{y,l,m-1}) = 0.$$

The conservation-law form of the Faraday equation is

$$\frac{\partial \mathbf{B}}{\partial t} + \operatorname{div}(\mathbf{vB} - \mathbf{Bv}) = 0.$$

This form is convenient for approximation if we integrate over the computational cell volume. On the other hand, if we want to solve the induction equation in terms of the components of the vector $\mathbf{b}$ defined at the finite volume interfaces, another integral form represented by Eq. (5.7.16) must be chosen. In acting so, for each surface S spanned over a contour L, using the Stokes theorem we can write out

$$\frac{\partial}{\partial t}\int_S \mathbf{b}\cdot d\mathbf{S} = \oint_L (\mathbf{v}\times\mathbf{B})\cdot d\mathbf{l}, \tag{5.7.17}$$

where $d\mathbf{l}$ is the element of the contour length oriented in accordance with the corkscrew rule with respect to the direction of the global normal to the cell interface.

The vector product $\mathbf{v}\times\mathbf{B}$ can be represented in terms of the components of a single axial vector $\boldsymbol{\Omega}$ as follows:

$$\Omega_x = vB_z - wB_y, \quad \Omega_y = wB_x - uB_z, \quad \Omega_z = uB_y - vB_x.$$

Thus, in the Cartesian coordinate system we obtain

$$\begin{aligned} \frac{\partial b_x}{\partial t} - \frac{\partial \Omega_z}{\partial y} + \frac{\partial \Omega_y}{\partial z} &= 0, \\ \frac{\partial b_y}{\partial t} + \frac{\partial \Omega_z}{\partial x} - \frac{\partial \Omega_x}{\partial z} &= 0, \\ \frac{\partial b_z}{\partial t} - \frac{\partial \Omega_y}{\partial x} + \frac{\partial \Omega_x}{\partial y} &= 0. \end{aligned} \tag{5.7.18}$$

The system for determining the components of the vector $\mathbf{b}$ is the following:

$$b^{k+1}_{x,l,m,n} = b^k_{x,l,m,n} + \frac{\Delta t}{\Delta y}(\bar{\Omega}_{z,l,m,n} - \bar{\Omega}_{z,l,m-1,n}) - \frac{\Delta t}{\Delta z}(\bar{\Omega}_{y,l,m,n} - \bar{\Omega}_{y,l,m,n-1}), \tag{5.7.19}$$

$$b^{k+1}_{y,l,m,n} = b^k_{y,l,m,n} + \frac{\Delta t}{\Delta z}(\bar{\Omega}_{x,l,m,n} - \bar{\Omega}_{x,l,m,n-1}) - \frac{\Delta t}{\Delta x}(\bar{\Omega}_{z,l,m,n} - \bar{\Omega}_{z,l-1,m,n}), \tag{5.7.20}$$

$$b^{k+1}_{z,l,m,n} = b^k_{z,l,m,n} + \frac{\Delta t}{\Delta x}(\bar{\Omega}_{y,l,m,n} - \bar{\Omega}_{z,l-1,m,n}) - \frac{\Delta t}{\Delta y}(\bar{\Omega}_{x,l,m,n} - \bar{\Omega}_{x,l,m-1,n}), \tag{5.7.21}$$

where

$$\begin{aligned} \bar{\Omega}_{x,l,m,n} &= \frac{1}{\Delta t \Delta x}\int_{t^k}^{t^{k+1}}\int_{x_{l-1/2}}^{x_{l+1/2}} \Omega_x(x,\, y_{m+1/2},\, z_{n+1/2})\,dt\,dx, \\ \bar{\Omega}_{y,l,m,n} &= \frac{1}{\Delta t \Delta y}\int_{t^k}^{t^{k+1}}\int_{y_{m-1/2}}^{y_{m+1/2}} \Omega_y(x_{l+1/2},\, y,\, z_{n+1/2})\,dt\,dy, \\ \bar{\Omega}_{z,l,m,n} &= \frac{1}{\Delta t \Delta z}\int_{t^k}^{t^{k+1}}\int_{z_{n-1/2}}^{z_{n+1/2}} \Omega_z(x_{l+1/2},\, y_{m+1/2},\, z)\,dt\,dz. \end{aligned}$$

Note that $\bar{\Omega}_{x,l,m,n}$, $\bar{\Omega}_{y,l,m,n}$, and $\bar{\Omega}_{z,l,m,n}$ are the time-averaged values of the respective components of the vector $\boldsymbol{\Omega}$ along the edges of the computational parallelepiped. We use the indices l, m, and n in their definition, though have in mind that they are always evaluated at the points $(x_l, y_{m+1/2}, z_{n+1/2})$, $(x_{l+1/2}, y_m, z_{n+1/2})$, and $(x_{l+1/2}, y_{m+1/2}, z_n)$, respectively. Thus, to solve the system (5.7.19)–(5.7.21), we need to evaluate the vector $\bar{\boldsymbol{\Omega}}$ on the grid edges. One of the natural ways (Ryu et al. 1998) to determine the averaged components of this vector is to use the magnetic fluxes (5.7.10)–(5.7.12) determined in the process of approximation the conservation equations (5.7.8). For example, if we take into account that $\Omega_x = vB_z - wB_y$, we obtain (compare with system 5.1.43)

$$\bar{\Omega}_{x,l,m,n} = \frac{1}{4}\left[\left(\bar{E}_{8,l,m+1/2,n} + \bar{E}_{8,l,m+1/2,n+1}\right) - \left(\bar{E}_{7,l,m,n+1/2} + \bar{E}_{7,l,m+1,n+1/2}\right)\right] \tag{5.7.22}$$

and similarly

$$\bar{\Omega}_{y,l,m,n} = \frac{1}{4}[(\bar{E}_{6,l,m,n+1/2} + \bar{E}_{6,l+1,m,n+1/2}) - (\bar{E}_{8,l+1/2,m,n} + \bar{E}_{8,l+1/2,m,n+1})], \tag{5.7.23}$$

$$\bar{\Omega}_{z,l,m,n} = \frac{1}{4}[(\bar{E}_{7,l+1/2,m,n} + \bar{E}_{7,l+1/2,m+1,n}) - (\bar{E}_{6,l,m+1/2,n} + \bar{E}_{6,l+1,m+1/2,n})]. \tag{5.7.24}$$

Another possible approach to evaluation of the vector $\boldsymbol{\Omega}$ on the cell edges was suggested by Dai and Woodward (1998). Its essence is also in averaging of the available data, though the method described above seems to be favorable due to its consistency with the upwind fluxes calculated at the cell interfaces (compare with Balsara and Spicer 1999).

Now we turn our attention to the calculation of the cell-centerd quantities. Since magnetic field occurs not only in the Faraday equation but also in the momentum and energy equations, we shall need the magnetic field vector components not only on the cell interfaces but also at their centers. This can be done by interpolation of $\mathbf{b}$ as follows:

$$(B_x)_{l,m,n} = \frac{1}{2}(b_{x,l,m,n} + b_{x,l-1,m,n}), \tag{5.7.25}$$

$$(B_y)_{l,m,n} = \frac{1}{2}(b_{y,l,m,n} + b_{y,l,m-1,n}), \tag{5.7.26}$$

$$(B_z)_{l,m,n} = \frac{1}{2}(b_{z,l,m,n} + b_{z,l,m,n-1}). \tag{5.7.27}$$

This interpolation procedure has the second order of accuracy for smooth functions and therefore can be used in the MHD codes based on TVD schemes of the same order of accuracy. It degrades, however, if the magnetic field vector is discontinuous, and the errors can be $O(1)$. This seems to be hardly avoidable in all methods based on the straightforward approximation of the Poisson equation and employing interpolation of discontinuous functions. On the other hand, we know that TVD schemes also degrade to the first order of approximation in the vicinity of discontinuities. The improvement lies in the application of the ENO schemes (see, e.g., Jiang and Wu 1999) and more sophisticated methods of interpolation. Worth mentioning is also a newly proposed procedure by Londrillo and Del Zanna (2000) combining the vector potential and staggered grid approaches.

5.8 Solar wind interaction with the magnetized interstellar medium

In Section 3.7 the problem on the solar wind–interstellar medium interaction was considered in a purely gas dynamic statement. The influence of the interstellar magnetic field can be important in the regions for which the value of magnetic pressure is comparable with dynamic pressure of the flow. The presence of magnetic field generally increases the maximum speed of small perturbations in the LISM flow, thus decreasing its effective Mach number (Fahr 1986; Fahr, Grzedzielski and Ratkiewicz 1988). The Mach number of the LISM calculated via the acoustic speed of sound is often assumed to be $M_\infty = 2$. At the same time, the magnitude of the LISM magnetic field, although not very well known, is estimated within 7×10^{-7} and 3×10^{-6} gauss (Axford 1972). This fact means (see, e.g., Pogorelov and Semenov 1997a) that magnetic pressure can exceed the value of thermal pressure. This is the reason to take MHD effects into account. In fact, if the LISM magnetic field becomes sufficiently strong, the flow can become submagnetosonic and the bow shock can disappear (see Fahr 1986). The stellar wind interaction with the magnetized interstellar medium was first studied numerically by Matsuda and Fujimoto (1993), although some of the results were disputed by Baranov and Zaitsev (1995). The last authors investigated the problem by the shock-fitting method only in the upwind part of the flow due to limitations of their numerical scheme. Pogorelov and Semenov (1997a) and Pogorelov and Matsuda (2000) presented the solution of the axisymmetric problem (the LISM magnetic field strength vector was assumed to be parallel to its velocity vector) of the SW–LISM interaction in the closed region around the Sun for various magnitudes of the LISM magnetic field.

Fahr et al. (1988) studied the shape of the heliopause on the basis of the Newtonian approach in the three-dimensional case of an arbitrary angle between the LISM magnetic field and velocity vectors. The authors found this shape approximately by equating the values of the total pressure on both sides of the heliopause. Three-dimensional problem was investigated by Washimi and Tanaka (1996), Pogorelov and Matsuda (1998a), Linde et al. (1998), Ratkiewicz et al. (1998), McNutt et al. (1999).

To illustrate the performance of high-resolution shock-capturing methods in application to multidimensional MHD equations (see the review by Pogorelov and Matsuda 1998b), in what follows we present the solution of the SW–LISM interaction in the closed region surrounding the Sun for various interstellar medium magnetic field vector $\mathbf{B}_\infty$ magnitudes and orientations with respect to the velocity vector $\mathbf{V}_\infty$. Two different numerical methods were applied. The first one is a high-resolution, second order of accuracy version of the TVD Lax–Friedrichs scheme (see Barmin et al. 1996). This method gives a substantial simplification of the numerical algorithm comparing with the methods based on the exact characteristic splitting of the Jacobian matrices of the MHD equations. The second one is based on the 8×8 version of the MHD CIR-type solver. We describe simple, but very effective numerical boundary conditions in the far field, which allow one to avoid the influence of spurious reflections from the subsonic outflow boundary. We include in our algorithm the procedure of satisfying the condition of source-free magnetic fields (the divergence-free condition), which is based to the approach of Aslan (1993) and Powell (1994).

While considering the SW–LISM interaction, we shall also pay attention to special irregular cases characterized by the presence of nonevolutionary shock waves. They can occur in both two- and three-dimensional cases. To clarify certain points of structural stability of nonevolutionary shocks we shall discuss the problem of a uniform supersonic and super-Alfvénic MHD flow over an infinitely conducting cylinder.

5.8.1 Statement of the problem. The flow pattern has already been shown schematically in Fig. 3.29. The system of governing equations for MHD flow of ideal, infinitely conducting, perfect plasma in the Cartesian coordinate system x, y, z (y-axis is perpendicular to the figure plane) is given by Eq. (5.7.7). This system contains an additional source term due to the procedure of the magnetic charge elimination. In the axisymmetric case ($\mathbf{B}_\infty \parallel \mathbf{V}_\infty$) it reduces to

$$\frac{\partial \mathbf{U}}{\partial t} + \frac{\partial \mathbf{E}}{\partial x} + \frac{\partial \mathbf{G}}{\partial z} + \mathbf{H}_1 + \mathbf{H}_2 = \mathbf{0}, \tag{5.8.1}$$

where

$$\mathbf{U} = \begin{bmatrix} \rho \\ \rho u \\ \rho w \\ e \\ B_x \\ B_y \\ B_z \end{bmatrix}, \quad \mathbf{E} = \begin{bmatrix} \rho u \\ \rho u^2 + p_0 - \dfrac{B_x^2}{4\pi} \\ \rho uv - \dfrac{B_x B_y}{4\pi} \\ \rho uw - \dfrac{B_x B_z}{4\pi} \\ (e + p_0)u - \dfrac{B_x}{4\pi}(\mathbf{v}\cdot\mathbf{B}) \\ 0 \\ uB_y - vB_x \\ uB_z - wB_x \end{bmatrix}, \quad \mathbf{G} = \begin{bmatrix} \rho w \\ \rho uw - \dfrac{B_x B_z}{4\pi} \\ \rho vw - \dfrac{B_y B_z}{4\pi} \\ \rho w^2 + p_0 - \dfrac{B_z^2}{4\pi} \\ (e + p_0)w - \dfrac{B_z}{4\pi}(\mathbf{v}\cdot\mathbf{B}) \\ wB_x - uB_z \\ wB_y - vB_z \\ 0 \end{bmatrix},$$

$$\mathbf{H}_1 = \frac{1}{x}\begin{bmatrix} \rho u \\ \rho(u^2 - v^2) - \dfrac{B_x^2 - B_y^2}{4\pi} \\ \rho uv - \dfrac{B_x B_y}{4\pi} \\ \rho uw - \dfrac{B_x B_z}{4\pi} \\ (e + p_0)u - \dfrac{B_x}{4\pi}(\mathbf{v}\cdot\mathbf{B}) \\ 0 \\ 0 \\ uB_z - wB_x \end{bmatrix}, \quad \mathbf{H}_2 = \begin{bmatrix} 0 \\ \dfrac{B_x}{4\pi} \\ \dfrac{B_y}{4\pi} \\ \dfrac{B_z}{4\pi} \\ \dfrac{\mathbf{v}\cdot\mathbf{B}}{4\pi} \\ u \\ v \\ w \end{bmatrix} \times \operatorname{div}\mathbf{B}.$$

This system is valid in the half-plane $x0z$ and can be obtained from (5.7.7) in the assumption of cylindrical symmetry. Its other form is

$$\frac{\partial x\mathbf{U}}{\partial t} + \frac{\partial x\mathbf{E}}{\partial x} + \frac{\partial x\mathbf{G}}{\partial z} + \tilde{\mathbf{H}}_1 + x\mathbf{H}_2 = \mathbf{0}, \tag{5.8.2}$$

where

$$\tilde{\mathbf{H}}_1 = \left[0,\ -p_0 - \rho v^2 + \tfrac{B_y^2}{4\pi},\ 0,\ 0,\ 0,\ -uB_y + vB_x,\ 0\right]^{\mathrm{T}}.$$

Though both presentations of the governing equations are mathematically equivalent, for numerical reasons it is often more convenient to use the former one, since it is supposed to give more stable results in the vicinity of the geometrical singularity $x = 0$. Note that system (5.8.1) can further be simplified if $v = B_y \equiv 0$ in the (x, z)-plane. The Euler gas dynamic equations can be obtained in different forms from systems (5.8.1) and (5.8.2) by assuming $\mathbf{B} \equiv \mathbf{0}$.

The above formulation implies that molecular and magnetic viscosities, heat conductivity, and anomalous transport effects are neglected. The charge-exchange processes between the neutral and plasma components of the winds are also disregarded in the context of this book, although they represent one of the important physical mechanisms governing the interaction (see the discussion in Section 3.7).

The specific heat ratio corresponding to the fully ionized plasma is $\gamma = 5/3$. The quantities of density, pressure, velocity, and magnetic field strength are normalized, respectively, by ρ_∞, $\rho_\infty V_\infty^2$, V_∞, and $V_\infty\sqrt{\rho_\infty}$, where the subscript ∞ marks the values in the uniform LISM flow. Time and the linear dimension are respectively related to L/V_∞ and L, where L is equal to 1 AU $= 1.5 \times 10^{11}$ m, that is, to the distance from the Sun to the Earth.

We perform calculations in the computational region between the inner and outer spherical surfaces. The flow from the Sun is supposed to be supersonic at the termination shock distance. For this reason we specify all parameter values on the inner boundary sphere. The uniform LISM flow is also supersonic, and we can fix all quantities on the inflow side of the outer boundary. Treatment of the outflow boundary can be more complicated. We use here the absorbing boundary conditions (Pogorelov and Semenov 1997a), which are the extension to MHD of the similar boundary conditions for pure gas dynamics (Sawada et al. 1986; Pogorelov and Semenov 1996b).

As initial values, the jump can be chosen between the SW and LISM parameters at a fixed distance R_{f} from the Sun smaller than the TS stand-off distance. For $R < R_{\mathrm{f}}$ the SW parameter distribution is specified. The magnetic field pressure in this region is supposed to be negligibly small in comparison with the SW hydrodynamic pressure; thus $\mathbf{B}_e = 0$. This is not actually true as far as the toroidal component of the solar magnetic field is concerned (see Washimi and Tanaka 1996; Linde et al. 1998). For $R > R_{\mathrm{f}}$, uniform distributions of the LISM pressure and density are adopted. If we solve the MHD problem, that is, if the LISM flow is magnetized, it is worth specifying the magnetic field strength distribution in a way such that the divergence-free condition is satisfied. For this purpose the magnetic and velocity fields in the LISM and SW flows are initially joined in the computational region in order that $\mathbf{B}$ conserves a constant angle with $\mathbf{v}$ and $\operatorname{div}\mathbf{B} = 0$. This is done by assuming

$$\begin{aligned}
U &= -|\mathbf{V}_\infty|\left[1 - \left(\frac{R_{\mathrm{f}}}{R}\right)^3\right]\cos\theta,\\
V &= 0, \qquad\qquad\qquad\qquad\qquad\qquad (5.8.3)\\
W &= |\mathbf{V}_\infty|\left[1 + \frac{1}{2}\left(\frac{R_{\mathrm{f}}}{R}\right)^3\right]\sin\theta
\end{aligned}$$

for $R > R_f$. Here U, V, and W represent the spherical components of the velocity vector in the directions R, φ, and θ, respectively. This distribution corresponds to the velocity distribution for an incompressible fluid flow over a sphere, directed along the z-axis (the θ count-off axis). The magnetic field is initialized in the same way, except that the field configuration is rotated about the y-axis so that the magnetic field vector is tilted with respect to the velocity vector by a desired angle.

The following solar wind and LISM parameters were chosen: $n_e = 7\,\mathrm{cm}^{-3}$, $V_e = 450\,\mathrm{km\,s}^{-1}$, $M_e = 10$, $n_\infty = 0.07\,\mathrm{cm}^{-3}$, $V_\infty = 25\,\mathrm{km\,s}^{-1}$, and $M_\infty = 2$. The dimensionless value of the magnetic field is specified via the Alfvén number $A_\infty = V_\infty/\sqrt{B_\infty/4\pi\rho_\infty}$.

5.8.2 Numerical algorithm.

Let us write out the numerical algorithm for solution of the SW–LISM interaction problem. Several aspects of it can be applied as well to various other MHD problems.

Finite-volume method

Finite-volume methods are known to be the most suitable for obtaining discontinuous solutions of hyperbolic differential equations. In these methods the governing equations in the conservation law form are integrated over finite control volumes of arbitrary complexity. To avoid the origin of a geometrical source term, which can appear in a curvilinear coordinate system (see Section 3.1), the integration of the system written for Cartesian vector components is widely adopted. In this case the particular coordinate system manifests itself not by the metric tensor components and Christoffel symbols but via the shape of the computational cell volumes and their surface areas.

To solve the system of (5.7.7) or (5.8.1), let us introduce a spherical mesh

$$
\begin{aligned}
&f^k_{l,m,n} = f(R_l, \varphi_m, \theta_n, t^k), \quad t^k = k\Delta t;\\
&R_l = R_{\min} + (l-1)\Delta R, \quad l = 1, 2, \ldots, L;\\
&\varphi_m = (m-3)\Delta\varphi, \quad m = 1, 2, \ldots, M;\\
&\theta_n = (n-2.5)\Delta\theta, \quad n = 1, 2, \ldots, N;\\
&\Delta R = (R_{\max} - R_{\min})/(L-1),\\
&\Delta\theta = \pi/(N-4), \quad \Delta\varphi = \pi/(M-5)
\end{aligned}
$$

with the origin at the Sun. Then, for each cell, the system (5.7.7) in the finite-volume formulation can be rewritten as follows:

$$
\begin{aligned}
&\frac{\mathbf{U}^{k+1}_{l,m,n} - \mathbf{U}^k_{l,m,n}}{\Delta t}\\
&+ \frac{1}{\Delta R}\left[\left(\frac{R_{l+1/2}}{R_l}\right)^2 \bar{\mathbf{E}}^k_{l+1/2,m,n} + \left(\frac{R_{l-1/2}}{R_l}\right)^2 \bar{\mathbf{E}}^k_{l-1/2,m,n}\right]\\
&+ \frac{1}{R_l \sin\theta_n \Delta\varphi\Delta\theta}\left[\Delta\theta\left(\bar{\mathbf{E}}^k_{l,m+1/2,n} + \bar{\mathbf{E}}^k_{l,m-1/2,n}\right)\right.\\
&\left.+ \Delta\varphi\left(\sin\theta_{n+1/2}\bar{\mathbf{E}}^k_{l,m,n+1/2} + \sin\theta_{n-1/2}\bar{\mathbf{E}}^k_{l,m,n-1/2}\right)\right] + \mathbf{H}^k_{l,m,n} = \mathbf{0}. \qquad (5.8.4)
\end{aligned}
$$

Here $\bar{\mathbf{E}}$ is the flux normal to the cell boundary, defined as

$$\bar{\mathbf{E}} = n_1\mathbf{E} + n_2\mathbf{F} + n_3\mathbf{G}, \tag{5.8.5}$$

where $\mathbf{n} = (n_1, n_2, n_3)$ is a unit outward vector normal to the cell surface.

Equation (5.8.4) has a time-discretized conservation law form for an individual computational cell. While the values of functions are given in the computational cell centers, to find numerical fluxes $\bar{\mathbf{E}}$ we need them also on the cell surfaces. Generally speaking, any particular method is specified by the formula for the calculation of this flux. The surface values of the functions can be found by interpolation of the center values. To attain the second order of accuracy in space, a piecewise-linear distribution of parameters inside computational cells is adopted. On the other hand, to avoid spurious oscillations near discontinuities, in the determination of the flow state vectors $\mathbf{U}^{\mathrm{L}}$ and $\mathbf{U}^{\mathrm{R}}$ on the left- and right-hand sides of each cell boundary we apply certain limitations on the derivatives of functions to be interpolated. Here we use the following reconstruction procedure in the radial direction (the indices m and n are omitted):

$$\begin{aligned}
&\mathbf{U}^{\mathrm{R}}_{l+1/2} = \mathbf{U}^k_{l+1} - \operatorname{minmod}(\Delta\mathbf{U}^k_{l+1/2}, \Delta\mathbf{U}^k_{l+3/2}, \tilde{\Delta}),\\
&\mathbf{U}^{\mathrm{L}}_{l+1/2} = \mathbf{U}^k_{l} + \operatorname{minmod}(\Delta\mathbf{U}^k_{l-1/2}, \Delta\mathbf{U}^k_{l+1/2}, \tilde{\tilde{\Delta}}),\\
&\tilde{\Delta} = 0.25\,(\Delta\mathbf{U}^k_{l+1/2} + \Delta\mathbf{U}^k_{l+3/2}),\\
&\tilde{\tilde{\Delta}} = 0.25\,(\Delta\mathbf{U}^k_{l-1/2} + \Delta\mathbf{U}^k_{l+1/2}),\\
&\operatorname{minmod}(x,y,z) = \operatorname{sgn}(x)\max\{0, \min[|x|, y\operatorname{sgn}(x), |z|]\}.
\end{aligned}$$

The above formulas are applied in the framework of the Lax–Friedrichs-type scheme to all primitive variables except for the thermal pressure. For the pressure we use a simple minmod reconstruction. This agrees with the discussion from Section 5.4.4. If the CIR-type method is used to determine fluxes through the computational cell interfaces, we apply the reconstruction procedure to the increments of characteristic variables. The more compressive limiters represented by the above formulas are applied in this case to the entropy and Alfvén characteristic waves, whereas the magnetosonic waves are the subject of the minmod reconstruction. The reconstruction procedure in the angular directions is similar.

The fluxes $\bar{\mathbf{E}}(\mathbf{U}^{\mathrm{R}}, \mathbf{U}^{\mathrm{L}})$ through the cell surfaces can be found by different methods. Wide recognition has been given to the total variation diminishing (TVD) shock-capturing methods based on the exact or some of the approximate solutions of the Riemann problem (the initial value problem on the disintegration of a jump between $\mathbf{U}^{\mathrm{R}}$ and $\mathbf{U}^{\mathrm{L}}$). Except for the one case, we shall present numerical results obtained with the help of the nonoscillatory extension of the Lax–Friedrichs scheme (Lax 1954) to the second order of accuracy (see Yee 1989). This scheme is much simpler than those based on the exact characteristic splitting of the Jacobian matrices of the MHD system and at the same time retains many of their positive properties.

The numerical flux (5.8.5) through the radial cell interface is calculated by the formula

$$\bar{\mathbf{E}}_{l+1/2,m,n} = \frac{1}{2}\Big[\bar{\mathbf{E}}\left(\mathbf{U}^{\mathrm{R}}_{l+1/2,m,n}\right) + \bar{\mathbf{E}}\left(\mathbf{U}^{\mathrm{L}}_{l+1/2,m,n}\right) - \hat{\mathbf{R}}_{l+1/2,m,n}\left(\mathbf{U}^{\mathrm{R}}_{l+1/2,m,n} - \mathbf{U}^{\mathrm{L}}_{l+1/2,m,n}\right)\Big].$$

Here $\hat{\mathbf{R}}_{l+1/2,m,n}$ is the diagonal matrix with the same elements on its diagonal equal to the spectral radius r (the maximum of eigenvalue magnitudes) of the Jacobian matrix $\partial\bar{\mathbf{E}}/\partial\mathbf{U}$:

$$
\begin{aligned}
&r = |U| + a_{\mathrm{f}}, \quad a_{\mathrm{f}}^2 = \frac{1}{2}\left((a^*)^2 + \sqrt{(a^*)^4 - 4a^2 b_R^2}\right), \\
&b_R = B_R/\sqrt{4\pi\rho}, \quad B^2 = B_R^2 + B_\varphi^2 + B_\theta^2, \\
&(a^*)^2 = (\gamma p + B^2/4\pi)/\rho, \quad a^2 = \gamma p/\rho,
\end{aligned}
$$

where U is the radial velocity component and B_R, B_φ, and B_θ are the radial and the angular components of the magnetic field strength vector, respectively. Similar formulas can be written for the fluxes in the angular directions. The matrix $\hat{\mathbf{R}}$ is evaluated by the approximate solution of the MHD Riemann problem (Pogorelov et al. 1995; Pogorelov and Semenov 1996a, 1997b, 1997c). Being accurate to the second order, the proposed scheme is much less dissipative than the original Lax–Friedrichs method and provides incomparably better shock resolution. Calculations were mainly performed in the spherical computational region with $R_{\min} = 24$ and $R_{\max} = 1200$. The mesh is $R \times \varphi \times \theta = 99 \times 21 \times 116$. In the axisymmetric cases, the computational region becomes circular as we omit the coordinate φ.

Far-field boundary conditions for MHD modelling in astrophysics

In this section we describe the approach to implementation of a nonreflecting boundary condition for hyperbolic systems like the Euler gas dynamic and MHD equations (see Pogorelov and Semenov 1997a). Its application for the Euler gas dynamic equations is presented in Section 3.7. Far-field boundary conditions are often encountered in astrophysical applications that are frequently characterized by the presence of very large distance scales.

Let us consider a local one-dimensional system

$$
\frac{\partial \mathbf{U}}{\partial t} + A\frac{\partial \mathbf{U}}{\partial x} = 0. \tag{5.8.6}
$$

The x-axis is directed perpendicular to the cell interface Γ.

Let us choose the following form of the unknown vector in Eq. (5.8.6):

$$
\mathbf{U} = [\rho,\ u,\ v,\ w,\ c_e,\ B_x,\ B_y,\ B_z]^{\mathrm{T}}, \tag{5.8.7}
$$

where, in addition to the purely gas dynamic case, the components occur of the magnetic field strength vector normal (B_x) and tangent (B_y and B_z) to Γ (see Eq. (1.3.24)).

For definiteness we consider the exit boundary ($u > 0$). The minimum eigenvalue in this case is $\lambda = u - a_{\mathrm{f}}$, where a_{f} is the largest of the two magnetosonic speeds a_{f} and a_{s} ($a_{\mathrm{f}} \geq c_e \geq a_{\mathrm{s}}$):

$$
\begin{aligned}
a_{\mathrm{f,s}} &= \frac{1}{2}\left[\left(c_e^2 + \frac{|\mathbf{B}|^2}{4\pi\rho} + \frac{c_e|B_x|}{\sqrt{\pi\rho}}\right)^{1/2} \pm \left(c_e^2 + \frac{|\mathbf{B}|^2}{4\pi\rho} - \frac{c_e|B_x|}{\sqrt{\pi\rho}}\right)^{1/2}\right], \\
|\mathbf{B}|^2 &= B_x^2 + B_y^2 + B_z^2.
\end{aligned} \tag{5.8.8}
$$

The eigenvector corresponding to this eigenvalue is

$$\mathbf{r} = \left[1,\ -\frac{a_\mathrm{f}}{\rho},\ \alpha B_y,\ \alpha B_z,\ \frac{(\gamma-1)c_e}{2\rho},\ 0,\ \beta B_y,\ \beta B_z\right]^\mathrm{T}, \tag{5.8.9}$$

$$\alpha = \frac{c_e a_\mathrm{s}\operatorname{sgn} B_x}{2\rho\sqrt{\pi\rho}(c_e^2 - a_\mathrm{s}^2)} = \sqrt{\frac{4\pi}{\rho}}\frac{a_\mathrm{s}(a_\mathrm{f}^2 - c_e^2)\operatorname{sgn} B_x}{c_e(B_y^2 + B_z^2)},$$

$$\beta = \frac{c_e^2}{\rho(c_e^2 - a_\mathrm{s}^2)} = \frac{4\pi(a_\mathrm{f}^2 - c_e^2)}{B_y^2 + B_z^2}.$$

If we seek the solution to (5.8.6) in the form of a simple wave $\mathbf{U}(x,t) = \mathbf{U}(\xi)$ with $\xi = x/t$, we obtain

$$(A - \lambda I)\,\mathbf{U}_\xi = \mathbf{0}, \quad \lambda = \xi,$$

where I is the identity matrix. Thus, $\mathbf{U}_\xi$ is proportional to the right eigenvector of A corresponding to $\lambda = \xi$. If we denote the proportionality factor as d, we can rewrite the system (5.8.6) in the form

$$\begin{aligned}
&\rho_\xi = d, \quad u_\xi = -\frac{a_\mathrm{f} d}{\rho}, \quad v_\xi = \alpha B_y d, \quad w_\xi = \alpha B_z d,\\
&(c_e)_\xi = \frac{(\gamma-1)c_e d}{2\rho}, \quad (B_x)_\xi = 0, \quad (B_y)_\xi = \beta B_y d,\\
&(B_z)_\xi = \beta B_z d, \quad u - a_\mathrm{f} = \xi.
\end{aligned} \tag{5.8.10}$$

Now in system (5.8.10) we substitute the equation for c_e by the equation for a_f, which can be easily obtained from Eq. (5.8.8) by direct differencing with respect to ξ and by using Eq. (5.8.10),

$$(a_\mathrm{f})_\xi = \frac{\vartheta d}{\rho}, \quad \vartheta = \rho\frac{\partial a_\mathrm{f}}{\partial \mathbf{U}}\cdot\mathbf{r}. \tag{5.8.11}$$

One can easily notice that the eigenvector (5.8.9) degenerates for $c_e = a_\mathrm{s}$. This happens if $B_y^2 + B_z^2 = 0$ and $c_e < a_\mathrm{A}$. In principle, we could choose a nondegenerate vector from Section 1.3.4, though the system obtained in that case would also require modifications. It is apparent from formulas (5.8.10) that the derivatives $(B_y)_\xi$ and $(B_z)_\xi$ become infinite at degeneration points. One can easily obtain, however, that

$$(B_{y,z}^2)_\xi = 8\pi(a_\mathrm{f}^2 - c_e^2)b_{y,z}^2\rho_\xi,$$

where $b_{y,z} = B_{y,z}/(B_y^2 + B_z^2)^{1/2}$. The right-hand side of this relation never degenerates if we assume, in accordance with Section 1.3.4, $b_y = b_z = 1/\sqrt{2}$ for $B_y^2 + B_z^2 \equiv 0$.

If we recall that

$$a_\mathrm{s} a_\mathrm{f} = c_e\frac{|B_x|}{2\sqrt{\pi\rho}},$$

the equations for the derivatives of the transverse components of the velocity vector can also be transformed to the form suitable for further application,

$$v_\xi = \frac{B_x}{4\pi\rho a_\mathrm{f}}(B_y)_\xi, \quad w_\xi = \frac{B_x}{4\pi\rho a_\mathrm{f}}(B_z)_\xi.$$

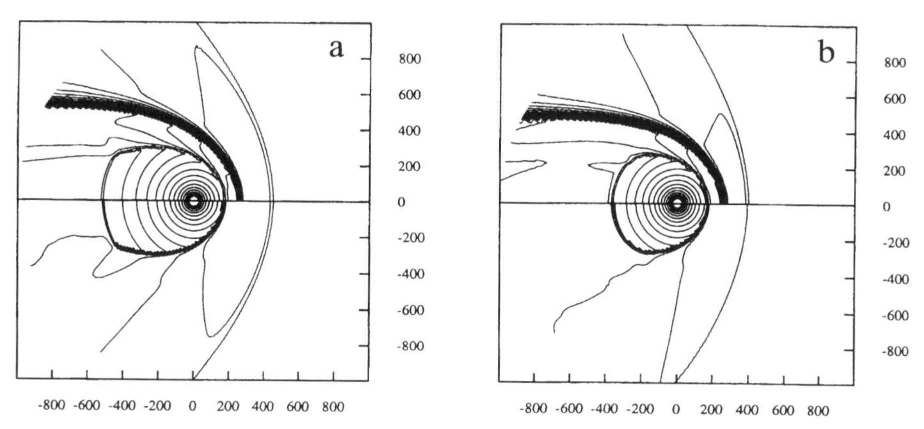

Figure 5.26 Pressure (below the symmetry line) and density logarithm isolines, $A = 10$ (a) and $A = 2$ (b): LF scheme.

Thus, by passing from d to ρ_ξ, we obtain the reduced system of equations that can be approximated similarly to Eq. (3.7.20):

$$
\begin{aligned}
&(a_{\rm f})_\Gamma = \left(\frac{u\vartheta + a_{\rm f}^2}{\vartheta + a_{\rm f}}\right)_0, \quad \rho_\Gamma = \rho_0\left(1 + \frac{u - a_{\rm f}}{\vartheta + a_{\rm f}}\right)_0, \\
&u_\Gamma = (a_{\rm f})_\Gamma, \quad (c_e)_\Gamma = (c_e)_0 + (\gamma - 1)\left(\frac{c_e}{2\rho}\right)_0 (\rho_\Gamma - \rho_0), \\
&(B_y^2)_\Gamma = B_{y0}^2 + 8\pi[(a_{\rm f}^2 - c_e^2)b_y^2]_0(\rho_\Gamma - \rho_0), \\
&(B_z^2)_\Gamma = B_{z0}^2 + 8\pi[(a_{\rm f}^2 - c_e^2)b_z^2]_0(\rho_\Gamma - \rho_0), \\
&(B_x)_\Gamma = (B_x)_0, \\
&v_\Gamma = v_0 + \left(\frac{B_x}{4\pi\rho a_{\rm f}}\right)_0 (B_{y\Gamma} - B_{y0}), \quad w_\Gamma = w_0 + \left(\frac{B_x}{4\pi\rho a_{\rm f}}\right)_0 (B_{z\Gamma} - B_{z0}).
\end{aligned}
\tag{5.8.12}
$$

Note that, in contrast to the purely gas dynamic case, the velocity components tangent to the boundary in the presence of the magnetic field, generally speaking, are different from their internal values.

5.8.3 Numerical results: axisymmetric case.

In this subsection we present the numerical results obtained for three different values of the Alfvén numbers $A_\infty = 10$, 2, and $\sqrt{2}$. In all considered cases, the y-components of vectors are assumed to be zero.

Regular configuration

The case with $A_\infty = 10$ corresponds to a very small magnetic field that only slightly influences the discontinuity pattern (see Fig. 5.26a). In the figure, the pressure logarithm (below the symmetry axis) and density logarithm isolines are presented.

In ideal MHD, magnetic field stresses are represented by tension along and by pressure perpendicular to **B**. For this reason, so long as the axisymmetric case is concerned, the increase of the LISM magnetic field results in the increase of the heliopause distance from the Sun, whereas its lateral distance to the symmetry axis becomes smaller. This effect can be seen in Fig.5.26b, which shows the similar isolines for the Alfvén number $A = 2$. This value corresponds to $p_{\text{magn}}/p_{\text{thermal}} \approx 0.83$ at infinity, that is, $B_\infty \approx 1.82$ (the dimensional value $\sim 1.6 \times 10^{-6}$ gauss), which is well below the estimated limits. Twenty equidistant isolines are shown in all isoline charts. In this case the bow shock is a fast MHD shock. Since $M_\infty = A_\infty = 2$, the maximum speed of the disturbance propagation parallel to the LISM streamlines remains the same as that in the absence of magnetic field. This magnetic field value is not sufficient to produce on the heliosheath effect comparable with that due to the presence of neutral atoms (Baranov and Zaitsev 1995). The size of the region between the bow shock and the heliopause is very important, however, in view of the charge exchange processes occurring inside it. Of course, stronger magnetic field (smaller Alfvén numbers) substantially affects the discontinuity pattern (Pogorelov and Semenov 1997a). Although magnetic field is absent in the region inside the heliopause, its action is revealed by the increased value of the total pressure at infinity. This leads to substantial decrease of the termination shock stand-off distance in the downstream region. Further increase in the magnetic field decreases the effective Mach number of the LISM flow, and the bow shock stand-off distance becomes smaller. The results shown in Fig. 5.26 were obtained with the help of the Lax–Friedrichs type scheme. For comparison, in Fig. 5.27 we show the discontinuity pattern for $A_\infty = 2$ obtained using the CIR-type scheme. Recall that numerical fluxes in such schemes are calculated via quantities obtained by simple arithmetic averaging of those on the left- and right-hand sides of the computational cell boundary. It is apparent that the resolution of discontinuities is considerably better in this case. On the other hand, it is worth admitting that we encountered certain difficulties in obtaining the steady-state solution in the vicinity of the stagnation point on the heliopause. This can be explained by the necessity to enforce the "entropy" condition in this region. For this purpose we applied the entropy fix procedure described in Section 2.10 to the linearly degenerate characteristic fields. Though the steady-state solution was finally obtained, this required larger number of time iteration cycles that in the case of applying the TVD Lax–Friedrichs scheme.

If $A_\infty < 1$, the bow shock must disappear. In this case the system (5.8.1) becomes elliptic. This justifies the development of the subsonic interaction models (Steinolfson 1994; Khabibrakhmanov and Summers 1996; Zank at al. 1996). The methods that should be applied to solution of the problem in this case lie outside the scope of this book.

Irregular interaction

If the magnetic field vector is parallel to the shock normal, we have a parallel shock. The behavior of parallel shocks depends on whether the Alfvén velocity ahead of them is smaller or larger than the acoustic speed of sound (see Section 5.3.3). In the former case, slow MHD shocks do not exist, while fast shocks are evolutionary and admissible for all their intensities if the flow is supersonic ahead of them. In the latter case, on the contrary, slow shocks are admissible for all their intensities if the flow is supersonic and sub-Alfvénic ahead of them, while fast shocks are admissible only in a certain range of parameters ahead of the shock, even if they correspond to a super-Alfvénic flow. Singular shocks are those for which the tangential component of magnetic field is equal to zero ahead of (behind) the shock and not

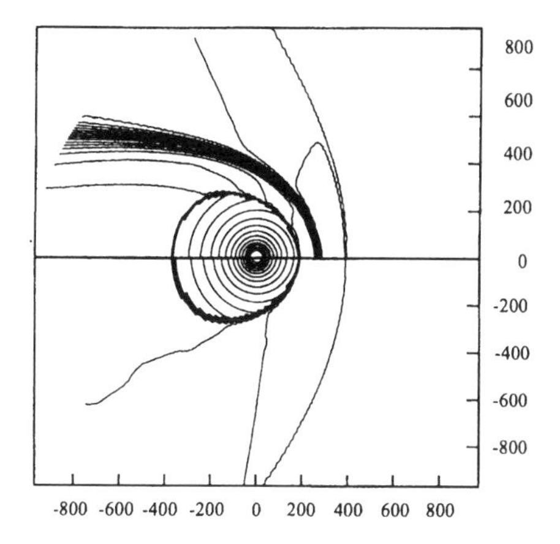

Figure 5.27 Pressure (below the symmetry line) and density logarithm isolines, $A = 2$: Roe-type scheme.

zero behind (ahead of) it. Such shocks are called switch-on (switch-off), as the tangential component of the magnetic field vector is switched on (off) at them. Switch-on shocks are always fast, while switch-off shocks are always slow, since the tangential component of the magnetic field always increases across fast and decreases across slow MHD shocks.

Let us consider what happens if we increase the value of the LISM magnetic field B_∞ with the rest of the LISM quantities being fixed. We have two dimensionless parameters relating the quantities ahead of the shock, namely, the Mach number $M_\infty = V_\infty/a_\infty$ (a_∞ is the acoustic speed of sound) and the Alfvén number $A_\infty = V_\infty/a_{A\infty}$. If $a_\infty > a_{A\infty}$ for $M_\infty > 1$, the forward point of the bow shock corresponds to a fast parallel shock that is always realizable. If we further increase B_∞, sooner or later $a_{A\infty}$ will become larger than a_∞ with $A_\infty > 1$. In this case the parallel shock, though remaining fast, will be still evolutionary until B_∞ acquires the value corresponding to the interval (see Section 5.3.3),

$$1 < A_\infty < \sqrt{\frac{(\gamma + 1)\, M_\infty^2}{2 + (\gamma - 1)M_\infty^2}}. \tag{5.8.13}$$

For A_∞ from the interval (5.8.13), the Alfvén number behind the shock is smaller than 1, thus resulting in a trans-Alfvénic shock that is not evolutionary. Occasionally, a singular (fast switch-on) shock becomes admissible exactly in this range of A_∞ (Section 5.3.3). On switch-on shocks $\mathbf{B} \parallel \mathbf{n}$ ahead of the shock but $\mathbf{B} \not\parallel \mathbf{n}$ behind it. That is, a tangential component of the magnetic field must appear at the forward point of the bow shock. On the other hand, this cannot occur due to geometrical reasons, since there is an infinite number of switch-on directions and all of them are equivalent. It could be fairly easily expected that the structure of the flow for the mentioned values of B_∞ must be different from that in the regular case.

The outlined problem is of fundamental importance in MHD modelling of astrophysical phenomena, regardless of the fact that we discuss it only in the application to the particular problem of the SW–LISM interaction. The MHD shock behavior is not only important

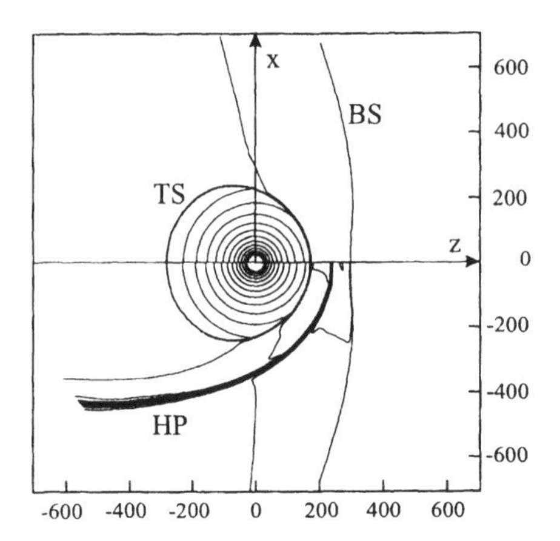

Figure 5.28 General configuration of the interaction: density (below the symmetry axis) and total pressure logarithm isolines.

for the interpretation of observational data in the physics of heliosphere, but it also must be quite clear to those creating numerical codes for MHD simulations. In contrast to pure gas dynamic problems, numerical solution of which can generally be obtained by direct solution of the discretized Euler system, more complicated physical phenomena, though also governed by hyperbolic systems, require deeper understanding of all mathematical aspects accompanying them (Kulikovskii et al. 1999).

The flow pattern originating at the forward point of the bow shock in the nonevolutionary interval was discussed by Steinolfson and Hundhausen (1990) and Matsuda and Fujimoto (1993). Baranov and Zaitsev (1995) showed impossibility of the stationary shock configuration presented in the latter paper but did not explain the reason of the phenomenon. Besides, they failed to obtain the solution in the questionable range of Alfvén numbers. Myasnikov (1997) obtained the solution similar to that of Matsuda and Fujimoto (1993), which turned out to be weakly nonstationary so that final solution could not be established. De Sterk et al. (1999) studied the plasma flow around an infinite cylinder and found out a variety of MHD shocks, some of them nonevolutionary.

In this section we present the results of high-resolution calculations in the two-dimensional axisymmetric and 2.5-dimensional statements. In the latter case we add a small axisymmetric rotational perturbation to the LISM velocity and magnetic field. As a result, only evolutionary shocks remain in the interaction region. This approach lies in the framework of the approach of Barmin et al. (1996), who studied the behavior of the nonevolutionary compound wave under action of rotational perturbations.

Let us choose such a strength of the magnetic field B_∞ for which $A_\infty = \sqrt{2}$, that is, we are within the interval (5.8.13). The dimensionless magnetic field in our statement is

$$\overline{B}_\infty = \frac{B_\infty}{V_\infty \sqrt{\rho_\infty}} = \frac{2\sqrt{\pi}}{A_\infty} = \sqrt{2\pi}.$$

This corresponds to $B_\infty \approx 2.3\ \mu$Gs.

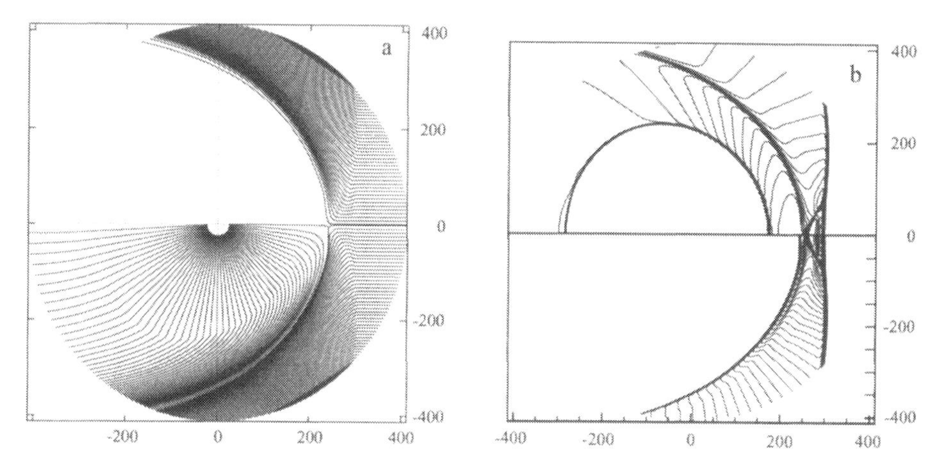

Figure 5.29 (a) Streamlines (below the symmetry axis) and magnetic field lines and (b) density (below the symmetry axis) and thermal pressure isolines.

Calculations for this value of the LISM magnetic field are performed in the half-circular region between $R_{\min} = 24$ AU and $R_{\max} = 1000$ AU. The number of cells is 501 in the radial and 504 in the angular direction. Such a mesh is necessary to resolve the fine structure of the SW–LISM interaction for the parameters corresponding to the interval (5.8.13)

In Fig. 5.28 we present a general view of the solar wind and interstellar medium interaction for the two-shock model (Baranov et al. 1971). The picture shows the chart of 19 equidistant isolines of the density logarithm (below the symmetry axis) and of the total pressure logarithm located between the minimum and maximum values. Here BS, TS, and HP are the bow shock, the termination shock, and the heliopause, respectively.

If we compare this figure with Figs. 5.26 and 5.27, it becomes clear that by increasing the space resolution and by avoiding transient solutions we obtained qualitatively different shape of the bow shock. It is now concave in the vicinity of the symmetry axis. In Fig. 5.29a we show the streamlines (below the axis) and the magnetic field lines. The prominent feature in the streamline behavior is that there exists a region around the symmetry axis where they decline toward it, in contrast with their usual inclination for larger Alfvén numbers (or weaker magnetic fields). Later they turn in the opposite direction at the additional, or secondary, shock, which is clearly seen in Fig. 5.29b. In this figure we present the thermal pressure isolines (20 isolines between the minimum and the maximum) and the density isolines (below the symmetry axis). In this case, to visualize the region of our interest, we show only isolines between 1 and 2.6 with the increment 0.04. Note that the thermal pressure suffers a jump at the heliopause everywhere except for the stagnation point. That is why the intensity of this jump decreases toward the axis until it degenerates into a compression fan. In the density isolines we can see a high-density layer. It is also called an entropy layer, since there is no jump of pressure across it. This layer originates owing to the difference in entropy between the streamlines crossing the secondary shock at right angles near the axis and at acute angles farther from it. This results in existence of the elongated proton density wall near the heliopause.

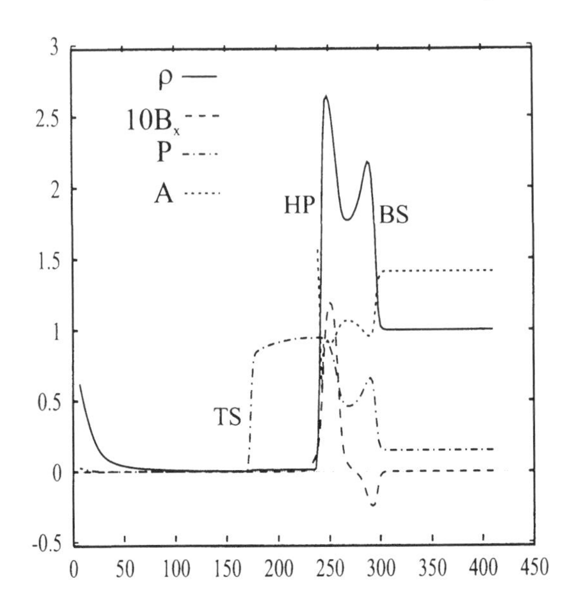

Figure 5.30 Parameter distributions along the line $x = 1.296$.

As far as the evolutionary shocks are concerned, we would like to look at the parameter distributions along the symmetry axis. We show the profiles of several variables along the line $x = 1.296$ in Fig. 5.30. We do not have computational points on the symmetry axis, and the above line is the closest to it available. It is clearly seen that the Alfvén number $A = 2\,|\mathbf{V}|\sqrt{\pi\rho}/|\mathbf{B}|$ decreases from the value above to the value below 1 across both the bow and the secondary shock. This means that these shocks are nonevolutionary in three dimensions. A small x-component of the magnetic field appears behind the bow shock. Later B_x smoothly tends to zero, changes its sign, then increases at the secondary shock, and disappears at the HP. The Alfvén number profile is abruptly interrupted, since A increases to infinity across the HP. The flow pattern presented in this section is qualitatively different from that obtained by Matsuda and Fujimoto (1993) and Myasnikov (1997). It also differs from the pattern obtained by Pogorelov and Semenov (1997a) for larger Alfvén numbers. Thus, in contrast with pure gas dynamics, the variation in the LISM parameters can cause substantial qualitative changes in the SW–LISM interaction pattern.

5.8.4 Numerical results: rotationally perturbed flow. It is obvious from the general theory of evolutionary shocks that the bow shock in the vicinity of the symmetry axis must be modified by a rotational perturbation of the flow. For this reason, remaining in the framework of the axisymmetric statement, we add the y-component (perpendicular to the xz-plane) to the magnetic field and velocity vectors, namely, $\overline{B}_y = 0.05$ and $\overline{V}_y = 0.05/\overline{B}_{z\infty}$. After that we solve the MHD system in the 2.5-dimensional formulation, that is, taking into account the variation of the y-components. The discontinuity pattern is shown in Fig. 5.31a, where the same density isolines are shown as in Fig. 5.29b. Though the picture has changed quantitatively, the general structure of the flow remains similar, since the secondary shock

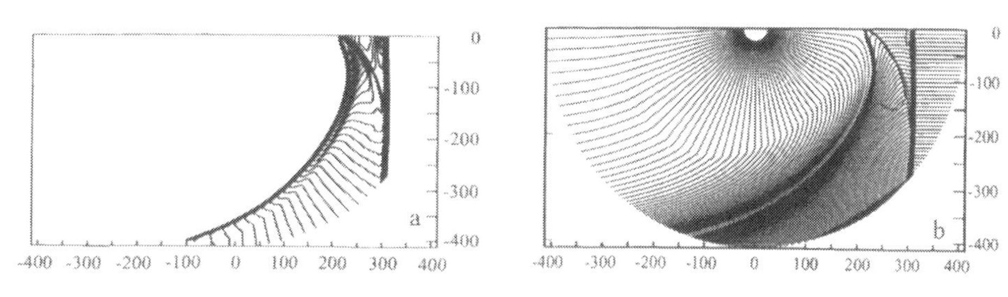

Figure 5.31 Density isolines (left) and streamlines and $A = 1$ lines: perturbed LISM (right).

and the density wall still exist. The streamlines pattern is also essentially the same, as seen from Fig. 5.31b, where they are shown together with the shocks and the isoline $A = 1$. We see that the secondary shock becomes a switch-off shock, since its surface coincides with the line $A = 1$ and the velocity (and magnetic field) vector looses its tangential component behind this shock.

There is also a very narrow region around the symmetry axis where $A < 1$. To look closer into this region, we present the parameter distributions along the line $x = 1.296$ (Fig. 5.32). It is clear that the Alfvén number becomes less than unity behind the shock. On the other hand, we see that, within a single density jump, the x-component of the magnetic field is switched on and then switched off without any space interval. This means that we have effectively two merged shocks near the axis. The secondary shock is well seen in the total pressure profile closer to the HP (compare it with the Alfvén number profile, which drops from 1 to a smaller value behind this shock). In principle, this merged shock is also nonevolutionary (Jeffrey and Taniuti 1964) and could have been destroyed if the rotational component of the magnetic field vector, owing to the boundary conditions, were not zero on the symmetry axis.

If x is approximately larger than 50, the switch-off shock in the pair of merged shocks disappears, as seen from Fig. 5.33 showing profiles along the line $x = 55.5$, and the bow shock becomes a single fast MHD shock with $A > 1$ behind it. As we move from $x = 0$ to $x = 55.5$, the intensity of the above switch-off shock gradually decreases.

As we pass from the concave part of the bow shock to its convex part, we meet the inflection point. At this point the secondary switch-off shock seems to approach the bow shock, though the density jump across it is rather small. If we look at the parameter profiles along the line $x = 153$ (Fig. 5.34) that passes in the vicinity of the point of the bow shock maximum convexity, we see that this shock again becomes a switch-on shock, since $A = 1$ behind it. The tangential component is switched, however, in the direction from the symmetry axis in this case. As we move farther from the mentioned point, the bow shock transforms into a fast MHD shock. Note that the y-component of the magnetic field acquires rather large values, though initial perturbation was very small. This means that the flow within $x < 200$ is sensitive to rotational perturbations. Farther from the symmetry axis the y-component becomes much smaller behind the bow shock than the x-component (a fraction of one percent).

We calculated the SW–LISM interaction problem for a controversial case of Alfvén

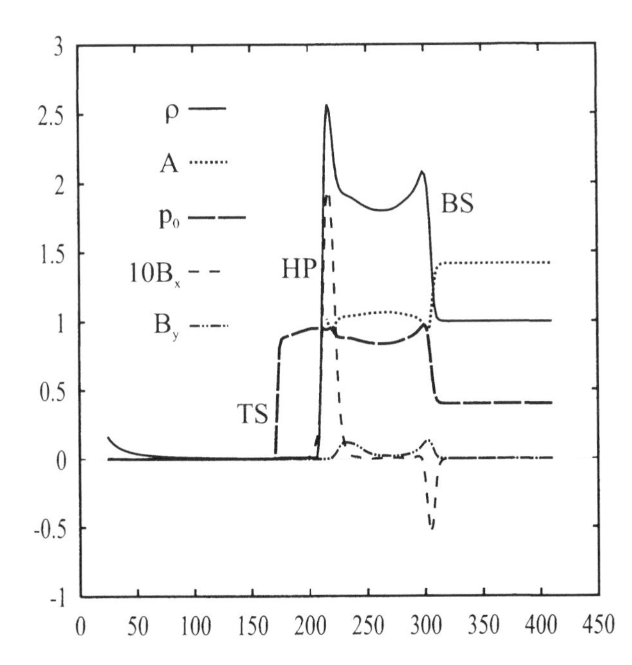

Figure 5.32 Parameter distribution along the line $x = 1.296$: perturbed LISM.

numbers close to unity. The flow structure in the region around the symmetry axis in this case is completely different as compared with that for larger values of A_∞. Besides, a high-entropy layer originates, which spreads over the heliopause up to $x \approx 300$ AU. The latter effect may be important in the interpretation of observational data.

We showed that nonevolutionary MHD shocks can be successfully avoided by adding out-of-plane perturbations. Such kinds of shock often originate if three-dimensional MHD equations are reduced to two-dimensional ones or even in 3D cases if plasma is not ideal and/or if out-of-plane perturbations are forcefully diminished. In this case nonevolutionary shocks can exist for a long time depending on the value of molecular and magnetic viscosities (see Barmin et al. 1996). No need to say that perturbations of any kind can be encountered in the LISM, since it can hardly be assumed absolutely uniform. The flow pattern that can be realized for other type of disturbances may be different, of course, quantitatively, although nonevolutionary shocks will either be absent or manifest themselves as time-dependent shock-like structures. The flow pattern obtained in the axisymmetric case can exist as a transient one. It is stationary only if three-dimensional disturbances do not act at all. This pattern, however, can be rather persistent to disturbances in the numerical treatment. This is due to the numerical viscosity and resistivity. In this case the disturbance must be either large enough or act for a sufficiently long time (have low frequency).

Presented numerical results show that the LISM magnetization can be of great importance for the interpretation of data obtained in the space experiment. On the other hand, the result of the magnetic field influence on the whole structure of the flow for $\mathbf{B}_\infty \not\parallel \mathbf{V}_\infty$ can be more complicated. It is clear that the flow pattern in this case is three-dimensional.

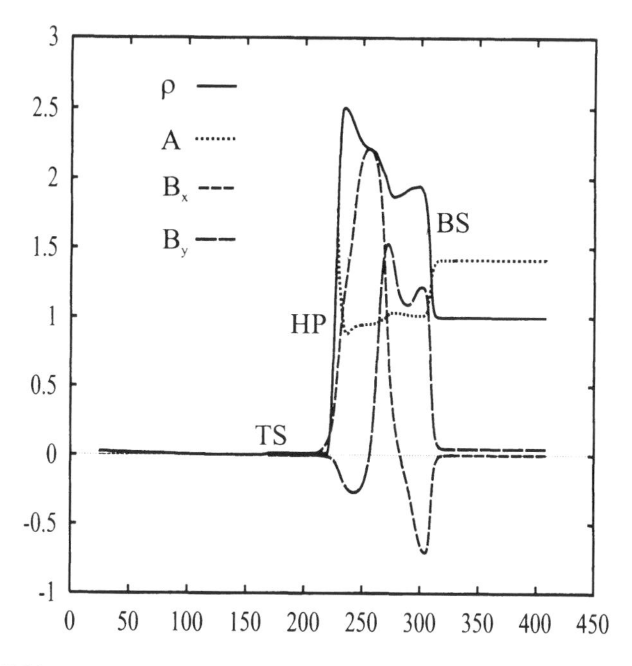

Figure 5.33 Parameter distribution along the line $x = 55.5$: perturbed LISM.

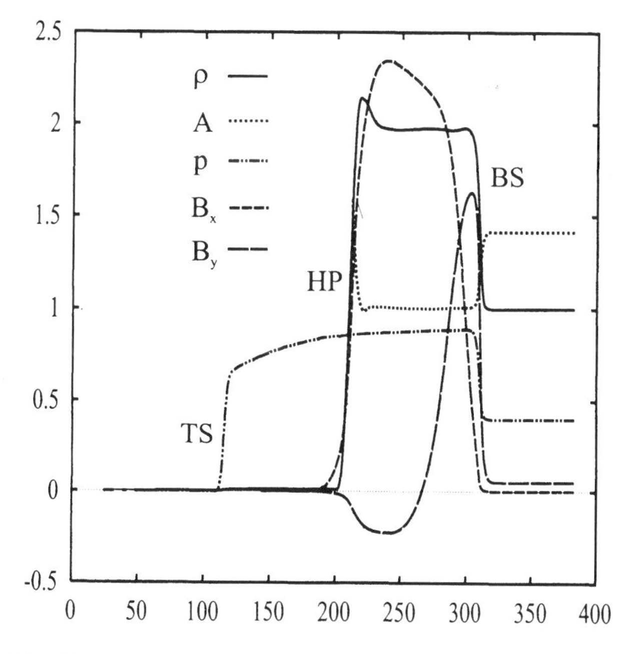

Figure 5.34 Parameter distribution along the line $x = 153$: perturbed LISM.

Besides, the velocity of propagation of fast magnetosonic disturbances along the LISM velocity vector can become much larger if the angle between $\mathbf{B}_\infty$ and $\mathbf{V}_\infty$ approaches to $\pi/2$. This means that the bow shock in the form of the fast magnetosonic shock is more likely to vanish.

The performance of the algorithm for implementation of the magnetic field divergence-free condition can be seen from Fig. 5.29a. The magnetic field lines remain parallel to the streamlines. One can also see that no magnetic field penetrates into the heliosphere. This result cannot be obtained without a special treatment of the divergence-free condition.

5.8.5 A note on the MHD flow over an infinitely conducting cylinder. We mentioned earlier the numerical results of parametric calculations of MHD plasma flow over an infinitely conducting cylinder and a sphere by De Sterk et al. (1999) (see also De Sterk, Low, and Poedts 1998; De Sterk and Poedts 1999; De Sterk 1999). Analysis of these results leads us to the conclusion that the necessity to solve the SW–LISM interaction problem as a whole may have resulted in the accuracy decrease in the vicinity of the complicated shock-wave structure occurring in the irregular regime of the flow. On the other hand, by solving a two-dimensional problem we can implement disturbances impossible under the restriction of axial symmetry. To clear out the point, we obtained the solution of the field-aligned ideal plasma flow over an infinitely conducting cylinder. The dimensionless parameters of the uniform flow are the Mach number $M_\infty = \sqrt{13.5}$, the Alfvén number $A_\infty = 1.5$, and the specific heat ratio $\gamma = 5/3$.

The steady-state pattern is shown in Fig. 5.35a, where the isolines of density logarithms are presented. It is easy to check that the parameters of the considered case belong to the interval (5.8.13). In contrast with our solution of the SW–LISM interaction problem, we managed to resolve two more shocks. It is clear that the triple point turned out to be detached from the bow shock. In Fig. 5.36 we show the enlarged picture of the zone of our interest. In addition, the magnetic field lines are presented in the same figure. The analysis shows that if we move upward from the point of the bow shock intersection with the symmetry axis, this shock is fast ($1 \rightarrow 2$) and evolutionary. Note that the bow shock is concave outward, in agreement with our solution of the SW–LISM interaction problem and the results by De Sterk et al. (1998). We see that further on this shock becomes convex outward, and we must have an inflection point on its profile. From this point the shock becomes nonevolutionary ($1 \rightarrow 3$). This lasts until we reach the extremum convex point where the $1 \rightarrow 3$ shock transforms into the switch-on ($1 \rightarrow 2 = 3$) shock. Further upward it becomes fast and evolutionary again. All the shocks lying between the bow shock and the body are nonevolutionary. One can also see several tangential discontinuities in the shock layer. Their presence becomes clear if we inspect the streamline behavior in Fig. 5.36.

Note that in our case the streamlines are everywhere parallel to the magnetic field lines. As noted by De Sterk (1999), the obtained solution is metastable if we introduce a small angle between $\mathbf{V}_\infty$ and $\mathbf{B}_\infty$. It is important that there is no necessity to pass to the three-dimensional formulation of the problem in order to destroy the obtained solution. As clear from the discussion given in Section 5.3.3, it suffices to introduce small magnetic field and velocity components parallel to the cylinder axis (preserving the flow alignment with the magnetic field). In this way, we let the magnetic field vector go out of the initial plane. Since in this case we distinguish one of the directions, it becomes the direction

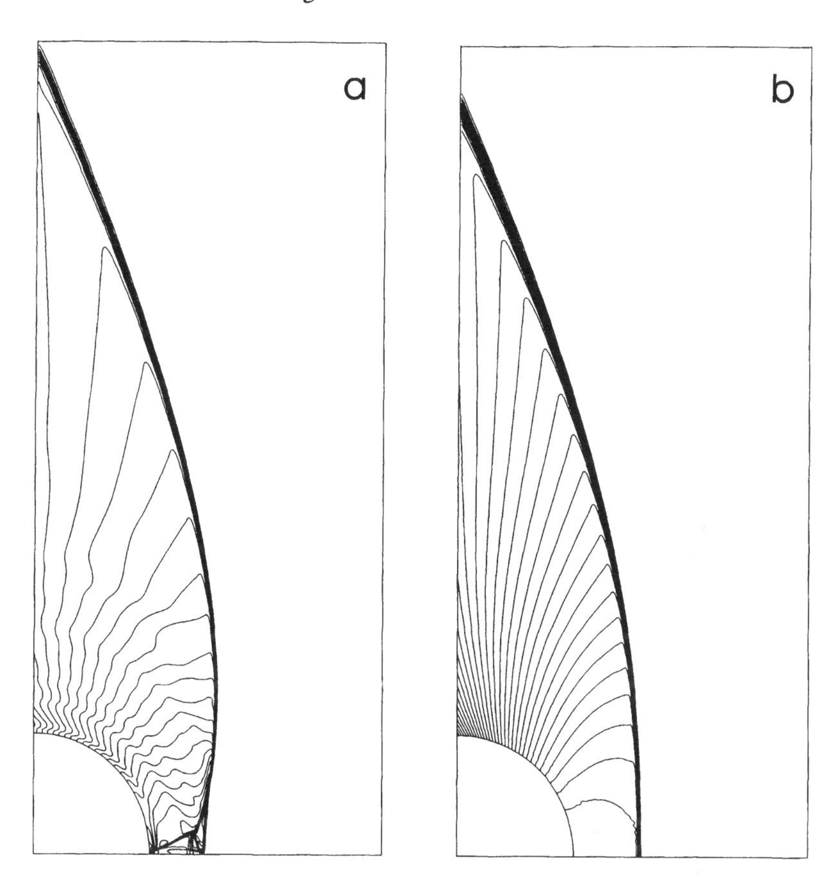

Figure 5.35 Density isolines for the purely planar (a) and perturbed (b) flows of ideal plasma over an infinitely conducting cylinder.

for switching on the magnetic field transversal component in the forward part of the bow shock, and no controversial irregular configurations are necessary to account for the flow symmetry. The steady-state solution is shown in Fig. 5.35b. The bow shock is everywhere fast; all additional discontinuities disappeared, and the solution in the plane perpendicular to the cylinder axis looks very much like those occurring in the regular flow. Note that by introducing the disturbance we in no way destroyed the invariability of the flow along the cylinder axis. Of course, the velocity and magnetic field components along it acquire fairly large values behind the bow shock. One may certainly argue that we introduced the disturbance of the uniform flow, which acted long enough to destroy the viscous structure of the initial solution, and that such a solution can exist for some time if the sign of the disturbance is eventually reversed. This is not as true as in the scenario of the evolution of plane trans-Alfvénic shocks described in Section 5.5. Recall that in that case, by reversing the perturbation, in certain cases we could preserve the presence of the nonevolutionary shock wave structure in our solution. Now, even if after some time we change the direction

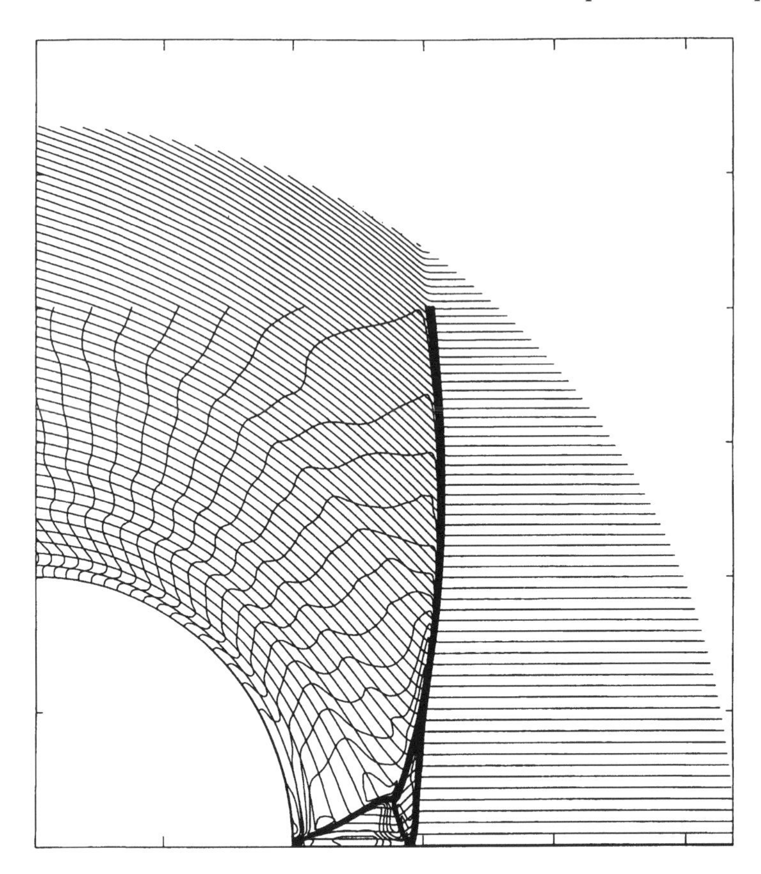

Figure 5.36 Density isolines and magnetic field lines for the purely planar cylinder.

of the magnetic field component parallel to the cylinder axis in the uniform flow, we will never eliminate the regions with high values of this component inside the shock layer. Thus, the solution of the first nonevolutionary kind will never appear again.

It is interesting to look at the three-dimensional bow shock obtained by De Sterk (1999) in the case of nonzero angle between $\mathbf{V}_\infty$ and $\mathbf{B}_\infty$ (Fig. 5.37). We see that various types of nonevolutionary shocks seem to be inevitable. The reason is in the fact that one can always choose a plane containing both the velocity and magnetic field vectors. It is apparent, however, that if we would like to destroy the nonevolutionary solution, it is necessary to perturb the symmetry plane. This means that if nonevolutionary shocks are expected to exist in the symmetric statement, the problem must be solved by disregarding the presence of this plane, and out-of-plane disturbances of $\mathbf{B}_\infty$ and $\mathbf{V}_\infty$ must be introduced.

5.8.6 Numerical results: three-dimensional modelling. In the previous subsection we described the influence of the magnetic field magnitude on the flow structure in the axisymmetric statement with $\mathbf{B}_\infty \parallel \mathbf{V}_\infty$. Our task now is to show that the variation of

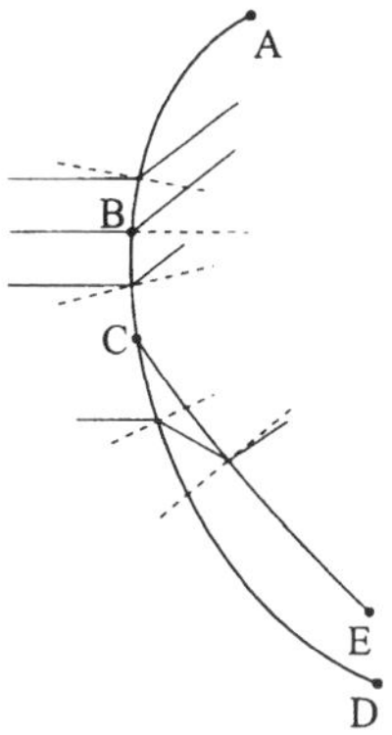

Figure 5.37 Three-dimensional bow shock pattern. B is the point where the external magnetic field is normal to the shock surface. AB , BC, and CD are fast, $1 \rightarrow 3$ trans-Alfvénic, and fast shocks, respectively. CE changes its type from $2 \rightarrow 4$ trans-Alfvénic to slow.

the angle between $\mathbf{B}_\infty$ and $\mathbf{V}_\infty$ can remarkably influence the shape of the heliopause and the whole flow structure even for $A_\infty = 2V_\infty\sqrt{\pi\rho_\infty}/B_\infty = 2$ (note that nonevolutionary shocks do not appear in this range of parameters). This result is due to the increase in the maximum speed of a small perturbation propagation in the direction parallel to the LISM velocity vector. In this case the Mach number calculated via the fast magnetosonic speed becomes smaller, resulting in a substantial increase of the bow shock stand-off distance. The intensity of this shock becomes smaller. In Figs. 5.38a and 5.38b the density decimal logarithm isolines are shown, respectively, in the xz plane, where both magnetic field and velocity vectors of the LISM at infinity are located, and in the "equatorial" plane (yz-plane) for the angle between $\mathbf{B}_\infty$ and $\mathbf{V}_\infty$ equal to $\pi/2$. Both planes are symmetry planes in this case. In Figure 5.39 the SW and LISM streamlines and the magnetic field lines are presented in the xz-plane. One can see that the heliopause is contracted in the equatorial plane in comparison with its profile in the xz plane as a result of the combined action of the magnetic field tension and pressure. Magnetic field lines become wrapped around the equatorial parts of the heliopause. In addition, magnetic pressure acting on the heliopause surface in the vicinity of the critical streamline pushes it closer to the Sun.

If magnetic field lines are directed at an angle $\pi/4$ to the LISM velocity at infinity, the contact discontinuity becomes tilted with respect to its position for $\mathbf{B}_\infty \parallel \mathbf{V}_\infty$ in a way consistent with the approximate consideration of Fahr et al. (1988). This effect is due to the highly anisotropic distribution of the magnetic pressure over the heliopause. The effective Mach number also changes its value along the bow shock, resulting in the strong variation of its stand-off distance. This can be seen in Figs. 5.40a and 5.40b, which show the density logarithm distribution in the symmetry (xz) and the equatorial (yz) planes, respectively. The streamlines and the magnetic field lines for this case are presented in Fig. 5.41. The wrapping behavior of the magnetic field lines can be qualitatively seen in Figure 5.42, which shows the lines starting from the plane slightly above the symmetry plane and parallel to it.

In Fig. 5.43 the density profiles along the z-axis are shown. The cases shown correspond

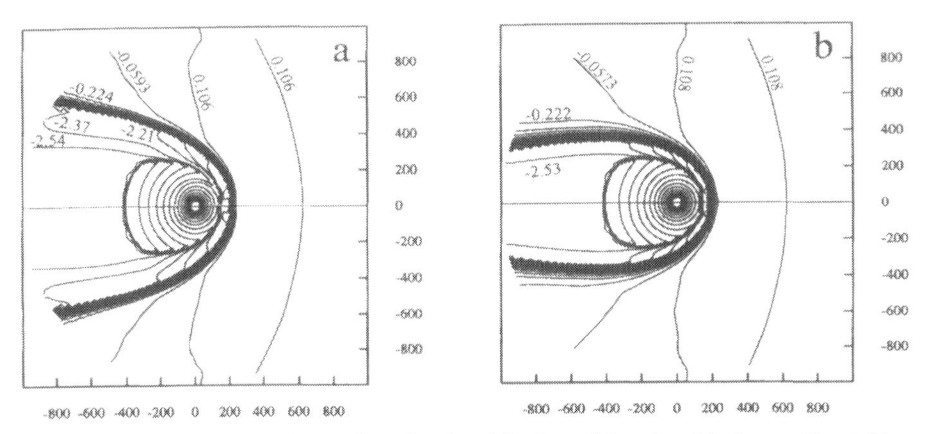

Figure 5.38 Density logarithm isolines for $A = 2$ in the xz (a) and yz (b) planes: $\mathbf{B}_\infty \perp \mathbf{V}_\infty$.

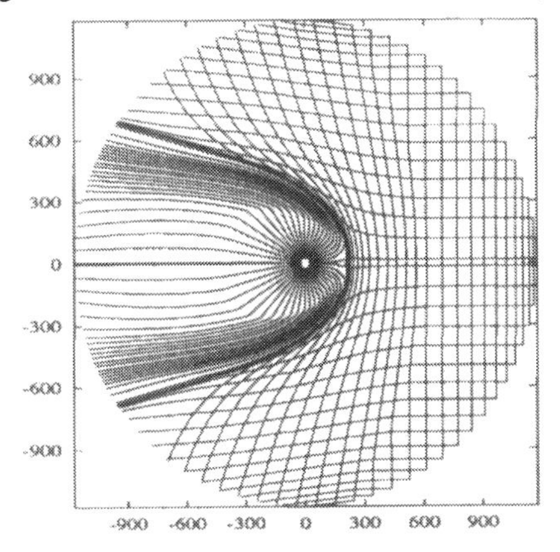

Figure 5.39 Streamlines and magnetic field lines for $A = 2$ in the xz plane: $\mathbf{B}_\infty \perp \mathbf{V}_\infty$.

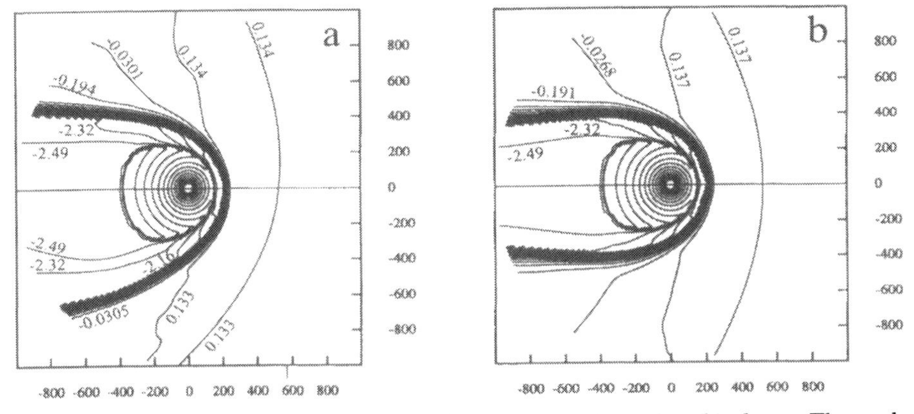

Figure 5.40 Density logarithm isolines for $A = 2$ in the xz (a) and yz (b) planes. The angle between $\mathbf{B}_\infty$ and $\mathbf{V}_\infty$ is equal to 45°.

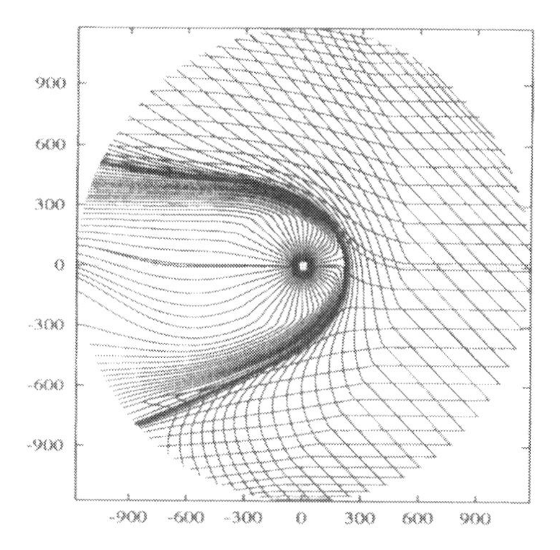

Figure 5.41 Streamlines and magnetic field lines for $A = 2$ in the xz plane. The angle between $\mathbf{B}_\infty$ and $\mathbf{V}_\infty$ is equal to 45°.

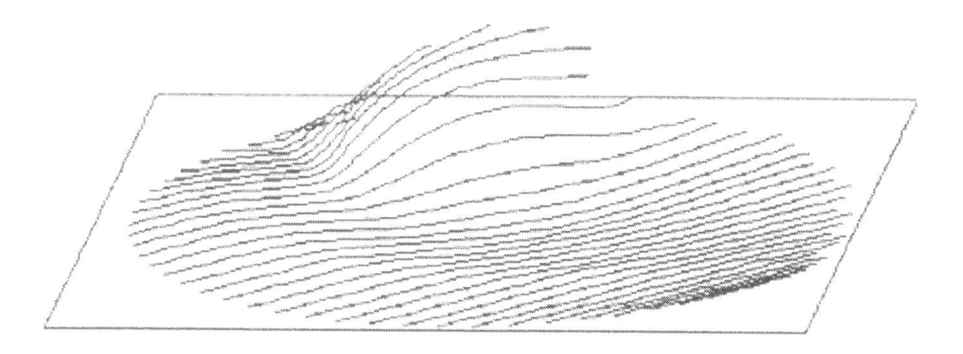

Figure 5.42 Magnetic field lines wrapping the heliopause.

to the absence of the magnetic field $\mathbf{B} = \mathbf{0}$ and to the angle α between $\mathbf{V}_\infty$ and $\mathbf{B}_\infty$ equal to 0, $\pi/2$, and $\pi/4$. The discontinuities inherent in the flow pattern are resolved with a reasonable accuracy, taking into account the small number of cells in one of the angular directions ($R \times \phi \times \theta = 99 \times 21 \times 116$), and without any spurious oscillations. Note that no artificial viscosity was added in the algorithm. The density jump across most of the shocks lies within 3–5% of the exact value. Somewhat larger errors can occur in the regions of vanishing density and pressure near the Mach disk in the rear side of the termination shock, although they cannot produce any substantial effect on the overall picture.

The results of calculations presented in this chapter show the peculiarities of the application of high-resolution numerical methods to complicated discontinuous MHD flows. It is quite clear that, although Godunov's methods show good performance close to that in pure gas dynamics, the presence of nonunique solutions to the MHD Riemann problem in its plane statement can cause the origin of transient solutions that are only due to the

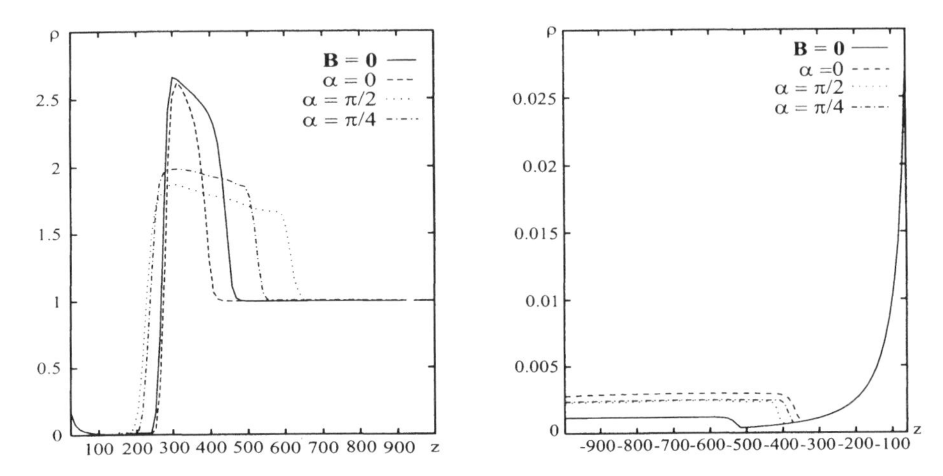

Figure 5.43 Density profiles along the z axis.

dissipation intrinsic in all numerical methods. These solutions have little connection with realistic phenomena occurring in plasma in the presence of physical dissipation. Moreover, if numerical dissipation is much larger that the physical one, extraneous nonevolutionary solutions can be obtained in the framework of application of shock-capturing numerical methods. Numerical schemes that automatically rule out these spurious solutions are still to be developed.

Dissipation processes are widely believed to provide the mechanism for selection the admissible solutions in ideal gas dynamics and MHD. This is caused by the fact that higher-order, small-scale, physical models for these fields of mechanics of continuous media contain second-order derivatives. On the contrary, some physical phenomena contain third-order derivatives in the small-scale models. This means that dispersion can be of major importance in the selection of admissible solutions in the framework of a large-scale model. Such phenomena will be discussed in Chapter 7.

Chapter 6
Solid Dynamics Equations

In this chapter we describe the construction of explicit numerical shock-capturing methods of the Godunov type for some systems of the solid dynamics equations. All these methods are based on the solution of a one-dimensional Riemann problem. We present CIR-type numerical schemes using the solution of a linearized Riemann problem solver and illustrate their specific features as applied to various solid dynamics systems.

Solid dynamics deals with the most general type of motion of deformable media, including, in particular, gas dynamic, shallow water, and magnetohydrodynamic flows; see Sedov (1971) and Maugin (1988). In our brief presentation we shall discuss two additional phenomena, elastic and plastic processes in isotropic media.

First we consider systems that generalize gas dynamic systems, see Section 3.1. The systems to be considered can include an arbitrary equation of state (EOS) and take into account elastic and plastic processes. For such systems, the problem arises to construct numerical schemes for hyperbolic systems written in nonconservative or partly conservative form. The CIR method described previously permits us to do this without difficulties. The developed numerical algorithms can adequately predict the dynamics of moving discontinuities and simulate monotone profiles of grid variables in the vicinity of discontinuities and in domains of large gradients of functions.

We consider also second-order hyperbolic systems that describe the dynamics of thin elastic shells, including the so-called Timoshenko-type systems. For particular systems we discuss their basic features. For example, some systems with special right-hand sides can have rapidly oscillating components of the solution. This fact may lead to development of instabilities in the numerical calculation. A special algorithm is developed for solution of such systems.

This chapter covers the basic approaches that may be applied in the investigation and construction of difference schemes for simulation of a number of other hyperbolic systems of solid dynamics equations. Solid dynamics is a rapidly developing branch of physics. For this reason the presentation of solid dynamics in a single chapter cannot be comprehensive. We provide proper references to the literature where particular issues are treated in more detail.

6.1 System of governing equations

Dozens of different forms of hyperbolic systems have been developed so far in solid dynamics; see, for example, Thomas (1961), Germain (1973), Godunov (1978), Kolarov, Baltov, and Boncheva (1979), Berdichevskii (1983), Kondaurov and Nikitin (1990), Bykovtsev and

Ivlev (1998), Godunov and Romenskii (1998), etc. This can be attributed to numerous types of motion of continuous deformable media and a large number of rheological models. Unlike gas dynamics, the choice of one or another rheological model in solid dynamics can substantially change the structure and behavior of the governing hyperbolic system. The variables required for the description of the motion of a medium, as well as their number, can be quite different, depending on the adopted model. The model itself can depend on the values of deformations, coordinate systems, and the processes and phenomena included in the model, e.g., linear and nonlinear elasticity, plasticity, electroplasticity and superelasticity, hardening and softening, fracture and cracking, viscosity, creep, brittleness, heredity, anisotropy, and others; for details, see Bell (1973), Lurie (1980), Zukas et al. (1982), Kobayashi (1987), Ionov and Selivanov (1987), Pluvinage (1989), Klyushnikov (1993), Bushman et al. (1993), Pavlov and Khokhlov (2000) and others. The consideration of the medium on the microlevel requires taking into account internal electromagnetic fields and related phenomena, such as magnetoelasticity (Ambartsumyan, Bagdasaryan, and Belubekyan 1977), piezoelectricity, elastic dielectrics, ferromagnetics, magnetothermoelasticity, and others (Maugin 1988; Godunov and Romenskii 1998).

The number of models used in solid dynamics is too large to be covered here. For this reason we restrict ourselves to the consideration of the following systems.

First we discuss those of solid dynamics systems that are simple generalizations of gas dynamic equations and describe, for some values of the parameters involved, usual ideal gas dynamic flows with one or another equation of state (EOS). Such solid dynamics systems arise when studying the motion of a nonideal medium in the case where the usual gas dynamic pressure p is insufficient to describe the internal contact interaction. In particular, the pressure p in ideal gas dynamics acts along the normal to the surface of the volume element and cannot take into account for friction forces, which are tangent to the surface.

Further, we consider systems of second-order equations that govern the dynamics of thin elastic shells. In particular, we discuss the dynamics equations for cylindrical and orthotropic shells, as well as the Klein–Gordon equation, which is the simplest case of the dynamics equations for thin elastic shells.

Consideration of these systems will allow us to demonstrate some specific features of solid dynamics equations and present approaches that may be helpful in constructing numerical methods for various solid dynamics equations.

6.1.1 Solid dynamics with an arbitrary EOS.

Consider the following time-dependent three-dimensional system of equations of isotropic solid dynamics in Eulerian form:

$$
\begin{aligned}
&\frac{\partial \rho}{\partial t} + \operatorname{div}(\rho \mathbf{v}) = 0,\\
&\frac{\partial \rho \mathbf{v}}{\partial t} + \operatorname{div}(\rho \mathbf{v}\mathbf{v} - \hat{\mathbf{P}}) = \mathbf{0},\\
&\frac{\partial e}{\partial t} + \operatorname{div}(e \mathbf{v} - \hat{\mathbf{P}} \cdot \mathbf{v}) = 0,\\
&\frac{D\hat{\mathbf{S}}}{Dt} + \tfrac{2}{3}\mu \operatorname{tr}(\hat{\mathbf{V}})\,\hat{\mathbf{I}} - 2\mu\hat{\mathbf{V}} + \theta(s)(\hat{\mathbf{S}} : \hat{\mathbf{V}})\,\hat{\mathbf{S}} = \hat{\mathbf{0}}.
\end{aligned}
$$

Here $\rho = \rho(t,x,y,z)$ is the density, t is time, $(x,y,z) = (x_1,x_2,x_3)$ are the Cartesian coordinates, $\mathbf{v} = \mathbf{v}(t,x,y,z) = [v_1, v_2, v_3]^{\mathrm{T}} = [u, v, w]^{\mathrm{T}}$ is the flow velocity,

$$e = \rho\varepsilon + \frac{1}{2}\rho\,(u^2+v^2+w^2) + \frac{1}{4\mu}(\hat{\mathbf{S}} : \hat{\mathbf{S}}) \tag{6.1.1}$$

is the total energy per unit volume for linearly elastic and ideal-plastic solids, $\varepsilon = \varepsilon(\rho, T)$ is the internal energy per unit mass, T is temperature, and $p = p(\rho, \varepsilon)$ is the pressure. The last term in Eq. (6.1.1) describes the shear elastic strain energy (Landau and Lifshitz 1986) and $\mu = \mu(\rho, \varepsilon)$ is one of the elastic Lamé coefficients, the so-called shear modulus. The above system generalizes the gas dynamic equations, see Section 3.1, by adding elastic and plastic processes to a solid under deformation. It is written out with the aid of the 3×3 symmetric Cauchy stress tensor $\hat{\mathbf{P}} = [P_{ij}]$, $P_{ij} = P_{ji}$.

The dot and colon stand for the contraction of tensors or vectors over one and two indices, respectively. In particular, $\hat{\mathbf{A}} \cdot \mathbf{b} = \sum_{p=1}^{3} A_{ip} b_p$ means the inner product (contraction) of a tensor $\hat{\mathbf{A}} = [A_{ij}]$ and a vector $\mathbf{b} = [b_1, b_2, b_3]^{\mathrm{T}}$; $\hat{\mathbf{A}} : \hat{\mathbf{B}} = \sum_{p=1}^{3}\sum_{q=1}^{3} A_{qp} B_{pq}$ denotes the contraction of tensors $\hat{\mathbf{A}} = [A_{ij}]$ and $\hat{\mathbf{B}} = [B_{ij}]$. Similarly, the dot product of two tensors, $\hat{\mathbf{A}} \cdot \hat{\mathbf{B}} = \sum_{p=1}^{3} A_{ip} B_{pj}$, can be defined.

We will use the representation

$$\hat{\mathbf{P}} \equiv \frac{\operatorname{tr}\hat{\mathbf{P}}}{3}\hat{\mathbf{I}} + \hat{\mathbf{S}} = -p\hat{\mathbf{I}} + \hat{\mathbf{S}}, \quad \operatorname{tr}\hat{\mathbf{S}} = 0,$$

where

$$\operatorname{tr}\hat{\mathbf{P}} = \sum_{i=1}^{3} P_{ii} = \hat{\mathbf{P}} : \hat{\mathbf{I}}$$

is the trace of the tensor $\hat{\mathbf{P}}$, and, under the assumption of isotropy,

$$p = -\tfrac{1}{3}\operatorname{tr}\hat{\mathbf{P}}.$$

The 3×3 symmetric tensor $\hat{\mathbf{S}} = [S_{ij}]$ is the deviator of the stress tensor $\hat{\mathbf{P}}$ and $\hat{\mathbf{I}} = \operatorname{diag}[1,1,1]$ is the identity tensor. Finally, we rewrite this system in the following closed form involving both conservative and nonconservative terms:

$$\frac{\partial \rho}{\partial t} + \operatorname{div}(\rho\mathbf{v}) = 0, \tag{6.1.2}$$

$$\frac{\partial \rho\mathbf{v}}{\partial t} + \operatorname{div}(\rho\mathbf{v}\mathbf{v} + p\hat{\mathbf{I}} - \hat{\mathbf{S}}) = \mathbf{0}, \tag{6.1.3}$$

$$\frac{\partial e}{\partial t} + \operatorname{div}[(e+p)\mathbf{v} - \hat{\mathbf{S}} \cdot \mathbf{v}] = 0, \tag{6.1.4}$$

$$\frac{D\hat{\mathbf{S}}}{Dt} + \tfrac{2}{3}\mu\operatorname{tr}(\hat{\mathbf{V}})\,\hat{\mathbf{I}} - 2\mu\hat{\mathbf{V}} + \theta(s)(\hat{\mathbf{S}} : \hat{\mathbf{V}})\,\hat{\mathbf{S}} = \hat{\mathbf{0}}. \tag{6.1.5}$$

In the general case the total energy per unit volume can be determined as

$$e = \rho\varepsilon + \tfrac{1}{2}\rho\,(u^2+v^2+w^2), \quad \varepsilon = \varepsilon(\rho, T, \hat{\mathbf{T}}^{\mathrm{e}}, \hat{\mathbf{T}}^{\mathrm{p}}),$$

where $\hat{\mathbf{T}} = \hat{\mathbf{T}}^{\mathrm{e}} + \hat{\mathbf{T}}^{\mathrm{p}}$ is the total strain; $\hat{\mathbf{T}}^{\mathrm{e}}$ and $\hat{\mathbf{T}}^{\mathrm{p}}$ are the elastic and plastic strains.

The notation $D\hat{\mathbf{A}}/Dt$ stands for the Jaumann derivative of a 3×3 tensor $\hat{\mathbf{A}} = [A_{ij}]$ with respect to time,

$$\frac{DA_{ij}}{Dt} = \frac{dA_{ij}}{dt} + \sum_{p=1}^{3}(A_{ip}\varphi_{pj} + A_{pj}\varphi_{pi}), \tag{6.1.6}$$

$$\varphi_{ij} = \frac{1}{2}\left(\frac{\partial v_i}{\partial x_j} - \frac{\partial v_j}{\partial x_i}\right),$$

where df/dt denotes the total (or substantive) derivative of a function f,

$$\frac{df}{dt} \equiv \frac{\partial f}{\partial t} + \sum_{p=1}^{3} v_p \frac{\partial f}{\partial x_p} = \frac{\partial f}{\partial t} + u\frac{\partial f}{\partial x} + v\frac{\partial f}{\partial y} + w\frac{\partial f}{\partial z}. \tag{6.1.7}$$

The tensor $\hat{\mathbf{V}} = [V_{ij}]$ is a symmetric strain rate tensor,

$$V_{ij} = \frac{1}{2}\left(\frac{\partial v_i}{\partial x_j} + \frac{\partial v_j}{\partial x_i}\right), \tag{6.1.8}$$

and the term that depends on $\theta(s)$ describes the plastic flow in accordance with the Prandtl–Reuss theory of plasticity. Specifically, $\theta(s) = 0$ if $s = \hat{\mathbf{S}} : \hat{\mathbf{S}} < 2k^2$ or for unloading, and $\theta(s) = \mu/k^2$ for loading otherwise, where k is the yield limit; see also the discussion below. The tensor $\hat{\mathbf{0}}$ is the 3×3 zero tensor.

A relation between ρ, p and ε in the form $p = p(\rho, \varepsilon)$ or $\varepsilon = \varepsilon(\rho, p)$, is called an equation of state. In particular, we can write out the equation of state (EOS) for an ideal perfect gas as

$$p = (\gamma - 1)\rho\varepsilon, \quad \varepsilon = \frac{p}{(\gamma - 1)\rho}. \tag{6.1.9}$$

The quantity $T = (\gamma - 1)\varepsilon$ gives the temperature in energy units; $\gamma > 1$ is the adiabatic index. A more complicated, two-term EOS has the form

$$p = (\gamma - 1)\rho\varepsilon + (\rho - \rho_0)c_0^2, \quad \varepsilon = \frac{p + \gamma p_0}{(\gamma - 1)\rho} - \frac{c_0^2}{\gamma - 1}, \tag{6.1.10}$$

where p_0 and c_0 are constants such that the relation $p = 0$ holds for $\varepsilon = 0$ and $\rho = \rho_0$; hence, $p_0 = \rho_0 c_0^2/\gamma$. Equation (6.1.9) can be obtained from (6.1.10) for $c_0 = 0$ and $p_0 = 0$; for more details see Section 3.1. For a solid, one may use the EOS in the form

$$\ln\frac{\rho}{\rho_0} = \frac{p}{K} - 3\alpha(T - T_0), \quad \varepsilon = c_V T, \quad K = \rho\left[\frac{\partial p}{\partial \rho}\right]_T, \quad \alpha = \frac{1}{3K}\left[\frac{\partial p}{\partial T}\right]_\rho, \tag{6.1.11}$$

where ρ_0 is the density of the initially undeformed material, T_0 is the initial temperature, K is the bulk modulus, α is the coefficient of linear expansion ($a = 3\alpha$ is the volumetric expansion coefficient), T is the temperature, and $c_V = [\partial\varepsilon/\partial T]_\rho$ is the specific heat at constant volume.

Equations (6.1.2), (6.1.3), and (6.1.4) express the conservation of the mass, momentum, and energy of the medium, respectively. Equation (6.1.5) describes the evolution of the stress tensor deviator by taking into account the elastic and plastic processes. Consider the equation of evolution of $\hat{\mathbf{S}}$ in more detail.

Hooke's law

For small deformations, Eq. (6.1.5) expresses Hooke's law. This law was suggested as early as 1678 to describe a linear relationship between stresses and strains in one-dimensional motions. Hooke's law in three dimensions defines a linear relation between the stress tensor $\hat{\mathbf{P}}$ and the tensor of small deformations $\hat{\mathbf{T}}$ with components

$$T_{ij} = \frac{1}{2}\left(\frac{\partial d_i}{\partial x_j} + \frac{\partial d_j}{\partial x_i}\right), \quad \text{or} \quad \hat{\mathbf{T}} = \frac{1}{2}[\nabla \mathbf{d} + (\nabla \mathbf{d})^{\mathrm{T}}], \tag{6.1.12}$$

where $\mathbf{d} = [d_1, d_2, d_3]$ is the displacement vector and $\nabla = [\partial/\partial x_1, \partial/\partial x_2, \partial/\partial x_3]^{\mathrm{T}}$. Hence, Hooke's law has the general form

$$\hat{\mathbf{P}} = \hat{\mathbf{C}} : \hat{\mathbf{T}} \quad \Longrightarrow \quad P_{ij} = \sum_{p,q=1}^{3} C_{ijpq} T_{qp}, \tag{6.1.13}$$

where $\hat{\mathbf{C}} = [C_{ijpq}]$ is a $3 \times 3 \times 3 \times 3$ tensor of elastic constants. In general, $\hat{\mathbf{C}}$ can depend on the stress tensor (Prager 1961). Using the specific symmetry of $\hat{\mathbf{C}}$ and adopting the isotropy condition, one can find that only two entries of $\hat{\mathbf{C}}$ out of $3^4 = 81$ are independent; see, for example, Bloch (1964) and Sedov (1971). In particular, it can be established that $C_{ijpp} = 0$ for $i \neq j$, $C_{iipp} = C_{ppii}$, $C_{1111} = C_{2222} = C_{3333} = \lambda + 2\mu$, and $C_{1122} = C_{1133} = C_{2233} = \mu$, where λ and μ are the so-called Lamé coefficients; μ is also referred to as the shear modulus. Then, relation (6.1.13) acquires a simpler form

$$P_{ij} = \lambda \, \mathrm{tr}(\hat{\mathbf{T}}) \, \delta_{ij} + 2\mu T_{ij}, \quad \text{or} \quad \hat{\mathbf{P}} = \lambda \, \mathrm{tr}(\hat{\mathbf{T}}) \, \hat{\mathbf{I}} + 2\mu \hat{\mathbf{T}}, \tag{6.1.14}$$

where δ_{ij} is the Kronecker delta. Note that, instead of μ and λ, other elastic coefficients can be used, for example, Young's modulus

$$E = \frac{(3\lambda + 2\mu)\,\mu}{\lambda + \mu} \tag{6.1.15}$$

and Poisson's ratio

$$\sigma = \frac{\lambda}{2(\lambda + \mu)}. \tag{6.1.16}$$

Let us rewrite Eq. (6.1.14) in terms of the deviator $\hat{\mathbf{S}} = [S_{ij}]$ of the stress tensor $\hat{\mathbf{P}}$. We have

$$S_{ij} = -\tfrac{2}{3}\mu \, \mathrm{tr}(\hat{\mathbf{T}}) \, \delta_{ij} + 2\mu T_{ij}, \quad \text{or} \quad \hat{\mathbf{S}} = -\tfrac{2}{3}\mu \, \mathrm{tr}(\hat{\mathbf{T}}) \, \hat{\mathbf{I}} + 2\mu \hat{\mathbf{T}}. \tag{6.1.17}$$

Note that Eq. (6.1.17) depends only on the coefficient $\mu = \mu(\rho, \varepsilon)$ and is independent of λ.

Let us differentiate Eq. (6.1.17) with respect to time and assume that $|d\mu/dt| \ll 1$. Then we arrive at the following equations for the evolution of $\hat{\mathbf{S}}$ in time in the elastic approximation:

$$\frac{DS_{ij}}{Dt} = -\tfrac{2}{3}\mu \, \mathrm{tr}(\hat{\mathbf{V}}) \, \delta_{ij} + 2\mu V_{ij}, \quad \text{or} \quad \frac{D\hat{\mathbf{S}}}{Dt} = -\tfrac{2}{3}\mu \, \mathrm{tr}(\hat{\mathbf{V}}) \, \hat{\mathbf{I}} + 2\mu \hat{\mathbf{V}}. \tag{6.1.18}$$

By the Stokes formula, $\hat{\mathbf{V}} = d\hat{\mathbf{T}}/dt$; hence, since

$$\frac{d\mathbf{d}}{dt} = \mathbf{v},$$

we obtain relation (6.1.8). The Jaumann derivative with respect to time, $D\hat{\mathbf{A}}/Dt$, is defined by formula (6.1.6). The Jaumann derivative takes into account the fact that the components of a tensor in Eulerian coordinates are not constant even if these components are constant in the system of coordinates instantaneously rotating together with a small neighborhood of the point under consideration. For details, see Thomas (1961), where higher derivatives with respect to time were considered, Prager (1961), Kolarov et al. (1979), and also the section on page 388.

Plastic and viscoplastic flows

So far we have dealt only with small elastic deformations described by Hooke's law. In the case of relatively large deformations, the elastic motion of the media transforms into a plastic flow. In the plastic flow the components of the stress tensor deviator are bounded from above. The Prandtl–Reuss theory of plasticity uses the Mises criteria for plastic flow; namely, if

$$s = \hat{\mathbf{S}} : \hat{\mathbf{S}} \geq 2k^2 \tag{6.1.19}$$

then plastic flow occurs, otherwise the motion of the medium is elastic. The constant $k = k(\rho, \varepsilon)$ is called the yield limit and the surface $\hat{\mathbf{S}} : \hat{\mathbf{S}} - 2k^2 = 0$ is the so-called Mises yield surface in the space of the deviator entries. The Mises surface is convex, continuous, and smooth.

Let us define the function $\theta(s)$ in Eq. (6.1.5) as

$$\theta(s) = \begin{cases} 0 & \text{if} \quad s = \hat{\mathbf{S}} : \hat{\mathbf{S}} < 2k^2, \\ 0 & \text{if} \quad s \geq 2k^2 \quad \text{for unloading,} \\ \dfrac{\mu}{k^2} & \text{if} \quad s \geq 2k^2 \quad \text{for loading} \end{cases} \tag{6.1.20}$$

and prove that this $\theta(s)$ permits one to satisfy the condition of nonincreasing deviator entries for plastic flow. Indeed, multiplying Eq. (6.1.5) by $\hat{\mathbf{S}}$ from the right and from the left and summing the two resulting equations, we obtain

$$\hat{\mathbf{S}} : \frac{D\hat{\mathbf{S}}}{Dt} + \frac{D\hat{\mathbf{S}}}{Dt} : \hat{\mathbf{S}} - 2\mu(\hat{\mathbf{S}} : \hat{\mathbf{V}} + \hat{\mathbf{V}} : \hat{\mathbf{S}}) + 2\theta(s)(\hat{\mathbf{S}} : \hat{\mathbf{V}})\,\hat{\mathbf{S}} : \hat{\mathbf{S}} = 0;$$

hence,

$$\frac{d\hat{\mathbf{S}} : \hat{\mathbf{S}}}{dt} = [2\mu - \theta(s)(\hat{\mathbf{S}} : \hat{\mathbf{S}})]2(\hat{\mathbf{S}} : \hat{\mathbf{V}}) = 0 \tag{6.1.21}$$

if $s = \hat{\mathbf{S}} : \hat{\mathbf{S}} \geq 2k^2$. Therefore in plastic flow the values of the deviatoric components cannot go beyond the yield surface $s = 2k^2$. Note that for plastic flows the last term in formula (6.1.1) describing the shear elastic strain energy is constant if $\mu =$ const and $k =$ const.

There are other criteria and models of nonelastic flow of medium; see, for example, Freudental and Geiringer (1958), Wilkinson (1960), Rabotnov (1988). In particular, the surface

$$s_{\mathrm{T}} = \tau(\hat{\mathbf{S}}) - k = 0 \tag{6.1.22}$$

is the so-called Tresca yield surface in the space of the deviator entries. This surface can be used in the criterion of plastic flow instead of the Mises yield surface. Here,

$$\tau(\hat{\mathbf{S}}) = \tfrac{1}{2} \max\left[|s_1 - s_2|, |s_2 - s_3|, |s_3 - s_1|\right],$$

where s_1, s_2, and s_3 are the eigenvalues of the deviator matrix $[S_{ij}]$. Tresca's surface is convex and piecewise continuous.

The yield limit k can suffer changes during plastic flow, in particular, due to the hardening process. In this case $k = k_0 + \beta\chi$, where k_0 is an initial yield limit, $\beta > 0$ is a constant, and χ is the hardening parameter depending on the work of the plastic strain,

$$d\chi = \hat{\mathbf{P}} : d\hat{\mathbf{T}}^{\mathrm{p}}, \tag{6.1.23}$$

where $\hat{\mathbf{T}} = \hat{\mathbf{T}}^{\mathrm{e}} + \hat{\mathbf{T}}^{\mathrm{p}}$ is the total strain, and $\hat{\mathbf{T}}^{\mathrm{e}}$ and $\hat{\mathbf{T}}^{\mathrm{p}}$ are the elastic and plastic strains. For $\beta < 0$ we have the process of softening.

In 1868 Maxwell suggested an elastoviscous model of media with relations

$$\frac{D\hat{\mathbf{S}}}{Dt} + \tfrac{2}{3}\mu\,\mathrm{tr}(\hat{\mathbf{V}})\,\hat{\mathbf{I}} - 2\mu\hat{\mathbf{V}} = -\frac{\hat{\mathbf{S}}}{\tau}, \tag{6.1.24}$$

where we see a new term, as compared with relations (6.1.18). The parameter τ is the stress relaxation time. In this case, the deviator entries are damped in time as

$$S_{ij} \sim \exp\left(-\frac{t}{\tau}\right).$$

Assume that Maxwell's damping occurs only in regions where plastic flow takes place. Such a flow is called elastoplastic–viscoplastic and elastoviscoplastic. In this case

$$\frac{D\hat{\mathbf{S}}}{Dt} + \tfrac{2}{3}\mu\,\mathrm{tr}(\hat{\mathbf{V}})\,\hat{\mathbf{I}} - 2\mu\hat{\mathbf{V}} + \theta(s)(\hat{\mathbf{S}} : \hat{\mathbf{V}})\,\hat{\mathbf{S}} = -\psi(s)\frac{\hat{\mathbf{S}}}{\tau}, \tag{6.1.25}$$

where

$$\psi(s) = \begin{cases} 0 & \text{if } s = \hat{\mathbf{S}} : \hat{\mathbf{S}} < 2k^2, \\ 2\dfrac{\sqrt{\hat{\mathbf{S}} : \hat{\mathbf{S}}} - \sqrt{2}\,k}{\sqrt{\hat{\mathbf{S}} : \hat{\mathbf{S}}}} & \text{if } s \geq 2k^2, \end{cases}$$

provided that $s = \hat{\mathbf{S}} : \hat{\mathbf{S}} \leq 2k^2$. Indeed, by analogy with relation (6.1.21) one can obtain

$$\frac{d}{dt}\left(\sqrt{\hat{\mathbf{S}} : \hat{\mathbf{S}}} - \sqrt{2}\,k\right) = -\frac{1}{\tau}\left(\sqrt{\hat{\mathbf{S}} : \hat{\mathbf{S}}} - \sqrt{2}\,k\right) \quad \text{if} \quad s = \hat{\mathbf{S}} : \hat{\mathbf{S}} \geq 2k^2.$$

Hence, $\sqrt{\hat{\mathbf{S}}:\hat{\mathbf{S}}}$ relaxes exponentially to $\sqrt{2}\,k$. This property permits one to bound the components of the stress tensor deviator. For $\theta(s) \neq 0$ the flow is elastoplastic–viscoplastic and for $\theta(s) \equiv 0$ is called elastoviscoplastic. In particular, the structure of elastoviscoplastic waves in Maxwell's medium was investigated by Godunov (1978).

In the three-dimensional case, system (6.1.2)–(6.1.5) has ten unknowns, for example, ρ, u, v, w, ε, S_{11}, S_{12}, S_{13}, S_{22}, and S_{23}. For S_{33} we have the relation $S_{33} = -S_{11} - S_{22}$. In nonconservative form this system can be rewritten as

$$\frac{d\rho}{dt} + \rho \operatorname{div} \mathbf{v} = 0, \tag{6.1.26}$$

$$\frac{d\mathbf{v}}{dt} + \frac{1}{\rho} \operatorname{div}(p\hat{\mathbf{I}} - \hat{\mathbf{S}}) = \mathbf{0}, \tag{6.1.27}$$

$$\frac{d\varepsilon}{dt} + \frac{p_s}{\rho} \operatorname{div} \mathbf{v} = 0, \qquad p_s = p + \frac{1}{4\mu}(\hat{\mathbf{S}}:\hat{\mathbf{S}}); \tag{6.1.28}$$

$$\frac{D\hat{\mathbf{S}}}{Dt} + \tfrac{2}{3}\mu\,\mathrm{tr}(\hat{\mathbf{V}})\,\hat{\mathbf{I}} - 2\mu\hat{\mathbf{V}} + \theta(s)(\hat{\mathbf{S}}:\hat{\mathbf{V}})\,\hat{\mathbf{S}} = \hat{\mathbf{0}}, \tag{6.1.29}$$

where we assume that all derivatives of μ are sufficiently small.

Let us consider the case of two space dimensions and write out system (6.1.26)–(6.1.29) in terms of the seven variables ρ, u, v, ε, S_{11}, S_{22}, and S_{12}. We have

$$\frac{d\rho}{dt} + \rho\left(\frac{\partial u}{\partial x} + \frac{\partial v}{\partial y}\right) = 0, \tag{6.1.30}$$

$$\frac{du}{dt} + \frac{1}{\rho}\left(\frac{\partial p}{\partial x} - \frac{\partial S_{11}}{\partial x} - \frac{\partial S_{12}}{\partial y}\right) = 0, \tag{6.1.31}$$

$$\frac{dv}{dt} + \frac{1}{\rho}\left(\frac{\partial p}{\partial y} - \frac{\partial S_{12}}{\partial x} - \frac{\partial S_{22}}{\partial y}\right) = 0, \tag{6.1.32}$$

$$\frac{d\varepsilon}{dt} + \frac{p_s}{\rho}\left(\frac{\partial u}{\partial x} + \frac{\partial v}{\partial y}\right) = 0, \tag{6.1.33}$$

$$\frac{DS_{11}}{Dt} - \frac{4}{3}\mu\frac{\partial u}{\partial x} + \frac{2}{3}\mu\frac{\partial v}{\partial y} + \theta(s)(\hat{\mathbf{S}}:\hat{\mathbf{V}})\,S_{11} = 0, \tag{6.1.34}$$

$$\frac{DS_{22}}{Dt} + \frac{2}{3}\mu\frac{\partial u}{\partial x} - \frac{4}{3}\mu\frac{\partial v}{\partial y} + \theta(s)(\hat{\mathbf{S}}:\hat{\mathbf{V}})\,S_{22} = 0, \tag{6.1.35}$$

$$\frac{DS_{12}}{Dt} - \mu\frac{\partial v}{\partial x} - \mu\frac{\partial u}{\partial y} + \theta(s)(\hat{\mathbf{S}}:\hat{\mathbf{V}})\,S_{12} = 0. \tag{6.1.36}$$

Here, t is time, $(x, y) = (x_1, x_2)$ are the plane Cartesian coordinates, $\rho = \rho(t, x, y)$ is the density, $\mathbf{v} = \mathbf{v}(t, x, y) = [v_1, v_2]^{\mathrm{T}} = [u, v]^{\mathrm{T}}$ is the flow velocity, ε is the internal energy per unit volume, $p = p(\rho, \varepsilon)$ is the pressure, and S_{11}, S_{22}, S_{12}, and $S_{33} = -S_{11} - S_{22}$ are the nonzero components of the deviator tensor. For plastic flow we have relation (6.1.20), where

$$s = \hat{\mathbf{S}}:\hat{\mathbf{S}} = S_{11}^2 + S_{22}^2 + (S_{11} + S_{22})^2 + 2S_{12}^2,$$

and

$$\hat{\mathbf{S}}:\hat{\mathbf{V}} = S_{11}\frac{\partial u}{\partial x} + S_{12}\left(\frac{\partial v}{\partial x} + \frac{\partial u}{\partial y}\right) + S_{22}\frac{\partial v}{\partial y}.$$

We have also

$$\frac{DS_{11}}{Dt} = \frac{dS_{11}}{dt} + \zeta S_{12}\left(\frac{\partial v}{\partial x} - \frac{\partial u}{\partial y}\right),$$
$$\frac{DS_{22}}{Dt} = \frac{dS_{22}}{dt} + \zeta S_{12}\left(\frac{\partial u}{\partial y} - \frac{\partial v}{\partial x}\right),$$
$$\frac{DS_{12}}{Dt} = \frac{dS_{12}}{dt} + \frac{1}{2}\zeta(S_{11} - S_{22})\left(\frac{\partial u}{\partial y} - \frac{\partial v}{\partial x}\right).$$

The parameter ζ is either 0 or 1, which allows one either to take into account or disregard the rotational terms in the Jaumann derivative for a more general analysis. For small rotational deformations these terms can be neglected.

Finally, we make some additional remarks concerning various versions and generalizations of the governing system of equations.

Strain tensors

In the above equations we used the tensor of small deformations, see Eq. (6.1.12). In some cases different strain tensors are used. They can be expressed in terms of the local linear transformation of the medium prescribed by the 3×3 matrix J. The matrix J is the Jacobian matrix of the transformation described by the formulas $x_i(t) = x_i(x_1^0, x_2^0, x_3^0, t)$, $i = 1, 2, 3$ with $x_i(t = 0) = x_i^0$. The coordinates x_1, x_2, and x_3 are the Eulerian, in particular Cartesian, coordinates, and x_1^0, x_2^0, and x_3^0 are the Lagrangian coordinates. The Jacobian matrix J defines the strain gradient tensor $\hat{\mathbf{J}}$ (Truesdell 1977):

$$\hat{\mathbf{J}} = [J_{ij}] = \left[\frac{\partial x_i}{\partial x_j^0}\right]. \tag{6.1.37}$$

In addition, the matrix $D = (J^{-1})^{\mathrm{T}}$ defines the so-called distortion tensor $\hat{\mathbf{D}}$ (Godunov 1978):

$$\hat{\mathbf{D}} = [D_{ij}] = \left[\frac{\partial x_i^0}{\partial x_j}\right]^{\mathrm{T}} = \left[\frac{\partial x_j^0}{\partial x_i}\right]. \tag{6.1.38}$$

We can diagonalize $\hat{\mathbf{J}}$ so that

$$\begin{aligned} &\hat{\mathbf{J}} = [J_{ij}] = U \operatorname{diag}[k_1, k_2, k_3] W, \\ &UU^{\mathrm{T}} = WW^{\mathrm{T}} = I, \quad \det U = \det W = 1, \end{aligned} \tag{6.1.39}$$

where the orthogonal matrices U and W describe the space rotation of an elementary volume and define the orthogonal tensors $\hat{\mathbf{U}}$ and $\hat{\mathbf{W}}$. The diagonal matrix $\operatorname{diag}[k_1, k_2, k_3]$ describes linear expansion of an elementary volume in three directions. If $k_1 = k_2 = k_3 = 1$, then the medium moves as a rigid body. By analogy,

$$\hat{\mathbf{D}} = [D_{ij}] = U \operatorname{diag}[1/k_1, 1/k_2, 1/k_3] W.$$

Then the Cauchy strain tensor becomes

$$\hat{\mathbf{T}}_{\mathrm{C}} = (\hat{\mathbf{J}} \cdot \hat{\mathbf{J}}^{\mathrm{T}})^{1/2} - \hat{\mathbf{I}} = U \operatorname{diag}[k_1 - 1, k_2 - 1, k_3 - 1] U^{\mathrm{T}}. \tag{6.1.40}$$

Along with the Cauchy strain tensor, other strain tensors are also used in solving particular problems, for example, the Finger strain tensor,

$$\hat{\mathbf{T}}_{\mathrm{F}} = \hat{\mathbf{I}} - (\hat{\mathbf{J}} \cdot \hat{\mathbf{J}}^{\mathrm{T}})^{-1/2} = U \operatorname{diag}[1 - 1/k_1, 1 - 1/k_2, 1 - 1/k_3] U^{\mathrm{T}}, \tag{6.1.41}$$

the Green strain tensor,

$$\hat{\mathbf{T}}_{\mathrm{G}} = \tfrac{1}{2}(\hat{\mathbf{J}} \cdot \hat{\mathbf{J}}^{\mathrm{T}} - \hat{\mathbf{I}}) = U \operatorname{diag}\left[\tfrac{1}{2}(k_1^2 - 1), \tfrac{1}{2}(k_2^2 - 1), \tfrac{1}{2}(k_3^2 - 1)\right] U^{\mathrm{T}}, \tag{6.1.42}$$

the Hencky strain tensor,

$$\hat{\mathbf{T}}_{\mathrm{H}} = U \operatorname{diag}[\ln k_1, \ln k_2, \ln k_3] U^{\mathrm{T}}, \tag{6.1.43}$$

and the Almansi strain tensor,

$$\begin{aligned} \hat{\mathbf{T}}_{\mathrm{A}} &= \tfrac{1}{2}[\hat{\mathbf{I}} - (\hat{\mathbf{J}}^{\mathrm{T}})^{-1} \cdot \hat{\mathbf{J}}^{-1}] = \tfrac{1}{2}[\hat{\mathbf{I}} - \hat{\mathbf{D}} \cdot \hat{\mathbf{D}}^{\mathrm{T}}] \\ &= U \operatorname{diag}\left[\tfrac{1}{2}(1 - 1/k_1^2), \tfrac{1}{2}(1 - 1/k_2^2), \tfrac{1}{2}(1 - 1/k_3^2)\right] U^{\mathrm{T}}, \end{aligned} \tag{6.1.44}$$

or

$$\hat{\mathbf{T}}_{\mathrm{A}} = \tfrac{1}{2}(\hat{\mathbf{I}} - \hat{\mathbf{g}}),$$

where $\hat{\mathbf{g}} = (J^{\mathrm{T}})^{-1} J^{-1} = D D^{\mathrm{T}}$ is the metric tensor.

For the case $k_1 = k_2 = k_3 = 1$ all these tensors are zero. The use of one or another tensor permits one to preserve some additional properties of the governing equations (Prager 1961; Godunov 1978).

From above strain tensors in assumption of small deformations the tensor (6.1.12) can be obtained. Indeed,

$$\begin{aligned} &\mathbf{d} = \delta\mathbf{x} - \delta\mathbf{x}^0 = (J - I)\delta\mathbf{x}^0, \\ &\nabla\mathbf{d} = (J - I)\nabla\mathbf{x}^0 = (J - I)J^{-1} \quad \Longrightarrow \quad J = (I - \nabla\mathbf{d})^{-1}. \end{aligned} \tag{6.1.45}$$

Hence, in particular, for the Almansi strain tensor we have

$$\begin{aligned} \hat{\mathbf{T}}_{\mathrm{A}} &= \tfrac{1}{2}[I - (J^{\mathrm{T}})^{-1} J^{-1}] = \tfrac{1}{2}[I - (I - \nabla\mathbf{d})^{\mathrm{T}}(I - \nabla\mathbf{d})] \\ &= \tfrac{1}{2}[\nabla\mathbf{d} + (\nabla\mathbf{d})^{\mathrm{T}} - \nabla\mathbf{d}\,(\nabla\mathbf{d})^{\mathrm{T}}] \approx \tfrac{1}{2}[\nabla\mathbf{d} + (\nabla\mathbf{d})^{\mathrm{T}}]. \end{aligned} \tag{6.1.46}$$

The Jaumann and other derivatives

Jaumann in 1905 suggested a derivative (6.1.6) that is an adequate measure of stress variations in Eulerian coordinates, since it possesses the following properties (Sedov 1960; Prager 1961; Kolarov et al. 1979):

- it preserves the Leibnitz differentiation rule for the contraction of two tensors,

$$D_t(\hat{\mathbf{A}} \cdot \hat{\mathbf{B}}) = D_t(\hat{\mathbf{A}}) \cdot \hat{\mathbf{B}} + \hat{\mathbf{A}} \cdot D_t(\hat{\mathbf{B}}),$$

where D_t stands here for the Jaumann derivative;

- $D_t\hat{\mathbf{A}}$ is linear in the components of $\hat{\mathbf{A}}$;
- the derivative of a 3×3 tensor is also a 3×3 tensor;
- if $D_t\hat{\mathbf{A}} = \hat{\mathbf{0}}$, then the derivatives of the invariants of $\hat{\mathbf{A}}$ are also zero;
- $D_t\hat{\mathbf{A}} = \hat{\mathbf{0}}$ if the point, with its small neighborhood moves as a rigid body with the angular velocity curl $\mathbf{v}$; and
- it preserves the symmetry of the tensor being differentiated.

In using different strain tensors and non-Eulerian coordinates, other types of derivatives arise, for example, Cotter–Rivlin, Truesdell, Hill, Green, Oldroyd, and Mandel derivatives. Unlike the Jaumann derivative, these derivatives do not satisfy all the above conditions; see Prager (1961) and Kolarov et al. (1979). These derivatives can be expressed as the Jaumann derivative $D_t\hat{\mathbf{A}}$ plus some quadratic form of $\hat{\mathbf{A}}$ and $\hat{\mathbf{L}} = [L_{ij}]$, where $L_{ij} = \partial v_i/\partial x_j$. A general representation for the mentioned derivatives was suggested by Mandel:

$$\widetilde{D}_t\hat{\mathbf{A}} = D_t\hat{\mathbf{A}} + a_1\hat{\mathbf{A}}\,\mathrm{tr}(\hat{\mathbf{L}}) + a_2\hat{\mathbf{A}}\cdot\hat{\mathbf{L}} + a_3\hat{\mathbf{L}}\cdot\hat{\mathbf{A}} + a_4\hat{\mathbf{L}}\,\mathrm{tr}(\hat{\mathbf{A}}),$$

where a_1, a_2, a_3, and a_4 are constants. In particular, the Truesdell derivative is defined as

$$D_t\hat{\mathbf{A}} - \hat{\mathbf{L}}\cdot\hat{\mathbf{A}} - \hat{\mathbf{A}}\cdot\hat{\mathbf{L}}^{\mathrm{T}} + \hat{\mathbf{A}}\,\mathrm{tr}(\hat{\mathbf{L}}).$$

Some of these derivatives depend on the coordinate basis in which the tensor is differentiated, as well as on the form of the strain tensor (Sedov 1960; Prager 1961). In particular, the Green derivative with respect to time arises in the relation between the Green strain tensor (6.1.42) and the corresponding strain rate tensor. In particular,

$$\begin{aligned}\frac{d\hat{\mathbf{T}}_{\mathrm{G}}}{dt} &= \frac{1}{2}\frac{d}{dt}\left(\sum_{p=1}^{3} J_{ip}J_{jp}\right) = \frac{1}{2}\frac{d}{dt}\left(\sum_{p=1}^{3}\frac{\partial x_i}{\partial x_p^0}\frac{\partial x_j}{\partial x_p^0}\right)\\ &= \frac{1}{2}\sum_{p=1}^{3}\left(\frac{\partial v_i}{\partial x_p^0}J_{jp} + J_{ip}\frac{\partial v_j}{\partial x_p^0}\right) = \frac{1}{2}\sum_{p,q=1}^{3}\left(\frac{\partial v_i}{\partial x_q}J_{qp}J_{jp} + J_{ip}\frac{\partial v_j}{\partial x_q}J_{qp}\right).\end{aligned}$$

For small deformations, see Eq. (6.1.45), one can find

$$\frac{d\hat{\mathbf{T}}}{dt} = V_{ij} + \sum_{p=1}^{3}\left(T_{ip}\frac{\partial v_j}{\partial x_p} + T_{jp}\frac{\partial v_i}{\partial x_p}\right),$$

where V_{ij} is the strain rate tensor of (6.1.8). Hence,

$$\frac{\widetilde{D}\hat{\mathbf{T}}}{\widetilde{D}t} = \hat{\mathbf{V}}, \quad \frac{\widetilde{D}T_{ij}}{\widetilde{D}t} \equiv \frac{dT_{ij}}{dt} - \sum_{p=1}^{3}\left(T_{ip}\frac{\partial v_j}{\partial x_p} + T_{jp}\frac{\partial v_i}{\partial x_p}\right),$$

where $\widetilde{D}/\widetilde{D}t$ is the Green derivative with respect to time.

The additional terms to the Jaumann derivative depend on the strain rate tensor and, hence, can also be interpreted as additional terms in the basic law describing the evolution of the stress tensor as a function of stresses and strain rates (Prager 1961).

Note that the symmetric Cauchy stress tensor $\hat{\mathbf{P}}$ is the stress tensor (per unit area) in the current configuration, x_i, and its symmetry follows from the equilibrium of the torques acting on the elementary volume; see, for instance, Sedov (1971) and Truesdell (1977). Unlike this, the nonsymmetric Piola (Piola–Kirchhoff) tensor is the stress tensor (per unit area) in the initial configuration of the body, x_i^0; the dynamics equations for this tensor use the Hill derivative with respect to time. To symmetrize the governing system, the modified symmetric Kirchhoff strain tensor is used. In this case, when formulating the dynamics equations in terms of this tensor, the Truesdell derivative with respect to time arises (Prager 1961).

Thus, derivatives with respect to time depend on the version of a strain tensor and a way of their presentation. These derivatives can be determined by the direct differentiation with respect to time the tensor in terms of the strain gradient tensor $\hat{\mathbf{J}}$ or the distortion tensor $\hat{\mathbf{D}}$ of (6.1.40)–(6.1.44). The nonuniqueness in the choice of derivatives with respect to time for the tensor of small deformations can be eliminated by using special variables, see Section 6.1.2.

The above remarks demonstrate the variety of representations of governing systems. But if the constitutive relations, conservation laws, and assumptions are the same, all these differential equations are equivalent and describe the same processes and flows for different ways of presentation. As far as the use of specific time derivatives is concerned, the choice of them depends on the system of coordinates and the basic tensors; see the discussion by Kondaurov and Nikitin (1990).

Some versions of governing equations

There are other forms of the model represented by (6.1.2)–(6.1.5). In particular, in the three-dimensional case, one can use the system of equations for the set of ten variables $\mathbf{v}$, e, and $\hat{\mathbf{P}}$:

$$\begin{aligned} &\frac{\partial \rho \mathbf{v}}{\partial t} + \operatorname{div}(\rho \mathbf{v}\mathbf{v} - \hat{\mathbf{P}}) = \mathbf{0}, \\ &\frac{\partial e}{\partial t} + \operatorname{div}(e\mathbf{v} - \hat{\mathbf{P}} \cdot \mathbf{v}) = 0, \\ &\frac{D\hat{\mathbf{P}}}{Dt} - \lambda \operatorname{tr}(\hat{\mathbf{V}})\,\hat{\mathbf{I}} - 2\mu\hat{\mathbf{V}} + \theta(s)(\hat{\mathbf{S}} : \hat{\mathbf{V}})\,\hat{\mathbf{S}} = \hat{\mathbf{0}}, \end{aligned} \tag{6.1.47}$$

where $\lambda = \lambda(\rho, \varepsilon)$ and $\mu = \mu(\rho, \varepsilon)$. The two systems, (6.1.2)–(6.1.5) and (6.1.47), have the same order, since we have five scalar equation for the components of $\hat{\mathbf{S}}$ (since $\operatorname{tr}\hat{\mathbf{S}} = 0$) and six equations for the components of $\hat{\mathbf{P}}$. To close (6.1.47) one can complement it by an EOS in the form $p = -\frac{1}{3}\operatorname{tr}\hat{\mathbf{P}} = p(\rho, \varepsilon)$, which can be used to determine ρ after $\hat{\mathbf{P}}$ has been calculated. From this EOS it can be found that

$$\frac{dp}{dt} + (\lambda + \tfrac{2}{3}\mu)\operatorname{div}\mathbf{v} = 0 \quad \Longrightarrow \quad \lambda = \rho\frac{\partial p}{\partial \rho} + \frac{p_s}{\rho}\frac{\partial p}{\partial \varepsilon} - \frac{2}{3}\mu, \quad p_s = p + \frac{1}{4\mu}(\hat{\mathbf{S}} : \hat{\mathbf{S}}).$$

Equations (6.1.47) are preferable to Eqs. (6.1.2)–(6.1.5) for the case of small density variations. The use of Eqs. (6.1.2)–(6.1.5) may diminish the accuracy of calculations in this case. In nonconservative form these equations become

$$\frac{d\mathbf{v}}{dt} - \frac{1}{\rho}\operatorname{div}\hat{\mathbf{P}} = \mathbf{0},$$

$$\frac{d\varepsilon}{dt} + \frac{p_s}{\rho}\,\mathrm{div}\,\mathbf{v} = 0, \tag{6.1.48}$$
$$\frac{D\hat{\mathbf{P}}}{Dt} - \lambda\,\mathrm{tr}(\hat{\mathbf{V}})\,\hat{\mathbf{I}} - 2\mu\hat{\mathbf{V}} + \theta(s)(\hat{\mathbf{S}} : \hat{\mathbf{V}})\,\hat{\mathbf{S}} = \hat{\mathbf{0}}.$$

For isothermal processes one can use Eqs. (6.1.2)–(6.1.5) for the variables ρ, $\mathbf{v}$, and $\hat{\mathbf{S}}$:

$$\frac{\partial\rho}{\partial t} + \mathrm{div}(\rho\mathbf{v}) = 0,$$
$$\frac{\partial\rho\mathbf{v}}{\partial t} + \mathrm{div}(\rho\mathbf{v}\mathbf{v} + p\hat{\mathbf{I}} - \hat{\mathbf{S}}) = \mathbf{0},$$
$$\frac{D\hat{\mathbf{S}}}{Dt} + \tfrac{2}{3}\mu\,\mathrm{tr}(\hat{\mathbf{V}})\,\hat{\mathbf{I}} - 2\mu\hat{\mathbf{V}} + \theta(s)(\hat{\mathbf{S}} : \hat{\mathbf{V}})\,\hat{\mathbf{S}} = \hat{\mathbf{0}},$$

where p is a function of only ρ.

In terms of ρ, $\mathbf{v}$, and $\hat{\mathbf{P}}$, the above system reads

$$\frac{\partial\rho\mathbf{v}}{\partial t} + \mathrm{div}(\rho\mathbf{v}\mathbf{v} - \hat{\mathbf{P}}) = \mathbf{0}, \tag{6.1.49}$$
$$\frac{D\hat{\mathbf{P}}}{Dt} - \lambda\,\mathrm{tr}(\hat{\mathbf{V}})\,\hat{\mathbf{I}} - 2\mu\hat{\mathbf{V}} + \theta(s)(\hat{\mathbf{S}} : \hat{\mathbf{V}})\,\hat{\mathbf{S}} = \hat{\mathbf{0}}.$$

Note that here $\mathrm{tr}\,\hat{\mathbf{P}} \neq 0$ and

$$\lambda = K - \frac{2}{3}\mu, \quad K = \rho\frac{dp}{d\rho}.$$

All these systems of equations are hyperbolic in a fairly wide range of parameters for physically relevant equations of state. In particular, the hyperbolicity of equations of nonlinear thermoelasticity was investigated in Godunov (1978), see also Godunov and Romenskii (1998). Note that the above equations can be enriched by taking into account the fracture processes; see, for example, Zukas et al. (1982), Atluri (1986), Rabotnov (1988), Chernikh (1996), and Kondaurov and Fortov (2000). Two simple methods of fracture simulation based on the strength limit and bulk porosity can be found in Section 3.1 and can be used in solid dynamics.

It should be noted that the solid dynamics equations that take into account the Prandtl–Reuss plasticity model have essentially nonconservative form, and this fact can be rigorously proved, see Kondaurov (1982a). This is due to the fact that the Prandtl–Reuss plasticity is not sensitive to strain rates. This fact leads to difficulties in obtaining the exact discontinuous solutions and in constructing adequate relations for the moving or stationary discontinuities that exist in the Prandtl–Reuss elastoplastic flows, as well as in constructing the exact Riemann problem solver. Let us mention an approach that allows one to construct jump relations for nonconservative equations (Sadovskii 1997). To obtain exact relations at discontinuities one should adopt additional assumptions or hypotheses, for example, in the form of nonclassical variational principles or inequalities such as those first applied by Haar and Karman (1909); see also dissipative inequalities in Ziegler (1963). In particular, the hypothesis of the existence of maximal energy dissipation at a discontinuity turns out to be very fruitful. To satisfy these variational inequalities a special permissible class of stresses is considered in the form of convex subsets in the stress space presented in parametric form. For this class of stresses, a closed system of discontinuity relations is obtained, including strong and weak elastic-plastic shock waves for different theories of plasticity and criteria (6.1.19) and (6.1.22); see details in Bykovtsev and Kretova (1972), Kamenyarz (1972), Druyanov (1982), Balashov (1993), Klyushnikov (1993), and Sadovskii (1997). However, this approach is not yet generally accepted today and is under investigation. Particularly, the following governing system of equations for $\mathbf{v}$ and $\hat{\mathbf{P}}$ under the assumption that $\rho = \text{const}$ and $\varepsilon = \text{const}$ can be used (Sadovskii 1997), see (6.1.49):

$$\rho \frac{d\mathbf{v}}{dt} - \operatorname{div} \hat{\mathbf{P}} = \mathbf{0},$$

$$\frac{d\hat{\mathbf{P}}}{dt} - \lambda \operatorname{tr}(\hat{\mathbf{V}})\, \hat{\mathbf{I}} - 2\mu \hat{\mathbf{V}} + \theta(s)(\hat{\mathbf{S}} : \hat{\mathbf{V}})\, \hat{\mathbf{S}} = \hat{\mathbf{0}}.$$

In our opinion, it is important to take into account the equation for the temperature or internal energy, since the elastic and plastic modulus depend on the density and temperature. Only such an approach could help one to generalize the exact solutions of gas dynamics. Another drawback of the above approach is the nonuniqueness in the parametrization of the class of permissible stresses. Moreover, by now there is no complete analysis of the structure and stability of the solutions obtained.

So far in governing equations we have dealt with the tensor $\hat{\mathbf{T}}$ describing small deformations, see (6.1.12). For nonsmall deformations one should use more complicated tensors with nonlinear terms, in particular, the Almansi strain tensor $\hat{\mathbf{T}}^{\mathrm{A}}$, see (6.1.46),

$$T_{ij}^{\mathrm{A}} = \frac{1}{2}\left(\frac{\partial d_i}{\partial x_j} + \frac{\partial d_j}{\partial x_i} - \sum_{k=1}^{3} \frac{\partial d_k}{\partial x_i}\frac{\partial d_k}{\partial x_j}\right), \tag{6.1.50}$$

or the Green strain tensor $\hat{\mathbf{T}}^{\mathrm{G}}$,

$$T_{ij}^{\mathrm{G}} \simeq \frac{1}{2}\left(\frac{\partial d_i}{\partial x_j} + \frac{\partial d_j}{\partial x_i} + \sum_{k=1}^{3} \frac{\partial d_k}{\partial x_i}\frac{\partial d_k}{\partial x_j}\right), \tag{6.1.51}$$

in the Eulerian or Lagrangian coordinates, respectively (Prager 1961; Kolarov et al. 1979; Zubchaninov 1990). These tensors are not linear in $\mathbf{d}$ and the generalization of Hooke's law for these tensors is not clear. Also, the problem still exists of writing the governing system in the conservative form. These problems can be overcome by considering the governing equations in terms of some special, more natural variables. This also permits one to eliminate the problem of choosing the tensor derivative with respect to time.

6.1.2 Conservative form of elastoviscoplastic solid dynamics.

Below we describe some conservative forms of solid dynamics equations for nonlinear thermoelasticity (Kondaurov 1981) and elastoviscoplasticity (Kondaurov 1982a, 1982b). This conservative representation turned out to be possible by using the strain gradient tensor $\hat{\mathbf{J}}$ of (6.1.37). We first discuss the basic properties of $\hat{\mathbf{J}}$.

At $t = 0$, consider a liquid volume G_0 with the density distribution $\rho_0 = \rho_0(x_1^0, x_2^0, x_3^0)$, where x_i^0 are the Lagrangian coordinates. Let the Eulerian coordinate system x_i coincide with x_i^0 at $t = 0$. Then the density distribution $\rho = \rho(t, x_1, x_2, x_3)$ in the liquid volume G_0 for other time instants can be determined from the mass conservation law:

$$\iiint_{G_0} \rho_0 \, dx_1^0 dx_2^0 dx_3^0 = \iiint_{G_0} \rho_0 \, dx_1 dx_2 dx_3 = \iiint_{G_0} \rho \, dx_1^0 dx_2^0 dx_3^0 = \iiint_{G_0} \rho \det \hat{\mathbf{J}} \, dx_1 dx_2 dx_3.$$

Hence,

$$\iiint_{G_0} \rho_0 \, dx_1 dx_2 dx_3 = \iiint_{G_0} \rho \det \hat{\mathbf{J}} \, dx_1 dx_2 dx_3,$$

or

$$\rho_0 = \rho \det \hat{\mathbf{J}}. \tag{6.1.52}$$

Let us determine the evolution of the strain gradient entries $[J_{ij}]$ in time. We have

$$\frac{dJ_{ij}}{dt} = \frac{d}{dt}\left(\frac{\partial x_i}{\partial x_j^0}\right) = \frac{\partial v_i}{\partial x_j^0} = \sum_{p=1}^{3} \frac{\partial v_i}{\partial x_p}\frac{\partial x_p}{\partial x_j^0} = \sum_{p=1}^{3} \frac{\partial v_i}{\partial x_p} J_{pj}, \tag{6.1.53}$$

and hence,

$$\frac{\partial J_{ij}}{\partial t} + \sum_{p=1}^{3}\left[v_p \frac{\partial J_{ij}}{\partial x_p} - \frac{\partial v_i}{\partial x_p} J_{pj}\right] = 0. \tag{6.1.54}$$

Differentiating relations (6.1.52) with respect to t yields

$$\frac{d\rho}{dt} = -\frac{\rho_0}{(\det \hat{\mathbf{J}})^2}\frac{d \det \hat{\mathbf{J}}}{dt} = -\frac{\rho}{\det \hat{\mathbf{J}}}\frac{d \det \hat{\mathbf{J}}}{dt}.$$

Using the Euler formula (Truesdell 1977)

$$\frac{1}{\det \hat{\mathbf{J}}}\frac{d \det \hat{\mathbf{J}}}{dt} = \operatorname{div} \mathbf{v}, \tag{6.1.55}$$

we obtain

$$\frac{d\rho}{dt} = -\rho \operatorname{div} \mathbf{v},$$

and, hence, Eq. (6.1.2). One can prove relation (6.1.55) by direct calculation:

$$\begin{aligned}
\frac{d \det \hat{\mathbf{J}}}{dt} &= \frac{d}{dt}\left(\sum_{i,j,k=1}^{3} e_{ijk} J_{1i} J_{2j} J_{3k}\right) \\
&= \sum_{i,j,k,p=1}^{3} e_{ijk}\left[\frac{\partial v_1}{\partial x_p} J_{pi} J_{2j} J_{3k} + \frac{\partial v_2}{\partial x_p} J_{1i} J_{pj} J_{3k} + \frac{\partial v_3}{\partial x_p} J_{1i} J_{2j} J_{pk}\right] \\
&= \sum_{i,j,k=1}^{3} e_{ijk}\left[\frac{\partial v_1}{\partial x_1} J_{1i} J_{2j} J_{3k} + \frac{\partial v_2}{\partial x_2} J_{1i} J_{2j} J_{3k} + \frac{\partial v_3}{\partial x_3} J_{1i} J_{2j} J_{3k}\right] = \det \hat{\mathbf{J}} \times \operatorname{div} \mathbf{v},
\end{aligned}$$

where e_{ijk} is the $3 \times 3 \times 3$ identity antisymmetric tensor (Korn and Korn 1968).

Let us sum Eq. (6.1.54) multiplied by ρ and Eq. (6.1.2) multiplied by J_{ij}. Then we obtain

$$\frac{\partial \rho J_{ij}}{\partial t} + \sum_{p=1}^{3}\left[\frac{\partial \rho v_p J_{ij}}{\partial x_p} - \frac{\partial \rho v_i J_{pj}}{\partial x_p}\right] = -\sum_{p=1}^{3} v_i \frac{\partial \rho J_{pj}}{\partial x_p}. \tag{6.1.56}$$

Let us prove the relations

$$\sum_{p=1}^{3} \frac{\partial \rho J_{pj}}{\partial x_p} = 0.$$

Suppose $D = J^{-1}$, then by definition

$$J_{ij} \equiv \frac{1}{\det D}\Delta_{ji} = \det \hat{\mathbf{J}}\, \Delta_{ji} = \frac{\rho_0}{\rho}\Delta_{ji},$$

where Δ_{ji} is the corresponding cofactor of the matrix D. Hence,

$$\rho J_{ij} = \rho_0 \Delta_{ji}.$$

And then one can find by direct calculations that

$$\sum_{p=1}^{3} \frac{\partial \rho J_{pj}}{\partial x_p} = \rho_0 \sum_{p=1}^{3} \frac{\partial \Delta_{jp}}{\partial x_p} \equiv 0. \tag{6.1.57}$$

In particular, for $j = 1$ we have

$$\sum_{p=1}^{3} \frac{\partial \Delta_{1p}}{\partial x_p} = \frac{\partial \Delta_{11}}{\partial x_1} + \frac{\partial \Delta_{12}}{\partial x_2} + \frac{\partial \Delta_{13}}{\partial x_3} = \frac{\partial}{\partial x_1}\left(\frac{\partial x_2^0}{\partial x_2}\frac{\partial x_3^0}{\partial x_3} - \frac{\partial x_2^0}{\partial x_3}\frac{\partial x_3^0}{\partial x_2}\right)$$
$$- \frac{\partial}{\partial x_2}\left(\frac{\partial x_2^0}{\partial x_1}\frac{\partial x_3^0}{\partial x_3} - \frac{\partial x_2^0}{\partial x_3}\frac{\partial x_3^0}{\partial x_1}\right) + \frac{\partial}{\partial x_3}\left(\frac{\partial x_2^0}{\partial x_1}\frac{\partial x_3^0}{\partial x_2} - \frac{\partial x_2^0}{\partial x_2}\frac{\partial x_3^0}{\partial x_1}\right) \equiv 0.$$

Hence, the right-hand side in Eq. (6.1.56) is equal to zero and the governing system of equations of isothermal elasticity has the conservative form

$$\frac{\partial \rho \mathbf{v}}{\partial t} + \operatorname{div}(\rho \mathbf{v}\mathbf{v} - \hat{\mathbf{P}}) = \mathbf{0},$$
$$\frac{\partial \rho J_{ij}}{\partial t} + \sum_{p=1}^{3} \frac{\partial}{\partial x_p}(\rho v_p J_{ij} - \rho v_i J_{pj}) = 0.$$

For determined $\hat{\mathbf{P}}$ one can use Hooke's laws for one of the strain tensors (6.1.40)–(6.1.44) that are dependent on J_{ij}. The density ρ can be found from (6.1.52). The number of variables of this system in the three-dimensional case is equal to 12. Note that these equations permits one to take into account arbitrary finite values of deformations.

The governing system of equations of thermoelasticity can be written out in the conservative form

$$\frac{\partial \rho \mathbf{v}}{\partial t} + \operatorname{div}(\rho \mathbf{v}\mathbf{v} - \hat{\mathbf{P}}) = \mathbf{0}, \tag{6.1.58}$$

$$\frac{\partial e}{\partial t} + \operatorname{div}(e\mathbf{v} - \hat{\mathbf{P}} \cdot \mathbf{v}) = 0, \tag{6.1.59}$$

$$\frac{\partial \rho J_{ij}}{\partial t} + \sum_{p=1}^{3} \frac{\partial}{\partial x_p}(\rho v_p J_{ij} - \rho v_i J_{pj}) = 0, \tag{6.1.60}$$

where

$$P_{ij} = \rho \sum_{p=1}^{3} J_{ip} \frac{\partial F}{\partial J_{jp}}, \quad S = -\frac{\partial F}{\partial T}, \quad \rho = \frac{\rho_0}{\det \hat{\mathbf{J}}}.$$

Here $F = \varepsilon - ST$ is the free energy per unit mass, $\varepsilon = \varepsilon(\rho, T, J_{ij})$ is the internal energy per unit mass, S is the entropy, T is the temperature, and $e = \rho\varepsilon + \frac{1}{2}\rho(u^2 + v^2 + w^2)$; see, for details, Kondaurov (1981, 1982a, 1982b), Bushman et al. (1993), and Godunov and Romenskii (1998). The number of variables of this system in the three-dimensional case is equal to 13. Equations (6.1.58)–(6.1.59) express the conservation momentum and energy of the medium, respectively. Equation (6.1.60) describes the continuity of media, in particular, the absence of cracks.

By analogy with Eqs. (6.1.58)-(6.1.60) the governing system of equations for the dynamics of elastoviscoplastic medium can be written out. A material is elastic inside the yield surface and plastic with hardening law (6.1.23) outside the surface. In this case the strain gradient tensor $\hat{\mathbf{J}}$ can be represented in the following form:

$$\hat{\mathbf{J}} = \hat{\mathbf{J}}^{\mathrm{e}} \cdot \hat{\mathbf{J}}^{\mathrm{p}}, \quad \hat{\mathbf{J}}^{\mathrm{e}} = [J_{ij}^{\mathrm{e}}], \quad \hat{\mathbf{J}}^{\mathrm{p}} = [J_{ij}^{\mathrm{p}}], \qquad \det[J_{ij}^{\mathrm{e}}] > 0, \quad \det[J_{ij}^{\mathrm{p}}] > 0.$$

Here, $\hat{\mathbf{J}}^{\mathrm{e}}$ is the elastic strain gradient tensor and $\hat{\mathbf{J}}^{\mathrm{p}}$ is the plastic strain gradient tensor, $\hat{\mathbf{J}}^{\mathrm{e}} \cdot \hat{\mathbf{J}}^{\mathrm{p}} \neq \hat{\mathbf{J}}^{\mathrm{p}} \cdot \hat{\mathbf{J}}^{\mathrm{e}}$.

Let us postulate that the independent variables describing the thermodynamic state of a medium are

$$\pi = \{J_{ij}^{\mathrm{e}}, J_{ij}^{\mathrm{p}}, \chi, T\}.$$

The variables J_{ij}^{p} and χ are assumed to satisfy the relations

$$\frac{dJ_{ij}^{\mathrm{p}}}{dt} = \Phi_{ij}(\pi), \quad \frac{d\chi}{dt} = \Phi_0(\pi) \ \text{ if } \ f(\pi) \geq 0,$$

and

$$\frac{dJ_{ij}^{\mathrm{p}}}{dt} = 0, \quad \frac{d\chi}{dt} = 0 \ \text{ if } \ f(\pi) < 0,$$

where $f(\pi) = 0$ is the yield surface, and Φ_{ij} and Φ_0 are determined by the rheology of the material.

The governing system of equations can be written out in the conservative form

$$\begin{aligned}
&\frac{\partial \rho \mathbf{v}}{\partial t} + \operatorname{div}(\rho \mathbf{v}\mathbf{v} - \hat{\mathbf{P}}) = \mathbf{0},\\
&\frac{\partial e}{\partial t} + \operatorname{div}(e\mathbf{v} - \hat{\mathbf{P}} \cdot \mathbf{v}) = 0,\\
&\frac{\partial \rho J_{ij}}{\partial t} + \sum_{p=1}^{3} \frac{\partial}{\partial x_p}(\rho v_p J_{ij} - \rho v_i J_{pj}) = 0,\\
&\frac{\partial \rho J_{ij}^{\mathrm{p}}}{\partial t} + \operatorname{div}(\rho \mathbf{v} J_{ij}^{\mathrm{p}}) = \rho \Phi_{ij},\\
&\frac{\partial \rho \chi}{\partial t} + \operatorname{div}(\rho \mathbf{v} \chi) = \rho \Phi_0,
\end{aligned} \tag{6.1.61}$$

where

$$P_{ij} = \rho \sum_{p=1}^{3} J^{\mathrm{e}}_{ip} \frac{\partial F}{\partial J^{\mathrm{e}}_{jp}}, \quad S = -\frac{\partial F}{\partial T}, \quad \rho = \frac{\rho_0}{\det \hat{\mathbf{J}}},$$

Here $F = \varepsilon - ST$ and $\varepsilon = \varepsilon(\rho, T, J^{\mathrm{e}}_{ij}, J^{\mathrm{p}}_{ij})$ is the internal energy per unit mass. The number of variables of this system in the three-dimensional case is equal to 20, since $\hat{\mathbf{J}}^{\mathrm{p}}$ is symmetric (Kondaurov 1982a, 1982b).

The conservative equations described above permit one to obtain the formal relations at discontinuities (Kondaurov 1981, 1982a, 1982b). This does not mean, however, that the obtained relations will adequately express the physical relations on discontinuities. Such relations can be obtained only from the divergent form equations of elastoviscoplasticity which are obtained from fundamental conservation laws of mass, momentum, energy, etc. Otherwise, the relations obtained on the basis of Eq. (6.1.61) will probably be valid only under the assumption of small plastic deformations.

The above system can be used to symmetrize the governing system of equations (Kondaurov 1981, 1982a, 1982b) and investigate the wave propagation (Bushman et al. 1993). However, this approach is not yet generally accepted today and is under investigation.

Note that the results of solving of the Riemann problem for plastic media by Bykovtsev and Kretova (1972), Kamenyarz (1972), Druyanov (1982), Balashov (1993) are based essentially on the existence of the hardening process. Therefore, their results are completely consistent with Eq. (6.1.61).

6.1.3 Dynamics of thin shells. Many modern machines and structures (buildings, ships, rockets, airplanes, and others) consist of thin metal and composite shells. Therefore, their investigation, analysis, and simulation are of great practical importance. There are a lot of different dynamics systems that describe the motion and vibration of thin shells; see Grigolyuk and Selezov (1973), Donnell (1976), Goldenveizer (1976), and Gorshkov and Tarlakovskii (1990). These systems were derived under various assumptions about kinematic and static constraints.

A thin shell is an elastic three-dimensional body bounded by two curvilinear surfaces such that the distance h between these surfaces is small as compared with the characteristic curvatures. In particular, a shell is usually regarded as *thin* if $\beta = h/R < 0.02$–0.05, where h is the shell thickness and R is the minimal radius of the characteristic shell curvatures. The fact that a shell is thin, under the assumption of elastic motion with $\rho = \text{const}$ and $\varepsilon = \text{const}$, permits one to reduce the three-dimensional solid dynamics equations to two-dimensional equations. Particularly, in constructing the governing equations in the theory of thin shells, the simplified Kirchhoff–Love and Love hypotheses are adopted; see Donnell (1976), Goldenveizer (1976), and Grigolyuk and Mamay (1997). This is the so-called classical shell theory, which describes most of the static problems and dynamics (or quasistatic) problems with smooth and relatively long-term loadings; see the review of mechanical experiments in Grigolyuk and Selezov (1973). However, the shell behavior under short pulse loading may be described inadequately by this theory. Besides, the dynamics equations of the classical shell theory are nonhyperbolic.

In 1921 Timoshenko suggested a fruitful modification of the classical theory by taking additionally into account the shear and rotational inertia of the shell cross-section.

The Timoshenko-type dynamics systems are hyperbolic, as well as the complete three-dimensional elasticity equations, and this permits one to describe adequately the effect of short pulse loading and the wave propagation in shells. Note that the derivation of Timoshenko-type systems requires taking into account more expansion terms than the classical theory and is, thus, more sophisticated.

To illustrate the difference between the classical theories of rods, plates, and shells and the Timoshenko theory, we consider as an example a simpler linear equation describing the vibration of a uniform rod (Grigolyuk and Selezov 1973), rather than a shell. The classical Bernoulli–Euler equation in this case becomes

$$\frac{\partial^2 w}{\partial t^2} + a_1 \frac{\partial^4 w}{\partial x^4} = 0,$$

the Rayleigh modification of this equation is given by

$$\frac{\partial^2 w}{\partial t^2} + a_1 \frac{\partial^4 w}{\partial x^4} - a_2^* \frac{\partial^4 w}{\partial t^2 \partial x^2} = 0,$$

and the Timoshenko-type equation has the form

$$\frac{\partial^2 w}{\partial t^2} + a_1 \frac{\partial^4 w}{\partial x^4} - a_2 \frac{\partial^4 w}{\partial t^2 \partial x^2} + a_3 \frac{\partial^4 w}{\partial t^4} = 0,$$

where a_1, a_2^*, a_2, and a_3 are some positive constants depending on the density of the rod, its geometrical parameters and elastic coefficients; w is the deflection. The last equation is hyperbolic (see Section 1.3.5). Experiments show that the Timoshenko equation adequately describe the wave propagation for shorter pulse loadings when the classical equations (Grigolyuk and Selezov 1973). This fact can be important in studying the specific features of vibrations of structures.

We will consider for simplicity only one-dimensional systems, which means that one of the radii of curvature is infinite, i.e., the shell is cylindrical. In general, the one-dimensional dynamics equations of thin shells with deformations, transverse displacements, and rotational inertia are hyperbolic systems of second-order equations of both mixed and conservative form. For example, such systems may have the conservative form

$$\mathbf{V}_{tt} + [\mathbf{G}(t, x, \mathbf{V}, \mathbf{V}_x)]_x = \mathbf{H}(t, x, \mathbf{V}, \mathbf{V}_x), \tag{6.1.62}$$

where $\mathbf{V} = [V_1, \ldots, V_M]^{\mathrm{T}}$, $\mathbf{G} = [G_1, \ldots, G_M]^{\mathrm{T}}$, and $\mathbf{H} = [H_1, \ldots, H_M]^{\mathrm{T}}$; $\mathbf{V} = \mathbf{V}(t, x)$ is the vector of displacements of the points of the shell midsurface and angles of rotation of the normal to the midsurface, t is time, and x is the space or angular coordinate. System (6.1.62) can be reduced to a quasilinear hyperbolic system of first order equations by introducing a new $3M$-dimensional vector of variables $\mathbf{U}$; see Section 6.3. Consider particular systems of equations.

The Klein–Gordon equation

The Klein–Gordon equation is the simplest case of a system of shell dynamics; it has the form

$$\frac{\partial^2 u}{\partial t^2} - \frac{\partial^2 u}{\partial x^2} + \frac{u}{\beta^2} = 0. \tag{6.1.63}$$

This equation is a special case of the one-dimensional wave equation of the general form

$$\frac{\partial^2 u}{\partial t^2} - c^2 \frac{\partial^2 u}{\partial x^2} + a\frac{\partial u}{\partial x} + bu = q(t,x). \tag{6.1.64}$$

The coefficient c^2 in Eq. (6.1.64), which is equal to unity in Eq. (6.1.63), defines the speed c of propagation of small perturbations. In particular, the Klein–Gordon equation is one of the basic equation of mathematical physics. It describes the dynamics of a free relativistic pseudo-particle (Vladimirov 1971), is used for the investigation of the dynamics of a thin vertical rod on an elastic foundation (Slepyan 1972), etc. In the two-dimensional case this equation describes, for example, the edge wave in a liquid under an elastic plate with a crack (Gol'dshtein, Marchenko, and Semenov 1994; Marchenko and Semenov 1994). Some details of numerical calculation of propagation of this wave can be found in Marchenko and Semenov (1995).

Equations of cylindrical shell dynamics

The dynamics equations for an infinite uniform cylindrical isotropic shell under the action of an external force independent of the axial coordinate can be written in the form

$$\begin{aligned} V_{tt} &= (T - \Psi Q)_x + \theta T + Q - \Psi q, \\ W_{tt} &= (Q + \theta T)_x + \Psi Q - T + q, \\ \Psi_{tt} &= -M_x - \theta M + 12\beta^{-2} Q, \end{aligned} \tag{6.1.65}$$

where

$$\theta = W_x - V, \quad Q = (\theta - \Psi)\delta_1, \quad T = V_x + W + \tfrac{1}{2}\theta^2, \quad M = -\Psi_x,$$
$$\beta = \frac{h}{R}, \quad \delta_1 = \tfrac{1}{2}(1-\sigma^3)\delta^2, \quad \delta^2 = 0.86;$$

see Ogibalov (1963) and Grigolyuk and Gorshkov (1976). In (6.1.65), h and R are the shell thickness and the midsurface radius, V and W are the tangential and normal displacements of the midsurface, Ψ is the angle of rotation of the normal to the midsurface, σ is Poisson's ratio of (6.1.16), $q = q(t,x)$ is the external force, and x is the polar angle.

The dynamics equations for an infinite uniform cylindrical isotropic shell for axisymmetric deformations can be written in the following form (Evseev and Semenov 1985, 1989):

$$\begin{aligned} V_{tt} &= (T - \Psi Q)_x - \Psi\theta Q - \Psi q, \\ W_{tt} &= (Q + \theta T)_x + \theta Q - T + q, \\ \Psi_{tt} &= -M_x + 12\beta^{-2} Q, \end{aligned} \tag{6.1.66}$$

where

$$\theta = W_x, \quad Q = (\theta - \Psi)\delta_1, \quad T = V_x + \tfrac{1}{2}\theta^2, \quad M = -\Psi_x,$$

and x is the axial coordinate.

The displacements in Eqs. (6.1.65)–(6.1.66) and the axial coordinate in Eq. (6.1.66) are normalized by R, the stresses are normalized by $B = Eh/(1-\sigma^2)$, where E is Young's modulus of (6.1.15), the torsional moment, time t, and the external force are normalized by $Bh^2/(12R)$, $R\sqrt{\rho h/B}$, and B/R, respectively, where ρ is the material density.

Dynamics equations of an orthotropic shell

The analysis of the motion of thin shells may be based on a simpler model, the so-called orthotropic shell model. For example, consider an orthotropic composite thin shell that consists of a linearly elastic isotropic stiff matrix reinforced by linearly elastic flexible filaments. The dynamics equations of thin shells for axisymmetric deformations in terms of dimensionless variables can be written (Teters 1969; Ricards and Teters 1974; Teters, Ricards, and Narusberg 1978; Banichuk, Kobelev, and Ricards 1988) as follows:

$$\begin{aligned} u_{tt} - a_1^2 u_{xx} &= b_1 w_x, \\ w_{tt} - a_2^2 w_{xx} &= b_2 \varphi_x - c_2 u_x - d_2 w + e_2 q, \\ \varphi_{tt} - a_3^2 \varphi_{xx} &= -b_3 w_x - c_3 \varphi, \end{aligned} \tag{6.1.67}$$

where

$$\begin{aligned} & a_1 = \gamma\beta^{-1}, \quad a_2 = \gamma\beta^{-1}\sqrt{\alpha}, \quad a_3 = \gamma\beta^{-1} \quad \Longrightarrow \quad \Delta t/\beta \le \Delta x, \\ & \alpha = (1 - \sigma_1\sigma_2)\frac{\mu_{12}}{E_1}, \quad \beta = \frac{h}{R}, \quad \gamma = \frac{R}{L}, \\ & x = \frac{\tilde{x}}{L}, \quad u = \frac{\tilde{u}}{L}, \quad w = \frac{\tilde{w}}{h}, \quad t = \tilde{t}\frac{ch}{R^2}, \\ & q = \tilde{q}\frac{(1 - \sigma_1\sigma_2)L}{hE_1}, \quad c = \sqrt{\frac{E_1}{(1 - \sigma_1\sigma_2)\rho}}, \\ & b_1 = O(\beta^{-1}), \quad b_2 = O(\beta^{-3}), \quad b_3 = O(\beta^{-1}), \\ & c_2 = O(\beta^{-3}), \quad c_3 = O(\beta^{-2}), \quad d_2 = \frac{E_1}{\beta^2 E_2} = O(\beta^{-2}), \quad e_2 = O(\beta^{-2}). \end{aligned}$$

In (6.1.67), h, R, and L are the shell thickness, the midsurface radius, and the shell length, respectively; u and w are the tangential and normal displacements of the midsurface, φ is the angle of rotation of the normal to the midsurface, $q = q(t, x)$ is the external force, and x is the axial coordinate. The tilde over a symbol denotes the corresponding dimensional variable; E_1, E_2, and σ_1 and σ_2 are Young's moduli and Poisson's ratios of the matrix and filaments, respectively; μ_{12} is the effective shear modulus and ρ is the material density.

6.2 CIR method for the calculation of solid dynamics problems

The solid dynamics equations can be represented in either nonconservative or mixed (conservative–nonconservative) forms. For each form, one can use special versions of the Courant–Isaacson–Rees (CIR) methods described previously in Section 2.3. The CIR methods are Godunov-type methods based on an approximate solution of a locally linearized hyperbolic system. This approach is a reliable instrument of numerical simulation. It permits one to simulate the flows of media with various moving discontinuities and obtain monotone profiles of grid variables in the vicinity of discontinuities or large gradients. The CIR-type methods are successfully used for simulation of various multidimensional problems of solid dynamics, taking into account various governing equations; see, for example, Petrov and Kholodov (1984a, 1984b), Korotin et al. (1987), Ivanov, Petrov, and Suvorova

(1989), Korotin, Petrov, and Kholodov (1989), Ivanov et al. (1990), Petrov, Tormasov, and Kholodov (1990), Petrov and Tormasov (1990), Ivanov and Petrov (1992), V. D. Ivanov et al. (1999), etc. There are more than 50 works devoted to the application of the CIR method in solid dynamic simulations.

In this section we will describe formulas that are necessary for practical implementation of the CIR class of numerical difference schemes.

Equations (6.1.30)–(6.1.36) in vector form read

$$\frac{\partial \mathbf{U}}{\partial t} + A_1 \frac{\partial \mathbf{U}}{\partial x} + A_2 \frac{\partial \mathbf{U}}{\partial y} = \mathbf{0}, \tag{6.2.1}$$

$$\mathbf{U} = [\rho, u, v, \varepsilon, S_{11}, S_{22}, S_{12}]^{\mathrm{T}}, \tag{6.2.2}$$

$$A_1 = \begin{bmatrix} u & \rho & 0 & 0 & 0 & 0 & 0 \\ z_0 & u & 0 & z_1 & -1/\rho & 0 & 0 \\ 0 & 0 & u & 0 & 0 & 0 & -1/\rho \\ 0 & a_1 & a_2 & u & 0 & 0 & 0 \\ 0 & b_1 & b_2 & 0 & u & 0 & 0 \\ 0 & c_1 & c_2 & 0 & 0 & u & 0 \\ 0 & \delta_1 & \delta_2 & 0 & 0 & 0 & u \end{bmatrix}, \tag{6.2.3}$$

$$z_0 = \frac{1}{\rho}\frac{\partial p}{\partial \rho}, \quad z_1 = \frac{1}{\rho}\frac{\partial p}{\partial \varepsilon}, \quad a_1 = \frac{p_s}{\rho}, \quad a_2 = 0,$$

$$b_1 = -\tfrac{4}{3}\mu + \theta S_{11}S_{11}, \quad b_2 = \zeta S_{12} + \theta S_{12}S_{11},$$

$$c_1 = \tfrac{2}{3}\mu + \theta S_{11}S_{22}, \quad c_2 = -\zeta S_{12} + \theta S_{12}S_{22},$$

$$\delta_1 = \theta S_{11}S_{12}, \quad \delta_2 = -\mu - \tfrac{1}{2}(S_{11} - S_{22})\zeta + \theta S_{12}S_{12}.$$

The matrix A_2 can be obtained from A_1 by interchanging $u \leftrightarrow v$, $S_{11} \leftrightarrow S_{22}$, as well as the second and third, the fifth and sixth rows and columns. However, we will show that it suffices to describe a diagonalization procedure for only A_1 (or A_2).

For the construction of a CIR-type scheme, one must diagonalize the matrix A_1 (or A_2) by representing it in the form

$$A = \Omega_{\mathrm{R}} \Lambda \Omega_{\mathrm{L}},$$

where A denotes A_1 (or A_2). The matrices Ω_{L} and Ω_{R} are composed of the left and right eigenvectors of A, respectively, and $\Lambda = \mathrm{diag}[\lambda_1, \ldots, \lambda_7]$ is the diagonal matrix of the real eigenvalues of A. One can find that all eigenvalues λ of A_1 satisfy the equation

$$(L^4 - L^2 r + q)L^3 = 0, \quad L = \lambda - u,$$

$$r = z_0\rho + a_1 z_1 - \frac{b_1 + \delta_2}{\rho} = \tilde{c}^2 - \frac{b_1 + \delta_2}{\rho}, \quad \tilde{c} = \sqrt{z_0\rho + a_1 z_1},$$

$$q = -\frac{r\delta_2}{\rho} - \frac{\delta_2^2}{\rho^2} - \frac{(b_2 - z_1 a_2 \rho)\delta_1}{\rho^2} = \frac{r^2}{4} - \left(\frac{r}{2} + \frac{\delta_2}{\rho}\right)^2 - \frac{(b_2 - z_1 a_2 \rho)\delta_1}{\rho^2},$$

where $\tilde{c}$ is the gas dynamic speed of sound (3.1.24). The matrix A_1 has seven eigenvalues, and hence the eigenvalue matrix Λ is expressed as

$$\Lambda = \mathrm{diag}[u, u, u, u - \lambda_{20}, u + \lambda_{20}, u - \lambda_{10}, u + \lambda_{10}],$$

$$\lambda_{20} = \sqrt{\tfrac{1}{2}r + \sqrt{\tfrac{1}{4}r^2 - q}}, \quad \lambda_{10} = \sqrt{\tfrac{1}{2}r - \sqrt{\tfrac{1}{4}r^2 - q}},$$

where

$$\frac{r^2}{4} - q = \left(\frac{r}{2} + \frac{\delta_2}{\rho}\right)^2 + \frac{(b_2 - z_1 a_2 \rho)\delta_1}{\rho^2}.$$

We assume that $r^2 \geq 4q \geq 0$ and $r \geq 0$. In this case the system has only real eigenvalues and $\lambda_{20} \geq \lambda_{10}$. We will assume also that $\delta_2 \neq 0$ and $q \neq 0$, whence $\lambda_{10}\lambda_{20} \neq 0$ and $r > 0$. The assumption $\delta_2 \neq 0$ for $\zeta = 0$ is equivalent to the condition of $\mu > 0$ and absence of the purely plastic shear flow with $S_{11} = S_{22} = 0$ and $S_{12} = k$.

For a one-dimensional elastic flow with $\theta(s) = 0$ and $\zeta = 0$ one can find

$$\lambda_{20} = \sqrt{\frac{\tilde{\lambda} + 2\mu}{\rho}}, \quad \lambda_{10} = \sqrt{\frac{\mu}{\rho}}, \tag{6.2.4}$$

where

$$\tilde{\lambda} = \rho\frac{\partial p}{\partial \rho} + \frac{p_s}{\rho}\frac{\partial p}{\partial \varepsilon} - \frac{2}{3}\mu = \rho\tilde{c}^2 - \frac{2}{3}\mu + \frac{p_s - p}{\rho}\frac{\partial p}{\partial \varepsilon}.$$

Here $\tilde{\lambda}$ and μ are the Lamé coefficients. For a one-dimensional isotropic plastic flow with $\theta(s) = \mu/k^2$ and, in particular, $S_{12} = 0$, $S_{11} = S_{22} = -\frac{1}{2}S_{33} = k/\sqrt{3}$, and $\zeta = 0$, we can find

$$\lambda_{20} = \sqrt{\frac{\tilde{\lambda} + \frac{4}{3}\mu}{\rho}}, \quad \lambda_{10} = \sqrt{\frac{\mu}{\rho}}. \tag{6.2.5}$$

The matrices Ω_L and Ω_R of A_1 have the form (Semenov 1988)

$$\Omega_L = \frac{1}{2[D,d]}\begin{bmatrix} -[a,\delta]/\rho & 0 & 0 & \delta_2 & 0 & 0 & -a_2 \\ -[b,\delta]/\rho & 0 & 0 & 0 & \delta_2 & 0 & -b_2 \\ -[c,\delta]/\rho & 0 & 0 & 0 & 0 & \delta_2 & -c_2 \\ -z_0 d_1 & \lambda_{20}d_1 & \rho\lambda_{20}D_1 & -z_1 d_1 & d_1/\rho & 0 & D_1 \\ -z_0 d_1 & -\lambda_{20}d_1 & -\rho\lambda_{20}D_1 & -z_1 d_1 & d_1/\rho & 0 & D_1 \\ -z_0 d_2 & \lambda_{10}d_2 & \rho\lambda_{10}D_2 & -z_1 d_2 & d_2/\rho & 0 & D_2 \\ -z_0 d_2 & -\lambda_{10}d_2 & -\rho\lambda_{10}D_2 & -z_1 d_2 & d_2/\rho & 0 & D_2 \end{bmatrix}, \tag{6.2.6}$$

$$d_1 = \rho\lambda_{20}^2 + \delta_2, \quad d_2 = \delta_1,$$

$$D_1 = \frac{(\delta_2 + \rho\lambda_{10}^2)d_1}{\rho\delta_1} = a_2 z_1 - \frac{b_2}{\rho}, \quad D_2 = \frac{(\delta_2 + \rho\lambda_{20}^2)d_2}{\rho\delta_1} = \frac{d_1}{\rho},$$

$$[f,g] = \det\begin{bmatrix} f_1 & f_2 \\ g_1 & g_2 \end{bmatrix} = f_1 g_2 - f_2 g_1;$$

$$\det\Omega_L = 2[D,d]\lambda_{20}\lambda_{10}\delta_2^2\rho q = 2[D,d]\delta_2^2\rho q^{3/2};$$

$$\Omega_R = \begin{bmatrix} \times & \times & \times & D_2\rho/\lambda_{20}^2 & D_2\rho/\lambda_{20}^2 & -D_1\rho/\lambda_{10}^2 & -D_1\rho/\lambda_{10}^2 \\ \times & \times & \times & -D_2/\lambda_{20} & D_2/\lambda_{20} & D_1/\lambda_{10} & -D_1/\lambda_{10} \\ \times & \times & \times & d_2/\rho\lambda_{20} & -d_2/\rho\lambda_{20} & -d_1/\rho\lambda_{10} & d_1/\rho\lambda_{10} \\ \times & \times & \times & E_2 & E_2 & -E_1 & -E_1 \\ \times & \times & \times & F_2 & F_2 & -F_1 & -F_1 \\ \times & \times & \times & G_2 & G_2 & -G_1 & -G_1 \\ \times & \times & \times & d_2 & d_2 & -d_1 & -d_1 \end{bmatrix}, \tag{6.2.7}$$

$$E_i = \frac{\rho a_1 D_i - a_2 d_i}{\rho\lambda_{i0}^2}, \quad F_i = \frac{\rho b_1 D_i - b_2 d_i}{\rho\lambda_{i0}^2}, \quad G_i = \frac{\rho c_1 D_i - c_2 d_i}{\rho\lambda_{i0}^2}, \qquad i = 1, 2.$$

Here the entries denoted by crosses within each of the first three columns $\mathbf{r}^1$, $\mathbf{r}^2$, and $\mathbf{r}^3$ are equal, respectively, to

$$\mathbf{r}^1 = \alpha_1\tilde{\mathbf{r}}^1 + \alpha_2\tilde{\mathbf{r}}^2 + \alpha_3\tilde{\mathbf{r}}^3, \quad \mathbf{r}^2 = \beta_1\tilde{\mathbf{r}}^1 + \beta_2\tilde{\mathbf{r}}^2 + \beta_3\tilde{\mathbf{r}}^3, \quad \mathbf{r}^3 = \gamma_1\tilde{\mathbf{r}}^1 + \gamma_2\tilde{\mathbf{r}}^2 + \gamma_3\tilde{\mathbf{r}}^3;$$
$$\tilde{\mathbf{r}}^1 = [z_1, 0, 0, -z_0, 0, 0, 0]^T, \quad \tilde{\mathbf{r}}^2 = [0, 0, 0, 1, \rho z_1, 0, 0]^T, \quad \tilde{\mathbf{r}}^3 = [0, 0, 0, 0, 0, 1, 0]^T,$$
$$\alpha_1 = \sigma = \frac{2[D, d]}{q}, \quad \alpha_2 = \frac{[b, \delta]\sigma}{\delta_2\rho^2}, \quad \alpha_3 = \frac{[c, \delta]\sigma}{\delta_2\rho},$$
$$\beta_1 = -\frac{\sigma}{z_1\rho}, \quad \beta_2 = -\frac{([a, \delta]z_1 + z_0\delta_2\rho)\sigma}{z_1\rho^2}, \quad \beta_3 = -\frac{[c, \delta]\sigma}{\delta_2\rho},$$
$$\gamma_1 = \gamma_2 = 0, \quad \gamma_3 = \frac{q\sigma}{\delta_2}.$$

The corresponding matrices Ω_L, Λ, and Ω_R of A_2 can be obtained similarly.

To use the CIR type methods of Section 2.3.1 one must calculate several vectors of type

$$\mathbf{Q} = \Omega_R \operatorname{diag}[\psi_1, \psi_1, \psi_1, \psi_4, \psi_5, \psi_6, \psi_7]\Omega_L \mathcal{P},$$
$$\mathbf{Q} = [Q_1, \dots, Q_7], \quad \mathcal{P} = \mathbf{U}_{m+1}^n - \mathbf{U}_m^n,$$

where $\psi_1 = \psi_2 = \psi_3$, since $\lambda_1 = \lambda_2 = \lambda_3 = u$. Note that the following representation can be used for the vector $\mathbf{Q}$ (see also gas dynamical formulas (3.4.4)):

$$\mathbf{Q} = \psi_1\mathcal{P} + \sum_{k=4}^{7}(\psi_k - \psi_1)\mathbf{R}_k, \tag{6.2.8}$$
$$\mathbf{R}_k = \Omega_R \operatorname{diag}_k[0, \dots, 1, \dots]\Omega_L\mathcal{P}, \qquad k = 4, 5, 6, 7;$$

where the diagonal matrix $\operatorname{diag}_k[0, \dots, 1, \dots]$ contains unity in the kth position. The vectors $\mathbf{R}_k$ can be directly calculated. In particular, for $k = 4$:

$$\mathbf{R}_4 = \frac{d_1}{2[D, d]} \begin{bmatrix} D_2\rho/\lambda_{20}^2 \\ -D_2/\lambda_{20} \\ d_2/\rho\lambda_{20} \\ E_2 \\ F_2 \\ G_2 \\ d_2 \end{bmatrix} \left(-z_0\mathcal{P}_1 + \lambda_{20}\mathcal{P}_2 + \frac{\rho\lambda_{20}D_1}{d_1}\mathcal{P}_3 - z_1\mathcal{P}_4 + \frac{1}{\rho}\mathcal{P}_5 + \frac{D_1}{d_1}\mathcal{P}_7\right).$$

Substituting

$$z_0\mathcal{P}_1 + z_1\mathcal{P}_4 \approx \frac{p^n_{m+1} - p^n_m}{\rho}$$

where p is the pressure, in many cases permits us to avoid the explicit usage of $\partial p/\partial\rho$ and $\partial p/\partial\varepsilon$. By analogy with (3.4.5) the compact algorithm of calculating (6.2.8) can be described. Approach (6.2.8) permits one to consider the case of $\lambda_{10} = 0$ without additional calculations. In this case $\lambda_1 = \lambda_2 = \lambda_3 = \lambda_6 = \lambda_7 = u$, and, hence, $\psi_1 = \psi_2 = \psi_3 = \psi_6 = \psi_7$, or

$$\mathbf{Q} = \psi_1\mathcal{P} + \sum_{k=4}^{5}(\psi_k - \psi_1)\mathbf{R}_k,$$
$$\mathbf{R}_k = \Omega_{\mathrm{R}}\,\mathrm{diag}_k[0,\ldots,1,\ldots]\Omega_{\mathrm{L}}\mathcal{P}, \qquad k = 4, 5;$$

Note that interchanging $u \leftrightarrow v$ (or $\mathcal{P}_2 \leftrightarrow \mathcal{P}_3$), $S_{11} \leftrightarrow S_{22}$ (or $\mathcal{P}_5 \leftrightarrow \mathcal{P}_6$), as well as interchanging $Q_2 \leftrightarrow Q_3$ and $Q_5 \leftrightarrow Q_6$ determines $\mathbf{Q}$ corresponding to the matrix A_2.

For the relation $\hat{\mathbf{S}} : \hat{\mathbf{S}} = 2k^2$ to be preserved in the region of plastic flow, one must correct the deviatory components S_{ij} in accordance with the rule

$$S_{ij} \to S_{ij}\frac{\sqrt{2}\,k}{\sqrt{\hat{\mathbf{S}} : \hat{\mathbf{S}}}}$$

at each time step.

One can use the above formulas for the matrices Ω_{L}, Λ, and Ω_{R} to construct a CIR-type numerical method for other systems of equations, in particular, for Eq. (6.1.48). Then, Eq. (6.1.48) in vector form reads

$$\frac{\partial\mathbf{W}}{\partial t} + \widetilde{A}_1\frac{\partial\mathbf{W}}{\partial x} + \widetilde{A}_2\frac{\partial\mathbf{W}}{\partial y} = \mathbf{0},$$
$$\mathbf{W} = [u, v, \varepsilon, P_{11}, P_{22}, P_{12}, P_{33}]^{\mathrm{T}},$$

where $\widetilde{A}_1 = M^{-1}A_1M$ and $\widetilde{A}_2 = M^{-1}A_2M$. The matrices $M(\mathbf{W}) = \partial\mathbf{U}/\partial\mathbf{W}$ and $M^{-1}(\mathbf{W}) = \partial\mathbf{W}/\partial\mathbf{U}$ are the Jacobian matrices of transformations $\mathbf{U} \to \mathbf{W}$ and $\mathbf{W} \to \mathbf{U}$, respectively. These Jacobian matrices can be easily determined from the formulas $\rho = \rho(p, \varepsilon)$, $S_{11} = P_{11} + p$, $S_{22} = P_{22} + p$, $S_{12} = P_{12}$, where $p = -\frac{1}{3}(P_{11} + P_{22} + P_{33})$, and the inverse formulas $P_{11} = S_{11} - p$, $P_{22} = S_{22} - p$, $P_{12} = S_{12}$, and $P_{33} = -3p(\rho, \varepsilon) - P_{11} - P_{22}$. The other variables, u, v, and ε, remain the same. The formulas for diagonalization of $\widetilde{A}_1$ can be found in Petrov and Kholodov (1984a) and Magomedov and Kholodov (1988).

By analogy, one can rewrite Eqs. (6.1.26)–(6.1.29) in terms of ρ, $\rho\mathbf{v}$, e, and $\hat{\mathbf{S}}$, which provides the conservative–nonconservative representation of the governing system of equations; see (6.1.2)–(6.1.5). Note that under assumption that $\mu = \text{const}$ the equation for the total energy, (6.1.4), can be rewritten in the familiar gas dynamic form

$$\frac{\partial\widetilde{e}}{\partial t} + \mathrm{div}[(\widetilde{e} + p)\mathbf{v}] = f_{\mathrm{s}}, \quad f_{\mathrm{s}} = \mathbf{v}\cdot\mathrm{div}\,\hat{\mathbf{S}} - (p_{\mathrm{s}} - p)\,\mathrm{div}\,\mathbf{v},$$

where $\widetilde{e} = \rho\varepsilon + \frac{1}{2}\rho\,(u^2 + v^2 + w^2)$ and $f_{\mathrm{s}} \to 0$ if $\hat{\mathbf{S}} \to \hat{\mathbf{0}}$.

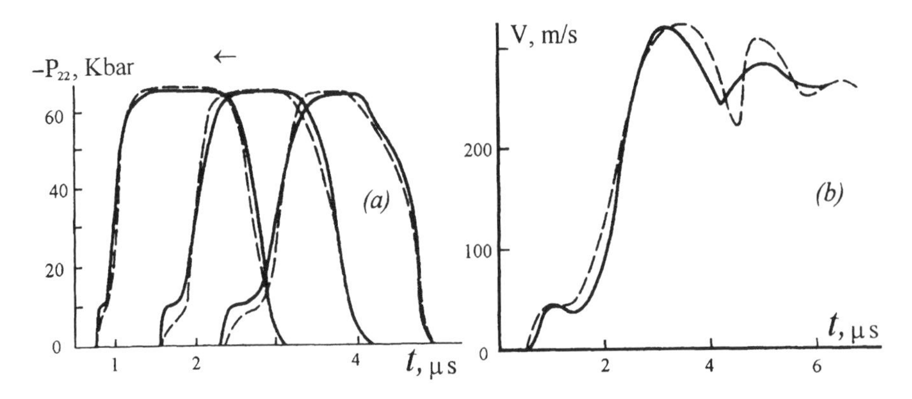

Figure 6.1 (a) The propagation of an elastoplastic wave resulting from the collision of two iron disks; (b) graphs of the velocity of the free surface of the target under the interaction with the wave.

In the general case of an arbitrary moving grid, in Eq. (6.2.1) one must diagonalize matrices $\alpha_1 A_1 + \alpha_2 A_2$, where α_1 and α_2 are coefficients depending on the local rotation of the coordinate system.

The three-dimensional equations of solid dynamics in vector form read

$$\frac{\partial \mathbf{U}}{\partial t} + A_1 \frac{\partial \mathbf{U}}{\partial x} + A_2 \frac{\partial \mathbf{U}}{\partial y} + A_3 \frac{\partial \mathbf{U}}{\partial z} = \mathbf{0}, \quad \mathbf{U} = [\rho, u, v, w, \varepsilon, S_{ij}]^{\mathrm{T}};$$

$$\text{or} \quad \frac{\partial \mathbf{W}}{\partial t} + \widetilde{A}_1 \frac{\partial \mathbf{W}}{\partial x} + \widetilde{A}_2 \frac{\partial \mathbf{W}}{\partial y} + \widetilde{A}_3 \frac{\partial \mathbf{W}}{\partial z} = \mathbf{0}, \quad \mathbf{W} = [u, v, w, \varepsilon, P_{ij}]^{\mathrm{T}}. \qquad (6.2.9)$$

To solve Eq. (6.2.9) by using an arbitrary moving grid, one must diagonalize 10×10 matrices of the form $\widetilde{A} = \alpha_1 \widetilde{A}_1 + \alpha_2 \widetilde{A}_2 + \alpha_3 \widetilde{A}_3$. Such a diagonalization in three-dimensional calculations (Petrov et al. 1990; Petrov and Tormasov 1990) is rather complicated. For this reason, in the three-dimensional cases, they determined, e.g., Ω_{R} for the matrix $\widetilde{A}$ analytically; however, the matrix Ω_{L} was calculated numerically.

As regards the equations in terms of strain gradient tensor $\hat{\mathbf{J}}$, for example, (6.1.61), note that this approach is not yet generally accepted today and is put through primary numerical investigation. In particular, within the framework of this approach, the CIR method was used for the one- and two-dimensional numerical simulation of elastoviscoplastic flows by Vorobiev et al. (1995), and Lomov and Kondaurov (1995, 1998).

6.2.1 Numerical simulation of spallation phenomena.

Consider some typical results of numerical simulation of solid dynamics problems with CIR schemes. These results were kindly granted to us by V. D. Ivanov; see for details Ivanov et al. (1990), Ivanov and Petrov (1992). All these results were obtained by using the nonconservative form of Eq. (6.1.48) in moving coordinates associated with the boundary of the body. In all simulations, the EOS of (6.1.11) was used.

Figures 6.1, 6.2, and 6.3 present the results of a one-dimensional simulation of elastoplastic flows accompanied by spallation processes. Figure 6.1a shows the propagation of an

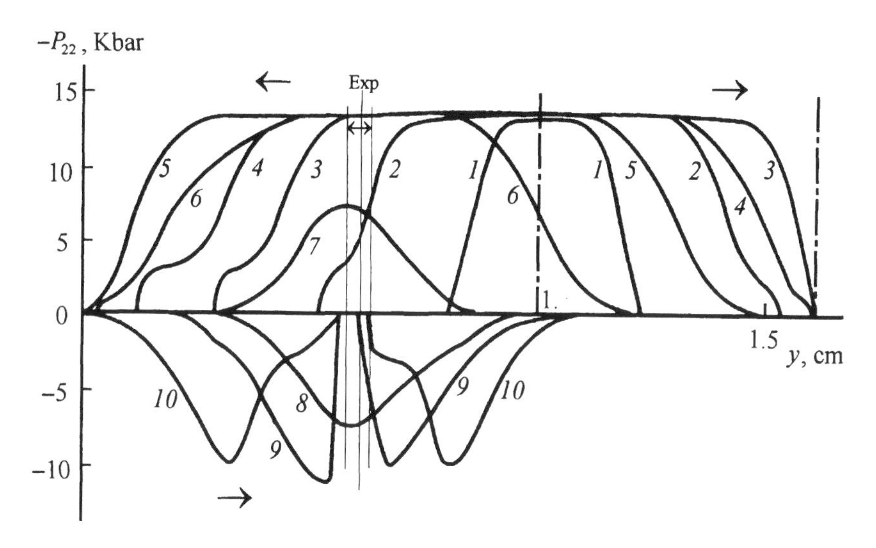

Figure 6.2 One-dimensional results of simulation of spallation experiments for two colliding disks, steel and aluminum.

elastoplastic wave in iron resulting from the collision of two Armco iron discs. This figure demonstrates the profiles of P_{22} at three different points of the target with thickness 10 mm. The solid lines represent the results of the numerical simulation and the dashed lines are the experimental results by Kanel' and Shcherban' (1980). Ahead of the plastic wave front moving with speed $\sqrt{(\tilde{\lambda} + \frac{4}{3}\mu)/\rho}$, one can see the elastic wave front moving with a faster speed, $\sqrt{(\tilde{\lambda} + 2\mu)/\rho}$; see Eqs. (6.2.4)–(6.2.5). It is apparent from Fig. 6.1a that a rarefaction wave begins to form on the right of the plastic wave. The experiments were carried out for projectile plates with thickness 2 and 5 mm for velocity 590 ± 10 m/s; the target plates had thicknesses 2, 10, and 15 mm.

Figure 6.1b shows the velocity profile v of the free surface of the target under the interaction with the elastoplastic wave. The first jump in the velocities is related to the time instant when the elastic wave front reaches the free surface. The further increase in the velocity is associated with the plastic wave. Here the dashed line represents the experimental data and the solid line corresponds to the numerical simulation. The time-oscillations of the velocity result from the propagation of the wave through a narrow spallation disk. This narrow plate arises due to splitting (spalling) of a portion of the material from the initial disc under the action of a positive stress (negative pressure) exceeding the Armco iron dynamic strength limit. Compare this with the preliminary gas dynamic investigation of spallation processes in Section 3.4.2. In this simulation it is assumed that the fracture of the material takes place whenever the positive stress exceeds the strength limit.

Figures 6.2 and 6.3 present the one-dimensional numerical results of simulation of spallation experiments for two colliding disks, steel and aluminum (Tarasov 1974; Ryibakov 1977). The line $x = 0$ in Fig. 6.2 is the free surface of the target plate and two dot-and-dash

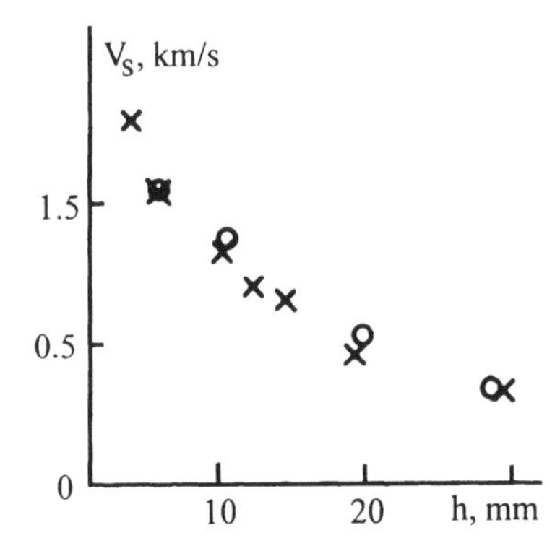

Figure 6.3 Graph of the velocity of the spall plate as a function of the thickness of the steel target.

vertical lines on the right are the boundaries of the projectile plate. Curves 1–7 show the profiles of P_{22} in the elastoplastic wave before the spallation and curves 8–10 with negative pressure correspond to the post-spall period. The thin vertical line labeled with "Exp" indicates the spallation boundary obtained in the experiments, and the other two thin lines bound the region of possible spallation determined from the numerical simulation.

In Fig. 6.3 the open circles show the values of the velocity V_s of the spall plate as a function of the thickness h of the steel target. The experimental results (Rybakov 1977) are shown by crosses. The experiments were carried out for steel projectile plates with thicknesses 1.06 and 1.52 mm and corresponding velocities 0.96 and 0.65 km/s; the steel target plates have thicknesses 3–50 mm. The experimental values of the dynamic strength limit were found to be 815 ± 47 kg/mm^2 and 750 ± 46 kg/mm^2 for experiments with projectile plates of thickness 1.06 mm and 1.52 mm, respectively.

Figures 6.4–6.7 illustrate some specific features of spallation phenomena in the two-dimensional case for cylindrical targets with a circular groove. The scheme of experiments (for details see Vovchenko, Krasyuk, and Semenov 1992) is presented in Fig. 6.4a–d. The arrows indicate the action of a laser pulse of intensity $\sim 10^{11}$ W/cm^2. In the experiments both non-grooved targets (Fig. 6.4a–b) and grooved targets (Fig. 6.4c–d) were used. The figure demonstrates the specific features of spallation processes for these two cases. Note that a groove may favor the process of cracking, and the spallation can originate for the same laser pulse only in the configuration in Fig. 6.4c but not in Fig. 6.4a.

Figure 6.5a–c shows the results of the laser tests of spallation processes in aluminum targets with a groove. The right photos correspond to different stages of spallation in experiments with laser pulse intensity 3.0×10^{11}, 3.5×10^{11}, and 4.3×10^{11} W/cm^2, respectively. The corresponding ablative pressures determined by the scaling law (Vovchenko, Krasyuk, Pashinin, and Semenov 1994) are 154, 187, and 247 Kbar; the respective durations of the laser pulses are 40, 10, and 40 ns.

The right photo in Fig. 6.5a corresponds to the case of incomplete spallation (Fig. 6.4c) and the right photo in Fig. 6.5c to the case of complete spallation (Fig. 6.4d). The left photos show the structure of melted aluminum that appears under the action of the laser pulse in

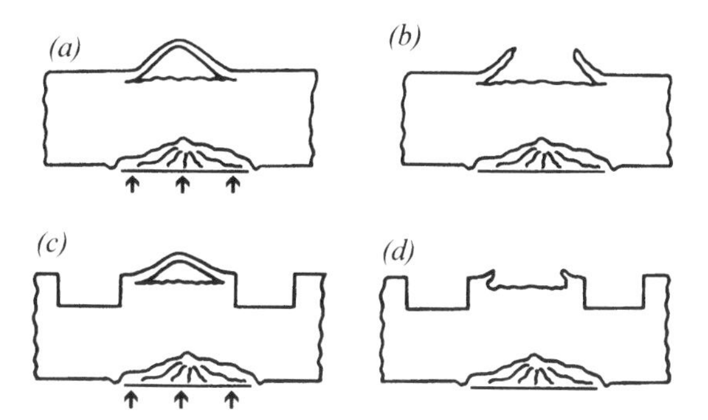

Figure 6.4 The scheme of the experiments for the spallation investigation.

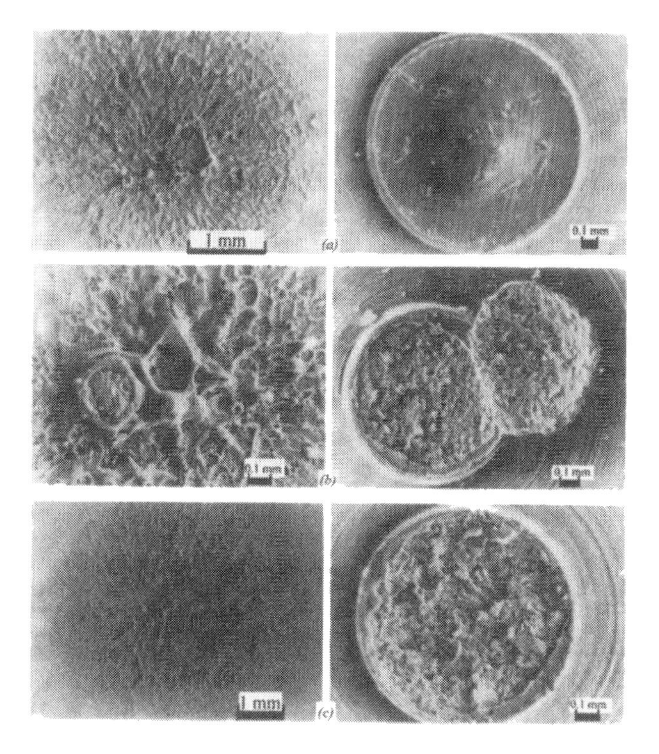

Figure 6.5 Photos of the results of the laser experiments of spallation processes in aluminum targets with groove.

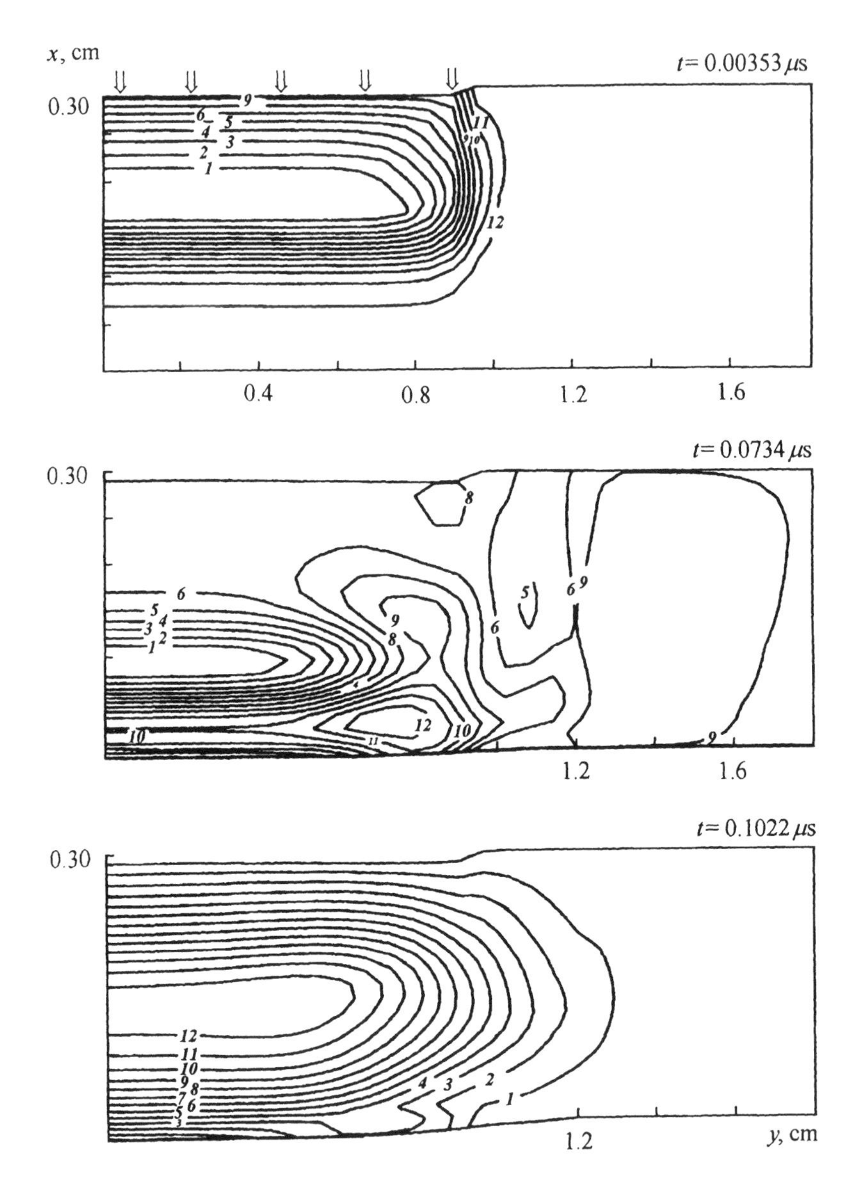

Figure 6.6 Results of 2D simulation of spallation phenomena for a non-grooved target for $t = 0.00353$, 0.0734, and 0.1022 μs.

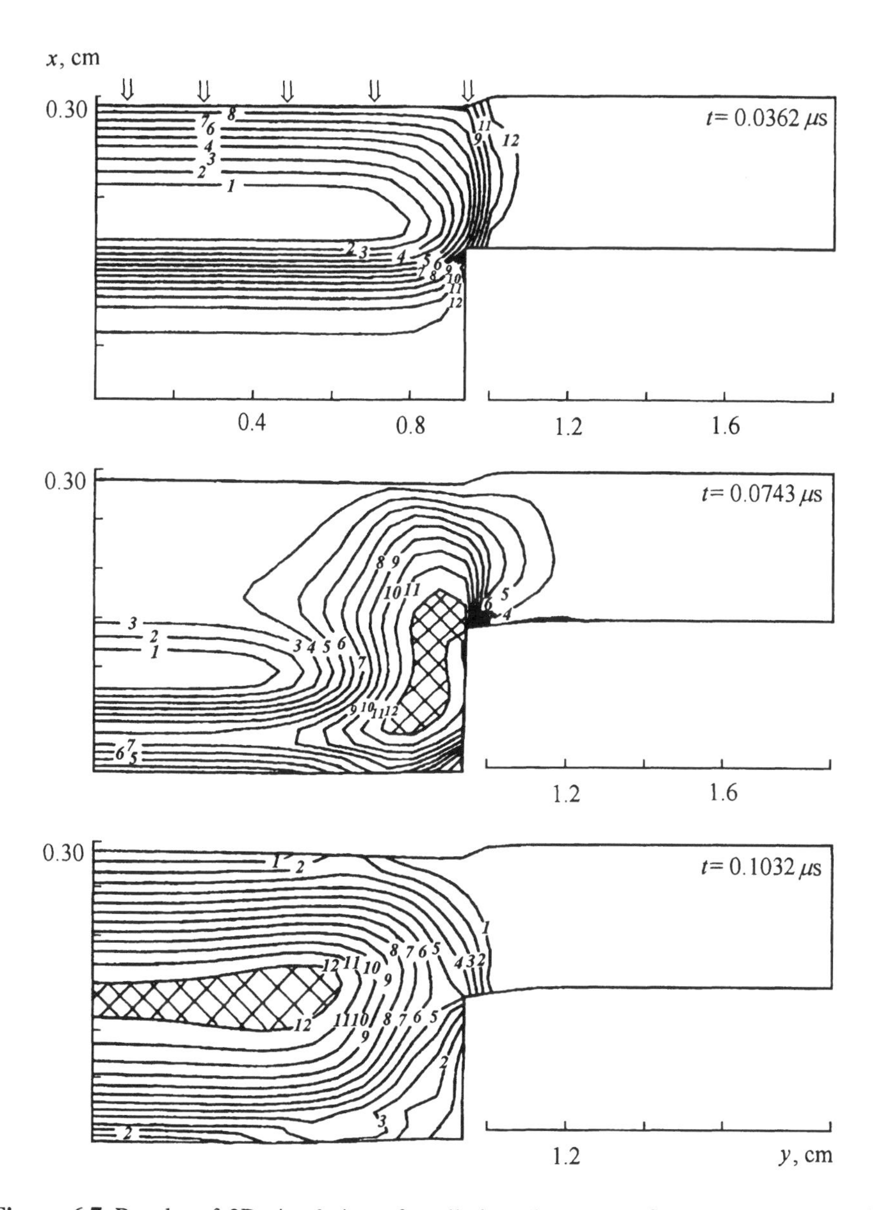

Figure 6.7 Results of 2D simulation of spallation phenomena for a grooved target for $t = 0.00362, 0.0743$, and $0.1032\ \mu s$.

the lower part of the targets.

Figures 6.6 and 6.7 present results of the two-dimensional numerical simulation of spallation phenomena for non-grooved (Fig. 6.6) and grooved (Fig. 6.7) targets in a moving system of cylindrical coordinates associated with the boundary of the target. These figures show isolines of maximal stresses for some time instants. (The greater the number of the isoline, the greater the stress value.)

Figures 6.6 and 6.7 correspond to the case where in experiment the spallation occurs only for grooved targets. In Fig. 6.7 the regions where a crack originates are shaded. A vertical crack arises at $t = 0.0743$ μs, see Fig. 6.7, and further a horizontal crack appears at $t = 0.1032$ μs. The most accurate results may be obtained by using a crack-fitting technique. In contrast, the target in Fig. 6.6 has no crack and spallation phenomena are not observed, which is supported by physical experiments; see Vovchenko, Krasyuk, and Semenov (1992).

6.3 CIR method for studying the dynamics of thin shells

In this section we describe the Godunov method for numerical solution of quasilinear hyperbolic systems of equations that govern the dynamics of thin shells under deformation. These systems are used for the investigation of transient wave processes under pulse loading (Ogibalov 1963; Ricards and Teters 1974; Grigolyuk and Gorshkov 1976).

First we describe methods for the general case and further discuss specific features of some particular systems, such as the Klein–Gordon equation and the systems of the dynamics of cylindrical and orthotropic shells.

One of the basic difficulties in the numerical investigation of such systems by explicit methods is the presence of the right-hand-side terms with large factors proportional to β^{-2}, where β is the ratio of the shell thickness to minimal radius of curvature of its middle surface, $\beta = 0.02$–0.001. In this case, rapidly oscillating and weakly damping components of the form $\exp(it/\beta)$, where $i = \sqrt{-1}$ and t is time, may arise in the solution. If such harmonics are not taken special care of, then in many cases this leads to fast development of numerical instabilities and requires one to carry out calculations with a time increment Δt much less than that admitted by the CFL condition and the condition of approximation of the right-hand sides. The problem may not be completely solved by using the techniques of increasing Δt based on the ideas of temporal averaging (Bazhenov and Chekmarev 1983; Abdukadyrov, Pinchukova, and Stepanenko 1984), the introduction of implicit approximation of the right-hand sides (Gur'yanov 1985; Ivanov 1987), and performing additional iterations (Panichkin 1981).

In the algorithm presented below, we use the ideas of the splitting method, which makes it possible to select rapidly oscillating components in a special manner so that the stability of explicit schemes is guaranteed, including even the cases where $\Delta t/\beta \gg 1$, where Δt is the time increment. To approximate the hyperbolic operator, a CIR-type difference scheme is applied; see Section 2.3.1. The method to be described below permits us to determine the limits of applicability of alternative numerical methods and the causes of their computational instability.

In general, the one-dimensional equations of the dynamics of thin shells with deformations, transverse displacements, and rotational inertia are hyperbolic systems of second-order equations in both mixed conservative–nonconservative and conservative form. For example, the system can have the conservative form (6.1.62). In (6.1.62) $\mathbf{V}$, $\mathbf{G}$, and $\mathbf{H}$ are M-dimensional vectors, $\mathbf{V} = \mathbf{V}(t,x)$ is the vector that includes the displacements of the shell's middle surface and the angles of deviation, t is time, and x is a spatial or angular coordinate.

Equations (6.1.62) can be reduced to a quasilinear hyperbolic system of first-order equations by introducing the following new $3M$ vector $\mathbf{U}$ (Evseev and Semenov 1985, 1989):

$$U_m = \frac{\partial V_m}{\partial t}, \quad U_{M+m} = \frac{\partial V_m}{\partial x}, \quad U_{2M+m} = V_m, \qquad m = 1,\dots,M.$$

In terms of the variables $\mathbf{U}$, Eq. (6.1.62) can be rewritten in the form

$$\begin{aligned} &\frac{\partial U_m}{\partial t} + \sum_{p=1}^{M}\left[\frac{\partial G_m}{\partial (V_p)_x}\frac{\partial U_{M+p}}{\partial x} + \frac{\partial G_m}{\partial V_p}U_{M+p}\right] = H_m - \frac{\partial G_m}{\partial x}, \\ &\frac{\partial U_{M+m}}{\partial t} - \frac{\partial U_m}{\partial x} = 0, \quad \frac{\partial U_{2M+m}}{\partial t} = U_m, \qquad m = 1,\dots,M. \end{aligned} \tag{6.3.1}$$

In vector form Eq. (6.3.1) becomes

$$\frac{\partial \mathbf{U}}{\partial t} + A(t,x,\mathbf{U})\frac{\partial \mathbf{U}}{\partial x} = \mathbf{F}(t,x,\mathbf{U}), \tag{6.3.2}$$

where A is a $3M \times 3M$ matrix, and the vectors $\mathbf{F}$ and $\mathbf{U} = \mathbf{U}(t,x)$ are $3M$ vectors.

To solve (6.3.2) we use the splitting method, which involves several stages. At the first stage, at each node x_j ($j = 1,\dots,N$) and on each time interval $[t^k, t^{k+1}]$, $t^{k+1} = t^k + \Delta t$, $k = 0, 1, \dots,$ we analytically solve the Cauchy problem for the system of ordinary differential equations (ODE),

$$\frac{d\mathbf{U}_j^0}{dt} = \mathbf{f}\left(t, x_j, \mathbf{U}_j^0\right), \quad \mathbf{U}_j^0 = \mathbf{U}_j^0(t), \qquad \mathbf{U}_j^0(t{=}t^k) = \mathbf{U}(t{=}t^k, x{=}x_j). \tag{6.3.3}$$

Here the right-hand side $\mathbf{f}$ is composed of the terms of $\mathbf{F}$ such that (i) $\mathbf{f}$ contains all terms with the large factor β^{-2} and (ii) problem (6.3.3) has an explicit solution $\mathbf{U}_j^0(t)$ that is the sum of terms of the form $R(x_j)\exp[(1+\delta)it/\beta]$, where $\delta \ll 1$. Analysis and investigation have shown that such a representation for $\mathbf{f}$ exists for all systems considered by Ogibalov (1963), Grigolyuk and Gorshkov (1976), Teters, Ricards, and Narusberg (1978), and some other researchers. This is a consequence of the fact that all terms in $\mathbf{F}$ containing the large factor β^{-2} are linear in $\mathbf{U}$ and justified by the mechanical sense of the problem. The use of the first-order system (6.3.2) facilitates the selection of $\mathbf{f}$. Some examples will be given below. Note that the selection of $\mathbf{f}$ is not unique. In some cases this allows one to obtain more compact expressions at this stage of the numerical algorithm.

At the second stage, at each node x_j on the same time interval, we solve the Cauchy problem for the system

$$\frac{d\mathbf{u}_j}{dt} = \mathbf{F}(t, x_j, \mathbf{u}_j) - \mathbf{f}(t, x_j, \mathbf{u}_j), \quad \mathbf{u}_j = \mathbf{u}_j(t), \qquad \mathbf{u}_j(t{=}t^k) = \mathbf{U}_j^0(t^{k+1}). \tag{6.3.4}$$

It can be solved at each node x_j by the standard explicit multilayer numerical procedure with the order of approximation $s \geq 2$ (Hall and Watt 1976; Hairer, Norsett, and Wanner 1987). In this case the approximation conditions lead to the constraint $(\Delta t/\beta)^s \leq q$ on the step size, where q is a small quantity determining the accuracy of integration at one step in time. Note that these stability conditions for (6.3.4) are less restrictive than the approximation conditions.

At the final stage, using the same time increment, we numerically integrate the following homogeneous hyperbolic system of first-order equations:

$$\frac{\partial \mathbf{U}}{\partial t} + A(t, x, \mathbf{U})\frac{\partial \mathbf{U}}{\partial x} = \mathbf{0}, \quad \mathbf{U}(t{=}t^k, x{=}x_j) = \mathbf{U}_j^k = \mathbf{u}_j(t^{k+1}). \tag{6.3.5}$$

The nonconservative system (6.3.5) can be integrated by an explicit Courant–Isaacson–Rees method of the first or second order of accuracy, see Sections 2.3.1 and 2.5.

For example, as a first simple step, one can use in numerical simulation the following hybrid difference scheme:

$$\frac{\mathbf{U}_j^{k+1} - \mathbf{U}_j^k}{\Delta t} + B_{j+1/2}^- \frac{\mathbf{U}_{j+1}^k - \mathbf{U}_j^k}{\Delta x} + B_{j-1/2}^+ \frac{\mathbf{U}_j^k - \mathbf{U}_{j-1}^k}{\Delta x} = \mathbf{0}, \quad \mathbf{U}_j^k = \mathbf{u}_j(t^{k+1}),$$

where the integer subscripts refer to the centers of difference cells, the half-integer subscripts refer to the corresponding boundaries, and Δx is the space mesh size. The superscripts k and $k+1$ are the numbers of the time layer. Since the equations of (6.3.5) are hyperbolic, we can write $A = \Omega_R \Lambda \Omega_L$. Then $B^{\pm} = \Omega_R \Lambda^{\pm} \Omega_L$, where $\Lambda^{\pm} = [\frac{1}{2}(1 \pm g_p)\lambda_p \delta_{pl}]$, and g_p is the hybridity coefficient chosen for each of the characteristic directions with the index p independently (see Eqs. (2.3.59)–(2.3.61) and (2.3.37) in Section 2.3.1).

In accordance with the mechanical sense of this model, the solution of the Timoshenko-type system cannot be discontinuous, and therefore the use of more sophisticated numerical methods, particularly those based on a mesh function reconstruction, may be unnecessary.

The investigation of stability conditions and approximations for (6.3.3)–(6.3.5) shows the relations $\Delta t = O(\beta)$ and $\Delta x = O(\beta)$ must be satisfied in computations based on schemes of first-order accuracy and $\Delta t = O(\sqrt{\beta})$ and $\Delta x = O(\sqrt{\beta})$ for schemes of second-order accuracy (Evseev and Semenov 1989). Therefore, a decrease in β must be accompanied by a decrease in Δt and Δx. This can be interpreted in the following way. When deriving the thin shell dynamics equations, a small parameter h, the shell thickness, is used. This parameter can be treated as the space mesh size in the direction normal to the shell surface. Therefore, to preserve the order of approximation in the numerical solution, Δt and Δx must satisfy the above relations.

In accordance with the final approximation of the splitting procedure, the method of (6.3.3)–(6.3.5) has the second order of accuracy in x on smooth solutions, while on non-smooth solutions it is first-order accurate. The described method preserves the monotonicity property. Numerical scheme (6.3.3)–(6.3.5) can easily be generalized to the systems of the dynamics of thin shells with two space variables; see Section 2.6.

Numerical calculations of model problems and problems of the dynamical behavior of cylindrical and orthotropic shells confirm the reliability and efficiency of the method. For illustration, consider some results of a numerical simulation of thin cylindrical shells governed by Eq. (6.1.65) (Semenov 1987). Figure 6.8 shows the dynamics of the normal

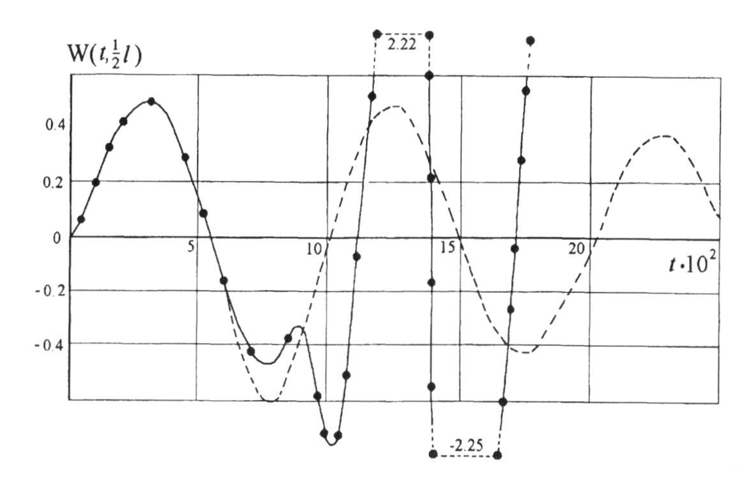

Figure 6.8 Graphs of $W(t, \frac{1}{2}l)$.

displacement $W(t, \frac{1}{2}l)$ at the middle of a shell with length l and $\beta^{-2} = 30,000$. The shell dynamics was investigated under the action of a left triangular pulse with duration 0.01 dimensionless time units, which is equal to unity in the scale of the figure. The pulse sets the shell into vibration and it vibrates further freely. The solid line shows the numerical results obtained using the described selection of rapidly oscillating terms, while the dashed line represents the numerical results obtained by a standard explicit–implicit difference scheme for systems of second-order equations. It is obvious that the latter scheme does not provide adequate results in simulating systems with large β^{-2}.

Figures 6.9 and 6.10 show $W(t, x)$ for six sequential time instants $t_1, \ldots, t_6$. From Fig. 6.8 one can see that the numerical results lose their stability for $t \geq 12$. However, from Fig. 6.10 it is apparent that the stability is lost earlier, for $t \geq 6.5$. The dashed line indicates the current solution for $t = t_6$. The loss of numerical stability in available methods for the analysis of thin shells was the reason for constructing a new version of numerical algorithms for simulating the dynamics of shells (Evseev and Semenov 1985, 1989, 1990).

It should be noted also that the reasons for the appearance of instabilities in previous methods can conditionally be divided into two groups. The first group is illustrated by the model equation

$$\frac{\partial U}{\partial t} = i\frac{U}{\beta}, \qquad i = \sqrt{-1}.$$

This equation has the exact solution

$$U(t) = U(0)\exp(it/\beta); \tag{6.3.6}$$

whence

$$|U(t)| = |U(0)|. \tag{6.3.7}$$

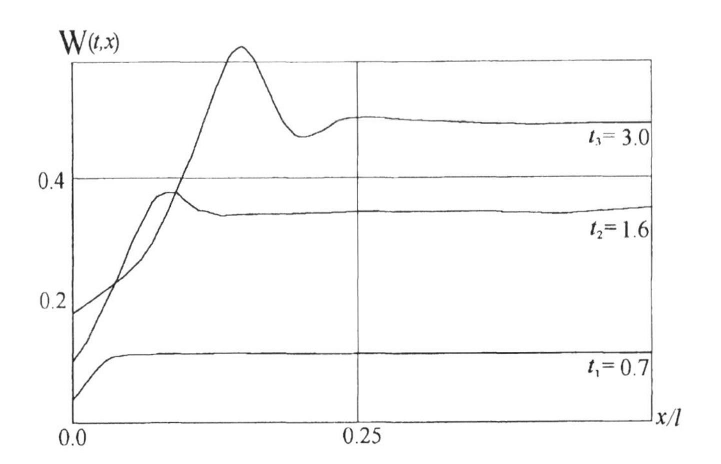

Figure 6.9 Graphs of $W(t, x)$ for $t_1 = 0.7$, $t_2 = 1.6$, and $t_3 = 3.0$.

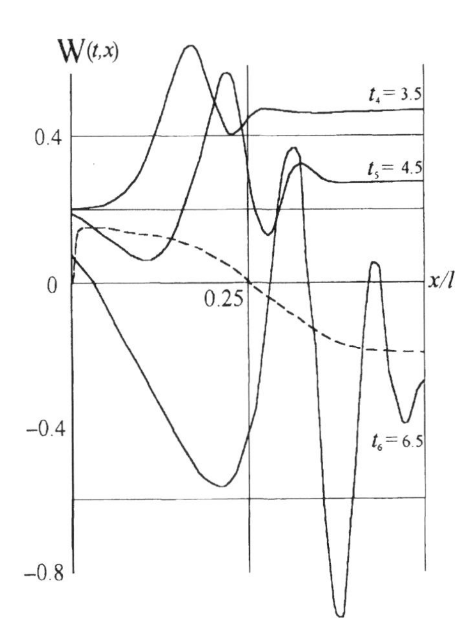

Figure 6.10 Graphs of $W(t, x)$ for $t_4 = 3.5$, $t_5 = 4.5$, and $t_6 = 6.5$.

The application of the explicit scheme

$$\frac{U^{k+1} - U^k}{\Delta t} = \frac{i}{\beta} U^k$$

leads to the numerical solution

$$U^{k+1} = \left(1 + \frac{i}{\beta}\Delta t\right) U^k;$$

whence

$$|U^{k+1}| = \left[1 + \left(\frac{\Delta t}{\beta}\right)^2\right] |U(0)| > |U(0)|.$$

Hence, in contrast to the approach of (6.3.6), for $\Delta t = O(\beta)$ there is an abrupt growth of U^{k+1} that disagrees with the properties of the exact equation.

The second group of instabilities arises in using implicit or temporally multilevel approximations of the right-hand side, which ensures the correct description of vibrations, although inadequate approximating of hyperbolic part of the system. This also leads to numerical instability. Note that the use of the implicit scheme

$$\frac{U^{k+1} - U^k}{\Delta t} = \frac{i}{2\beta}(U^{k+1} + U^k)$$

permits one to preserve the property (6.3.7) in the numerical solution. Unfortunately, no simple generalization of this scheme to an arbitrary ODE (6.3.3)–(6.3.4) has been made so far.

Thus, we have presented an explicit two-layer conditionally stable method for numerical integration of the dynamics equations of thin shells. The method is based on the reduction of the dynamics equations to a quasilinear hyperbolic system of first-order equations and precise selection of the rapidly oscillating components in the solution. The method makes it possible to carry out calculations in a broad range of variation of the determining parameters without loss of numerical stability and without the necessity to invoke any smoothing procedures. The method also permits one to determine the limits of applicability and establish the reasons for the loss of stability of some known algorithms. Further, we consider the key features of implementing of method in numerical practice.

6.3.1 The Klein–Gordon equation. Consider the Klein–Gordon equation

$$\frac{\partial^2 u}{\partial t^2} - \frac{\partial^2 u}{\partial x^2} + \frac{u}{\beta^2} = 0. \tag{6.3.8}$$

Let us perform the transformation

$$U_1 = \frac{\partial u}{\partial t}, \quad U_2 = \frac{\partial u}{\partial x}, \quad U_3 = u. \tag{6.3.9}$$

Thus, we reduce Eq. (6.3.8) to a hyperbolic system of first-order equations,

$$\frac{\partial U_1}{\partial t} - \frac{\partial U_2}{\partial x} = -\frac{1}{\beta^2} U_3, \quad \frac{\partial U_2}{\partial t} - \frac{\partial U_1}{\partial x} = 0, \quad \frac{\partial U_3}{\partial t} = U_1,$$

or, in vector form,

$$\frac{\partial \mathbf{U}}{\partial t} + A\frac{\partial \mathbf{U}}{\partial x} = \mathbf{F}, \tag{6.3.10}$$

$$\mathbf{U} = [U_1, U_2, U_3]^{\mathrm{T}}, \tag{6.3.11}$$

$$\mathbf{F} = [F_1, F_2, F_3]^{\mathrm{T}}, \quad F_1 = -\frac{U_3}{\beta^2}, \quad F_2 = 0, \quad F_3 = U_1;$$

$$A = \begin{bmatrix} 0 & -1 & 0 \\ -1 & 0 & 0 \\ 0 & 0 & 0 \end{bmatrix}. \tag{6.3.12}$$

System (6.3.10)–(6.3.12) is hyperbolic, and the coefficient matrix A of it has only real eigenvalues and a complete set of eigenvectors. Hence, the matrix A can be written in the equivalent form

$$A = \Omega_{\mathrm{R}}\Lambda\Omega_{\mathrm{L}}, \tag{6.3.13}$$

where

$$\Omega_{\mathrm{R}} = \begin{bmatrix} \frac{1}{2} & \frac{1}{2} & 0 \\ \frac{1}{2} & -\frac{1}{2} & 0 \\ 0 & 0 & 1 \end{bmatrix}, \quad \Lambda = \mathrm{diag}[-1, 1, 0], \quad \Omega_{\mathrm{L}} = \begin{bmatrix} 1 & 1 & 0 \\ 1 & -1 & 0 \\ 0 & 0 & 1 \end{bmatrix}. \tag{6.3.14}$$

6.3.2 Dynamics equations of cylindrical shells. The dynamics equations for an infinite uniform isotropic cylindrical shell subjected to the action of external forces independent of the axial coordinate can be represent in the form (6.1.65).

The dynamics equations for an infinite uniform isotropic cylindrical shell subjected to the action of external forces that depend on the axial coordinate x can be written in the form (6.1.66).

By the change of variables

$$\begin{aligned} &U_1 = \frac{\partial V}{\partial t}, \quad U_2 = \frac{\partial W}{\partial t}, \quad U_3 = \frac{\partial \Psi}{\partial t}, \\ &U_4 = \frac{\partial V}{\partial x}, \quad U_5 = \frac{\partial W}{\partial x}, \quad U_6 = \frac{\partial \Psi}{\partial x}, \\ &U_7 = V, \quad U_8 = W, \quad U_9 = \Psi, \end{aligned} \tag{6.3.15}$$

Eqs. (6.1.65) and (6.1.66) can be reduced to the form (6.3.10) with

$$A = \begin{bmatrix} 0 & 0 & 0 & -1 & -\alpha & 0 & 0 & 0 & 0 \\ 0 & 0 & 0 & -\theta & -\gamma & 0 & 0 & 0 & 0 \\ 0 & 0 & 0 & 0 & 0 & -1 & 0 & 0 & 0 \\ -1 & 0 & 0 & 0 & 0 & 0 & 0 & 0 & 0 \\ 0 & -1 & 0 & 0 & 0 & 0 & 0 & 0 & 0 \\ 0 & 0 & -1 & 0 & 0 & 0 & 0 & 0 & 0 \\ 0 & 0 & 0 & 0 & 0 & 0 & 0 & 0 & 0 \\ 0 & 0 & 0 & 0 & 0 & 0 & 0 & 0 & 0 \\ 0 & 0 & 0 & 0 & 0 & 0 & 0 & 0 & 0 \end{bmatrix}, \tag{6.3.16}$$

where

$$\alpha = \theta - \delta_1 \Psi, \quad \gamma = T + \theta^2 + \delta_1. \tag{6.3.17}$$

The vector $\mathbf{F} = [F_1, \ldots, F_9]^{\mathrm{T}}$ on the right-hand side of system (6.1.65) is defined by its components as

$$\begin{aligned} F_1 &= Q + U_5 + \theta T - \alpha U_4 + (\delta_1 U_9 - Q) U_6 - q U_9, \\ F_2 &= Q U_9 - T + \theta U_5 - \gamma U_4 - \delta_1 U_6 + q, \\ F_3 &= 12\beta^{-2} Q - \theta M, \\ F_4 &= F_5 = F_6 = 0, \quad F_7 = U_1, \quad F_8 = U_2, \quad F_9 = U_3, \end{aligned} \tag{6.3.18}$$

where, in accordance with Eqs. (6.1.65), (6.3.15), and (6.3.17),

$$\begin{aligned} &\theta = U_5 - U_7, \quad Q = (\theta - U_9)\delta_1, \quad T = U_4 + U_8 + \tfrac{1}{2}\theta^2, \\ &M = -U_6, \quad \alpha = \theta - \delta_1 U_9, \quad \gamma = T + \theta^2 + \delta_1. \end{aligned} \tag{6.3.19}$$

The components of the right-hand-side vector of system (6.1.66) in the quasilinear form are given by

$$\begin{aligned} F_1 &= -\theta Q U_9 + \delta_1 U_6 U_9 - Q U_6 - q U_9, \\ F_2 &= \theta Q - T - \delta_1 U_6 + q, \\ F_3 &= 12\beta^{-2} Q, \\ F_4 &= F_5 = F_6 = 0, \quad F_7 = U_1, \quad F_8 = U_2, \quad F_9 = U_3, \end{aligned} \tag{6.3.20}$$

where, in accordance with Eqs. (6.1.66) and (6.3.15),

$$\theta = U_5, \quad Q = (\theta - U_9)\delta_1, \quad T = U_4 + \tfrac{1}{2}\theta^2, \quad M = -U_6. \tag{6.3.21}$$

The matrices Λ, Ω_{R}, and Ω_{L} corresponding to A of (6.3.16) can be found in Evseev and Semenov (1985, 1989). Specifically,

$$\begin{aligned} \Lambda &= \operatorname{diag}[-1, -a, -b, b, a, 1, 0, 0, 0], \\ a &= \frac{1}{\sqrt{2}} \sqrt{1 + \gamma + \sqrt{(1-\gamma)^2 + 4\alpha\theta}}, \\ b &= \frac{1}{\sqrt{2}} \sqrt{1 + \gamma - \sqrt{(1-\gamma)^2 + 4\alpha\theta}}, \end{aligned} \tag{6.3.22}$$

and Ω_{L}, Ω_{R} have the block matrix form

$$\Omega_{\mathrm{L}} = \begin{bmatrix} B & 0 \\ 0 & I \end{bmatrix}, \quad \Omega_{\mathrm{R}} = \begin{bmatrix} B^{-1} & 0 \\ 0 & I \end{bmatrix},$$

where $I = \operatorname{diag}[1, 1, 1]$ is the 3×3 identity matrix, and the 6×6 matrix B has the following structure:

$$B = \begin{bmatrix} 0 & 0 & 1 & 0 & 0 & 1 \\ \times & \times & \times & \times & \times & \times \\ \times & \times & \times & \times & \times & \times \\ \times & \times & \times & \times & \times & \times \\ \times & \times & \times & \times & \times & \times \\ 0 & 0 & 1 & 0 & 0 & -1 \end{bmatrix}. \tag{6.3.23}$$

Here the entries denoted by crosses within each row are proportional, respectively, to either

$$\lambda^2-\gamma, \quad \alpha, \quad 0, \quad -(\lambda^2-\gamma)\lambda, \quad -\lambda\alpha, \quad 0 \tag{6.3.24}$$

or

$$\theta, \quad \lambda^2-1, \quad 0, \quad -\lambda\theta, \quad -(\lambda^2-1)\lambda, \quad 0, \tag{6.3.25}$$

where $\lambda=-a,-b,b$, and a. In the case $a=1$, we have two pairs of multiple eigenvalues; see Eq. (6.3.22). For the matrix B to be nonsingular in this case, in (6.3.23) we choose the entries marked by crosses so that

$$B=\begin{bmatrix} 0 & 0 & 1 & 0 & 0 & 1 \\ a^2-\gamma & \alpha & 0 & (a^2-\gamma)a & a\alpha & 0 \\ \theta & b^2-1 & 0 & b\theta & (b^2-1)b & 0 \\ \theta & b^2-1 & 0 & -b\theta & (1-b^2)b & 0 \\ a^2-\gamma & \alpha & 0 & (\gamma-a^2)a & -a\alpha & 0 \\ 0 & 0 & 1 & 0 & 0 & -1 \end{bmatrix}, \tag{6.3.26}$$

$$B^{-1}=\frac{1}{2e}\begin{bmatrix} 0 & (b^2-1)ab & -ab\alpha & -ab\alpha & (b^2-1)ab & 0 \\ 0 & -\theta ab & (a^2-\gamma)ab & (a^2-\gamma)ab & -\theta ab & 0 \\ e & 0 & 0 & 0 & 0 & e \\ 0 & (b^2-1)b & -a\alpha & a\alpha & (1-b^2)b & 0 \\ 0 & -\theta b & (a^2-\gamma)a & (\gamma-a^2)a & \theta b & 0 \\ e & 0 & 0 & 0 & 0 & -e \end{bmatrix}, \tag{6.3.27}$$

where

$$e=(a^2b^2-\gamma b^2-a^2+\gamma-\alpha\theta)ab=(a^2-b^2)(b^2-1)ab.$$

The fact that all eigenvalues in (6.3.22) are real ensures the hyperbolicity of the systems (6.1.65) and (6.1.66). The requirements that all eigenvalues of the system must be real determines the domain of applicability for the above mathematical models describing the dynamics of thin shells.

6.3.3 Dynamics equations of orthotropic shells.

The dynamics equations of orthotropic thin shells for the case of axisymmetric deformations can be represented in the form (6.1.67). By introducing the vector

$$\mathbf{U}=[U_1,\dots,U_9]^{\mathrm{T}}=[u_t,w_t,\varphi_t,u_x,w_x,\varphi_x,u,w,\varphi]^{\mathrm{T}}, \tag{6.3.28}$$

we can transform Eq. (6.1.67) to system (6.3.10) with the 9×9 matrix $A=[A_{ij}]$ whose nonzero entries are given by

$$A_{14}=-a_1^2, \quad A_{25}=-a_2^2, \quad A_{36}=-a_3^2, \quad A_{41}=A_{52}=A_{63}=-1. \tag{6.3.29}$$

For this A one can find that

$$\Omega_L = \begin{bmatrix} 1 & 0 & 0 & -a_1 & 0 & 0 & 0 & 0 & 0 \\ 0 & 1 & 0 & 0 & -a_2 & 0 & 0 & 0 & 0 \\ 0 & 0 & 1 & 0 & 0 & -a_3 & 0 & 0 & 0 \\ 1 & 0 & 0 & a_1 & 0 & 0 & 0 & 0 & 0 \\ 0 & 1 & 0 & 0 & a_2 & 0 & 0 & 0 & 0 \\ 0 & 0 & 1 & 0 & 0 & a_3 & 0 & 0 & 0 \\ 0 & 0 & 0 & 0 & 0 & 0 & 1 & 0 & 0 \\ 0 & 0 & 0 & 0 & 0 & 0 & 0 & 1 & 0 \\ 0 & 0 & 0 & 0 & 0 & 0 & 0 & 0 & 1 \end{bmatrix}, \tag{6.3.30}$$

$$\Lambda = \mathrm{diag}[a_1, a_2, a_3, -a_1, -a_2, -a_3, 0, 0, 0], \tag{6.3.31}$$

$$\Omega_R = \frac{1}{2} \begin{bmatrix} 1 & 0 & 0 & 1 & 0 & 0 & 0 & 0 & 0 \\ 0 & 1 & 0 & 0 & 1 & 0 & 0 & 0 & 0 \\ 0 & 0 & 1 & 0 & 0 & 1 & 0 & 0 & 0 \\ -\alpha_1 & 0 & 0 & \alpha_1 & 0 & 0 & 0 & 0 & 0 \\ 0 & -\alpha_2 & 0 & 0 & \alpha_2 & 0 & 0 & 0 & 0 \\ 0 & 0 & -\alpha_3 & 0 & 0 & \alpha_3 & 0 & 0 & 0 \\ 0 & 0 & 0 & 0 & 0 & 0 & 2 & 0 & 0 \\ 0 & 0 & 0 & 0 & 0 & 0 & 0 & 2 & 0 \\ 0 & 0 & 0 & 0 & 0 & 0 & 0 & 0 & 2 \end{bmatrix}, \tag{6.3.32}$$

$$\alpha_1 = \frac{1}{a_1}, \quad \alpha_2 = \frac{1}{a_2}, \quad \alpha_3 = \frac{1}{a_3}.$$

In addition, the components of the right-hand-side vector $\mathbf{F}$ are expressed as

$$F_1 = b_1 U_5, \quad F_2 = b_2 U_6 - c_2 U_4 - d_2 U_8 + e_2 q(t, x), \quad F_3 = -b_3 U_5 - c_3 U_9, \tag{6.3.33}$$
$$F_4 = F_5 = F_6 = 0, \quad F_7 = U_1, \quad F_8 = U_2, \quad F_9 = U_3.$$

6.3.4 Selection of rapidly oscillating components.

Below we illustrate the procedure of selecting the rapidly oscillating components (6.3.3) for Eqs. (6.1.63), (6.1.65), (6.1.66), and (6.1.67).

The Klein–Gordon equation

For the Klein–Gordon equation (6.1.63) the Cauchy problem of (6.3.3) reads

$$\frac{dU_1}{dt} = -\beta^{-2} U_3, \quad \frac{dU_2}{dt} = 0, \quad \frac{dU_3}{dt} = U_1, \tag{6.3.34}$$
$$U_1(t{=}0) = U_{10}, \quad U_2(t{=}0) = U_{20}, \quad U_3(t{=}0) = U_{30}.$$

There is an exact solution of Eq. (6.3.34),

$$U_1^0(t) = U_{10} \cos(t/\beta) - U_{30} \beta^{-1} \sin(t/\beta),$$
$$U_2^0(t) = U_{20},$$
$$U_3^0(t) = U_{30} \cos(t/\beta) + U_{10} \beta \sin(t/\beta).$$

This is the solution of the first stage (6.3.3). For $t = \Delta t$ this solution provides the initial data for the problem given by (6.3.4). Note that the stage (6.3.34) is stable for arbitrary Δt. For the approximation of the right-hand sides to be adequate, we must require at least that

$$\frac{\Delta t}{\beta} < \frac{\pi}{2}.$$

There is no need to perform the second stage (6.3.4) for the Klein–Gordon equation, since $\mathbf{f} = \mathbf{F}$. Nevertheless, we shall illustrate this stage by adding, for example, a term γU_3^2 to the first equation in (6.3.34). In this case, the stage (6.3.4) for (6.3.34) becomes

$$\begin{aligned}
&\frac{du_1}{dt} = \gamma u_3^2, \quad \frac{du_2}{dt} = 0, \quad \frac{du_3}{dt} = 0, \\
&u_1(t{=}0) = U_1^0(\Delta t), \quad u_2(t{=}0) = U_2^0(\Delta t), \quad u_3(t{=}0) = U_3^0(\Delta t).
\end{aligned} \tag{6.3.35}$$

Hence,

$$u_1(t) = U_1^0(\Delta t) + \gamma t[U_3^0(\Delta t)]^2, \quad u_2(t) = U_2^0(\Delta t), \quad u_3(t) = U_3^0(\Delta t).$$

For simplicity, in the case of small β, one can neglect the terms $O(\beta)$ in $U_i^0(\Delta t)$.

Dynamics of cylindrical shells

In terms of $\mathbf{U}$ (6.3.15), the components of $\mathbf{F}$ for Eq. (6.1.65) (see also Eqs. (6.3.18) and (6.3.19)) are expressed as

$$\begin{aligned}
F_1 &= Q + U_5 + \theta T - \alpha U_4 + (\delta_1 U_9 - Q)U_6 - qU_9 \\
&= (U_5 - U_7 - U_9)\delta_1 + U_5 + \theta T - \alpha U_4 + (\delta_1 U_9 - Q)U_6 - qU_9, \\
F_2 &= QU_9 - T + \theta U_5 - \gamma U_4 - \delta_1 U_6 + q, \\
F_3 &= 12\beta^{-2}Q - \theta M = (U_5 - U_7 - U_9)12\beta^{-2}\delta_1 - \theta M, \\
F_4 &= F_5 = F_6 = 0, \quad F_7 = U_1, \quad F_8 = U_2, \quad F_9 = U_3.
\end{aligned}$$

Stage (6.3.3) for this $\mathbf{F}$ becomes

$$\begin{aligned}
&\frac{dU_1}{dt} = (U_5 - U_7 - U_9)\delta_1 + U_5, \\
&\frac{dU_3}{dt} = (U_5 - U_7 - U_9)12\beta^{-2}\delta_1, \\
&\frac{dU_5}{dt} = 0, \quad \frac{dU_7}{dt} = U_1, \quad \frac{dU_9}{dt} = U_3
\end{aligned} \tag{6.3.36}$$

with the initial data $U_i(t{=}0) = U_{i0}$, $i = 1, 3, 5, 7, 9$. Substituting $U_1 = dU_7/dt$, $U_3 = dU_9/dt$, and $U_5 = U_{50}$ into the first two equations of (6.3.36) yields the following system of two ordinary differential equations:

$$\begin{aligned}
\frac{d^2U_7}{dt^2} &= (1 + \delta_1)U_{50} - (U_7 + U_9)\delta_1, \\
\frac{d^2U_9}{dt^2} &= 12\beta^{-2}\delta_1 U_{50} - (U_7 + U_9)12\beta^{-2}\delta_1.
\end{aligned}$$

By summing these two equations, we obtain

$$\frac{d^2(U_9+U_7)}{dt^2} = (a^2+1)U_{50} - (U_9+U_7)a^2, \quad a = \sqrt{12\beta^{-2}\delta_1 + \delta_1} \sim \beta^{-1}.$$

It follows that

$$f(t) = U_9(t) + U_7(t) = (U_{90}+U_{70})\cos(at) + (U_{30}+U_{10})a^{-1}\sin(at) + (1+a^{-2})[1-\cos(at)]U_{50},$$

and, hence,

$$\frac{dU_1}{dt} = -\delta_1 f(t) + (1+\delta_1)U_{50}, \quad \frac{dU_7}{dt} = U_1,$$
$$\frac{dU_3}{dt} = [U_{50} - f(t)](a^2-\delta_1), \quad \frac{dU_9}{dt} = U_3.$$

In particular,

$$\begin{aligned} U_1(t) &= U_{10} + [U_{50}t - (U_{90}+U_{70})a^{-1}\sin(at) \\ &\quad + (U_{30}+U_{10})a^{-2}\cos(at) - (t - a^{-1}\sin(at))(1+a^{-2})U_{50}]\delta_1 \\ &\quad + U_{50}t - (U_{30}+U_{10})a^{-2}\delta_1, \\ U_3(t) &= [U_{50}t - (U_{90}+U_{70})a^{-1}\sin(at) \\ &\quad + (U_{30}+U_{10})a^{-2}\cos(at) - (t-a^{-1}\sin(at))(1+a^{-2})U_{50}](a^2-\delta_1) \\ &\quad + (U_{30}+U_{10})a^{-2}\delta_1 - U_{10}. \end{aligned}$$

For $\beta \to 0$ we have $U_1(t) = U_7(t) = U_9(t) = O(1)$ and $U_3(t) = O(\beta^{-1})$.

The above selection of rapidly oscillating terms is not unique; neither is it the simplest one. For example, consider another approach. Instead of the first equation in (6.3.36), one can use the more simple equation $dU_1/dt = 0$. Then, the equations for U_1, U_3, U_5, U_7, and U_9 acquire the form

$$\frac{dU_1}{dt} = 0, \quad \frac{dU_3}{dt} = (U_5 - U_7 - U_9)a^2, \quad \frac{dU_5}{dt} = 0,$$
$$\frac{dU_7}{dt} = U_1, \quad \frac{dU_9}{dt} = U_3; \qquad a = \beta^{-1}\sqrt{12\delta_1};$$

whence,

$$\begin{aligned} &U_1(t) = U_{10}, \quad U_7(t) = U_{70} + U_{10}t, \quad U_5(t) = U_{50}, \\ &U_3(t) = c + (U_{30} - c)\cos(at) - (U_{90} - b)a\sin(at), \\ &U_9(t) = b + ct + (U_{30} - c)a^{-1}\sin(at) + (U_{90} - b)\cos(at); \\ &b = U_{50} - U_{70}, \quad c = -U_{10}. \end{aligned}$$

For $\beta \to 0$ we have $U_1(t) = U_7(t) = U_9(t) = O(1)$ and $U_3(t) = O(\beta^{-1})$.

For system (6.1.66) with the right-hand side $\mathbf{F}$ of (6.3.20)–(6.3.21) represented in terms of $\mathbf{U}$, the equations for stage (6.3.3) can also be found. We have

$$\frac{dU_3}{dt} = (U_5 - U_9)a^2, \quad \frac{dU_5}{dt} = 0, \quad \frac{dU_9}{dt} = U_3; \qquad a = \beta^{-1}\sqrt{12\delta_1}.$$

The exact solution of these equations is given by

$$U_3(t) = U_{30}\cos(at) - (U_{90} - U_{50})a\sin(at), \quad U_5(t) = U_{50},$$
$$U_9(t) = U_{50} + U_{30}a^{-1}\sin(at) + (U_{90} - U_{50})\cos(at).$$

For $\beta \to 0$ we have $U_5(t) = U_9(t) = O(1)$ and $U_3(t) = O(\beta^{-1})$.

Note that the values of the velocities U_1, U_2, and U_3 can be large. But this cannot lead to instabilities in the numerical stages (6.3.4) and (6.3.5), since the right-hand sides F_i, $i = 1, \dots 6$, of the hyperbolic system, see (6.3.18) and (6.3.20), do not depend on these functions. Furthermore, F_7, F_8, and F_9 correspond to the zero eigenvalues and do not affect the stage (6.3.5).

Dynamics of orthotropic shells

For system (6.1.67) with the right-hand side **F** of (6.3.33) in terms of **U**, we can solve the corresponding system of ordinary differential equations,

$$\frac{dU_i}{dt} = F_i, \quad U_i(t{=}0) = U_{i0}, \qquad i = 1, \dots, 9,$$

exactly. Provided that there is no external loading $q = q(t, x)$, this solution is

$$U_1(t) = U_{10} + b_1 t U_{05},$$
$$U_2(t) = U_{20}\cos(Dt) - (U_{80} - \alpha_8)D\sin(Dt),$$
$$U_3(t) = U_{30}\cos(Ct) - (U_{90} - \alpha_9)C\sin(Ct),$$
$$U_4(t) = U_{40}, \quad U_5(t) = U_{50}, \quad U_6(t) = U_{60},$$
$$U_7(t) = U_{70} + U_{10}t + \tfrac{1}{2}t^2 b_1 U_{05},$$
$$U_8(t) = \alpha_8 + (U_{80} - \alpha_8)\cos(Dt) + \frac{\sin(Dt)}{D}U_{20},$$
$$U_9(t) = \alpha_9 + (U_{90} - \alpha_9)\cos(Ct) + \frac{\sin(Ct)}{C}U_{30};$$
$$D = \sqrt{d_2}, \quad C = \sqrt{c_3}, \quad \alpha_8 = \frac{b_2 U_{60} - c_2 U_{40}}{d_2}, \quad \alpha_9 = -\frac{b_3 U_{50}}{c_3}.$$

For $\beta \to 0$ and $t/\beta \le \Delta x$ we have $U_1(t) = U_4(t) = U_5(t) = U_6(t) = U_7(t) = hU_8(t) = U_9(t) = O(1)$ and $hU_2(t) = U_3(t) = O(\beta^{-1})$.

Chapter 7
Nonclassical Discontinuities and Solutions of Hyperbolic Systems

Earlier, in Chapter 1, we studied the behavior of discontinuities and constructed solutions of one-dimensional gas dynamic equations and some other systems of equations. In the case of small-amplitude waves described by arbitrary hyperbolic systems of equations, which represent conservation laws, the generalization of the standard classical behavior of the Riemann waves, discontinuities, and self-similar solutions is ensured by the Lax theorems (Lax 1957, see Chapter 1).

A discontinuity in the solution of a hyperbolic system of first-order equations will be called classical if

(i) the hyperbolic systems are the same on either side of the discontinuity;

(ii) the system represents conservation laws whose number n is equal to the number of equations;

(iii) only conditions resulting from the conservation laws must be satisfied on the discontinuity.

Such discontinuities were considered in Chapter 1. We showed that the set of evolutionary discontinuities (shock waves) can be divided into n types. For weak shock waves there is a one-to-one correspondence between the shocks and Riemann waves. In small-amplitude discontinuities of each type the change in quantities is similar to that in the corresponding Riemann waves.

The best known nonclassical discontinuity is the combustion front in a combustible gas mixture (see, for example, Williams 1965). Besides conservation laws, an additional relation that ensures the existence of the front structure and determines its propagation velocity must be satisfied on this front. Similar discontinuities are also well known in other problems with phase transitions. Clearly, the additional relations and, consequently, the behavior of the discontinuities depend on some processes inside the structure.

In physical situations to be considered in this chapter we shall study the set of admissible (i.e., having continuous structure) discontinuities and show that discontinuities with additional relations can also exist in the absence of anything resembling phase transitions. Moreover, if an oscillatory process induced by dispersion takes place inside the discontinuity structure, then the discontinuities with additional relations may be of different types (see Sections 7.5, 7.6, and 7.7) with a discrete set of the velocities. Thus, the set of admissible discontinuities acquires a complex structure, as it were to break down into parts. The number of these parts becomes greater when the dispersion processes dominate over the dissipative ones inside the discontinuity structure. In other physical situations, discontinuities with a few (more than one) additional relations can exist (Section 7.8).

The set of admissible discontinuities determines solutions of the problems. In particular, we can mention such an important problem as the Riemann problem of disintegration of an arbitrary discontinuity. The solution of the problem may be nonunique. Nonuniqueness of solutions can also take place if all admissible discontinuities are classical. This was observed in gases with fairly complex equations of state. In this case there exist shock waves which, in accordance with the conservation laws, can disintegrate into a system of waves propagating with different velocities (Galin 1958; Gardner 1959). Similar situations were also noted in other physical problems in which finite (non-small) discontinuities arise. One of the possible physical situations with similar nonuniqueness associated with the presence of a shock wave capable of disintegration is described in Section 7.4, where nonlinear small-amplitude waves in a weakly anisotropic elastic medium are considered for a general relation between stresses and strains.

However, the possibility of the disintegration of a shock wave does not imply that the shock wave does disintegrate and that the shock waves that can disintegrate do not exist. Discontinuities capable of disintegration can be called metastable (Truskinovskii 1994). As is well known, under certain conditions metastable states can exist a long time. To clarify the point of physical realizability of metastable discontinuities and investigate the uniqueness of solutions of the problems, it is necessary to perform a more detailed consideration of the smaller-scale problem in which the discontinuities are "smeared" and to study the interaction between the corresponding continuous waves. In Section 7.4, in particular, we outline the results concerning the interaction between viscoelastic waves. On the basis of numerical experiments it is found that metastable waves are fairly stable with respect to finite perturbations. This makes it possible to consider these metastable waves as physically realizable. For a certain class of the elasticity theory problems it is possible to formulate a rule that allows one to determine without calculations which of the solutions is realized in the nonunique case.

The choice of examples to be considered in this chapter does not pretend to be complete or cover all qualitative effects and is dictated by personal tastes of the authors. Nevertheless, this choice demonstrates a fairly large number of various qualitative peculiarities of solutions. These peculiarities must be taken into account in developing numerical methods for continuum mechanics.

Prior to considering particular problems, we give three sections of the general character. In Section 7.1 the evolutionary conditions for nonclassical fronts are discussed. In Section 7.2 the correlation between the evolutionary conditions for discontinuities and the existence of their structure is studied. The boundary conditions on a discontinuity are obtained from the requirement of existence of their structure. Section 7.3 deals with the behavior of classical discontinuities near Jouget points on the shock adiabatic curves. Under certain conditions the presence of the Jouget points on the shock adiabatic curve may lead to the absence or nonuniqueness of self-similar solutions.

Since the main aim of this chapter is to consider discontinuous solutions, and taking into account the fact that in many cases the behavior of solutions can be assumed to be one-dimensional in a small vicinity of a point belonging to the smooth surface of the discontinuity, the problems with one independent space variable are considered.

7.1 Evolutionary conditions in nonclassical cases

We shall study solutions of one-dimensional hyperbolic systems of first-order partial differential equations in the presence of interfaces or fronts—in three-dimensional cases they are surfaces—that separate regions in which these solutions are constructed. We will consider the generic case in which these systems and the number of equations can be different on each side of the front. The surfaces separating substances with different properties (e.g., solids and liquids, conducting and nonconducting media) can be examples of such fronts.

In order for initial boundary value problems to be well posed, as in the case of classical discontinuities (Chapter 1), we must specify a certain number of boundary conditions in accordance with the equations valid on both sides of the boundary. These boundary conditions relate the quantities ahead of and behind the discontinuity and its velocity.

Let the motion of the front be governed by the equation $x = X(t)$, where the x-axis of the Cartesian coordinate system is directed along the normal to the front. Then the velocity of the front is determined from the relation $W = dX/dt$. We denote the number of equations (order of the system) ahead of ($x > X(t)$) and behind the front ($x < X(t)$) by n^{R} and n^{L}, respectively. Characteristic velocities evaluated for the states at the points immediately ahead of and behind the front will be denoted by $\lambda_1^{\mathrm{R}}, \lambda_2^{\mathrm{R}}, \ldots, \lambda_{n^{\mathrm{R}}}^{\mathrm{R}}$ and $\lambda_1^{\mathrm{L}}, \lambda_2^{\mathrm{L}}, \ldots, \lambda_{n^{\mathrm{L}}}^{\mathrm{L}}$.

The evolutionary conditions (Lax 1957; Landau and Lifshitz 1987) on the front allow us to find a solution of the linearized initial boundary value problem of the interaction of the front with small disturbances. The boundary conditions and equations on either side of the front are linearized independently. These evolutionary conditions impose constraints on the boundary conditions to be satisfied on the front.

For the hyperbolic systems the unknown quantities to be determined from the linearized boundary conditions are amplitudes of weak outgoing waves and the disturbance of the front velocity. For this reason the evolutionary conditions on the front can be stated as follows: in the (x, t)-plane the total number of outgoing characteristics must be equal to $N - 1$, where N is the number of boundary conditions.

The number of outgoing characteristics is determined by the number of the inequalities $\lambda_i^{\mathrm{R}} - W > 0$, $\lambda_j^{\mathrm{L}} - W < 0$ to be satisfied (see Section 1.4.4).

Let us enumerate the characteristic velocities so that

$$\lambda_1^{\mathrm{R}} \le \lambda_2^{\mathrm{R}} \le \cdots \le \lambda_{n^{\mathrm{R}}}^{\mathrm{R}}, \qquad \lambda_1^{\mathrm{L}} \le \lambda_2^{\mathrm{L}} \le \cdots \le \lambda_{n^{\mathrm{L}}}^{\mathrm{L}}$$

and let the inequalities

$$\lambda_l^{\mathrm{R}} < W < \lambda_{l+1}^{\mathrm{R}}, \quad \lambda_k^{\mathrm{L}} < W < \lambda_{k+1}^{\mathrm{L}}, \quad 0 \le l \le n^{\mathrm{R}}, \quad 0 \le k \le n^{\mathrm{L}} \tag{7.1.1}$$

be satisfied. To avoid misinterpreting, we must put here

$$\lambda_0^{\mathrm{R}} = -\infty, \quad \lambda_0^{\mathrm{L}} = -\infty, \quad \lambda_{n^{\mathrm{R}}+1}^{\mathrm{R}} = \infty, \quad \lambda_{n^{\mathrm{L}}+1}^{\mathrm{L}} = \infty$$

Here, $n^{\mathrm{R}} - l$ and k are the numbers of different outgoing characteristics directed to the right and to the left from the front, respectively. The evolutionary condition requires the equality

$$k + n^{\mathrm{R}} - l = N - 1 \tag{7.1.2}$$

to be satisfied.

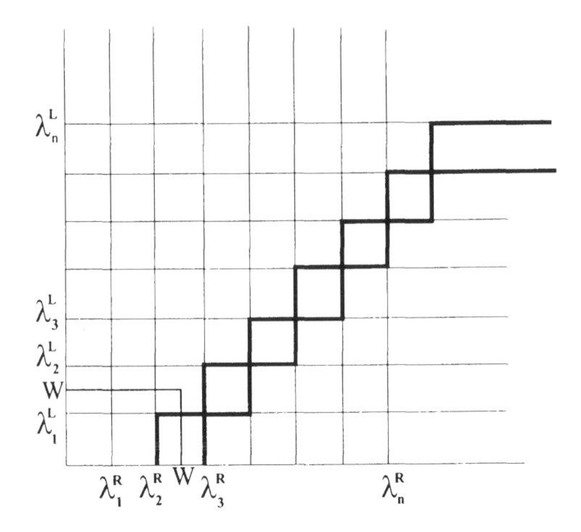

Figure 7.1 Evolutionary diagram.

The total number N of the boundary conditions on discontinuity is not always known *a priori*. Suppose we know that M conditions ($M \leq N$) are certainly satisfied (for example, these conditions result from the conservation laws). In what follows, these M relations will be called *basic relations*. Then inequalities (7.1.1) to be satisfied together with the equality

$$k + n^{\mathrm{R}} - l = M - 1 \tag{7.1.3}$$

will be called *a priori evolutionary conditions*.

As in the case of classical discontinuities, a discontinuity can be represented by a point in the evolutionary diagram (see Fig. 7.1) in which λ_i^{R}, W and λ_i^{L}, W are plotted on the horizontal and vertical axes, respectively. If the quantities with the superscript "R" are known, then on the horizontal axis the velocities can be plotted in the real scale, whereas on the vertical axis the corresponding velocities are arbitrarily scaled with retaining the inequalities between them. For a given N, the discontinuities represented in the diagram by points lying inside the sequence of rectangular regions satisfying inequalities (7.1.1) turn out to be evolutionary. The outlined regions go up to the right and are attached to each other by corners (Fig. 7.1). For brevity, these rectangular regions will be called *evolutionary rectangles*. Since N can be different for different discontinuities, the evolutionary rectangles do not always form the above sequence. Fronts belonging to the same or different evolutionary rectangles will be called fronts of the same or different types, respectively.

A priori evolutionary discontinuities (for which equality (7.1.3) must be satisfied for M = const) correspond to points of the sequence of rectangular regions (*a priori* evolutionary rectangles) going up to the right. If $M = N$, i.e., all the N boundary conditions coincide with the basic ones, the number of the conditions necessary for discontinuities to be evolutionary and that on *a priori* evolutionary discontinuities coincide, as well as the rectangles corresponding to these discontinuities. The condition $N > M$ means that besides the M basic boundary conditions there are $N - M$ *additional conditions*. In this case the evolutionary rectangles are displaced upward with respect to the sequence of *a priori* evolutionary discontinuities shown in Fig. 7.1.

For the rectangles lying lower than the sequence of *a priori* evolutionary rectangles, the number of boundary conditions is *greater than that necessary for discontinuities to be evolutionary*. Adding new boundary conditions cannot make these discontinuities be evolutionary. As a rule, these nonevolutionary fronts represent two or more merged evolutionary discontinuities propagating with identical velocities. They can acquire different velocities under the action of small disturbances and the initial discontinuity disintegrates.

Similar to classical discontinuities, in certain cases the linearized system of relations on a discontinuity *can split off into two (or more) independent subsystems*. Each of the small perturbation wave amplitudes (on either side of the front) and the disturbance δW enter only into one of them. The evolutionary condition represents the requirement of resolvability of each subsystem. This means that the number of the relations in each subsystem must be equal to the number of unknown quantities. However, we can require the satisfaction of condition (7.1.2) for the complete system and for one of the subsystems. Thus, there exists an additional evolutionary condition resulting from the resolvability of the subsystem. In this case there are two sets of rectangular regions in the evolutionary diagram, and those discontinuities for which W is located in *overlapping rectangles* are evolutionary.

As in the case of classical discontinuities, the front whose velocity coincides with one of the characteristic velocities ahead of or behind the front is considered to be evolutionary, though we imply that the interaction of such a discontinuity with small disturbances requires special analysis. Thus, in what follows inequalities (7.1.1) will be considered as nonstrict,

$$\lambda_l^{\mathrm{R}} \leq W \leq \lambda_{l+1}^{\mathrm{R}}, \qquad \lambda_k^{\mathrm{L}} \leq W \leq \lambda_{k+1}^{\mathrm{L}}. \tag{7.1.4}$$

If one of the inequalities in relations (7.1.4) transforms into the strict equality for a noninfinitesimal discontinuity, this equality is called *Jouget condition* and the corresponding front is called *Jouget front* similarly to the terminology adopted in detonation theory. The Jouget fronts correspond to points on the boundaries of the evolutionary rectangles. The Jouget condition can be satisfied both behind and ahead of the front.

7.2 Structure of fronts. Additional boundary conditions on the fronts

In this section we will study some consequences resulting from an analysis of the structures of fronts and discontinuities. Hyperbolic equations can be considered as an asymptotic form of equations describing large-scale phenomena (Gelfand 1959). These phenomena are governed by a *more complete system of equations*. From this viewpoint the fronts and discontinuities correspond to relatively narrow zones of sharp changes in the solutions of the complete system. If we neglect the width of these zones, they can be considered as discontinuities. Relations on the discontinuities must relate parameters on both sides of these narrow zones and can be obtained by constructing solutions of the complete system in these narrow zones. These solutions are called solutions of the discontinuity structure problem. The requirement for these solutions to exist provides certain relations between quantities on either side of the transition zone.

In some cases these relations coincide with the basic relations, known in advance, that result from the conservation or other known physical laws. In other cases, constraints in the form of inequalities arise. The hypothesis implying that only those fronts that have structure can exist and be used to construct solutions leads to the uniqueness of solutions of a number of problems for hyperbolic equations (Galin 1959; Oleinik 1959; Kalashnikov 1959; Rozhdestvenskii and Yanenko 1983) whose solutions are nonunique without this hypothesis. However, as will be, in particular, seen from what follows in the subsequent sections, the above-mentioned hypothesis does not always lead to the uniqueness of solutions in the hyperbolic approximation, and the uniqueness of solutions of the hyperbolic systems is not a physical requirement.

The requirement of the existence of the structure very often shows that there are discontinuities on which additional boundary conditions in the form of equalities must be satisfied. Without additional conditions these discontinuities would be nonevolutionary and, consequently, should be rejected. The combustion fronts mentioned above are among these discontinuities. The fronts with additional relations play an important part in various fields of mechanics and physics including the examples considered below.

This section is primarily devoted to estimation of the number of the additional relations on discontinuities resulting from the existence of the structure. The results can be briefly formulated as follows. Under some natural and general assumptions formulated below in Sections 7.2.1 and 7.2.2 it is shown that the total number of additional and basic relations should ensure the evolutionarity of the discontinuity (Kulikovskii 1968). Moreover, it is obvious that the condition of the entropy nondecrease is satisfied if the discontinuity has a structure. Therefore, the requirement of existence of a structure represents a fairly rigorous rule for selection of discontinuities.

We will assume that the discontinuity structure can be described by a system of ordinary differential equations resulting from the complete system if the solution depends on a single variable $\xi = -x + Wt$, where x is the space variable directed along the normal to the wave front, W is the velocity of the discontinuity, and t is time. To prove the above statement one must analyze two sets of solutions of ordinary differential equations describing the structure, namely, the solutions tending to constant values as $\xi \to \infty$ or $\xi \to -\infty$. Each of these sets depends not only on the velocity W of the discontinuity and the quantities at $\xi = \infty$ or $\xi = -\infty$, but also on arbitrary constants C_i^{L} or C_j^{R} characterizing the process of variation of the quantities inside the structure. The number of these constants is estimated in Section 7.2.3. To obtain a solution describing the structure and valid for any ξ we should match the solutions going from $\xi = \infty$ and $\xi = -\infty$ at a finite value of ξ, i.e., inside the structure. The simplest variant of such matching lies in the requirement for the solution to be continuous.

Eliminating C_i^{L} and C_i^{R} from the matching conditions, we obtain relations between the quantities specifying the solution at $\xi = \pm\infty$ and W and thereby giving conditions on discontinuity for the hyperbolic systems of equations describing large-scale phenomena. Evaluation of the number of the relations obtained (see Section 7.2.4) generally leads to the result formulated above. Both the result obtained and peculiar possible situations in which this can be invalid (for example, if the number of the basic relations on discontinuity is greater than that necessary for the discontinuity to be evolutionary) are discussed in Section 7.2.5. The possibility for the structure to have internal discontinuities, which was previously excluded by the assumption made in Section 7.2.1, is analyzed in Section 7.2.5.

7.2.1 Equations describing the discontinuity structure.

First, let us consider the complete system of equations describing the discontinuity structure. Sometimes the structure can be described by the initial system of hyperbolic equations admitting discontinuous solutions. This happens for discontinuities occurring in solutions of linear equations and for discontinuities corresponding to Riemann waves that propagate, preserving their shapes. But in order to solve the problem of the discontinuity structure we must generally use a system of equations that differs from the initial hyperbolic system.

Equations describing the discontinuity structure must be based on physical processes inside the discontinuity. On the other hand, the complete system describing the structure must be consistent with the large-scale equations whose discontinuous solutions are studied.

For this purpose we will assume that for large characteristic time T and length L the equations describing the structure transform into a hyperbolic system or two systems if the systems are different on either side of the discontinuity. The latter can occur if we admit that inside the discontinuity structure there can exist a surface (for simplicity, one surface) on which certain coefficients of the equations describing the structure are discontinuous or become identically equal to zero.

Thus, we will assume that a complete system of equations describing the discontinuity structure is defined everywhere and this system tends to the corresponding hyperbolic systems far from the discontinuity, where the solution, in accordance with our assumption, is characterized by large scales of T and L. Let us represent the complete system in the form of a system of first-order equations

$$A_{mj}\frac{\partial v_j}{\partial t} + B_{mj}\frac{\partial v_j}{\partial x} + F_m = 0, \qquad m, j = 1, 2, \ldots, Q. \tag{7.2.1}$$

The coefficient matrices A_{mj}, B_{mj}, and F_m are assumed to be functions of v_j. The rank of the matrix $[A_{mj}]$ can be less than R. In particular, system (7.2.1) can be parabolic. We assume that the equations

$$F_m(v_k) = 0, \qquad m = 1, 2, \ldots, Q \tag{7.2.2}$$

are mutually dependent and their solutions depend on n independent variables $U_1, U_2, \ldots, U_n$, $n < Q$. This means that system (7.2.1) can be written out in the form such that the first n equations do not involve F_m, that is $F_m \equiv 0$ for $m = 1, 2, \ldots n$.

In studying phase transitions, inside the structure there frequently exists a critical surface on each side of which the number of independent equations (7.2.2) is different, say $n = n^{\mathrm{R}}$ and $n = n^{\mathrm{L}}$ on the right- and left-hand sides, respectively. Accordingly, the set of variables U_m, generally speaking, can be different on both sides of the critical surface: U^{R}_m, $m = 1, 2, \ldots, n^{\mathrm{R}}$ and U^{L}_j, $j = 1, 2, \ldots, n^{\mathrm{L}}$.

For example, in studying such fronts in an electromagnetic field (Section 7.8) on which the conductivity of the medium becomes nonzero (or, on the contrary, vanishes) the Ohm law is used in the form (in the one-dimensional case):

$$\frac{c}{4\pi}\frac{\partial \mathbf{B}}{\partial x} = \sigma\left(\mathbf{E} + \frac{\mathbf{v}}{c} \times \mathbf{B}\right),$$

where the conductivity σ depends on the temperature T and $\sigma \equiv 0$ at temperatures lower than some critical value T^* (T^* can depend on the density and other parameters). If T acquires the value T^* inside the structure, the surface $T = T^*$ is critical. In the nonconducting region the quantities $\mathbf{E}$, $\mathbf{v}$, and $\mathbf{B}$ are independent unknown variables, whereas in the conducting region (for large-scale phenomena) they obey the relation $\mathbf{E} = -\mathbf{B} \times \mathbf{v}/c$. Similarly, if an elastic solid experiences a reversible phase transition into the liquid state (or *vice versa*) the elastic coefficient responsible for the shear elasticity vanishes (or, respectively, becomes nonzero).

While considering phenomena characterized by slow variations in space and time, in particular, phenomena occurring far from the transition zone (structure), we can neglect the terms with space and time derivatives in those of equations (7.2.1) in which F_m are present. This makes it possible to express v_j, $j = 1, 2, \ldots, Q$, in terms of U_m^α, $m = 1, 2, \ldots, n^\alpha$, $\alpha =$ R, L, and substitute them in the remaining n^α equations. In this case we obtain the system of n^{R} equations for n^{R} unknown functions U_m^{R} and the system of n^{L} equations for n^{L} unknowns U_m^{L}. Both systems will be assumed to be hyperbolic and identified with the hyperbolic systems describing the states on either side of the transition zone of the discontinuity. Thus, we can write out the system governing large-scale phenomena in the form

$$a_{mj}^\alpha \frac{\partial U_j^\alpha}{\partial t} + b_{mj}^\alpha \frac{\partial U_j^\alpha}{\partial x} = 0, \qquad j, m = 1, 2, \ldots, n^\alpha, \qquad \alpha = \mathrm{R}, \mathrm{L}, \tag{7.2.3}$$

where a_{mj}^α and b_{mj}^α depend on U_k^α. In what follows systems (7.2.3) are called *simplified* systems (as compared with system (7.2.1)).

Let us make some additional assumptions concerning the complete system (7.2.1). The first assumption relates to the behavior of solutions of the complete linearized system (7.2.1). Linearizing system (7.2.1) in the vicinity of constant quantities v_j^0 satisfying relations (7.2.2), we obtain the system with constant coefficients. We shall seek its solutions in the form

$$v_j = v_j^0 + v_j^* e^{i(kx - \omega t)},$$

where v_j^* are constants and k is the wave number. From the linearized system (7.2.1), we obtain

$$(-i\omega A_{mj}^0 + ikB_{mj}^0 + F_{mj}^0)v_j^* = 0,$$

where A_{mj}^0, B_{mj}^0 and F_{mj}^0 are the values of A_{mj}, B_{mj} and $\partial F_m / \partial v_j$ for $v_j = v_j^0$. If

$$\det[F_{mj}^0 - i\omega A_{mj}^0 + ikB_{mj}^0] = 0 \tag{7.2.4}$$

the system has a nonzero solution v_j^*.

For given v_j^0 Eq. (7.2.4) relates ω to k and is called the *dispersion relation*. The dispersion relation makes it possible to find the multivalued analytical functions $\omega(k)$ or $k(\omega)$. The number of their branches is determined by the highest powers of ω and k in the dispersion relation (7.2.4).

We shall assume that the following *dissipation condition* is satisfied: if ω and k obey the dispersion relation, then

$$\operatorname{Im} \omega < 0, \qquad \text{if} \quad \operatorname{Im} k = 0, \quad k \neq 0. \tag{7.2.5}$$

Thus, the system is dissipative in some domain of v_j^0 space if any sinusoidal in x solution of the system (7.2.1) linearized in the vicinity of every point $v_j^0 = \text{const}$ of the domain exponentially decreases with time for any finite wave length. The value $k = 0$ can correspond to ω with $\operatorname{Im}\omega = 0$.

One more assumption ensures the continuity of the solution of the problem of the discontinuity structure. By passing to the moving coordinate system

$$x' = x - Wt, \qquad W = \text{const},$$

we can transform the system (7.2.1) to the form

$$A_{mj}\frac{\partial v_j}{\partial t} + B'_{mj}\frac{\partial v_j}{\partial x'} + F_m = 0, \qquad B'_{mj} = B_{mj} - WA_{mj}. \tag{7.2.6}$$

In what follows we will assume that

$$\det[B'_{mj}] \neq 0 \tag{7.2.7}$$

over the variation range of v_j and W (in what follows W is the velocity of the wave representing the discontinuity structure). Assumption (7.2.7) is made only for convenience. As shown in Section 7.2.6, if this assumption is rejected the structure can have internal discontinuities but the number of additional relations on the discontinuities occurring in solutions of the large-scale equations remains the same.

It is obvious that if system (7.2.1) is dissipative, the same holds for the system (7.2.6). In fact, the solutions of the linearized system (7.2.1) in the form $\exp[i(kx - \omega t)]$ and the solutions of the linearized system (7.2.6) in the form $\exp[i(k'x' - \omega' t)]$ are the identical solutions written in different coordinate systems. If certain solutions decrease with time, the other also decrease. The coincidence of exponential arguments results in $\omega' = \omega(k) - kW$ for $k' = k$, where the function $\omega(k)$ must satisfy the dispersion relation (7.2.4).

7.2.2 Formulation of the structure problem and additional assumptions.

Let us investigate the existence of a steady-state structure of a discontinuity in the form of a traveling wave:

$$v_m \to v_m(\xi), \qquad \xi = -x + Wt = -x', \tag{7.2.8}$$

with $v_m \to v_m^{\mathrm{R}}$ as $\xi \to -\infty$ and $v_m \to v_m^{\mathrm{L}}$ as $\xi \to \infty$.

Clearly, the quantities v_m^{R} and v_m^{L} must satisfy the system (7.2.2) and, consequently, the sets of variables $U_1^{\mathrm{R}}, U_2^{\mathrm{R}}, \ldots, U_{n^{\mathrm{R}}}^{\mathrm{R}}$ and $U_1^{\mathrm{L}}, U_2^{\mathrm{L}}, \ldots, U_{n^{\mathrm{L}}}^{\mathrm{L}}$ can be assigned to them.

Let us find out the number of relations between the variables W, U_m^{R}, and U_m^{L} allowing us to obtain the solution in the form (7.2.8). When studying shock structures, we will assume that the velocity W of the discontinuity does not coincide with any of characteristic velocities of the simplified hyperbolic systems (7.2.3) for $U_m = U_m^{\mathrm{R}}$, $m = 1, 2, \ldots, n^{\mathrm{R}}$ and $U_j = U_j^{\mathrm{L}}$, $j = 1, 2, \ldots, n^{\mathrm{L}}$. In what follows, we will consider this generic case, which allows us to determine how many boundary conditions is required for a discontinuity to be evolutionary.

Without loss of generality, we assume that only a single critical surface can exist inside the structure. On this surface some of the functions A_{mj}, B_{mj}, F_m can be discontinuous or identically equal to zero on one side. Such surfaces can correspond to physical transitions occurring in the medium, and their locations should be determined by expressions relating the variables v_j that specify the medium state

$$\Phi_0(v_j) = 0. \tag{7.2.9}$$

Equations describing the steady-state structure of a discontinuity have the form

$$B'_{mj}\frac{dv_j}{d\xi} = F_m, \tag{7.2.10}$$

where B'_{mj} can be determined from equalities (7.2.6).

Owing to condition (7.2.7) every solution must be everywhere continuous and uniquely continued with respect to the variable ξ (F_m are assumed to be bounded and differentiable). Besides, on the critical surface all v_j are assumed to be continuous. This gives us R conditions relating v_j. If we take into account relation (7.2.9), the total number of the conditions on the critical surface is $R + 1$. In what follows, these $R + 1$ conditions will be called *matching conditions*. We construct the solution on either side of the discontinuity starting from the states corresponding to $\xi = \pm\infty$ and extend them to the matching point. If there is no real critical surface, expression (7.2.9) relating v_j must be specified formally and we must require continuity of v_j on the surface defined by this relation. Thus, the cases with no critical surfaces do not formally differ from the general case of their presence.

Thus, the following assumptions are made:

• after simplification, the complete systems of equations describing large-scale phenomena become hyperbolic and we must determine the conditions on discontinuities arising in its solutions;

• dissipation conditions (7.2.5) must be satisfied;

• relation (7.2.7) ensuring the continuity of solutions of the equations describing the structure must be satisfied—in Section 7.2.6 we consider the case without this assumption;

• both ahead of and behind a discontinuity its velocity W does not coincide with the corresponding characteristic velocities.

In addition, $(R + 1)$ matching conditions must be satisfied. This seems so natural that we do not consider this as an assumption.

7.2.3 Behavior of the solution as $\xi \to \pm\infty$.

Let us linearize system (7.2.6) in a vicinity of homogeneous states attained by the solution as $\xi \to \pm\infty$ and investigate some consequences of assumptions (7.2.5) and (7.2.7) (Lyubarskii 1961). From these equations it follows that for $\operatorname{Im}\omega' > 0$ the linearized system (7.2.6) has no solutions in the form $\exp[i(k'x' - \omega't)]$ with real k'. Consider the dispersion relation corresponding to the linearized system (7.2.6). Let p and q be the numbers of its roots $k'(\omega')$ (with their multiplicities) located in the upper and lower halves of the complex plane k' for $\operatorname{Im}\omega' > 0$.

From condition (7.2.7) it follows that $p + q = Q$. If the parameters v_j characterizing the state about which the linearization is performed and W vary continuously, the numbers p and q do not change when $\operatorname{Im}\omega$ remains in the upper half-plane ω. In fact, due to condition (7.2.5) none of the roots k' can intersect the real k'-axis for such change in the parameters. In accordance with condition (7.2.7), $|k| < \infty$ for finite values of ω and the sign of $\operatorname{Im} k$ cannot change due to the passage of the root k through infinity.

We will assume that there exists an intersection or a contact of the set of all possible states v_s^{R}, $s = 1, 2, \ldots, Q$, corresponding to $\xi = -\infty$ and depending on U_r^{R}, where $r = 1, 2, \ldots, n^{\mathrm{R}}$, and the set of all possible states v_j^{L}, $j = 1, 2, \ldots, Q$, corresponding to $\xi = \infty$ and depending on U_s^{L}, where $s = 1, 2, \ldots, n^{\mathrm{L}}$. For example, when the surface of discontinuity separates conducting and nonconducting media, in the conducting medium the quantities **E**, **B**, and **v** are connected by the relation $\mathbf{E} = \mathbf{B} \times (\mathbf{v}/c)$. This set of the quantities is a subset of the quantities **E**, **B**, and **v**, which are independent in the nonconducting medium. Thus, if the set of the states v_s^{R} intersects the set of the states v_j^{L} the quantities p and q remain the same in the union of these sets. We take this for granted in what follows. Since for $\operatorname{Im}\omega' > 0$ in a dissipative system all disturbances decay in the direction of their propagation, the quantities p and q are equal to the numbers of small linear disturbances described by the complete system and travelling to the right and to the left, respectively (Briggs 1964; Sobolev 1958).

Let us evaluate the number of roots $k_j'(\omega)$ of the dispersion relation for the linearized system (7.2.6) such that $\operatorname{Im} k' > 0$ for $\omega' = 0$. As $\omega' \to 0$ so that $\operatorname{Im}\omega' > 0$, some of the roots k' also tend to zero from the upper half-plane and vanish in the limit. Let us find the number of these roots. Note that if ω' and k' tend to zero simultaneously, then the time and space scales of the corresponding solution tend to infinity. Therefore, the appropriate exponential solution must be described by the simplified hyperbolic system (7.2.3). The quantities $\omega = \omega' + Wk'$ and $k = k'$ must satisfy the dispersion relation of the system obtained by linearizing Eq. (7.2.3). Since the system is hyperbolic and does not involve nondifferential terms, the solution of the dispersion relation has the form

$$k^\alpha = \omega/\lambda_j^\alpha, \qquad j = 1, 2, \ldots, n^\alpha; \qquad \alpha = \mathrm{R}, \mathrm{L},$$

where λ_j^α are the characteristic velocities of the simplified system (7.2.3).

Accordingly, we obtain expressions for the asymptotic behavior of the roots k_j' such that $k_j' \to 0$ as $\omega' \to 0$,

$$k_j'^\alpha = \omega'/\lambda_j'^\alpha, \qquad \lambda_j'^\alpha = \lambda_j^\alpha - W, \qquad j = 1, 2, \ldots, n^\alpha. \tag{7.2.11}$$

In accordance with the above assumption, all $\lambda_j'^\alpha$ are nonzero. From this it follows that the complete linearized system (7.2.6) has exactly n^α roots $k'(\omega)$ tending to zero as $\omega' \to 0$ and the number of roots k_j' which tend to zero from the upper half-plane is equal to the number of the characteristic velocities of the simplified system (7.2.3) satisfying the inequalities $\lambda_j^\alpha - W > 0$. In what follows, we will denote the number of such characteristic velocities by r^α.

Clearly, r^α depends on W and the values of $v_j^{\mathrm{R,L}}$ in whose vicinity we linearized the system. In the lower half-plane k' we have l^α roots tending to zero as $\omega' \to 0$, where $l^\alpha = n^\alpha - r^\alpha$. Thus, if $\omega' = 0$, $p - r^\alpha$ and $q - l^\alpha$ roots of the dispersion relation of the system (7.2.6) remain in the upper and lower half-planes, respectively.

We shall construct a solution of the problem of the steady-state shock structure that is a solution of system (7.2.10) starting from $\xi = \pm\infty$. For large negative ξ and fixed values of v_j^{R} the solution should only slightly differ from v_j^{R} and, therefore, the corrections to v_j^{R} can be obtained from the linearized equations. Generally, solutions of these equations depend exponentially on ξ, and it is convenient to obtain them from the solutions in the form $\exp[i(k'x' - \omega' t)]$, where $x' = x - Wt = -\xi$, by passing to the limit $\omega' \to 0$.

Since solutions of the linearized equations should tend to zero as $\xi \to -\infty$, the exponential functions occurring in the solution can correspond only to those values of $k' = k_m^{\mathrm{R}}$ that remain in the upper half-plane after ω' vanishes. From this we can conclude that unknown solutions depend on $p - r^{\mathrm{R}}$ arbitrary constants C_m^{R} that are the factors multiplying the exponential functions $\exp(-ik_m^{\mathrm{R}}\xi)$. As ξ increases, the solutions can be uniquely continued even if the solution becomes nonlinear. This is ensured by condition (7.2.7).

Thus, the set of solutions of the system (7.2.10) tending to constant values as $\xi \to -\infty$ for a given W is determined by $p - r^{\mathrm{R}}$ arbitrary constants C_m^{R} and the state at $\xi = -\infty$, which, in its turn, is determined by n^{R} quantities $U_1^{\mathrm{R}}, U_2^{\mathrm{R}}, \ldots, U_{n^{\mathrm{R}}}^{\mathrm{R}}$. Similarly, for the same W the set of solutions tending to constant values as $\xi \to \infty$ is determined by $q - l^{\mathrm{L}}$ arbitrary constants C_j^{L} (which are the factors of the exponential functions decreasing with increase in ξ) and by n^{L} quantities $U_1^{\mathrm{L}}, U_2^{\mathrm{L}}, \ldots, U_{n^{\mathrm{L}}}^{\mathrm{L}}$. We recall that $p - r^{\mathrm{R}}$ and $q - l^{\mathrm{L}}$ are equal to the differences of the numbers of small disturbances governed by the complete and simplified systems and propagating from the discontinuity upstream ($x > 0$) and downstream ($x < 0$), respectively.

7.2.4 Additional relations on discontinuities.

As mentioned above, in order to solve the problem on the discontinuity structure it is necessary to satisfy $Q + 1$ matching conditions. The matching conditions can be considered as conditions relating parameters determining the solution on either side of the matching point,

$$\Theta_a(U_1^{\mathrm{R}}, \ldots, U_{n^{\mathrm{R}}}^{\mathrm{R}}, C_1^{\mathrm{R}}, \ldots, C_{p-r^{\mathrm{R}}}^{\mathrm{R}}, U_1^{\mathrm{L}}, \ldots, U_{n^{\mathrm{L}}}^{\mathrm{L}}, C_1^{\mathrm{L}}, \ldots, C_{q-l^{\mathrm{L}}}^{\mathrm{L}}, W) = 0,$$
$$a = 1, 2, \ldots, Q + 1. \tag{7.2.12}$$

The quantity W is added to the arguments of the functions Θ_a since the solutions matched depend on W. Generally, we can eliminate $p - r^{\mathrm{R}} + q - l^{\mathrm{L}} = Q - r^{\mathrm{R}} - l^{\mathrm{L}}$ quantities C_i^{R} and C_j^{L}. As a result, we have $Q + 1 - (Q - r^{\mathrm{R}} - l^{\mathrm{L}}) = r^{\mathrm{R}} + l^{\mathrm{L}} + 1$ relations of the form

$$\Phi_b(U_1^{\mathrm{R}}, \ldots, U_{n^{\mathrm{R}}}^{\mathrm{R}}, U_1^{\mathrm{L}}, \ldots, U_{n^{\mathrm{L}}}^{\mathrm{L}}, W) = 0, \quad b = 1, 2, \ldots, r^{\mathrm{R}} + l^{\mathrm{L}} + 1, \tag{7.2.13}$$

where Φ_b are certain functions. We recall that r^{R} and l^{L} are the numbers of outgoing characteristics of the simplified system (7.2.3) directed to the right and to the left from the discontinuity. This means that the sum $r^{\mathrm{R}} + l^{\mathrm{L}}$ is equal to the total number of different outgoing characteristics. Thus, the number of relations obtained from the requirement of the existence condition for the discontinuity structure is equal to the number of the evolutionary conditions for the discontinuity occurring in the simplified hyperbolic system describing large-scale phenomena.

Let us make a few remarks on the results obtained. If Eq. (7.2.10) describes a steady-state motion of a continuous medium, the mass, momentum, and energy conservation laws (and, possibly, other laws corresponding to the problem under consideration) must be their consequences. The first integrals of the system considered correspond to the conservation laws which makes it possible to relate parameters $U_i^{\mathrm{R}}, U_i^{\mathrm{L}}$, and W. Clearly, these relations are involved in the set of formulas (7.2.13). However, the relations on discontinuity that express the conservation or other universal laws are independent of the processes occurring inside the structure (if the processes do not contradict to these laws) and, as a rule, are known in advance. We called them *basic relations* on discontinuity. Besides the basic relations, the system (7.2.13) can involve *additional relations* on discontinuity that depend on processes occurring inside the structure, i.e., on the choice of the complete system (7.2.1).

When deriving relations (7.2.13), we assumed that the solution of the problem on the discontinuity structure problem exists. When studying small-amplitude waves we can expand a solution in a series in the wave amplitude. From the complete system of fairly general equations for any number of the conservation laws (or in their absence) we can obtain the Burgers equation possessing transitional solutions that can be considered as discontinuity structures, the same equations holding on either side of the discontinuity. These solutions give $n = n^{\mathrm{R}} = n^{\mathrm{L}}$ relations between U_m^{R}, U_m^{L}, and W and an inequality determining the direction of their variation in the wave.

The existence condition for the solution of the discontinuity structure can generally impose certain constraints on u_r^{L} and u_s^{R}, which have the form of inequalities:

$$\Psi_q(u_r^{\mathrm{L}}, u_s^{\mathrm{R}}, W) \geq 0. \tag{7.2.14}$$

Such inequalities must involve, for example, the frequently used condition of the entropy nondecrease on discontinuities without heat supply. This condition must obviously be satisfied on discontinuities having structure.

There is one more important point. When deriving relations (7.2.13) from (7.2.12), we use a part of equalities (7.2.12) to find the quantities C_m^{R} and C_j^{L} and then substitute them into the remaining relations. These quantities can be found if there exists a nonzero minor of the Jacobian obtained by differentiating the left-hand sides of relations (7.2.12) with respect to these quantities. If this condition is not satisfied, some of the quantities C_m^{R} and C_j^{L} do not occur in relations (7.2.12) or occur in them not independently. In this case elimination of C_m^{R} and C_j^{L} from (7.2.12) results in relations (7.2.13) whose number is greater than $r^{\mathrm{R}} + l^{\mathrm{L}} + 1$. Obviously, this case corresponds to nonevolutionary discontinuities.

If this occurs, the above-mentioned constants C_m^{R} and C_j^{L} cannot be uniquely determined. As a result, the solution representing the structure of such a nonevolutionary discontinuity, if it exists, is nonuniquely determined and depends on one or more parameters. As any degeneration, the degeneration of all the minors of the above Jacobian can be considered as an exception. However, if the initial and final points of the structure are such that the number of the basic relations is greater than $r^{\mathrm{R}} + l^{\mathrm{L}} + 1$, the above-mentioned degeneration is predetermined. Examples of such nonevolutionary discontinuities are known in magnetohydrodynamics (Germain 1959, see also Section 5.5 of this book).

7.2.5 Main result and its discussion.

The above considerations make it possible to formulate the final result as follows. If the number of basic relations on discontinuity is less than it is necessary for the discontinuity to be evolutionary, then from the existence condition for the solution structure one can find as many additional relations as necessary for the discontinuity to be evolutionary. The particular form of additional relations depends on the equations describing the shock structure.

If a discontinuity is *a priori* evolutionary, or, in other words, if the number of the basic relations is equal to the number of evolutionary conditions, no additional relations of the equality type originate from our investigation of the discontinuity structure. However, certain inequalities can arise granting that the solution representing the discontinuity structure exists when they are satisfied.

If the number of the basic relations is greater than that necessary for a discontinuity to be evolutionary, the structure of such nonevolutionary discontinuity, if it exists, is nonunique (the solution includes arbitrary parameters). We recall that usually such discontinuities disintegrate nonlinearly under the action of small disturbances.

Note that in addition to nontrivial discontinuities with not all C_m^{R} and C_j^{L} simultaneously equal to zero, there exist situations in phase transition fronts in which all these quantities vanish. This occurs on continuous fronts separating the states governed by different simplified systems of equations. Such continuous fronts are known, for example, in magnetohydrodynamics (Butler 1965) when the conductivity is assumed to be nonzero only at fairly high temperatures $T > T_*$.

In conclusion we should note that we considered only one-dimensional and steady-state discontinuity structures. Real structures, however, can be unsteady and multidimensional, i.e., both temporal oscillations and multidimensional flows are possible inside the narrow zone describing the discontinuity. The solution representing a one-dimensional steady-state structure can be unstable or not exist at all. Moreover, as mentioned above, the transition zone can expand in time remaining, however, narrow as compared with the external scale of the problem.

The existence of a steady-state discontinuity structure does not guarantee the realizability of the discontinuity. To be sure, we must be convinced that the continuous solution obtained is stable. This is a difficult and, unfortunately, not very frequently considered problem. Strictly speaking, the absence of the steady-state structure does not mean that the corresponding discontinuity does not exist, since it can have unsteady or multidimensional structure. In the coordinate system attached to the discontinuity the structure can be periodic in time or with respect to the coordinates in the wave front surface. For such discontinuities the problem of obtaining additional relations was considered by Kulikovskii (1988).

7.2.6 A remark on deriving additional relations when condition (7.2.7) is not satisfied.

When considering particular problems on the discontinuity structure we can encounter situations in which the complete system of equations does not satisfy condition (7.2.7) ensuring the continuity of the shock structure. In most cases such a system is a result of "oversimplification" of dissipative or other smearing mechanisms. If we assume that the structure is described by a fairly complete set of dissipative mechanisms, condition (7.2.7) can be met, and the transition to a simpler system of equations for which (7.2.7) is not satisfied can be performed by tending a part of dissipative coefficients to zero. In the limit discontinuities can arise inside the structure. The omitted dissipative coefficients are essential only in a small vicinity of these discontinuities. If relations on these internal discontinuities are known or can be obtained by the above mentioned passage to the limit, they must be used for construction of the discontinuity structure as a whole and obtaining the additional relations.

Let us consider the behavior of the solution of the problem on the discontinuity structure when assumption (7.2.7) is not satisfied, i.e., for some values of W and v_j the equality

$$\det[B_{mj} - WA_{mj}] = 0, \quad m, j = 1, 2, \ldots, Q \tag{7.2.15}$$

holds. Let this equality be satisfied for a certain set of values $W = \lambda_j^*$ that are functions of v_s and assume, for simplicity, that they are simple roots of Eq. (7.2.15). The quantities

λ_j^* are characteristic velocities of the complete system (7.2.1). Since the system can be not hyperbolic, the number of the roots can differ from the order Q of the system.

If ω and k tend to infinity being of the same order, the dispersion relation (7.2.4) yields the following asymptotic relation between ω and k:

$$k_j = \omega/\lambda_j^*$$

for a chosen characteristic velocity λ_j^*.

For real k, $\operatorname{Im}\omega$ is assumed to be finite and from the last equality we can conclude that all λ_j^* are real, since, both λ_j^* and its complex conjugate satisfy Eq. (7.2.15).

In the coordinate system moving with an arbitrary velocity W, as $k' \to \infty$ and $\omega' \to \infty$ we obtain

$$k_j' = \frac{\omega'}{\lambda_j^{*\prime}}, \qquad \lambda_j^{*\prime} = \lambda_j^* - W, \tag{7.2.16}$$

where $\lambda_j^{*\prime}$ for the system (7.2.6) can be determined similarly to λ^* in (7.2.1).

As can be seen from equality (7.2.16), if for $\operatorname{Im}\omega' > 0$ the difference $\lambda_j^* - W$ passes through zero from negative to positive values the corresponding root k_j passes from the lower to upper half-plane of the complex plane k' passing through infinity.

This allows us to determine the numbers of the roots $k'(0)$ lying, for $\omega = 0$, in the upper and lower half-planes for various W and v_j. These numbers coincide with the numbers of constants $C_m^{\mathrm{R,L}}$ in the asymptotic representation of the solution of the problem on the discontinuity structure as $\xi \to \infty$ and $\xi \to -\infty$, respectively (see Section 7.2.3). It is usually not difficult to determine these numbers for very large W exceeding all characteristic velocities of both the simplified and complete systems. Then we can follow their change as W decreases.

Thus, in accordance with the previous considerations and the results obtained in Section 7.2.3, the number of roots k_j' lying in the upper and lower half-planes k' changes when W passes through the characteristic velocities of the complete (7.2.1) or simplified (7.2.3) systems, respectively. In this case, in accordance with Section 7.2.3, if W becomes less than some characteristic velocity of the simplified system (one of the differences $\lambda_k - W$ changes sign from "+" to "–"), one root $k_j'(0)$ disappears from the upper half-plane and appears in the lower one. When W becomes less than a characteristic velocity λ_j^* of the complete system, then in accordance with the preceding considerations the opposite situation occurs. Namely, the number of the roots $k_j'(0)$ in the upper half-plane increases by one, while one root $k_m'(0)$ disappears from the lower half-plane (the root passes through infinity).

Without considering the general situation, we shall analyze a particular case in which only one of the differences $\lambda_j^* - W$ changes sign inside the structure passing from negative (ahead of the front) to positive (behind the front) values. Comparing this case with that in which the differences $\lambda_m^* - W$ are definite (positive or negative) and the signs of all the differences $\lambda_r^* - W$ and $\lambda_l^{\mathrm{R}} - W$ are the same in the initial state at $\xi = -\infty$, we can say that if $\lambda_j^* - W$ is not of fixed sign the asymptotic representation of the solution for $\xi \to \infty$ has one arbitrary constant less than for the case of a definite (positive or negative) difference $\lambda_m^* - W$. However, in this case an evolutionary discontinuity on which $\lambda_m^* - W$ is discontinuous and changes sign must be present inside the structure. Conditions under which internal discontinuities can occur in the shock wave structure were studied by

O. I. Dementii and S. V. Dementii (1978). This discontinuity can be characterized by a single arbitrary parameter, the shock amplitude. Generally, after the matching procedure we have a sufficient number of parameters for obtaining relations in the form (7.2.13) satisfying the evolutionary conditions.

When one of the differences $\lambda_j^* - W$ changes sign, from plus ahead of the discontinuity to minus behind it, the asymptotic form of the solution for $\xi \to \pm\infty$ contains one constant more than in the case of definite signs of the differences $\lambda_m^* - W$. However, when $\lambda_j^* - W$ vanishes, the system (7.2.10) has a singular point and the condition ensuring that the integral curve passes through it gives the necessary condition to find the additional constant.

7.2.7 Hugoniot manifold. It is often convenient to specify some quantities on one side of the discontinuity, for example U_k^{R}, and study the set of the states arising on the other side of the discontinuity, i.e., the set of points in the U_m^{L}-space that satisfy the boundary conditions on the discontinuity. Initially, we shall consider the set of values of U_m^{L} that are distinguished only by the basic boundary conditions, calling this set (by analogy with gas dynamics and magnetohydrodynamics) the shock adiabatic manifold or the shock adiabatic curve, if its dimension is equal to unity. In what follows, we will also call this set the Hugoniot manifold or the Hugoniot curve, respectively. When all the quantities U_k^{R} are fixed and the number of basic relations is equal to n^{L} (i.e., to the number of variables U_m^{L}), as in gas dynamics, magnetohydrodynamics, and elasticity theory, the shock adiabatic (or Hugoniot) manifold is a curve in the U_m^{L}-space and the velocity of the discontinuity is a function assigned on this curve.

The set of points U_m^{L} that are the limiting points of the solution of the problem on the shock structure for given initial values of U_k^{R} will be called the admissible part of the Hugoniot manifold. In certain cases, when studying the Hugoniot manifold it is convenient to specify only a part of the quantities U_k^{R}.

By analyzing the discontinuities corresponding to the evolutionary points of the Hugoniot manifold, we can distinguish three types of domains on it: (i) domains I in which the evolutionary conditions are satisfied due to the basic relations, i.e., the number of outgoing long-wave perturbations described by the simplified systems is one less than the number of the basic relations; (ii) domains II in which the number of the basic relations is less than that required for the discontinuity to be evolutionary; (iii) domains III in which the number of the basic relations is greater than that required for the discontinuity to be evolutionary.

In accordance with the statements proved above, the complete system (7.2.13) of boundary conditions on discontinuity corresponding to domain I does not involve additional boundary conditions and the requirement for the existence of the solution can lead only to appearance of constraints in the form of inequalities. In this case only a part of domain I belongs to the admissible part of the Hugoniot manifold.

In domain II the requirement of the existence a solution of the problem of the shock structure gives additional boundary conditions that make the discontinuity evolutionary. In this case a subdomain of a lower dimension can be separated from domain II on the Hugoniot manifold. This subdomain belongs to the admissible part of the Hugoniot manifold.

Points belonging to domain III correspond to nonevolutionary discontinuities, since there are too many basic relations for a discontinuity to be evolutionary and the situation cannot be improved. As shown above in Section 6.2.4, the structure of such discontinuities, if it exists, depends on arbitrary parameters. In certain cases, such discontinuities can consist of several evolutionary discontinuities moving at the same velocity. Obviously, small disturbances can initiate a difference between the relative velocities of these discontinuities. In this case the initial nonevolutionary discontinuity disintegrates into evolutionary discontinuities moving at different velocities.

It is interesting that in the classical cases—in particular, in gas dynamics and magnetohydrodynamics—it is sufficient to specify quantities on one side of a discontinuity and one more parameter (for example, the velocity of the discontinuity) to determine parameters on the other side using the complete set of boundary conditions. Such discontinuities can be called one-parameter discontinuities. From this viewpoint deflagration fronts are zero-parameter discontinuities. In the generic case, in which the evolutionary conditions are satisfied, multiparametric discontinuities can exist. For such discontinuities the number of the parameters can depend on the choice of the side of the discontinuity on which we specify the parameters. The discontinuities with "the negative number of parameters" can occur if the state on one side of the discontinuity cannot be arbitrary and due to the complete set of boundary conditions on the discontinuity the parameters turn out to be related. Such relations can exist in solutions of certain mechanical problems if the discontinuity can induce appropriate forward-propagating disturbances that modify the initial state U_m^{R}. In what follows, some examples of such discontinuities will be given (see Section 7.8).

It is natural to treat multiparametric discontinuities as a result of merging of one-parameter discontinuities. In this case for the resulting discontinuity to be evolutionary the number of parameters characterizing it must be equal to the number of parameters characterizing merged discontinuities.

As noted above, in many cases small-amplitude shock wave structures are described by the Burgers equation. We indicated that the corresponding discontinuities are one-parameter ones and the appearance of discontinuities differing from one-parameter discontinuities, as well as their splitting and merging, can be expected if the discontinuity has finite intensity or changes the properties of the medium.

In what follows we consider a few problems of continuum mechanics and study a discontinuity structure to obtain the complete system of boundary conditions on discontinuities. An analysis of some solutions of the problems, in particular, the results on the existence of the solution for various values of determining parameters, the uniqueness of the solutions and their dependence on the initial and boundary conditions allow us to make conclusions on the correctness of the models chosen to construct large-scale solutions.

7.3 Behavior of the Hugoniot curve in the vicinity of Jouget points and nonuniqueness of solutions of self-similar problems

In this section we will consider classical shock waves for which the same hyperbolic systems of equations are valid on either side of the shock and relations between the quantities on the discontinuity result from the conservation laws. We will describe features of the behavior of the Hugoniot curve in the vicinity of Jouget points at which the velocity of the shock wave coincides with one of the characteristic velocities behind the shock wave.

In the case considered, in accordance with Chapter 1, all shock waves can be divided into n types and the k-shock can be associated with its own characteristic velocity λ_k. Accordingly, we will distinguish "proper" and "improper" Jouget points. At the proper Jouget points the velocity of the k*th* shock is equal to its own k*th* characteristic velocity $W = \lambda_k^{\mathrm{L}}$ and at the improper Jouget points the equality $W = \lambda_{k-1}^{\mathrm{L}}$ holds.

We will consider the generic case in which for given U_i^{R} the relations on discontinuities resulting from the conservation laws (see Chapter 1)

$$\{F_i(U_k)\} = W\{U_i\} \quad i, k = 1, 2, \ldots, n \tag{7.3.1}$$

determine a curve in the U_i-space. Here, as usual, the braces denote a jump of the embraced quantities, that is,

$$\{U_i\} = U_i - U_i^{\mathrm{R}}, \quad \{F_i(U_k)\} = F_i(U_k) - F_i(U_k^{\mathrm{R}}),$$

where U_i^{R} and U_i correspond to the states ahead of and behind the discontinuity, respectively.

Let us prove that (i) in the generic case the shock velocity W has an extremum at a Jouget point on the Hugoniot curve and, vice versa, if the shock velocity has an extremum at some point on the Hugoniot curve, this point is a Jouget point (Hanyga 1976); (ii) the presence of improper Jouget points on the Hugoniot curve can lead to nonuniqueness or nonexistence of solutions to self-similar problems (Kulikovskii and Sveshnikova 1996).

Let us differentiate equality (7.3.1) assuming that $U_i^{\mathrm{R}} = \text{const}$, i.e., considering the variation of the quantities along the Hugoniot curve

$$(F_{ij} - W\delta_{ij})dU_j = (U_i - U_i^{\mathrm{R}})dW, \quad F_{ij} = F_{ij}(U_k) = \frac{\partial F_i(U_k)}{\partial U_j}. \tag{7.3.2}$$

The matrix $[F_{ij} - W\delta_{ij}]$ is the characteristic matrix of the initial system of equations

$$\frac{\partial U_i}{\partial t} + \frac{\partial}{\partial x}(F_i(U_k)) = 0.$$

For nonzero dU_j the equality $dW = 0$ can be valid only if $\det[F_{ij} - W\delta_{ij}] = 0$. This means that $W = \lambda$, where λ is one of the characteristic velocities at the point at which differentiation is performed, i.e., behind the shock wave. In other words,

$$\frac{dW}{d\sigma} = 0,$$

where $d\sigma$ is an element of the arc along the Hugoniot curve.

On the other hand, if $W = \lambda$ at a certain point on the Hugoniot curve, then multiplying Eq. (7.3.2) by the left eigenvector $\mathbf{l}$ of the matrix $[F_{ij} - W\delta_{ij}]$ and summing, we obtain

$$(U_i - U_i^{\mathrm{R}})l_i dW = 0.$$

If

$$(U_i - U_i^{\mathrm{R}})l_i \neq 0 \tag{7.3.3}$$

then $dW = 0$ for nonzero dU_j. This means that, in general, the derivative of the shock velocity with respect to the arc length of the Hugoniot curve is equal to zero at the Jouget point.

In order to clarify the behavior of the shock velocity in the vicinity of the Jouget point and understand the consequences of violation of condition (7.3.3), we will introduce a local coordinate system at the Jouget point with the basis vectors aligned with the eigenvectors of the matrix $[F_{ij}]$, i.e., aligned with the tangents to the integral curves of the Riemann waves. In this coordinate system equalities (7.3.2) can be written in the form

$$(\lambda_i - W)\delta u_i = u_i \delta W, \qquad i = 1, 2, \ldots, n. \tag{7.3.4}$$

Here, no summation is performed over the repeated index i, and δu_i and u_i are the components of $\delta\mathbf{U}$ and $\mathbf{U} - \mathbf{U}^{\mathrm{R}}$ in the new axes.

Assume that the difference $\lambda_1 - W$ is equal to zero and changes sign at the point considered, while the other differences $\lambda_i - W$ are nonzero at it. The quantity u_1 is nonzero and proportional to the scalar product of the vector $(\mathbf{U} - \mathbf{U}^{\mathrm{R}})$ by the left eigenvector of the matrix $[F_{ij}]$ and, therefore, is proportional to the left-hand side of (7.3.3). If inequality (7.3.3) holds, then from the first equality of (7.3.4) we can see that W has an extremum (maximum or minimum depending on the sign of u_1) at the considered Jouget point.

If condition (7.3.3) is not satisfied and u_1 vanishes and changes sign at the Jouget point, then W is monotone along the Hugoniot curve. If the matrix $[F_{ij}]$ is symmetric then its left and right eigenvectors coincide and condition (7.3.3) means that the tangent to the Hugoniot curve is not orthogonal to the vector $\mathbf{U} - \mathbf{U}^{\mathrm{R}}$ at the Jouget point. In what follows, we will assume that inequality (7.3.3) is satisfied. This corresponds to the generic case.

One more important property of the Hugoniot curve consists in the fact that it is tangent to one of the eigenvectors of the matrix $[F_{ij}]$ at the Jouget points and only at them. This results from equalities (7.3.4), if we write out them at a certain point on the Hugoniot curve. If we move this point along the Hugoniot curve to the Jouget point at which $W = \lambda_1$, then for a given length of the vector $\delta\mathbf{U}$ the quantity δW tends to zero together with all δu_j except for δu_1. Thus, the Hugoniot curve and the integral curves of the corresponding Riemann wave have a common tangent at the Jouget point.

From the above it follows that if we put λ_i^{R} and W on the horizontal axis of the evolutionary diagram in their real scales, the curve $W(\sigma)$ representing the shock adiabat has vertical tangents at the points of its intersection horizontal lines $W = \lambda_i^{\mathrm{L}}$. Thus, $W(\sigma)$ acquires extrema at these points (Fig. 7.2).

The physical reason of such behavior of the Hugoniot curve at the Jouget points is very simple. If the Jouget condition $W - \lambda^{\mathrm{L}} = 0$ is satisfied, then alongside with the initial discontinuity we can consider a succession of the initial discontinuity and a small perturbation propagating behind the discontinuity with the same velocity. This can be interpreted as a combined discontinuity that has the same velocity as the initial one. Hence it follows that $dW = 0$ at the considered point for nonzero dU_j. On the other hand, if the Hugoniot curve has two close points corresponding to discontinuities moving with the same velocity, then the discontinuity corresponding to a jump from one of these points to another can be considered as a small discontinuity moving with the same velocity as the jump from the initial point into the state ahead of the introduced small discontinuity. This means that

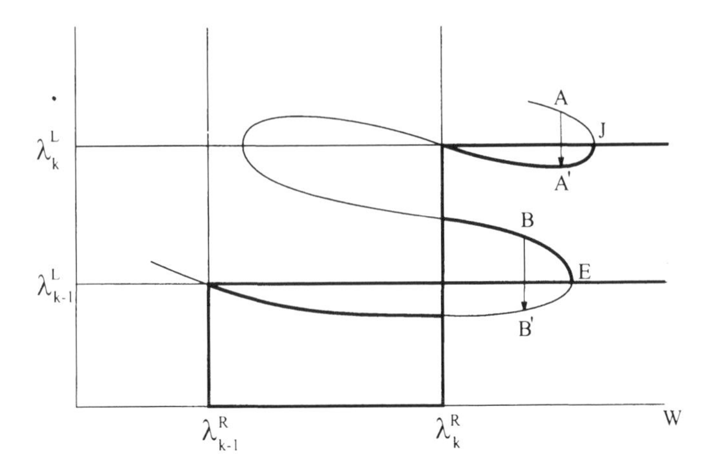

Figure 7.2 Weak shock waves in the vicinity of Jouget points on the evolutionary diagram.

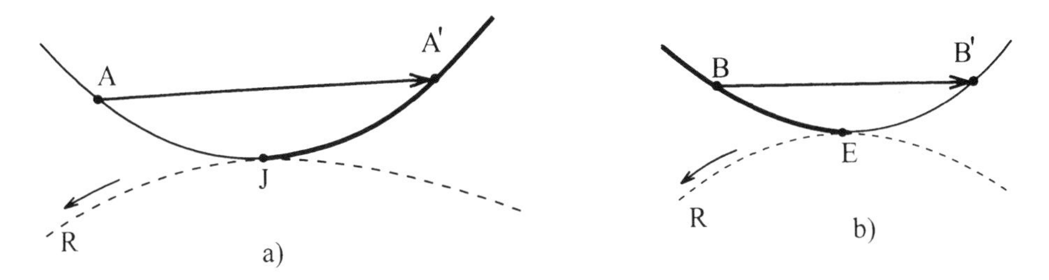

Figure 7.3 Mutual arrangement of the evolutionary part of a Hugoniot curve and the integral curve of an expanding Riemann wave in the vicinity of "proper" (a) and "improper" (b) Jouget points.

the Jouget condition $W = \lambda^L$ must be satisfied if these close points coincide and the small discontinuity disappears. It is apparent that this occurs at the extremum point of $W(\sigma)$.

From the proved proposition it follows that the Jouget shock wave may contact with a Riemann wave and their velocities coincide at the contact point.

Let us consider an evolutionary fragment of the Hugoniot curve in the U_i-space terminated at the Jouget point and an integral curve of the corresponding Riemann wave passing through this point.

In Figs. 7.3a and 7.3b the continuous curves represent arcs of the Hugoniot curve in the U_i-space in the vicinity of the Jouget points: the proper point J and the improper point E. The dotted curves represent the integral curves of the Riemann waves passing through these points. The bold lines represent the evolutionary fragments of the Hugoniot curve corresponding to the points lying inside the evolutionary rectangles shown in Fig. 7.2.

From what we have said it follows that these curves have a common tangent at the Jouget point. Let us find the direction of variation of the quantities in the expanding Riemann wave, the characteristic velocity of which decreases from the leading to the rear front. For this purpose we shall use the diagram shown in Fig. 7.2. In Fig. 7.3 we have shown the enlarged portions of the Hugoniot curve represented in Fig. 7.2 in the vicinity of the proper and improper Jouget points.

Since the velocity $W(\sigma)$ of the discontinuity has extrema at the Jouget points, we can find a pair of points corresponding to the same value of the shock velocity, and this value is close to one in that extremum point. Clearly, the jump from one point to another point in this pair satisfies all the conservation laws, and the velocity of this shock will be equal to that in the jumps from the initial point to each of these points. For example, in Fig. 7.2 we have shown small jumps (shock waves) $A \to A'$ and $B \to B'$. The velocity of each jump is close to the corresponding characteristic velocity λ.

A small discontinuity propagating with a velocity close to the characteristic velocity λ is evolutionary if $W > \lambda$ ahead of the shock and $W < \lambda$ behind the shock. This means that small discontinuities are evolutionary if the jump occurs downward across the horizontal straight line $W = \lambda^L$ in Fig. 7.2. In Fig. 7.3 the weak evolutionary jumps $A \to A'$ and $B \to B'$ are shown with arrows. In accordance with the Lax theorem (see Chapter 1), the variation of U_i in the expanding Riemann wave is opposite to that in the shock. This determines the direction of variation of the quantities in expanding Riemann waves shown by bent arrows in Fig. 7.3.

Thus, if the Hugoniot curve intersects *the upper boundary of the evolutionary region*, i.e., the proper Jouget point J is realized, then *the integral curve* in the U_i-space that corresponds to an expanding Riemann wave of the same type is directed opposite to the direction of the parameter variation in the shock and *is a continuation of the evolutionary part of the Hugoniot curve* with the common tangent at the Jouget point (Fig. 7.3a).

On the contrary, if the Jouget point is improper (point E) and corresponds to the intersection of the Hugoniot curve with *the lower boundary of the evolutionary region* shown in Fig. 7.2, then in the U_i-space this point is the origin of a fragment of the integral curve that corresponds to an expanding Riemann wave of the type different from the type of the shock wave. This fragment of the integral curve and the evolutionary part of the Hugoniot curve form a curve with a cusp at the Jouget point J (Fig. 7.3b).

Let us consider the solution of the self-similar problem with independent variable x/t. Let one of the discontinuities be the Jouget discontinuity or the one close to it. These solutions consist of Riemann waves and discontinuities separated by regions of uniform distribution.

As shown in Chapter 1, in the standard situations the self-similar solutions contain n waves and can be constructed by fitting the amplitudes of these waves. For a given state ahead of the Riemann or shock wave, the state behind the wave belongs to a certain curve in the U_i-space, namely, to the integral curve of the Riemann wave or the Hugoniot curve. In order to determine accurately the variation of all quantities in the wave it suffices to specify a single quantity called the wave amplitude. Determination of a self-similar solution reduces to connecting the points, which specify the states on different sides of the wave system, using a polygonal line composed of integral curves of expanding Riemann waves or Hugoniot curves corresponding to evolutionary shocks. In acting so it is necessary to satisfy the inequalities between the wave velocities; namely, the velocity of the leading front of a wave propagating in the wake of another must not be higher than that of the rear front of the preceding wave. A solution exists in the vicinity of a certain point $U_i = U_i^*$ if it exists at

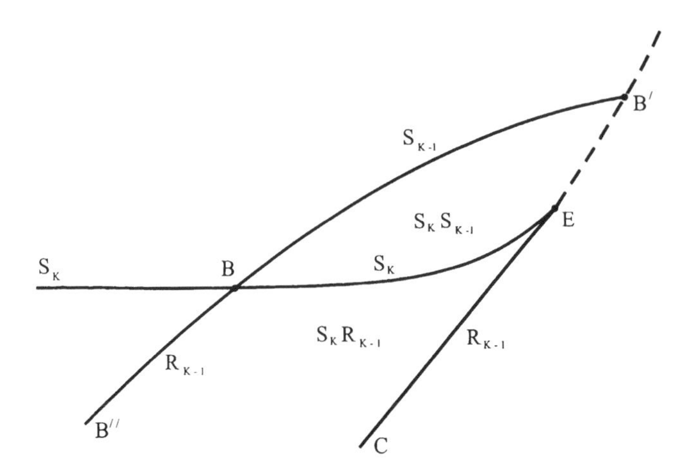

Figure 7.4 Domains in the parameter space in which solutions containing the shock wave S_k in the vicinity of an "improper" Jouget point take place.

this point itself and the Jacobian $\det[\partial U_i/\partial a_j]$ is not zero. Here, a_i are the amplitudes of the waves that compose the solution.

We now consider the solution of a self-similar problem in which one of the discontinuities is described by a point close to the proper Jouget point J or coinciding with it (see Fig. 7.2). As shown above, in this case the evolutionary segment of the Hugoniot curve corresponding to the k-shock in the U_i space can be smoothly continued beyond the Jouget point by a segment of the integral curve of the expanding k-Riemann wave.

The segment of the Hugoniot curve corresponding to the k-shock together with its continuation in the form of the k-Riemann wave can be considered as a single curve of the k-waves in the U_i-space, a single parameter (amplitude) varying along the curve. This does not prevent us from finding the wave amplitude, since there are no particular reasons for the Jacobian to become zero.

Another situation occurs in the vicinity of the improper Jouget point E at which the equality $W = \lambda_{k-1}^{\mathrm{L}}$ is satisfied for the k-shock (see Fig. 7.2) In this case the segment of the Hugoniot curve corresponding to the k-shock is tangent at the Jouget point to the integral curve of the $(k-1)$-Riemann wave. This segment together with the integral curve corresponding to the expanding $(k-1)$-Riemann wave form a curve with a cusp at the Jouget point, see curve BEC in Fig. 7.4, in which S_k and R_{k-1} denote corresponding parts of the Hugoniot curve and the self-similar expanding Riemann wave; the dotted curve EB' corresponds to the nonevolutionary segment of the Hugoniot curve.

If in a small vicinity of the Jouget point E we replace the Hugoniot curve S_k and the Riemann wave R_{k-1} by their common tangent, then the variation in the parameters characteristic of these waves clearly results in a motion along the same ray. Hence it follows that, under variation of the amplitudes of n different waves used in the construction of the solution, the points U_i^* corresponding to the state behind the wave system occupy not the entire n-dimensional vicinity of the Jouget point but only an $(n-1)$-dimensional hypersurface. This means that the Jacobian $\det[\partial U_i/\partial a_j]$ vanishes at the Jouget point and in the linear approximation adopted the solution does not exist in its vicinity.

In order to construct a self-similar solution in the nonlinear approximation we now consider a simpler case in which by a special choice of initial and/or boundary conditions, the solution is made to consist of only two waves (namely, of the k- and $(k-1)$-waves). Note that each point lying on the evolutionary segment S_k of the Hugoniot curve (for example, point B in Fig. 7.4) corresponds to the state behind the k-discontinuity. Either an evolutionary $(k-1)$-discontinuity S_{k-1} or a continuous self-similar expanding $(k-1)$-Riemann wave R_{k-1} can propagate in the wake of the preceding wave. In the U_i-space, the state behind these waves belongs to the curves composed of the evolutionary part BB' of the Hugoniot curve S_{k-1} and its continuation, represented by the integral curve BB'' of the Riemann wave R_{k-1}. When the point B moves along S_k, the combined curve $B''BB'$ covers a two-dimensional surface. For points lying on this surface we can construct a solution consisting of the sequence of an evolutionary k-discontinuity and a $(k-1)$-wave (continuous or discontinuous) that is close to the solution corresponding to the Jouget point E. This solution will be called solution I.

Let us show that the domain of definition for solution I is a surface with the edge on the other side of which there is no self-similar solution of type I. For this purpose we need to find the curve representing this edge. If a sufficiently small evolutionary $(k-1)$-discontinuity propagates through the state B behind an evolutionary k-discontinuity propagating at a velocity W_B, then the velocity of the former is lower then W_B, since the velocity of the $(k-1)$-discontinuity is close to λ_{k-1} and, in accordance with the evolutionary conditions (see Chapter 1), $W_B - \lambda_{k-1}^{\mathrm{L}} > 0$ behind the k-discontinuity. If we increase the amplitude of this $(k-1)$-discontinuity for a fixed state B, then its velocity will also increase and can become equal to W_B when the point representing the state behind this discontinuity arrives at a state B' (Fig. 7.4). In the physical space, the k- and $(k-1)$-discontinuities will merge and form a single nonevolutionary discontinuity corresponding to the point B'.

Further increase in the amplitude of the $(k-1)$-discontinuity does not correspond to any solution of the self-similar problem since the $(k-1)$-discontinuity cannot overtake the k-discontinuity. In the space of variables U_i this means that the domain in which a solution S_kS_{k-1} consisting of two discontinuities of kth and $(k-1)$th types exists represents a surface bounded by the curve consisting of points similar to B and B', that is, by the Hugoniot curve with the point B lying on its evolutionary part and the point B' on the nonevolutionary part on the other side from the Jouget point E.

As mentioned above, on the other side of the evolutionary segment of the Hugoniot curve S_k we can construct a self-similar solution consisting of a k-discontinuity S_k and a $(k-1)$-Riemann wave R_{k-1}. We shall denote this solution by S_kR_{k-1}. The two-dimensional surface corresponding to the solutions S_kS_{k-1} and S_kR_{k-1} is bounded by a curve that consists of the nonevolutionary part of the Hugoniot curve (the arc EB' in Fig. 7.4) and the integral curve of the Riemann wave (the arc EC) tangent to the Hugoniot curve at the point E.

On the other side of this edge CEB' of the two-dimensional surface *the self-similar solution does not exist or, if it exists, must have another structure (solution of type II)*. Two cases are possible.

Case (a). The domain in which a solution of type II exists is bounded by the same edge. Thus, in the vicinity of the point E solutions exist and are unique, though their structures are different on either side of the separating curve $B'EC$. For this to occur it is necessary that the second solution, as well as the first solution considered above, be genetically associated with the nonevolutionary part of the Hugoniot curve (the arc EB' in Fig. 7.4). This is possible, however, if there exists another combination of two discontinuities, different from that considered above, that merge into a single discontinuity as they approach the curve representing the nonevolutionary part of the Hugoniot curve from the other side. Precisely this situation can be realized if there exists another branch of the Hugoniot curve corresponding to the velocity interval containing W_E. If there exists a proper Jouget point J, the inequality $W_J > W_E$ must hold, i.e., the point J lies to the right from the point E (see Fig. 7.2). In this case for velocities somewhat lower than W_E there are two different shock waves corresponding to the kth evolutionary rectangle (Fig. 7.4). These shocks together with a $(k-1)$-shock can compose two combinations corresponding to a nonevolutionary discontinuity close to the shock transition into the point E.

Note that in the vicinity of the nonevolutionary part of the Hugoniot curve, near the point E the two self-similar solutions can be interpreted as a result of disintegration of a nonevolutionary shock under the action of small perturbations. These solutions are not close in the sense of absolute value of their difference, but they remain close if we use other measures of their difference, for example, those taking into account the dimension of the domain in the physical space, where the solutions differ strongly. In such a norm the solutions I and II differ slightly. This is associated with the fact that the velocities of the shock waves formed as a result of the nonevolutionary shock disintegration are close to the velocity of the initial nonevolutionary shock.

Case (b). If W is close to W_E, there are no other k-shocks. Then there is only a single (considered above) combination of two waves that merge into a nonevolutionary discontinuity. For example, this would hold if $W_J < W_E$, see Fig. 7.2. There are two radically different situations. In the first of them there exist some self-similar solutions everywhere in the neighborhood of point E. In this case the other solution II is valid on the other side of the edge $B'EC$ (Fig. 7.4). This solution is not directly associated with the Hugoniot curve in the neighborhood of the point E. The second solution is generally valid in the entire neighborhood of the point E and considerably differs from the solution I in its domain of existence. Generally, the boundary of the domains of existence of solution II is not located close to the point E. This means that there are two different self-similar solutions in the half-neighborhood of the point E. They are represented by solutions I and II. The second situation is the total absence of the solution in the domain, where solution I does not exist (in Fig. 7.4 to the right of curve $B'EC$).

Thus, if an improper Jouget point exists on the Hugoniot curve, in the case (a) the solution changes its type when crossing the nonevolutionary segment of the Hugoniot curve and in the case (b) the solution is either nonunique or does not exist for certain values of the determining parameters.

The above considerations treat the case in which there are only two waves with nonzero amplitudes (k- and $(k-1)$-waves). However, we can readily verify that all the conclusions will also be valid when these waves are a part of a more complex wave sequence.

The behavior of the solution in the presence of an improper Jouget point on the Hugoniot curve was first investigated by Kulikovskii and Sveshnikova (1985a, 1985b, 1995) in their study of quasitransverse waves in an elastic anisotropic medium. They showed that the situation described in the case (a) occurs for $W_J > W_E$ and the nonuniqueness of solution described in the case (b) occurs if $W_J < W_E$. The latter case will be further considered in Section 7.4.

7.4 Nonlinear small-amplitude waves in anisotropic elastic media

In this section we briefly outline the results of recent investigations of nonlinear small-amplitude waves in elastic media in which nonlinearity and anisotropy interact with each other (for details see Kulikovskii and Sveshnikova 1995). In the presence of nonlinearity and anisotropy a number of classical problems, such as the piston problem or the Riemann problem, have more than one solution. We consider the question of their admissibility in viscoelastic media under vanishing viscosity. The case of vanishing anisotropy is also discussed.

7.4.1 Basic equations. We will consider the motion of an elastic medium in the form of plane waves. This means that solutions depend on a single Cartesian coordinate and time. For continuous motion the governing equations can be written in the form (Bland 1969)

$$\rho_0 \frac{\partial v_i}{\partial t} = \frac{\partial}{\partial x}\frac{\partial \Phi}{\partial u_i}; \quad \frac{\partial u_i}{\partial t} = \frac{\partial v_i}{\partial x}; \quad \frac{\partial S}{\partial t} = 0; \quad i = 1, 2, 3. \tag{7.4.1}$$

Here, we used the Lagrangian approach in which $x_1, x_2, x_3 \equiv x$ are the Cartesian coordinates in the initial state of the medium. The unknown quantities are v_i and u_i, where $v_i = \partial w_i/\partial t$ are the velocity components and $u_i = \partial w_i/\partial x$ are the derivatives of the components of the displacement vector. These quantities, as well as the entropy S, are considered to be functions of x and time t. In (7.4.1), $\Phi = \rho_0 U$, where U is the internal energy per unit of mass and ρ_0 is the density in the initial state. The internal energy is assumed to depend on the strain tensor

$$\varepsilon_{ij} = \frac{1}{2}\left(\frac{\partial w_i}{\partial x_j} + \frac{\partial w_j}{\partial x_i} + \frac{\partial w_k}{\partial x_i}\frac{\partial w_k}{\partial x_j}\right).$$

Owing to the geometry of the problem, the derivatives $\partial w_\beta/\partial x_\alpha$; α, $\beta = 1, 2$ (here and in what follows, the Latin and Greek indices take the values 1, 2, and 3 and 1 and 2, respectively) are independent of time. Without loss of generality, the derivatives $\partial w_3/\partial x_\alpha$ can be disregarded in the plane waves. This does not lead to loss of generality since, in fact, we can make them equal to zero by rotating the body as a whole. In addition, Φ depends both on the entropy S and some internal invariable parameters including those determining anisotropy (Lokhin and Sedov 1963). In general, these parameters can depend on the Lagrangian coordinates. However, in the problem considered below we assume them to be constant. Thus, the internal energy depends on u_i and S (the entropy S can change its value on shocks, where the unknown quantities are discontinuous). The system is completed by specifying the function $\Phi(u_i, S)$. Note that this function is determined up to addition of a linear function of u_i without changing of equations (7.4.1). For incompressible media $u_3 =$ const and $v_3 =$ const and the system (7.4.1) contains only the equations with $i = 1$ and 2.

For stable media Eq. (7.4.1) constitutes a hyperbolic system of partial differential equations. They include the second Newton's law, the equation governing the variation of strains under the action of displacements, and the equation accounting for reversibility of the processes under consideration.

The relations

$$\left\{\frac{\partial \Phi}{\partial u_i}\right\} = -\rho_0 W\{v_i\}, \quad \{v_i\} = -W\{u_i\}, \quad \{\Phi\} = \frac{1}{2}\left(\frac{\partial \Phi^{\mathrm{L}}}{\partial u_k} + \frac{\partial \Phi^{\mathrm{R}}}{\partial u_k}\right)\{u_k\} \quad (7.4.2)$$

must be satisfied on discontinuities (shock waves) in the absence of external actions. Here, the superscripts "R" and "L" denote the quantities ahead of and behind the discontinuity, respectively; the braces denote the jump of the embraced quantities. For example, $\{\Phi\} = \Phi^{\mathrm{L}} - \Phi^{\mathrm{R}} = \Phi(u_i^{\mathrm{L}}, S^{\mathrm{L}}) - \Phi(u_i^{\mathrm{R}}, S^{\mathrm{R}})$. In addition, $W = dx/dt$ is the velocity of the discontinuity calculated via the Lagrangian variable. The first two groups of Eq. (7.4.2) express the conservation of momentum and the continuity of displacements. The last of Eq. (7.4.2) results from the energy conservation across the discontinuity (the velocity is eliminated from this equation by means of the first of Eq. (7.4.2)). The discontinuity is thermodynamically admissible if $\{S\} \geq 0$.

If the medium is isotropic in the (x_1, x_2)-plane (we shall call this the wave isotropy condition), the arguments u_1 and u_2 enter into Φ as the combination $u_1^2 + u_2^2$. Thus, in the case of the wave isotropy Φ depends on $u_1^2 + u_2^2$, u_3, and S. We shall consider the cases in which the wave anisotropy is small and Φ differs only slightly from a function of the above-mentioned arguments, that is,

$$\Phi(u_i, S) = F(u_1^2 + u_2^2, u_3, S) + qp(u_1,\ u_2,\ u_3), \quad (7.4.3)$$

where q is a small parameter and $p(u_i)$ is a certain function "of the order of unity." Actually, in what follows we will assume that $p(u_i)$ can be expanded in a power series in u_i and the first coefficients of the expansion are of the same order as those in the expansion of the function F.

A number of detailed studies of nonlinear waves have been performed in the case of the wave isotropy by Bland (1969), Hanyga (1976), and Lenskii (1981, 1982, 1983a, 1983b).

Bland (1969) considered small-amplitude waves propagating through an isotropic initially strainless medium. It is obvious that this case corresponds to $q = 0$ in equality (7.4.3). The expression for F was taken as the sum of a few first terms in the expansion of the internal energy in a series in components of the strain tensor ε_{ij} and S. It was taken into account that F depends on ε_{ij} only in terms of invariant of this tensor. The following representation is generally valid:

$$F = \frac{1}{2}\mu\left(u_1^2 + u_2^2\right) + \left(\frac{1}{2}\lambda + \mu\right)u_3^2 + au_3^3 + bu_3\left(u_1^2 + u_2^2\right) - \quad (7.4.4)$$
$$-\frac{h}{4}\left(u_1^2 + u_2^2\right)^2 + \rho_0 T_0(S - S_0).$$

Here, μ, λ, a, b, and h are constant coefficients that determine the elastic properties of the medium (in the isotropic case μ and λ are the Lamé coefficients), ρ_0, T_0, and S_0 are the density, the temperature, and the entropy in a certain state chosen as the initial one. In

Eq. (7.4.4) we retain only those terms of the fourth power of u_i that affect the principal nonlinear effects occurring in the waves. Among the terms containing the entropy we retain only the principal term, since the change in the entropy $S - S_0$ turns out to be very small in the waves considered below. For incompressible media there are no terms containing u_3 in expression (7.4.4).

Hanyga (1976) and Lenskii (1981, 1982, 1983a, 1983b) considered finite-amplitude nonlinear isotropic waves ($q = 0$) in elastic media for different choice of the function F. We should also mention the investigation of magnetohydrodynamic waves (see Kulikovskii and Lyubimov (1965) and references therein), since magnetohydrodynamics of perfectly conducting media is a particular case of elasticity. In all these cases the behavior of the waves manifests certain common features. In particular, in the case of the wave isotropy there exist so-called rotational waves in which $u_1^2 + u_2^2 = \text{const}$. These waves can be both continuous, when they propagate preserving their shape, and discontinuous with $\{S\} = 0$. These discontinuities propagate at the same velocity as continuous waves. The presence of these waves determines a special form of evolutionary conditions imposed on the shocks. Moreover, some peculiarities take place in solutions of the structure problems. Below, we shall investigate these problems by comparing the cases with $q \neq 0$ and $q = 0$.

Consider small-amplitude waves in the presence of small anisotropy and choose equalities (7.4.3) and (7.4.4) as the equations of state. The function $p(u_k)$ occurring in (7.4.3) can also be expanded in a power series. For small q we can restrict our attention to the quadratic terms. By rotating the axes u_1 and u_2, we can reduce the dependence of the anisotropic part of the internal energy on u_1 and u_2 to the form

$$p = \frac{1}{2}(u_2^2 - u_1^2). \tag{7.4.5}$$

In isotropic media linear waves can be divided into longitudinal and transverse waves. Only u_3 is variable in the longitudinal waves, whereas the transverse waves exhibit variation only in u_1 and u_2. Thus, for small nonlinearity and anisotropy the waves can be subdivided into "quasilongitudinal" and "quasitransverse" waves.

7.4.2 Quasilongitudinal waves.

Though in quasilongitudinal waves u_1 and u_2 are variable together with u_3, the ratios of variations of u_1 and u_2 to that of u_3 are not greater than $\max(\varepsilon, g)$ by the order of magnitude. Here, $\varepsilon = \max(u_1, u_2)$. Quasilongitudinal waves propagating in the positive (or negative) direction of x correspond to a single family of characteristics. Taking into account only the principal order of nonlinearity, one can describe the variation of quantities in quasilongitudinal small-amplitude waves propagating through a homogeneous background by the equations

$$\frac{\partial u_3}{\partial t} + \lambda_3(u_3)\frac{\partial u_3}{\partial x} = 0, \quad \lambda_3^2(u_3) = \frac{1}{\rho_0}\left(\frac{1}{2}\lambda + \mu + 6au_3\right), \tag{7.4.6}$$

where λ, μ, and a are the expansion coefficients in (7.4.4).

In quasilongitudinal waves the quantities u_1 and u_2 can be expressed in terms of u_3. By passing to a moving coordinate system and choosing an appropriate scale for u_3, the coefficient of $\partial u_3/\partial x$ can be made equal to u_3. This reduces Eq. (7.4.6) to the standard form. The results obtained for quasilongitudinal small- and finite-amplitude waves are outlined in the studies performed by Bland (1969), Hanyga (1976), Burenin and Chernyshov (1978), and Lenskii (1982, 1983a, 1983b) for $q = 0$. Sveshnikova (1982) considered small-amplitude waves for $q \neq 0$. In addition to continuous waves, solutions can involve discontinuities (shock waves). In the small-amplitude approximation the latter obey the Lax theorems (see Chapter 1).

7.4.3 Quasitransverse waves. Quasitransverse waves corresponding to two families of characteristics with close velocities (they coincide in the absence of nonlinearity and anisotropy) are of major interest. Complex interaction of nonlinearity and anisotropy manifests itself most clearly in the behavior of the quasitransverse waves. For this reason, the subsequent consideration is completely devoted to these waves. To describe the interaction of two quasitransverse waves moving with close velocities in the positive direction of the x-axis in the case of small nonlinearity and anisotropy, under fairly general conditions on the basis of the complete system (7.4.1) of elasticity equations one can derive the following simplified system of two equations (Kulikovskii 1986; Kulikovskii and Sveshnikova 1995):

$$\frac{\partial u_\alpha}{\partial t} + \frac{\partial}{\partial x}\left(\frac{\partial H}{\partial u_\alpha}\right) = 0, \qquad \alpha = 1, 2; \tag{7.4.7}$$

$$H = \frac{1}{2} f\left(u_1^2 + u_2^2\right) - \frac{1}{4} k\left(u_1^2 + u_2^2\right)^2 + \frac{1}{2} g\left(u_2^2 - u_1^2\right),$$

$$f = \sqrt{\frac{\mu}{\rho_0}}, \qquad k = \frac{1}{\sqrt{\mu\rho_0}}\left(\frac{b^2}{\lambda + \mu} - 2h\right), \qquad g = \frac{q}{2\sqrt{\rho_0\mu}},$$

where k is the coefficient associated with nonlinearity, g is a small parameter characterizing anisotropy, and f is the characteristic velocity in the absence of nonlinearity and anisotropy (i.e., for $g = 0$ and $k = 0$). Only the principal terms responsible for nonlinearity and anisotropy are retained in system (7.4.7). The term corresponding to anisotropy is proportional to g in the expression for $H(u_1, u_2)$. It originates from the similar terms in (7.4.3) and (7.4.5). In the case of the wave anisotropy ($g = 0$) the system (7.4.7) was obtained and used in magnetohydrodynamics (Rogister 1971).

One of the reasons accounting for the wave anisotropy can be the presence of a small preliminary deformation of the isotropic medium. In this case the quantity g turns out to be proportional to the difference $\varepsilon_{22} - \varepsilon_{11}$ between the initial strains in the wave plane.

In deriving the system (7.4.7), we used equalities (7.4.3) and (7.4.4). Assuming that nonzero initial conditions correspond to waves associated only with the two families of the characteristics mentioned above, we approximately solved the equations for perturbations on other characteristics and substituted the result obtained in the remaining two equations for quasitransverse waves. In particular, there is an approximate relation that expresses u_3 in terms of u_1 and u_2:

$$u_3 = -\frac{b}{\lambda + \mu}\left(u_1^2 + u_2^2\right) + \text{const}. \tag{7.4.8}$$

The system (7.4.7) defines the following relations on shock:

$$\left\{\frac{\partial H}{\partial u_\alpha}\right\} = W\{u_\alpha\}, \qquad \alpha = 1, 2, \tag{7.4.9}$$

where $W = dx/dt$ is the velocity of the shock. These equations express the conservation of momentum across shocks. Since the simplified system (7.4.7) does not involve entropy or any other thermodynamic variable, the condition of the entropy nondecrease on the discontinuity must be expressed as the condition of nonincrease in the mechanical energy

$$\frac{1}{2}\left(\left(\frac{\partial H}{\partial u_\alpha}\right)^{\mathrm{L}} + \left(\frac{\partial H}{\partial u_\alpha}\right)^{\mathrm{R}}\right)\{u_\alpha\} \leq 0. \tag{7.4.10}$$

On quasitransverse shocks the relation between u_3 and u_1 and u_2 is given by the same relation (7.4.8) as in continuous quasitransverse waves.

As seen from the expression for H, the coefficient g can have any sign; for example, it can be made positive for an appropriate choice of numbering for the quantities u_1 and u_2. By passing to a moving coordinate system, we can eliminate f. Also by varying the scales of the dependent and independent variables, we can make the quantities g and $|k|$ to be equal to unity. In this case the system of equations and the relations on the discontinuity preserve their form and the function H acquires one of two canonical forms for $k < 0$ and $k > 0$, respectively,

$$H = \frac{1}{2}(u_2^2 - u_1^2) \pm \frac{1}{4}(u_1^2 + u_2^2)^2.$$

In what follows, however, when writing the equations, we shall retain the coefficients g and k for the sake of convenience of studying the dependence of solutions on these physically meaningful parameters. The possibility of reducing H to one of the two canonical forms indicates that there are only two qualitatively different cases of the wave behavior occurring for $k > 0$ or $k < 0$.

Although the majority of the results given below were initially obtained by means of approximate solutions of the complete system (7.4.1), (7.4.2) of elasticity equations, their presentation will be carried out on the basis of the simplified system (7.4.7), (7.4.9).

7.4.4 Riemann waves.

This name is attributed to the solutions in the form

$$u_i = u_i(\theta(x, t)), \tag{7.4.11}$$

where θ is an unknown function of x and t. Substituting (7.4.11) in (7.4.7), we obtain

$$(H_{\alpha\beta} - \lambda\delta_{\alpha\beta})\frac{du_\beta}{d\theta} = 0, \qquad \lambda = -\frac{\partial\theta/\partial t}{\partial\theta/\partial x}, \qquad H_{\alpha\beta} = \frac{\partial^2 H}{\partial u_\alpha \partial u_\beta}. \tag{7.4.12}$$

Here, α, $\beta = 1, 2$, $\delta_{\alpha\beta}$ is the Kronecker delta, and λ is the characteristic velocity. As seen from Eq. (7.4.12), the characteristic velocity must be equal to one of the eigenvalues of the matrix $[H_{\alpha\beta}]$ enumerated so that $\lambda_1 \leq \lambda_2$. Since λ_1 and λ_2, being the eigenvalues of the matrix $[H_{\alpha\beta}]$, are functions of u_γ, from the second of equations (7.4.12) we obtain the following equation for θ

$$\frac{\partial\theta}{\partial t} + \lambda_\alpha(u_\gamma(\theta))\frac{\partial\theta}{\partial x} = 0, \tag{7.4.13}$$

which shows that θ and, consequently, $u_\gamma(\theta)$ are constant on each of the characteristics represented by the straight lines in the (x, t)-plane. They obey the formula

$$\frac{dx}{dt} = \lambda_\alpha(u_\gamma(\theta)). \tag{7.4.14}$$

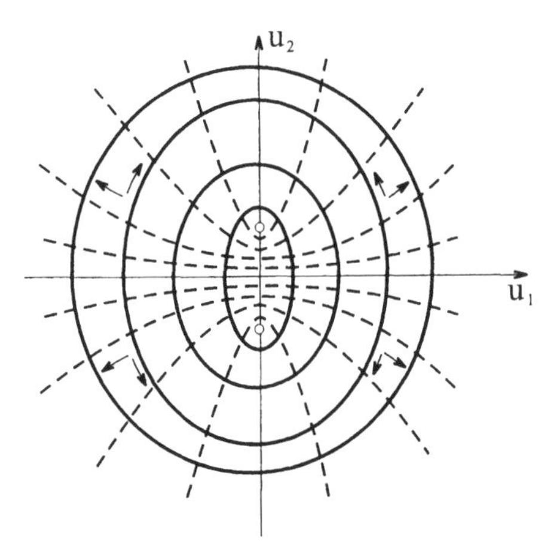

Figure 7.5 Integral curves of quasitransverse Riemann waves in a weakly anisotropic elastic medium.

In the (u_1, u_2)-plane the equality $\lambda_1 = \lambda_2$ is satisfied only at the points $u_1 = 0$ and $u_2 = \pm\sqrt{G}$, $G = 4g/k$ (here we assume that $g/k > 0$, since sign of g can be changed by renumbering the variables u_1 and u_2). The integral curves of the Riemann waves are tangent to the eigenvectors of the matrix $H_{\alpha\beta}$ and form two orthogonal families in the (u_1, u_2)-plane (Fig. 7.5) (Sveshnikova 1982). The axes u_1 and u_2 are the axes of symmetry for the set of integral curves. The solutions corresponding to λ_1 and λ_2 are called slow and fast quasitransverse simple waves, respectively.

Consider Eqs. (7.4.12) for $\lambda = \lambda_1$ and $\lambda = \lambda_2$. The points at which $\lambda_1 = \lambda_2$ are singular points of two families of their integral curves. For $k > 0$ ellipse-like curves correspond to the fast waves ($\lambda = \lambda_2$) and hyperbola-like curves correspond to the slow waves ($\lambda = \lambda_1$). For $k < 0$ the situation is opposite. In the approximation adopted the entire pattern of integral curves of the quasitransverse waves independently of sign of k is always the same to within a scale determined by $\sqrt{G}$. As $g \to 0$ or $u_1^2 + u_2^2 \to \infty$ the integral curves transforms into rays and circles. This corresponds to the wave isotropy case considered by Bland (1969). In Fig. 7.5 the arrows indicate the direction of decrease in λ along the integral curves for $k > 0$. If $k < 0$, λ decreases in the opposite direction. As far as the wave is considered with the variation of quantities described by those fragments of the integral curves along which λ decreases, it is expanding in time, since the leading front of the wave moves faster than the rear front.

7.4.5 Shock waves.

From relation (7.4.9) we can obtain the formulas (Kulikovskii and Sveshnikova 1980, 1982, 1995)

$$(u_1^2 + u_2^2 - R^2)(U_1 u_2 - U_2 u_1) + 2G(u_1 - U_1)(u_2 - U_2) = 0, \tag{7.4.15}$$
$$G = 4g/k, \quad R^2 = U_1^2 + U_2^2$$

and determine the shock velocity W. In (7.4.15) U_1 and U_2 correspond to the quantities ahead of the discontinuity. Behind the discontinuity the variables are still denoted as u_1 and

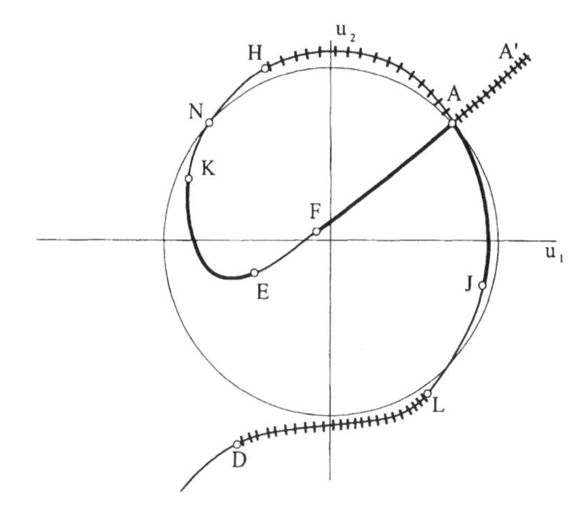

Figure 7.6 Hugoniot curve of a quasitransverse shock: continuous and hatched elements correspond to evolutionary fragments for $k > 0$ and $k < 0$, respectively.

u_2.

The Hugoniot curve in the (u_1, u_2)-plane is defined by Eq. (7.4.15) for fixed values of U_1 and U_2. As can be seen from Eq. (7.4.15) the Hugoniot curve is independent of sign of k, since by our choice $\operatorname{sgn} g = \operatorname{sgn} k$. In Fig. 7.6 we have plotted one of the possible variants of this curve. It corresponds to U_1 and U_2 substantially greater than $\sqrt{G}$. The condition of nonincreasing mechanical energy (7.4.10) or nondecreasing entropy can be written in the form

$$k(u_1^2 + u_2^2 - R^2)\{(u_1 - U_1)^2 + (u_2 - U_2)^2\} \leq 0. \tag{7.4.16}$$

Here, the left-hand side of the inequality is equal (with the opposite sign) to the quantity proportional to the entropy production in the shock. The circle $u_1^2 + u_2^2 = R^2$, where the jump in the entropy changes its sign, will be called the entropy circle. Note that the Hugoniot curve intersects the entropy circle at points with the coordinates $(-U_1, U_2)$ and $(U_1, -U_2)$. The first of them is the point N in Fig. 7.6.

In the case considered, the *a priori* evolutionary conditions for shocks can be reduced to one of the systems of inequalities

$$\begin{cases} \lambda_2^R \leq W, \\ \lambda_1^L \leq W \leq \lambda_2^L, \end{cases} \qquad \begin{cases} \lambda_1 \leq W \leq \lambda_2^R, \\ W \leq \lambda_1^L \end{cases} \tag{7.4.17}$$

defining fast and slow shocks, respectively. Inequalities (7.4.17) can be graphically represented in the evolutionary diagrams as segments belonging to the outlined rectangles. In Figs. 7.7a and 7.7b we have reproduced some variants of the Hugoniot curves in the cases with $k > 0$ and $k < 0$, respectively.

As mentioned in Section 7.1, along the horizontal axis of the evolutionary diagram the shock velocity is always reproduced in the real scale. Along the vertical coordinate the inequalities between W, λ_1^L, and λ_1^L are met only qualitatively. In Figs. 7.6 and 7.7 identical letters refer to the same points. In the shock velocity diagram the initial point A

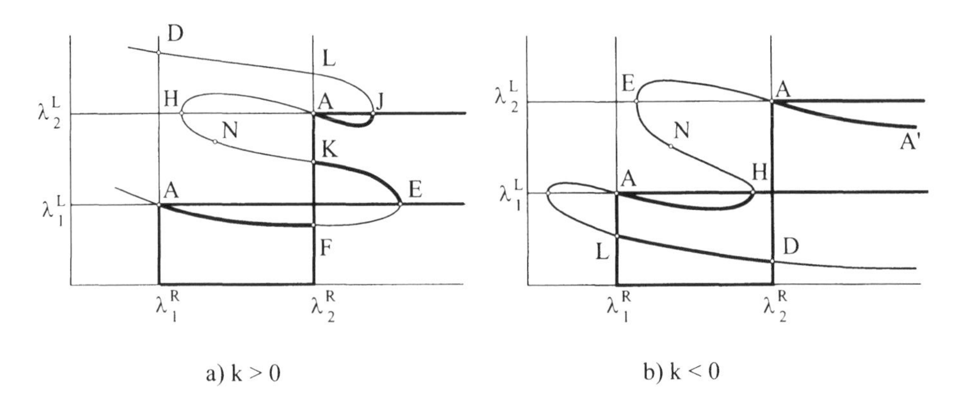

Figure 7.7 Evolutionary diagrams.

$(u_1 = U_1,\ u_2 = U_2)$ is represented by two points corresponding to its location on the two intersecting branches of the Hugoniot curve. In Fig. 7.6 the evolutionary segments of the Hugoniot curve are shown by bold curves for $k > 0$ and by transverse hatching for $k < 0$.

If we consider the Hugoniot curve as a function of $u_1/\sqrt{G}$ and $u_2/\sqrt{G}$, only two parameters, namely, the coordinates of the initial point, remain in Eq. (7.4.15). The evolutionary parts of the Hugoniot curve also depend on sign of k. The variants represented in Figs. 7.7a and 7.7b correspond to the case with fairly large $u_1/\sqrt{G}$ and $u_2/\sqrt{G}$ (more than 2). If we diminish these quantities, the point E in Fig. 7.7a moves to the left and the evolutionary segment KE disappears. Simultaneously, the point H in Fig. 7.7b moves to the right and an additional separate evolutionary segment appears in the fast wave region. More complete description of the shape of the Hugoniot curve was given by Kulikovskii and Sveshnikova (1982, 1995).

In Section 7.1 the points such as F, E, K, J, D, L, and H representing the points at which the shock velocity coincides with one of the characteristic velocities ahead of or behind the shock were called Jouget points. The shocks corresponding to the endpoints of the evolutionary intervals play an important role in constructing self-similar solutions. It is worth noting that for the elastic waves considered here the requirement of the *a piori* evolutionarity implies satisfaction of the condition of nonnegativity of the entropy production.

Along with the evolutionary segments of the Hugoniot curve, the nonevolutionary segments can also be important. For example, Kulikovskii and Sveshnikova (1984) showed that the points of the segment FE (Fig. 7.7a) correspond to one or two combinations of two shocks moving with the same velocities. The nonevolutionary segments adjoining the point A correspond to the states ahead of the shock wave, the state behind which is represented by the point A.

As $g \to 0$ for fixed U_1 and U_2, the Hugoniot curve in the $(u_1,\ u_2)$-plane approaches a circle with center at the origin and a straight line passing through the origin and the initial point. In the limit the evolutionary segment EK of the Hugoniot curve (Fig. 7.7a) occupies a part of the circular arc and the segment of the straight line from the circle to the mid-radius.

It is worth noting that for $g = 0$, first, the *entire* circle corresponds to evolutionary discontinuities whose velocity coincides with a characteristic velocity on either side of the discontinuity front and, second, the *entire* ray issuing from the origin in the direction opposite to that toward the initial point corresponds to nonevolutionary discontinuities. The reason for this qualitative difference between the evolutionary segments of the Hugoniot curve in the limiting case $g \to 0$ and those for $g = 0$ will be discussed below.

7.4.6 Self-similar problems and nonuniqueness of solutions.

Solutions of self-similar problems can serve as a justification test for investigation of simple waves and shocks.

Let the state of the elastic half-space $x > 0$ be given by constant values $u_1 = U_1$ and $u_2 = U_2$ at $t = 0$ and the boundary conditions are given by $u_1 = u_1^*$ and $u_2 = u_2^*$ for $x = 0$ and $t > 0$. This is the so-called "piston" problem. It is necessary to find a solution of Eq. (7.4.7) as a function of $x/t > 0$. The solution should consist of Riemann waves and discontinuities separated by regions of uniform states. In what follows, we assume that only *a priori* evolutionary shocks studied above are admissible. The motivation is given below.

If we go back from the simplified system (7.4.7) to the complete system (7.4.1) of elasticity equations, in the solution of the problem under consideration we encounter two cases. In one of them u_3 is negligibly small both in the initial and boundary conditions. Another one requires us to solve the problem of constructing a system of quasitransverse waves remaining after departure of the longitudinal wave. The solution of this problem was investigated by Kulikovskii and Sveshnikova (1985a, 1985b, 1995) for $k > 0$ and $k < 0$.

In Fig. 7.8 we assumed that $k > 0$ and reproduced one of the variants of the solution for given values of U_1 and U_2 and arbitrary values of u_1^* and u_2^*. The plane of the variables u_1^* and u_2^* is divided into domains in which we indicated the waves constituting the solution. The following notation is used: R_1 and R_2 are slow and fast Riemann waves; S_1 is a slow shock; S_2 and S_2^* are fast shocks corresponding to the segments AJ and EK of the Hugoniot curve, respectively; S_{2J} and S_{2K} are Jouget shocks of the same types as the shocks $A \to J$ and $A \to K$, respectively.

The possibility to construct two different solutions for points belonging to the hatched domain $Q'QPETE'$ (see Fig. 7.8) seems to be of major interest. Each of the two solutions can be continuously continued into one of the neighboring domains. For example, in the domain QPE (see Fig. 7.8) we have the solution $S_2^*S_1$ continuously continued into the domain $QKFP$ across the boundary QP. This solution contains two shocks that separate zones with constant values of u_1 and u_2 in the (x, t)-plane. Moreover, in the domain QEP we have a solution $S_2^*R_2S_{2K}S_1$ continuously continued across the boundary PET. This solution consists of the system of fast waves $S_{2J}R_2S_{2K}$ that directly follow one after the other without separation and of a slow shock S_1 separated from the above system by a zone of constant quantities. Both solutions coincide on the curve QP but are different on the curve PEE'. The variant of the solution represented in Fig. 7.8 corresponds to the Hugoniot curve plotted in Fig. 7.6 and occurs for $k > 0$ and U_1 and U_2 sufficiently greater than $\sqrt{G}$. Note that the nonuniqueness described here originates from the presence of an improper Jouget point on the shock adiabatic curve. It was considered in the previous section from the general viewpoint.

It is of interest that for given U_1 and U_2 as $G \to 0$ the domain of nonuniqueness tends to the interior of an angle with vertex at the origin. However, for $G = 0$ the solution turns out

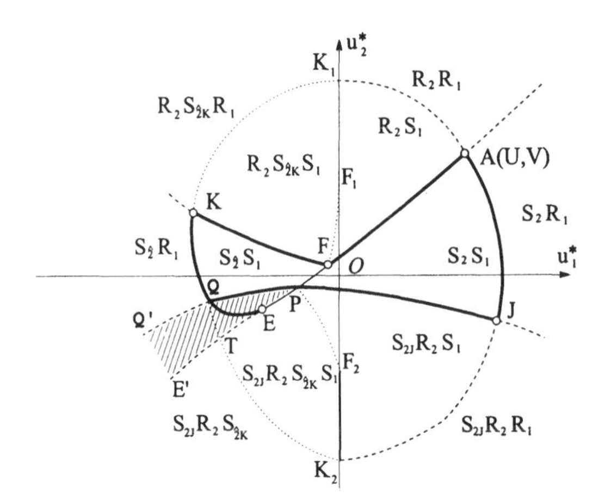

Figure 7.8 Solution of the "piston" problem for $k > 0$: dependence of the solution on the boundary values $u_i|_{x=0} = u_i^*$.

to be unique. An explanation of this illusive contradiction will be given below in Section 7.4.8.

Along with the piston problem, the Riemann problem is also of great interest. This is due to the fact that the solution to this problem is frequently used to calculate numerical fluxes through computational cell volumes. In this case the solution of the problem is sought in the entire space for the following initial conditions:

$$t = 0, \qquad x > 0: \qquad u_1 = U_1, \qquad u_2 = U_2;$$
$$t = 0, \qquad x < 0: \qquad u_1 = u_1^*, \qquad u_2 = u_2^*.$$

This problem can be reduced to solution of two piston problems in the regions $x > 0$ and $x < 0$ with conditions at $x = 0$ unknown in advance. The investigation of this problem (Kulikovskii and Sveshnikova 1988, 1995) also leads to the conclusion on the possibility of the solution nonuniqueness. The necessity of solving the Riemann problem with nonunique solutions can arise in the interaction of quasitransverse waves (both overtaking and colliding ones). This means that the elasticity equations does not allow us to predict uniquely the solution of naturally arising problems. The requirement of smoothness for initial conditions cannot prevent the formation of shocks, the collision of which leads to nonunique continuation of the solution in time. Collision of more than two shocks or any other formation of an arbitrary discontinuity way corresponds to special initial or boundary conditions. These conditions represent, in a certain sense, a set of zero extent among all initial or boundary conditions. Therefore, in order to construct unique solutions of the problems corresponding to smooth initial data in the generic case, it is sufficient to have a rule for selection of solutions to the Riemann problem for a discontinuity arising as a result of interaction of two shock waves.

7.4.7 Waves in viscoelastic media, vanishing viscosity.

To select admissible discontinuities and find unique solutions, we consider the discontinuity structure and the problems similar to those discussed above for a viscoelastic medium (within the framework of the Kelvin–Voigt model). The presence of viscosity leads to appearance of additional terms in Eq. (7.4.7),

$$\frac{\partial u_\alpha}{\partial t} + \frac{\partial}{\partial x}\left(\frac{\partial H}{\partial u_\alpha}\right) = \nu \frac{\partial^2 u_\alpha}{\partial x^2}, \qquad \alpha = 1, 2. \tag{7.4.18}$$

Here, the coefficient ν is assumed to be constant and proportional to the viscosity coefficient. Note that the viscosity coefficient ν occurring in (7.4.18) can be made equal to unity by changing the scales of independent variables without variation of the quantities f and g in expression (7.4.7) for H. By considering an elastic medium as the limit of a viscoelastic medium as $\nu \to 0$ ("vanishing viscosity"), we would like to obtain a selection principle for unique solutions of the problems whose solutions are nonunique in the absence of viscosity.

The shock structures were investigated by Kulikovskii and Sveshnikova (1987) for $k > 0$ and Chugainova (1991) for $k < 0$. Solutions of viscoelastic equations were sought in the form of travelling waves, that is, functions depending on $\xi = -x + Wt$ in which u_α tends to finite limits as $\xi \to \pm\infty$. Clearly, as $\nu \to 0$ the width of the transition zone, where the solution substantially differs from the limiting values, tends to zero and this solution transforms into a discontinuity representing a shock satisfying the conservation laws (7.4.9).

For $k > 0$ the existence of the solutions corresponding to shocks belonging to the segment *EK* of the Hugoniot curve was proved analytically. In a series of numerical calculations it was shown that there always exists a shock structure for all the other evolutionary shocks. Thus, the investigation does not disregard any part of the shock waves as "unreal" or "spurious." At the same time, no discontinuities with structure were found among *a priori* nonevolutionary discontinuities.

A number of time-dependent non-self-similar problems for viscoelastic media were solved numerically by Chugainova (1988, 1993a). For these problems the self-similar solutions considered above represent asymptotics as $t \to \infty$. Consider the half-space $x > 0$, in which the quantities $u_1 = U_1$ and $u_2 = U_2$ are constant at $t = 0$. Let us change the boundary values of u_1 and u_2 at $x = 0$ either instantaneously or within a finite time interval τ so that u_1 and u_2 acquire constant values u_1^* and u_2^* for $t > \tau$. Consider large times for which the characteristic linear dimension of the problem is also large. In this case we can expect that the effect of viscosity and variability of the boundary conditions within the interval τ weakens and the solution will tend to a certain, although not known in advance, self-similar solution of the inviscid problem. This solution may depend on the functions specifying the boundary values of u_1 and u_2 at $0 \leq t \leq \tau$.

A series of calculations showed that the simpler solutions S_2S_1 or S_2R_1 are usually realized at large t in the domain $Q'QPETE'$ in the (u_1^*, u_2^*)-plane (Fig. 7.8). It was possible to obtain complex solutions $S_{2J}R_2S_{2K}S_1$ or $S_{2J}R_2S_{2K}R_1$ only if on a time interval τ_1 we maintained the boundary condition at $x = 0$ in the domain $E'TEPF_2K_2$ (Fig. 7.8). This domain is adjacent to the nonuniqueness domain from below, where only the complex solution exists. This solution might develop during the time τ_1. If τ_1 is sufficiently large and we change the boundary conditions in a way that the point (u_1^*, u_2^*) transfers to the domain of nonuniqueness, perturbations originating from the boundary cannot change the type of the solution.

In addition, Chugainova (1993a) carried out a series of calculations of the interaction between shocks. As a shock here we understand the solution possessing a structure. Two variants of the problem were considered. In the first variant, by satisfying appropriate boundary conditions, the wave transformed with time into the solution representing the shock structure in the region $x > 0$. Then, the boundary conditions were changed and the second shock with its own structure and a higher velocity (that is overtaking the first shock) was similarly created. The final state corresponding to the state behind the second shock was chosen so that it belonged to the domain of nonuniqueness. As a result of interaction between these shocks, once the second wave had overtaken the first one, the solution approached simpler asymptotics represented by the wave patterns S_2S_1 or S_2R_1.

In the second variant, collision of shocks moving in opposite directions was considered. In order to solve the simplified system (7.4.7) instead of the complete system of viscoelasticity equations, the following considerations were used. The phenomena studied are associated with the interaction of the effects of nonlinearity, anisotropy, and viscosity. If the amplitudes of the interacting waves are fairly small, the effects of nonlinearity and viscosity (assumed to be of the same order of magnitude) have no time to manifest themselves during the interaction of the shock structure. Thus, the interaction between shocks moving in opposite directions can be considered as the collision of linear inviscid waves. After collision each wave will move through a new surrounding, as it would have climbed up the other wave. However, in this case the states ahead of and behind this wave will not satisfy relations (7.4.9) on discontinuity, and the action of small effects of nonlinearity, anisotropy, and viscosity finally leads to disintegration of this wave. The smaller the corresponding terms in Eq. (7.4.18), the longer the disintegration time.

In the calculations the parameters of the colliding waves were intentionally chosen so that the problem had *a priori* unknown asymptotics. The calculations showed that in all considered collisions of shocks moving in opposite directions the solution on each side of the collision position also approached simpler asymptotics containing two waves.

As a result of the described numerical experiments with viscoelastic quasitransverse waves we can make the following conclusions on the asymptotic behavior of solutions for cases in which the corresponding inviscid problem has more than one solution. First, by setting initial or boundary conditions (for example, by suitably "smearing" the time variation of the boundary conditions in the piston problem or the space variation of the initial conditions in the Riemann problem), we can attain solutions tending to any of the two asymptotics. Second, the simpler asymptotic consisting of two waves in some sense has larger "domain of attraction," since the solution tends to precisely this asymptotic if we do not apply measures to enforce special initial or boundary conditions. Third, in all calculated cases of interaction of two shocks (overtaking or colliding ones), the solution tends to the simpler asymptotic.

These calculations allow us to introduce a hypothesis according to which the simpler asymptotic must always develop in the interaction of two shock waves.

Therefore, if an elastic medium is the limit of a viscoelastic medium as viscosity tends to zero, the initial conditions specified for all x are continuous and, moreover, correspond to the "generic case," then the suggested hypothesis makes it possible to construct a unique solution.

However, we should note that this hypothesis still requires a proof or more detailed numerical tests. In addition, all the situations considered in this section refer to viscoelastic media. The solution selection principles can be different if other dissipation mechanisms exist in the medium (see the subsequent sections of this chapter).

7.4.8 Role of the wave anisotropy and passage to the limit $g \to 0$. We noted above that only a single, more complex solution must be retained in the region $E'TEPF_2K_2$ adjoining the domain of nonuniqueness from below in the (u_1^*, u_2^*)-plane. Disappearance of one of two different solutions, when the point (u_1^*, u_2^*) crosses the boundary $PETE'$ of these domains from above downward, results in a certain specific behavior of the solution that can be called catastrophic restructuring. Note that on the upper boundary $QQ'P$ of the domain of nonuniqueness, self-similar solutions do not differ one from another and such a phenomenon does not occur.

Catastrophic restructuring can occur (Kulikovskii 1989), for example, in the case with $k > 0$ for small values of $\Gamma = G/(u_1^2 + u_2^2)$ in the process of evolution of a fast simple wave represented in the (u_1, u_2)-plane by an integral curve of the ellipsoidal shape. This integral curve is close to a circle for small Γ (see Fig. 7.5). As a result of this evolution, a shock wave S_2 arises; the smaller Γ, the slower this process. If the point representing the state behind the shock wave S_2 moves along the Hugoniot curve approaching the point E, the shock S_2 generates a slow simple wave R_1 that lags behind S_2. When the point representing S_2 arrives at the point E along the Hugoniot curve, the wave S_{2E} disintegrates into the combination $S_{2J}R_2S_{2K}S_1$, the shock S_1 moving ahead of the wave R_1 generated earlier. For small Γ the profile of the slow compound wave, i.e., the combination S_1R_1, has a characteristic shape. The amplitude of thus generated slow waves is of the order of $\max(u_1, u_2)$. As $\Gamma \to 0$, it tends to $\sqrt{U_1^2 + U_2^2}/2$; the effective width of the profile and its duration tend to zero. This phenomenon was simulated numerically by Chugainova (1993b).

Both this phenomenon and the nonuniqueness of the solution occur only for $g \neq 0$, whereas for $g = 0$ they are absent. In this connection we shall consider the difference between the cases $g = 0$ and very small g in more detail. We start from the question about evolutionarity of the shock waves. As noted above, a part of the Hugoniot curve tends to coincide with the isentropic circle as $g \to 0$. In this limit the velocity of the corresponding shocks coincides with the fast (for $k > 0$) characteristic velocities on either side of the discontinuity. As a result, as $g \to 0$, the decay rate of nonevolutionary waves corresponding to points on the Hugoniot curve belonging to the vicinity of the isentropic circle diminishes. For $g = 0$ this process never occurs and the entire circle corresponds to evolutionary discontinuities.

A part of the segment EK of the Hugoniot curve lying on the straight line AO is evolutionary for $g \neq 0$. We shall now explain why it becomes nonevolutionary in the limit as $g \to 0$. Let us consider the interaction between the shock S_2 and small fast perturbations. For small g fast perturbations correspond to a displacement of the point (u_1, u_2) approximately along a circle with center at the origin, whereas slow perturbations correspond to approximately radial displacements. As can be seen from Fig. 7.8, if for small g a small fast perturbation overtakes the shock wave S_2 corresponding to a point fairly distant from the isentropic circle, this leads to some displacement of the point representing the shock along the Hugoniot curve and generation (reflection) of a slow wave whose amplitude increases with $|dr/d\theta|$. Here, $r = r(\theta)$ is the formula describing the Hugoniot curve in polar coordinates. At $g = 0$ this derivative becomes indefinitely large. This corresponds to an infinitely large reflection coefficient, that is, to nonevolutionarity. At the same time, one must note another circumstance. Consider the region of the quantities u_1^* and u_2^* in which we construct the solution using shocks belonging to the segment *EK* located fairly far from the isentropic circle. Then if we solve the self-similar problem, the size of this region decreases as $g \to 0$ and is of zero extent for $g = 0$.

Let us consider the behavior of two different solutions of the same problem when the solution is nonunique as $g \to 0$. For this purpose we shall consider the situation in which the point (u_1^*, u_2^*) is fixed and lies inside the nonuniqueness domain. Passing to the limit, the velocities of the fast waves involved in the two solutions tend to the same value equal to the velocity of a rotational wave. The states behind the fast shock S_2 and behind the combination of the fast waves $S_{2J}R_2S_{2K}$ also tend to coincide. Therefore, for small g the domain of the (x, t)-plane in which the solutions differ by a finite value is a narrow sector with the vertex at the origin. The vertex angle of this sector tends to zero as $g \to 0$. At $g = 0$ the solutions coincide and, consequently, the self-similar problem has a unique solution.

7.4.9 Final conclusions.

Let us summarize the features of the problems associated with quasitransverse waves in anisotropic elastic media. First, solutions of even very simple self-similar problems are nonunique. Second, solutions of these problems do not depend continuously on the determining parameters and admit the possibility of sudden restructuring accompanied by disintegration of shocks under the action of small perturbations. We can also note that the above-mentioned phenomena occur even for small-amplitude waves in elastic media of the most general equation of state. The properties described are not only intrinsic in the fairly complex system of elasticity equations but also in the simple second-order system (7.4.7) describing the propagation of quasitransverse waves in the same direction. The case $g = 0$, in which there is no nonuniqueness, seems to be a degenerate one.

Note that these properties were obtained under the assumption that only *a priori* evolutionary shock waves can be used to construct the solutions. This is true if the elastic medium is considered as a limit of a viscoelastic medium.

It is also worth noting that nonuniqueness of solutions must be treated as "physically" unremovable. In fact, a non-self-similar solution for the system with nonzero viscosity can tend with time to one of the two possible asymptotics representing self-similar solutions of the inviscid problem. The resulting solution depends on details in setting the initial or boundary conditions. These details relates to small time (or space) intervals that tend to zero under vanishing viscosity. However, as can be concluded on the basis of the performed numerical experiments, the asymptotics of solutions to the problems on the interaction of shock waves is always associated with a single (simpler) type.

7.5 Electromagnetic shock waves in ferromagnets

In this section we consider plane nonlinear electromagnetic waves in nonconducting ferromagnets at rest. For long waves whose dispersion and dissipation are negligible, the governing equations constitute a hyperbolic system (system of the Maxwell equations). It can be reduced to the form coinciding with the nonlinear elasticity equations with the similar relations on discontinuities. Below we consider the cases in which the magnetic field is inclined at a small angle to the shock normal. In such cases all the relations on electromagnetic shock waves coincide with those on sufficiently weak shocks in elastic media. Thus, all the results obtained in the previous section for the Riemann waves, Hugoniot curves, and *a priori* evolutionary shocks turn out to be valid for the electromagnetic waves.

In order to describe shorter waves and, in particular, the structure of the electromagnetic waves, we use the Landau–Lifshitz equation describing the variation of the magnetization vector. In this case the structure of discontinuities and the set of admissible discontinuities differ significantly from those occurring in the case of viscoelasticity equations considered in the previous section. For the electromagnetic waves considered here, this set has a complicated structure and, in particular, contains *a priori* nonevolutionary discontinuities. Nonlinear electromagnetic waves in nonconducting ferromagnets have earlier been studied from various viewpoints by Bogatyrev (1974), Kataev (1963), and Gaponov, Ostrovskii, and Freidman (1967). It is essential that in what follows we simultaneously take into account nonlinearity, dispersion, dissipation, and anisotropy of the medium (Gvozdovskaya and Kulikovskii 1997). The waves for which the magnetic field is inclined at finite angle to the wave normal were studied by Gvozdovskaya and Kulikovskii (1999b).

7.5.1 Long-wave approximation. Elastic analogy.

The basis of the long-wave approximation is given by the Maxwell equations and the equilibrium relation between the magnetic field strength **H** and the magnetic induction **B**. Assume that the medium is not polarizable, that is, the electric field strength **E** coincides with the electric displacement vector **D** and there are no electric charges. Then for plane waves the Maxwell equations yield

$$\frac{1}{c}\frac{\partial E_1}{\partial t}+\frac{\partial H_2}{\partial x}=0,\quad \frac{1}{c}\frac{\partial E_2}{\partial t}-\frac{\partial H_1}{\partial x}=0,\quad \frac{\partial E_3}{\partial t}=0,\quad \frac{\partial E_3}{\partial x}=0,$$

$$\frac{1}{c}\frac{\partial B_2}{\partial t}+\frac{\partial E_1}{\partial x}=0,\quad \frac{1}{c}\frac{\partial B_1}{\partial t}-\frac{\partial E_2}{\partial x}=0,\quad \frac{\partial B_3}{\partial t}=0,\quad \frac{\partial B_3}{\partial x}=0. \tag{7.5.1}$$

Here, we used the standard notation and the Gaussian system of units. The waves are assumed to propagate along the $x_3 \equiv x$ axis of Cartesian coordinate system. From these equations it follows that $E_3 = \text{const}$ and $B_3 = \text{const}$. These quantities must be found from external conditions. In order to clarify the analogy with the elasticity theory, we change the scale of t or x so that $c = 1$ in the new system of units. In addition, we set $E_2 = \varepsilon_1$ and $E_1 = -\varepsilon_2$. Then the equations acquire the form

$$\frac{\partial \varepsilon_\alpha}{\partial t}=\frac{\partial H_\alpha}{\partial x},\qquad \frac{\partial B_\alpha}{\partial t}=\frac{\partial \varepsilon_\alpha}{\partial x},\qquad \alpha=1,2. \tag{7.5.2}$$

These equations are invariant under simultaneous multiplication of the units for the quantities $\varepsilon_\alpha, H_\alpha$, and B_α by the same factor. We choose these units so that the values of u_α, H_α, and B_α are less than the Gaussian ones by the factor of $\sqrt{4\pi}$. In this case the total energy per unit volume of the field and medium can be written in the form

$$U = \frac{1}{2}\left(\varepsilon_1^2 + \varepsilon_2^2\right) + \Phi(B_\alpha) + \rho_0 T_0 (S - S_0). \tag{7.5.3}$$

Note that the magnetic part of the energy Φ also depends parametrically on B_3. The last term of Eq. (7.5.3) represents the thermal energy per unit volume of the medium. The thermal energy is represented by the additive term, as it was in the case of small-amplitude nonlinear elastic waves. This is associated with the fact that in what follows we will consider either the waves with S = const or with a very small change in S. As well known, H_i can be expressed in terms of derivatives of the magnetic energy Φ with respect to B_i. In the chosen system of units these expressions can be written in the form

$$H_\alpha = \frac{\partial \Phi}{\partial B_\alpha}. \tag{7.5.4}$$

Equations (7.5.2) and (7.5.4) coincide with the first two differential equations of (7.4.1) describing elastic waves in incompressible media. In this case the role of the quantities v_α and u_α will be played by the quantities ε_α and B_α, respectively.

Moreover, if the process of magnetization is reversible (in what follows we adopt this for fairly slow processes), the equation

$$\frac{\partial S}{\partial t} = 0 \tag{7.5.5}$$

must be added to Eqs. (7.5.2) and (7.5.4).

In many cases the magnetic energy of the medium can be represented in the form suitable for both equilibrium and nonequilibrium states. We assume the energy to be a function of two independent arguments

$$\Phi_m = \Phi_m(B_i, M_k), \qquad M_i = B_i - H_i, \qquad i, k = 1, 2, 3. \tag{7.5.6}$$

Here, the components M_i of the magnetization vector are defined so that their values are less than those in the Gaussian system of units by the factor of $\sqrt{4\pi}$. In equilibrium states (in particular, those ahead of and behind a discontinuity) the magnetic energy can be represented as

$$\Phi(B_i) = \min_{M_k} \Phi_m(B_i, M_k), \qquad i, k = 1, 2, 3 \tag{7.5.7}$$

and for given values of B_i the quantities M_k can be found from the equations

$$\frac{\partial \Phi_m}{\partial M_k} = 0. \tag{7.5.8}$$

For magnetic fields exceeding the saturation threshold, the magnitude of the magnetization vector can be assumed to be constant: $|\mathbf{M}| = M$ = const (Landau and Lifshitz 1984; Maugin 1988; Gurevich and Melkov 1994). In what follows, the quantities M_α, $\alpha = 1, 2$

are chosen as independent variables. It is obvious that $M_3 = \sqrt{M^2 - M_1^2 - M_2^2}$. If the energy of the magnetic field is mainly determined by the mutual orientation of the vectors **B** and **M** and depends only slightly on the mutual orientation of the vector **M** with the crystal lattice (this is a typical case), then by taking into account the equalities $B_3 = \text{const}$ and $M = \text{const}$, the expression for the internal energy can be written, to within a constant term, in the form (Landau and Lifshitz 1984; Maugin 1988)

$$\Phi_m(B_\alpha, M_\alpha) = \frac{1}{2}\left(B_1^2 + B_2^2\right) - B_\alpha M_\alpha - B_3\sqrt{M^2 - M_1^2 - M_2^2} + g\varphi(M_\alpha). \qquad (7.5.9)$$

Here, the form of the function $\varphi(M_\alpha)$ is determined by the properties of the medium and g is a small parameter. In what follows, we neglect the terms of the order of g^2. Two terms in the middle of the right-hand side of (7.5.9) represent the scalar product $\mathbf{M} \cdot \mathbf{B}$. In order to compare these expression with the standard form, note that $\mathbf{M} \cdot \mathbf{B} = \mathbf{M} \cdot \mathbf{H} + \text{const}$ since $\mathbf{B} = \mathbf{H} + \mathbf{M}$ and $M^2 = \text{const}$.

Taking into account the first equality of (7.5.6), from (7.5.9) and (7.5.8) we obtain

$$\frac{\partial \Phi_m}{\partial B_\alpha} = \frac{\partial \Phi}{\partial B_\alpha} = B_\alpha - M_\alpha = H_\alpha, \qquad (7.5.10)$$

that is, relations (7.5.4) are satisfied.

Since g is small, taking into account equality (7.5.7), from (7.5.9) we obtain

$$\Phi(B_\alpha) = \frac{1}{2}\left(B_1^2 + B_2^2\right) - M\sqrt{B_3^2 + B_1^2 + B_2^2} + g\varphi^*(B_\alpha). \qquad (7.5.11)$$

Here, φ^* is the value acquired by $\varphi(M_\alpha)$ if the vectors **B** and **M** are parallel (terms of the order higher than or equal to g^2 are neglected).

In what follows, we will consider the case in which the vector **B** is inclined at a small angle to the x_3-axis and B_α are small. Then the function Φ determined by equality (7.5.11) can be expanded in a power series in terms of B_α,

$$\Phi = \frac{1}{2}\left(1 - \frac{M}{B_3}\right)\left(B_1^2 + B_2^2\right) + \frac{M}{8B_3^3}\left(B_1^2 + B_2^2\right)^2 + \frac{g_1}{2}\left(B_1^2 - B_2^2\right). \qquad (7.5.12)$$

Since g is small, in the last term obtained from the expansion of $g\varphi^*$ we retain only the terms quadratic in B_α (linear terms that do not affect the behavior of discontinuities were omitted). The term B_1B_2 was eliminated by rotation of the B_1 and B_2 coordinate axes. In addition, the term containing $B_1^2 + B_2^2$ was added to the first terms of (7.5.12).

There are materials (for example, ferromagnetic cubic crystals) for which $\varphi^*(B_\alpha)$ is a quadratic form with respect to B_1 and B_2, regardless of smallness of B_α (Gurevich and Melkov 1994).

Along with continuous waves, the Maxwell equations admit discontinuities. Relations on them result from the integral representation of these equations. In the chosen system of units they read

$$W\{\varepsilon_\alpha\} + \{H_\alpha\} = 0, \quad W\{B_\alpha\} + \{\varepsilon_\alpha\} = 0, \quad \{B_3\} = 0, \quad \{E_3\} = 0, \qquad (7.5.13)$$

where W is the velocity of discontinuity divided by the speed of light c. These relations coincide with the relations on discontinuities in elastic media.

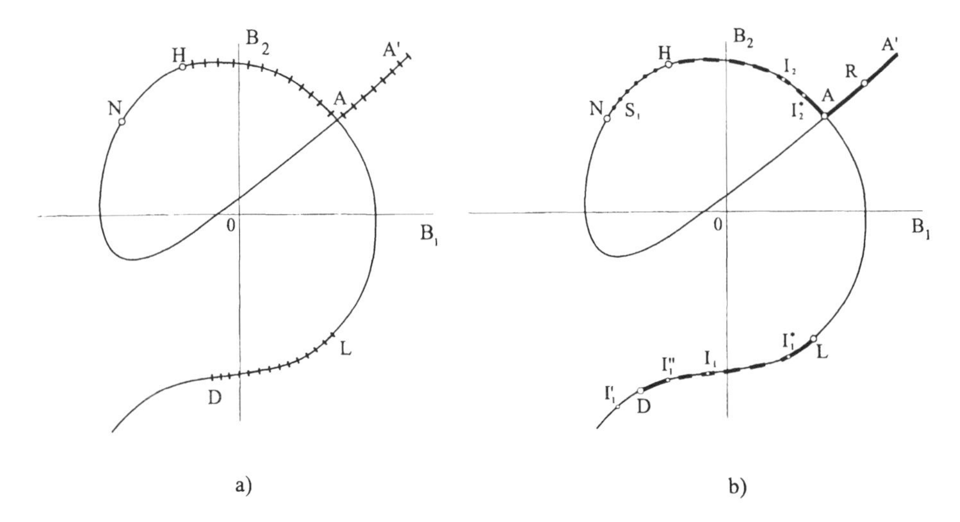

Figure 7.9 (a) *A priori* evolutionary fragments (hatched) of a Hugoniot curve. (b) Admissible part of a Hugoniot curve for waves in anisotropic ferromagnets: bold curves and black dots.

Note that in the framework of the above-mentioned analogy with elastic waves, expansion (7.5.12) shows that electromagnetic waves correspond to $k < 0$. Accordingly, to facilitate the comparison with the results of the preceding Section 6.4 we changed the numeration of the axes x_1 and x_2. This can be seen from the last term of Eq. (7.5.12). All the results obtained for this case in Section 7.4 concerning the Riemann waves, Hugoniot curves, and *a priori* evolutionary shocks can be extended as they are to the electromagnetic waves.

Let us recall the shape of the Hugoniot curve and its mapping onto the evolutionary diagram for $k < 0$ (Figs. 7.9a and 7.10). In Fig. 7.9a the *a priori* evolutionary segments (AA' corresponds to fast waves, and AH and LD to slow waves) are reproduced. The Hugoniot curve is mapped into the evolutionary diagram in Fig. 7.10.

7.5.2 Structure of electromagnetic shock waves. In the ferromagnets the variation of the magnetization vector is governed by the Landau–Lifshitz equation (Landau and Lifshitz 1984),

$$\frac{\partial \mathbf{M}}{\partial t} = \gamma(\mathbf{M} \times \mathbf{H}_{ef}) - \lambda \mathbf{H}_{ef}. \tag{7.5.14}$$

The vector $\mathbf{H}_{ef}$ is the gradient of the function $\Phi_m(B_i, M_k)$ with respect to the variables M_k for fixed B_α. In the approximation considered $|\mathbf{M}| = M = \text{const}$. Thus, the gradient is calculated on the surface of the sphere $M_1^2 + M_2^2 + M_3^2 = M^2$ and lies in a plane tangent to the sphere. The parameters γ and λ are determined by the properties of material. In various materials $\lambda/(\gamma M)$ ranges from 10^{-2} to 5×10^{-5}. The last term of the Landau–Lifshitz equation is usually written in the form $\lambda \mathbf{M} \times (\mathbf{M} \times \mathbf{H}_{ef})/M^2$ coinciding with that written above, since $\mathbf{M} \cdot \mathbf{H}_{ef} = 0$.

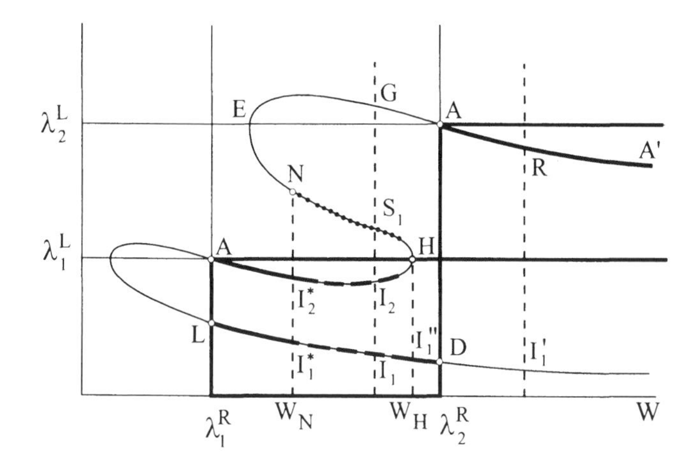

Figure 7.10 Admissible part of a Hugoniot curve on the evolutionary diagram. The notation is the same as in Fig. 7.9.

The Landau–Lifshitz equation has been successfully used for description of rapidly moving waves by Gurevich and Melkov (1994). There exists an experimental confirmation of the Landau–Lifshitz damping in rigid magnets (Landau and Lifshitz 1984). In the isotropic case without dissipation the behavior of solitons was investigated on the basis of these equations by Nakata (1991). More complicated systems of equations were used for description of the motion of magnetic domain walls or slow waves by Maugin (1988) and Kosevich, Ivanov, and Kovalev (1990). In this case the dependence of Φ_m on the derivatives of the magnetic field strength was also taken into account. For this reason the terms with second-order derivatives $\partial^2 M_\alpha/\partial x^2$ additionally appeared in Eq. (7.5.14). In order to neglect these effects, we must require that the wavelength l satisfy the condition $l = cW/\gamma H_{ef} \gg l_*$, where W is the wave velocity and l_* is the thickness of the domain wall ($l_* \sim 10^{-6}$ cm). This condition is assumed to be satisfied.

Let us chose the variables M_α as curvilinear coordinates on the sphere $|\mathbf{M}| = M$. Then, using Eq. (7.5.9) and projecting Eq. (7.5.14) on the M_1- and M_2-axes, we obtain

$$\frac{\partial M_1}{\partial t} = -\lambda \frac{M^2 - M_1^2 - M_2^2}{M^2 - M_2^2} \left(\frac{\partial \Phi_m}{\partial M_1} \right)_{B_\alpha} - \gamma M_3 \left(\frac{\partial \Phi_m}{\partial M_2} \right)_{B_\alpha}, \tag{7.5.15}$$

$$\frac{\partial M_2}{\partial t} = \gamma M_3 \left(\frac{\partial \Phi_m}{\partial M_1} \right)_{B_\alpha} - \lambda \frac{M^2 - M_1^2 - M_2^2}{M^2 - M_1^2} \left(\frac{\partial \Phi_m}{\partial M_2} \right)_{B_\alpha}. \tag{7.5.16}$$

Multiplying these equations by $(\partial \Phi_m/\partial M_1)_B$ and $(\partial \Phi_m/\partial M_2)_B$, respectively, and summing, we obtain

$$\left(\frac{\partial \Phi_m}{\partial M_\alpha} \right)_{B_\alpha} \frac{\partial M_\alpha}{\partial t} \tag{7.5.17}$$

$$= -\lambda (M^2 - M_1^2 - M_2^2) \left(\frac{1}{M^2 - M_2^2} \left(\frac{\partial \Phi_m}{\partial M_1} \right)_{B_\alpha}^2 + \frac{1}{M^2 - M_1^2} \left(\frac{\partial \Phi_m}{\partial M_2} \right)_{B_\alpha}^2 \right) \leq 0.$$

Note that the left-hand side of Eq. (7.5.17) is equal to the total derivative of the magnetic energy for $\mathbf{B} = \text{const}$. This quantity represents the energy dissipation associated with the variation of the magnetic moment, which is proportional (with the opposite sign) to the entropy production.

As follows from Eqs. (7.5.15)–(7.5.16), the left-hand side of inequality (7.5.17) is equal to zero only if $\partial M_1/\partial t = 0$ and $\partial M_2/\partial t = 0$.

We will seek the solution of the problem of the electromagnetic shock wave structure in the form of a travelling wave in which all the quantities depend on the variable $\xi = Wt - x$. The values of $\xi = -\infty$ and $\xi = \infty$ correspond to the states ahead of and behind the discontinuity, and W is the dimensionless velocity of the wave under consideration. The solution can be determined from the differential equations (7.5.2), (7.5.15), and (7.5.16) in which $\partial/\partial t$ and $\partial/\partial x$ should be substituted by $Wd/d\xi$ and $-d/d\xi$, respectively. In this case Eq. (7.5.2) can be integrated. By virtue of the equality $B_\alpha = H_\alpha + M_\alpha$ and the conditions for $\xi \to -\infty$ we obtain

$$B_\alpha(\xi) = B_\alpha^{\mathrm{R}} + \frac{1}{1 - W^2}\left(M_\alpha(\xi) - M_\alpha^{\mathrm{R}}\right), \tag{7.5.18}$$

where B_α^{R} and M_α^{R} are the values of B_α and M_α at $\xi = -\infty$. This makes it possible to introduce a function $\Phi^*(M_1, M_2)$ such that

$$\Phi^*(M_1, M_2) = -\frac{(M_1 + P_1)^2 + (M_2 + P_2)^2}{2(1 - W^2)} - \tag{7.5.19}$$

$$-B_3\sqrt{M^2 - M_1^2 - M_2^2} + g\varphi(M_1, M_2),$$

$$P_\alpha = M_\alpha^{\mathrm{R}} - (1 - W^2)B_\alpha^{\mathrm{R}} = M_\alpha^{\mathrm{R}}\left(1 - \frac{1 - W^2}{M_3^{\mathrm{R}}}B_3\right).$$

This function is independent of current values of B_1 and B_2. By virtue of relations (7.5.18) we can write

$$\left(\frac{\partial \Phi_m}{\partial M_\alpha}\right)_{B_\gamma} = \frac{\partial \Phi^*(M_1, M_2)}{\partial M_\alpha}. \tag{7.5.20}$$

Equalities (7.5.20) make it possible to represent the equations describing the electromagnetic shock wave structure and relation (7.5.17) in the form

$$W\frac{dM_1}{d\xi} = -\lambda_1\frac{\partial \Phi^*}{\partial M_1} - \gamma M_3\frac{\partial \Phi^*}{\partial M_2}, \tag{7.5.21}$$

$$W\frac{dM_2}{d\xi} = \gamma M_3\frac{\partial \Phi^*}{\partial M_1} - \lambda_2\frac{\partial \Phi^*}{\partial M_2}, \tag{7.5.22}$$

$$\lambda_\alpha = \lambda\frac{M^2 - M_1^2 - M_2^2}{M^2 - M_\alpha^2}, \qquad \alpha = 1, 2,$$

$$W\frac{d\Phi^*}{d\xi} = -\lambda_1\left(\frac{\partial \Phi^*}{\partial M_1}\right)^2 - \lambda_2\left(\frac{\partial \Phi^*}{\partial M_2}\right)^2 \le 0. \tag{7.5.23}$$

Since in accordance with (7.5.18) the derivatives of M_α with respect to ξ are proportional to the derivatives of B_α and taking into account that for $\xi \to \pm\infty$ all the derivatives in the solution of the shock structure problem must vanish, the states at $\xi = \pm\infty$ may correspond only to singular points of the system (7.5.21)–(7.5.22).

The singular points of the system (7.5.21)–(7.5.22) coincide with the critical points of the function $\Phi^*(M_1, M_2)$ at which its partial derivatives are equal to zero. By virtue of equalities (7.5.20) the derivatives $\partial\Phi(M_\alpha, B_\beta)/\partial M_\gamma$ also vanish at these points. In accordance with Eq. (7.5.8) this means that the singular points of the system (7.5.21)–(7.5.22) correspond to equilibrium states.

If we eliminate the electric field components from the relations on discontinuity (7.5.13), then the equality $M_i = B_i - H_i$ allows us to represent them in the form

$$B^{\mathrm{L}}_\alpha = B^{\mathrm{R}}_\alpha + \frac{1}{1 - W^2}\left(M^{\mathrm{L}}_\alpha - M^{\mathrm{R}}_\alpha\right). \tag{7.5.24}$$

Equalities (7.5.24) differ from (7.5.18) only by the presence of the superscript "L." To each variable M_α we can assign the variable B_α using equalities (7.5.18). Then transition from one of the singular points of the system (7.5.21)–(7.5.22) to another ensures satisfaction of the relations on electromagnetic shock waves with thermodynamically equilibrium states ahead of and behind the discontinuity.

Relations (7.5.24) show that for a given value of W the arrangement of the singular points of the system (7.5.21)–(7.5.22) in the (M_1, M_2)-plane differs from that of the points characterizing the possible states ahead of and behind the discontinuity in the (B_1, B_2)-plane only by scale and translation. This makes it possible to use the results of Section 7.4 for determining the arrangement of the singular points.

The left-hand side of inequality (7.5.23) (which coincides with the left-hand side of (7.5.17)) represents dissipation of the magnetic energy. In the approximation adopted it coincides with the entropy production to within the multiplier $\rho_0 T_0$. Thus, the relations $\{\Phi^*\} > 0$ and $\{\Phi^*\} < 0$ are satisfied on the Hugoniot curve inside and outside the entropy circle, respectively. In particular, $\{\Phi^*\} = 0$ at point N with the coordinates $(-B^{\mathrm{R}}_1,\ B^{\mathrm{R}}_2)$.

In accordance with inequality (7.5.23) the function $\Phi^*(M_1, M_2)$ decreases along all the integral curves of the system (7.5.21)–(7.5.22) intersecting the level lines of this function in the (M_1, M_2) plane. In this case the smaller the ratio λ/γ, the smaller the angles between integral curves and level lines.

Note that at $\lambda = 0$, that is, in the absence of dissipation, the integral curves coincide with the level lines of the function $\Phi^*(M_\beta)$. In addition, closed level lines of the function $\Phi^*(M_1, M_2)$ that do not approach the critical points correspond to solutions of the system (7.5.21)–(7.5.22) periodic in ξ, that is, to periodic undamped waves. The level lines whose two endpoints arrive at a singular point correspond to solitary waves.

Owing to relation (7.5.23), types of the singular points of the system (7.5.21)–(7.5.22) are obviously associated with types of the corresponding critical points of the function $\Phi^*(M_\alpha)$. If $\Phi^*(M_\alpha)$ has a maximum, then by virtue of (7.5.23) the entire vicinity of this point is occupied by integral curves issuing from it with increasing ξ. Consequently, the singular point is either a node or a focus. For fairly small $\lambda/\gamma M$ (this is precisely the case considered) the singular point is a focus with issuing integral curves. If $\Phi^*(M_\alpha)$ has a minimum, then for small $\lambda/\gamma M$ the singular point is a focus with arriving integral curves. Similarly, we can readily obtain that a saddle point of the function $\Phi^*(M_1, M_2)$ corresponds to a saddle point of the system (7.5.21)–(7.5.22).

By varying the parameters determining the function Φ^* (below, we consider only variation in W) we cannot change the type of a critical point until it is isolated. This can occur if critical points merge. One of the principal coefficients of the quadratic form representing the function Φ^* in the vicinity of each critical point vanishes in the merging. Prior to merging, the corresponding coefficients have opposite signs at the considered critical points. In the generic case the second coefficient remains nonzero and definite (positive or negative) and has the same sign at the two points. This means that one of the merging points is a saddle point and the other is a node. After merging the points can either disappear or change places.

In the process of merging the intensity of the corresponding discontinuity tends to zero and at the moment of merging the velocity W of the discontinuity is equal to the characteristic velocity calculated for the state corresponding to this point. If the shock velocity W coincides with one of the values of the characteristic velocity at the point A when it merges with another singular point, the initial point A with given coordinates $M_1^{\mathrm{R}}, M_2^{\mathrm{R}}$ changes its type. Consider another example. Let two critical points disappear at the point H (Fig. 7.9, 7.10) with increase in W. In this case the velocity W is equal to the characteristic velocity behind the discontinuity. If W is fairly close to unity and M_1^{R} and M_2^{R} are fixed, from (7.5.19) it is clear that the initial point A represents the maximum point of the function $\Phi^*(M_1, M_2)$.

From what we have discussed it follows that for all $W > \lambda_2^{\mathrm{R}}$ in the $(M_1^{\mathrm{R}}, M_2^{\mathrm{R}})$-plane the initial point A is a maximum point of the function $\Phi^*(M_1, M_2)$ and the system (7.5.21)–(7.5.22) has unstable focus. The point R (corresponding to the point R in Fig. 7.10) is a saddle point of the function $\Phi^*(M_1, M_2)$ and of the system (7.5.21)–(7.5.22). The point I_1' corresponding to the same velocity W (see Fig. 7.10) is a stable focus of system (7.5.21)–(7.5.22).

In Fig. 7.11 we have reproduced the arrangement of the singular points (cf. Fig. 7.10) and level lines of the function $\Phi^*(M_1, M_2)$. Since for $W > \lambda_2^{\mathrm{R}}$ the vicinity of the point A is occupied by the integral curves issuing from this point, then, obviously, there always exists an integral curve arriving at the saddle point R as $\xi \to \infty$. The nonevolutionary shock wave $A \to I_1'$ (with too many boundary conditions) corresponds to a set of the integral curves depending on a single arbitrary parameter and connecting A and I_1'. In the case considered all but one integral curves issuing from A arrive at I_1'. This agrees with the general consideration of the structure of similar nonevolutionary discontinuities outlined in Section 7.2. The shock waves $A \to I_1'$ have no importance since they must nonlinearly disintegrate into evolutionary discontinuities.

Consider now the interval $W_H < W < \lambda_2^{\mathrm{R}}$. In this case the pattern of the level lines and the behavior of the integral curves are qualitatively the same as above. We must only interchange the letters A and R in Fig. 7.11. There exist two integral curves connecting the points A and I_1' (separatrices of the saddle point A). Each of these integral curves represents the structure of a slow shock wave existing for any W on the above velocity interval.

Of major interest is the velocity interval $W_E < W < W_H$ on which we have five singular points of the system (7.5.21)–(7.5.22). Among these points are the initial saddle point A, two stable foci I_1 and I_2, one other saddle point S_1, and an unstable focus G. For $W_N < W < W_H$ we have $\Phi^*(A) > \Phi^*(S_1)$ and for certain values of W an integral curve can exist that connects the saddle points $(A \to S_1)$. For this case in Fig. 7.12 we have reproduced the pattern of the level lines $\Phi^*(M_1, M_2) = \text{const}$. The integral curve $A \to S_1$ is shown in Fig. 7.13.

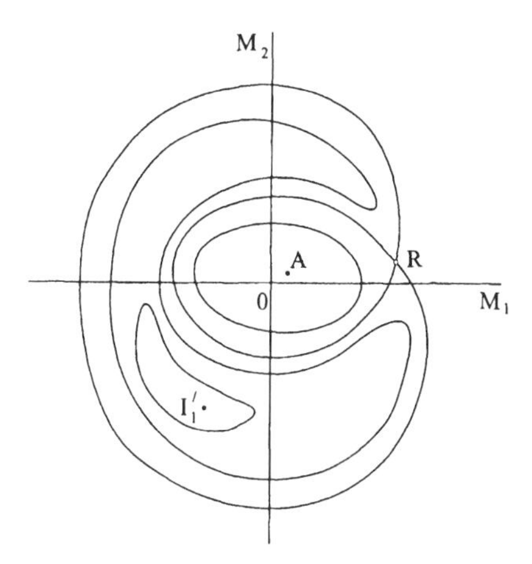

Figure 7.11 Level lines of the function $\Phi^*(M_1, M_2)$ (see formula (7.5.19)) for $W > \lambda_2^R$.

As $\xi \to \infty$, each of the two integral curves issuing from the point A must arrive at one of the singular points I_1, I_2, or S_1. Arrival of the integral curve at the saddle point S_1 is exceptional. This is possible for a specific value of the parameter W. At some point (say T) the integral curve intersects the level line that passes through the point S_1 and has the figure eight shape. Before the integral curve issuing from the point A arrives at the figure eight curve, it rotates several times around this curve. The number of rotations is determined by the difference $\Phi^*(A) - \Phi^*(S_1)$ and the angle between the integral curve and the level line. As mentioned above, this angle is of the order of $\lambda/\gamma M$. If the difference $\Phi^*(A) - \Phi^*(S_1)$ is finite and $\lambda/\gamma M$ is small, the number of rotations is large. In this case a small, of the order of $\lambda/\gamma M$, variation of W suffices to add or subtract one rotation until the integral curve intersects the figure eight curve. Under this variation of W the point T runs through the entire figure eight curve and arrives at the saddle point S_1 twice, each time from different sides. Thus, the values of W for which integral curves connect the points A and S_1 are separated by small intervals of length about $\lambda/\gamma M$. Each of these values corresponds to a point on the Hugoniot curve. This point belongs to the *a priori* nonevolutionary segment *HN* of the Hugoniot curve. On the other hand, since W acquires an isolated value at this point, we must treat this as the existence of an additional relation on the discontinuity. Thus, the discontinuity $A \to S_1$ turns out to be evolutionary.

As we vary W, there appear intervals of W on which both integral curves issuing from the point A arrive at one of the singular points I_1 or I_2. This means that only one of the slow shock waves has structure. On the segments I_2^*H and $I_1^*I_1''$ of the Hugoniot curve corresponding to slow shock waves the set of admissible discontinuities can be represented as a broken line consisting of individual strokes. In Figs. 7.9b and 7.10 the strokes are located so that in the generic case for any W at least one of the slow shock waves has structure.

If W lies on the interval $W_E < W < W_N$, the level line $\Phi^*(M_1, M_2) = \Phi^*(A)$ has the shape of the figure eight curve with the point A at the center. The integral curves issuing from A cannot now arrive at the saddle point S_1 since $\Phi^*(S_1) > \Phi^*(A)$. In this case each

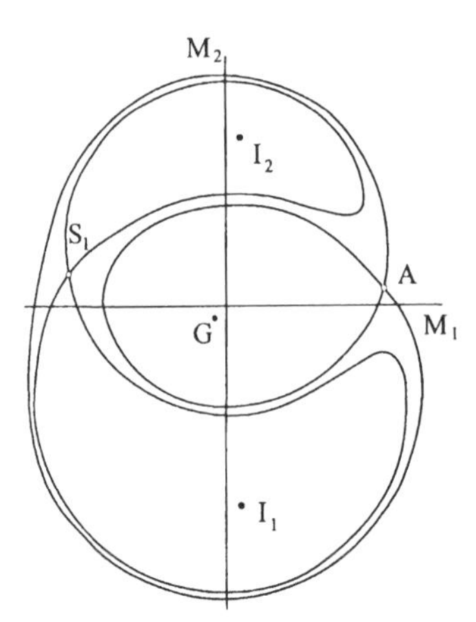

Figure 7.12 Level lines of the function $\Phi^*(M_1, M_2)$ for $W_N < W < W_H$.

separatrix of the point A issuing in the direction of decreasing Φ^* and going to the minimum points I_1 and I_2 inevitably arrives at these points. Thus, for any W the two slow shock waves have structure on this velocity interval.

On the interval $\lambda_1^R < W < W_E$ there are three singular points A, I_1, and I_2. As above, the point A is connected by integral curves with the points I_1 and I_2. This ensures the existence of structure for both slow shock waves.

7.5.3 The set of admissible discontinuities.

The investigation performed above makes it possible to distinguish a set of points on the Hugoniot curve that correspond to admissible discontinuities, that is, the admissible part of the Hugoniot curve. The admissible part of the Hugoniot curve contains both the entire fast branch AA' and the segments AI_2^*, LI_1^*, and $I_1''D$ (Figs. 7.9b and 7.10) corresponding to slow shock waves.

The slow shock segment of the Hugoniot curve corresponding to $W_* < W < W_H$ may have gaps, i.e., intervals on which the shock waves have no structure. At a given point, for example, I_2, this occurs if the two separatrices issuing from A arrive at another singular point (I_1). As shown above, in this case the lengths of these intervals and the lengths of the admissible intervals between them are of the order of $\lambda/\gamma M \ll 1$. Moreover, the *a priori* nonevolutionary segment NH of the Hugoniot curve contains a set of isolated points corresponding to admissible discontinuities; the distance between these points is of the order of $\lambda/\gamma M$. In Figs. 7.9b and 7.10 we have shown the set of admissible discontinuities by bold segments of the Hugoniot curve and by isolated points. The Hugoniot curve and admissible elements on it were obtained numerically by Gvozdovskaya and Kulikovskii (1999b).

7.5.4 Nonuniqueness of solutions.

Summarizing our study of the discontinuity structure, we can supplement the large-scale model formulated in Section 7.5.1 with the requirement that all the discontinuities used to construct the solutions of the problems must belong to the set of admissible discontinuities.

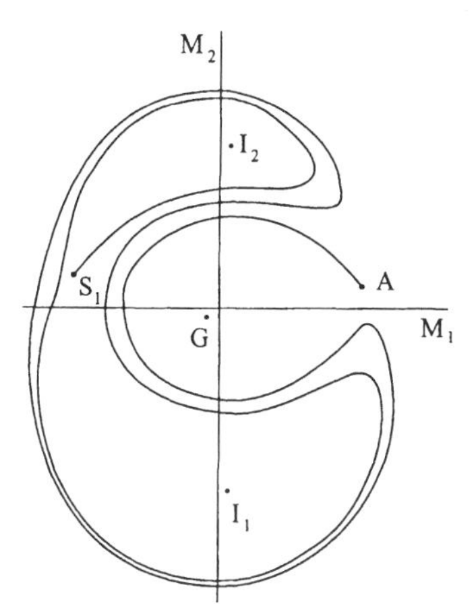

Figure 7.13 Integral curve connecting A and S_1.

However, we can show that, even if this restriction is adopted, the large-scale model does not guarantee the uniqueness of our solution. The nonuniqueness arises due to the presence of the set of isolated points on the Hugoniot curve with specific (distinguished) values of the shock velocity. These points lie inside a diagonal rectangle $\lambda_1^R \leq W < \lambda_2^R$, $\lambda_1^L \leq W \leq \lambda_2^L$ of the evolutionary diagram (Fig. 7.10). The corresponding discontinuities will be called intermediate discontinuities.

In order to be convinced of the nonuniqueness, we can consider the following example of the self-similar wave problem in the region $x > 0$ with the initial ($t = 0$) and boundary ($x = 0$) conditions specified by points A and P in the (M_1, M_2)-plane (Fig. 7.14). The solution is a function of x/t. In the self-similar problems, Riemann waves or moderate-amplitude shocks can propagate ahead of and behind an intermediate discontinuity (fast waves propagate ahead of the discontinuity and slow waves behind it). In Fig. 7.14 we have plotted the curves C_1C_1', C_2C_2', ..., C_nC_n' drawn by the isolated points $C_1, C_2, \ldots, C_n$ belonging to the Hugoniot curve when the point for which the Hugoniot curve is constructed moves along the curve AA' that corresponds to states behind fast shock waves starting from the point A. We also plotted the integral curve $P'P$ of a slow Riemann wave propagating behind the intermediate discontinuities. In accordance with Section 7.4 the characteristic velocity λ_2 decreases along the integral curve with decrease in B_2 and, consequently, in M_2. One of the solutions of the problem formulated above has the following structure. Initially, a fast shock wave $A \to A_i''$ propagating through the state A transfers this state to such a point that the state behind an intermediate discontinuity is characterized by a point C_i'' lying on the integral curve of the Riemann wave passing through the point P. This Riemann wave expanding with time corresponds to the integral curve $C_i''P$ and propagates behind the intermediate discontinuity. The state behind the Riemann wave is characterized by the point D and satisfies the boundary condition at $x = 0$. Obviously, there are as many such solutions as there are isolated points C_i on the Hugoniot curve corresponding to the point A and there exists a finite domain of the boundary values for which the solution of the self-similar problem is not unique.

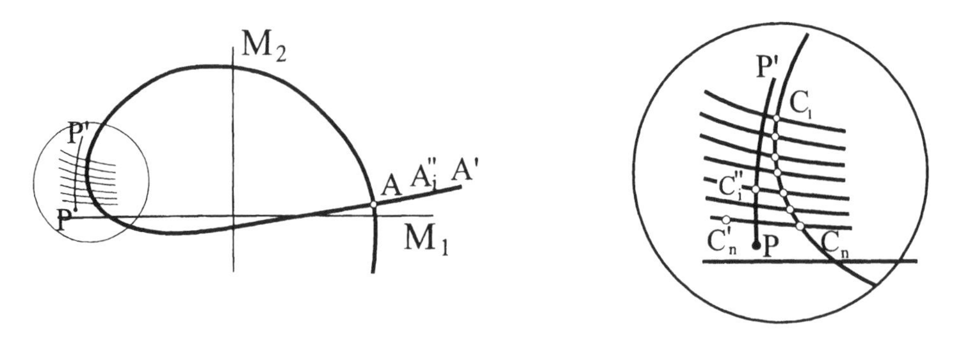

Figure 7.14 Nonuniqueness of the solution of a self-similar initial boundary-value problem for electromagnetic waves in ferromagnets.

Thus, even if we supplement the large-scale model for nonlinear electromagnetic waves with the requirement of admissibility of discontinuities, this does not allow us to find unique solutions of the problems, in particular, self-similar ones.

7.6 Shock waves in composite materials

In this section we consider nonlinear quasitransverse waves in composite materials, that is, in elastic media possessing an internal structure associated with the periodicity of their properties. If we consider fairly long waves, the behavior of the medium is similar to that of a certain averaged medium with elastic properties. The elasticity coefficient can be calculated if the structure of the initial periodic medium is known (Bakhvalov and Panasenko 1989). Thus, in the long-wave approximation the processes can be described by the elastic model. In particular, elastic media resulting from the averaging can be nonlinear and weakly anisotropic and, consequently, can be described by the model used in Section 7.4.

For smaller-scale phenomena the averaged description of motions of elastic media turns out to be still possible; however, certain additional terms leading to the wave dispersion can arise in the equations to be solved. On the other hand, there always exist certain dissipative mechanisms that will be considered below as additional viscous stresses. Taking dispersion into account necessitates proper consideration of the discontinuity structure to select admissible discontinuities. This investigation is a necessary element to construct a new large-scale model describing the averaged medium under consideration.

7.6.1 Basic equations and the discontinuity structure. As mentioned above, the large-scale equations coincide with Eq. (7.4.1). As in Section 7.4, we will consider small-amplitude waves in weakly anisotropic media. Thus, relations (7.4.3)–(7.4.5) for Φ still hold.

For the sake of simplicity, we will consider incompressible media for which the system of governing equations is as follows:

$$\rho_0 \frac{\partial v_i}{\partial t} = \frac{\partial}{\partial x}\frac{\partial \Phi}{\partial u_i}, \qquad \frac{\partial u_i}{\partial t} = \frac{\partial v_i}{\partial x}, \qquad \frac{\partial S}{\partial t} = 0, \qquad i = 1, 2. \tag{7.6.1}$$

The expression for Φ has the form

$$\Phi = \frac{1}{2}\mu\left(u_1^2 + u_2^2\right) - \frac{h}{4}\left(u_1^2 + u_2^2\right)^2 + \frac{1}{2}q\left(u_2^2 - u_1^2\right) + \rho_0 T_0(S - S_0). \qquad (7.6.2)$$

Note that the coefficients h and q characterize the nonlinearity of the medium and the wave anisotropy, respectively. Their role is the same as that of the corresponding coefficients k and g in the simplified system (7.4.7). It can readily be shown that all the subsequent basic conclusions are also valid for compressible media.

Passing to the complete system of equations in which dispersion and dissipation are taken into account, we additionally assume that the dispersion and viscous terms are small and are of the order of the nonlinear terms of the considered solutions. The dispersion terms involve higher-order derivatives of u_i with respect to x. Assuming that these terms and nonlinearity are small, we can consider all coefficients as constants. This means that the dispersion terms are taken into account only in the linear approximation. Moreover, we retain only the lowest-order derivatives with respect to x in the dispersion terms, since in the long-wave solutions the values of derivatives decrease with their order. The waves are assumed to be long as compared with the inner geometric scale of the medium. The long-wave nature, or the space large scale, of solutions is due to the small nonlinearity adopted. This property will be obvious in our study of the problem on the discontinuity structure.

As follows from the results obtained by Bakhvalov and Panasenko (1989) and Bakhvalov and Eglit (2000), the dispersion terms can be introduced by adding terms with the third or second derivatives of u_i with respect to x in the first group of Eq. (7.6.1). The second-order terms can occur in the equations of motion as $b_{ij}u_j''$, where $u_j'' = \partial^2 u_j/\partial x^2$ and b_{ij} is an antisymmetric 2×2 matrix. It is precisely this form of the matrix b_{ij} that ensures dispersion in the absence of dissipation. The expressions $b_{ij}u_j''$ must represent in equations of motion the components of a two-dimensional vector. For antisymmetric matrices b_{ij} this vector can be written in the form $\mathbf{u}'' \times \mathbf{b}$, where $\mathbf{u} = u_1\mathbf{e}_1 + u_2\mathbf{e}_2$, and $\mathbf{b}$ is a pseudo-vector directed along the x-axis. Note that only if $\mathbf{b}$ is a pseudo-vector, the expression $\mathbf{u}'' \times \mathbf{b}$ is a vector.

Thus, if we want to have terms with second-order derivatives in the equations of motion it is necessary to involve a pseudovector in the formulation of the problem. It is precisely this situation takes place when nonlinear electromagnetic waves propagate in ferromagnets (Section 7.5). In this case the initial magnetic field can be considered as a pseudovector.

In what follows, we will consider the case with no pseudovectors involved in the formulation of the problem so that the dispersion terms are represented by the third derivatives $m_{ij}\partial^3 u_j/\partial x^3$ with $m_{ij} = \text{const}$. Since in the isotropic case $m_{ij} = m\delta_{ij}$, the dispersion terms can be written in the form $m\partial^3 u_i/\partial x^3$.

For example, similar terms with $m > 0$ must arise on the left-hand side of the equation of motion if a homogeneous elastic easily-deformable medium involves uniformly distributed rigid rods of fairly high flexural rigidity that are either parallel or nearly parallel to the x-axis.

In addition, we assume that the dissipation processes are represented by the viscous terms $\eta(\partial^2 v_i/\partial x^2)$, with $\eta = \text{const} > 0$ occurring on the left-hand side of the first group of equations (7.6.1) with the minus sign. The viscous terms are assumed to be small. Therefore, we can choose a linear isotropic form of them. Thus, the complete system of equations suitable for describing the discontinuity structure can be written as

$$\rho_0 \frac{\partial v_i}{\partial t} - \frac{\partial}{\partial x}\left(\frac{\partial \Phi}{\partial u_i}\right) + m\frac{\partial^3 u_i}{\partial x^3} - \eta \frac{\partial^2 v_i}{\partial x^2} = 0, \qquad \frac{\partial u_i}{\partial t} = \frac{\partial v_i}{\partial x}. \tag{7.6.3}$$

7.6.2 Discontinuity structure; admissible discontinuities.

For $m = 0$, that is, if there is no dispersion and only viscosity is taken into account, the shock structure was considered in Section 7.4 and it was shown that all the *a priori* evolutionary discontinuities have structure, i.e., all these discontinuities are admissible.

Let us consider the effect of dispersion on the shock structure and determine the set of admissible discontinuities. We will seek bounded solutions of the system (7.6.3) that are functions of $\xi = x - Wt$. Integrating the first two equations with respect to x and eliminating v_i, we can represent the system (7.6.3) in the form

$$mu_i'' + \eta W u_i' = -\frac{\partial P}{\partial u_i}, \qquad i = 1, 2; \tag{7.6.4}$$

$$P = \frac{1}{2}(\rho_0 W^2 - \mu)\left(u_1^2 + u_2^2\right) - \frac{1}{2}q\left(u_2^2 - u_1^2\right) + \frac{h}{4}\left(u_1^2 + u_2^2\right)^2 + A_1 u_1 + A_2 u_2,$$

where A_i are integration constants chosen by accounting for the condition $\partial P/\partial u_i = 0$ in the state $u_i = u_i^{\mathrm{R}}$ ahead of the discontinuity. Primes here denote derivatives with respect to ξ.

Multiplying Eq. (7.6.4) by u_i' and summing, we obtain

$$\left(\frac{m}{2}u_i' u_i' + P\right)' = -\eta W u_i' u_i'.$$

Note that $u_i' u_i' = u_1'^2 + u_2'^2 > 0$.

The right-hand side of this equation represents a dissipative function taken with the opposite sign. Therefore, the total jump $\{P\}$ across the discontinuity structure is of the sign opposite to the entropy production, that is $\{P\} = 0$ if the entropy production is equal to zero.

Equations (7.6.4) have the form similar to that of equations of motion for a material point of mass m under the action of a potential force specified by the potential energy P and the friction proportional to the velocity u_i'. In this case the variable ξ plays the role of time. Below, we will use this analogy to obtain certain qualitative conclusions on the shock structure and the set of admissible discontinuities. More detailed investigation necessitates the use of numerical methods.

For a given function P and fixed W the pattern of integral curves of the system (7.6.4) in the (u_1, u_2, u_1', u_2')-space and, consequently, the existence of the shock structure, is determined by the ratio $\eta/\sqrt{m}$, which characterizes the ratio of the viscous and dispersion terms in the equations. For the sake of simplicity, we shall restrict our attention to consideration of small g such that $g/hR^2 \ll 1$, where $R^2 = (u_1^{\mathrm{R}})^2 + (u_2^{\mathrm{R}})^2$.

7.6.3 Case $h > 0$.

If we set $q = 0$ and $A_i = 0$, then for $\rho_0 W^2 < \mu$ the graph of the function $P(u_1, u_2)$ has the form of an annular groove whose outer walls go indefinitely

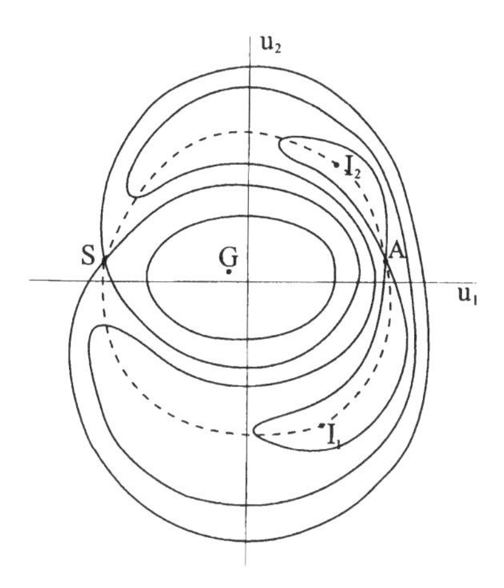

Figure 7.15 Level lines of the function $P(u_1, u_2)$, see formula (7.6.4).

upward and a local maximum is located at the point $u_1 = 0, u_2 = 0$. If $q \neq 0$ and $A_i = 0$, then the depth of the groove is not constant and the groove has two local minima separated by two saddle points. The local maximum at the origin is conserved provided that $\rho_0 W^2 < \mu$. For sufficiently small A_i the number and the types of the singular points remain the same (see Figs. 7.15 and 7.16 in which the groove is shown by the dotted curve). The qualitative form of the function $P(u_1, u_2)$ is preserved.

Along with the equalities $u_i' = 0$, the stationary points of the function $P(u_1, u_2)$ determine singular points of the system (7.6.4) and may correspond to the states at $\xi = \pm\infty$. The situations in which five singular points exist are of major interest. For example, this can occur for $W_H < W < \lambda_2^R$, see Fig. 7.17a, in which the vertical straight lines $W =$ const are shown by dotted lines. In the (u_1, u_2)-plane the points I_1, I_2, S, and G on the Hugoniot curve correspond to the stationary points of the function P. In addition, there is a stationary point A corresponding to the initial state. In the case considered, the point G is a maximum point, I_1 and I_2 are minimum points, and S and A are saddle points of the function $P(u_1, u_2)$. In the four-dimensional space (u_1, u_2, u_1', u_2') the projections of the integral curve on the (u_1, u_2)-plane represent trajectories of a mass point moving with friction in this plane under the action of the force determined by the potential energy $P(u_1, u_2)$. The structure of a discontinuity with the upstream state given by the point A corresponds to a trajectory arriving at the point A as $\xi \to \infty$ from one of the critical points I_1, I_2, G, or S. The departing velocity is assumed to be zero, that is, $u_1' = 0$, $u_2' = 0$. In this case the departing point represents the state downstream the discontinuity.

Clearly, the integral curves issuing from the minima of the potential energy (I_1 or I_2) at zero velocity $u_1' = 0$, $u_2' = 0$ cannot exist, i.e., there is no structure of the nonevolutionary shock waves $A \to I_1$ and $A \to I_2$.

Let us consider the set of trajectories issuing from the point G (maximum) at $u_1' = 0$, $u_2' = 0$. This set represents a node in the (u_1, u_2)-plane and, consequently, depends on a single parameter. As $\xi \to \infty$ all the trajectories must terminate at one of the points A, I_1, I_2, or S. Obviously, the sets of trajectories arriving at the points I_1 and I_2 (minima) also depend

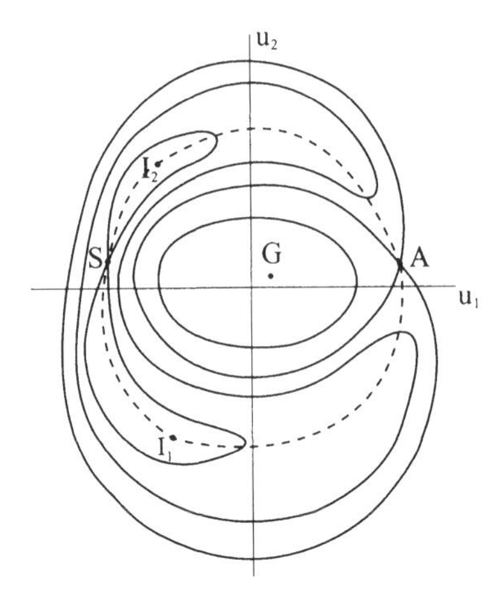

Figure 7.16 Level lines of the function $P(u_1, u_2)$.

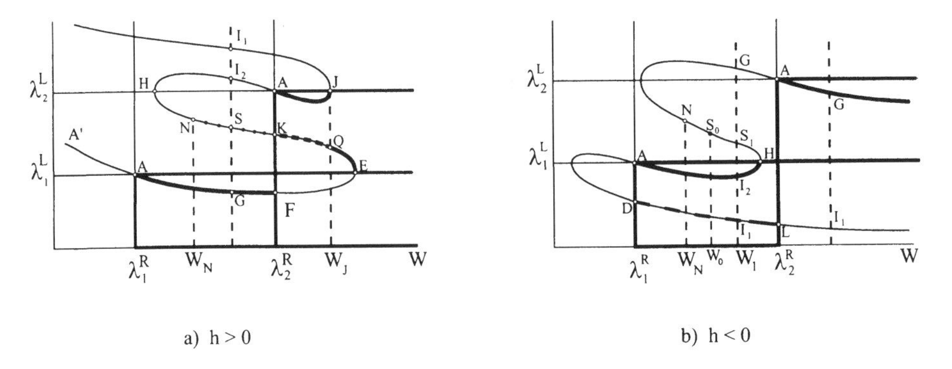

Figure 7.17 Evolutionary diagrams and admissible discontinuities in composite materials.

on a single parameter. This means that there exist integral curves coming into the points A and S that separate these flows. Thus, a structure of the shock wave $A \to G$, which is a slow shock wave, exists (Fig. 7.17a).

Let us clarify the point on the existence of an integral curve going from S to A as ξ increases. Clearly, the point S must be located higher than the point A on the graph of the function P (Fig. 7.15). This holds if $W_N < W < \lambda_2^R$ (the point N is symmetric to the point A about the u_2-axis). Taking into account the above-mentioned relation between $\{P\}$ and the entropy production, we can conclude that the equality $P(S) = P(A)$ is satisfied provided that the point S coincides with the point N (see Section 7.4). If $W < W_N$, then the point A turns out to lie higher than the point S and there is no integral curve going from S to A and, consequently, there is no shock wave structure $A \to S$ (Fig. 7.16).

Let us now consider the situation represented in Fig. 7.15 for $P(S) > P(A)$. There are two oppositely directed trajectories issuing from the point S with $u_1' = 0$, $u_2' = 0$. Heavy mass points, departing from the saddle point S, move mainly along the groove and generally terminate their motion at one of the minima I_1 or I_2. The effect of friction is characterized by the ratio $\eta/\sqrt{m}$. If it is sufficiently small, then in its motion a mass point performs some oscillations along the groove, each time passing through the point A. We can expect that under variation of the parameters of Eq. (7.6.4) (for example, under variation of W), the number of the oscillations will also be changing. In addition, we can indicate alternating intervals on the W axis corresponding to the termination of motion at the points I_1 or I_2. Clearly, these intervals are separated by the values of W such that the mass point considered tends neither to I_1 nor I_2 as $\xi \to \infty$. In this case there is a single possibility, namely, as $\xi \to \infty$ the mass point tends to the unstable (saddle) singular point A. This means that for these values of W the shock wave $A \to S$ has structure. As can be seen from Fig. 7.17a, the shock $A \to S$ is *a priori* nonevolutionary. This shock becomes evolutionary provided that the shock velocity is equal to the distinguished values at which the shock structure exists. This equality must be considered as an additional condition on the discontinuity. In this sense this shock front is similar to the combustion front in a gas.

For small values of the friction coefficient the number of oscillations that the heavy mass point performs along the groove is very large, and a small variation in W suffices to change the stopping point from I_1 by I_2 or vice versa. In this case W passes through the value W_0 corresponding to the stopping point at A. This means that for small $\eta/\sqrt{m}$ the differences between the values of W for which a shock wave structure $A \to S$ exists are small. These values are located on the interval $W_N < W < \lambda_2^R$ as shown in Fig. 7.17a.

On the interval $\lambda_2^R < W < W_J$, the level line pattern qualitatively conserves the shape shown in Fig. 7.15. However, the letters A and I_2 must interchange their position in this case. It is obvious, that the structure of the fast shock $A \to I_2$ always exists (I_2 belongs to the segment AJ of the Hugoniot curve). The structure of the discontinuity $A \to I_1$ does not exist. The structure of the discontinuity $A \to G$, where G belongs to the interval FE in Fig. 7.17a, is not unique. This is typical of metastable nonevolutionary discontinuities.

The following is valid for the shock $A \to S$ structure on the interval $\lambda_2^R < W < W_J$ (S belongs to the segment KQ in Fig. 7.17a). Each of the trajectories issuing from the point S at zero initial velocities in the opposite directions may terminate at the point A (point I_2 in Fig. 7.15). This trajectory represents the structure of a fast shock wave corresponding to the segment QK of the Hugoniot curve. However, the oscillatory character of the motion along the groove has an important consequence. For each trajectory the values of W can be subdivided into two intervals. For W belonging to one of them the trajectory terminates either at A or at I_1. The smaller $\eta/\sqrt{m}$, the smaller the length of these intervals. The shock wave structure $A \to S$ does not exist if both integral curves issuing from S arrive at the point I_1. We can expect that this situation occurs for fairly small η if W is close to λ_2^R. In fact, in this case the points I_1 and A are close each other and the depression of the groove is small in the vicinity of the point A. In this case we can also expect that the lengths of the W-axis intervals corresponding to arrival of each trajectory at A is smaller than those of the intervals for I_2 and for the two trajectories the intervals do not overlap the entire W axis.

For $W > W_J$ the groove has only a single hollow with the deepest point at A. In this case both integral curves issuing from the point S arrive at the point A. Thus, the shock wave structure $A \to S$ exists.

Summarizing the above consideration, in Figs. 7.17a and 7.18a we have reproduced the set of all admissible discontinuities in the case $h > 0$ by means of bold segments and

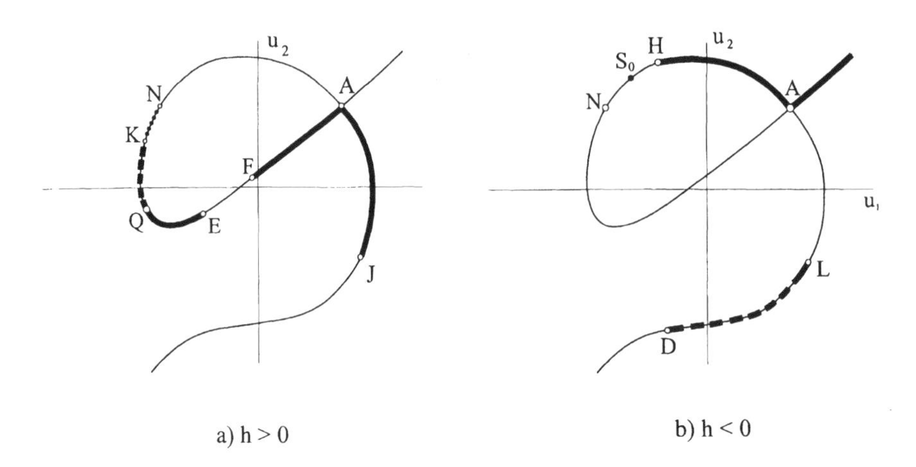

Figure 7.18 Arrangement of the set of admissible discontinuities in composite materials on Hugoniot curves in the u_1, u_2 plane.

isolated points. The number of these isolated points and short segments, as mentioned above, depends on the parameter $\eta/\sqrt{m}$, which determines the structure correct to a scale.

7.6.4 Case $h < 0$.

In this case the graph of the function $P(u_1, u_2)$ for small q and A_i resembles the surface of a volcano crater. If $W = W_1$ (see the vertical dotted straight line in Fig. 7.17b) and $\lambda_1^R < W < \lambda_2^R$, there are five stationary points of the function $P(u_1, u_2)$. The pattern of the level lines remains the same as that represented in Figs. 7.15 and 7.16. However, the direction of variation in the function P from one line to the other is opposite. In this case the point G is a minimum point, the points I_1 and I_2 are maximum points, and A and S are saddle points.

For small values of q and A_i the height of the crater walls varies within a narrow range. In this case in the entire set of the trajectories departing from the smaller maximum (in Figs. 7.15 and 7.16 this is the point I_2) we can distinguish two beams of trajectories going in the opposite directions approximately along the ridge (the upper edge) of the crater (see dotted curve in Figs. 7.15 and 7.16). When a trajectory of a heavy particle deflects at fairly large distances from the ridge, the particle is rather strongly accelerated in the transverse direction and goes aside, and its trajectory leaves the beam considered. Each of these beams can collide with an obstacle. This corresponds to the second maximum. As noted above, the trajectories of each beam are separated and one part of the trajectories goes inside, while the other goes outside the crater. In this case, there exists a separating trajectory in each of the beams that remains on the ridge infinitely long. Obviously, these trajectories must terminate at the points A and S, respectively. The first of them represents the structure of the slow shock wave $A \to I_2$.

Let us now consider the trajectories issuing from the higher maximum (I_1). As in the previous case, for small q we can distinguish two beams of the trajectories going approximately along the ridge. In each beam there is a trajectory that must terminate on the ridge at the points A or S after performing some oscillations along the ridge and passing (for fairly small $\eta/\sqrt{m}$) through the lower maximum. For fairly small $\eta/\sqrt{m}$ we can expect that there exist intervals of the values of W on which the two trajectories terminate at the point S. In this case on the segment DL of the Hugoniot curve there are intervals not corresponding to admissible shock waves (Figs. 7.17b and 7.18b). In any case, this is true in its part adjoining the point D. Both the lengths of these intervals and the distances between them tend to zero as $\eta/\sqrt{m} \to 0$.

Let us now investigate the existence of structure for an intermediate shock wave $A \to S$. Obviously, there is no such structure for $P(S) < P(A)$ (Fig. 7.15), that is, for $W < W_N$. As W increases, then depending on U_1 and U_2 we can find a value $W_0 > W_N$ such that for $W > W_0$ the trajectory issuing from the point S in the direction of the crater (this trajectory is unique) surmounts the declined part of the ridge in the vicinity of the point A and goes out into the outer region. From continuity considerations it is clear that for the separating value $W = W_0$ there exists a trajectory $S \to A$ corresponding to the intermediate shock wave structure $A \to S_0$.

Thus, we obtain a set of admissible discontinuities reproduced in the evolutionary diagram (Fig. 7.17b) by bold segments and the isolated point S_0.

In conclusion, in Fig. 7.18b we have reproduced the Hugoniot curve with the distinguished set of admissible discontinuities in the (u_1, u_2)-plane.

As in the case of electromagnetic shock waves, the set of admissible intermediate shocks (which includes many points, for $h > 0$) exhibits nonuniqueness of solutions of the problem.

7.7 Longitudinal nonlinear waves in elastic rods

In this section we will consider a very simple example of the nonclassical behavior of discontinuities in the solutions of large-scale equations exhibiting complex structure of the set of admissible discontinuities. The simplicity of the large-scale model is due to the fact that only a single type of perturbations can propagate through the rod in each direction along the characteristics. However, to obtain the nonclassical behavior of the shocks in this case inside the shock wave structure not only dispersion but also a more complex nonlinearity must exist. This nonlinearity must ensure the existence of at least two Jouget points on the Hugoniot curve. In the previous sections, two Jouget points existed in the generic case. This was granted by the presence of two families of characteristics with the same direction.

7.7.1 Large-scale model. While considering very long longitudinal waves in a homogeneous rectilinear rod, we assume that the rod is elastic and the total pressure in its cross-section is defined by the formula

$$P = P(u), \qquad u = \frac{\partial w}{\partial x}, \tag{7.7.1}$$

where w is the displacement of the rod material along the rod axis aligned with the x-axis. Here we assumed that the strain u varies at distances much larger than the rod diameter so that we can put $w = w(x, t)$. The second assumption made in formula (7.7.1) implies that we neglect the dependence of P on the entropy S. This assumption was made to simplify

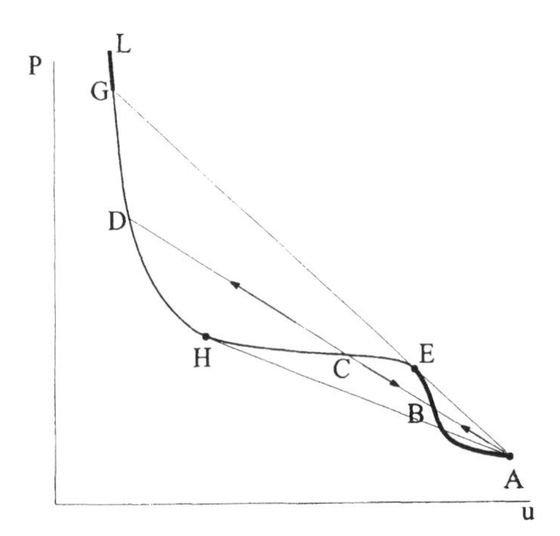

Figure 7.19 Total pressure P in a rod as a function of the strain u. The Rayleigh lines and the state ahead of a discontinuity (A) and possible states behind the discontinuity (B, C, D) are shown.

the problem and it can readily be omitted. In what follows we will consider the dependence of P on u represented by the curve $ABECHDGL$ in Fig. 7.19.

If (7.7.1) is satisfied, the equation of motion for the homogeneous linear rod can be represented in the form

$$\frac{\partial^2 u}{\partial t^2} = -\frac{\partial^2 P(u)}{\partial x^2}. \tag{7.7.2}$$

Here, x is the Lagrangian coordinate. We use the scale of the variable x such that the coefficient on the left-hand side of the equation is equal to unity.

Equation (7.7.2) has hyperbolic type for $\partial P/\partial u \leq 0$. In this case both continuous and discontinuous solutions must be considered. As determined in Section 7.2, the basic relations on discontinuities result from the integral form of the basic equations. In the case considered it is represented by the conservation of the x-component of the momentum. Equation (7.7.2) expresses this conservation for smooth solutions. The relation on discontinuities has the form

$$P(u^{\mathrm{R}}) - W^2(u - u^{\mathrm{R}}) = P(u). \tag{7.7.3}$$

Here, W is the velocity of the discontinuity.

The superscript "R" corresponds to the quantities ahead of the discontinuity and the quantities without superscript correspond to the state behind the discontinuity. Equations (7.7.2) and (7.7.3) determine the large-scale model. These equations coincide with the corresponding relations describing one-dimensional motions of a gas if P is its pressure, $1 + u = V$ is the specific volume, and we neglect any dependence on entropy.

The satisfaction of (7.7.3) means that we have the following expressions for the velocity of the discontinuity W and the velocity λ of small perturbations:

$$W^2 = \frac{P(u) - P(u^{\mathrm{R}})}{u^{\mathrm{R}} - u}, \qquad \lambda^2 = -\frac{dP}{du}. \tag{7.7.4}$$

From the viewpoint of the momentum conservation, any pair of points on the curve $P(u)$ in the (u, P)-plane may be chosen as states ahead of and behind the discontinuity if $\partial P/\partial u < 0$ and $u^{\mathrm{L}} \leq u^{\mathrm{R}}$. Thus, in the case considered the Hugoniot curve is the curve $P(u)$ with a distinguished point $(u^{\mathrm{R}}, P(u^{\mathrm{R}}))$.

Using equalities (7.7.4), we can readily verify that in Fig. 7.19 the segment AE and the part of the Hugoniot curve lying above the point H are *a priori* evolutionary segments of the Hugoniot curve for which the inequality $\lambda^{\mathrm{R}} < W < \lambda^{\mathrm{L}}$ is satisfied.

For given W and state $(u^{\mathrm{R}}, P(u^{\mathrm{R}}))$ ahead of the discontinuity the states behind it can be represented in the (u, P)-plane by points of intersection of the curve $P(u)$ with the straight line defined by the left-hand side of Eq. (7.7.3). This straight line is called the Rayleigh line. For a fairly complex function $P(u)$ several intersection points can exist (see the points B, C, and D in Fig. 7.19).

If we assume that shock waves corresponding to all these points exist, it is obvious that solutions of the problems of motion of the rod turn out to be nonunique. We must additionally select only physically admissible discontinuities among those satisfying relation (7.7.3). As shown by Galin (1959), who investigated shock waves in gases with complicated equations of state, the "false" nonuniqueness of the solutions can exist for the Hugoniot curves of the form shown in Fig. 7.19 even if we require that the entropy does not decrease across the discontinuity and the discontinuity is *a priori* evolutionary. Note that we assumed that P is independent of S. In this case the Hugoniot curve is defined by $P(u)$.

To obtain a unique solution Galin (1959) suggested retaining only the admissible discontinuities that have structure when the gas viscosity and heat conductivity are taken into account. In this case evolutionary discontinuities described by a jump from the initial point to a neighboring point on the Rayleigh line turn out to be admissible. For the initial state A, see Fig. 7.19, admissible states can be represented by points belonging to the segments AE and GL on the graph of the function $P(u)$. The segments are shown by bold curves in Fig. 7.19. Here, E is the point of tangency between the curve $P(u)$ and the straight line that issues from A. There are no discontinuities with additional boundary conditions on them. Oleinik (1959) arrived at similar conclusions when studying first-order equations with analogous nonlinearity.

7.7.2 Model for moderate-scale motions.

Equations (7.7.2) account for neither dispersion nor dissipation phenomena. These phenomena could be important in description of the discontinuity structure. The presence of dispersion is associated with finite transverse dimensions of the rod. If we consider longitudinal linear waves whose length l satisfies the inequality $d \ll l \ll L$, where L is the characteristic external dimension of the problem and d is the characteristic transverse dimension of the rod, then the basic contribution of dispersion effects can be represented by the term $\beta^2 \partial^4 u/\partial x^2 \partial t^2$ with $\beta = \mathrm{const}$ (Love 1892). This term should be added to the right-hand side of Eq. (7.7.2). For a circular rod the equality $\beta = \sigma r$ must be satisfied, where σ is Poisson's ratio and r is the radius of the rod. If nonlinearity is small, for moderately long waves the dispersion term preserves the form with $\beta = \mathrm{const}$. If the dissipative processes in the rod are neglected, due to the presence of nonlinearity solitons and nonlinear periodic waves can propagate along the rod (Potapov 1985).

In what follows we will assume that the medium is viscoelastic. Thus, we must add the term $\mu \partial^2 u/\partial x^2$, $(\mu > 0)$, which describes dissipation, to the right-hand side of Eq. (7.7.2).

As a result, for moderately long waves ($l \gg r$) we obtain the equation

$$\frac{\partial^2 u}{\partial t^2} = -\frac{\partial^2 P(u)}{\partial x^2} + \beta^2 \frac{\partial^4 u}{\partial t^2 \partial x^2} + \mu \frac{\partial^3 u}{\partial t \partial x^2}. \tag{7.7.5}$$

As $l \to \infty$, the two additional terms become infinitely small as compared with the terms of Eq. (7.7.2), since they involve higher-order derivatives.

As will be seen from the equation describing the discontinuity structure (see the next subsection), the smaller the difference between $P(u)$ and a linear function, the larger the characteristic linear scale l of change in the solution. The condition $l \gg r$ ensures the applicability of the approximate equation (7.7.5). To satisfy this condition the nonlinearity and variations in initial and boundary conditions must be small.

The important difference between Eq. (7.7.5) and the equations used by Galin (1959) and Oleinik (1959) is in the presence of the dispersion term in Eq. (7.7.5) for $\beta^2 \neq 0$.

7.7.3 Equations describing the discontinuity structure. Let us seek the solution of the problem of the discontinuity structure in the form of a travelling wave (W is the velocity of this wave)

$$u = u(\xi), \qquad \xi = -(x - Wt), \qquad \lim_{\xi \to -\infty} u = u^{\mathrm{R}}, \qquad \lim_{\xi \to \infty} u = u^{\mathrm{L}}.$$

From Eq. (7.7.5) it follows that

$$\beta^2 W^2 u'' + \mu W u' = F(u) \tag{7.7.6}$$

with

$$F(u) = W^2(u - u^{\mathrm{R}}) - P(u^{\mathrm{R}}) + P(u).$$

Here, primes denote the derivatives with respect to ξ. In addition, we used the fact that the derivatives of u with respect to ξ vanish at $u = u^{\mathrm{R}}$.

Note that Eq. (7.7.6) coincides, up to notation, with the equation used earlier by Kulikovskii (1984) for investigation of structure of discontinuities occurring in the solution of the first-order equation

$$\frac{\partial u}{\partial t} + \frac{\partial \varphi(u)}{\partial x} = 0. \tag{7.7.7}$$

This equation was considered as the one describing a large-scale model for the equation of the form

$$\frac{\partial u}{\partial t} + \frac{\partial \varphi(u)}{\partial x} = \mu \frac{\partial^2 u}{\partial x^2} - m \frac{\partial^3 u}{\partial x^3}. \tag{7.7.8}$$

The latter equation gives the equation of the type (7.7.6) for investigation of the discontinuity structure.

Equation (7.7.5) describes the waves propagating in both directions. If the nonlinearity is small, the interaction between these waves is also small. In this case we can investigate these waves independently and obtain an individual equation in the form of Eq. (7.7.8) for each wave. In essence, Eq. (7.7.8) is the general form of the equation describing the wave associated with a single characteristic family with small but general nonlinearity. Unlike the well-known Korteweg–de-Vries–Burgers equation, the nonlinearity specified in Eq. (7.7.8) by the function $\varphi(w)$ can be conceptually more complicated. In the Korteweg–de-Vries–Burgers equation $\varphi(w) = w^2/2$. However, here we are not going to derive an equation in the form (7.7.8) from Eq. (7.7.5).

7.7.4 Admissible discontinuities. Consider the situation represented in Fig. 7.19. Here, the initial point A and the value of W^2 are chosen so that in addition to the point A there are three points B, C, and D of intersection of the Rayleigh line with the plot of the function $P(u)$. This means that the function $F(u)$ given by expression (7.7.6) vanishes for the values of its argument u equal to $u_A = u^{\mathrm{R}}$, u_B, u_C, and u_D. It can occur that one of the values u_B, u_C, or u_D coincides with u^{L}, where $u^{\mathrm{L}} = \lim_{\xi\to\infty} u(\xi)$, for $u_A = \lim_{\xi\to-\infty} u(\xi)$. In fact, for the solution describing the discontinuity structure all the derivatives of u with respect to ξ vanish as $\xi \to \infty$. This means that the equality $F(u^{\mathrm{L}}) = 0$ must be satisfied. Clearly, if the equalities $F(u^{\mathrm{R}}) = 0$ and $F(u^{\mathrm{L}}) = 0$ hold, the basic relation on discontinuity (7.7.3) is satisfied. As a result, all three values u_B, u_C, and u_D must satisfy this relation.

As mentioned above, if we assume that Eq. (7.7.3) is the only relation on discontinuity, then the jumps $u^{\mathrm{R}} \to u_B$ and $u^{\mathrm{R}} \to u_D$ are evolutionary, whereas the discontinuity $u^{\mathrm{R}} \to u_C$ is not evolutionary.

For further qualitative investigation of the solution representing the discontinuity structure let us note that for given u the function $F(u)$ is the difference between the ordinates of the curve $P(u)$ and the Rayleigh line. The form of Eq. (7.7.6) necessitates that the form of the solution, i.e., the form of the function $u(\xi)$ defined to within a change in the scale of ξ depends only on the combination β/μ but not on β and μ separately. In fact, if we change the scale along the x-axis, the coefficients $\beta^2 W^2$ and μW are also changing, while the ratio β/μ remains constant. In particular, the value of u^{L}, that is, the choice of u^{L} equal to u_B, u_C, or u_D depends on the same combination.

In order to simplify the subsequent analysis we choose the scale of the variable ξ such that Eq. (7.7.6) reduces to the form

$$\beta^2 u'' + \mu u' = F(u) \tag{7.7.9}$$

and the previous expression (7.7.6) for $F(u)$ remains valid. The quantity W occurs now only in this expression.

If $\beta = 0$ and $\mu \neq 0$, then in accordance with (7.7.9) we obtain $du/d\xi = F(u)/\mu$. The direction of variation of u is determined by the sign of $F(u)$. In Fig. 7.19 this direction is shown by arrows on the Rayleigh line. We see now that the discontinuity terminated at the point B nearest to A has structure under the condition that this point lies to the left of u^{R} in Fig. 7.19. The set of the points with the coordinates $(u, P(u))$, which can represent the state behind the discontinuity for a fixed suitable W, is outlined in bold in Fig. 7.19. In the case considered all the discontinuities with structure are *a priori* evolutionary but not all *a priori* evolutionary discontinuities have structure. The evolutionary discontinuities with the state behind them belonging to the segment HG of the curve $F(u)$ have no structure. The points H and E are points of tangency of the curve $F(u)$ with straight lines that issue from the point A. No additional relations appear. The set of u representing the final states u^{L} in the discontinuity structure consists of the two intervals: the interval $[u^{\mathrm{R}}, u_E]$ and the interval adjoining the point $u = u_G$ on the right. Consequently, the constraints imposed on the states behind an admissible discontinuity can be written as inequalities. These results agree with those obtained by Galin (1959) and Oleinik (1959).

Note that in mathematical literature the mentioned inequalities are referred to as the Oleinik conditions. Sometimes, these conditions are also called the entropy condition

(Harten, Hyman, and Lax 1976), although this is not the condition of nondecreasing entropy and for $\beta \neq 0$ there exist other discontinuities with positive entropy production. This will be clear from what follows.

Consider now the case with $\beta > 0$ and $\mu > 0$ (Kulikovskii and Gvozdovskaya 1998). Then Eq. (7.7.9) can be considered as an equation of motion for a body of mass β^2 under the action of the force $F(u)$ in the presence of friction specified by the expression $\mu du/d\xi$. Here ξ plays the role of time. If we specify the initial state at A (see Fig. 7.18) and change W within a certain interval, there appear four equilibrium states in which $F(u) = 0$.

The analogy between solutions of Eq. (7.7.9) and oscillatory motions of a body makes it possible, by varying W, to investigate qualitatively the structure of the set u^L. In Fig. 7.20 we have shown a set of u^L, i.e., the admissible part of the Hugoniot curve for a fixed moderate value of β/μ, by means of bold segments and isolated points. We see that it consists of the intervals $AR, R'J', JK, K'M', MN, N'P', PQ$, and $Q'L$ represented by bold curves and the isolated points $C_1, C_2, \ldots, C_n$, (in the case shown in the figure we chose $n = 7$). For a given function $F(u)$ the number of the isolated points and intervals depends on the ratio β/μ and increases indefinitely with this quantity.

In order to prove this, note that for small W corresponding to points on the interval AR (see Fig. 7.20) the mass point acquires speed moving along the interval AB (Fig. 7.19) on which $F < 0$ and for fairly large β/μ gets through the equilibrium point B falling into the domain with $F > 0$. The point B can be attained monotonically for smaller β/μ. If W is fairly small, the mass point cannot get on the interval with $F > 0$, returns to the point B which lies on the segment AR of the plot of $P(u)$ and, after performing oscillations in the vicinity of this point, stops at B under the action of friction.

If we increase W, the force F decreases on the interval BC (Fig. 7.19) and the length of this interval also decreases. Simultaneously, the acceleration of the mass point increases along the interval AB. Therefore, for fairly small β/μ there exists such a value of $W = W_1$ that the mass point getting through the stable equilibrium state B approaches an unstable equilibrium point of the type C (point C_1 in Fig. 7.20) as $t \to \infty$. At $W = W_1$ this point corresponds to the state behind an admissible discontinuity.

If W further increases, the mass point passes through the unstable point of the type C and appears in the zone of attraction of a stable equilibrium point of the type D, where it stops after some oscillations. It is easy to understand that there exists an interval of W for which the states behind an admissible shock occur to such points (in Fig. 7.20 they occupy the interval $R'J'$).

If we further increase W, for fairly large β/μ there exists such a value of $W = W_J$ that the mass point accelerated on the interval AB passes through the points C and D from the right to the left with such a velocity that in the reverse motion occurring as $t \to \infty$ the mass point indefinitely approaches an unstable equilibrium point of the type C from the left. This point C_2 corresponds to the admissible discontinuity.

For $W > W_J$ the mass point passes through the unstable point of the type C from the left to the right and tends to a stable equilibrium point of the type B as $t \to \infty$. These solutions corresponds to points belonging to the interval JK (see Fig. 7.20).

Similar considerations can be further continued. At very high velocities W solutions will obviously terminate at stable points of the type D lying on the interval $Q'L$. The number of admissible intervals and isolated points on the Hugoniot curve is determined by

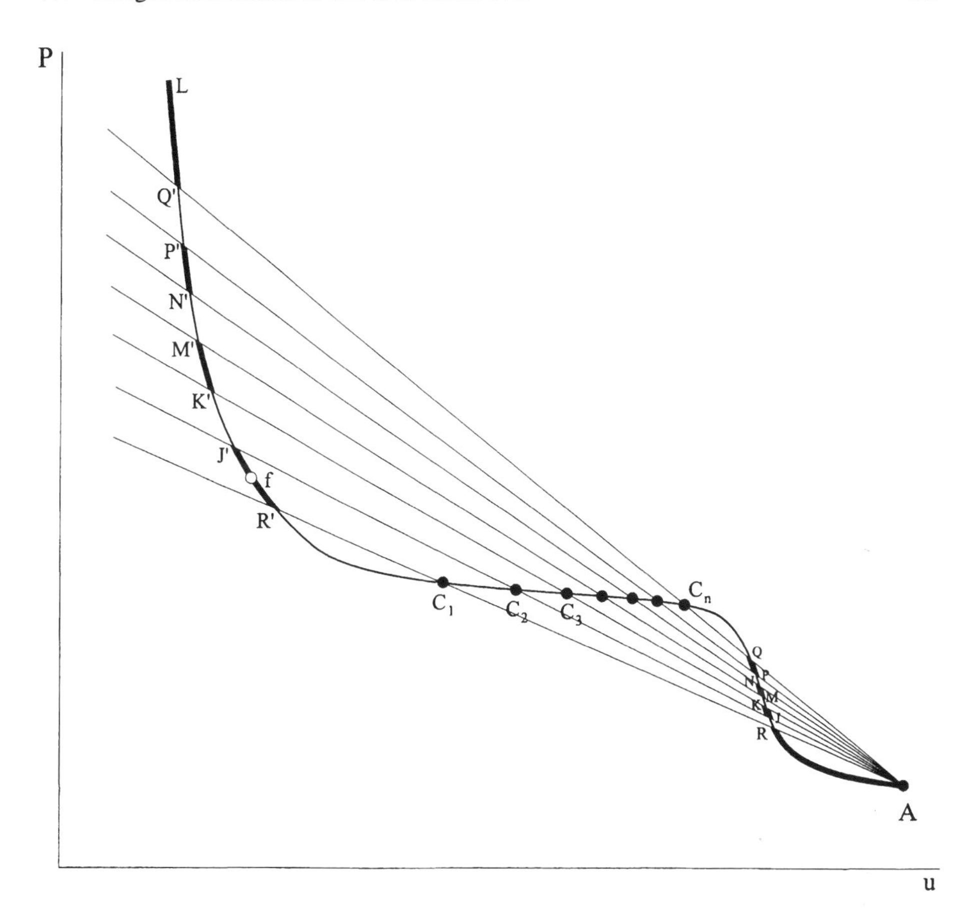

Figure 7.20 Admissible states behind discontinuities for moderate value of β/μ: bold curves and black dots.

the maximum possible number of oscillations near the unstable equilibrium points of the type C for given β/μ (the maximum is taken with respect to W). Clearly, if β = const the friction also decreases with μ and the maximum possible number of these oscillations increases.

The above considerations are based on the analogy between Eq. (7.7.9) and the equation of motion of a mass point. We can also qualitatively investigate the field of the integral curve in the plane of the variables u and $p = m(du/d\xi)$ and the dependence of this field on W. As an illustration, in Fig. 7.21 we have reproduced the above-mentioned field of integral curves corresponding to the case with the termination point being the stable point of the type D.

The isolated points C_i belong to the admissible part of the Hugoniot curve and correspond to certain values of the velocity W of the discontinuity. The equality of W to one of these values represents an additional boundary condition. In other words, this is an additional

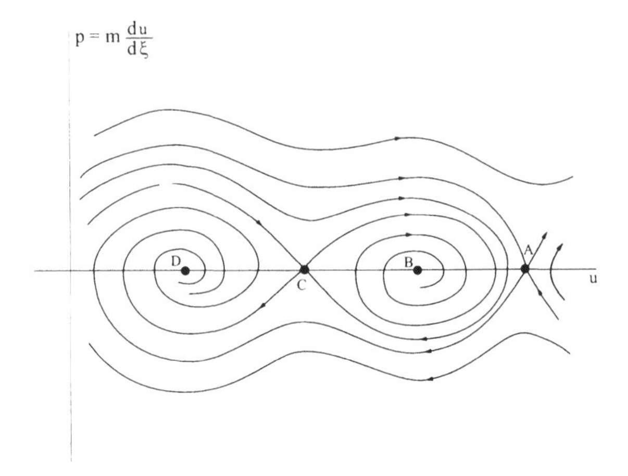

Figure 7.21 Integral curves in the (u, p)-plane, where $p = \beta^2 \dfrac{du}{d\xi}$.

relation on the discontinuity that makes it evolutionary. This relation originates if the discontinuity has structure.

If β and μ tend to zero with $\beta/\mu = \text{const}$ or if the characteristic linear dimension of the problem tends to infinity together with the length scale, then the solution representing the discontinuity structure transforms into a discontinuity admissible for given $F(u)$, β/μ, and u^{R}. If $\beta/\mu \to \infty$, the number of the admissible intervals and isolated points on the Hugoniot curve tends to infinity.

7.7.5 More precise large-scale model. Nonuniqueness.

To state a more precise model for large-scale phenomena, we shall require that all the discontinuities used for constructing the solutions must be admissible, that is, have structure. We are not going to consider solutions containing such discontinuities for large β/μ in detail and only note an obvious nonuniqueness of solutions to self-similar problems that can be obtained within the framework of this statement. It is clear that we disregard in this case unsteady events occurring on lengths of the order of the shock structure width.

As an example, we consider a self-similar "piston" problem with the initial state at $t = 0$ and $x > 0$ given by the point A and with the state at $x = 0$ and $t > 0$ specified by a point f belonging to the interval $J'R'$ (Fig. 7.20). The solution must consist of self-similar continuous waves of the form $u = u(x/t)$ for $x/t > 0$, discontinuities with $W > 0$, and zones with constant distributions of u. Clearly, there exists a solution consisting of a single discontinuity $A \to f$. In addition, there exist solutions consisting of one of the discontinuities $A \to C_i$, $i = 2, 3, \ldots, n$, and another discontinuity $C_i \to f$ moving at some lower velocity behind the former one. It can readily be seen that the discontinuities $C_i \to f$ have structure. The nonuniqueness can also occur for other boundary conditions different from those specified by the point f.

Although this has not been specially investigated, it is fairly clear that the self-similar solutions constructed above may represent asymptotics of solutions of non-self-similar problems for Eq. (7.7.4) as $t \to \infty$. The formulations of non-self-similar problems can differ due to various setting of the initial and/or boundary conditions for small values of x or t determined by the width l of the shock wave structure and time l/W. These quantities are infinitesimal from the viewpoint of the large scale. But precisely these vanishingly small intervals in the formulation of the problem determine one or other asymptotics of the solution as $t \to \infty$.

7.8 Ionization fronts in a magnetic field

As one more example of the nonclassical behavior of fronts and discontinuities, in this section we will study fronts propagating through a gas with the conductivity changing from zero to infinity. The presence of a magnetic field determines the existence of the fronts of different types and is significant for the phenomenon to be considered. The systems of equations describing the behavior of the electromagnetic fields and the medium are different on either side of these discontinuities. Thus, the fronts considered are fronts of phase transition.

7.8.1 Large-scale model. If a fairly strong shock wave propagates through a non-conducting gas, the increase in temperature can initiate the gas conductivity. In this case the gas starts to interact with the magnetic field. There is no interaction between the electromagnetic field and the gas ahead of the front. Hence the electromagnetic field can be described by the Maxwell equations, which in the absence of electric currents and charges have the form

$$\frac{\partial \mathbf{B}}{\partial t} + c \operatorname{curl} \mathbf{E} = \mathbf{0}, \qquad \operatorname{div} \mathbf{B} = \mathbf{0}, \tag{7.8.1}$$

$$\frac{\partial \mathbf{E}}{\partial t} - c \operatorname{curl} \mathbf{B} = 0, \qquad \operatorname{div} \mathbf{E} = 0. \tag{7.8.2}$$

The behavior of the gas is described by the gas dynamic equations

$$\frac{d\rho}{dt} + \rho \operatorname{div} \mathbf{v} = 0, \quad \rho \frac{d\mathbf{v}}{dt} = -\operatorname{grad} p, \quad \frac{d\varepsilon}{dt} + \frac{p}{\rho} \operatorname{div} \mathbf{v} = 0. \tag{7.8.3}$$

Here, $\varepsilon = \varepsilon(p, \rho)$ is the internal energy of unit of mass of the gas. For a perfect gas we have

$$\varepsilon = \frac{p}{\rho(\gamma - 1)}.$$

The other notation is conventional. Equations (7.8.1)–(7.8.3) disregard dissipation processes, that is, they correspond to the large-scale approximation.

In this large-scale approximation the conducting gas, the magnetic and electric fields behind an ionization front can be described by the magnetohydrodynamic equations (see Chapter 5)

$$\frac{d\rho}{dt} + \rho \operatorname{div} \mathbf{v} = 0, \quad \rho \frac{d\mathbf{v}}{dt} = -\operatorname{grad} p + \frac{1}{4\pi} \operatorname{curl} \mathbf{B} \times \mathbf{B}, \tag{7.8.4}$$

$$\frac{d\varepsilon}{dt} + \frac{p}{\rho} \operatorname{div} \mathbf{v} = 0, \quad \frac{\partial \mathbf{B}}{\partial t} - \operatorname{curl}(\mathbf{v} \times \mathbf{B}) = \mathbf{0}, \quad \mathbf{E} = -\frac{\mathbf{v}}{c} \times \mathbf{B}, \tag{7.8.5}$$

$$\operatorname{div} \mathbf{B} = 0, \quad \mathbf{j} = \frac{c}{4\pi} \operatorname{curl} \mathbf{B}. \tag{7.8.6}$$

The last equalities in Eqs. (7.8.5) and (7.8.6) determine the electric field **E** and the electric current density **j** which do not occur in the other differential equations. On the contrary, in a nonconducting medium the vector **E** is the independent quantity to be determined from the Maxwell equations.

Note that the Maxwell equations have the same form both in nonconducting and conducting media. This is clear from the last two equations of (7.8.5).

From expression (7.8.5) for **E** we see that in conducting media $|\mathbf{E}|$ is a small quantity of the order of $|\mathbf{v}||\mathbf{B}|/c$. In what follows, we assume that the order of $|\mathbf{E}|$ is everywhere the same.

On the ionization front that separates regions occupied by conducting and nonconducting gases certain relations (called basic relations in Section 7.2) must be satisfied. They can be obtained from the integral form of the governing equations. The system of the basic relations on a discontinuity can be written in the form (Germain 1959; see also Kulikovskii and Lyubimov 1965)

$$\left(\frac{B_1 V}{4\pi} - B_0 v_1\right)^{\mathrm{L}} + E_0 = 0, \qquad \left(\frac{B_2 V}{4\pi} - B_0 v_2\right)^{\mathrm{L}} = 0, \tag{7.8.7}$$

$$\{v_1 - B_0 B_1\} = 0, \qquad \{v_2 - B_0 B_2\} = 0, \qquad \left\{p + m^2 V + \frac{B_1^2 + B_2^2}{8\pi}\right\} = 0,$$

$$\left\{\varepsilon(p, V) - \frac{m^2 V^2 + v_1^2 + v_2^2}{2} - \frac{V(B_1^2 + B_2^2)}{8\pi} - E_0 B_1 + B_0(B_1 v_1 + B_2 v_2) + VJ\right\} = 0.$$

In the last equality J denotes the quantity $p + m^2 V + (B_1^2 + B_2^2)/8\pi$ which is the same on either side of the front.

In writing (7.8.7), we used a Cartesian coordinate system attached to the discontinuity with the x_3-axis normal to the surface of the discontinuity. Thus, its velocity is equal to zero, and the directions of the axes x_1 and x_2 are such that ahead $E_1 = 0$. Taking into account the last equality of (7.8.5) we see that $v_1 = 0$ and $B_2 = 0$ behind the discontinuity. We also used the mass conservation law in the form

$$v_3^{\pm} = mV^{\pm}, \qquad V = \frac{1}{\rho}, \tag{7.8.8}$$

where m is the mass flux. In addition, we have introduced the notations $B_0 = B_3/4\pi m$ and $E_0 = cE_2/4\pi m$. The quantities m, B_0, and E_0 are continuous on the discontinuity. As usual, the superscripts "R" and "L" denote the quantities ahead of and behind the discontinuity, respectively, and the braces denote a jump of the embraced quantity, that is,

$$\{a\} = a^{\mathrm{L}} - a^{\mathrm{R}}.$$

As will be shown below, the relations on ionization fronts are not always limited to the basic boundary conditions (7.8.7). In order to obtain additional relations we should consider a moderate-scale model and investigate the structure of the ionization fronts.

7.8.2 Moderate-scale model. Let us consider the case in which the ionization of gas is everywhere small enough (Kulikovskii and Lyubimov 1959; Taussig 1967; Barmin and Kulikovskii 1968a) to neglect its contribution to the equation of state. In what follows the gas is considered to be perfect. The magnetic viscosity $\eta_m = c^2/4\pi\sigma$ is assumed to be the greatest dissipation coefficient. Barmin and Kulikovskii (1971) reviewed studies of the ionization fronts under other assumptions. Such fronts were also considered in the monograph by Liberman and Velikovich (1986).

In the adopted approximation the model for moderate-scale phenomena reduces to the system of MHD equations with finite electric conductivity in the absence of viscosity and heat conductivity. Equations (7.8.4) and (7.8.6) preserve their form and equations (7.8.5) should be replaced by

$$\frac{d\varepsilon}{dt} + \frac{p}{\rho}\operatorname{div}\mathbf{v} = \frac{\eta_m}{4\pi\rho}(\operatorname{curl}\mathbf{B})^2, \qquad \frac{\partial \mathbf{B}}{\partial t} - \operatorname{curl}(\mathbf{v}\times\mathbf{B}) + \operatorname{curl}(\eta_m \operatorname{curl}\mathbf{B}) = 0, \quad (7.8.9)$$

$$\mathbf{j} = \sigma\left(\mathbf{E} + \frac{\mathbf{v}}{c}\times\mathbf{B}\right), \qquad \eta_m = \frac{c^2}{4\pi\sigma}.$$

The right-hand side of the first of Eqs. (7.8.9) represents the action of the Joule heat. The relation for $\mathbf{j}$ is the simplest form of Ohm's law (without taking the Hall effect into account). As $\sigma \to \infty$, Ohm's law transforms into the last of relations (7.8.5). The second of Eqs. (7.8.9) is a consequence of the Maxwell equations and Ohm's law.

To investigate the discontinuity structure we assume that all the quantities are functions of a single Cartesian coordinate $x = x_3$, i.e., we shall consider the discontinuity structure in the coordinate system attached to the discontinuity. Then the system describing the variation of the quantities in the structure can be written in the following form:

$$\begin{cases} \dfrac{dB_1}{dx} = \dfrac{4\pi m}{\eta_m}\left(\dfrac{B_1 V}{4\pi} - B_0 v_1 + E_0\right), \\ \dfrac{dB_2}{dx} = \dfrac{4\pi m}{\eta_m}\left(\dfrac{B_2 V}{4\pi} - B_0 v_2\right), \end{cases} \quad (7.8.10)$$

$$\begin{cases} v_1 - B_0 B_1 = 0, \qquad v_2 - B_0 B_2 = 0, \\ p + m^2 V + \dfrac{B_1^2 + B_2^2}{8\pi} = J, \qquad J = \text{const}, \\ \varepsilon(p, V) - \dfrac{m^2 V^2 + v_1^2 + v_2^2}{2} - \dfrac{V(B_1^2 + B_2^2)}{8\pi} - \\ -E_0 B_1 + B_0(B_1 v_1 + B_2 v_2) + VJ = \mathcal{E}, \qquad \mathcal{E} = \text{const}. \end{cases} \quad (7.8.11)$$

Equations (7.8.10) express the fact that the tangential components of the electric field are constant. This is a consequence of the first of Eqs. (7.8.1) for time-independent solutions. In writing (7.8.10), we took into account Ohm's law and the last equality of Eqs. (7.8.6). Equalities (7.8.11) express conservation of the momentum and energy fluxes and have the structure similar to Eqs. (7.8.7). Equations (7.8.10) and (7.8.11) are written in the coordinate system moving in the discontinuity plane in a way that the fluxes of the components of momentum along the x_1- and x_2-axes are equal to zero at any x = const. The fluxes of the x-component of momentum and energy per unit area of the plane x = const are denoted by J and $\mathcal{E}$, respectively.

Under these assumptions the quantities cannot be continuous in the nonconducting region. This follows from Eqs. (7.8.10) and (7.8.11), since the right-hand sides of Eqs. (7.8.10) vanish for $\sigma = 0$.

Since the magnetic viscosity is the only dissipation coefficient present in the system, the magnetic field must be continuous, while the remaining quantities can be discontinuous inside the ionization front structure. Clearly, these discontinuities represent gas dynamic shock waves.

To describe an ionization front we must specify some switching conditions for conductivity. In the simplest case we can assume that σ is a function of the thermodynamic state of the gas, for example, $\sigma = \sigma(T)$, where T is the temperature, so that $\sigma \equiv 0$ at $T \leq T^*$ and $\sigma > 0$ at $T > T^*$. In this case the conductivity is switched on if $T \leq T^*$ ahead of the front and $T > T^*$ behind it. However, as follows from system (7.8.10), parameters governing the conductivity coefficient are not very important. All we need to know is on which fronts the conductivity actually switches on. This makes it possible to consider cases with nonequilibrium conductivity.

7.8.3 The set of admissible discontinuities.

Equations (7.8.11) represent four equalities that relate the quantities v_1, v_2, p, V, B_1, and B_2. These relations make it possible to express v_1, v_2, p, and V in terms of B_1 and B_2. As follows from the above considerations, for a perfect gas these functions are two-valued and these possible values may correspond, respectively, to the states ahead of and behind a gas dynamic shock that is at rest in the chosen coordinate system.

To be specific, let us consider the (V, B_1, B_2)-space. The equation relating V, B_1, and B_2 can be found from relations (7.8.11). For a perfect gas it has the form

$$\frac{\gamma+1}{\gamma-1}m^2V^2 - 2\frac{\gamma}{\gamma-1}\left(J - \frac{B_\tau^2}{8\pi}\right)V + 2\mathcal{E} - B_0^2B_\tau^2 + 2E_0B_1 = 0, \tag{7.8.12}$$

where

$$B_\tau{}^2 = B_1^2 + B_2^2.$$

The two-valued function $V(B_1, B_2)$ that can be determined from Eq. (7.8.12) constitutes a surface Σ in the (B_1, B_2, V)-space. We assume that it is composed of two parts. The greater and smaller values of V correspond to the supersonic and subsonic parts Σ_1 and Σ_2, respectively. The surfaces Σ_1 and Σ_2 are in contact along the curve on which the x-component of the velocity component $v_3 = mV$ is equal to the speed of sound $c_e = \sqrt{\gamma p V}$.

The state ahead of the front ($x = -\infty$) can be specified by an arbitrary point on the part of the surface Σ, where $\sigma = 0$. In this case, as mentioned above, the right-hand sides of Eqs. (7.8.10) are equal to zero. In the state behind the ionization front ($x = \infty$, $\sigma \neq 0$) the expressions in the parentheses on the right-hand sides of Eqs. (7.8.10) vanish. Thus, this state is described by singular points of the system (7.8.10) which can be written in the form

$$\begin{aligned}
\frac{dB_1}{dx} &= \frac{4\pi m}{\eta_m}\left\{B_1\left(\frac{V}{4\pi} - B_0^2\right) + E_0\right\}, \\
\frac{dB_2}{dx} &= \frac{4\pi m}{\eta_m}B_2\left(\frac{V}{4\pi} - B_0^2\right).
\end{aligned} \tag{7.8.13}$$

Here, $V = V(B_1, B_2)$ is the introduced two-valued function which specifies the surface Σ. When writing Eqs. (7.8.13), we used the first two relations of (7.8.11).

The singular points A_i of the system (7.8.13) correspond to one-dimensional MHD flows. They have been studied by Germain (1959) who considered the MHD shock structure for $\sigma \neq 0$. There can be not more than four singular points A_1, A_2, A_3, and A_4 at which the following inequalities

$$mV(A_4) < a_s, \quad a_s < mV(A_3) < a_A, \qquad (7.8.14)$$
$$a_A < mV(A_2) < a_f, \quad a_f < mV(A_1)$$

must be satisfied. Here, $a_s \leq a_A \leq a_f$ are the characteristic velocities of the MHD equations and $mV = v_3$ is the gas velocity with respect to the wave. Fast and slow magnetohydrodynamic shock waves correspond to the transitions $A_1 \to A_2$ and $A_3 \to A_4$, respectively. If these shock waves are infinitely weak, the velocities of fast and slow shock waves are equal to a_f and a_s, respectively. The rotational (Alfvén) Riemann wave that preserves its shape and the corresponding reversible rotational discontinuity propagate in a conducting gas with the velocity a_A (see Section 7.5).

A solution representing the structure of the ionizing shock wave should consist of a jump from Σ_1 to Σ_2 with constant magnetic field and segments of magnetohydrodynamic integral curves of the system (7.8.13) going into one of the singular points A_i.

We restrict ourselves to consideration of ionization fronts propagating with a supersonic velocity through a given nonconducting state. In this case the gas may become conducting only if the ionization front structure starts from a discontinuity (gas dynamic shock) behind which the conductivity is nonzero.

Subsonic ionization fronts can also exist (Barmin and Kulikovskii 1968a). In this case the state ahead of the front must belong to the curve separating the conducting and nonconducting states on Σ_2. This situation occurs since a gas dynamic shock can propagate ahead of the subsonic ionization front. This shock transfers the gas to the state that separates the domains on the surface Σ with $\sigma > 0$ and $\sigma \equiv 0$. Thus, the subsonic ionization fronts require information on the boundary between the conducting and nonconducting states. For the sake of simplicity, we disregard this case.

Typically the surface Σ is a closed convex surface whose cross-section by the plane $B_2 = 0$ is shown in Fig. 7.22 (curve a). All the singular points of the system lie in this plane on the intersection of the surface Σ with the hyperbola

$$B_1 \left(\frac{V}{4\pi} - B_0^2 \right) = -E_0. \qquad (7.8.15)$$

The point A_1 always lies on the upper supersonic part of the surface Σ, while the point A_4 always lies on the lower subsonic part. This follows from inequalities (7.8.14), since $a_f > c_e$ and $a_s < c_e$. Depending on the flow parameters, the points A_2 and A_3 can lie both on Σ_2 and Σ_1.

Since the supersonic ionization front structure starts with a gas dynamic shock (in which B_1 and B_2 are continuous) occurring from the upper part of the surface Σ_1 to the lower part Σ_2, we consider the integral curves of the system (7.8.13) belonging to Σ_2 in projection on the (B_1, B_2)-plane. In Fig. 7.23 we have qualitatively shown the integral curves when the

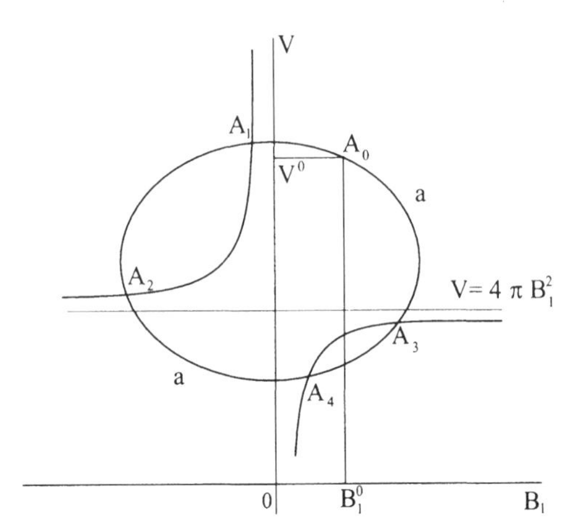

Figure 7.22 State ahead of an ionization front (A_0) and possible states behind it (A_2, A_3, A_4).

point A_2 lies on Σ_2. The arrows denote the direction of variation of quantities for decreasing x, i.e., along the gas stream.

We shall consider different types of supersonic ionization fronts and distinguish the types of the fronts by the type of the singular points corresponding to the state behind the front, i.e., depending on which of inequalities (7.8.14) is satisfied. In Fig. 7.24 we have shown the evolutionary diagram of the fronts, the corresponding singular points A_1, A_2, A_3, and A_4 being indicated in the corresponding rectangles. On the horizontal axis we have plotted the gas dynamic speed of sound c_e. In the rectangles we also indicated the number K of additional relations on the discontinuities necessary to make it evolutionary.

Slow ionization front (I^-) corresponds to the state A_4 behind the front. For I^- the first of inequalities (7.8.14) is satisfied. The integral curves arrive at the singular point A_4, which is a node (Fig. 7.23). These integral curves occupy a domain Q in the (B_1, B_2)-plane bounded by curves l issuing from A_2 and arriving at A_3, which is a saddle point. Since this front has a fixed final state, the initial state must lie on Σ_1 above the domain Q. The solution representing the discontinuity structure starts with a gas dynamic shock and then the quantities vary continuously on the surface Σ_2 along the integral curves considered above.

The requirement of existence of the structure imposes no additional conditions, except for inequality-type conditions. This agrees with the evolutionary conditions. If instead of the final state we fix the velocity of ionizing front W and the initial state, the set of the final states will generally be two-dimensional. Taking into account the possibility of variation in W, we can conclude that the admissible part of the shock adiabatic manifold corresponding to slow ionization fronts is three-dimensional.

Intermediate ionization front ($I^\sim$) corresponds to the state A_3 behind the front. The second of inequalities (7.8.14) is satisfied. If the singular point A_3 lies on Σ_2 (Fig. 7.23), then this point is a saddle point. Two above-mentioned integral curves l arrive at this point. The point representing the initial state must lie on Σ_1 above the curves l. On crossing the gas dynamic shock preceding the ionization front structure, the point representing the state of the medium falls on one of the integral curves l and arrives at A_3 along this curve.

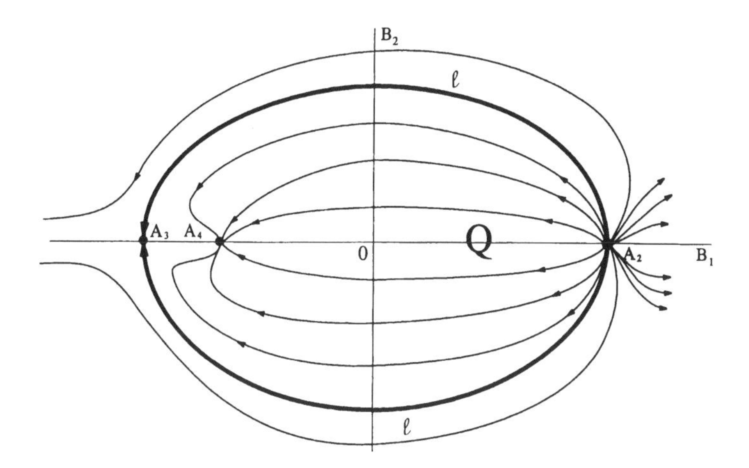

Figure 7.23 Integral curves describing variation of the magnetic field across ionization front structures.

If the point A_3 lies on Σ_1, only the point A_4 remains on Σ_2 and all the integral curves arrive at this point. Therefore, if a jump occurs from the initial state to Σ_2, we cannot arrive at A_3.

Thus, if the intermediate ionization front has structure, a single restriction on the set of the initial states arises, namely, B_1^R and B_2^R should belong to the curve l. This restriction must be treated as a single additional relation on the ionization front. This corresponds to the evolutionary conditions. Thus, for a given state ahead of the front and a fixed W, the set of states behind the front is one-dimensional. Hence, if we take into account the variation of W, the admissible part of the Hugoniot manifold corresponding to intermediate ionization fronts is a two-dimensional set.

Fast ionization front (I^+) corresponds to the state A_2 behind the front. The third of inequalities (7.8.14) is satisfied. If the point A_2 lies on Σ_2, this point is a node with integral curves issuing from it. In this case the point A_2 can be reached from the initial point only through a jump corresponding to a gas dynamic shock. This leads to two additional relations $\{B_1\} = 0$ and $\{B_2\} = 0$ and the changes in all the quantities across the fast front coincide with those across the gas dynamic shock. The part of the Hugoniot manifold that corresponds to these fronts is one-dimensional.

If the point A_2 lies on Σ_1, this point is a saddle point. The integral curves that arrive at this point lie in the plane $B_2 = 0$. On the curve separating Σ_1 and Σ_2 the integral curve arriving at A_2 changes its direction. Thus, there are no integral curves arriving at the point A_2, which lies on Σ_1. As a result, the fast ionization front with supersonic velocity behind the front has no structure. Besides, there are no ionization fronts such that the downstream state corresponds to the point A_1.

Note that presenting the results of this subsection we disregarded the electric field. In the conducting region the electric field can be determined from relation (7.8.5), while in the nonconducting region the tangential components of the electric field can be found from the condition of their continuity on the front in the attached coordinate system. If we add the

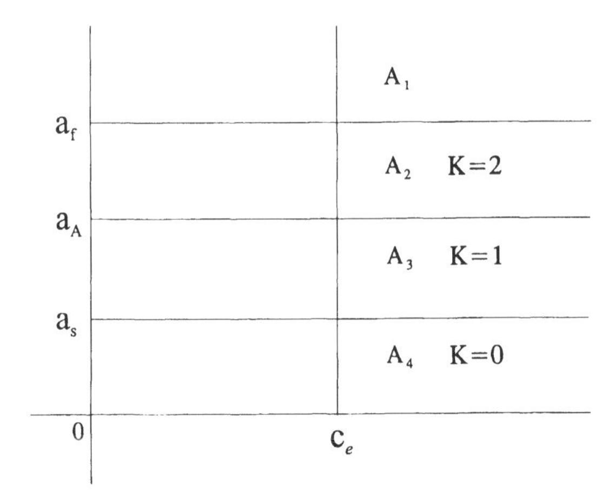

Figure 7.24 Evolutionary diagrams of ionization fronts.

electric field ahead of the discontinuity to the parameters to be determined, we should add two additional relations on the ionization fronts for the tangential components of the electric field. The Hugoniot manifolds of the ionization fronts considered above were obtained by specifying all the quantities ahead of the front, except for E_1 and E_2, which remained arbitrary. However, since the electromagnetic field does not interact with the gas in the nonconducting region, under certain conditions the problem of determining the electric field strength is separated from that for the other quantities. This justifies application of the approach developed above.

However, if along with the quantities ρ, p, $\mathbf{B}$, and $\mathbf{v}$ we also specify E_1 and E_2 ahead of the ionization front, then it turns out that, for example, fast ionization fronts can exist not for arbitrary combinations of these quantities. An electromagnetic wave radiated ahead of the front changes the electromagnetic field, thus ensuring the possibility of realizing the ionization fronts.

7.8.4 The simplest self-similar problem. Let us now consider *the piston problem*. Let a nonconducting gas occupy the half-space $x > 0$ at $t = 0$. Assume now that initially the gas is at rest, its density and pressure are constant, and the magnetic and electric fields $\mathbf{B}$ and $\mathbf{E}$ are uniform with $E^2 \ll B^2$.

An infinitely conducting plane piston starts moving at $t = 0$ and preserves its speed for $t > 0$. Suppose an ionization front develops in this case. If an infinitely conducting gas is in contact with the surface of an infinitely conducting body, all three velocity components must be continuous on the contact surface. This follows from the nonpenetration condition and the continuity of tangential components of the electric field in the coordinate system attached to the piston.

Thus, we can fix three velocity components on the piston surface. This agrees with the fact that infinitely conducting gas and magnetic field frozen in the gas constitute an anisotropic elastic medium (Landau and Lifshitz 1984).

Thus formulated problem is self-similar and its solution must depend on x/t. Only electromagnetic waves can propagate ahead of the ionization front, which is assumed to be supersonic. This is true if the normal component of the piston velocity is fairly high and directed into the gas. As is well known, variations of magnetic and electric fields in electromagnetic waves are equal in absolute value in nonconducting regions. Behind and immediately ahead of the ionization front $E = O(vB/c)$. Therefore, if the value of the initial electric field E is fairly small, its variation and, consequently, variations of B turn out to be small. In what follows, we neglect the variation of B in the electromagnetic wave. This means that we can fix the magnetic field ahead of the ionization front and the boundary condition determining the change in the electric field across the ionization front is satisfied. Thus, we can disregard this boundary condition. Note that the change in electric field fully determines an electromagnetic wave. Note also that the quantity B_3 is independent of x and t since the problem is one-dimensional.

The solution of the problem must consist of an ionization front and, possibly, successive MHD waves. If the ionization front is slow, then from the first of inequalities (7.8.14) it follows that MHD waves cannot propagate behind the ionization front, as their velocities are higher than its velocity. Since the part of the Hugoniot manifold corresponding to this type of the ionization fronts is three-dimensional, it is possible to choose the state behind the front such that the gas and piston velocity vectors coincide. Thus, we can expect that if a slow ionization front is formed, then all quantities are constant behind it and the gas and piston velocities coincide.

If the ionization front is intermediate, then in accordance with the second inequality of (7.8.14) the self-similar solution can incorporate a slow MHD wave travelling behind the front. This wave may be either a shock or a rarefaction wave. Variation of quantities in this wave is characterized by a single parameter. Variation of the quantities in the ionization front is characterized by two parameters. Using the three parameters, we can satisfy three boundary conditions on the piston.

If the ionization front is fast, then in accordance with the third of inequalities (7.8.14), two waves, namely, a slow magnetohydrodynamic wave and a rotational discontinuity whose velocity with respect to gas is equal to a_A, can propagate behind the front. Variation of quantities in the fast ionization front and in each of the two successive waves is characterized by a single parameter. Using the three parameters, we can satisfy the conditions on the piston.

7.8.5 Variation of the gas velocity across ionization fronts.

To avoid considering both subsonic ionization fronts and complex phenomena arising when the state of the gas turns out to be close to the ionization threshold, we shall restrict our attention to the investigation of velocity variation in ionization fronts in which increase in the density and pressure exceeds certain values. Velocity variation in ionization fronts without this restriction and the piston problem were investigated by Barmin and Kulikovskii (1968b).

We start from consideration of an ionization front with such propagation velocity W and with the large normal magnetic field component B_3 so that the inequality $V^0 < 4\pi B_0^2$ is satisfied. Here, $B_0 = B_3/4\pi m$, $m = W/V^0$, and V^0 is the specific volume of the gas ahead of the front, i.e., in the undisturbed state. This inequality means that the front velocity W is less than the Alfvén velocity $a_A = B_3\sqrt{V^0/4\pi}$ calculated via the initial density. The

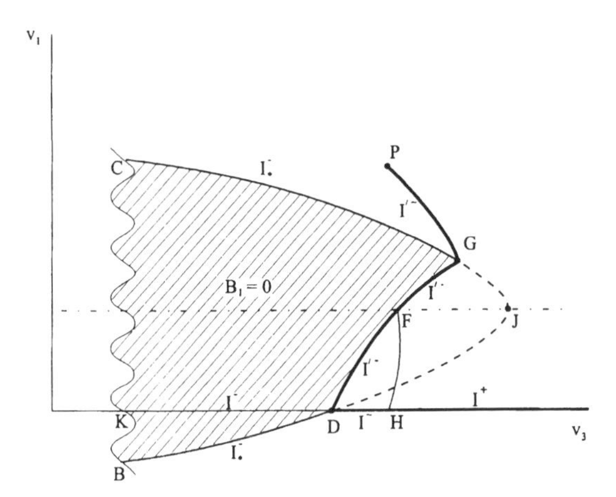

Figure 7.25 Projection of a Hugoniot manifold in the (v_3, v_1)-plane for plane ionization fronts.

equality $V = 4\pi B_0^2$ corresponds to the horizontal asymptote of the hyperbola (7.8.15). The quantity $4\pi B_0^2$ decreases for increasing W and m.

We shall assume that the ionization fronts under consideration propagate through a certain state of a gas at rest with $V = V^0$, $p = p^0$, $\mathbf{v} = 0$, and $\mathbf{B} = B_3\mathbf{e}_3 + B_1^0\mathbf{e}_1$. For the sake of simplicity, we shall consider the situation in which both the piston velocity and the initial magnetic fields lie in the (x_1, x_3)-plane (recall that $x_3 \equiv x$).

Among the ionization fronts of any type considered in Section 7.8.3 there are fronts with invariable magnetic field components B_1 and B_2. Change of the other quantities across these fronts is the same as in gas dynamic shock waves. Accordingly, only the velocity component v_3 does change and in the (v_3, v_1)-plane the state behind these fronts corresponds to the v_3-axis. In these ionization fronts the state behind the gas dynamic shock represents one of the singular points A_i of the system (7.8.13).

As mentioned above, we consider discontinuities with fairly large variations in quantities, i.e., ionization fronts behind which $v_3 > v_3(K)$, where K is a certain point on the v_3-axis (Fig. 7.25) and $v_3(K) > 0$ is the value of v_3 at this point.

For the given W and sufficiently large values of B_3, the mutual location of the curve a and the horizontal asymptote of the hyperbola (7.8.15) for fronts with invariable value of B_1 is shown in Fig. 7.26. The state ahead of the front is denoted by A_0. The state behind the front is represented by the point A_4.

As W increases, the mutual location of the curve a, the point A_0 and the hyperbola (7.8.15) changes and is qualitatively shown in Figs. 7.27 and 7.28. In this connection the final point becomes A_3, then A_2. Accordingly, in Fig. 7.25 the segments corresponding to slow (KD), intermediate (DH), and fast ionization fronts (H, ∞) are arranged on the v_3-axis from the left to the right.

Let us consider now the cases in which B_1 is not constant. The velocity component v_1 varies simultaneously with B_1 in accordance with the first equality of (7.8.11). If we vary E_0, the point A_4 in Fig. 7.26 can move from the position $B_1 = B_1^0$ in any direction. In this

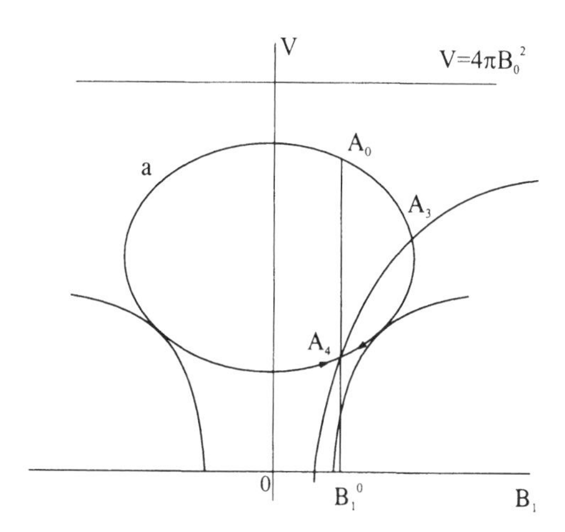

Figure 7.26 States ahead of and behind an ionization front (A_0 and A_4) and the location of Jouget points in the case of $V^{\circ} \ll 4\pi B_0^2$.

case the ionization front structure consists of a gas dynamic shock from the initial point to the lower part of the curve a, which occurs at the fixed B_1, and a part of the curve a specified by the first of Eqs. (7.8.10) from the state behind the gas dynamic shock to the point A_4. Position of the point A_4 with respect to the position $B_1 = B_1^0$ can be changed by varying E_0 (the curve a depends on E_0, but variations of E_0 only slightly affect its shape).

In slow ionization fronts I^- under consideration the maximum variations of B_1 and v_1 occurs if the hyperbola (7.8.15) touches the curve a from the right or the left (the positions of the hyperbola are shown in Fig. 7.26). At the points of tangency, the points A_3 and A_4 merge and, therefore, the Jouget condition is satisfied behind the discontinuity, i.e., the downstream gas velocity with respect to the front is equal to the slow magnetosonic velocity a_s (the velocity of an infinitesimal slow MHD shock). In particular, the fronts with $B_1 = 0$ are located between these ionizing Jouget fronts denoted by I_*^-. In Fig. 7.25 we have reproduced the range of the velocity components v_1 and v_3 denoted by I^-, which corresponds to variation of the velocity in slow ionization fronts. This domain is bounded from above and below by curves marked by I_*^-. Each point of these curves corresponds to a slow ionization front for which the Jouget condition $W = a_s$ is satisfied downstream.

Obviously, at the point D (see Fig. 7.25) the Jouget condition is satisfied. A comparison of Figs. 7.22 and 7.27 shows that at $v_3 = v_3(D)$ the Jouget point is located at $B_1 = B_1^0$. This corresponds to the movement of the Jouget curve across the v_3-axis (see Fig. 7.25) from the bottom to the top for increasing v_3. In Fig. 7.25 the Jouget curve for $v_3 > v_3(D)$ is continued by the dotted line.

Let us consider possible changes in v_1 for $v_3(D) < v_3 < v_3(H)$. Points on the segment DH correspond to intermediate ionization fronts $I^{\sim}$. As seen from Fig. 7.27, a complex discontinuity composed of a gasdynamic shock $A_0 \to A_3$ and a slow MHD shock wave $A_3 \to A_4$ that propagate with the same velocities is also possible. In what follows, this slow magnetohydrodynamic shock wave will be denoted by S^-. The directions of variation of B_1 is given by the first of Eqs. (7.8.13). It is shown by arrows in the figures. One can easily

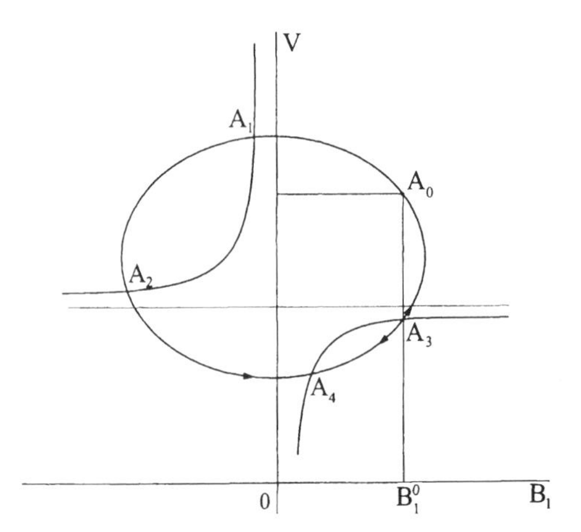

Figure 7.27 Intermediate ionization front ($A_0 \to A_3$) for continuous magnetic field. The limiting location of the state behind the slow ionization front (A_4).

observe that the integral curve cannot arrive at the point A_4 if the point A_3 lies to the left of the straight line $B_1 = B_1^0$. Accordingly, the extreme right position of the point A_4 in Fig. 7.27 corresponds to the presence of the above-mentioned double discontinuity $A_0 \to A_3 \to A_4$ denoted by $I^{\sim}S^-$. We denote this slow ionization front as $I'^- \equiv I^{\sim}S^-$. Obviously, the Jouget point at which $B_1 > 0$ lies between the points A_3 and A_4. In Fig. 7.25 the curve DF corresponds to I'^- and the dotted curve DJ to the Jouget points. In the (v_3, v_1)-plane the curve DF represents the lower boundary of the domain I^-. As earlier, the upper boundary is represented by the Jouget curve I_*^- corresponding to $B_1 < 0$.

If $E_0 = 0$, the curve I'^- issuing from the point D arrives at the point F at which $B_1 = 0$. The horizontal asymptote of hyperbola (7.8.15) passes in this case through the point that represents the state behind the gas dynamic shock starting from the point A_0. When E passes through zero, the points A_2 and A_3 representing the state behind the gas dynamic shock exchange their positions. Thus, the jump S^- into the point F occurs from the point H of the v_3-axis that separates the segments corresponding to the discontinuities $I^{\sim}$ and I^+. At $E_0 = 0$ the curve a becomes symmetric about the V-axis (see Eq. (7.8.12)). In this case the jump from the state H to the point with $B_1 = -B_1^0$ and $V = V_H$ that is symmetric about the V-axis is also possible. This is a rotational (or Alfvén) discontinuity. It propagates at the velocity $a_A = B_3\sqrt{V/4\pi}$ with respect to the gas ahead of and behind the discontinuity. In what follows, we denote it as A. In the plane case considered, the tangential component $\mathbf{B}_\tau = \mathbf{B} - B_3\mathbf{e}_3$ of the magnetic field rotates by 180°. The rotational discontinuity is known to have no steady-state structure. But at large t the width of A increases slower than the length scale in self-similar solution. This makes it possible to consider A as a discontinuity.

In Fig. 7.25 the point P denotes the state behind the combination of the discontinuities I_H and A moving with identical velocities. Here, I_H stands for the ionization front. The state behind this front corresponds to the point H at which $\{B_1\} = 0$.

Along with intermediate ionization fronts corresponding to the interval DH of the v_3-axis, there are other intermediate fronts. Their structure consists of a gas dynamic shock

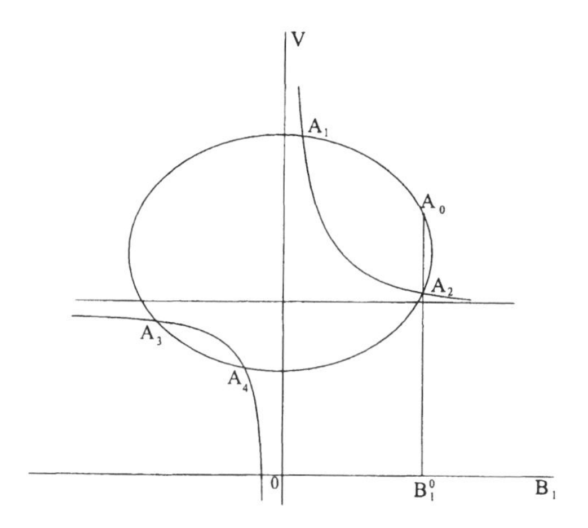

Figure 7.28 Fast ($A_0 \to A_2$) and intermediate ($A_0 \to A_3$) ionization fronts.

$A_0 \to A_2$ and the integral curve l (see Fig. 7.23) from A_2 to A_3. Clearly, the velocities of these fronts coincide with the velocity of the discontinuity $A_0 \to A_2$ representing a fast ionization front I^+. The state behind I^+ belongs to the v_3-axis to the right of the point H (Fig. 7.25). The fronts $A_0 \to A_2 \to A_3$ will be denoted by $I'^{\sim}$. In the (v_3, v_1)-plane the set of the fronts $I'^{\sim}$ corresponds to a curve (this is the curve PG in Fig. 7.25) that issues from the point P.

It is worth noting that there exists a slow ionization front consisting of the above-mentioned front $I'^{\sim}$ (transition $A_0 \to A_2 \to A_3$) and a slow shock wave S^- moving with the same velocity. In Fig. 7.28 this corresponds to the transition $A_3 \to A_4$. These ionization fronts will be denoted by I'^-, since they consist of the intermediate ionization front $I'^{\sim}$ and a slow MHD shock wave S^- moving with the same velocity. The set of possible states behind such fronts represents a curve FG that is a continuation of the curve DF in Fig. 7.25.

If $V > 4\pi B_0^2$ at points behind the gas dynamic discontinuity, i.e., these points lie above hyperbola's asymptote, then, as follows from the direction of variation in B_1, in order for an integral curve to terminate at the point A_4 it is necessary that the point A_2 lies to the right of the initial point in Fig. 7.28. With increase in $|E_0|$ the point A_4 goes along the curve a from its position represented in Fig. 7.28 (this is the state behind the front I'^-) to the Jouget point. In Fig. 7.28 the point A_3 corresponding to the state behind the front $I'^{\sim}$ lies to the left of the Jouget point. Accordingly, in the (v_3, v_1)-plane the curve corresponding to the fronts $I'^{\sim}$ lies above the Jouget curve I_*^-, while the curve corresponding to points of type of A_4 lies below the Jouget curve. This can be seen from Fig. 7.28.

As the velocity of the discontinuity increases, the horizontal asymptote of the hyperbola passes lower and the points A_3 and A_4 representing the states behind $I'^{\sim}$ and I'^- approach each other (see Fig. 7.28). At a certain velocity of the discontinuity these points merge and at a higher velocity they disappear. Accordingly, as v_3 increases, the curves $I'^{\sim}$, I_*^-, and I'^- in the v_3, v_1 plane initially approach each other and then terminate at a single point (point G in Fig. 7.25).

Thus, in the (v_3, v_1)-plane (Fig. 7.25) the set of states behind all possible ionization fronts propagating through the state with given $p = p^0$, $V = V^0$, $\mathbf{v} = 0$, and $\mathbf{B} = B_3\mathbf{e}_3 + B_1^0\mathbf{e}_1$ consists of the following fragments: (i) a part of the plane bounded by the closed curve *KBDFGC* whose points represent the states behind I^-; (ii) the interval *DH* on the v_3-axis and the curve *PG* whose points represent the states behind $I^\sim$ and $I'^\sim$; (iii) points of the v_3-axis to the right of the point *H* that represent the states behind I^+. The set (i)+(ii)+(iii) is the admissible part of the Hugoniot manifold in the (v_3, v_1)-plane. More exactly, it is the projection of the plane-polarized part of the Hugoniot manifold to this plane. Recall that by our assumptions B_2 and hence v_2 are equal to zero behind the fronts.

If we do not restrict ourselves to consideration of plane-polarized fronts, that is, if we also consider the fronts with $v_2 \neq 0$, then the slow ionization fronts (I^-) will correspond to a three-dimensional domain in the (v_1, v_3, v_2)-space, the intermediate ionization fronts $(I^\sim)$ will correspond to a surface intersecting the (v_3, v_1)-plane along the curves *DH* and *PG*, and the fast ionization fronts (I^+) will correspond to a ray that belongs to the v_3-axis and goes from the point *H* to infinity.

7.8.6 Constructing the solution of the piston problem. The Hugoniot manifold constructed above for plane-polarized fronts in the v_3, v_1 plane, on the other hand, represents the solution of the piston problem for piston's velocity belonging to this Hugoniot manifold. In this case the solution consists of a single ionization front. Let us consider the consequences of the presence of other waves in the solution of the piston problem.

One of these waves is the slow Riemann wave (R^-). Without detailed consideration of variation of the physical quantities in this wave, we can note that the signs of variations of the specific volume V (and, consequently, v_3) and the magnetic field component B_1 in a slow Riemann wave are opposite to those in a slow MHD shock wave (Landau and Lifshitz 1984). Hence we can conclude that in the expanding slow Riemann wave v_3 decreases and $|v_1|$ increases. Note, however, that in order to avoid the necessity of introducing recombination fronts (Butler 1965; Kulikovskii 1968; Barmin 1970) we must not proceed too far along the integral curves describing slow Riemann waves expanding with time.

Note that a slow Riemann wave can propagate behind a slow ionization front only if the latter is a Jouget front (I_*^-). This makes it possible to construct the solution of the piston problem for domains that adjoin the curves I_*^- bounding the domain I^- from the outside. In Fig. 7.29 these domains are denoted as $I_*^-R^-$.

Slow Riemann and shock waves can propagate behind fast and intermediate ionization fronts. In Fig. 7.29 the corresponding velocity domains are denoted as $I^\sim R^-$, $I^\sim S^-$, I^+R^-, and I^+S^-. Note that the domain $I^\sim S^-$ adjoins the domain I^- along the curve *DF*. On this curve the velocities of the waves $I^\sim$ and S^- become identical. The solution $I^\sim S^-$ ceases to exist above the curve *DF* due to the fact that the velocity of S^- becomes higher then the velocity behind $I^\sim$ (reverse relationship between the velocities of the fronts). The domain corresponding to the solution I^+S^- is bounded by the curve *FF'* on which $B_1 = 0$.

There are also domains *P'PGQQ'* and *PMFG* adjoining the curve *PG*. In the former and in the latter domains our solution has the form $I'^\sim R^-$ and $I'^\sim S^-$, respectively. On the curve *FG* the velocities of $I'^\sim S^-$ and I'^- coincide and below this curve the solution $I'^\sim S^-$ ceases to exist due to the reverse relationship between the velocities of the discontinuities.

On the curve *PQQ''* the solution has the form I^+A. The curve itself can be characterized by the fact that $B_1 = -B_1^0$ on it. In the domains *Q''QPMFF'* and *P'PQQ''* the solution has

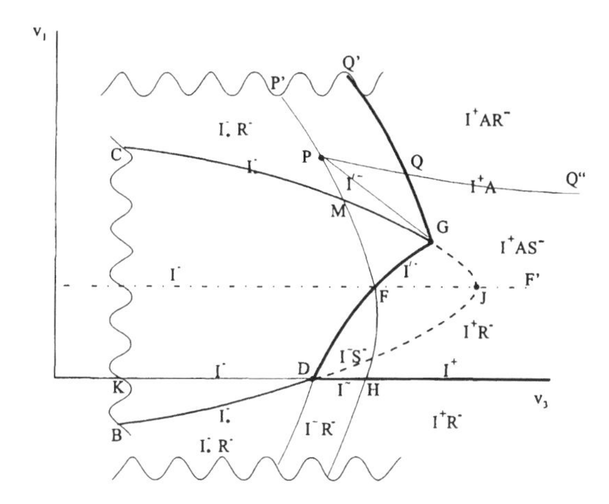

Figure 7.29 Solution of the piston problem in the presence of an ionization front as a function of the piston velocity.

the form I^+AS^- and I^+AR^-, respectively.

Thus, in the (v_3, v_1)-plane there exists domain $P'PMFGQQ'$ in which there are several solutions of the piston problem. In the other considered part of the (v_3, v_1)-plane the solution exists and is unique.

Let us outline the solutions in the domain of nonuniqueness. They are represented by $I_*^-R^-$, $I'^{\sim}R^-$, and I^+AR^- in the domain $P'PQQ'$; by $I_*^-R^-$, $I'^{\sim}R^-$, and I^+AS^- in the domain PQG; by $I_*^-R^-$, $I'^{\sim}S^-$, and I^+AS^- in the domain PGM; and by I^-, $I'^{\sim}S^-$, and I^+AS^- in the domain MGF.

Note that in considering the piston problem with three nonzero velocity components of the piston the solutions containing I^-, $I^{\sim}$, and I^+ correspond to three-dimensional domains. In this case, if the ionization front is slow, then this front is a single wave in the solution of the problem; if the ionization front is intermediate, then there is also a slow wave (Riemann or shock wave); and, finally, if the ionization front is fast, then in addition to the above-mentioned waves, the solution contains a rotational discontinuity.

If we pass from one into another domain of the values of **v**, the number of the waves composing the solutions changes. A typical situation is splitting-off or merging of discontinuities on the boundaries between the domains. For example, on one side of the curve DF the solution consists of a single wave I^-, whereas on the other side the solution consists of two waves $I^{\sim}S^-$. On the curve DF itself the velocities of the waves $I^{\sim}$ and S^- coincide. When merging or splitting, the discontinuities preserve the number of the free parameters characteristic of these discontinuities.

7.9 Discussion

In this chapter we presented several examples of unusual behavior of discontinuities occurring in solutions of hyperbolic systems and considered problems for which the solutions of self-similar problems are not unique. Certain possible reasons of the nonuniqueness were indicated.

Primarily, the presented results clearly show that in the generic case the hyperbolic systems themselves and the corresponding conservation laws (in other words, large-scale models) do not determine unique solutions. Already in 1961, Godunov proposed an example of the equation whose admissible discontinuities depend on the form of functions specifying dissipation coefficients. In Sections 7.4, 7.5, and 7.6 we considered several physical problems with the same large-scale models. However, the set of admissible discontinuities, that is, discontinuities consistent with the corresponding small-scale models, turns out to be different, since the models describing the discontinuity structure are different.

If the small-scale model leads to oscillations in the discontinuity structure, the set of admissible discontinuities can have a specific dispersive configuration described in Sections 7.5, 7.6, and 7.7. In such cases the solutions that can be constructed for self-similar problems by using only admissible discontinuities are explicitly nonunique. Solutions of self-similar problems can also be nonunique both in the case of the classical behavior of discontinuities, for example, in elastic media (Sections 7.4), and in the presence of discontinuities with phase transitions (Sections 7.8). Hugoniot manifolds can be multidimensional, and the number of additional relations on discontinuity can be greater than unity.

The nonuniqueness of solutions of self-similar problems means that the large-scale model (consisting of a hyperbolic system and a set of admissible solutions) is not complete. If we use a more complete system of equations (small-scale model) for which the solutions are continuous, we can expect the solutions be continuous. Consequently, when studying large-scale phenomena, in addition to the choice of admissible discontinuities it is necessary to know how to make a proper selection among solutions of self-similar problems when the need arises. One of the selection rules is formulated on the basis of numerical experiments for the solution of the Riemann problem of the interaction of two shock waves in elastic media (Section 7.4). Generally, we cannot formulate more or less universal rules to determine the type of the self-similar solution.

All these facts are of great importance for developing numerical methods for solving the problems governed by hyperbolic systems and admitting discontinuous solutions. Apparently, when analyzing large-scale phenomena in the cases similar to those considered in this chapter, it is necessary to carry out at least local calculations in narrow space zones for short time intervals in more detail using the systems suitable for description of small-scale phenomena. It is crucial that the calculations in the narrow zones must take into account the physics of the processes occurring inside them. Otherwise, we can make an error even in estimating the set of admissible discontinuities. In particular, as follows from Section 7.7, similar errors can arise in the attempt to apply certain numerical methods developed for physical situations without dispersion to the calculation of processes with oscillations developing in the shock wave structure.

Bibliography

Abarbanel, S., Gottlieb, D. (1973) Higher order accuracy finite difference algorithms for quasi-linear, conservation law hyperbolic systems, *Math. Comput.* **27**, 505–523.

Abarbanel, S., Zwas, G. (1971) Third and fourth order accurate schemes for hyperbolic equations of conservation law form, *Math. Comput.* **25**, 229–236.

Abarbanel, S., Gottlieb, D., Turkel, E. (1975) Difference schemes with fourth order accuracy for hyperbolic equations, *SIAM J. Appl. Math.* **29**, 329–351.

Abdukadyrov, S.A., Pinchukova, N.I., Stepanenko, M.V. (1984) An algorithm of numerical solution of dynamic equations of elastic media and structures, *Fiziko-Tekhn. Problemy Razrab. Poleznyh Iskopaemyh [Soviet Mining Sci.]*, No. 6, 34–41 [in Russian].

Abramovich, G.N. (1976) *Applied Gas Dynamics*, Nauka, Moscow [in Russian].

Akhiezer, A.I., Lyubarskii, G.Ya., Polovin, R.V. (1959) On the stability of shock waves in magnetohydrodynamics, *Soviet Physics–JETP* **8**, 507–511.

Akhiezer, A.I., Akhiezer, I.A., Polovin, R.V., Sitenko, A.G., Stepanov, K.N. (1975) *Plasma Electrodynamics*, Pergamon Press, London.

Alalykin, G.V., Godunov, S.K., Kireeva, I.L., Pliner, L.A. (1970) *Solution of One-Dimensional Gas Dynamic Problems on Moving Grids*, Nauka, Moscow [in Russian].

Ambartsumyan, S.A., Bagdasaryan, G.E., Belubekyan, M.V. (1977) *Magnetoelasticity of Thin Shells and Plates*, Nauka, Moscow [in Russian].

Anderson, D.A. (1974) A comparison of numerical solutions to the inviscid equations of fluid motion, *J. Comput. Phys.* **15**, 1–20.

Anderson, W.K., Thomas, J.L., van Leer, B. (1985) A comparison of finite volume flux vector splittings for Euler equations, *AIAA Paper*, No. 85-0122.

Andreev, A.A., Kholodov, A.S. (1989) On three-dimensional supersonic flow past blunt bodies when interference is present, *U.S.S.R. Comput. Maths Math. Phys.* **29**, No. 1, 103–107.

Aslan, N. (1993) *Computational Investigations of Ideal MHD Plasmas with Discontinuities*, Ph.D. Thesis, Nuclear Engineering Department, University of Michigan.

Aslan, N. (1996a) Numerical solutions of one-dimensional MHD equations by fluctuation approach, *Int. J. Numer. Meth. Fluids* **22**, 569–580.

Aslan, N. (1996b) Two-dimensional solutions of MHD equations with an adapted Roe method, *Int. J. Numer. Meth. Fluids* **23**, 1211–1222.

Aslan, N. (1999) MHD-A: A fluctuation splitting wave model for planar magnetohydrodynamics, *J. Comput. Phys.* **153**, 437–466.

Atluri, S.A. (Ed.) (1986) *Computational Methods in the Mechanics of Fracture*, Computational Methods in Mechanics **2**, Elsevier, New York.

Axford, W.I. (1972) The interaction of the solar wind with the interstellar medium, in *Solar Wind*, NASA SP-308, 609–658.

Azarenok, B.N., Ivanenko, S.A. (1999) Moving grid technology for finite volume methods in gas dynamics, in *Finite Volumes for Complex Applications II - Problems and Perspectives*, R. Vilsmaeier,

F. Benkhaldoum, and D. Hanel (Eds.), 795–802, Hermes, Paris.

Azarenok, B.N., Ivanenko, S.A. (2000) Application of adaptive grids in numerical analysis of time-dependent problems in gas dynamics, *Comput. Maths Math. Phys.* **40**, No. 9, 1330–1349.

Azarova, O.A., Vlasov, V.V., Grudnitskii, V.G., Popov, N.A., Rygalin, V.N. (1993a) Difference schemes of minimal stencil and its application in shock-fitting algorithms, in *Algorithms for Numerical Simulation of Discontinuous Flows*, V.M. Borisov (Ed.), 9–55, Computing Center, Russian Academy of Sciences, Moscow [in Russian].

Azarova, O.A., Vlasov, V.V., Grudnitskii, V.G., Rygalin, V.N. (1993b) Simulation of one-dimensional nonstationary gas dynamic flows with fitting of shock waves and contact surfaces, in *Algorithms for Numerical Simulation of Discontinuous Flows*, V.M. Borisov (Ed.), 56–79, Computing Center, Russian Academy of Sciences, Moscow [in Russian].

Babenko, K.I., Ivanova, V.N., Kosorukov, A.L., Radvogin, Yu.B. (1980) *Supersonic Flow near Smooth Bodies Including Chemical Nonequilibrium*, Preprint No. 54, Keldysh Institute of Applied Mathematics, USSR Academy of Sciences, Moscow [in Russian].

Baker, T.J. (1999) Delaunay–Voronoï methods, Chapter 16 in *Handbook of Grid Generation*, J.F. Thompson, B.K. Soni, and N.P. Weatherill (Eds.), CRC, Boca Raton, FL.

Bakhvalov, N.S., Eglit, M.E. (2000) Effective dispersive equations for wave propagation in periodic media, *J. Russian Acad. Sci. Math. Doklady* **61**, No. 1, 1–4.

Bakhvalov, N.S., Panasenko, G.P. (1989) *Homogenization. Averaging Processes in Periodic Media*, Kluwer, Dortrecht.

Baklanovskaya, V.F., Pal'tsev, B.V., Chechel, I.I. (1979) Boundary value problems for St. Venant's equations on the plane, *U.S.S.R. Comput. Maths Math. Phys.* **19**, No. 3.

Balashov, D.B. (1993) Decay of discontinuity in linearly strain-hardening elastic-plastic medium, *Mech. Solids* **28**, No. 2, 113–120.

Balsara, D.S. (1998a) Linearized formulation of the Riemann problem for the adiabatic and isothermal magnetohydrodynamics, *Astrophys. J. Suppl.* **116**, 119–131.

Balsara, D.S. (1998b) Total variation diminishing scheme for adiabatic and isothermal magnetohydrodynamics, *Astrophys. J. Suppl.* **116**, 133–153.

Balsara, D.S., Spicer, D.S. (1999) A staggered mesh algorithm using high order Godunov fluxes to ensure solenoidal magnetic fields in magnetohydrodynamic simulations, *J. Comput. Phys.* **149**, 270–292.

Banichuk, N.V., Kobelev, V.V., Ricards, R.B. (1988) *Optimization of Elements of Structures Made of Composite Materials*, Mashinostroenie, Moscow [in Russian].

Baranov, V.B. (1990) Interaction of the solar wind with the external plasma, in *Physics of the Outer Heliosphere*, 287–297, Pergamon Press, New York.

Baranov, V.B., Krasnobaev, K.V. (1977) *Hydrodynamic Theory of Cosmic Plasma*, Nauka, Moscow [in Russian].

Baranov, V.B., Malama, Yu.G. (1993) Model of the solar wind interaction with the local interstellar medium: Numerical solution of self-consistent problem, *J. Geophys. Res.* **98**, A15157–A15163.

Baranov, V.B., Zaitsev, N.A. (1995) On the problem of the solar wind interaction with magnetized interstellar plasma, *Astron. Astrophys.* **304**, 631–637.

Baranov, V.B., Zaitsev, N.A. (1998) On the problem of the heliospheric interface response to cycles of the solar activity, *Geophys. Res. Lett.* **25**, 4051–4054.

Baranov, V.B., Ermakov, M.K., Lebedev, M.G. (1981) A three-component model of the solar wind–interstellar medium interaction: some numerical results, *Sov. Astron. Lett.* **7**, No. 3, 206–208.

Baranov, V.B., Krasnobaev, K.V., Kulikovskii, A.G. (1971) A model of solar wind–ISM interaction, *Soviet Physics Doklady* **15**, 791–793.

Baranov, V.B., Lebedev, M.G., Ruderman, M.S. (1979) Structure of the region of solar wind–interstellar medium interaction and its influence on H atoms penetrating the solar wind, *Astrophys. Space Sci.* **66**, 441–451.

Barkhudarov, E.M., Mdvinishvili, M.O., Taktakishvili, M.I., Tsintsadze, N.L., Chelidze, T.Ya. (1987) Interferometric study of the passage of a shock wave through a laser spark, *Sov. Tech. Phys.* **32**, 1409.

Barmin, A.A. (1962) A study of discontinuity surfaces with energy release (absorption) in magnetohydrodynamics, *J. Appl. Math. Mech.* **26**, 801–810.

Barmin, A.A. (1970) Recombination fronts for arbitrary magnetic field orientation, *Fluid Dynamics* **5**, No. 3, 359–363.

Barmin, A.A., Gogosov, V.V. (1960) Piston problem in magnetohydrodynamics, *Doklady Akademii Nauk SSSR [Soviet Physics Doklady]* **134**, No. 5, 1041–1043 [in Russian].

Barmin, A.A., Kulikovskii, A.G. (1968a) On shock waves ionizing gas in a magnetic field, *Doklady Akademii Nauk SSSR [Soviet Physics Doklady]* **178**, 55–58 [in Russian].

Barmin, A.A., Kulikovskii, A.G. (1968b) Gas velocity variation in ionizing shock waves. The problem of the conductive piston, *J. Appl. Math. Mech.* **32**, 506–510.

Barmin, A.A., Kulikovskii, A.G. (1969) Gas-ionizing shock waves in an arbitrary oriented magnetic field, in *Problems of Hydrodynamics and Continuum Mechanics*, SIAM, Philadelphia, 39–54.

Barmin, A.A., Kulikovskii, A.G. (1971) Ionization and recombination fronts in electromagnetic field, *Itogi Nauki. Hydromechanics* **5**, 5-31, VINITI, Moscow [in Russian].

Barmin, A.A., Kulikovskii, A.G., Pogorelov, N.V. (1996) Shock-capturing approach and nonevolutionary solutions in magnetohydrodynamics, *J. Comput. Phys.* **126**, 77–90.

Barth, T.J., Jespersen, D.C. (1989) The design and application of upwind schemes on unstructured meshes, *AIAA Paper*, No. 89-0366.

Batani, D., Koenig, M., Benuzzi, A., Krasyuk, I.K., Pashinin, P.P., Semenov, A.Yu., Lomonosov, I.V., Fortov, V.E. (1999a) Problems in the optical measurement of dense plasma heating in laser shock wave compression, *Plasma Phys. Control. Fusion* **41**, 93–103.

Batani, D., Koenig, M., Benuzzi, A., Krasyuk, I.K., Pashinin, P.P., Semenov, A.Yu., Lomonosov, I.V., Fortov, V.E. (1999b) Problems of measurement of dense plasma heating in laser shock-wave compression, *Laser Part. Beams* **17**, No. 2, 265–273.

Bayliss, A., Turkel, E. (1982) Far field boundary conditions for compressible flows, *J. Comput. Phys.* **48**, 182–199.

Bazhenov, V.G., Chekmarev, D.T. (1983) An approach of finite-difference approximation of functions and their derivatives in numerical solution of problems of the Timoshenko theory of plates and shells, in *Applied Problems of Strength and Plasticity. Algorithmization and Automation of Solution of the Problems of Elasticity and Plasticity*, No. 25, 15–22, Gor'kii State University, Gor'kii [in Russian].

Beam, R.M., Warming, R.F. (1976) Upwind second-order difference schemes and applications in aerodynamic flows, *AIAA J.* **14**, 1241–1249.

Belikov, V.V., Semenov, A.Yu. (1985) Godunov's type method with the Kolgan modification for numerical solution of 2D shallow water equations, in *Proc. 10th Conf. MPTI Young Sci., March 23–Apr. 7, 1985*, Pt. 1, No. 5983-85, 179–214, VINITI, Moscow [in Russian].

Belikov, V.V., Semenov, A.Yu. (1988a) *An Explicit Numerical Method for Simulation of Shallow Water Equations Based on Exact Solution of the Riemann Problem*, Preprint No. 42, General Physics Institute, USSR Academy of Sciences, Moscow [in Russian].

Belikov, V.V., Semenov, A.Yu. (1988b) Godunov's type method for numerical solution of the two-dimensional shallow water equations, in *Proc. 17th Session Sci. Methodol. Seminar on Ship*

Hydrodynamics, Varna, Oct. 17-22, 1988 **2**, 56/1–56/6.

Belikov, V.V., Semenov, A.Yu. (1997a) New non-Sibson interpolation on arbitrary system of points in Euclidean space, in *15th IMACS World Congress on Scientific Computation, Modelling and Applied Mathematics, Berlin, August 1997*, **2**: *Numerical Mathematics*, A. Sydow (Ed.), 237–242, Wissenshaft und Technik Verlag, Berlin.

Belikov, V.V., Semenov, A.Yu. (1997b) A Godunov-type method based on an exact solution to the Riemann problem for the shallow-water equations, *Comput. Maths Math. Phys.* **37**, No. 8, 974–986.

Belikov, V.V., Semenov, A.Yu. (1997c) The Godunov-type method for numerical modelling of shallow water flows, in *Computational Fluid Dynamics of Natural Flows*, Trudy IOFAN [Proc. Gen. Phys. Inst., Russian Acad. Sci.], **53**, 5–43, Nauka/Fizmatlit, Moscow [in Russian].

Belikov, V.V., Semenov, A.Yu. (1998) A Godunov's type method based on an exact solution to the Riemann problem for the shallow water equations, in *Proc. 4th European Comput. Fluid Dyn. Conf., Athens, Sept. 7-11, 1998*, K.D. Papailiou et al. (Eds.), **1**, No. 1, 310–315, John Wiley, Chichester, U.K.

Belikov, V.V., Semenov, A.Yu. (2000) Non-Sibsonian interpolation on arbitrary system of points in Euclidean space and adaptive isolines generation, *Appl. Numer. Math.* **32**, No. 4, 371–387.

Belikov, V.V., Ivanov, V.D., Kontorovich, V.K., Korytnik, S.A., Semenov, A.Yu. (1997) The non-Sibsonian interpolation: A new method of interpolation of the values of a function on an arbitrary set of points, *Comput. Maths Math. Phys.* **37**, 9–15.

Bell, J.F. (1973) *Encyclopedia of Physics. Mechanics of Solids I*, Volume VIa/1, C.A. Truesdell (Ed.), Springer, New York.

Bell, J.B., Colella, P., Trangenstein, J.A. (1989) Higher order Godunov methods for general systems of conservation laws, *J. Comput. Phys.* **82**, 362–397.

Belotserkovskii, O.M., Grudnitskii, V.G. (1980) Investigation of non-stationary gas flows with a complex internal structure by methods of integral relations, *U.S.S.R. Comput. Maths Math. Phys.* **20**, No. 6.

Belotserkovskii, O.M., Kholodov, A.S. (1999) Majorizing schemes on unstructured grids in the space of indeterminate coefficients, *Comput. Maths Math. Phys.* **39**, 1730–1747.

Belotserkovskii, O.M., Grudnitskii, V.G., Prokhorchuk, Yu.A. (1983) Difference scheme of second-order of accuracy on the minimal pattern for hyperbolic equations, *U.S.S.R. Comput. Maths Math. Phys.* **23**, No. 1, 81–86.

Belotserkovskii, O.M., Grudnitskii, V.G., Rygalin, V.N. (1983) The isolation of discontinuities in the calculation of one-dimensional nonsteady gas flows, *Soviet Physics Doklady* **28**, 695–697.

Belotserkovskii, O.M., Gushchin, V.A., Kon'shin, V.N. (1987) The splitting method for investigating flows of a stratified liquid with a free surface, *U.S.S.R. Comput. Maths Math. Phys.* **27**, No. 2, 181–191.

Belotserkovskii, O.M., Kholodov, A.S., Turchak, L.I. (1986) Grid-characteristic methods in multidimensional problems of gas-dynamics, in *Current Problems in Computational Fluid Dynamics*, O.M. Belotserkovskii and V.P. Shidlovsky (Eds.), 125–189, Mir, Moscow.

Belotserkovskii, O.M., Golovachev, Yu.P., Grudnitskii, V.G. et al. (1974a) *Numerical Investigation of Modern Problems in Gas Dynamics*, O.M. Belotserkovskii (Ed.), Nauka, Moscow [in Russian].

Belotserkovskii, O.M., Osetrova, S.D., Fomin, V.N., Kholodov, A.S. (1974b) The hypersonic flow of a radiating gas over blunt body, *U.S.S.R. Comput. Maths Math. Phys.* **14**, No. 4, 168–179.

Belotserkovskii, O.M., Demchenko, V.V., Kosarev, V.I., Kholodov, A.S. (1978) Numerical simulation of some problems of shells compression by laser, *U.S.S.R. Comput. Maths Math. Phys.* **18**, No. 2.

Belov, N.A., Myasnikov, A.V. (1999) On the stability of the contact surface separating two hypersonic source flows, *Fluid Dynamics* **34**, 379–387.

Belytshko, T., Krongauz, Y., Organ, D., Fleming, M., Krysl, P. (1996) Meshless methods: An overview and recent developments, *Comput. Meth. Appl. Mech. Eng.* **139**, No. 1-4, 3–47.

Ben-Artzi, M., Falkovitz, J. (1984) A second-order Godunov-type scheme for compressible fluid dynamics, *J. Comput. Phys.* **55**, 1–32.

Benz, W. (1988) Applications of smooth particle hydrodynamics (SPH) to astrophysical problems, *Comput. Phys. Comm.* **48**, 97–105.

Berdichevskii, V.L. (1983) *Variational Principles of Mechanics of Continuous Media*, Nauka, Moscow [in Russian].

Bland, D.R. (1969) *Nonlinear Dynamic Elasticity*, Blaisdell, Waltham, MA.

Blazhevich, Y.V., Ivanov, V.D., Petrov, I.B., Petviashvili, J.V. (1999) High impact modeling with smooth particle hydrodynamics, *Matematicheskoe Modelirovanie [Math. Modeling]* **11**, No. 1, 88-100 [in Russian].

Bloch, V.I. (1964) *Theory of Elasticity*, Khar'kov State University, Khar'kov [in Russian].

Blum, P.W., Fahr, H.J. (1969) Solar wind tail and the anisotropic production of fast hydrogen atoms, *Nature* **223**, 936–937.

Bogatyrev, Yu.K. (1974) *Pulse Tools with Nonlinear Distributed Parameters*, Sovetskoe Radio, Moscow [in Russian].

Bol'shov, L.A., Burdonskii, I.N., Velikovich, A.L. et al. (1987) Acceleration of foils by a pulsed laser beam, *Soviet Physics–JETP* **65**, 1160–1169.

Bondarenko, Yu.A., Burdonskii, I.N., Gavrilov, V.V. et al. (1981) Experimental study of the acceleration of thin foils by high-power laser pulses, *Soviet Physics–JETP* **54**, 85–89.

Boris, J.P., Book D.L. (1973) Flux-corrected transport. I. SHASTA, a fluid transport algorithm that works, *J. Comput. Phys.* **11**, 38–69.

Boris, J.P., Book, D.L. (1976) Flux-corrected transport. III. Minimal-error FCT algorithms, *J. Comput. Phys.* **20**, 397–431.

Boris, J.P., Book, D.L., Hain, K. (1975) Flux-corrected transport. II. Generalizations of the method, *J. Comput. Phys.* **18**, 248–283.

Borrel, M., Montagne, J.L. (1985) Numerical study of a non-centered scheme with application to aerodynamics, *AIAA Paper*, No. 85-1497.

Bowyer, A. (1981) Computing Dirichlet tesselations, *Comput. J.* **24**, No. 2, 162–166.

Brackbill, J.U., Barnes, D.C. (1980) The effect of nonzero $\nabla \cdot \mathbf{B}$ on the numerical solution of the magnetohydrodynamic equations, *J. Comput. Phys.* **35**, 426–430.

Brackbill, J.U., Monaghan, J.J. (Eds.) (1988) *Particle Methods in Fluid Dynamics and Plasma Physics*, Comput. Phys. Comm. **48**, No. 1.

Brandt, J.C. (1970) *Introduction to the Solar Wind*, W.H. Freeman, New York.

Briggs, R.J. (1964) *Electron Stream Interactions with Plasma*, MIT Press, Cambridge, MA.

Brio, M., Wu, C.C. (1988) An upwind differencing scheme for the equations of ideal magnetohydrodynamics, *J. Comput. Phys.* **75**, 400–422.

Bronshtein, I.N., Semendyayev, K.A. (1973) *A Guide Book to Mathematics; Fundamental Formulas, Tables, Graphs, Methods*, Springer, New York.

Brueckner, K.A., Jorna, S. (1974) Laser-driven fusion, *Revs. Mod. Phys.* **46**, 325–367.

Burenin, A.A., Chernyshov, A.D. (1978) Shock waves in an isotropic elastic medium under plane finite strain, *J. Appl. Math. Mech.* **42**, 758–765.

Burges, M.D.J., Dragila, R., Luther-Davies, B. et al. (1985) Characterization of plasmas produced by a laser line focus, *Phys. Rev. A: Gen. Phys.* **32**, 2899–2908.

Burstein, S.Z., Mirin, A.A. (1970) Third order difference methods for hyperbolic equations, *J. Comput. Phys.* **5**, 547–571.

Bushman, A.V., Fortov, V.E. (1983) Models of equations of state, *Usp. Fiz. Nauk* **140**, No. 2, 177–232 [in Russian].

Bushman, A.V., Lomonosov, I.V., Fortov, V.E. (1992) *Equations of State for Metals at High Density of Energy*, Inst. Chem. Phys., Russian Academy of Sciences, Chernogolovka [in Russian].

Bushman, A.V., Kanel, G.I., Ni, A.L., Fortov, V.E. (1993) *Intense Dynamic Loading of Condensed Matter*, Taylor and Francis, London.

Butler, D.S. (1965) One-dimensional flow in an ionizing gas, *J. Fluid Mech.* **23**, 1–21

Bykovtsev, G.I., Ivlev, D.D. (1998) *Theory of Plasticity*, Dalnauka, Vladivostok [in Russian].

Bykovtsev, G.I., Kretova, L.D. (1972) Shock wave propagation in elastic-plastic media, *J. Appl. Math. Mech.* **36**, 94–104.

Cargo, P., Gallice, G. (1995) Un solveur de Roe pour les équations de la magnétohydrodynamique, *C. R. Acad. Sci. Paris* **320**, Serie 1, 1269–1272.

Cargo, P., Gallice, G. (1997) Roe matrices for ideal MHD and systematic construction of Roe matrices for systems of conservation laws, *J. Comput. Phys.* **136**, 446–466.

Chakravarthy, S.R., Osher, S. (1983) Numerical experiments with the Osher upwind scheme for the Euler equations, *AIAA J.* **21**, 1241–1248.

Champney, J.H., Chaussee, D.S., Kutler, P. (1982) Computation of blast wave–obstacle interactions, *AIAA Paper*, No. 82-0227.

Chan, W.M. (1999) Hyperbolic methods for surface and field grid generation, Chapter 5 in *Handbook of Grid Generation*, J.F. Thompson, B.K. Soni, and N.P. Weatherill (Eds.), CRC, Boca Raton, FL.

Chao, J.K., Lyu, L.H., Wu, B.H. et al. (1993) Observation of an intermediate shock in interplanetary space, *J. Geophys. Res.* **98**, 17443–17450.

Charakhch'yan, A.A. (1979) A version of the completely conservation scheme for the gasdynamic equations, *U.S.S.R. Comput. Maths Math. Phys.* **19**, No. 1, 272–276.

Charakhch'yan, A.A (1992) Application of moving regular grids to computation of gasdynamic flows with interfaces, in *Modern Problems in Computational Aerohydrodynamics*, A.A. Dorodnicyn and P.I. Chushkin (Eds.), 189–210, CRC Press, London.

Charakhch'yan, A.A. (1997a) Stability of shaped-charge jets generated under pulse action on conical targets, *J. Appl. Mech. Techn. Phys.* **38**, No. 3, 337–340.

Charakhch'yan, A.A. (1997b) Numerical investigation of circular cumulative jets compressing deuterium in conical targets, *Plasma Phys. Control. Fusion* **39**, No. 2, 237–247.

Charakhch'yan, A.A. (2000a) On algorithms for computing the Riemann problem by means for Godunov's scheme, *Comput. Maths Math. Phys.* **40**, 746–760.

Charakhch'yan, A.A. (2000b) Richtmyer–Meshkov instability of an interface between two media due to passage of two successive shocks, *J. Appl. Mech. Techn. Phys.* **41**, No. 1, 23–31.

Charakhch'yan, A.A., Krasyuk, I.K., Pashinin, P.P., Semenov, A.Yu. (1999) On mechanism of deuterium heating in laser experiments with conical targets, *Laser Part. Beams* **17**, 749–752.

Chernikh, K.F. (1996) *Introduction to the (Physical and Geometrical) Nonlinear Theory of Fracture*, Nauka/Fizmatlit, Moscow [in Russian].

Chow, E., Monaghan, J.J. (1997) Ultrarelativistic SPH, *J. Comput. Phys.* **134**, 296–305.

Chu, C.K., Taussig, R.T. (1967) Numerical experiments of magnetohydrodynamic shocks and the stability of switch-on shocks, *Phys. Fluids* **10**, 249–256.

Chugainova, A.P. (1988) The formation of a self-similar solution for the problem of non-linear waves in an elastic half-space, *J. Appl. Math. Mech.* **52**, 541–545.

Chugainova, A.P. (1991) *Peculiarities of Nonlinear Waves Behaviour in Elastic and Viscoelastic Media*, Ph.D. Thesis, Moscow State University, Moscow [in Russian]

Chugainova, A.P. (1993a) The interaction of nonlinear waves in slightly anisotropic viscoelastic medium, *J. Appl. Math. Mech.* **57**, 375–381.

Chugainova, A.P. (1993b) Transformation of a nonlinear elastic wave in a medium with weak anisotropy, *Mech. Solids* **28**, 71–83.

Clifton, R.J. (1967) A difference method for plane problems in dynamic elasticity, *Quart. Appl. Math* **25**, 97–116.

Cohen, R., Kulsrud, R. (1974) Nonlinear evolution of parallel-propagating hydromagnetic waves, *Phys. Fluids* **17**, 2215–2225.

Colella, P., Woodward, P.R. (1984) The piecewise parabolic method (PPM) for gas dynamical simulations, *J. Comput. Phys.* **54**, 174–201.

Courant, R., Friedrichs, K.O. (1976) *Supersonic Flow and Shock Waves*, Applied Mathematical Sciences **21**, Springer, New York.

Courant, R., Lax, P. (1949) On nonlinear partial differential equations with two independent variables, *Comm. Pure Appl. Math.* **2**, 255–273.

Courant, R., Friedrichs, K., Lewy, H. (1928) Über die partiellen Differenzengleichungen der mathematischen Physik, *Math. Annalen* **100**, H. 1/2, 32–74.

Courant, R., Isaacson, E., Rees, M. (1952) On the solution of nonlinear hyperbolic differential equations by finite differences, *Comm. Pure Appl. Math.* **5**, 243–255.

Dai, W., Woodward, P.R. (1994a) An approximate Riemann solver for ideal magnetohydrodynamics, *J. Comput. Phys.* **111**, 354–372.

Dai, W., Woodward, P.R. (1994b) Extension of the piecewise parabolic method to multidimensional ideal magnetohydrodynamics, *J. Comput. Phys.* **115**, 485–514.

Dai, W., Woodward, P.R. (1998) On the divergence-free condition and conservation laws in numerical simulations for supersonic magnetohydrodynamic flows, *Astrophys. J.* **494**, 317–335.

Dementii, O.I., Dementii, S.V. (1978) Phase velocities and discontinuous structure of a shock wave, *Prikladnaya Matematika Teoreticheskaya Fizika [Applied Mathematics and Theoretical Physics]*, No. 6, 25–32 [in Russian].

De Neef, T., Moretti, G. (1980) Shock fitting for everybody, *Computers and Fluids* **8**, 327–334.

Denus, S., Fiedorowicz, H., Nagraba, S. et al. (1983) Time-delayed filaments of prolonged durability on laser irradiated microspheres, *Opt. Commun.* **47**, No. 2, 127–130.

De Sterk, H. (1999) *Numerical Simulation and Analysis of Magnetically Dominated MHD Bow Shock Flows with Applications in Space Physics*, Ph.D. Thesis, Katholieke Universiteit Leuven and National Center for Atmospheric Research.

De Sterk, H., Poedts, S. (1999) Field-aligned magnetohydrodynamic bow shock flows in the switch-on regime. Parameter study of the flow around a cylinder and results for the axi-symmetrical flow over a sphere, *Astron. Astrophys.* **343**, 641–649.

De Sterk, H., Low, B.C., Poedts, S. (1998) Complex magnetohydrodynamic bow shock topology in field-aligned low-β flow around a perfectly conducting cylinder, *Phys. Plasmas* **5**, 4015–4027.

De Sterk, H., Deconinck, H., Poedts, S., Roose, D. (1999) A bow shock flow containing (almost) all types of ("exotic") MHD discontinuities, in *Hyperbolic Problems: Theory, Numerics, Application*, International Series on Numerical Mathematics **129**, 195–204, Birkhäuser Verlag, Basel, Switzerland.

De Vore, C.R. (1991) Flux-corrected transport techniques for multidimensional compressible magnetohydrodynamics, *J. Comput. Phys.* **92**, 142–160.

Dhareswar, L.J., Naik, P.A., Nandwana, P.D., Pant, H.C. (1987) Characteristics of plasma flow from laser irradiated planar thin foil targets, *J. Appl. Phys.* **61**, 4458–4463.

Di Giacinto, M., Valorani, M. (1989) Shock detection and discontinuity tracing for unsteady flows, *Computers and Fluids* **17**, 61–84.

Dolezal, A., Wong, S.S.M. (1995) Relativistic hydrodynamics and essentially non-oscillatory shock capturing schemes, *J. Comput. Phys.* **120**, 266-277.

Donnell, L.H. (1976) *Beams, Plates, and Shells*, McGraw-Hill, New York.

Druyanov, B.A. (1982) Generalized solutions of the dynamical theory of plasticity and thermoplasticity, *Soviet Physics Doklady* **27**, 1026–1027.

Duderstadt, J.J., Moses, G.A. (1982) *Inertial Confinement Fusion*, John Wiley, New York.

Durlofsky, L., Engquist, B., Osher, S. (1992) Triangle based adaptive stencils for the solution of hyperbolic conservation laws, *J. Comput. Phys.* **98**, 64–73.

Einfeldt, B., Munz, C.-D., Roe, P.L., Sjörgreen, B. (1991) On Godunov-type methods near low densities, *J. Comput. Phys.* **92**, 273–295.

Emery, A.F. (1968) An evaluation of several differencing methods for inviscid flow problems, *J. Comput. Phys.* **2**, 306–331.

Engquist, B., Majda, A. (1977) Absorbing boundary conditions for the numerical simulation of waves, *Math. Comput.* **31**, 629–651.

Engquist, B., Osher, S. (1981) One-sided difference approximations for nonlinear conservation laws, *Math. Comput.* **36**, 321–351.

Eulderink, A., Mellema, G. (1995) General relativistic hydrodynamics with Roe solver, *Astron. Astrophys. Suppl.* **110**, 587–623.

Evans, R.G. (1981) Radiation cooling instabilities in laser-heated plasma, *J. Phys. D: Appl. Phys.* **14**, No. 10, L173–L177.

Evans, C.R., Hawley, J.F. (1988) Simulation of magnetohydrodynamic flows: A constrained transport method, *Astrophys. J.* **332**, 659–677.

Evseev, E.G. (1985) The method of solution of nonlinear equations of thin shell dynamics, in *Nonlinear Theory of Thin Structures and Biomechanics, Proc. First All-Union Symp.*, Tbilisi, 191-193 [in Russian].

Evseev, E.G., Semenov, A.Yu. (1985) Reduction of thin shell dynamics (Timoshenko type) equations to the first order hyperbolic system and a hybrid numerical method for its solution, in *Proc. 10th Conf. MPTI Young Sci., 1985, March 23–Apr. 7*, Part 1, No. 5983-85, 169-178, VINITI, Moscow [in Russian].

Evseev, E.G., Semenov, A.Yu. (1989) *Numerical Method for Solution of the Thin Shell Dynamics System*, Preprint No. 20, General Physics Institute, USSR Academy of Sciences, Moscow [in Russian].

Evseev, E.G., Semenov, A.Yu. (1990) A method for numerical solution of the equations for the dynamics of thin shells based on separating out rapidly oscillating components, *Soviet Math. Doklady* **41**, No. 1, 118–121.

Fahr, H.J. (1986) Is the heliospheric interface submagnetosonic? Consequences for the LISM presence in the heliopause, *Adv. Space Res.* **6**, 13–25.

Fahr, H.J., Grzedzielski, S., Ratkiewicz, R. (1988) Magnetohydrodynamic modeling of the 3-dimensional heliopause using the Newtonian approximation, *Annales Geophysicae* **6**, 337–354.

Farin, G. (1990) Surfaces over Dirichlet tessellations, *Comput. Aided Geom. Design* **7**, 281–292.

Falle, S.A.E.G., Komissarov, S.S. (1996) An upwind numerical scheme for relativistic hydrodynamics with a general equation of state, *Month. Notices Royal Astron. Soc.* **278**, 586–602.

Fedorenko, R.P. (1962) The application of difference schemes of high accuracy to the numerical solution of hyperbolic equations, *U.S.S.R. Comput. Maths Math. Phys.* **2**, 1355–1365.

Fedotova, Z.I. (1978) On the application of an invariant finite-difference scheme to fluid oscillation in a basin, *Chislennye Metody Mekhaniki Sploshnoi Sredy [Numer. Meth. Mech. Contin.*

Media] **9**, No. 3, 137–146 [in Russian].

Fortov, V.E., Goel, B., Munz, C.-D. et al. (1996) Numerical simulations of nonstationary fronts and interfaces by the Godunov method in moving grids, *Nuclear Science and Engineering* **123**, 169–189.

Freudental, A.M., Geiringer, H. (1958) *The Mathematical Theories of the Inelastic Continuum*, Springer, Berlin.

Friedrichs, K.O. (1948) On the derivation of the shallow water theory: Appendix to the formation of breakers and bores by J.J. Stoker, *Comm. Pure Appl. Math.* **1**, 81–85.

Fritts, M.J., Crowley, W.P., Trease, H. (1988) The Free-Lagrange Method, *Lect. Notes Phys.* **238**, Springer, Berlin.

Fukuda, N., Hanawa, T. (1999) Gravitational and parametric instabilities of the interstellar medium in which the Alfén wave travels, *Astrophys. J.* **517**, 226–241.

Galin, G.Ya. (1958) Disturbance propagation in media with nonlinear dependence of stresses on strains and temperature, *Doklady Akademii Nauk SSSR [Soviet Physics Doklady]* **120**, No. 4, 730–734 [in Russian].

Galin, G.Ya. (1959) Theory of shock waves, *Soviet Physics Doklady* **4**, 757–760.

Gallice, G. (1996) Résolution numérique des équations de la magnétohydrodynamique idéale bidimensionnelle, in *Actes du workshop Méthode Numérique pour la M.H.D., CMAP, Ecole Polytechnique, Fevrier 12–13, 1996*, 101–132.

Gaponov, A.V., Ostrovskii, L.A., Freidman, G.I. (1967) Electromagnetic shock waves, *Izvest. Vuzov. Radiophys.* **10**, No. 9–10, 1376–1413 [in Russian]

Gardner, C.S. (1963) Comment on "Stability of Step Shocks," *Phys. Fluids* **6**, 1366–1367.

Garsia-Navarro, P., Hubbard, M.E., Priestley, A. (1995) Genuinely multidimensional upwinding for the 2D shallow water equations, *J. Comput. Phys.* **121**, 79–93.

Gelfand, I.M. (1959) Some problems of quasilinear equations theory, *Uspekhi Matematicheskikh Nauk [USSR Mathematical Surveys]* **14**, No. 9, 87–158 [in Russian].

George, P.-L., Borouchaki, H. (1998) *Delaunay Triangulation and Meshing. Application to Finite Elements*, Hermes, Paris.

Germain, P. (1959) Contribution a la théorie des ondes de choc en magnétodynamique des fluides, *Office National d'Etudes et de Recherches Aeronautiques*, Publ. 97, Paris.

Germain, P. (1973) *Cours de Mecanique des Milieux Continus*, **1**, Masson, Paris.

Givoli, D. (1991) Nonreflecting boundary conditions, *J. Comput. Phys.* **94**, 1–29.

Glaister, P. (1988a) Flux difference splitting for the Euler equations in one spatial co-ordinate with area variation, *Int. J. Numer. Meth. Fluids* **8**, 97–119.

Glaister, P. (1988b) An approximate linearized Riemann solver for the Euler equations for real gas, *J. Comput. Phys.* **74**, 382–408.

Glaister, P. (1991) A Riemann solver for "barotropic" flow, *J. Comput. Phys.* **93**, 477–480.

Glaister, P. (1995) A weak formulation of Roe's approximate Riemann solver applied to the St. Venant equations, *J. Comput. Phys.* **116**, 189–191.

Godlewski, E., Raviart, P.-A. (1996) *Numerical Approximation of Hyperbolic Systems of Conservation Laws*, Springer, New York.

Godunov, S.K. (1959) A finite difference method for the computation of discontinuous solutions of the equations of fluid dynamics, *Mat. Sbornik [Math. Notes]* **47(89)**, No. 3, 271–306 [in Russian].

Godunov, S.K. (1961) Nonunique blurring of discontinuities in solutions to quasilinear systems, *Doklady Akademii Nauk SSSR [Soviet Physics Doklady]* **136**, No. 2, 272–273 [in Russian].

Godunov, S.K. (1978) *Elements of Mechanics of Continuous Media*, Nauka, Moscow [in Russian].

Godunov, S.K., Romenskii, E.I. (1998) *Elements of Mechanics of Continuous Media and Conservation Laws*, Nauchnaya Kniga, Novosibirsk [in Russian].

Godunov, S.K., Ryabenkii, V.S. (1987) *Difference Schemes: An Introduction to the Underlying Theory*, North-Holland, Amsterdam.

Godunov, S.K., Zabrodin, A.V., Prokopov, G.P. (1961) A computational scheme for two-dimensional non-stationary problems of gas dynamics and calculation of the flow from a shock wave approaching a stationary state, *U.S.S.R. Comput. Maths Math. Phys.* **1**, 1187–1219.

Godunov, S.K., Zabrodin, A.V., Pliner, L.A. et al. (1968) Method of calculation of two-dimensional nonstationary gas dynamics problems in regions with complex geometry, *Report of Institute of Applied Mathematics*, USSR Academy of Sciences, Moscow [in Russian].

Godunov, S.K., Deribas, A.A., Zabrodin, A.V., Kozin, N.S. (1970) Hydrodynamic effects on colliding solids, *J. Comput. Phys.* **5**, No. 3, 517–539.

Godunov, S.K., Zabrodin, A.V., Ivanov, M.Ya., Kraiko, A.N., Prokopov, G.P. (1979) *Résolution Numérique des Problèmes Multidimensionnels de la Dynamique des Gaz*, Mir, Moscou.

Godwal, B.K., Ng, A., Da Silva, L. et al. (1990) Shock melting and Hugoniot calculation for gold, *Phys. Rev. Lett.* **144**, 26–30.

Gogosov, V.V. (1961) Resolution of an arbitrary discontinuity in magnetohydrodynamics, *J. Appl. Math. Mech.* **25**, 148–170.

Golant, V.E., Zhilinsky, A.P., Sakharov, I.E. (1980) *Fundamentals of Plasma Physics*, John Wiley, New York.

Goldenveizer, A.L. (1976) *Theory of Elastic Thin Shells*, Nauka, Moscow [in Russian].

Gol'din, V.Ya., Kalitkin, N.N., Shishova, T.V. (1965) Non-linear difference schemes for hyperbolic equations, *U.S.S.R. Comput. Maths Math. Phys.* **5**, No. 5, 229–239.

Gol'dshtein, R.V., Marchenko, A.V., Semenov, A.Yu. (1994) Boundary waves in liquid under an elastic plate with crack, *J. Russian Acad. Sci. Physics–Doklady* **39**, No. 11, 813–815.

Golubov, B.I., Vitushkin, A.G. (1977) Function variation, in *Mathematical Encyclopedia* **1**, 606-608, Soviet Encycl., Moscow [in Russian].

Goncharov, S.F., Semenov, A.Yu. (1995) Numerical simulation of jet-like structures in laser plasma, in *Laser Interaction and Related Plasma Phenomena, 12 Int. Conf., April 1995, Osaka*, S. Nakai and G.H. Miley (Eds.), Amer. Inst. Phys. Conference Proc. **369**, Pt. I, 499–504, Woodbury, NY.

Goncharov, S.F., Serov, R.V., Yanovskii, V.P. (1987) Data processing of interferograms of laser plasma with abrupt gradient of density profile, *Sov. Phys. Lebedev Inst. Brief Comm.*, No. 7, 24–26 [in Russian].

Goncharov, S.F., Semenov, A.Yu., Serov, R.V., Yanovskii, V.P. (1992) Large-scale jet-like structures in the coronal region of plasma produced under laser pulse interaction with planar solids, in *Investigation of Physical Processes in Planar and Conical Targets*, Trudy IOFAN [Proc. Gen. Phys. Inst., Russian Acad. Sci.] **36**, 228–246, Nauka, Moscow [in Russian].

Gorbachev, Yu.E., Zhmakin, A.I., Fursenko, A.A. (1985) Numerical modeling of processes in a relaxing gas during the rapid application of energy, *J. Appl. Mech. Techn. Phys.* **26**, No. 2, 169–177.

Gorshkov, A.G., Tarlakovskii, D.V. (1990) *Transient Aerohydroelasticity of Spherically-Shaped Bodies*, Nauka, Moscow [in Russian].

Göttelmann, J. (1999) A spline collocation scheme for the spherical shallow water equations, *J. Comput. Phys.* **148**, 291–298.

Gressier, J., Moschetta, J.-M. (1998) On the pathological behaviour of upwind schemes, *AIAA Paper*, No. 98-0110.

Grigolyuk, E.I., Gorshkov, A.G. (1976) *Interaction of Elastic Structures with a Fluid*, Sudostroenie, Leningrad [in Russian].

Grigolyuk, E.I., Mamay, V.I. (1997) *Nonlinear Deformation of Thin Constructions*, Nauka/Fizmatlit, Moscow [in Russian].

Grigolyuk, E.I., Selezov, I.T. (1973) *Nonclassical Theories of Vibration of Rods, Plates, and Shells*, Advances in Science and Technology. Mechanics of Solids **5**, VINITI, Moscow [in Russian].

Grin', V.T., Kraiko, A.N., Slavyanov, N.N. (1981) Solution to the problem of starting a nozzle mounted at the end of a shock tube, *Fluid Dynamics* **16**, 897–903.

Grudnitskii, V.G., Prokhorchuk, Yu.A. (1977) An approach to construction of difference schemes with arbitrary order of accuracy for partial differential equations, *Doklady Akademii Nauk SSSR* **234**, No. 6, 1249–1252 [in Russian].

Gurevich, A.G., Melkov, G.A. (1994) *Magnetic Oscillations and Waves*, Nauka, Moscow [in Russian].

Gur'yanov, A.A. (1985) Numerical solution of dynamic problems in the theory of shells by the method of flux-corrected transport, *VINITI Manuscript*, No. 2832-85, VINITI, Moscow [in Russian].

Gvozdovskaya, N.I., Kulikovskii, A.G. (1997) On electromagnetic shock waves and their structure in anisotropic ferromagnets, *J. Appl. Math. Mech.* **61**, 135–143.

Gvozdovskaya, N.I., Kulikovskii, A.G. (1999a) Quasitransverse shock waves in elastic media with an internal structure, *J. Appl. Mech. Tech. Phys.* **40**, 341–346.

Gvozdovskaya, N.I., Kulikovskii, A.G. (1999b) Investigation of electromagnetic shock-wave structure in anisotropic ferromagnets with easy axis, *Wave Motion* **29**, 23–34.

Haar, A., Karman, T. (1909) in *Nachr. von der Königlichen Gesellshaft der Wissenshaften zu Göttingen, Math-phys. K.*, No. 2, 204.

Hada, T. (1994) Evolutionary conditions in the dissipative MHD system: Stability of intermediate MHD shock waves, *Geophys. Res. Lett.* **21**, 2275–2278.

Haines, M.G. (1974) An electron thermal instability in a resistive non-equilibrium fully ionized plasma, *J. Plasma Phys.* **12**, 1–14.

Haines, M.G. (1981) Thermal instability and magnetic field generated by large heat flow in a plasma, especially under laser-fusion conditions, *Phys. Rev. Lett.* **47**, 917-920.

Hairer, E., Norsett, S.P., Wanner, G. (1987) *Solving Ordinary Differential Equations I. Nonstiff Problems*, Springer, Berlin.

Hall, G., Watt, J.M. (Eds.) (1976) *Modern Numerical Methods for Ordinary Differential Equations*, Clarendon Press, Oxford.

Hanawa, T., Nakajima, Y., Kobuta, K. (1994) *Extensions of Roe's Approximate Riemann Solver for General Equation of State and Magnetohydrodynamics*, Preprint DPNU-94-34, Department of Physics, Nagoya University.

Hanyga, A. (1976) On the solution to the Riemann problem for arbitrary hyperbolic system of conservation laws, *Polish Acad. Sci. Publications of Geophysics*, A-1 (98), Panstvowe wydavnitstvo naukowe, Warszawa.

Harten, A. (1978) The artificial compression method for computation of shocks and contact discontinuities: III. Self-adjusting hybrid schemes, *Math. Comput.* **32**, 363–389.

Harten, A. (1983) High resolution schemes for hyperbolic conservation laws, *J. Comput. Phys.* **49**, 357–393.

Harten, A. (1987) ENO schemes with subcell resolution, *J. Comput. Phys.* **83**, 148–184.

Harten, A., Hyman, J.M. (1983) Self-adjusting grid methods for one-dimensional hyperbolic conservation laws, *J. Comput. Phys.* **50**, 235–269.

Harten, A., Osher, S. (1987) Uniformly high-order accurate nonoscillatory schemes. I, *SIAM J. Numer. Anal.* **24**, 279–309.

Harten, A., Zwas, G. (1972a) Self-adjusting hybrid schemes for shock computations, *J. Comput. Phys.* **9**, 568–583.

Harten, A., Zwas, G. (1972b) Switched numerical Shuman filters for shock calculations, *J. Engrg. Math.* **6**, No. 2, 207–216.

Harten, A., Hyman, J.M., Lax, P.D. (1976) On finite difference approximations and entropy conditions for shocks, *Comm. Pure Appl. Math.* **29**, 297–322.

Harten, A., Osher, S., Engquist, B., Chakravarthy, S.R. (1986) Some results on uniformly high-order accurate essentially nonoscillatory schemes, *Appl. Numer. Math.* **2**, No. 3-5, 347–377.

Harten, A., Engquist, B., Osher, S., Chakravarthy, S.R. (1987) Uniformly high-order nonoscillatory schemes. III, *J. Comput. Phys.* **71**, 231–303.

Hedstrom, G.W. (1979) Nonreflecting boundary conditions for nonlinear hyperbolic systems, *J. Comput. Phys.* **30**, 222–237.

Henshaw, W.D. (1987) A scheme for the numerical solution of hyperbolic systems of conservation laws, *J. Comput. Phys.* **68**, 25–47.

Hirao, A., Ogasawara, M. (1981) Magnetic field generating thermal instability including the Nernst effect, *J. Phys. Soc. Jap.* **50**, 668–672.

Hirsch, C. (1990) *Numerical Computation of Internal and External Flows* **1**, **2**, John Wiley, Chichester, U.K.

Hora, H. (1981) *Physics of Laser Driven Plasmas*, John Wiley, New York.

Hornung, K., Malama, Yu.G., Thoma, K. (1996) Modeling of the very high velocity impact process with respect to in-situ ionization measurements, *Adv. Space Res.* **17**, No. 12, (12)77–(12)86.

Hubbard, M.E., Baines, M.J. (1997) Conservative multidimensional upwinding for the steady two-dimensional shallow water equations, *J. Comput. Phys.* **138**, 419–448.

Huynh, H.T. (1995) Accurate upwind schemes for Euler equations, *AIAA Paper*, No. 95-1737.

Ikeuchi, S., Spitzer, L. (1984) Scattering of shock waves by a spherical cloud, *Astrophys. J.* **283**, 825–832.

Ilyushin, A.A. (1990) *Mechanics of Continuous Media*, Moscow State University, Moscow [in Russian].

Imshennik, V.S., Bobrova, N.A. (1997) *Dynamics of Collision Plasma*, Energoatomizdat, Moscow [in Russian].

Ionov, V.N., Selivanov, V.V. (1987) *Dynamics of Fracture of Deformable Body*, Mashinostroenie, Moscow [in Russian].

Ivanenko, S.A. (1999) Harmonic mappings, Chapter 8 in *Handbook of Grid Generation*, J.F. Thompson, B.K. Soni, and N.P. Weatherill (Eds.), CRC, Boca Raton, FL.

Ivanov, I.E., Kryukov, A.I. (1996) High resolution monotone method for computation internal and jet inviscid flows, *Matematicheskoe Modelirovanie [Math. Modeling]* **8**, No. 6, 47–55 [in Russian].

Ivanov, I.E., Kryukov, A.I., Pogorelov, N.V. (1999) Grid adaptation for gas dynamic and astrophysical flows, in *Proc. 8th International Meshing Roundtable, Lake Tahoe, Oct. 10–13, 1999*, Sandia Report 99-2288, 313–319, Albuquerque, NM.

Ivanov, M.Ya., Koretskii, V.V., Kurochkina, N.Ya. (1980a) Analysis of the properties of first-order accurate front-capturing schemes, *Chislennye Metody Mekh. Sploshnoi Sredy [Numer. Meth. Mech. Contin. Media]* **11**, No. 1, 81–110 [in Russian].

Ivanov, M.Ya., Koretskii, V.V., Kurochkina, N.Ya. (1980b) Analysis of the properties of second-order accurate front-capturing schemes, *Chislennye Metody Mekh. Sploshnoi Sredy [Numer. Meth. Mech. Contin. Media]* **11**, No. 2, 41–63 [in Russian].

Ivanov, M.Ya., Koretskii, V.V., Kurochkina, N.Ya. (1980c) Analysis of the properties of high-order accurate front-capturing schemes, *Chislennye Metody Mekh. Sploshnoi Sredy [Numer. Meth. Mech. Contin. Media]* **11**, No. 4, 83–103 [in Russian].

Ivanov, V.D., Petrov, I.B. (1992) Numerical simulation of deformation and fracture under laser interaction with targets, in *Investigation of Physical Processes in Planar and Conical Targets*, Trudy IOFAN [Proc. Gen. Phys. Inst., Russian Acad. Sci.] **36**, 247–266, Nauka, Moscow [in Russian].

Ivanov, V.D., Petrov, I.B., Suvorova, Yu.V. (1989) Numerical solutions of two-dimensional dynamic problems of hereditary viscoelastic theory, *Mechanics of Composite Materials*, No. 3, 419–424 [in Russian].

Ivanov, V.D., Kondaurov, V.I., Petrov, I.B., Kholodov, A.S. (1990) Calculation of dynamic deformation and fracture of elastic-plastic body by grid-characteristic methods, *Matematicheskoe Modelirovanie [Math. Modeling]* **2**, No. 11, 11–29 [in Russian].

Ivanov, V.D., Petrov, I.B., Tormasov, A.G., Kholodov, A.S., Pashutin, R.A. (1999) Grid-characteristic method of dynamic deformation calculation on irregular grids, *Matematicheskoe Modelirovanie [Math. Modeling]* **11**, No. 7, 118–127 [in Russian].

Ivanov, V.L. (1987) Approximation of systems of hyperbolic equations that contain large parameters in the non-differential terms, *U.S.S.R. Comput. Maths Math. Phys.* **27**, No. 5, 73–77.

Jeffrey, A. (1976) *Quasilinear Hyperbolic Systems and Waves*, Research Notes in Mathematics, Pitman, London.

Jeffrey, A., Taniuti, T. (1964) *Nonlinear Wave Propagation*, Academic Press, New York.

Jiang, G.-S., Wu, C.C. (1999) A high-order WENO finite difference scheme for equations of ideal magnetohydrodynamics, *J. Comput. Phys.* **150**, 561–594.

Kalashnikov, A.S. (1959) Construction of generalized solutions of quasilinear equations of first order without convexity conditions as limits of solutions of parabolic equations with a small parameter, *Doklady Akademii Nauk SSSR [Soviet Physisc Doklady]* **127**, No. 1, 27–30 [in Russian].

Kamenetskii, V.F., Semenov, A.Yu. (1989) *Construction of Hybrid Difference Schemes Based on Characteristic Relations*, Commun. Appl. Math., Computing Center, USSR Academy of Sciences, Moscow [in Russian].

Kamenetskii, V.F., Semenov, A.Yu. (1993) *Self-Adjusting Shock Fitting in Shock-Capturing One- and Two-Dimensional Gas Dynamic Calculations*, Preprint No. 36, General Physics Institute, USSR Academy of Sciences, Moscow [in Russian].

Kamenetskii, V.F., Semenov, A.Yu. (1994) Self-consistent isolation of discontinuities in continuous calculations of gas-dynamic flows, *Comput. Maths Math. Phys.* **34**, 1287–1297.

Kamenyarz, Ya.A. (1972) Simple waves and collapse of discontinuity in elastic-plastic medium with Mises condition, *J. Appl. Math. Mech.* **36**, 296–305.

Kanel', G.I., Shcherban', V.V. (1980) Plastic deformation and cleavage rupture of Armco iron in a shock wave, *Combustion, Explosion and Shock Waves* **16**, No. 4, 439–446.

Kanel, G.I., Razorenov, S.V., Utkin, A.V., Fortov, V.E. (1996) *Shock-Wave Phenomena in Condensed Media*, Janus-C, Moscow [in Russian].

Kantrovitz, A., Petschek, H. (1966) MHD characteristics and shock waves, in *Plasma Physics in Theory and Applications*, W.B. Kunkel (Ed.), 148–206, McGraw-Hill, New York.

Kataev, I.G. (1963) *Electromagnetic Shock Waves*, Sovetskoe Radio, Moscow [in Russian].

Kennel, C.F., Blandford, R.D., Wu, C.C. (1990) Structure and evolution of small-amplitude intermediate shock waves, *Phys. Fluids B* **2**, 253–269.

Khabibrakhmanov, I.K., Summers, D. (1996) A new model of interaction between the solar wind and the local interstellar medium, *J. Geophys. Res.* **101**, A7609–A7618.

Kholodov, A.S. (1978) On construction of difference schemes with positive approximation for hyperbolic equations, *U.S.S.R. Comput. Maths Math. Phys.* **18**, No. 6.

Kholodov, A.S. (1980) The construction of higher order accurate difference schemes for hyperbolic equations, *U.S.S.R. Comput. Maths Math. Phys.* **20**, No. 6.

Kholodov, A.S. (1984) The construction of difference schemes with positive approximation for equations of parabolic type, *U.S.S.R. Comput. Maths Math. Phys* **24**, No. 5, 41–48.

Kholodov, A.S., Petrov, I.B. (1984) Regularization in discontinuous numerical solutions of equations of hyperbolic type, *U.S.S.R. Comput. Maths Math. Phys.* **24**, No. 4, 128–138.

Kilpio, A.V., Kochiev, D.G., Malyutin, A.A. et al. (1992) "Kamerton", a one-channel laser installation, in *Investigation of Physical Processes in Planar and Conical Targets*, Trudy IOFAN [Proc. Gen. Phys. Inst., Russian Acad. Sci.] **36**, 202-212, Nauka, Moscow [in Russian].

Kim, J., Ryu, D., Jones, T.W., Hong, S.S. (1999) A multidimensional code for isothermal magnetohydrodynamic flows in astrophysics, *Astrophys. J.* **514**, 506–519.

Klyushnikov, V.D. (1993) Plasticity theory: Current state and development prospects, *Mech. Solids* **28**, No. 2, 93–105.

Knorr, G., Mond, M. (1980) The representation of shock-like solutions in an Eulerian mesh, *J. Comput. Phys.* **38**, 212–226.

Kobayashi, A.S. (Ed.) (1987) *Handbook of Experimental Mechanics*, Prentice-Hall, Englewood Cliffs, NJ.

Kolarov, D., Baltov, A., Boncheva, N. (1979) *Mechanics of Plastic Media*, Mir, Moscow [in Russian].

Koldoba , A.V., Kuznetsov, O.A., Ustyugova, G.V. (1992) *Quasi-Monotone Difference Schemes of a Higher-Order Approximation for the MHD Equations*, Preprint No. 69, Keldysh Institute of Applied Mathematics, Russian Academy of Sciences, Moscow [in Russian].

Kolgan, V.P. (1972) Application of the minimum-derivative principle in the construction of finite-difference schemes for numerical analysis of discontinuous solutions in gas dynamics, *Uchenye Zapiski TsAGI [Sci. Notes of Central Inst. of Aerohydrodynamics]* **3**, No. 6, 68–77 [in Russian].

Kolgan, V.P. (1975) Finite-difference schemes for computation of two-dimensional discontinuous solutions of nonstationary gas dynamics, *Uchenye Zapiski TsAGI [Sci. Notes of Central Inst. of Aerohydrodynamics]* **6**, No. 1, 9–14 [in Russian].

Kolmogorov, A.N., Fomin, S.V. (1975) *Introductory Real Analysis*, Dover, New York.

Komissarov, S.S. (1999) A Godunov-type scheme for relativistic magnetohydrodynamics, *Month. Notices Royal Astron. Soc.* **303**, 343–366.

Kondaurov, V.I. (1981) Conservation laws and symmetrization of the equations of the nonlinear theory of thermoelasticity, *Soviet Physics Doklady* **26**, No. 2, 234–236.

Kondaurov, V.I. (1982a) On conservation laws of elastoviscoplastic medium with finite deformations, *Mech. Solids* **17**, No. 6.

Kondaurov, V.I. (1982b) Equations of elastoviscoplastic medium with finite deformations, *J. Appl. Mech. Techn. Phys.* **23**, No. 4, 584–591.

Kondaurov, V.I., Fortov, V.E. (2000) *Fundamentals of Thermomechanics of Condensed Media*, Moscow Inst. Phys. Tech., Moscow [in Russian].

Kondaurov, V.I., Nikitin, L.V. (1990) *Theoretical Basis of Rheology of Geomaterials*, Nauka, Moscow [in Russian].

Kon'shin, V.N. (1985) *Numerical Modelling of Wave Motion of Liquid*, Ph.D. Thesis, Moscow Inst. Phys. Tech., Moscow [in Russian].

Korn, G.A., Korn, T.M. (1968) *Mathematical Handbook for Scientists and Engineers*, McGraw-Hill, New York.

Korobeinikov, V.P. (1991) *Problems of Point-Blast Theory*, The American Institute of Physics, New York.

Korotin, P.N., Petrov, I.B., Kholodov, A.S. (1989) Numerical simulation of elastic and elastoplastic solids under the action of intensive energy beams, *Matematicheskoe Modelirovanie [Math. Modeling]* **1**, No. 7, 1–12 [in Russian].

Korotin, P.N., Petrov, I.B., Pirogov, V.B., Kholodov, A.S. (1987) On numerical solution of related problems of supersonic flow over deformable shells of finite thickness, *U.S.S.R. Comput. Maths Math. Phys.* **27**, No. 4, 181-188.

Kosevich, A.M., Ivanov, B.A., Kovalev, A.S. (1990) Magnetic solitons, *Phys. Reports* **194**, 3–4.

Kostrykin, V.S., Fomin, V.N., Kholodov, A.S. (1976) Three-dimensional radiating flow around blunted cones and ellipsoids of revolution, *U.S.S.R. Comput. Maths Math. Phys.* **16**, No. 2, 166–174.

Kotchine, N.E. (1926) Sur la théorie des ondes de choc dans un fluide, *Rendicotti del Circolo Matematico di Palermo* **50**, 305–344.

Kraiko, A.N., Makarov, V.E., Tillayeva, N.I. (1980) The numerical construction of shock wave fronts, *U.S.S.R. Comput. Maths Math. Phys.* **20**, No. 3.

Krasyuk, I.K., Semenov, A.Yu. (1992) Numerical modeling of gas dynamic phenomena in conical targets, in *Investigation of Physical Processes in Planar and Conical Targets*, Trudy IOFAN [Proc. Gen. Phys. Inst., Russian Acad. Sci.] **36**, 83–111, Nauka, Moscow [in Russian].

Krasyuk, I.K., Pashinin, P.P., Semenov, A.Yu. (1994a) New similarity relations based on statistical analysis of laser thermonuclear fusion experiments, *J. Russian Acad. Sci. Physics-Doklady* **39**, No. 5, 330–332.

Krasyuk, I.K., Pashinin, P.P., Semenov, A.Yu. (1994b) New scalings based on statistical analysis of laser fusion experiments, *Laser Physics* **4**, 532–537.

Krebs, J., Hillebrandt, W. (1983) The interaction of supernova shockfronts and nearby interstellar clouds, *Astron. Astrophys.* **128**, 411–419.

Krivtsov, V.M., Naumova, I.N., Shmyglevskii, Yu.D., Zubov, V.I. (1992) Computation of the interaction of a laser radiation beam with an aluminium vessel and its vapor, in *Modern Problems in Computational Aerohydrodynamics*, A.A. Dorodnicyn and P.I. Chushkin (Eds.), 165–188, CRC Press, London.

Kröner, D. (1997) *Numerical Schemes for Conservation Laws*, Wiley and Teubner, Chichester, U.K.

Kulikovskii, A.G. (1968) Surfaces of discontinuity separating two perfect media of different properties. Recombination waves in magnetohydrodynamics, *J. Appl. Math. Mech.* **32**, 1145–1152.

Kulikovskii, A.G. (1979) Properties of shock adiabats in the neighborhood of Jouget points, *U.S.S.R. Fluid Dynamics* **14**, No. 2, 317–320.

Kulikovskii, A.G. (1984) A possible effect of oscillations in the structure of a discontinuity on the set of admissible discontinuities, *Soviet Physics Doklady* **29**, 283–285.

Kulikovskii, A.G. (1986) Equations describing the propagation of non-linear quasitransverse waves in a weakly non-isotropic elastic body, *J. Appl. Math. Mech.* **50**, 455–461.

Kulikovskii, A.G. (1988) Strong discontinuities in continuum flows and their structure, *Trudy MIAN [Proc. Steklov Inst. Math.]* **182**, No. 4, 261–291 [in Russian].

Kulikovskii, A.G. (1991) Peculiarities of the behavior of nonlinear quasitransverse waves in an elastic medium with a small anisotropy, *Proc. Steklov Inst. Math.*, Issue 1, 153–160.

Kulikovskii, A.G., Gvozdovskaya, N.I. (1998) The effect of dispersion on the set of admissible discontinuities in continuum mechanics, *Proc. Steklov Inst. Math.* **223**, 55–65 [in Russian].

Kulikovskii, A.G., Lyubimov, G.A. (1959) On magnetohydrodynamic gas ionizing shock waves, *Soviet Physics Doklday* **129**, 52–55.

Kulikovskii, A.G., Lyubimov, G.A. (1965) *Magnetohydrodynamics*, Addison Wesley, Reading, MA.

Kulikovskii, A.G., Sveshnikova, E.I. (1980) On shock waves propagation in stressed isotropic non-linearly elastic media, *J. Appl. Math. Mech.* **44**, 367–374.

Kulikovskii, A.G., Sveshnikova, E.I. (1982) Investigation of the shock adiabat of quasitransverse waves in a prestressed elastic medium, *J. Appl. Math. Mech.* **46**, 667–683.

Kulikovskii, A.G., Sveshnikova, E.I. (1984) Some properties of the shock adiabat of quasitransverse elastic waves, *J. Appl. Math. Mech.* **48**, 575–578.

Kulikovskii, A.G., Sveshnikova, E.I. (1985a) A self-similar problem on the action of a sudden load on the boundary of an elastic half-space, *J. Appl. Math. Mech.* **49**, 214–220.

Kulikovskii, A.G., Sveshnikova, E.I. (1985b) Nonlinear waves arising when changing stresses on an elastic half-space boundary, in *Problems of Nonlinear Mechanics of Continuous Media*, Volgus, Tallinn [in Russian].

Kulikovskii, A.G., Sveshnikova, E.I. (1987) On the structure of quasitransverse elastic shock waves, *J. Appl. Math. Mech* **51**, 711–716.

Kulikovskii, A.G., Sveshnikova, E.I. (1988) The decay of an arbitrary initial discontinuity in an elastic medium, *J. Appl. Math. Mech.* **52**, 786–790.

Kulikovskii, A.G., Sveshnikova, E. (1995) *Nonlinear Waves in Elastic Media*, CRC Press, Boca Raton, FL.

Kulikovskii, A.G., Sveshnikova, E.I. (1996) The existence and uniqueness of self-similar solutions involving Jouguet points on the shock adiabatic curve, *J. Appl. Math. Mech.* **60**, 61–65.

Kulikovskii, A.G., Pogorelov, N.V., Semenov, A.Yu. (1999) Mathematical aspects of numerical solution of hyperbolic systems, in *Hyperbolic Problems: Theory, Numerics, Application, International Series on Numerical Mathematics* **130**, 589–598, Birkhäuser Verlag, Basel, Switzerland.

Kupriyanova, T.V., Mikhailov, Yu.Ya., Chinilov, A.Yu. (1991) Construction of an approximate conservative problem of the decay of an explosion in the case of an equilibrium gas, *Comput. Maths Math. Phys.* **31**, No. 4, 50–56.

Kuropatenko, V.F. (1992) Equations of state for mathematical modeling in physics and mechanics, *Matematicheskoe Modelirovanie [Math. Modeling]* **4**, No. 12, 112–136 [in Russian].

Kutler, P., Lomax, H., Warming, R. (1972) Computation of space shuttle flow fields using noncentered finite-difference schemes, *AIAA Paper*, No. 72-193.

Kutler, P., Lomax, H., Warming, R.F. (1973) Second- and third-order noncentered difference schemes for nonlinear hyperbolic equations, *AIAA J.* **11**, 189–196.

Kvitov, S.V., Bushman, A.V., Kulish, M.I., Lomonosov, I.V., Polishchuk, A.Ya., Semenov, A.Yu., Ternovoi, V.Ya., Filimonov, A.S., Fortov, V.E. (1991) Measurements of the optical emission of a dense bismuth plasma during its adiabatic expansion, *JETP Letters* **53**, No. 7, 353–357.

Landau, L.D., Lifshitz, E.M. (1984) *Electrodynamics of Continuous Media*, Pergamon Press, Oxford.

Landau, L.D., Lifshitz, E.M. (1986) *Theory of Elasticity*, Pergamon Press, New York.

Landau, L.D., Lifshitz, E.M. (1987) *Fluid Mechanics*, Pergamon Press, Oxford.

Latter, R. (1955) Similarity solution for a spherical shock wave, *J. Appl. Phys.* **26**, 954–960.

Lax, P.D. (1954) Weak solutions of nonlinear hyperbolic equations and their numerical computation, *Comm. Pure Appl. Math.* **7**, 159–193.

Lax, P.D. (1957) Hyperbolic systems of conservation laws II, *Comm. Pure Appl. Math.* **10**, 537–566.

Lax, P.D. (1972) *Hyperbolic Systems of Conservation Laws and the Mathematical Theory of Shock Waves*, Conf. Board Math. Sci. Regional Conference Series in Applied Mathematics **11**, SIAM, Philadelphia.

Lax, P.D., Wendroff, B. (1960) Systems of conservation laws, *Comm. Pure Appl. Math.* **13**, 217–237.

Lax, P.D., Wendroff, B. (1964) Difference schemes for hyperbolic equations with high order of accuracy, *Comm. Pure Appl. Math.* **17**, 381–398.

Lenskii, E.V. (1981) Shock adiabat of a plane longitudinal-shear discontinuity, *Moscow Univ. Mech. Bull.* **36**, No. 1–2, 24–27.

Lenskii, E.V. (1982) Propagation of plane waves of a 2-component strain state in a non-linear compressible medium, *Moscow Univ. Mech. Bull.* **37**, No. 6, 101–106.

Lenskii, E.V. (1983a) Plane compression/shear waves in a nonlinear-elastic incompressible medium, *Mech. Solids* **18**, No. 6, 85–94.

Lenskii, E.V. (1983b) Simple waves in a nonlinear-elastic medium, *Moscow Univ. Mech. Bull.* **38**, No. 3, 11–18.

Le Veque, R.J. (1992) *Numerical Methods for Conservation Laws*, Birkhäuser Verlag, Basel, Switzerland.

Liberman, M.A., Velikovich, A.L. (1986) *Physics of Shock Waves in Gases and Plasmas*, Springer, New York.

Liepmann, H.W., Roshko, A. (1957) *Elements of Gasdynamics*, John Wiley, New York.

Liewer, P.C., Karmesin, S.R., Brackbill, J.U. (1996) Hydrodynamic instability of the heliopause driven by plasma-neutral charge-exchange interactions, *J. Geophys. Res.* **101**, A17119–A17127.

Lighthill, M.L. (1957) Dynamics of dissociating gas. Part I. Equilibrium flow, *J. Fluid Mech.* **2**, 1–32.

Lin, S.C., Teare, J.D. (1963) Rate of ionization behind shock waves in air. II. Theoretical interpretation, *Phys. Fluids* **6**, 355–375.

Linde, T.J., Gombosi, T.I., Roe, P.L., Powell, K.G., DeZeeuw, D.L. (1998) Heliosphere in the magnetized local interstellar medium: Results of a three-dimensional MHD simulation, *J. Geophys. Res.* **103**, A1889–A1904.

Liou, M.-S. (1996) A sequel to AUSM: AUSM$^+$, *J. Comput. Phys.* **129**, 364–382.

Liou, M.-S., Steffen, C.J., Jr. (1993) A new flux splitting scheme, *J. Comput. Phys.* **107**, 23–39.

Liu, X.D., Osher, S., Chan, T. (1994) Weighted essentially non-oscillatory schemes, *J. Comput. Phys.* **115**, 200–212.

Lokhin, V.V., Sedov, L.I. (1963) Nonlinear tensor functions of several tensor arguments, *J. Appl. Math. Mech* **27**, 597–629.

Lomov, I.N., Kondaurov, V.I. (1995) Application of Godunov-type methods for the solution of condensed matter problem, in *Shock Compression of Condensed Matter–1995*, S.C. Schmidt and W.C. Tao (Eds.), 259–262, The American Institute of Physics.

Lomov, I.N., Kondaurov, V.I. (1998) Fracture of brittle material with initial porosity under high energy density flows, in *Shock Compression of Condensed Matter–1997*, S.C. Schmidt et al. (Eds.), 247–250, The American Institute of Physics.

Londrillo, P., Del Zanna, L. (2000) High-order upwind schemes for multidimensional magneto-hydrodynamics, *Astrophys. J.* **530**, 508–524.

Love, E.H. (1892) *A Treatise on the Mathematical Theory of Elasticity*, Cambridge University Press.

Lurie, A.I. (1980) *Nonlinear Theory of Elasticity*, Nauka, Moscow [in Russian].

Lyatkher, V.M., Militeev, A.N. (1978) A numerical study of flow geometry in tailraces, *Gidrotekhn. Stroitelstvo [Hydrotech. Construction]*, No. 6, 27–32.

Lyatkher, V.M., Militeev, A.N., Mishuev, A.V., Sladkevich, M.S. (1986) A numerical study of tsunamis approaching shores, in *Tsunami Waves Research: Generation and Propagation of Tsunami Waves in Oceans* **1**, 110–119, Joint Geophys. Committee, Moscow [in Russian].

Lyubarskii, G.Ya. (1961) On the shock wave structure, *Prikladnaya Matematika i Mekhanika [J. Appl. Math. Mech.]* **25**, No. 6, 1041–1049 [in Russian].

Lyubarskii, G.Ya., Polovin, R.V. (1960) The disintegration of unstable shock waves in magnetohydrodynamics, *Soviet Physics–JETP* **9**, 902–906.

Lyubimov, A.N., Rusanov, V.V. (1970) *Gas Flows Past Blunt Bodies*, **1-2**, Nauka, Moscow [in Russian] (see also NASA-TT-F 715, Feb. 1973)

MacCormack, R.W. (1969) The effect of viscosity in hypervelocity impact cratering, *AIAA Paper*, No. 69-354.

Magomedov, K.M. (1971) Grid-characteristic method for numerical solution of gas dynamics problems, in *Proc. Sect. Numer. Meth., 2nd Int. Colloq. Implosion Gas Dyn. React. Systems, Novosibirsk, Aug. 19-23, 1969* **1**, 328–356, Computing Center, USSR Academy of Sciences, Moscow [in Russian].

Magomedov, K.M., Kholodov, A.S. (1969) The construction of difference schemes for hyperbolic equations based on characteristic relations, *U.S.S.R. Comput. Maths Math. Phys.* **9**, No. 2, 158–176.

Magomedov, K.M., Kholodov, A.S. (1988) *Grid-Characteristic Numerical Methods*, Nauka, Moscow [in Russian].

Makarenko, N.I. (1985) A validation of three-dimensional and two-layer models of shallow water, in *Nonlinear Problems in the Theory of Surface and Internal Waves*, L.V. Ovsiannikov and V.N. Monakhov (Eds.), 78–96, Nauka, Novosibirsk [in Russian].

Makhanov, S.S., Semenov, A.Yu. (1994) A stable non-negative numerical method for calculating the flow of a liquid in an open channel, *Comput. Maths Math. Phys.* **34**, 85–95.

Makhanov, S.S., Semenov, A.Yu. (1995) Modeling of motion of fluid with free surface under level recession, *Water Resources* **22**, No. 4, 357–362.

Makhanov, S.S., Semenov, A.Yu. (1996) A two-dimensional non-negative algorithm for calculating the flow of a liquid in an open channel, *Comput. Maths Math. Phys.* **36**, 501–507.

Makhanov, S.S., Semenov, A.Yu. (1998) Numerical methods for non-linear parabolic boundary-value problems with a priory bounded solution, in *Proc. 4th Europ. Comput. Fluid Dyn. Conf., Athens, Sept. 7-11, 1998*, K.D. Papailiou et al. (Eds.), **1**, No. 1, 78–82, John Wiley, Chichester, U.K.

Malama, Yu.G., Kestenboim, Kh.S., Hornung, K. (2000) *Numerical Simulation of High-Speed Impact of Dust Particles with the Target by Second-Order Godunov's Scheme*, Preprint No. 664, Institute for Problems in Mechanics, Russian Academy of Sciences, Moscow [in Russian].

Malyshev, A.P. (1996) A monotone difference scheme of increased accuracy for the computer simulation of wave processes, *Comput. Maths Math. Phys* **36**, 1295–1998.

Marchenko, A.V., Semenov, A.Yu. (1994) Edge waves in a shallow fluid beneath a fractured elastic plate, *Fluid Dynamics* **29**, 589–592.

Marchenko, A.V., Semenov, A.Yu. (1995) Computing the definite integrals of the Wiener–Hopf method by summing powers of residues, *Comput. Maths Math. Phys.* **35**, 357–362.

Marchuk, A.G., Chubarov, L.B., Shokin, Yu.I. (1983) *Numerical Simulation of Tsunamis*, Nauka, Novosibirsk [in Russian].

Marconi, F., Salas, M. (1973) Computation of three-dimensional flows about aircraft configurations, *Computers and Fluids* **1**, 185–195.

Marconi, F., Salas, M.D., Yaeger, L.S. (1976) Steady super/hypersonic inviscid flow around real configurations, *NASA CR-2675*.

Markovskii, S.A. (1998a) Oscillatory disintegration of nonevolutionary magnetohydrodynamic discontinuities, *J. Exper. Theor. Phys.* **86**, No.2, 340-347.

Markovskii, S.A. (1998b) Nonevolutionary discontinuous magnetohydrodynamic flows in a dissipative medium, *Phys. Plasmas* **5**, 2596–2604.

Markovskii, S.A., Skorokhodov, S.L. (2000a) Oscillatory disintegration of a trans-Alfvénic shock: A magnetohydrodynamic simulation, *Phys. Plasmas* **7**, 158–165.

Markovskii, S.A., Skorokhodov, S.L. (2000b) Disintegration of trans-Alfvénic shocks due to variable viscosity and resistivity, *J. Geophys. Res.* **105**, No. A6, 12705–12712.

Marquina, A., Martí, J.M., Ibáñez, J.M., Miralles, J.A., Donat, R. (1992), Ultrarelativistic hydrodynamics: High-resolution shock-capturing methods, *Astron. Astrophys.* **258**, 566–571.

Martí, J.M., Müller, E. (1994) The analytical solution of the Riemann problem in relativistic hydrodynamics, *J. Fluid Mech.* **258**, 317–333.

Martí, J.M., Müller, E. (1996) Extension of the piecewise parabolic method to one-dimensional relativistic hydrodynamics, *J. Comput. Phys.* **123**, 1–14.

Martí, J.M., Müller, E. (1999) Numerical hydrodynamics in special relativity, *Living Reviews in Relativity* **2**, 1999-3 [http:/www.livingreviews.org/Articles/Volume2/1999-3marti].

Matsuda, T., Fujimoto, Y. (1993) MHD interaction between the solar wind and local interstellar medium, in *Proc. 5th Int. Symp. on Comput. Dyn., Sendai, Aug. 31–Sept. 3, 1993* **2**, 186–193.

Matsuda, T., Fujimoto, Y., Shima, E., Sawada, K., Inaguchi, T. (1989) Numerical simulations of interaction between stellar wind and interstellar medium, *Progr. Theor. Phys.* **81**, 810–822.

Matsuno, K. (1999) Higher-order upwind method for hyperbolic grid generation, *Computers and Fluids* **28**, 825–851.

Maugin, G.A. (1988) *Continuum Mechanics of Electromagnetic Solids*, North-Holland, New York.

McNutt, R.L., Jr., Lyon, J., Goodrich, C.C., Wiltberger, M. (1999) 3D MHD simulations of the heliosphere–VLISM interaction, in *Solar Wind Nine*, S.R. Habbal et al. (Eds.), 823–826, CP471, The American Institute of Physics.

McRae, D.S., Lafli, K.R. (1999) Dynamic grid adaption and grid quality, Chapter 34 in *Handbook of Grid Generation*, J.F. Thompson, B.K. Soni, and N.P. Weatherill (Eds.), CRC, Boca Raton, FL.

Men'shov, I.S. (1990) Increasing the order of approximation of Godunov's scheme using solutions of the generalized Riemann problem, *U.S.S.R. Comput. Maths Math. Phys* **30**, No. 5, 54–65.

Men'shov, I.S. (1991) Generalized problem of breakup of an arbitrary discontinuity, *J. Appl. Math. Mech.* **55**, 67–74.

Men'shov, I.S. (1992) Increasing the accuracy of the Godunov scheme for calculating stationary supersonic gas flows based on the solution of generalized Riemann problem, *Comput. Maths Math. Phys* **32**, No. 2, 257–263.

Mileshin, V.I., Tillyaeva, N.I. (1982) Comparison of calculated and experimental data on flow past axisymmetric intakes in regimes with an ejected shock wave, *Uchenye Zapiski TsAGI [Sci. Notes of Central Inst. of Aerohydrodynamics]* **13**, No. 2, 135–141 [in Russian].

Mishuev, A.V., Sladkevich, M.S., Silchenko, A.S. (1984) Analysis of experimental results and a numerical method for long waves approaching a slope, in *Conference on Tsunamis, Gor'kii, USSR, 1984, Sept. 18-21*, 121–123, USSR Academy of Sciences, Gor'kii [in Russian].

Monaghan, J.J. (1989) On the problem of penetration in particle methods, *J. Comput. Phys.* **82**, 1–15.

Monaghan, J.J. (1997) SPH and Riemann solvers, *J. Comput. Phys.* **136**, 298–307.

Monaghan, J.J., Gingold, R.A. (1983) Shock simulation by the particle method SPH, *J. Comput. Phys.* **52**, 374–389.

Moretti, G. (1963) Three-dimensional supersonic flow computations, *AIAA J.* **1**, 2192-2193.

Moretti, G. (1974) Three-dimensional, supersonic, steady flows with any number of embedded shocks, *AIAA Paper*, No. 74-10.

Moretti, G. (1979) The λ-scheme, *Computers and Fluids* **7**, 191–205.

Moretti, G. (1987) A technique for integrating two-dimensional Euler equations, *Comput. Fluids* **15**, 59–75.

Moretti, G., Abbett, M. (1966) A time dependent computational method for blunt body flows, *AIAA J.* **4**, 2136-2141.

Moretti, G., Bleich, G. (1967) Three-dimensional flow around blunt bodies, *AIAA Paper*, No. 67-222.

Mulder, W.A., van Leer, B. (1983) Implicit upwind methods for the Euler equations, *AIAA Paper*, No. 83-1930.

Myasnikov, A.V. (1997) *On the Problem of the Solar Wind Interaction with Magnetized Interstellar Medium*, Preprint No. 195, Institute for Problems in Mechanics, Russian Academy of Sciences, Moscow.

Myong, R.S., Roe, P.L. (1997) Shock waves and rarefaction waves in magnetohydrodynamics. Part 2. The MHD system, *J. Plasma Phys.* **58**, 521–552.

Myong, R.S., Roe, P.L. (1998) On Godunov-type schemes for magnetohydrodynamics. 1. A model system, *J. Comput. Phys.* **147**, 545–567.

Nakajima, Y., Hanawa, T. (1996) Formation and evolution of filamentary molecular clouds with oblique magnetic field, *Astrophys. J.* **467**, 321–333.

Nakata, I. (1991) Nonlinear electromagnetic waves in a ferromagnet, *J. Phys. Soc. Japan* **60**, 77–81.

Naumova, I.N., Shmyglevskii, Yu.D. (1978) Algorithm and program of the Riemann problem solver for an nonperfect gas, *Inform. Bull. Algorithms and Programs*, No. 3, П002809, All-Union Sci. Tech. Inform. Center, Moscow [in Russian].

Nazhestkina, E.I., Rusanov, V.V. (1980) The approximation of the boundary conditions in difference schemes, *U.S.S.R. Comput. Maths Math. Phys.* **20**, No. 6.

Nessyahu, H., Tadmor, E. (1988) Non-oscillatory central differencing for hyperbolic conservation laws, *ICASE Report*, No. 88-51.

Nessyahu, H., Tadmor, E. (1990) Non-oscillatory central differencing for hyperbolic conservation laws, *J. Comput. Phys.* **87**, 408–463.

Nestor, O.H., Olsen, H.N. (1960) Numerical methods for reducing line and surface probe data, *SIAM Rev.* **2**, No. 3, 200–207.

Neugebauer, M. (1999) The three-dimensional solar wind at solar activity minimum, *Rev. Geophys.* **37**, 107–126.

Ng, A., Parfeniuk, D., Da Silva, L. (1985a) Hugoniot measurements for laser-generated shock wave in aluminum, *Phys. Rev. Lett.* **54**, No. 24, 2604–2607.

Ng, A., Parfeniuk, D., Da Silva, L. (1985b) Measurement of shock heating in laser-irradiated solids, *Opt. Commun.* **53**, No. 6, 389–393.

Ng, A., Chin, G., Da Silva, L. et al. (1989) Laser-produced shock wave in aluminum-gold targets for the study of Hugoniot melting in gold, *Opt. Commun.* **72**, No. 5, 297–301.

Ogasawara, M., Hirao, A., Ohkubo, H. (1980) Hydrodynamic effects on field-generating thermal instability in laser-heated plasma, *J. Phys. Soc. Jap.* **49**, 322–326.

Ogibalov, P.M. (1963) *Problems of the Dynamics and Stability of Shells*, Moscow State University, Moscow [in Russian].

Oleinik, O.A. (1959) On the uniqueness and stability of generalized solution of the Cauchy problem for quasilinear equation, *Uspekhi Matematicheskikh Nauk [USSR Mathematical Surveys]* **14**, No. 2(86), 159–164 [in Russian].

Oran, E.S., Boris, J.P. (1987) *Numerical Simulation of Reactive Flows*, Elsevier, New York.

Ortega, J.M., Rheinboldt, W.C. (1970) *Iterative Solution of Nonlinear Equations in Several Variables*, Academic Press, New York.

Osher, S. (1981) Numerical solution of singular perturbation problems and hyperbolic systems of conservation laws, in *North Holland Mathematical Studies* **47**, 179–205.

Osher, S. (1984) Riemann solvers, the entropy condition, and difference approximations, *SIAM J. Numer. Anal.* **21**, No. 2, 217-235.

Osher, S., Shu, C.-W. (1988) Efficient implementation of essentially non-oscillatory shock-capturing schemes, *J. Comput. Phys.* **77**, 439–471.

Osher, S., Solomon, F. (1982) Upwind difference hyperbolic systems of conservation laws, *Math. Comput.* **38**, 339–374.

Ostapenko, V.V. (1998) On strong monotonicity of nonlinear difference schemes, *Comput. Maths Math. Phys.* **37**, 1119–1133.

Ovsiannikov, L.V. (1985) The Lagrangian approximation in waves theory, in *Nonlinear Problems in the Theory of Surface and Internal Waves*, L.V. Ovsiannikov and V.N. Monakhov (Eds.), 10–77, Nauka, Novosibirsk [in Russian].

Pandolfi, M. (1975) Numerical experiments of free surface water motion with bores, *Lect. Notes Phys.* **35**, 304–312, Springer, Heidelberg.

Pandolfi, M. (1984) A contribution to the numerical prediction of unsteady flows, *AIAA J.* **22**, 602–610.

Pandolfi, M., D'Ambrosio, D. (1998) Upwind methods and carbuncle phenomenon, in *Computational Fluid Dynamics 98* **1**, 126–131, John Wiley, Chichester, U.K.

Panichkin, V.I. (1981) Axially symmetric wave processes in shell structures with dampers, in *Applied Problems of Strength and Plasticity. Statics and Dynamics of Deformable Systems*, No. 18, 74-84, Gor'kii State University, Gor'kii [in Russian].

Papa, L. (1984) Application of the Courant–Isaacson–Rees method to solve the shallow-water hydrodynamic equations, *Appl. Math. Comput.* **15**, 85–92.

Parker, E.N. (1961) Stellar wind regions, *Astrophys. J.* **134**, 20–27.

Parshikov, A.N. (1999) Application of a solution to the Riemann problem in the SPH method, *Comput. Maths Math. Phys.* **39**, 1173–1182.

Pauls, H.L., Zank, G.P. (1996) Interaction of a nonuniform solar wind with the local interstellar medium, *J. Geophys. Res.* **101**, A17081–A17092.

Pavlov, P.V., Khokhlov, A.F. (2000) *Solid-State Physics*, Vysshaya Shkola, Moscow [in Russian].

Petrov, I.B., Kholodov, A.S. (1984a) Numerical study of some dynamic problems of the mechanics of a deformable rigid body by the mesh-characteristic method, *U.S.S.R. Comput. Maths Math. Phys.* **24**, No. 3, 61–73.

Petrov, I.B., Kholodov, A.S. (1984b) Regularization of discontinuous numerical solutions of equations of hyperbolic type, *U.S.S.R. Comput. Maths Math. Phys.* **24**, No. 4, 128-138.

Petrov, I.B., Tormasov, A.G. (1990) About numerical solution of three-dimensional collision problem, *Matematicheskoe Modelirovanie [Math. Modeling]* **2**, No. 2, 58–72 [in Russian].

Petrov, I.B., Tormasov, A.G., Kholodov, A.S. (1990) On the use of hybrid grid-characteristic schemes for the numerical solution of three-dimensional problems in the dynamics of the deformable solid, *U.S.S.R. Comput. Maths Math. Phys.* **30**, No. 4, 191-196.

Petrovskii, I.G. (1991) *Lectures on Partial Differential Equations*, Dover, New York.

Pluvinage, G. (1989) *Mecanique elastoplastique de la rupture (critères d'amorçage)*, Cepadues-Editions, Toulouse.

Pogorelov, N.V. (1987) Diagonalization and simultaneous symmetrization of matrices in the stationary gas dynamic equations, *Soviet Physics Doklady* **32**, 552–553.

Pogorelov, N.V. (1988a) Supersonic chemically reacting flows near bodies flying at high angle of attack, *Modelirovanie v Mekhanike [Modelling in Mechanics]* **2**, No. 3, 116–125 [in Russian].

Pogorelov, N.V. (1988b) *Investigation of Supersonic Nonequilibrium Reacting Airflow past Blunt Bodies at High Angles of Attack*, Preprint No. 350, Institute for Problems in Mechanics, USSR Academy of Sciences, Moscow [in Russian].

Pogorelov, N.V. (1990) Three-dimensional nonequilibrium reaction airflow around a body penetrating into an equilibrium heated zone, *U.S.S.R. Fluid Dynamics* **25**, 918–925.

Pogorelov, N.V. (1993) Numerical investigation of a nonstationary gas dynamic stellar wind–interstellar medium interaction, in *Proc. 5th Int. Symp. on Comput. Fluid Dyn., Sendai, August 31–September 3, 1993*, **3**, 7–12.

Pogorelov, N.V. (1995) Periodic stellar wind / interstellar medium interaction, *Astron. Astrophys.* **297**, 835–840.

Pogorelov, N.V. (1997) Numerical simulation of nonstationary gasdynamic interaction of the solar wind with the local interstellar medium, *Comput. Fluid Dyn. J.* **6**, No. 2, 213–222.

Pogorelov, N.V. (1998) Computational fluid dynamics methods for astrophysical applications, in *Computational Fluid Dynamics '98* **2**, 815–820, John Wiley, Chichester, U.K.

Pogorelov, N.V. (2000) Nonstationary phenomena in the solar wind and interstellar medium interaction, *Astrophys. Space Sci.*, to appear.

Pogorelov, N.V., Matsuda, T. (1998a) Influence of the interstellar magnetic field direction on the shape of the global heliopause, *J. Geophys. Res* **103**, A237–A245.

Pogorelov, N.V., Matsuda, T. (1998b) Application of numerical methods to modeling the stellar wind and interstellar medium interaction, in *CFD Review 1998* **2**, 932–962, World Scientific, Singapore.

Pogorelov, N.V., Matsuda, T. (2000) Nonevolutionary MHD shocks in the solar wind and interstellar medium interaction, *Astron. Astrophys.* **354**, 697–702.

Pogorelov, N.V., Semenov, A.Yu. (1996a) Peculiarities of numerical modelling of discontinuous MHD flows, in *Numerical Methods in Engineering 96*, 1022–1027, John Wiley, Chichester, U.K.

Pogorelov, N.V., Semenov, A.Yu. (1996b) A modification of non-reflecting boundary conditions for gasdynamic simulation in astrophysics, *Comput. Maths Math. Phys.* **36**, No. 3, 395–404.

Pogorelov, N.V., Semenov, A.Yu. (1997a) Solar wind interaction with the magnetized interstellar medium: Shock-capturing modelling, *Astron. Astrophys.* **321**, 330–337.

Pogorelov, N.V., Semenov, A.Yu. (1997b) A family of approximate solutions to the MHD Riemann problem retaining the shock relations, *Comput. Maths Math. Phys.* **37**, No. 3, 320–328.

Pogorelov, N.V., Semenov, A.Yu. (1997c) A family of shock-fitted approximate solutions to the problem of the magnetic gasdynamic discontinuity, *J. Russian Acad. Sci. Physics-Doklady* **42**, 37–42.

Pogorelov, N.V., Shevelev, Yu.D. (1981) On the application of nonorthogonal curvilinear coordinate systems in the calculation of supersonic blunt body flows at high angle of attack, *Chislennye Metody Mekhaniki Sploshnoi Sredy [Numer. Meth. Mech. Contin. Media]* **12**, No. 6, 65–80 [in Russian].

Pogorelov, N.V., Ohsugi, Y., Matsuda, T. (2000) Towards steady-state solutions for supersonic wind accretion on to gravitating objects, *Month. Notices Royal Astron. Soc.* **313**, 198–208.

Pogorelov, N.V., Barmin, A.A., Kulikovskii, A.G., Semenov, A.Yu. (1995) Approximate Riemann solvers and valid solutions of MHD calculations, in *6th Int. Symp. on Comput. Fluid. Dyn., Collection Tech. Papers, Lake Tahoe, 1995*, **2**, 952–957.

Polishchuk, A.Ya., Semenov, A.Yu., Ternovoi, V.Ya., Fortov, V.E. (1991) *Numerical Modeling of the Shock-Induced Luminescence of Free Metal Surface*, Preprint No. 25, General Physics Institute, Russian Academy of Sciences, Moscow [in Russian].

Polovin, R.V., Cherkasova, K.P. (1962) Disintegration of nonevolutionary shock waves, *Soviet Physics–JETP* **14**, 190–191.

Polovin, R.V., Demutskii, V.P. (1960) Shock adiabat in magnetohydrodynamics, *Ukrainskii Fizicherskii Zhurnal [Ukraine Phys. J.]* **5**, No. 3, 3–11 [in Russian].

Potapkin, A.V. (1983) The use of adaptive meshes in calculations of flow with large gradients, *Chislennye Metody Mekh. Sploshnoi Sredy [Numer. Meth. Mech. Contin. Media]* **14**, No. 3, 126-139 [in Russian].

Potapov, A.I. (1985) *Nonlinear Strain Waves in Rods and Plates*, Gor'kii State University, Gor'kii [in Russian].

Powell, K.G. (1994) An approximate Riemann solver for magnetohydrodynamics, *ICASE Report No. 94-24*, ICASE NASA Langley Research Center, Hampton, VA.

Powell, K.G., Roe, P.L., Myong, R.S., Gombosi, T., De Zeeuw, D. (1995) An upwind scheme for magnetohydrodynamics, in *AIAA 12th Comput. Fluid Dyn. Conf., San Diego, CA, June 19–22, 1995*, 661-680.

Powell, K.G., Roe, P.L., Linde, T.J., Gombosi, T.I., De Zeeuw, D. (1999) A solution-adaptive upwind scheme for ideal magnetohydrodynamics, *J. Comput. Phys.* **154**, 284–309.

Prager, W. (1961) *Einfürung in die Kontinuumsmechanik*, Birkhäuser Verlag, Basel, Switzerland.

Prokopov, G.P. (1999) *Investigation of Relaxation Processes in the Three-Temperature Model of Nonstationary Flows of Heat Conductive Gas*, Preprint No. 66, Keldysh Institute of Applied Mathematics, Russian Academy of Sciences, Moscow [in Russian].

Prokopov, G.P. (2000) *Approximation of Equations of State for the Three-Temperature Model of Heat Conductive Gas Nonstationary Flows*, Preprint No. 33, Keldysh Institute of Applied Mathematics, Russian Academy of Sciences, Moscow [in Russian].

Quirk, J.J. (1994) A contribution to the great Riemann problem debate, *Int. J. Numer. Meth. Fluids.* **18**, 555–574.

Rabotnov, Yu.N. (1988) *Mechanics of Deformable Solids*, Nauka, Moscow [in Russian].

Ratkiewicz, R., Barnes, A., Molvik, G.A. et al. (1998) Effect of varying strength and orientation of local interstellar magnetic field on a configuration of exterior heliosphere: 3D MHD simulations, *Astron. Astrophys.* **335**, 363–369.

Ricards, R.B, Teters, G.A. (1974) *Stability of Shells Made of Composite Materials*, Zinatne, Riga [in Russian].

Richardson, D.J. (1964) The solution of two-dimensional hydrodynamic equations by the method of characteristics, in *Methods in Computational Physics* **3**, 295–318, Academic Press, New York.

Richardson, J., Wang, C. (1999) The global nature of solar cycle variations of the solar wind dynamic pressure, *Geophys. Res. Lett.* **26**, 561–564.

Richtmyer, R.D., Morton, K.W. (1967) *Difference Methods for Initial-Value Problems*, Interscience, New York.

Riemann, B. (1860) Über die Fortpflanzung ebener Luftwellen von endlicher Schwingungsweite, *Göttinger Abhandlungen* **8**.

Rizzi, A. (1978) Numerical implementation of solid-body boundary conditions for the Euler equations, *ZAMM* **58**, T301–T304.

Rizzi, A. (1982) Damped Euler-equation method to compute transonic flow around wing–body combinations, *AIAA J.* **20**, 1321–1328.

Rizzi, A., Eriksson, L.-E. (1984) Computation of the flow around wings based on the Euler equations, *J. Fluid. Mech.* **148**, 45–71.

Roache, P.J. (1976) *Computational Fluid Dynamics*, Hermosa, Albuquerque, NM.

Rodionov, A.V. (1987a) Monotonic scheme of the second order of approximation for the continuous calculation of non-equilibrium flows, *U.S.S.R. Comput. Maths Math. Phys.* **27**, No. 2, 175–180.

Rodionov, A.V. (1987b) Methods of increasing the accuracy of Godunov's scheme, *U.S.S.R. Comput. Maths Math. Phys.* **27**, No. 6, 164-169.

Rodionov, A.V. (1996) A numerical method for solving Euler's equations which retains the approximation on a deformed net, *Comput. Maths Math. Phys.* **36**, No. 3, 379–389.

Roe, P.L. (1981) Approximate Riemann problem solvers, parameter vectors and difference scheme, *J. Comput. Phys.* **43**, 357–372.

Roe, P.L. (1985) Some contributions to the modelling of discontinuous flows, in *Lectures in Applied Mathematics* **22**, 163–193, AMS, Providence, RI.

Roe, P.L. (1986) Characteristic-based schemes for the Euler equations, in *Ann. Rev. Fluid Mech.* **18**, 337–365.

Roe, P.L. (1989) Remote boundary conditions for unsteady multidimensional aerodynamic computations, *Computers and Fluids* **17**, 221–231.

Roe, P.L., Balsara, D.S. (1996) Notes on the eigensystem of magnetohydrodynamics, *SIAM J. Appl. Math.* **56**, 57.

Roe, P.L., Pike, J. (1984) Efficient construction and utilization of approximate Riemann solutions, in *Computing Methods in Applied Sciences and Engineering, VI*, Proc. 6th Symp. on Comput. Meth. in Appl. Sci. and Eng., Versailles, France, 1983, Dec. 12-16, R.G. Glowinski and J.-L. Lions (Eds.), 499–518, North-Holland, Amsterdam.

Rogister, A. (1971) Parallel propagation of nonlinear low-frequency waves in high-β plasma, *Phys. Fluids* **14**, 2733–2739.

Ross, M., Young, D.A. (1993) Theory of the equation of state at high pressure, in *Ann. Rev. Phys. Chem.* **44**, 61–87.

Rozhdestvenskii, B.L., Yanenko, N.N. (1983) *Systems of Quasilinear Equations and Their Applications to Gas Dynamics*, Amer. Math. Soc. Transl. of Math. Monographs, **55**, AMS, Providence, RI.

Rusanov, V.V. (1961) The calculation of the interaction of non-stationary shock waves and obstacles, *U.S.S.R. Comput. Maths Math. Phys.* **1**, No. 2, 304–320.

Rusanov, V.V. (1968) Difference schemes of the third order of accuracy for shock-capturing computation of discontinuous solutions, *Doklady Akademii Nauk SSSR [Soviet Math. Doklady]* **180**, No. 6, 1303-1305 [in Russian].

Rusanov, V.V. (1970) On difference schemes of third order of accuracy for non-linear hyperbolic systems, *J. Comput. Phys.* **5**, 507–516.

Ryibakov, A.P. (1977) Spalling in steel produced by explosion of a sheet charge and collision of a plate, *J. Appl. Mech. Techn. Phys.* **18**, No. 1, 131–134.

Ryu, D., Jones, T.W., Frank, A. (1995) Numerical magnetohydrodynamics in astrophysical algorithm and test for multidimensional flow, *Astrophys. J.* **452**, 785–796.

Ryu, D., Miniati, F., Jones, T.W., Frank, A. (1998) A divergence-free upwind code for multidimensional magnetohydrodynamic flows, *Astrophys. J.* **509**, 244–255.

Sadovskii, V.M. (1997) *Discontinuous Solutions in the Problems of Elastic-Plastic Dynamics*, Nauka/Fizmatlit, Moscow [in Russian].

Salas, M. (1976) Shock-fitting method for complicated two-dimensional supersonic flows, *AIAA J.* **14**, 583–588.

Samsonov, A.M., Deiden, G.V., Porubov, A.V., Semenova, I.V. (1997) Longitudinal strain solitons in a nonlinear elastic rod, in *Russian Science: Stand and Revitilize*, 33–40, Nauka, Moscow [in Russian]

Saurel, R., Larini, M., Loraud, J.C. (1994) Exact and approximate Riemann solvers for real gases, *J. Comput. Phys.* **112**, 126–137.

Sawada, K., Shima, E., Matsuda, T., Inaguchi, T. (1986) The Osher upwind scheme and its application to cosmic gas dynamics, *Mem. Fac. Eng. Kyoto Univ.* **48**, 240–264.

Sazonov, I.A., Semenov, A.Yu. (1993) Joint algorithms of entropy correction in gas dynamical ENO-calculations, in *Applied Problems of Aeromechanics and Geocosmic Physics*, T.V. Kondranin (Ed.), 63–67, Moscow Inst. Phys. Tech., Moscow [in Russian].

Schneider, V., Katschner, U., Rischke, D.H., Waldhauser, B., Maruhn, J.A., Munz, C.-D. (1993) New algorithms for ultra-relativistic numerical hydrodynamics, *J. Comput. Phys.* **105**, 92–107.

Sedov, L.I. (1946) The flow of air under intense explosion, *Doklady Akademii Nauk SSSR* **52**, No. 1, 17–20 [in Russian].

Sedov, L.I. (1960) Different definitions of the rate of change of a tensor, *J. Appl. Math. Mech.* **24**, 579–586.

Sedov, L.I. (1971) *A Course in Continuum Media*, Wolters-Nordhoff, Groningen, Netherlands.

Sedov, L.I. (1993) *Similarity and Dimensional Methods in Mechanics*, 10th ed., CRC, Boca Raton, FL.

Semenov, A.Yu. (1983) Usage of Godunov's method for shallow water equations with bottom relief, in *Proceed. 8th Conf. MPTI Young Sci., 1983, March 27 - Apr. 7*, Part 1, No. 5927-83, 150–157, VINITI, Moscow [in Russian].

Semenov, A.Yu. (1984) A method of constructing hybrid difference schemes for hyperbolic systems, *Soviet Math. Doklady* **30**, 602–605.

Semenov, A.Yu. (1987) *Construction of Hybrid Difference Schemes for Numerical Solution of Hyperbolic Systems*, Ph.D. Thesis, Moscow Inst. Phys. Tech. / Gen. Phys. Inst., USSR Acad. Sci., Moscow [in Russian].

Semenov, A.Yu. (1988) *Characteristic Method for Construction of Hybrid Difference Schemes. Application to Gas Dynamics*, Preprint No. 32, General Physics Institute, USSR Academy of Sciences, Moscow [in Russian].

Semenov, A.Yu. (1992) Algorithms for monotone nonoscillatory reconstruction of mesh function, in *The 3rd Russia-Japan Joint Symp. on Comput. Fluid Dynamics*, Vladivostok, Russia, 1992, Aug. 25-30, Abstracts, **2**, 165–166.

Semenov, A.Yu. (1995a) The limiting monotonic reconstruction and its application in gas dynamic calculation, in *6th Int. Symp. on Comput. Fluid. Dyn., Collection Tech. Papers, Lake Tahoe, 1995*, **3**, 1093–1098.

Semenov, A.Yu. (1995b) Modelling of the shock-induced luminescence of free metal surface, in *Laser Interaction and Related Plasma Phenomena, 12 Int. Conf., Osaka, April 1995*, E. Nakai and G.H. Miley (Eds.), Amer. Inst. of Physics, AIP Conference Proc. **369**, Pt. I, 434–442, Woodbury, NY.

Semenov, A.Yu. (1995c) Marching generation of orthogonal grids in a plane and on surfaces, *Comput. Maths Math. Phys.* **35**, 1359–1371.

Semenov, A.Yu. (1995d) Noniterative marching generation of orthogonal contour-fitted grids, in *6th Int. Symp. on Comput. Fluid. Dyn., Collection Tech. Papers, Lake Tahoe, 1995*, **3**, 1087–1092.

Semenov, A.Yu. (1996) Marching noniterative generation of orthogonal contour-fitted grids, in *Numerical Grid Generation in Computational Field Simulations*, B.K. Soni et al. (Eds.), Proc. of the 5th Int. Conf., April 1-5, 1996, **1**, 117–125, Mississippi State University.

Semenov, A.Yu. (1997) A modified Courant–Isaacson–Rees method for gas dynamics with an arbitrary equation of state, *Comput. Maths Math. Phys.* **37**, 1334–1340.

Serre, D. (1996) *Systèmes de Lois de Conservation*, Diderot, Paris.

Shokin, Yu.I. (1973) Method of the first differential approximation in theory of finite-difference schemes for hyperbolic system of equations, *Trudy MIAN [Proc. Steklov Inst. Math.]* **122**, No. 2, 66-84 [in Russian].

Shokin, Yu.I., Yanenko, N.N. (1985) *The Method of Differential Approximation: Application to Gas Dynamics*, Nauka, Novosibirsk [in Russian].

Shu, S.-W. (1987) TVB uniformly high-order shemes for conservation laws, *Math. Comput.* **49**, 105–121.

Shu, S.-W., Osher, S. (1988) Efficient implementation of essentially non-oscillatory shock-capturing schemes I, *J. Comput. Phys.* **77**, 439–471.

Shu, S.-W., Osher, S. (1989) Efficient implementation of essentially non-oscillatory shock-capturing schemes II, *J. Comput. Phys.* **83**, 45–71.

Shugaev, F.V. (1983) *Shock Wave Interaction with Disturbances*, Moscow State University, Moscow [in Russian].

Sibson, R. (1980) A vector identity for the Dirichlet tessellation, *Math. Proc. Cambridge Philos. Soc.* **87**, 151–155.

Sibson, R. (1981) A brief description of the natural neighbour interpolant, in *Interpreting Multivariate Data*, 21–36, John Wiley, Chichester, U.K.

Slepyan, L.I. (1972) *Nonstationary Elastic Waves*, Sudostroenie, Leningrad [in Russian].

Sobolev, S.L. (1958) On mixed problems for partial differential equations with two independent variables, *Doklady Akademii Nauk SSSR* **122**, No. 4, 555-558 [in Russian].

Sod, A.G. (1978) Review. A survey of several finite difference methods for systems of nonlinear hyperbolic conservation laws, *J. Comput. Phys.* **27**, 1–31.

Sofronov, I.L. (1993) Conditions for complete transparency on the sphere for the three-dimensional wave equation, *J. Russian Acad. Sci. Math. Doklady* **46**, 397.

Sofronov, I.L. (1998) Artificial boundary conditions of absolute transparency for two- and three-dimensional external time-dependent scattering problems, *Europ. J. Appl. Math.* **9**, 561–588.

Soo, S.L. (1967) *Fluid Dynamics of Multiphase Systems*, Blaisdell, London.

Spotz, W.F., Taylor, M.A., Swarztrauber, P.N. (1998) Fast shallow-water equation solvers in latitude-longitude coordinates, *J. Comput. Phys.* **145**, 432–444.

Srinivas, K., Gururaja, J., Krishra, K.P. (1976) An assessment of the quality of selected finite difference schemes for time dependent compressible flows, *J. Comput. Phys.* **20**, 140–159.

Stamper, J.A., Gold, S.H., Obenschain, S.P., Mclean, E.A., Sica, L. (1981) Dark-field study of rear-side density structure in laser-accelerated foils, *J. Appl. Phys.* **52**, 6562–6566.

Steger, J.L., Chaussee, D.S. (1980) Generation of body-fitted coordinates using hyperbolic partial differential equations, *SIAM J. Sci. Stat. Comput.* **1**, 431–437.

Steger, J.L., Warming, R.F. (1981) Flux vector splitting of the inviscid gasdynamic equations with application to finite difference methods, *J. Comput. Phys.* **40**, 263–293.

Steinolfson, R.S. (1994) Termination shock response to large-scale solar wind fluctuations, *J. Geophys. Res.* **99**, A13307–A13314.

Steinolfson, R.S., Hundhausen, A.J. (1990) Coronal mass ejection shock fronts containing the two types of intermediate shocks, *J. Geophys. Res.* **95**, 20693–20699.

Steinolfson, R.S., Pizzo, V.J., Holzer, T. (1994) Gasdynamic models of the solar wind/interstellar medium interaction, *Geophys. Res. Lett.* **21**, 245–248.

Stellingwerf, R.F., Wingate, C.A. (1993) Impact modeling with smooth particle hydrodynamics, *Int. J. Impact Eng.* **14**, 707–718.

Stoker, J.J. (1948) The formation of breakers and bores, *Comm. Pure Appl. Math.* **1**, 1–87.

Stoker, J.J. (1957) *Water Waves: The Mathematical Theory with Applications*, Interscience, New York.

Strang, G. (1963) Accurate partial difference methods I: Linear Cauchy problems, *Arch. Ration. Mech. Anal.* **12**, No. 5, 392–402.

Stulov, V.P. (1969) Similarity law for supersonic flow past blunt bodies, *U.S.S.R. Fluid Dynamics* **4**, No. 4, 93–96.

Sukumar, N., Moran, B., Semenov, A.Yu., Belikov, V.V. (2000) Natural neighbor Galerkin methods, *Int. J. Numer. Meth. Eng.* **49**, to appear.

Sun, M.T., Wu, S.T., Dryer, M. (1995) On the time-dependent numerical boundary conditions for magnetohydrodynamic flows, *J. Comput. Phys.* **116**, 330–342.

Sveshnikova, E.I. (1982) Simple waves in non-linear elastic medium, *J. Appl. Math. Mech.* **46**, 509–512.

Sweby, P.K. (1984) High resolution schemes using flux limiters for hyperbolic conservation laws, *SIAM J. Numer. Anal.* **21**, 995-1011.

Syrovatskii, S.I. (1959) The stability of shock waves in magnetohydrodynamics, *Soviet Physics–JETP* **8**, 1024–1028.

Tadmor, E. (1997) *Approximate Solutions of Nonlinear Conservation Laws*, CAM Report 97-51, Univ. Calif., Los Angeles.

Tanaka, T. (1994) Finite-volume schemes on an unstructured grid system for three-dimensional MHD simulation of inhomogeneous systems including strong background potential field, *J. Comput. Phys.* **111**, 381–389.

Tanaka, T., Washimi, H. (1999) Solar cycle dependence of the heliospheric shape deduced from a global MHD simulation of the interaction process between a nonuniform time-dependent solar wind and the local interstellar medium, *J. Geophys. Res.* **104**, A12605–A12616.

Taniuti, T., Nishihara, K. (1983) *Nonlinear Waves*, Pitman Monographs and Studies in Mathematics **15**, Pitman, London.

Tarasov, B.A. (1974) Resistance of plates to fracture under shock loading, *Problemy Prochnosti [Strength of Materials]* **6**, No. 3, 121–122 [in Russian].

Taussig, R.T. (1967) Comparison of oblique, normal and transverse ionizing shock waves, *Phys. Fluids* **10**, 1145–1161.

Taylor, G.I. (1950) The formation of a blast by a very intense explosion, *Proc. Roy. Soc.* **A201**, No. 1065, 159–186.

Taylor, M. (1996) *Partial Differential Equations*, Applied Mathematical Sciences **117**, Springer, New York.

Taylor, T.D., Ndefo, E., Masson, B.S. (1972) A study of numerical methods for solving viscous and inviscid flow problems, *J. Comput. Phys.* **9**, 99–119.

Teters, G.A. (1969) *Nonsimple Loading and Stability of Shells Made of Polymer Materials*, Zinatne, Riga [in Russian].

Teters, G.A., Ricards, R.B., Narusberg, V.L. (1978) *Optimization of Shells Made of Multilayer Composites*, Zinatne, Riga [in Russian].

Thiell, G., Meyer, B. (1985) Thermal instabilities as an explanation of jet-like structures observed on laser irradiated thin planar targets at 1.06 and 0.35 μm wavelenghts, *Laser Part. Beams* **3**, No. 1, 51–58.

Thomas, T.Y. (1961) *Plastic Flow and Fracture in Solids*, Academic Press, New York.

Thompson, K.W. (1987) Time-dependent boundary conditions for hyperbolic systems, *J. Comput. Phys.* **68**, 1–24.

Thompson, K.W. (1990) Time-dependent boundary conditions for hyperbolic systems II, *J. Comput. Phys.* **89**, 439–461.

Thompson, J.F., Warsi, Z.U.A., Mastin, C.W. (1985) *Numerical Grid Generation, Foundations and Applications*, North-Holland, New York.

Tidman, D.A., Shanny, R.A. (1974) Field-generating thermal instability in laser-heated plasmas, *Phys. Fluids* **17**, 1207–1210.

Tillyaeva, N.I. (1986) The generalization of the modified Godunov scheme for arbitrary continuous grids, *Uchenye Zapiski TsAGI [Sci. Notes of Central Inst. of Aerohydrodynamics]* **17**, No. 2,

18–26 [in Russian].

Todd, L. (1964) Evolution of the trans-Alfvénic normal shocks in a gas of finite electrical conductivity, *J. Fluid Mech.* **18**, 321–336.

Todd, L. (1965) The evolution of trans-Alfvénic shocks in gases of finite electrical conductivity, *J. Fluid Mech.* **21**, 193–209.

Todd, L. (1966) Evolution of switch-on and switch-off shocks in a gas of finite electrical conductivity, *J. Fluid Mech.* **24**, 597–608.

Toro, E.F. (1992) Viscous flux limiters, in *Notes Numer. Fluid. Dyn.* **35**, 592–600, Vieweg.

Toro, E.F. (1997) *Riemann Solvers and Numerical Methods for Fluid Dynamics. A Practical Introduction*, Springer, Berlin.

Tóth, G., Odstrčil, D. (1996) Comparison of some flux corrected transport and total variation diminishing numerical schemes for hydrodynamic and magnetohydrodynamic problems, *J. Comput. Phys.* **128**, 82–100.

Trainor, R.J., Lee, Y.T. (1982) Analytic models for the design of laser-generated shock-wave experiments, *Phys. Fluids* **25**, 1898–1907.

Truesdell, C.A. (1977) *A First Course in Rational Continuum Mechanics*, Academic Press, New York.

Truskinovsky, L. (1994) Transition to detonation in dynamic phase changes, *Arch. Ration. Mech. Analysis* **125**, 375–397.

Tsynkov, S.V. (1995) An application of nonlocal external conditions to viscous flow computations, *J. Comput. Phys.* **116**, 212–225.

Tsynkov, S.V. (1998) Numerical solution of problems on unbounded domains. A review, *Appl. Num. Math.* **27**, 465.

Turkel, E. (1977) Symmetric hyperbolic difference schemes and matrix problems, *Linear Algebra and Appl.* **16**, No. 2, 109–129.

van Albada, G.D., van Leer, B., Roberts, W.W. (1982) A comparative study of computational methods to cosmic gas dynamics, *Astron. Astrophys.* **108**, 76–84.

Vankeirsbilck, P., Deconinck, H. (1992) Solution of the compressible Euler equations with higher order ENO-schemes on general unstructured mesh, in *Computational Fluid Dynamics'92* **2** 843–850, Elsevier, Amsterdam.

van Leer, B. (1973) Towards the ultimate conservative difference scheme. I. The quest of monotonicity, *Lect. Notes Phys.* **18**, No. 1, 163–168.

van Leer, B. (1974) Towards the ultimate conservative difference scheme. II. Monotonicity and conservation combined in a second-order scheme, *J. Comput. Phys.* **14**, 361–370.

van Leer, B. (1977a) Towards the ultimate conservative difference scheme. III. Upstream-centered finite-difference schemes for ideal compressible flow, *J. Comput. Phys.* **23**, 263–275.

van Leer, B. (1977b) Towards the ultimate conservative difference scheme. IV. A new approach to numerical convection, *J. Comput. Phys.* **23**, 276–299.

van Leer, B. (1979) Towards the ultimate conservative difference scheme. V. A second-order sequel to Godunov's method, *J. Comput. Phys.* **32**, 101–136.

van Leer, B. (1982) Flux-vector splitting for the Euler equations, *ICASE Report* 82-30.

Vázquez-Cendón, M.E. (1999) Improved treatment of source terms in upwind schemes for the shallow water equations in channels with irregular geometry, *J. Comput. Phys.* **148**, 497–526.

Velichko, S.A., Lifshitz, Yu.B., Solntsev, I.A. (1999) Computation of unsteady flows with a scheme of improved accuracy, *Comput. Maths Math. Phys.* **34**, 817–831.

Vinokur, M. (1974) Conservation equations of gas dynamics in curvilinear coordinate systems, *J. Comput. Phys.* **14**, 105–125.

Vitushkin, A.G. (1955) *Multidimensional Variations*, GosIzdat Tekn.-Teor. Lit., Moscow [in Russian].

Vladimirov, V.S. (1971) *Equations of Mathematical Physics*, A. Jeffrey (Ed.), Marcel Dekker, New York.

Volchenkov, V.Ya., Belikov, V.V., Tsypin, V.T., Paich, Yu.L., Pisarev, Yu.V., Semenov, A.Yu. (1985) Development of a technique for determination of deformation and optimal dimensions of reinforcements at the exits of hillside culverts, *Tech. Report of Central Research Institute for Construction*, P-XI-5-84/85 (sect. 1), No. 02.85.0082641, Moscow [in Russian].

Voltsinger, N.E., Pyaskovskii, R.V. (1977) *Shallow Water Theory. Oceanology Problems and Numerical Methods*, Gidrometeoizdat, Leningrad [in Russian].

von Neumann, J., Richtmyer, R.D. (1950) A method for the numerical calculation of hydrodynamic shocks, *J. Appl. Phys.* **21**, No. 3, 232–237.

Vorobiev, O.Yu., Lomov, I.N., Shutov, A.V., Kondaurov, V.I., Ni, A.L., Fortov, V.E. (1995) Application of schemes on moving grids for numerical simulation of hypervelocity impact problems, *Int. J. Impact Eng.* **17**, 891–902.

Vovchenko, V.I., Krasyuk, I.K., Semenov, A.Yu. (1992) Ablation and dynamic characteristics of laser action on plane targets, in *Investigation of Physical Processes in Planar and Conical Targets*, Trudy IOFAN [Proc. Gen. Phys. Inst., Russian Acad. Sci.] **36**, 129–201, Nauka, Moscow [in Russian].

Vovchenko, V.I., Krasyuk, I.K., Pashinin, P.P., Prokhorov, A.M., Semenov, A.Yu., Fortov, V.E. (1992) Gas compression and heating in conical targets, in *Investigation of Physical Processes in Planar and Conical Targets*, Trudy IOFAN [Proc. Gen. Phys. Inst., Russian Acad. Sci.], **36**, 5–82, Nauka, Moscow [in Russian].

Vovchenko, V.I., Krasyuk, I.K., Pashinin, P.P., Semenov, A.Yu. (1994) Wide-range dependence of ablation pressure on laser radiation intensity, *J. Russian Acad. Sci. Physics-Doklady* **39**, 633–634.

Voynovich, P.A., Zhmakin, A.I., Fursenko, A.A. (1988) Numerical simulation of the interaction of shock waves in spatially inhomogeneous gases, *Sov. Tech. Phys.* **33**, 748-753.

Wallis, M.K. (1971) Shock-free deceleration of the solar wind? *Nature Phys. Sci.* **233**, 23–25.

Wallis, M.K. (1975) Local interstellar medium, *Nature* **254**, 202–203.

Wallis, M.K., Dryer, M. (1976) Sun and comets as sources of an external flow, *Astrophys. J.* **205**, 895–899.

Warming, R.F., Beam, R.M., Hyett, B.J. (1975) Diagonalization and simultaneous symmetrization of the gas dynamic matrices, *Math. Comput.* **29**, 1037–1045.

Washimi, H., Tanaka, T. (1996) 3-D magnetic field and current system in the heliopause, *Space Sci. Rev.* **78**, 85–94.

Weibel, E.S. (1959) Spontaneously growing transverse waves in a plasma due to an anisotropic velocity distribution, *Phys. Rev. Lett.* **2**, No. 3, 83–84.

Weyl, H. (1949) Shock waves in arbitrary fluids, *Comm. Pure Appl. Math.* **2**, No. 2-3, 103–122.

Wilkins, M.L. (1980) Use of artificial viscosity in multidimensional fluid dynamic calculations, *J. Comput. Phys.* **36**, 281–303.

Wilkinson, W.L. (1960) *Non-Newtonian Fluid*, Pergamon Press, London.

Willi, O., Rumsby, P.T. (1981) Filamentation on laser irradiated spherical targets, *Opt. Commun.* **37**, No. 1, 45–48.

Willi, O., Rumsby, P.T., Sartang, S. (1981) Optical probe observations of nonuniformities in laser-produced plasmas, *IEEE J. Quant. Electron.* **17**, 1909–1917.

Willi, O., Rumsby, P.T., Hooker, C., Raven, A., Lin, Z.Q. (1982) Observations of instabilities in the corona of laser produced plasma, *Opt. Commun.* **41**, No. 2, 110–114.

Williams, F.A. (1965) *Combustion Theory*, Addison Wesley, Reading, MA.

Witham, G.B. (1974) *Linear and Nonlinear Waves*, Pure and Applied Mathematics, Wiley Interscience, New York.

Woodward, P., Colella, P. (1984) The numerical simulation of the two-dimensional fluid flow with strong shocks, *J. Comput. Phys.* **54**, 115–173.

Wu, C.C. (1990) Formation, structure, and stability of MHD intermediate shocks, *J. Geophys. Res.* **95**, 987–990.

Wu, C.C., Kennel, C.F. (1992a) Structure and evolution of time-dependent intermediate shocks, *Phys. Rev. Lett.* **68**, 56–59.

Wu, C.C., Kennel, C.F. (1992b) Evolution of small-amplitude intermediate shocks in a dissipative and dispersive system, *J. Plasma Phys.* **47**, 85–102.

Wu, C.C., Kennel, C.F. (1993) The small amplitude MHD Riemann problem, *Phys. Fluids B* **5**, 2877–2886.

Xu, Z.Z., Lee, P.H.Y., Lin, L.H. et al. (1987) Filamentation and jets in linefocused laser-produced plasmas, *Opt. Commun.* **61**, No. 3, 199–202.

Xu, Z.Z., Jiang, Z.M., Lin, L.H. et al. (1988) Large-scale jet structures in laser-produced plasmas, *Opt. Commun.* **69**, No. 1, 49–53.

Xu, Z.Z., Chen, S.S., Lin, L.H. et al. (1989) Characteristics and evolution of plasma-jet-like structures in line-focused laser-produced plasmas, *Phys. Rev. A: Gen. Phys.* **39**, 808–815.

Yamamoto, Y., Karashima, K. (1980) Floating shock fitting for three-dimensional inviscid supersonic flows. Part I. General description, *Trans. Jap. Soc. Aeronaut. Space Sci.* **23**, No. 59, 1–17.

Yee, H.C. (1987) Construction of explicit and implicit symmetric TVD schemes and their applications, *J. Comput. Phys.* **68**, 151–179.

Yee, H.C. (1989) *A Class of High-Resolution Explicit and Implicit Shock-Capturing Methods*, von Karman Institute for Fluid Dynamics Lecture Series 1989-04 (NASA TM-101088).

Yee, H.C., Sandham, N.D., Djomehri, M.J. (1999) Low-dissipative high-order shock-capturing methods using characteristic-based filters, *J. Comput. Phys.* **150**, 199–238.

Zabrodin, A.V., Prokopov, G.P. (1998) Algorithm of numerical modelling of two-dimensional nonstationary flows of a heat conductive gas in the three-temperature approach, *Voprosy Atomnoi Nauki i Tehniki [Problems of Atomic Science and Engineering]*, Ser.: Math. Model. Phys. Processes, No. 3, 3–16 [in Russian].

Zachary, A., Colella, P. (1992) A higher-order Godunov method for the equations of ideal magnetohydrodynamics, *J. Comput. Phys.* **99**, 341–347.

Zachary, A.L., Malagoli, A., Colella, P. (1994) A higher-order Godunov method for multidimensional ideal magnetohydrodynamics, *SIAM J. Sci. Comput.* **15**, 263–284.

Zaitsev, N.A., Radvogin, Yu.B. (1990) *Numerical Method for Solving the Problem of Supersonic Flow around a Spherically Symmetric Source*, Preprint No. 86, Keldysh Institute of Applied Mathematics, USSR Academy of Sciences, Moscow [in Russian].

Zalesak, S.T. (1979) Fully multidimensional flux-corrected transport algorithms for fluids, *J. Comput. Phys.* **31**, 335–362.

Zank, G.P., Pauls, H.L., Williams, L.L., Hall, D.T. (1996) Interaction of the solar wind with the local interstellar medium: A multifluid approach, *J. Geophys. Res.* **101**, A21639–A21655.

Zegeling, P.A. (1999) Moving grid techniques, Chapter 37 in *Handbook of Grid Generation*, J.F. Thompson, B.K. Soni, and N.P. Weatherill (Eds.), CRC, Boca Raton, FL.

Zel'dovich, Ya.B., Raizer, Yu.P. (1967) *Physics of Shock Waves and High-Temperature Hydrodynamic Phenomena*, **1–2**, Academic Press, New York.

Zheleznyak, M.B., Mnatsakanyan, A.Kh., Pervukhin, S.V. (1986) Unsteady and nonequilibrium airflow near a stagnation line, *U.S.S.R. Fluid Dynamics* **22**, 993–996.

Zhukov, A.I. (1960) Application of the method of characteristics for numerical solution of the one-dimensional gas dynamic problems, *Trudy MIAN [Proc. Steklov Inst. Math.]* **58** [in Russian].

Ziegler, H. (1963) *Some Extremum Principles in Irreversible Thermodynamics with Application to Continuum Mechanics*, North-Holland, Amsterdam.

Zubchaninov, V.G. (1990) *Basics of Theory of Elasticity and Plasticity*, Vysshaya Shkola, Moscow [in Russian].

Zubov, V.I., Krivtsov, V.M., Naumova, I.N., Shmyglevskii, Yu.D. (1980) A numerical comparison of different models of metal vaporization, *U.S.S.R. Comput. Maths Math. Phys.* **20**, No. 6.

Zubov, V.I., Krivtsov, V.M., Naumova, I.N., Shmyglevskii, Yu.D. (1986) A numerical comparison of different models of metal vaporization, *U.S.S.R. Comput. Maths Math. Phys.* **26**, No. 6, 88–91.

Zubov, V.I., Krivtsov, V.M., Naumova, I.N., Shmyglevskii, Yu.D. (1993) Interaction of a laser beam with an aluminum vessel and the aluminum vapor so generated, *Math. Model. Comput. Exper. (MMCE)* **1**, No. 3, 255–260.

Zukas, J.A., Nicholas, T., Swift, H.F., Greszczuk, L.B., Curran, D.R. (1982) *Impact Dynamics*, John Wiley, New York.

Index